4th EDITION

Agriscience

FUNDAMENTALS & APPLICATIONS

4th EDITION

Agriscience

FUNDAMENTALS & APPLICATIONS

L. DeVere Burton
and
Elmer L. Cooper

THOMSON

DELMAR LEARNING

Australia Canada Mexico Singapore Spain United Kingdom United States

Agriscience: Fundamentals & Applications, 4th Edition
L. Devere Burton and Elmer L. Cooper

Vice President, Career Education Strategic Business Unit:
Dawn Gerrain

Director of Editorial:
Sherry Gomoll

Acquisitions Editor:
David Rosenbaum

Developmental Editor:
Gerald O'Malley

Editorial Assistant:
Christina Gifford

Director of Production:
Wendy A. Troeger

Production Manager:
JP Henkel

Production Editor:
Kathryn B. Kucharek

Technology Project Manager:
Sandy Charette

Director of Marketing:
Wendy Mapstone

Marketing Specialist:
Gerard McAvey

Cover Images:
Getty Images

Cover Design:
essence of 7

For permission to use material from this text or product, submit a request online at http://www.thomsonrights.com
Any additional questions about permissions can be submitted by email to thomsonrights@thomson.com

Library of Congress Cataloging-in-Publication Data

Burton, L. DeVere
   Agriscience : fundamentals & applications / L. DeVere Burton Elmer L. Cooper.—4th ed.
      p. cm.
   Cooper's name appears first on the earlier editions.
   ISBN 1-4018-5962-3
      1. Agriculture. I. Cooper, Elmer L. II. Title.

S495.B97 2005
630-dc22   2004062908

NOTICE TO THE READER

Contents

Section 1

Agriscience in the Information Age 2

Unit 1	The Science of Living Things 4
Unit 2	Better Living through Agriscience 18
Unit 3	Biotechnology 45

Section 2

You and the New Millennium 58

Unit 4	Career Options in Agriscience 60
Unit 5	Supervised Agricultural Experience 77
Unit 6	Leadership Development in Agriscience 101

Section 3

Natural Resources Management 122

Unit 7	Maintaining Air Quality 124
Unit 8	Water and Soil Conservation 140
Unit 9	Soils and Hydroponics Management 161
Unit 10	Forest Management 195
Unit 11	Wildlife Management 220
Unit 12	Aquaculture 238

Section 4

Integrated Pest Management 252

| Unit 13 | Biological, Cultural, and Chemical Control of Pests .... 254 |
| Unit 14 | Safe Use of Pesticides .... 272 |

Section 5 — Plant Sciences 296

Unit 15	Plant Structures and Taxonomy	298
Unit 16	Plant Physiology	316
Unit 17	Plant Reproduction	333

Section 6 — Crop Science 358

Unit 18	Home Gardening	360
Unit 19	Vegetable Production	377
Unit 20	Fruit and Nut Production	399
Unit 21	Grain, Oil, and Specialty Field-Crop Production	414
Unit 22	Forage and Pasture Management	437

Section 7 — Ornamental Use of Plants 454

Unit 23	Indoor Plants	456
Unit 24	Turfgrass Use and Management	476
Unit 25	Trees and Shrubs	498

Section 8 — Animal Sciences 516

Unit 26	Animal Anatomy, Physiology, and Nutrition	518
Unit 27	Animal Health	537
Unit 28	Genetics, Breeding, and Reproduction	554
Unit 29	Small Animal Care and Management	571
Unit 30	Dairy and Livestock Management	591
Unit 31	Horse Management	616

Section 9 — Food Science and Technology 636

| Unit 32 | The Food Industry | 638 |
| Unit 33 | Food Science | 657 |

Section 10

Communications and Management in Agriscience 684

Unit 34 Marketing in Agriscience 686
Unit 35 Agribusiness Planning 709
Unit 36 Entrepreneurship in Agriscience 727
Glossary 741
Appendix A: Developing a Personal Budget 779
**Appendix B: Plan Supervised Agricultural
                Experience Programs** 781
Index 788

Preface

Welcome to the agriscience world of the 21st century! *Agriscience: Fundamentals and Applications, Fourth Edition,* is about a new century of agricultural and agriscience developments.

The first class to use this textbook is a generation of students whose lives span across two different centuries and two different millennia. It is interesting to consider that in all of the ages of time since humans engaged in agricultural pursuits, nearly all of the agricultural innovations and technologies that have ever been known to humankind have evolved in less than 100 years.

It is to the agriscience students of the new millennium that this textbook is dedicated, for the agriculturists, scientists, and innovators of tomorrow are today's high school students. The "millennium generation" will be called on to feed the world as the human population nearly doubles to 10 billion people. To do this, they will need to learn more than any other generation has ever learned, and they will need to discover more new ways to increase food production than any other generation has ever discovered. They must accomplish this using marginal land, as many of our fertile farms are swallowed up to build cities and towns. *Agriscience: Fundamentals and Applications, Fourth Edition* is the modern agriscience textbook that will introduce the "millennium generation" to agricultural careers. This generation will also lead the industry that the people of the United States depend on to feed and clothe our citizens and to export surplus agricultural products to other regions of the world.

The fourth edition of this book expands on the original text and the ideas of earlier editions. The science component has been strengthened, and new examples of agricultural applications of science and technology have been added. The book is intended for introductory level agriscience classes at the ninth and tenth grades.

NEED FOR AN INTRODUCTORY TEXTBOOK

This book is an introductory textbook in a series of modern secondary agricultural textbooks published by Delmar, a division of Thomson Learning. It addresses the most basic levels of agriscience using language and examples that are matched to the needs of beginning students in the natural science career pathway.

The revisions in this new edition are the work of current Delmar agriscience author L. DeVere Burton. He is also the author of three other textbooks in the agriscience series: *Agriscience & Technology,*

Second Edition; Fish and Wildlife; Principles of Zoology and Ecology; and *Introduction to Forestry Science.* Each of these works, including this edition of *Agriscience: Fundamentals and Applications, Fourth Edition* reflects the premise on which agricultural education was founded—that most students learn best as they apply the principles of science and agriculture to real-life problems.

ORGANIZATION

This edition of *Agriscience: Fundamentals and Applications* is organized into 10 sections and 36 units. Each section introduces the subjects that will be covered in the individual units. The text and illustrations for each section have been revised. Each unit begins with a stated objective and a list of competencies to be developed. Important terms are listed at the beginning of each unit and highlighted in the text. They are also included in the glossary at the end of the book. Each unit contains profiles on science, careers, and agriculture, and concludes with student activities and a section on self evaluation. The book concludes with a complete and thorough index.

NEW FEATURES AND ENHANCED CONTENT

The science content of this edition has been significantly strengthened. Each unit includes a feature called "Hot Topics in Agriscience." Each of these features describes a recent scientific discovery for which an agricultural application has been identified. "Suggested Class Activities" is another feature that is found at the beginning of each unit. New photographs and illustrations have been added throughout the book. They are intended to bring a sharper focus to the agriscience emphasis of the text. Internet icons are featured throughout the textbook. They include key words for Internet searches on the topics of discussion. This feature will help students to explore agriscience topics beyond the boundaries of this textbook.

- "Suggested Class Activities" in each unit give both the student and the instructor an innovative way to become actively involved with the content of each unit.

- "Hot Topics in Agriscience" is a standard unit feature that describes recent scientific discoveries for which an agricultural application has been identified.

- Internet icons are placed throughout each unit. These icons include key search terms that will help students and instructors to explore agriscience topics beyond the scope of the textbook. (NOTE: Web site URLs were accurate at time of publication. Because of the fluid nature of the Internet, Delmar cannot guarantee that all Web site addresses will remain current for the duration of this edition.)

- Broad applications to science, math, agriculture, natural resources, and the environment provide the appropriate balance for the evolving agriscience curriculum.

- The addition of many new, full-color photos and illustrations will help to enhance learning.

EXTENSIVE TEACHING/ LEARNING MATERIALS

A complete supplemental package is provided together with this textbook. It is intended to assist teachers as they plan their teaching strategies by providing materials that are up-to-date and efficiently organized. These materials are also intended to assist students who wish to explore beyond the confines of the textbook. They include the following resources:

Instructor's Guide to Text

ISBN: 1-4018-5970-4

The Instructor's Guide provides answers to the end-of-chapter questions and additional materials to assist the instructor in the preparation of lesson plans.

Lab Manual

ISBN: 1-4018-5964-X

CD-ROM ISBN: 1-4018-7766-4

This comprehensive lab manual reinforces the text content. It is recommended that students complete each lab to confirm understanding of essential science content. Great care has been taken to provide instructors with low-cost, strongly science-focused labs.

Lab Manual Instructor's Guide

ISBN: 1-4018-5968-2

The Instructor's Guide for the lab manual provides answers to lab manual exercises and additional guidance for the instructor.

Classmaster CD-ROM

ISBN: 1-4018-5966-6

This new technology supplement provides the instructor with valuable resources to simplify the planning and implementation of the instructional program. It includes the Teacher's Resource Guide; transparency masters; motivational questions and activities; answers to questions in the text; and lesson plans to provide the instructor with a cohesive plan for presenting each topic. Also included is a computerized test bank of questions, giving the instructor an expanded capability to create tests.

Acknowledgments

The author and publisher wish to express their appreciation to the many individuals, FFA associations, and organizations that have supplied photographs and information necessary for the creation of this text. A very special thank you goes to all the folks at the National FFA organization and the USDA photo libraries, who provided many of the excellent photographs found in this textbook. Because of their efforts, this is a better book.

The author and publisher also gratefully acknowledge the unique expertise provided by the contributing authors to the text. Their work provides a special assurance that the content in each unit is at the cutting edge of the respective topic and is particularly appropriate for the level of the text. The contributing authors are:

Robert S. DeLauder, Agriscience Instructor, Damascus, Maryland;

Thomas S. Handwerker, Ph.D., Department of Agriculture, University of Maryland at Princess Anne;

Curtis F. Henry, Business Manager, College of Agriculture, University of Maryland at College Park;

Dr. David R. Hershey, Assistant Professor, Department of Horticulture, University of Maryland at College Park;

Robert G. Keenan, Agriscience Instructor, Landsdowne High School, Baltimore, Maryland;

J. Kevin Mathias, Ph.D., Institute of Applied Agriculture, University of Maryland at College Park;

Renee Peugh, Biological Science Consultant, Boise, Idaho;

Regina A. Smick, Ed.D., Academic Advisor and Instructor, College of Agriculture, Virginia Tech, Blacksburg; and

Gail P. Yeiser, Instructor, Institute of Applied Agriculture, University of Maryland at College Park.

A special thank you is also extended to the reviewers of this Fourth Edition. Their content expertise and suggestions for updates and improvements greatly enhanced the overall text.

Sarah Osborn Welty
Walkersville High School
Walkersville, MD

Russell Moore
Roane County High School
Spencer, WV

Douglas Brown
Cential Columbia High School
Bloomsburg, PA

Ken Moncrief
Ceres High School
Ceres, CA

Terry Bowlds
Franklin County High School (retired)
Frankfort, KY

Ed Winkle
Clermont Northeastern
Batavia, OH

Larin Crossley
Preston High School
Preston, ID

Juston Lamb
Pekin High School
Packwood, IA

4th EDITION

Agriscience

FUNDAMENTALS & APPLICATIONS

L. DeVere Burton
and
Elmer L. Cooper

Agriscience in the Information Age

Better Living through Research

Science and technology are modern miracles that have opened up the door to areas of research, turning the dreams of humankind into realities. Space station research, new frontiers to investigate, and our never ending quest for knowledge have exploded into many new and exciting careers.

You could become one of the people growing plants or animals in a space station high above the Earth. Or, you might be an engineer who designs the animal or plant-growing module of the space station, or a molecular geneticist or plant breeder designing new plants to grow well in low gravity, or a food scientist developing packaging for space-grown produce.

One fast-growing career area is in plant science. As you will learn, plants are "green machines" that capture, package, and store energy from the sun through photosynthesis. They supply food and fiber for animals and humans to help sustain life. But, human knowledge and energy are required to help plants function in the overall "green machine" that constitutes our food, fiber, and natural resources system.

Students of the 21st century will also become the agricultural professionals of the 21st century. They will become the agricultural producers, processors, marketers, and scientists who will discover new ways to feed the citizens of the United States and the world. This will be accomplished by conducting basic research and applying it to the agricultural food system.

Whether you choose a career in plant or animal science, sales and marketing, mechanics, or processing, it is sure to be rewarding. By studying agriscience you are opening the door to exciting educational programs and careers that contribute to better living conditions for people everywhere. What role will you play in the challenging task of producing the food and fiber that will be required by future generations?

Biosphere 2 *(Courtesy of Global Ecotechnics Corporation; photographer Gill Kenny)*

The Science of Living Things

OBJECTIVE

To recognize the major sciences that explain the development, existence, and improvement of living things.

MATERIALS LIST

✓ writing materials

✓ newspapers and magazines

✓ encyclopedias

✓ Internet connection

COMPETENCIES TO BE DEVELOPED

After studying this unit, you should be able to:

- define agriscience.
- discover agriscience in the world around us.
- relate agriscience to agriculture, agribusiness, and renewable natural resources.
- state the major sciences that support agriscience.
- describe basic and applied sciences that relate to agriscience.

SUGGESTED CLASS ACTIVITIES

1. Invite a retired farmer to be a guest speaker on the topic of improvements or advances in the science and technology of agricultural production that he or she has experienced during his or her career. Have the students make a list of the agricultural technologies that are discussed. Speculate on new agricultural technologies that the students may experience during their careers.

2. Obtain a copy of the application process for the National FFA Agriscience Student Award. This award offers excellent scholarship opportunities to students who plan and carry out agriscience research projects. Discuss some local agricultural problems that might be addressed by students who express interest in planning a research project in agriscience.

3. In groups of four or five students, research "organic farming" on the Internet. Compare and contrast organic farming and traditional farming. In your search, you may include factors such as cost versus yield, consumer demand, or any other factor that drives production. Prepare a 5-minute presentation on your findings and present it to the class.

TERMS TO KNOW

agriscience	industrial technology	integrated pest management	entomology
agriculture	high technology	organic food	agronomy
agriculture/agribusiness and	aquaculture	water resources	horticulture
renewable natural	agricultural engineering	environment	ornamentals
resources	animal science technology	turf	animal sciences
agribusiness	crop science	biology	agricultural economics
renewable natural resources	soil science	chemistry	agricultural education
technology	biotechnology	biochemistry	

Life in the United States and throughout the world is changing every moment of our lives. The space we occupy, as well as the people we work and play with, may be constant for a brief time. However, these are quick to change with time and circumstances. The things we need to know and the resources we have to use are constantly shifting as the world turns around us.

Humans have the gift of intelligence—the ability to learn and to know (Figure 1–1). This permits us to compete successfully with the millions of other creatures that share the earth with us (Figure 1–2). In ages past, humans have not always fared well in this competition. Wild animals had the advantages of speed, strength, numbers, hunting skills, and superior senses over humans. These superior senses of sight, smell, hearing, heat sensing, and reproduction all helped certain animals, plants, and microbes to exercise control over humans to meet their own needs.

The cave of the cave dweller, lake of the lake dweller, and cliff of the cliff dweller indicate early human reliance on natural surroundings for basic needs (food, clothing, and shelter) (Figure 1–3). Those early homes gave humans some protection from animals and unfavorable weather.

FIGURE 1–1 Humans have the gift of intelligence—the ability to learn and to know. *(Courtesy USDA/ARS K10087-1)*

FIGURE 1–2 The gift of intelligence has permitted humans to compete with and benefit from animals even though most animals are superior to humans in other ways. *(Courtesy USDA/ARS K7102-12)*

FIGURE 1–3 Early humans had to rely on natural settings to shield them from danger and the elements.

FIGURE 1–4 Americans spend only 11 percent of their incomes on food. *(Courtesy Getty Images)*

However, they were still exposed to disease, the pangs of hunger, the sting of cold, and the oppression of heat.

The world of agriscience has changed the comfort, convenience, and safety of people today. In the United States, we spend only 11 percent of our wages to feed ourselves (Figure 1–4). People in many nations spend more than half of their incomes on food. We are fortunate that our scientists have discovered new ways to produce greater amounts of food and fiber (such as cotton) from each acre of agricultural land. They have done this by finding ways to stimulate growth and production of animals and plants and to reduce losses from diseases, insects, and parasites. We have also learned to preserve our food from one production cycle until the next without excessive waste. The agriscience, agribusiness, and renewable natural resources of the nation provide materials for clothing, housing, and industry at an equally attractive price.

AGRISCIENCE DEFINED

Agriscience is a relatively new term that you may not find in your dictionary. **Agriscience** is the application of scientific principles and new technologies to agriculture. **Agriculture** is defined as the activities involved with the production of plants and animals and related supplies, services, mechanics, products, processing, and marketing (Figure 1–5). Actually, modern agriculture covers so many activities that a simple definition is not possible. Therefore, the U.S. Department of Education has used the phrase **agriculture/agribusiness and renewable natural resources** to refer to the broad range of activities in agriculture.

Agriculture generally has some tie-in or tie-back to animals or plants. However, production agriculture, or farming and ranching, accounts for only one-fifth of the total jobs in agriculture (Figure 1–6). The other four fifths of the jobs in agriculture are nonfarm and nonranch jobs, such as sales of farm equipment and supplies, plant and animal research,

Hot Topics in Agriscience

AGRICULTURAL RESEARCH: FEEDING A HUNGRY WORLD

It is estimated that the world's population will reach 7 billion people by 2010 and 9 billion in 2030 in comparison with approximately 6 billion today. During the same period, the amount of land and fresh water per person will decrease. Food production will have to become much more efficient if the people of the world are to have enough food to eat. During the past 50 years, food production has increased at a rate that is greater than the increases in population. Agricultural research has resulted in greater productivity of food, plants, and animals, and new technologies have made it possible for farmers to perform their work with greater efficiency. The key to an adequate food supply for the growing human population in the new millennium is agricultural research. New agricultural technologies that lead to the development of more efficient plants and animals and more efficient agricultural machinery will be needed. In addition, we will need to discover new food sources and maintain a healthy environment as the population approaches 10 billion people.

Environment refers to all the conditions, circumstances, and influences surrounding and affecting an organism or group of organisms.

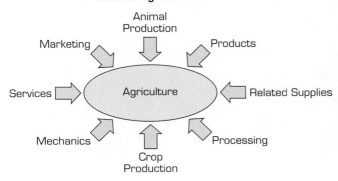

What Is Agriculture?

Animal Production

Marketing Products

Services → Agriculture ← Related Supplies

Mechanics Processing

Crop Production

FIGURE 1–5 Agriculture consists of all of the steps involved in producing a plant or animal and getting the plant and animal products to the people who consume them.

FIGURE 1–6 Farming and ranching account for approximately one-fifth of the agricultural jobs in the United States. *(Courtesy Getty Images)*

INTERNET TIPS:

Forming your search into a question will narrow the results. Example: What is agriscience?

INTERNET KEY WORDS:

renewable, natural resources

processing of agricultural products (Figures 1–7 and 1–8), agricultural education, and maintaining the health of plants and animals. **Agribusiness** refers to commercial firms that have developed with or stemmed out of agriculture (Figure 1–9).

Renewable natural resources are the resources provided by nature that can replace or renew themselves. Examples of such resources are wildlife, trees, and fish (Figure 1–10). Some occupations in renewable natural resources are game trapper, forester, and fisher (someone who harvests fish, oysters, and other seafood).

Technology is defined as the application of science to solve a problem. The application of science to an industrial use is called **industrial technology**. Hence, agriscience was coined to describe the application of high technology to agriculture. **High technology** refers to the use of

FIGURE 1–7 Agricultural education teachers and agricultural extension educators are among those whose careers are related to agriculture.

FIGURE 1–8 Veterinarians and veterinary technicians are people whose careers relate to agriculture in the field of animal health.

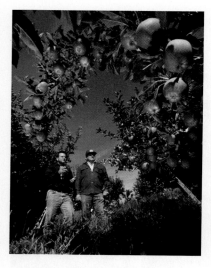

FIGURE 1–9 Agribusinesses are important to the life of most communities. *(Courtesy USDA/ARS K7614-6)*

FIGURE 1–10 Mature trees provide renewable sources for wood products. Young trees of most species begin to grow after mature trees are harvested, allowing sunlight to penetrate the forest floor. *(Courtesy Getty Images)*

electronics and state-of-the-art equipment to perform tasks and control machinery and processes (Figure 1–11). It plays an important role in the industry of agriculture.

Agriscience includes many endeavors. Some of these are aquaculture, agricultural engineering, animal science technology, crop science, soil science, biotechnology, integrated pest management, organic foods, water resources, and environment. **Aquaculture** means the growing and management of living things in water, such as fish. **Agricultural engineering** consists of the application of mechanical and other engineering principles in agricultural settings. **Animal science technology** refers to the use of modern principles and practices for animal growth and management (Figure 1–12). **Crop science** refers to the use of science principles in growing and managing crops. **Soil science** refers to the study of the properties and

FIGURE 1–11 Technology is used in agriscience for such purposes as testing feeds for nutrient content and testing food for purity from chemicals. *(Courtesy USDA/ARS #K-3396-6)*

FIGURE 1–12 Veterinarians use animal sciences to help keep our pets and production animals healthy. *(Courtesy USDA/ARS K5441-1)*

management of soil to grow plants. **Biotechnology** refers to the management of the genetic characteristics transmitted from one generation to another and its application to our needs. It may be defined as the use of cells or components of cells to produce products and processes (Figure 1–13).

The phrase **integrated pest management** refers to the many different methods used together to control insects, diseases, rodents, and

Agri-Profile

CAREER AREA:

Agriscientist

In recent years science has played an increasing role in the lives of plants and animals and the people around them. These living bodies include plants ranging in size from microscopic bacteria to the huge redwood and giant sequoia trees. They include animals from the one-celled amoeba to elephants and whales.

Only recently has science identified the nature of viruses and permitted humans to observe the submicroscopic world in which they exist. The electron microscope, radioactive tracers, computers, electronics, robots, and biotechnology are just a few of the developments that have revolutionized the world of living things. We call this the world of science. Agriscience is a part of this world. Through agriscience, humans can control their destinies better than at any time in known history.

Agriscience spans many of the major industries of the world today. Some examples are food production, processing, transportation, selling, distribution, recreation, environmental management, and professional services. Study of and experiences in a wide array of basic and applied sciences are appropriate preparations for careers in agriscience.

Students experience the wonder of living things. *(Courtesy USDA/ARS K5304-17)*

FIGURE 1–13 Genetic engineering and other forms of biotechnology have developed into one of the most important priorities in research today. *(Courtesy USDA/ARS #K-5011-19)*

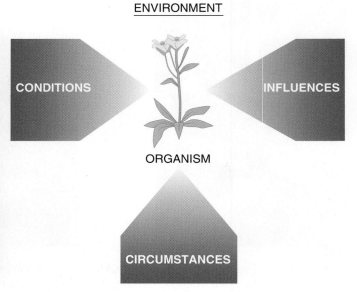

FIGURE 1–14 The term *environment* refers to all the conditions, circumstances, and influences surrounding and affecting an organism or group of organisms.

other pests. **Organic food** is a term used for foods that have been grown without the use of chemical pesticides. **Water resources** cover all aspects of water conservation and management. Finally, **environment** refers to all the conditions, circumstances, and influences surrounding and affecting an organism or group of organisms (Figure 1–14). This generally means air, water, and soil, but it may also include such things as temperature, presence of pollutants, intensity of light, or other influences.

AGRISCIENCE AROUND US

Agriscience and technologies have helped humans change their living conditions from dependence on hand labor to a highly mechanized society. In the process, food and fiber production have become much more efficient. Many members of U.S. society have become free to pursue new careers in business, industry, or the arts, because they are no longer required to spend all of their time finding or producing food for themselves and their families. Fewer than 2 percent of the people in the United States are farmers. On average, each farmer produces enough food for approximately 135 people. In fact, the large surplus of food that is produced in the United States is shipped to many other countries in the world.

Whether you live in the city, town, or country, you are surrounded by the world of agriscience. Plants use water and nutrients from the soil and release water and oxygen into the air. Animals provide companionship as pets and assistance with work. Both plants and animals are sources of food. Many microscopic plants and animals are silent garbage disposals (Figure 1–15). They assist in the process of decay of the unused plant and animal residue around us. This process returns

FIGURE 1–15 Composting is a process that uses bacteria to break down plant residue. The material that remains is used to provide nutrients to crops and gardens.

FIGURE 1–16 Turf has become an important crop, especially in areas near large cities where it is used to establish new lawns. *(Courtesy USDA/ARS #CS-0311)*

nutrients to the soil and has many other benefits to our environment and well-being.

Agriscience encompasses the wildlife of our cities and rural areas, and the fish and other life in streams, ponds, lakes, and oceans. Plants are used extensively to decorate homes, businesses, shopping malls, buildings, and grounds. When one crop is used less, another takes its place. This occurs even where land changes from farm use to suburban and urban uses.

Corn has long been referred to as king among crops in the United States. Yet, in one state, turf grass recently replaced corn as the number one crop. **Turf** is grass that is used for decorative, as well as soil-holding, purposes. This change has occurred as more land is being used for roads, housing, businesses, institutions, recreation, and other nonfarm uses (Figure 1–16).

Agriculture and the agriscience activities that support it extend far beyond the borders of the United States. Many nations throughout the world depend on agriscience to improve the production of their crop and livestock industries. Agriculture is a global industry, and although the United States exports many of its agricultural products, it also imports many agricultural products from other parts of the world. Many of the flowers used by florists in the United States come from Colombia, South America, and other foreign countries (Figure 1–17). Bulbs come from Holland and meat products are imported from Argentina. Lumber is shipped from the United States to Japan, only to return in the form of plywood and other processed lumber commodities. A decrease in the price of sow bellies or an unexpected change in the price of grain futures in Chicago can affect business and investment around the world (Figure 1–18).

The great water-control projects on the Colorado River have permitted the transformation of the American Southwest from a desert to

FIGURE 1–17 Flowers often are imported to the United States during seasons of the year when local florists are unable to produce them at competitive prices.

FIGURE 1–18 Marketing in agriscience has become big business. *(Courtesy National FFA; FFA # 18)*

INTERNET KEY WORDS:

agriculture
journals
biology
chemistry

irrigated lands (Figure 1–19). This is now an area of intensive crop production that has stimulated national population shifts. Water management has transformed the great dust bowl of the American West into the "bread basket" of the world.

Agriscience enterprises extend beyond farming to such fields as journalism and communications. Agricultural publications such as magazines, journals, and newspapers provide information to farmers, helping to make farm production more efficient. Radio and television programs provide similar services to agriculture. They provide a communications link among such people as agricultural specialists, agricultural extension educators, wildlife biologists, and others to communicate the latest information to farmers and other managers of natural resources. Such subjects as plants, animals, wildlife, market reports, gardening, and lawn care are popular "Saturday morning" topics.

FIGURE 1–19 The great Southwestern desert has been transformed into highly productive land, using irrigation water from the huge dams on the Colorado River.

AGRISCIENCE AND OTHER SCIENCES

INTERNET KEY WORDS:

entomology

agronomy

animal science

Agriscience is really the application of many sciences. Colleges of agriculture and life sciences perform a dual role of conducting research and teaching students in these sciences. **Biology** is one of the three basic sciences. It derives from two Greek words: *bios,* meaning "life," and *logy* meaning "to study." It is the science that studies all living things (organisms) and the environment in which these organisms live. Biology emphasizes the structures, functions, and behaviors of all living organisms. It focuses on the traits that organisms have in common as well as their differences. Having an understanding of biology is important to you, the agriscience student. New biological discoveries can affect many areas of your life. Examples include the choice to plant a new breed of flower in the front yard, changes in the way food is processed that results in a fresher product with a long shelf life, and new medical treatments that can help keep you healthy.

Chemistry is another basic science. It is the branch of science that studies the nature and characteristics of elements or simple substances. Chemists study the changes substances undergo when mixed with different substances. These changes, or reasons as scientists call them, are responsible for compounds that have been used by humans for thousands of years, whereas other compounds like the artificial sweetener are a relatively new discovery. Chemistry currently benefits us in many ways; for example, it is responsible for the creation of new medicines, textiles, fuel, and fertilizers.

Biochemistry is the last of the three basic sciences. It is a combination of biology and chemistry. Recall the definition of *bio,* which means life, when added to the word *chemistry,* it is easy to see that the science of biochemistry is the study of the chemical activities or processes of living organisms. These chemical activities take place in the cells and molecules of living organisms. This science is responsible for explaining things such as brain function, how genetic traits are passed from one generation to another, and how cells communicate and work together inside an organism.

Applied science uses the basic sciences in practical ways. For instance, **entomology** is the science of insects, the most abundant species on the planet. They account for more than 3 million human deaths per year, they transmit diseases, and they are our principal competitors for food. Insects, however, are required for pollination by half the plants on earth to produce seeds. It is important to find ways to help control problematic insects safely without causing secondary problems such as halting pollination of plants.

There are many other applied sciences. Knowledge of biology, chemistry, and biochemistry is fundamental in entomology and to the other applied sciences listed below.

Agronomy is the science of soil management and crops. Its focus is on the growth, management, and improvement of field crops such as wheat and corn. The goal is to increase food production and quality while maintaining a healthy environment. **Horticulture** is the science and art involved in the cultivation, propagation, processing, and marketing of flowers, turf, vegetables, fruits, nuts, and ornamental plants. **Ornamentals** are plants grown for their appearance or beauty. Examples are flowers, shrubs, trees, and grasses. Horticulture is a unique science because it also incorporates the art of plant design.

Science Connection

THE AGRISCIENCE PROJECT

The scientific method is an excellent and widely used method for systematic inquiry and documentation of new findings. The agriscience student is encouraged to learn and use the scientific method for classroom, laboratory, and field studies. The following procedures will guide you in your quest for new knowledge in agriscience.

Step 1. Identify the Problem

Decide precisely and specifically what it is that you wish to find out. For example, "How much nitrogen fertilizer is needed to grow healthy corn plants?" Be careful to limit your topic to a single researchable objective. Your teacher can suggest other topics that could be researched.

Step 2. Review the Literature

Reviewing the literature simply means reading up on and becoming well informed about the topic. See what is already known about it. Magazines, newspapers, reference books, encyclopedias, trade journals, computer information systems, television, cooperative extension meetings, teleconferences, and personal interviews may be sources of appropriate information. Be sure to seek information on appropriate ways to conduct research on the type of problem you have chosen.

The written report completes the research project and enables others to benefit from the new knowledge. *(Courtesy USDA/ARS K8329-2)*

Step 3. Form a Hypothesis

After learning as much as you can about the topic, develop a hypothesis or statement to be proven or disproven, which will solve the problem. For example, "Trout grown in 60° F water will grow faster than trout grown in 75° F water."

The **animal sciences** are applied sciences that involve animal growth, care, and management of domestic livestock. They include veterinary medicine, animal nutrition, animal reproduction, and animal production and care. Animal scientists work to discover science principles related to animals. Scientific principles are then applied to animal management plans to improve animal health and production.

Economics is the study of how societies use available resources to meet the needs of people. **Agricultural economics** is about factors that affect the management of agricultural resources, including farms and agribusinesses, to meet the needs of the human population. Farm policy and international trade are important components of agricultural economics. **Agricultural education** is one of the most unique programs available to students. It offers organized instruction, supervised agricultural experience (SAE), FFA, and extension education activities. Agricultural communications, journalism, and community development are also components of agricultural education. These and other disciplines are part of the dynamic study known as agriscience.

Step 4. Prepare a Project Proposal

Prepare a proposal outlining how you think the project should be done. Include the timelines, facilities, and equipment required, as well as anticipated costs and a description of how you will do the project.

Step 5. Design the Experiment

Considering the information gathered in Step 2, develop a plan for carrying out the project so you can test the hypothesis. This is the most critical step in your research project. If this is not done correctly, you may invest considerable time, work, and expense and end up with incorrect or invalid conclusions. The method or procedure should be carefully thought out and discussed with your teacher or other research authorities. This is to ensure that your design will actually measure what you are testing.

Step 6. Collect the Data

In this phase you actually conduct the experiment. Here you test and/or observe what takes place and record what you measure or observe.

Step 7. Draw Conclusions

Summarize the results. Make all appropriate calculations. Determine if the information allows you to accept or reject the hypothesis, or if the information is inconclusive.

Step 8. Prepare a Written Report

The written report provides you, your teacher, and other interested parties with a permanent record of your research. From this you can report to your peers, get course credit, and possibly apply for awards. Perhaps you can use a computer and hone your word-processing skills. For scientists, the written report becomes a permanent document that is kept by the research institution and becomes the basis for articles in research publications for all the world to see. The results become part of the "literature" on the topic.

FIGURE 1–20 There are plenty of good jobs for agricultural graduates whether at the technical degree, bachelor degree, or graduate degree levels of education. *(Courtesy Getty Images)*

A PLACE FOR YOU IN AGRISCIENCE

What about career opportunities in agriscience? By 2005, the nation's agricultural colleges experienced the strongest demand for graduates in a decade. A U.S. Department of Agriculture (USDA) study group had forecasted a national shortage of 4,000 agricultural and life sciences graduates per year. The shortage became reality in the early 1990s. Employers are offering higher salaries and more job variety to agriscience college graduates than ever before. Career opportunities continue to be strong in these fields, and they also extend into technology as it relates to agricultural systems. Consequently, high school agricultural, horticultural, or other agriscience program participants have opportunities to obtain good jobs and have rewarding careers (Figure 1–20). These opportunities are described in later units in this text. By studying agriscience, you open the door to exciting educational programs and careers that contribute to professional satisfaction and prosperity.

STUDENT ACTIVITIES

1. Write the Terms to Know and their meanings in your notebook.

2. List examples of animals that have better senses than do humans. Indicate the sense(s) along with the animals.

3. Write a paragraph or two on (1) cave, (2) lake, and (3) cliff dwellers. Explain how the types and locations of their homes provided protection from (1) animals and (2) unfavorable weather. An encyclopedia would be a good resource for this activity.

4. Ask your teacher to assign you to a small discussion group to talk about the responses to Activity 3.

5. Place a map of your school community on a bulletin board. Insert a colored map pin in every location of a farm, ranch, or agribusiness in your school community.

6. Talk to your County Extension Agent or another agricultural leader regarding the importance and role of agriscience, agribusiness, and renewable natural resources in your county.

7. Select one of the sciences mentioned in this unit. Prepare a written report on the meaning and nature of that science. Report to the class.

SELF EVALUATION

A. Multiple Choice

1. Humans have the ability to learn and know. This is known as
   a. achievement.
   b. intelligence.
   c. intuition.
   d. spontaneity.

2. The percentage of an average U.S. worker's pay that is used for food is
   a. 11 percent.
   b. 14 percent.
   c. 50 percent.
   d. 74 percent.

3. The best term to describe the application of scientific principles and new technologies to agriculture is
   a. agribusiness.
   b. renewable natural resources.
   c. farming.
   d. agriscience.

4. Harmful insects, rodents, and diseases are all referred to as
   a. animals.
   b. plants.
   c. pests.
   d. parasites.

5. Agriscience encompasses
   a. wildlife and fish.
   b. ornamental plants and trees.
   c. farms and agribusinesses.
   d. all of the above and more.

6. Irrigated lands are generally used for
   a. intensive crop production.
   b. wildlife refuges.
   c. forests.
   d. boating and fishing.

7. An example of a basic science is
   a. agronomy.
   b. aquaculture.
   c. horticulture.
   d. chemistry.

8. An example of an applied science is
   a. animal science.
   b. biochemistry.
   c. biology.
   d. chemistry.

9. One relationship of agriscience with many other sciences is that
   a. agriscience is the application of many other sciences.
   b. agriscience is entirely different from all other sciences.
   c. agriscience is an old term and an old science.
   d. agriscience is a narrow science and is easily defined.

10. The career and job outlook in agriscience is
    a. a strong demand for college graduates.
    b. a shortage of 4,000 trained workers per year.
    c. higher salaries are being offered.
    d. all of the above.

B. Matching

_____ 1. Aquaculture
_____ 2. Renewable resource
_____ 3. Agribusiness
_____ 4. Chemistry
_____ 5. High technology
_____ 6. Biology
_____ 7. Organic food
_____ 8. Environment

a. Commercial firms in agriculture
b. Electronics and ultramodern equipment
c. Growing in water
d. Basic science of plants and animals
e. Can replace itself
f. Characteristics of elements
g. Space and mass around us
h. Grown without chemical pesticides

C. Completion

1. Integrated pest management refers to the application of many different methods used together to _____.

2. The transformation of the American Southwest from desert to irrigated lands was made possible, in part, by water-control projects on the _____ River.

3. By studying agriscience, you open the door to exciting educational programs that may lead to _____.

Better Living through Agriscience

MATERIALS LIST

✓ paper
✓ pen or pencil

✓ current newspaper
✓ encyclopedias

✓ agriscience magazines
✓ Internet connection

COMPETENCIES TO BE DEVELOPED

After studying this unit, you should be able to:

- describe the conditions of desirable living spaces.
- discuss the influence of climate on our environment.
- compare the influences of humans, animals, and plants on the environment.
- examine the problems of an inadequate environment.
- identify some significant world population trends.
- identify significant historical developments in agriscience.
- state practices used to increase productivity in agriscience.
- identify important research achievements in agriscience.
- describe future research priorities in agriscience.

SUGGESTED CLASS ACTIVITIES

1. Discuss ways that farm work has changed during the last 100 years. Identify several important tasks that must be done by farmers. Describe how each of those tasks was done 100 years ago. Describe how farmers perform each of those tasks today. What scientific discoveries have contributed to greater efficiency in doing farm work today?

2. Investigate ways that new and modern farming methods have contributed to opportunities in career fields other than agriculture. How have efficient farming methods benefitted all the citizens of the United States?

3. Using library materials, the Internet, or other scientific sources, learn about climates in the latitudes of 90, 60, and 30 degrees. Using the information you have obtained about these regions, determine what kinds of crops might be raised at each of these latitudes.

TERMS TO KNOW

sewage system	insect	legume	genetic engineering
safe water	immune	tofu	monoclonal antibody
polluted	reaper	Katahdin	mastitis
condominium	combine	BelRus	coccidiosis
townhouse	moldboard plow	Russet	Impatiens
famine	cotton gin	aerosol	hybrid
contaminate	corn picker	Beltsville Small White	deficiency
urine	barbed wire	Green Revolution	laser
feces	milking machine	feedstuff	
parasite	tractor	selective breeding	

INTERNET KEY WORDS:

best places to live

Living conditions in the world vary extensively. In all countries, there are some very desirable places to live and work (Figure 2–1), yet even in highly developed countries, there are pockets of poverty. How do you explain the differences in living conditions from one place to another? Why do living conditions vary from one community to another? From one neighborhood to another? From one house to another? The wealth and preferences of individuals explain some of the differences. Yet, the environment or the area around us has much to do with the quality of life. It also has much to do with the way we feel about ourselves and others.

VARIETY IN LIVING CONDITIONS

The population of the world today is approximately 6.5 billion people (Figure 2–2). How long will it take for the population to reach 7 billion? Somewhere, a new child will have the distinction of being the 7 billionth human being living on the planet Earth (Figure 2–3). What will the home and community be like where "Baby 7 Billion" is born? Will there be

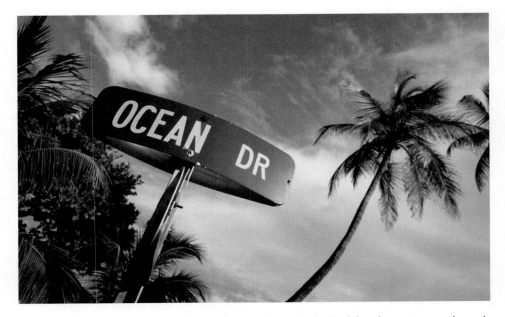

FIGURE 2–1 People throughout the world seek desirable places to work and live. *(Courtesy Getty Images)*

FIGURE 2–2 The population of the world is expected to approach 7 billion by 2010. *(Courtesy of D.C. Committee to Promote Washington)*

FIGURE 2–3 What kind of life on planet Earth will "Baby 7 Billion" have? *(Courtesy NASA)*

adequate food? Will that child be warm, but not too warm or too cold? Will there be sufficient food? Will the child be kept free from serious illness? Will his or her family have a house or good living space they call home? Will they have clothing to permit them to live and work outside the home in relative comfort? What will be the quality of life of others around the 7 billionth human being? Will that child survive, and will he or she go on to live a happy life? Positive answers to these questions would indicate a good environment for a person. These same questions should be asked for the rest of humankind.

The Homes We Live in

Homes of the Very Poor   Homes of the poorest people range in size and quality from nothing to a piece of cardboard or a scrap of wood on the ground. Many survive the freezing winters with only a tattered blanket on the warm sidewalk grates of our modern cities. For others, housing may take the form of a grass hut or a shack made of wood, cardboard, plastic, or scraps of sheet metal. These people often depend on the outdoors to provide water and washing areas and as a receptacle for human waste. Large families often share such homes with pets, poultry, or other livestock (Figure 2–4).

In cities, the poor frequently live in old buildings that are in bad condition and with plumbing that does not work. Drugs, crime, poisonous lead paint, and disease are typical hazards for these people. The steamy streets and sidewalks provide little relief from the oppressive summer heat.

Homes of the Less Fortunate   People with modest sources of income may have homes that are simple, but provide basic protection from the elements. Such homes may be of wood, stone, masonry blocks, sheet metal, brick, or other fairly permanent material. The presence of windows and doors may provide protection from the elements and some privacy.

FIGURE 2–4 The poor of the world live in substandard housing, and it is not uncommon for homes to be shared with pets, livestock, or poultry. *(Photo courtesy of Elmer Cooper)*

INTERNET KEY WORDS:

sewage treatment

These people frequently have access to water that is safe to drink, but bathrooms may be nonexistent or toilets may not have safe sewage-disposal systems. A **sewage system** receives and treats human waste (Figure 2–5). To be regarded as safe, a sewage system must decompose human waste and release by-products that are free from harmful chemicals and disease-causing organisms (Figure 2–6). In most countries of the world, people rely on creeks or rivers to supply their drinking water, bathe the family, wash the clothes, and carry away the human waste.

People with low or modest incomes may live in housing with bathrooms and running water. However, maintenance of the systems may be poor, and ignorance of the users may cause conditions that are health hazards.

FIGURE 2–5 Safe sewage disposal is essential to good health.

FIGURE 2–6 A modern sewage treatment plant in a community is one of the advantages of modern living.

In the undeveloped countries, the lower classes are fortunate if there is a source of safe water (water that is free of harmful chemicals and disease-causing organisms) at the village center (Figure 2–7). Modest and simple running-water facilities are generally the first evidence of community development in many such areas. Even a single faucet with unpolluted water is a major step forward for many communities. Communities that do not have a source of safe water must use whatever water they have available to them. Too often, water is polluted or unsafe to drink because it contains waste materials, chemicals, or unhealthful organisms.

Homes of the Middle and Upper Classes The middle- and upper-class people of the world can afford and enjoy housing that is clean, safe, and convenient. Such living spaces are often found as single houses. They are found in both rural and urban areas. In towns, villages, and urban areas, homes may also be in the form of townhouses, condominiums, or apartment buildings. A condominium is a building with many individually owned living areas or units. All living space of a

FIGURE 2–7 For much of the world's population, even a single community source of safe water is a luxury.
(Photo courtesy of Elmer Cooper)

single unit is generally on one floor. A townhouse is one of a row of houses connected by common side walls. Each unit is generally two or three stories high, giving the occupants more variety of living space.

Food

INTERNET KEY WORDS:

world food supply

Until the 1970s, much of the world went to bed hungry. Only a few countries had sufficient food for their people. Most countries had problems with distribution. Not everyone had food of sufficient amount and quality for proper nutrition. Today, major famines (widespread starvation) are still a fact of life and death. Even in the modern world, serious famines have occurred in various countries.

United Nations scientist John Tanner concluded that, in theory, the world could feed itself; but in practice it could not. It is estimated that nearly a billion people are not getting enough food for an active working life. Although some countries enjoy an adequate food supply from their own production and imports, most nations have many individuals who do not receive proper nutrition.

Family

Family life may well be the dominant force that shapes the environment for most individuals. The family has control of the household activities and sets the priorities of its members. The family has considerable influence over maintaining attractive surroundings and promoting warm relations among individuals.

For some, the family chooses the neighborhood and community where they live. A wise choice, however, is based on having the knowledge of better opportunities and the necessary resources to move to a better living environment. For most of the world's population, the communities where individuals are born are the communities where they are raised and spend their lives.

Neighborhood and Community

The neighborhood and community have substantial influence on the environment in which we live. Some communities have tree-lined country roads with attractive fields, pastures, or woodlands to provide variety in the landscape. Other communities may have the advantage of attractive homes, businesses, or community centers (Figure 2–8). Urban areas may

FIGURE 2–8 Good communities provide nice places to live and work. *(Courtesy Getty Images)*

Hot Topics in Agriscience

PRESERVING THE ENVIRONMENT USING INTENSIVE FARMING PRACTICES

When the first colonists arrived in America from Europe, there were abundant land resources available for food production. Trees were removed to make way for the farms. Midway through the 21st century, as the world's population approaches 10 billion people, the forests of the world will again become endangered unless we can increase the food production of current farmland. We will need to increase the efficiency of our existing farmlands to produce an adequate food supply for a growing world population. If we fail to increase the food production of our land, forests will probably be converted to farms, because more land will be required to produce the additional food that will be needed.

Responsible use of intensive farming practices is likely to play a big role in preserving our forest lands. Applications of agricultural chemicals to crops contribute to high production of food by controlling weeds and insects. The application of fertilizers to farmland is also a proven method for sustaining high levels of production. However, good judgment must always be exercised in the application of fertilizers and chemicals to ensure that they are used safely and that they do not pollute the environment. It is an interesting paradox to consider that farm fertilizers and pesticides may be our best hope for preserving the forests and other natural environments in the world. As many foreign governments can attest, preservation of the environment becomes a low priority to people who are starving.

boast high-rise buildings for work and residence. These provide beautiful vistas of city lights or harbor scenes of commerce and recreation.

Neighborhoods and villages are parts of larger communities. These communities are influenced greatly by the families who live in the immediate area. If families work together toward common goals, they can shape the character, education, religious activities, social outlets, employment opportunities, and other broad aspects of their environment.

Climate and Topography

INTERNET KEY WORDS:

climate prediction
causes of climate

Climate and topography are also important factors affecting our environment. Climate is the average yearly temperature and precipitation for a region. Unlike topography, which is the physical shape of the earth, the climate of a given area is shaped by many factors. The movement of heat by wind and ocean currents, the amount of heat absorbed from the sun, latitude, and the amount of precipitation received are all factors that influence climate. The climate affects what kinds of crops can be raised. The tropical areas of the world produce crops such as pineapples and bananas, whereas the more mild climates found in the U.S. mainland are better adapted for crops such as corn and wheat (Figure 2–9).

Average annual temperatures are very high near the equator. Yet people living near ocean waters, even in tropical areas, enjoy a moderate climate with cool breezes most of the time. Inland, the inhabitants are likely to experience hot, humid weather with high rates of rainfall. The high rainfall, in turn, stimulates heavy plant growth, resulting in jungle conditions. Similarly, sea-level elevations may create balmy 80° F temperatures, whereas a short trip to the top of a nearby mountain may reveal snow on its peak (Figure 2–10).

Climate and Latitude

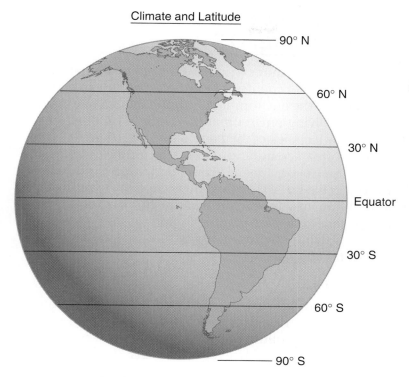

FIGURE 2–9 The distance from the equator is measured by degrees with zero being the equator and 90 degrees North being the north pole. Distance from the equator affects the climate of a region.

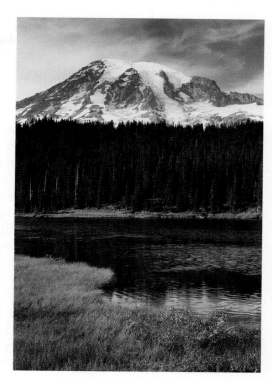

FIGURE 2–10 Topography is an important factor that affects the temperature of the environment. *(Courtesy Getty Images)*

Northern areas, such as Alaska, may border on the Arctic Circle and have long, frigid winters. Yet those same latitudes enjoy summers suitable for short-season crops (Figure 2–11). People inhabit most areas of the Earth, so the climate and topography where they find themselves create environmental conditions that influence their quality of life.

FIGURE 2–11 The climate near Anchorage, Alaska is influenced by a short growing season; however, the number of hours of daylight per day is high, contributing to excellent growth for some kinds of food crops.

FACTORS INFLUENCING LIVING ENVIRONMENTS

Humans are the only living creatures who have the ability to make choices that affect their living environments. For example, people can choose to protect or even to clean up the environments around them. They may also choose to damage their living environments. Living in an environment that is free of pollution is a choice of the people who live there.

People have developed scientific processes to remove waste products, poisons, and disease organisms from food, water, and the air (Figure 2–12). However, the body is limited in its capacity to remove poisons and harmful organisms. To remain healthy, humans and animals must limit their exposure to disease organisms and poisons. One important reason to protect living environments is to keep pollutants from entering food and water supplies.

FIGURE 2–12 Processes have been developed that will remove poisons and harmful organisms from water. *(Courtesy of PhotoDisc)*

Humans and Animals

Some human activities can be very damaging to living environments. For example, exhaust fumes from cars are known to cause acid precipitation. Leaky gasoline tanks pollute the soil, which in turn pollutes the drinking water supply. Improper use of chemicals (lawns, gardens, farms, and factories) pollutes rivers, streams, and lakes. Poor soil management results in erosion to the soil and contamination of streams, lakes, and reservoirs with silt.

A major problem for humans and animals is to avoid **contaminating** (adding material that will change the purity or usefulness of a substance) food and water with secretions from their own bodies. **Urine** and **feces** are liquid and solid body wastes. They are serious contaminants of food and water. Diseases are often spread by body contact, by eating impure food and water, or by breathing contaminated air.

There are serious animal diseases that spread from animal to animal by contact with body wastes. If animals have plenty of living space, this generally does not cause serious problems. But, as with humans, when animals are concentrated, health hazards increase. Fortunately, most diseases are spread among a given species of animal and not from one species to another. For instance, most diseases of dogs do not spread to cats. Similarly, most diseases of animals do not infect humans. However, there are some animal disorders that cause human sickness. Internal **parasites** are organisms that live on or inside other organisms with no benefit to the hosts. An internal parasite is an organism that lives inside of another organism called a host. The parasite is an unwelcome guest, because it always causes some kind of harm to its host by feeding on it. Brucellosis is an example of an animal disease that may be transferred from animals to humans, creating enormous health problems.

Insects

Insects impact heavily on our environment. Some cause damage to our living environment and others help to improve it. The cockroach is an unwelcome guest in many households of the world (Figure 2–13). Some cockroaches feed on human waste and then on the food of humans. They transmit disease from waste material to food and water. In poor housing conditions, they can move from household to house-

Hot Topics in Agriscience

UNDERSTANDING MAD COW DISEASE

Mad cow disease is a disease of cattle that has been known to pass to humans when they eat the brain or spinal cord tissues of an affected cow. This disease has devastated the cattle herds in Europe and Asia since it was found in the United Kingdom in 1986. Cattle herds that had diseased animals have been destroyed and burned to prevent the spread of the disease. The organism responsible for the disease eats holes in the brain of the cow, causing it to be unable to walk or function properly. The disease is spread when cows are fed infected nerve tissue, such as brain or spinal cord tissue from diseased cows, that has been processed into animal feed.

The human form of the disease is called variant of Creutzfeldt–Jakob disease. It is known to have caused human deaths in Europe. It is not known to spread from one human to another human. Researchers continue to seek additional information on the spread of the disease. Meat that does not contain infected brain or spinal cord tissue is believed to be safe for human consumption. The disease was detected in the United States for the first time in a Washington state dairy cow that had become crippled when its calf was born. The cow was butchered, and the tissue was tested in keeping with government regulations. Because the disease was confirmed, crippled or "downer cows" may no longer be used for human food.

hold. In doing so, they often leave disease organisms and illnesses in their wake.

Insects cause damage to the environment in other ways. They damage and kill trees and other plants. A population of harmful insects can expand quickly, and damage to trees can be extensive unless steps are taken to control them. For example, pine beetles are capable of killing entire populations of pine trees (Figure 2–14). Imagine the damage to the environment of a community when many of the trees in parks, yards, and streets are of a single variety that is susceptible to a highly destructive insect pest. When a tree or other plant is not affected by a harmful insect, it is sometimes described as being **immune** to the pest or organism. Many plants in the environment are immune to some insects, but they are often vulnerable to other insect species.

Not all insects are destructive to the environment. Some insects, such as bees and other pollinators, are useful to living environments. Without these insects to carry pollen from one flower to another, many

FIGURE 2–13 Cockroaches are serious household pests that may contaminate food or beverages. *(Courtesy USDA/ARS K7233-6)*

FIGURE 2–14 Pine beetles have destroyed a forest of pine trees that were first weakened by drought, and then infested with beetles.

FIGURE 2–15 Chemicals are needed in our modern society, but they may threaten the health of animals and people if misused or abused. *(Courtesy Getty Images)*

plants could not reproduce. Some insects, such as the lady bird beetle, prey on harmful insects. These insects play important roles in keeping populations of harmful insects in check. Insects do have profound effects on the living environments that surround us.

Chemicals

Chemicals can have both helpful and harmful effects on living environments (Figure 2–15). Oil spills and industrial chemical discharges have caused serious problems in oceans, lakes, rivers, and streams (Figure 2–16). Chemical pesticides continue to threaten wildlife, fish, shellfish, beneficial insects, microscopic organisms, plants, animals, and humans. In the middle 1960s, American biologist Rachel Carson shocked the world with her book, *Silent Spring*. This was one of the first books to provide convincing evidence of environmental damage being done by pesticides.

In 1972, DDT (dichloro-diphenyl-trichloroethane) was banned in the United States because of its damaging effects on the environment. This insecticide had been used to control mosquitoes, which carried the dreaded malaria organism. DDT was also a very effective chemical used against flies, and it enjoyed widespread use in homes, on farms and ranches, and wherever flies were a problem. Yet, because it was determined that DDT was responsible for interfering with the reproduction of birds by weakening the egg shells, it had to be discontinued; safer substitutes have been found.

It has been reported that over 10,000 different pesticides are registered for use in the states that surround one of our major coastal bays. Needless to say, careful management and control of so many different chemicals is absolutely essential. It requires the utmost care to avoid unacceptable damage to our environment.

FIGURE 2–16 Spills of petroleum or chemicals in water environments cause severe damage to populations of wild animals and plants. *(Courtesy Getty Images)*

OUR SHARED LIVING ENVIRONMENT

INTERNET KEY WORDS:

human population growth

All of the members of the plant and animal kingdoms, including humans, must share the living environments that are available on Earth. Some living environments are more friendly to the inhabitants than others, and plant and animal life tends to be concentrated in the warm, temperate regions of the world. Other environments, such as the Arctic and Antarctic regions, do not support the variety of plants and animals that are found in other places. The same is true of the high altitude, mountainous areas. One thing is certain: the living environments of the Earth will never grow any larger. In fact, the habitable regions may actually decline for many species of plants and animals as humans pave the Earth and create cities. Only a few species are able to adapt to the less favorable environments as they are crowded from their natural ranges.

As we ponder the life of "Baby 7 Billion" (see earlier), we must wonder if we are doing our part to preserve and enhance the environment. Plants, animals, insects, soil, water, and air must be kept in reasonable balance or all will suffer. Plants are generally considered to improve living environments, but excessive plant growth can infringe on the space for humans and animals. Excessive populations of humans and animals can damage a plant species until it is unable to adequately reproduce itself. Too many animals can compete excessively with humans for food, water, and space. Some species of insects are regarded as harmful by people because they feed on desirable crops or bother humans or livestock. However, many species of insects are beneficial to plants, animals, or humans.

Humans and animals tend to consume or remove plants, which hold soil in place and prevent erosion from wind and water (Figure 2–17). For instance, during the 1960s, most of the forests of China were cut and not replanted. Rapid and alarming soil erosion followed. The government then placed a high priority on reforestation and reversed the trend. Soil is needed to hold nutrients until plants require them. Similarly, we need the soil to filter and store clean water for plant growth and human and animal consumption. Plants take water from the soil and release water and oxygen to the air, which benefits humans and animals.

FIGURE 2–17 Plants are necessary to conserve our soil, especially in steep, highly erodable locations. *(Courtesy Getty Images)*

World Population Growth

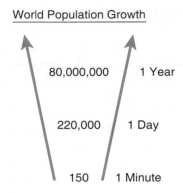

80,000,000 ⟋ 1 Year

220,000 / 1 Day

150 / 1 Minute

FIGURE 2–18 Earth's population is expected to reach 7 billion by 2010 and 8 billion by 2022.

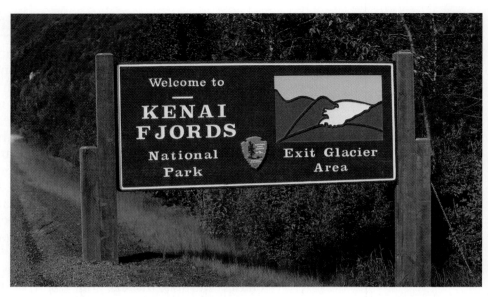

FIGURE 2–19 National parks and areas that are set aside as wilderness are intended to preserve the environment and the plant and animal species that live within it. *(Courtesy of PhotoDisc)*

The number of human beings in the world is growing at the rate of 150 every minute; 220,000 a day; 80 million a year (Figure 2–18). At this rate, the Earth's population will reach 7 billion by 2010 and 8 billion by 2022. Can Earth sustain such population growth? Will humans find enough to eat? Will we learn to protect our environment or will we destroy the system that supports life itself? Will we survive the competition of such population growth, but sacrifice our quality of life? Might we, in fact, improve our quality of life by using our intelligence to improve our environment?

Humans are the only living creatures who can choose to improve the living environments for themselves and other living organisms. The motivation to improve the environment is seldom present when people are hungry, however. As food production becomes more efficient, less of the total land area is required for the production of food. This allows some land areas and the living environments in the region to be preserved in their natural condition. For example, the national system of parks and monuments includes large areas where living environments support native plants and animals (Figure 2–19). Vast regions have been designated as "wilderness," and the type of human activity in these areas is strictly controlled to favor wild creatures. Humans also intervene in damaged environments to clean and restore them. The favorable economic conditions that are generally found in the United States have increased our ability and motivation to restore natural living environments and to preserve others.

AGRISCIENCE IN OUR GROWING WORLD

The keys to a prosperous future, indeed the bottom line for survival of the world's population, can be found in agriscience. Agriscience is the science of food production, processing, and distribution. It is the system that supplies fiber for building materials, rope, silk, wool, cotton, and

Agri-Profile

CAREER AREA:

Environmental Management

Management of the environment requires the attention of consumers, as well as professionals. However, specialists in air and water quality, soils, wildlife, fire control, automotive emissions, and factory emissions all help maintain a clean environment against tremendous population pressures in many localities. Helicopter, airplane, and satellite crews gather important data for scientific analysis to help monitor the quality of our environment.

Individuals in environmental careers may work indoors or outdoors; in urban or rural settings; or in boats, planes, factories, laboratories, or parks. Careers range from laborer to professional. Environmental concerns are high on global agendas today as nations attempt to reduce global hunger and pollution.

Environmental management requires skills in observation, analysis, and interpretation.
(Courtesy USDA/ARS #K-5184-1)

medicines. It provides the grasses and ornamental trees and shrubs that beautify our landscapes, protect the soil, filter out dust and sound, and supply oxygen to the air.

Agriscience accounts for 20 percent of jobs in the United States. It is the mechanism that permits the United States and other developed countries of the world to enjoy high standards of living. It is the system that nondeveloped countries are using in their efforts to feed and clothe their bulging populations. People look to agriscience for the necessary technology to compete on a par with other nations in the 21st century. We must look to agriscience to maintain and improve our quality of life.

The United States is a major world supplier of food. It is also a major supplier of fiber for clothing and of trees for lumber, posts, piling, paper, and wood products. The use of ornamental plants and acreage devoted to recreation has never been greater in the history of our country.

CHANGING POPULATION PATTERNS

The United Nations organization has reported that more children than ever before are surviving to adulthood. It also indicated that adults are living longer (Figure 2–20). Together these trends mean more population growth and more pressure on the environment. Advancements in medical science and services have made good health and longer lives a reality, but only for those who can afford good nutrition and modern health services. Similarly, through agriscience, we have made substantial gains in providing food, fiber, and shelter for the world (Figure 2–21). At the same time, the environment has stayed reasonably clean, considering the impact of bulging populations.

In the past, individuals younger than 25 years constituted the world's largest population group. This occurred because children were valued for the help they provided in making the family living. Children and young adults were engaged in a nation's labor force and provided

FIGURE 2–20 Worldwide child mortality rates and life expectancy estimates. *(Courtesy United Nation Fund for Population Activities)*

the manpower for armies. In most countries, the young respected their elders and provided for the needs of the elderly within the family.

The age profiles of people in developed countries are quite different from those of developing countries. Honduras has the traditional population pattern, with its largest number of citizens younger than 5 years. The number per age group then decreases to the smallest number, which occurs in the age group older than 80 years. When the Honduras population groups are displayed by sex in a bar graph, the graph takes the shape of a pyramid (Figure 2–22). Canada's pattern is slightly different. Its greatest population group is around the 20-year mark. Its graph reminds you of a Christmas tree, with its narrow bottom and cone appearance. Sweden, a country known for its excellent health services and high survival rate, has age brackets that are about equal. The population graph for that country resembles a column.

China has about one fifth of the world's population, yet it has been reasonably successful at feeding its population by keeping about 70 percent of its work force on farms. In contrast, less than 2 percent of the work force in the United States is necessary to operate the nation's farms.

In the mid-1970s, China implemented a policy whereby each couple was limited to one child. They called it the 4-2-1 policy. This means that extended families consisted of four grandparents, two parents, and one child. What would be the outcome if such a policy were strictly enforced for several generations? You would expect the pyramidal shape of

#1.
LABOR REQUIRED TO PRODUCE WHEAT, CORN, AND COTTON (IN HOURS)

	1800	1935-39	1955-59	1980-84	1990 or Later
Wheat (100 bu.)	373	67	17	7	7
Corn (100 bu.)	344	108	20	3	2.88
Cotton (1 bale)	601	209	74	5	5

#2.
YIELDS PER ACRE OF WHEAT, CORN, AND COTTON

	1800	1940	1960	1985-86	1990 (Prelim)
Wheat (bu.)	15	15	20	34	39.5
Corn (bu.)	25	29	55	118	118.5
Cotton (lb.)	154	253	446	630	640.0

FIGURE 2–21 Changes in agriscience productivity.

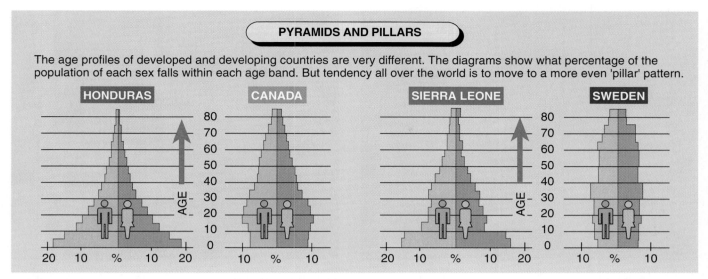

PYRAMIDS AND PILLARS

The age profiles of developed and developing countries are very different. The diagrams show what percentage of the population of each sex falls within each age band. But tendency all over the world is to move to a more even 'pillar' pattern.

HONDURAS | **CANADA** | **SIERRA LEONE** | **SWEDEN**

FIGURE 2–22 Age profiles and population patterns for developing and developed countries. *(Courtesy United Nations Fund for Population Activities)*

China's population graph to change to the shape of a Christmas tree, and, in time, to an upside-down pyramid! What would be the implications of feeding a nation with a population of mostly elderly people?

IMPACT OF AGRISCIENCE

History records little progress in agriculture for thousands of years. Then, starting in the early 1800s, the use of iron spurred inventions that revolutionized agriculture in the United States, British Isles, and northern Europe (Figure 2–23). However, for most of the world, progress has been much slower. In some nations, government leaders are slow to implement agriscience, because the nation initially would experience massive unemployment as machines displaced human labor.

Progress through Agricultural Engineering

Mechanization through inventive engineering was an important factor in the United States's agricultural development. The change from 90 percent to less than 2 percent of the workers being farmers evolved over a 200-year period. Machines helped make this possible. The old saying that "necessity is the mother of invention" suggests the relationship between an inventor's problem and the use of previously acquired skills to solve that problem. The solution is frequently a new device, machine, or process.

One of the most significant technologies to increase the efficiency of farm production was the generation and distribution of electricity to rural farming areas. Many of the labor-intensive jobs that were performed by hand 75 years ago are now performed by machines that are powered by electricity. Some examples include grain augers, milking machines, water pumps, fans, conveyor belts, power tools, and many other machines (Figure 2–24). The electric motor has revolutionized the world.

FIGURE 2–23 The inventions of the 1800s brought revolutionary changes in agriculture in the United States and Europe. *(Courtesy USDA 01di1473)*

FIGURE 2–24 Electric grain elevator.

FIGURE 2–25 Cyrus McCormick's reaper led to the development of the modern grain combine. *(Photo Courtesy of Elmer Cooper)*

INTERNET KEY WORDS:

rural electricity

American Inventors The United States has provided the inventors of many of the world's most important agricultural machines (Figure 2–25). In 1834, Cyrus McCormick invented the **reaper**, a machine to cut small grain. Later, a threshing device was added to the reaper, and the new machine was called a combine. The reaper cut and bundled the grain in the field. Today, grain is harvested with a machine called a **combine**, which cuts and threshes in a single operation. One modern combine operator can cut and thresh as much grain in 1 day as 100 individuals could cut and bundle in the 1830s.

Thomas Jefferson's invention of an iron plow to replace the wooden plow of the time was of great significance. Later, in 1837, a blacksmith named John Deere experienced the frustration of prairie soil sticking to the cast-iron plows of the time. It became apparent that Jefferson's invention would not work in the rich prairie soils of the Midwest. Through numerous attempts at shaping and polishing a piece of steel cut from a saw blade, the steel **moldboard plow** evolved. That plow permitted plowing of the rich, deep prairie soils of the great American West for agricultural production and launched the beginning of the John Deere Company.

In 1793, Eli Whitney invented the cotton gin. The **cotton gin** removed the cotton seeds from cotton fiber. This paved the way for an expanded cotton and textile industry. In 1850, Edmund W. Quincy invented the mechanical **corn picker**, which removed ears of corn from the stalks. During the same era, Joseph Glidden developed **barbed wire**, with sharp points to discourage livestock from touching fences. This effective fencing permitted establishment of ranches with definite boundaries. In 1878, Anna Baldwin invented a **milking machine** to replace hand milking (Figure 2–26). In 1904, Benjamin Holt invented the **tractor**, which became the source of power for belt-driven machines as well as for pulling.

INTERNET KEY WORDS:

history, agriculture, mechanized agriculture

Formation of Machinery Companies Many of the early inventors worked alone or with one or two partners. They were all workers in the

FIGURE 2-26 The invention of the mechanical milking machine has greatly reduced the amount of human labor required to care for dairy animals.

area of agricultural mechanics and, as such, in agriscience. By the early 1900s, the inventors or other enterprising people had formed companies to produce agricultural machinery or process agricultural products. This made invention a continuing process. Successive inventions were used to improve earlier inventions and to develop new equipment and supplies to meet the needs of a changing agricultural industry.

The development of mechanical cotton pickers and corn harvesters greatly expanded the output per farm. Significant expansion of U.S. agriculture also resulted from the development of irrigation technology. Since the end of World War II, the mechanization of U.S. agriculture has moved at a breathtaking pace (Figure 2–27).

Mechanizing Undeveloped Countries In the undeveloped countries of the world, many engineers, teachers, and technicians have sought simple, tough, reliable machines to improve agriculture. In such countries, the United States's highly developed, complex, computerized, and expensive machinery does not work for long. Most countries do not have people trained for the variety of agriculture mechanics jobs that are needed to support U.S. agriculture.

A machine with rubber tires is useless if a tire is damaged and repair services are not available. Similarly, failure of an electronic device may cause a $100,000 machine to become junk in the hands of an unskilled person in a country without appropriate repair facilities. This is the case in most undeveloped countries in Central and South America, Asia, and Africa. For the undeveloped nations of the world, other aspects of agriscience must become the vehicles for advancing agricultural productivity.

FIGURE 2-27 Since World War II, U.S. agriscience has progressed at a breathtaking pace. *(Courtesy USDA/ARS K5061-3)*

Improving Plant and Animal Performance

Humans have improved on nature's support of plant and animal growth since they discovered that the loosening of soil and planting of seeds could result in new and better plants. Even before that discovery, they probably aided plant growth by keeping animals away from them until fruit or other plant parts edible to humans were harvested.

The human touch has permitted plants and animals to increase production and performance to the point where less and less people are needed to produce the food supply for the United States. Surplus food is exported to many other nations.

One of the remarkable occurrences of the 20th century was the mechanization of agriculture. The many technologies that were developed for the agricultural industry have caused farms to become larger. Many people have been displaced from their family farms because they did not adopt the new technologies and farming practices that were needed to make their farms more efficient in a timely manner. Many of these people have learned trades other than farming and have become productive citizens in other industries. Without the farming revolution of the last 60 years, our citizens would not be free to pursue other occupations. The U.S. space program is possible only because our scientists do not have to produce their own food. The efficiency of U.S. farms has contributed to the freedom of our citizens to engage in many new and exciting occupations. These include the development of computers and other technologies that have resulted in the current "information age."

Bio-Tech Connection

AGRISCIENCE AND BIOTECHNOLOGY

Agriscience is heavily impacted by biotechnology, which addresses the continuation of life. In ornamental horticulture one uses a myriad of plants to beautify the interiors of homes, businesses, and institutions. Such plants also consume the carbon dioxide gas and supply oxygen for humans, animals, insects, and other living matter. In outdoor settings, ornamental trees, shrubs, and turfgrass grace our yards, streets, parks, and other public areas. Our highways rely on plants to screen off on-coming traffic, provide living hedges, absorb sound, prevent soil erosion, and create a stimulating environment to keep motorists alert. Further, fruits, vegetables, grains, and forage crops of gardens, ranches, and farms provide the backbone of the world's

Biotechnology provides valuable mechanisms to improve life. *(Courtesy USDA/ARS #K-5011-14)*

food supply. And, of course, trees provide wood, paper, and other fiber products. Plants reproduce both sexually and asexually through many different processes. The productive capability of these and other plants have been greatly improved through the efforts of scientists, technicians, and growers.

Similarly, many animal species have been modified over the centuries by humans through domestication, selection, breeding, and care. Currently, biotechnology is improving the productivity of plants and animals and providing new foods and medicines to enrich our lives. Genetic engineering enables humans to modify and utilize microorganisms in our fight against harmful insects and other pests. Today, the earth is providing food, shelter, habitat, health care, and other essentials to more people than at any time in history. However, there is much malnutrition and starvation in the world. The causes tend to be rooted in deficiencies in government, national infrastructure, poverty, and lack of education. The technology of food production is believed to be adequate to meet the world food needs if it could be applied at maximum levels.

Improving Life through Agriscience Research

Unlocking the Secrets of the Soybean Americans have long appreciated the extensive research on the peanut done by the American scientist George Washington Carver. Carver is credited with finding more than 300 uses for the peanut. These include food for humans, feed for livestock, cooking fats and oils, cosmetics, wallboard, plastics, paints, and explosives.

Less known are the secrets of the soybean. The Chinese have known for centuries that the soybean is a versatile plant with many uses. Calling it the "yellow jewel," the Chinese are said to have grown the soybean 3,000 years ago. The strong flavor of the soybean itself is not appealing, but the bean is a legume and is nutritious. A legume is a plant that hosts nitrogen-fixing bacteria. These bacteria convert nitrogen from the air to a form that can be used by plants. Legume plants are excellent sources of protein for humans and animals.

A Chinese scholar is believed to have first made tofu from soybeans in 164 BC. Tofu is a popular Chinese food made by boiling and crushing soybeans, coagulating the resulting soy milk, and pressing the curds into desired shapes. Today, tofu is a major food in the diet of China's huge population. It provides a reasonably healthful diet. Tofu can be fermented; marinated; smoked; steamed; deep-fried; sliced; shredded; made into candy; or shaped into loaves, cakes, or noodles.

Soy oil is the world's most plentiful vegetable oil. It is first extracted from the soybean, and the material that is left is processed into a protein-rich livestock feed known as soybean meal. The components of the soybean are used for hundreds of items. These range from food products, to lubricants, paper, chalk, paint, printing ink, and plastics (Figure 2–28). As early as 1940, Henry Ford evidently shocked journalists with an unusual demonstration. He slammed an ax into a Ford automobile trunk door made from highly resilient plastic. The new plastic was made from soybeans.

Baked Potatoes Many improvements in our way of life can be traced to agriscience research. For instance, the U.S. Department of Agriculture developed many pest-resistant varieties of potatoes. A case in point is the work with the Katahdin, a popular potato variety of the 1930s. From the Katahdin, scientists developed the BelRus, a superior baking variety bred to grow well in the Northeast. In a similar manner, the Russet potato grown in the volcanic soils of Idaho and the Northwest has been improved through research. Selection of parent stock has increased its resistance to diseases and insects, resulting in greater yields.

The Common Aerosol Before World War II, death from malaria was commonplace in the tropics. The deaths of U.S. soldiers from malaria triggered intensive research on the control of mosquitoes, the carrier of the malaria-causing organism. Development of the "bug bomb" resulted. Our present-day aerosol (a can with contents under pressure) resulted from that early research.

Turkey for the Small Family In your grandparents' time, Thanksgiving was probably observed by having all the relatives visit to consume the typical 30-lb turkey. As families became smaller and more scattered, the need for such large birds decreased; but even people with small families

INTERNET KEY WORDS:

soybean science

FIGURE 2–28 The soybean is the world's most important source of vegetable oil, and it provides the basic materials for hundreds of products. *(Courtesy of the American Soybean Association)*

INTERNET KEY WORDS:

biotechnology, agriculture

FIGURE 2–29 The Beltsville Small White turkey was developed to meet the needs of small families. *(Courtesy USDA/ARS)*

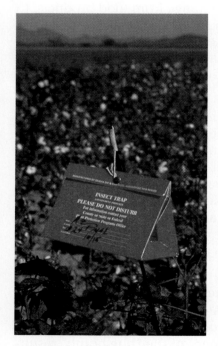

FIGURE 2–30 Attractants are valuable nonpolluting chemicals that lure insects to traps, bait, or systems of control through sterility. *(Courtesy USDA/ARS #K- 4113-1)*

liked turkey. The 30-lb bird was too much, so the problem was to develop a breed of turkey that weighed 8 to 12 lb at maturity (Figure 2–29). A solution was the Beltsville Small White turkey, named after the Beltsville Agricultural Research Center in Maryland, where the breed was developed. Further research and development has yielded meat animals with high yields of lean meat and less fat.

The Green Revolution During the 1950s, starvation was rampant in many countries of the world. A major question was: "Could the world's agriculture sustain the new population growth?" The solution was partly in the development of new, higher yielding, disease- and insect-resistant varieties of small grain for developing countries. The result was the Green Revolution, a process whereby many countries became self-sufficient in food production in the 1960s by using improved plant varieties and proven management practices.

Cultivated Blueberries Wild blueberries were enjoyed in early times when people had time to pick the tiny berries growing in the wild. But labor costs became too high to harvest such berries for sale. The solution was development of high-quality, large-fruited blueberry varieties from the wild. This started the new and valuable cultivated-blueberry industry of today.

Nutritional Values Until recently, feeding of animals and human nutrition were based on poor methods of feed and food analysis. The problem was: How can one recommend what to feed or what to eat if the content of food for humans and crops for livestock cannot be accurately determined? The solution was to develop detergent chemical methods for determining the nutritional value of feedstuff (any edible material used for animals). The procedures are now used widely throughout the world in both human and animal nutrition.

Biological Attractants The use of chemical pesticides provided short-term solutions to many insect-control problems. However, it has become apparent that chemicals have some disadvantages and additional means of control must be found. A partial solution was to discover chemicals that insects produce and give off to attract their mates (Figure 2–30). These are produced in the laboratory. Laboratory production of these chemicals has permitted mass trapping of insects to survey insect populations for integrated pest-management programs.

Recent Breakthroughs in Agriscience

Genetically Engineered Tomato Calgene, an agricultural biotechnology firm in Davis, California, developed a bio-engineered tomato that resists rotting. The new tomato was developed by turning off the gene that causes the tomato to soften and rot. The new tomato lasts longer on the shelf at the grocery store, retains its flavor longer, and tastes how a tomato should taste.

Natural Rubber Production Scientists at the USDA Western Regional Research Center at Albany, California, have modified a scrubby bush called "guayule" by genetically engineering methods to produce up to 1,000 kilograms of rubber per hectare from the plant (an increase from 200 kg/hectare from native plants). The plant looks like sagebrush.

Hot Topics in Agriscience

GENETICALLY MODIFIED FOODS

Humans have been modifying the genes of plants and animals for centuries using a technique called selective breeding. Plants or animals that exhibit desired genetic characteristics are selected as the parents of the next generation. For example, if a scientist wanted a wheat plant that was resistant to a disease, he or she would plant a field of wheat and expose it to the disease. Then the researchers would find the plants that survived the disease and use them to parent the next generation. This process is then repeated until a variety is found that has a high resistance to the disease. This process is difficult and can take years.

Genetic engineering is a technique that allows scientists to physically put certain genes into the cells of a plant or animal without the rigorous process of selective breeding. As a result, genetic engineering is much faster than selective breeding. This technique has revolutionized agriculture. One common genetically modified crop is herbicide-resistant wheat. A gene, which is resistant to the deadly effects of plant-killing herbicides, is positioned into the DNA of the wheat. As a result, a farmer can spray an entire field with an herbicide, killing only the weeds and not the valuable crops. Genetic modification is a powerful tool that has resulted in the improvement of a number of agricultural products.

This new advancement does not come without controversy. The United Kingdom, as well as other countries, has placed bans on genetically modified crops. Activist groups, together with media, have spurred the governments of these countries to halt the sale of genetically modified products because of perceptions that the products are not safe for people to use. The controversy over genetically modified food is not isolated to Europe. In the United States, from 2001 to 2002, almost 160 pieces of legislation relating to genetic engineering were introduced in 39 states. Oregon voters turned down a bill that would require all genetically modified food to be reported on the food labels. In 2002, Maryland placed a ban on all genetically engineered fish. This controversy may become the biggest struggle faced by the agriscience community in this and future decades.

INTERNET KEY WORDS:

mastitis, animals
genetically modified food

This new technology makes it possible to produce a domestic supply of natural rubber in the United States.

Bio-engineered Designer Foods By altering the genetic structure of food products, scientists have created new foods such as crispy vegetables, sweeter carrots, leaner meats, high-protein milk, longer lasting melons, and healthier cooking oils.

Monoclonal Antibodies in Goats' Milk Monoclonal antibodies are natural substances in blood that fight diseases and infections. Transgenic goats have been developed by inserting a gene into the goats' DNA, causing them to produce and secrete up to 4 grams of the monoclonal antibody in each liter of milk. This level of antibody production is 10 to 100 times greater than traditional methods of production from cell cultures. This early work with transgenic goats produced an anti-cancer antibody.

Bio-diesel from Animal Fat Excess animal fat (tallow) that is trimmed from the carcasses of meat animals is a low-value by-product of the meat processing industry. A process has been developed that converts tallow to bio-diesel, a product very much like the diesel fuel extracted from crude oil (Figure 2–31). The fat is heated to a liquid, followed by a purification process. The purified fat product is then mixed with methyl alcohol and a chemical catalyst. The bio-diesel that results from

FIGURE 2–31 A form of bio-diesel made from animal fat shows promise as a fuel for diesel engines.

the process has approximately the same heating value and power potential as traditional diesel fuel, and it will burn in an ordinary diesel engine.

Mastitis Reduced The mastitis organism has always been a serious problem for dairy farmers. Mastitis is an infection of the milk-secreting glands of cattle, goats, and other milk-producing animals. The resulting loss of milk production adds millions of dollars yearly to the cost of milk in the United States. Recent research efforts resulted in the development of abraded plastic loops for insertion into cow udders. The procedure resulted in a reduction in clinical mastitis of 75 percent. The reduction in infections resulted in increased milk production, averaging nearly 4 lb of milk per cow per day.

Human Nutrition Studies in human nutrition have demonstrated the benefits of decreasing the amount of animal fat in the diet, and increasing the proportion of fat from vegetable sources. This practice reduces high blood pressure and risk for heart attack. Much of the progress in human nutrition has grown out of research on animals and plants by agriscientists. Research on human nutrition has yielded new recommendations for healthful eating.

INTERNET KEY WORDS:

fire ant control

Fire Ant Control Fire ants infest 230 million acres in the southern areas of the United States. Their presence in the warmer climates of the world is a constant threat to the well-being of humans and livestock. A new synthetic control for fire ants increases the ratio of nonproductive drone ants to worker ants. This ratio change gradually weakens the colony and causes it to die.

New Hope for Coccidiosis Control Coccidiosis is a disease that costs poultry growers nearly $300 million per year in the United States alone. Recently, the USDA and private industry genetically engineered a para-

FIGURE 2–32 Infrared photos taken from satellites help diagnose problems such as ant infestations in fields and pastures. *(Courtesy USDA/ARS #K-3957-11)*

INTERNET KEY WORDS:

hybrid rice production

INTERNET KEY WORDS:

agricultural discoveries

site constituent that stimulates birds to develop immunity to coccidiosis. Hopefully, this will be the first step in the process of developing an effective vaccine against this persistent pest.

Exotic Flowers Horticulturists, gardeners, and hobbyists will be delighted with the new varieties of Impatiens (a popular, easy-to-grow, summer-flowering plant). Plant explorers introduced exotic new germ plasm, and plant breeders developed a new technique called ovuleculture to develop hybrids and new kinds of Impatiens. A hybrid is the offspring of a plant or animal derived from the crossing of two different species or varieties.

Satellites and Nitrogen Gas Lasers Nutrient deficiencies in growing corn and soybean crops are not easy to detect from the ground. A deficiency occurs when a nutrient is not available in the amounts that are needed for optimum growth. New technology now permits the monitoring of deficiencies of iron, nitrogen, potassium, and other nutrients using nitrogen gas lasers (devices used to determine wavelengths given off by the plants) from satellites. These wavelengths indicate the levels of various nutrients in plants (Figure 2–32).

Sugar Beet and Rice Hybrids The development of new varieties is a technique that has been used in agriscience for many decades to improve plant performance. A recent breakthrough has provided a sugar-beet hybrid with a high ratio of taproot weight to leaf weight. The hybrid yields about 15 percent more sugar per acre than previous varieties.

On the other side of the world, Chinese agronomists developed hybrid rice. Hybrid rice is capable of yielding up to 40 percent more rice per acre than traditional varieties. The combination of hybrid semi-dwarf rice varieties, improved irrigation, and chemical fertilizers has increased world rice production from 240 million metric tons in the 1960s to more than 500 million metric tons in the 1990s.

A study of USDA publications reveals great numbers of improved varieties, new products, and superior processes that have been discovered or developed through agriscience research.

AGRISCIENCE AND THE FUTURE

By 2005, the average American farmer was capable of producing enough food and fiber for approximately 140 people. Agriscience will become even more important in the next 100 years. As the world's population increases, it will require a highly sophisticated agriscience industry to provide the food, clothing, building materials, ornamental plants, recreation areas, and open-space needs for the world's billions. Americans will have to work more in the international arena, as more countries become highly competitive in agriscience and trade barriers are removed. Research and development will continue to play a dominant role as they lead the way in agriscience expansion in the future.

The USDA has developed the following mission statement to guide the agency: To enhance the quality of life for the American people by supporting production of agriculture:

* ensuring a safe, affordable, nutritious, and accessible food supply;
* caring for agricultural, forest, and range lands;

- supporting sound development of rural communities;
- providing economic opportunities for farm and rural residents;
- expanding global markets for agricultural and forest products and services; and
- working to reduce hunger in the United States and throughout the world.

The new century and millennium bring a new set of challenges to U.S. and world agriculture. The international business economy is a dominant factor in marketing agricultural products. A major share of U.S. agricultural production is now consumed in foreign countries, and this trend is expected to increase the volume of agricultural commodities that are sold outside of U.S. borders. Canada and Mexico have become two of our biggest markets. They have also become successful competitors with the United States for a world market share in some agricultural commodity markets.

The future of U.S. agriculture will require that farmers become even more efficient in the production of food and fiber crops. Animal agriculture will depend on scientific improvements in production methods and in the genetic superiority of food animals to improve the quality of animal products. The efficiency with which they are produced must also improve for animal products to continue to demand a strong market share of the food supply in a world crowded with humans.

STUDENT ACTIVITIES

1. Write the Terms to Know and their meanings in your notebook.
2. Develop a bulletin board that illustrates the components of our environment.
3. Collect newspaper articles that describe environmental problems in your community.
4. Prepare a two- or three-page paper describing a good environment in which to live. Include factors such as home, community, air, water, cleanliness, wildlife, plants, and animals.
5. Ask your teacher to invite a public health official to your class to discuss health problems in the community and how they could be reduced by improving the environment.
6. Draw a chart that illustrates some relationships among plants, animals, trees, soil, water, air, and people.
7. Look up three prominent U.S. inventors and describe the events that led to the inventions that made them famous.
8. Make a model of one of the machines that strongly influenced agriscience development.
9. Make a collage depicting some important discoveries, inventions, and developments in agriscience.
10. Assume that a bar graph depicting China's population for 1975 was pyramid-shaped, similar to that of Honduras. Make a bar graph representing what the population pattern will look like after two generations of 4-2-1 families.
11. Ask your teacher to arrange a field trip to study the variety of living environments in your community.
12. In groups of three or four students, develop an idea for a new piece of farm equipment that will improve the harvest or production of a locally grown crop, farming technique, or ranching practice.

SELF EVALUATION

A. Multiple Choice

1. To be regarded as safe, a sewage system must
  a. be connected to a city system.
  b. be constructed from concrete block.
  c. decompose human waste.
  d. discharge into a stream or river.

2. Safe water is
  a. any water pumped from wells.
  b. water collected from a roof.
  c. free of harmful chemicals and organisms.
  d. water taken from free-flowing rivers.

3. Apartments on one level in large buildings and owned by the residents are called
  a. condominiums.
  b. single houses.
  c. townhouses.
  d. villas.

4. The world's food production capability indicates that
  a. it could feed itself in theory, but not in practice.
  b. it could probably never keep up with population growth.
  c. food supplies outpace demand, and reduced production is recommended.
  d. widespread famine could not be helped by better distribution.

5. Contaminants of food and water include
  a. registered pesticides.
  b. contact by cockroaches.
  c. feces and urine.
  d. all of the above.

6. Although plants cannot move to food and water, they survive because of
  a. their capacity to reproduce.
  b. their ability to survive without food and water.
  c. their roots, which extract water from any material.
  d. parasites that convert water to nutrients.

7. The world's population is projected to increase to 7 billion people by
  a. 2000.
  b. 2010.
  c. 2022.
  d. 2034.

8. Agriscience in the United States
  a. accounts for 20 percent of the jobs.
  b. is likely to diminish in importance.
  c. reduces our standard of living.
  d. is being replaced by biotechnology.

9. Agricultural development would be described as
  a. rapid until about 1850.
  b. slow in most countries.
  c. very rapid in the United States in recent years.
  d. rapid in the past, but decreasing today.

10. The inventor of the iron plow was
  a. Cyrus McCormick.
  b. John Deere.
  c. Joseph Glidden.
  d. Thomas Jefferson.

11. The combine is a combination of the reaper and a
  a. corn picker.
  b. cotton gin.
  c. cotton picker.
  d. threshing device.

12. "Yellow jewel" is the name given by the Chinese to
    a. a special type of horse.   c. garden peas.
    b. a very young emperor.   d. soybeans.

13. Beltsville Small White is a
    a. breed of rabbit.   c. type of building.
    b. breed of turkey.   d. variety of soybean.

14. The great advance in world food production in the 1960s was called the
    a. biological attractants.   c. Greening of America.
    b. Green Revolution.   d. Great Leap Forward.

B. Matching Group 1

_____ **1.** Neighborhood   a. Release oxygen into the air
_____ **2.** Starvation   b. A priority in China
_____ **3.** Parasite   c. Registered pesticides in one bay area
_____ **4.** Immune   d. Part of a community
_____ **5.** DDT   e. Not harmed by
_____ **6.** 10,000   f. Famine
_____ **7.** Cockroach   g. Lives on another organism
_____ **8.** Reforestation   h. Banned insecticide
_____ **9.** Plants   i. Feeds on human waste and food

Matching Group 2

_____ **1.** Aerosol   a. Sense nutrient deficiencies
_____ **2.** Barbed wire   b. "Bug bomb"
_____ **3.** Coccidiosis   c. Nitrogen fixation
_____ **4.** Cotton gin   d. Disease of poultry
_____ **5.** Impatiens   e. Sharp points to discourage livestock
_____ **6.** Laser   f. Colorful flower
_____ **7.** Legume   g. Eli Whitney
_____ **8.** Low-input agriculture   h. Curd from soybeans
_____ **9.** Mastitis   i. John Deere
_____ **10.** Milking machine   j. New research objective
_____ **11.** Steel moldboard plow   k. Cyrus McCormick
_____ **12.** Reaper   l. Infection of milk-secreting glands
_____ **13.** Tofu   m. Anna Baldwin

UNIT
3
Biotechnology

OBJECTIVE

To examine elements of biotechnology.

MATERIALS LIST

✓ paper
✓ pencil or pen

✓ encyclopedias
✓ agriscience magazines

✓ Internet access

COMPETENCIES TO BE DEVELOPED

After studying this unit, you should be able to:

- define biotechnology, DNA, and other related terms.
- compare methods of plant and animal improvement.
- discuss historic applications of biotechnology.
- explain the concept of genetic engineering.
- describe applications of biotechnology in agriscience.
- state some safety concerns and safeguards in biotechnology.

SUGGESTED CLASS ACTIVITIES

1. Introduce this unit by serving one or more small servings of bread, yogurt, or cheese to class members. Explain to the class that each of these foods is a product of biotechnology, because living organisms act on the food to preserve it. Name as many food products as you can that are products of biotechnology.

2. Describe the controversy concerning the use of biotechnology methods to improve the productivity of food plants and animals. Identify some Web sites on the Internet where students may learn more about the production of food by plants and animals that have been improved using genetic engineering techniques.

3. Scientists have a working knowledge of the cloning process. Recently, scientists have cloned equine. Now that it is possible to clone a mammal by replicating its chromosomes, conduct a class discussion of ethics in biotechnology.

45

TERMS TO KNOW

bio	progeny	Cytosine (C)	clone
improvement by selection	deoxyribonucleic acid	Thymine (T)	genetic engineering
selective breeding	(DNA)	mapping	ice-minus
genetics	nucleic acid	gene splicing	X-Gal
heredity	base	recombinant DNA	bovine somatotropin (BST)
gene	Adenine (A)	technology	porcine somatotropin (PST)
generation	Guanine (G)	gene mapping	insulin

Biotechnology has become an important tool in agriscience. It promises unprecedented advancements in plant and animal improvement, pest control, environmental preservation, and life enhancement. However, there are real dangers that this new power over life processes can lead to unmanageable consequences in careless, uninformed, or criminal hands. Therefore, governments, scientists, agencies, corporations, and individuals have moved cautiously in the pursuit of new benefits through biotechnology.

Bio means life or living, therefore, biotechnology is the application of living processes to technology. Although many definitions abound for biotechnology, one of the more popular definitions is the use of microorganisms, animal cells, plant cells, or components of cells to produce products or carry out processes with living organisms.

HISTORIC APPLICATIONS OF BIOTECHNOLOGY

Living organisms have been used for centuries to alter and improve the quality and types of food for humans and animals. Examples include the use of yeast to make bread rise, bacteria to ferment sauerkraut, bacteria to produce dozens of types of cheeses and other dairy products, and microorganisms to transform fruit and grains into alcoholic beverages (Figure 3–1). Similarly, green grasses and grains have been stored in airtight spaces and containers, such as silos, where bacteria convert sugars and starches into acids. The acids provide a desirable taste and protect the feed from spoilage by other microorganisms. The converted feed is called silage (Figure 3–2).

IMPROVING PLANT AND ANIMAL PERFORMANCE

Humans have improved on nature's support of plant and animal growth since they discovered that the loosening of soil and planting of seeds could result in new plants. Even before that discovery, they probably aided plant growth by keeping animals away from plants until they yielded fruit or other plant parts that were edible by humans.

FIGURE 3–1 Many foods owe their texture and taste to micro-organisms such as yeast. *(Courtesy USDA/ARS #K36072-2)*

FIGURE 3–2 Silage consists of grains and/or green plant material preserved by the action of bacteria in an airtight environment. *(Courtesy National FFA; FFA #167)*

FIGURE 3–3 Improvement by selection means picking the best specimens for breeding purposes. *(Courtesy National FFA; FFA #162)*

INTERNET KEY WORDS:

Mendel, genetics

Improvement by Selection

History documents the domestication of the dog, horse, sheep, goat, ox, and other animals thousands of years ago. Improvement by selection soon followed. **Improvement by selection** means picking the best plants or animals for producing the next generation (Figure 3–3). As people bought, sold, bartered, and traded, they were able to get animals that had desirable characteristics, such as speed, gentleness, strength, color, size, milk production, and the like. By mating animals with characteristics that humans preferred, the offspring of those animals would tend to exhibit the characteristics of the parents and further intensify the desired characteristics. By accident, the owner was practicing **selective breeding**, or the selection of parents to get desirable characteristics in the offspring.

The chariot armies of the Egyptians and Romans, the might of the Chinese emperors, the speed of the invading barbarians into northern Europe, the strength of mounts carrying armored knights into battle, and the evasive Arabians of the desert, all provide convincing testimony to early successes at breeding horses for specific purposes.

Improvement by Genetics

An Austrian monk named Gregor Johann Mendel is credited with discovering the effect of genetics on plant characteristics. **Genetics** is the science of heredity. **Heredity** is the transmission of characteristics from an organism to its offspring through genes in reproductive cells. **Genes** are components of cells that determine the individual characteristics of living things. Mendel experimented with garden peas. He observed there was definitely a pattern in the way different characteristics were passed down from one generation to another. **Generation** refers to the offspring, or **progeny**, of common parents.

In 1866, Mendel published a scientific article reporting the results of his experiments. He had discovered that certain characteristics occurred in pairs, for example, short and tall in pea plants. Furthermore, he observed that one of those characteristics seemed to be dominant over

INTERNET KEY WORDS:

genetic engineering

Agri-Profile

CAREER AREAS:

Genetic Engineering

Genetic engineering cuts across many fields of endeavor. Procedures for genetic modification of organisms have been developing for more than a decade. Biologists, microbiologists, plant breeders, and animal physiologists are some examples of specialists who might use genetic engineering in their work. The work settings include the field, laboratory, classroom, and commercial operations.

People involved in genetic engineering usually have advanced degrees and they are highly specialized in a narrow area of research, such as cellular biology. Others may work in applied research in areas such as weevil control in small grains, nutrient requirements of small grains, or reproductive problems in dairy cattle. Others may be crop, animal, or pest control technicians who help manage the plants or animals that are the subjects of research. Still others may work in laboratories and devote most of their time to analysis and observation.

Because genetic engineering is a relatively new field and the applications are so numerous, the opportunities will expand as the field develops.

Using a DNA probe, an animal physiologist examines film showing gene patterns of various animals. *(Courtesy of USDA/ARS #K-1968-13)*

FIGURE 3–4 Mendel's extensive experimentation, observation, and record keeping provided the foundation for the modern science of genetics. *(Courtesy National FFA; FFA #199)*

the other. If tall was the dominant characteristic, then tall plants crossed with tall or short plants produced mostly tall plants; but some plants would be short. It was observed that the short characteristic could be hidden in tall plants in the form of a recessive gene. Such recessive genes could not express themselves in the form of a short plant unless both genes in the plant cells were the recessive genes for shortness. He also observed that short plants crossed with short plants always had short plants as offspring. This happened because there were no tall characteristics in either parent to dominate the characteristic of the offspring.

Mendel's work provides an excellent example of the power of the written word. His discoveries and conclusions would have been lost if they had not been recorded. The usefulness of his discoveries was not recognized until long after his death. In 1900, other scientists reviewed his writings and built upon the observations and conclusions he had reported. Today, biologists credit his work as being the foundation for the scientific study of heredity (Figure 3–4). Principles of heredity apply to animals as well as plants. More information on the principles of heredity may be found in subsequent units of this text.

DNA—GENETIC CODE OF LIFE

Of the estimated 300,000 kinds of plants and more than 1 million kinds of animals in the world, all are different in some ways. Conversely, plants and animals have certain similar characteristics that lend themselves to classification and permit prediction of characteristics of offspring by viewing

Science Connection

SCIENCE CONNECTION

Trait Predictability

A *trait* is another word used to describe a characteristic of an organism. Mendel discovered that parents pass their traits to their offspring. In this type of reproduction, each parent contributes one half of the genes. The result of parental combinations can be shown and predicted in a Punnett square. The Punnett square represents the results of some of Mendel's experiments. An allele is the different form in which genes can occur. In this example, there are three possible alleles. (1) "TT" represents a pea plant that is tall. Homozygous means that an organism has the same alleles for a given trait. (2) "tt" represents a homozygous pea plant that appears short. (3) "Tt" represents a heterozygous pea plant that is tall. The gene for "tall" is expressed because "T" is dominant over "t." Because of this dominance, both "TT" and "Tt" will result in a tall plant. "tt" will result in a short plant. Heterozygous means that an organism has different alleles for a given trait. Because the alleles for this trait are heterozygous, or different, this organism is called a hybrid. The Punnett square cross is done one box at a time; by taking the parental alleles to the left and top of each box, the offspring allele is found inside the four boxes.

Tall pea plant

		T	t
Tall pea plant	T	**TT** Tall pea plant Homozygous	**Tt** Tall pea plant Heterozygous
	t	**Tt** Tall pea plant Heterozygous	**tt** Short pea plant Homozygous

the parents. That is, the individual fertilized cell, called the embryo, contains coded information that determines what that cell and its successive cells will become. The coded material in a cell is called DNA. DNA is an acronym for deoxyribonucleic acid.

It is believed that a universal chemical language unites all living things. It was observed in the early 1800s that all living organisms are composed of cells, and that cells of microscopic organisms, as well as larger plants and animals, are basically the same. Then, in 1867, Friedrich Meischer observed that the nuclei of all cells contain a slightly acidic substance. He named the substance nucleic acid. Later, the name was expanded to deoxyribonucleic acid, or DNA. DNA in all living cells is similar in structure, function, and composition, and is the transmitter of hereditary information. A gene is a small section of DNA that is responsible for a trait. Chromosomes are rod-like structures made of DNA and other substances that hold genes.

DNA occurs in pairs of strands intertwined with each other and connected by chemicals called bases. The pairs of DNA strands may be likened to the two sides of a wire ladder. The bases may be likened to the rungs of that wire ladder. The different bases are: (1) Adenine, (2) Guanine, (3) Cytosine, and (4) Thymine. The first letters of each name of the bases—A, G, C, and T,—have become known as the genetic alphabet of the language of life (Figure 3–5).

If one end of the wire ladder is held, whereas the other end is twisted, the resulting shape would be called a double helix. This is the shape of DNA strands in a cell. Two strands of DNA and the bases between the strands compose a specific gene (Figure 3–6). The order or sequence of the bases between the DNA strands is the code by which a gene controls

DNA Strand   DNA Strand

Bases

FIGURE 3–5 The components of DNA may be likened to a ladder with very close rungs.

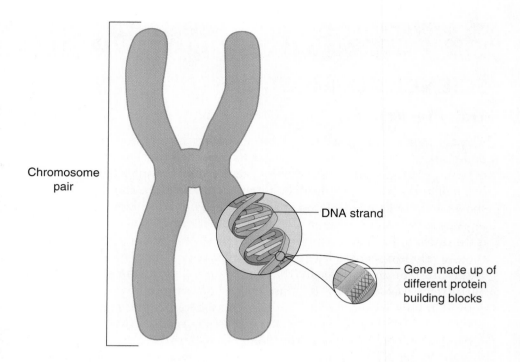

Chromosome pair

DNA strand

Gene made up of different protein building blocks

FIGURE 3–6 A gene is a small segment of DNA located at a specific site on a chromosome that influences or controls a hereditary trait.

a specific trait. Therefore, each rung with its accompanying side pieces of DNA constitutes a gene containing the genetic code to a single trait (Figure 3–7). The genetic material in the cells of a given microbe, plant, animal, or human can be isolated and observed. The trait or traits that a given gene controls can be identified and/or the combination of genes that influence a single trait can be determined.

Some examples of individual traits are hair color, tendency for baldness in humans, height of plants at maturity, and tendency of females to have twin offspring.

As cells divide, the DNA strands separate from each other and create duplicate strands to go to the new cells. Therefore, the genetic codes are duplicated and passed on from old cells to new cells as growth occurs and individuals reproduce (Figure 3–8). The process of identifying the location of a specific gene on a chromosome is called **mapping**.

Scientists can now remove individual genes carrying certain genetic information and replace them with genes containing other genetic instructions. By doing so, a given characteristic or performance can be altered in the microorganism or other plant or animal. For instance, plants that are susceptible to being eaten by certain insects may be altered so they will have resistance to that insect. The process of removing and inserting genes into DNA is called **gene splicing**, or **recombinant DNA technology**. The process of finding and recording the location of genes is called **gene mapping**.

Scientists have a working knowledge of how genetic information is stored in a cell, duplicated, and passed on from cell to cell as they divide and the organism forms. Furthermore, the process of transmitting genetic codes from parents to offspring and from parent to clone is common scientific knowledge. A **clone** is an exact duplicate of something. A major breakthrough was made in the early 1980s as scientists developed the process of genetic engineering. **Genetic engineering** refers to the movement of genetic information in the form of genes from one cell to another.

INTERNET KEY WORDS:

DNA, genetic code

DNA

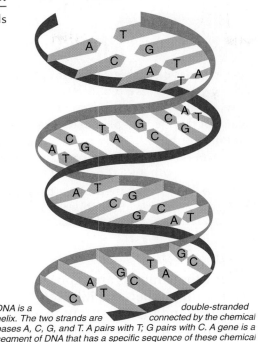

DNA is a double-stranded helix. The two strands are connected by the chemical bases A, C, G, and T. A pairs with T; G pairs with C. A gene is a segment of DNA that has a specific sequence of these chemical base pairs.

FIGURE 3–7 A chromosome is a structure that holds the genetic information of a cell. The DNA is wound tightly with other substances to form the chromosome. A gene is made of DNA. There are thousands of genes on each strand of DNA.

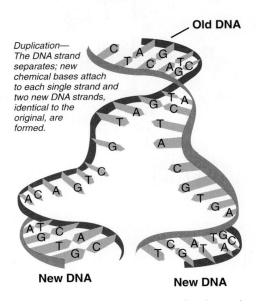

Duplication— The DNA strand separates; new chemical bases attach to each single strand and two new DNA strands, identical to the original, are formed.

Old DNA

New DNA **New DNA**

FIGURE 3–8 DNA strands divide and bases attach themselves to the new strands to form identical genes for new cells.

IMPROVING PLANTS AND ANIMALS

Scientists have learned to improve plants, animals, and microbes by manipulating the genetic content of cells. This procedure permits more choices for the researcher and more rapid observation of results. With the manipulation of cellular material, the scientist can now alter the characteristics of microorganisms, as well as those of larger plants and animals. This new capability has some amazing implications for human efforts to improve the quality of life.

In 1988, California scientists made the first outdoor tests of a product called ice-minus. **Ice-minus** are bacteria that have been genetically altered to retard frost formation on plant leaves. Synthetic chemicals are now available to protect fruit crops when temperatures fall 4 to 6 degrees below what would normally damage the fruiting process. Similarly, in 1988, genetically altered bacteria were injected into elm trees in an effort to control the deadly Dutch Elm disease. Furthermore, bacteria were genetically engineered so they turn a brilliant shade of blue in the presence of a compound called **X-Gal**. Such bacteria can be easily detected and traced in experimental or real-life situations. The ability to mark microorganisms has made biotechnology a safer and more manageable enterprise.

In animal science, the hormone **bovine somatotropin (BST)** has long been known for its stimulation of increased milk production in cows. However, it was not available for commercial use until bacteria were altered to produce the hormone. Another example of hormone production by genetically altered bacteria is an animal hormone called **porcine somatotropin (PST)**, which increases meat production in swine.

Hot Topics in Agriscience

TRANSGENIC ANIMALS—A NEW KIND OF FARMING

Transgenic animals have been modified using genetic engineering methods to contain genes that are not naturally found in the animal. Because the new gene is inserted into the chromosomes of the transgenic animal, it may be passed on to its offspring. Insulin and growth hormone are some of the first products for human use to be produced in this way. Most of the medical products produced by transgenic animals consist of proteins that are separated from milk or blood. These products include Human Protein C, an anticlotting protein that dissolves blood clots in humans. Other products include hemoglobin, a blood substitute produced in transgenic pigs, and Factors VIII and XI, which cause clotting in human blood to stop severe bleeding disorders such as hemophilia. All of these drugs are tested to ensure they are safe. The transgenic animal becomes a "living drug factory" that produces human proteins in milk or blood.

Any time that animals are exposed to a disease, there are individual animals that do not become infected. Sometimes an entire population of animals is found to be resistant to a disease that is highly contagious to other animal populations of the same species. In some instances, the disease resistance is because of a gene that has mutated or changed. It is now possible to identify the location of a resistant gene on a chromosome and to isolate it. This new genetic material can be transferred successfully to the chromosomes of an animal that is susceptible to the disease. Such a genetically altered animal is capable of passing the disease resistance to its offspring.

It is also possible to alter the genetic material in plants to improve them. For example, the Colorado potato beetle is a highly destructive insect that can completely destroy the plants in a potato field. A gene has been identified that causes plants to produce a substance in the leaves that is toxic to the potato beetle. New potato varieties were created by inserting this gene into the DNA of commercial potato varieties. The new potato variety kills the beetles when they eat the leaves of the plants. No risk to humans has been found or demonstrated.

It is now evident that genetic engineering and other forms of biotechnology hold great promise in controlling diseases, insects, weeds, and other pests. The plants and animals that we nurture for food, fiber, recreation, and preservation, can benefit from biotechnology as well as humans. The environment will be enhanced by less frequent use of chemical pesticides and greater use of biological controls (Figure 3–9).

INTERNET KEY WORDS:

potato beetle
genetic engineering

SOLVING PROBLEMS WITH MICROBES

Microscopic plants and animals lend themselves well to genetic engineering. Microbes reproduce quickly and can be genetically engineered to produce products needed by other plants, animals, and humans. One of the first commercial products made by genetic engineering was insulin. Insulin is the chemical used by people with diabetes to control their blood sugar levels. Previously, insulin was available only from animal pancreas tissue. It was in short supply, and was very expensive.

PLANT GENETIC ENGINEERING

Regenerated whole plant and its progeny carry the antibiotic resistant trait. Though antibiotic resistant plants are not commercially attractive, this kind of genetic engineering system could be used to make plants resistant to drought, salty soil, or insects.

FIGURE 3–9 Plant genetic engineering is becoming commonplace in the battle against diseases, insects, weeds, and other pests.

However, a bacterium called *Escherica coli* was genetically engineered to produce insulin much like cows produce milk and bees produce honey (Figure 3–10).

WASTE MANAGEMENT

Environmental pollution and the elimination of waste products from home, business, industry, utility, government, military, and other sources has become a major problem throughout the world. Landfills are becoming full and causing problems of leakage into the groundwater, old dump sites are creating new problems, waste is piling up, and sewage and chemical disposal is a constant problem. Environmental laws mandate reductions in solid waste disposal of 40 percent or more in comparison with 1990.

Biotechnology is being used to help solve waste disposal problems. Genetically altered bacteria are used to feed on oil slicks and spills to transform this serious pollutant into less harmful products. Similarly, bacteria have been developed to decompose or deactivate dioxin,

INTERNET KEY WORDS:

clone, Dolly

HUMAN INSULIN PRODUCTION

FIGURE 3–10 As a result of genetic engineering of bacteria, insulin is now readily available and relatively inexpensive.

Bio-Tech Connection

CLONING OF "DOLLY"

One of the most remarkable scientific accomplishments of the 20th century was the successful cloning of "Dolly" the sheep. Dolly was cloned in the laboratory from a single cell obtained from the udder of another sheep. A clone is an exact genetic copy of another living organism. Dolly is genetically identical to the sheep from which the cell was obtained. Scientists treated the cell in the laboratory to make it react like a fertilized egg cell. It was placed in the uterus of a female sheep where it developed into a lamb. The cell divided and developed just as a fertilized egg would do in a normal pregnancy. What is the significance of this event? It is now possible to produce a new generation of sheep (or other mammals), with each animal identical to the most productive animal in the herd. Imagine having a whole herd of cows just like your best cow. Confusing, perhaps? Even their spots would be the same, and you would have to give each cow a permanent microchip (or other identification) just to identify her.

The first successful cloning of a large animal was done using a cell from the udder of another female sheep. *(Courtesy of USDA/ARS #K-1968-13)*

polychlorinated biphenyl (PCB), insecticides, herbicides, and other chemicals in our rivers, lakes, and streams. Bacteria are capable of converting solid waste from humans and livestock into fuel that is used to generate electricity or to heat buildings.

Although great progress has been made in pollution reduction in some areas, pollution is still one of the world's greatest problems. Biotechnology has brought some spectacular breakthroughs in use of or decomposition of waste materials (Figure 3–11).

FIGURE 3–11 Biotechnology has been used to modify bacteria, allowing them to break down crude oil. These bacteria have become important in reducing the harmful effects on the environment when oil spills occur. *(Courtesy USDA #CS 336)*

SAFETY IN BIOTECHNOLOGY

INTERNET KEY WORDS:

safe products, biotechnology

Federal and state governments monitor biotechnology research and development very closely. Much fear has been expressed about the perceived dangers of genetically modified organisms. Therefore, appropriate policies, procedures, and laws have had to be developed as biotechnology has evolved. Research priorities and initiatives require discussion and interaction by scientists, government agencies, and other authorities. Products are tested in laboratories, greenhouses, and other enclosures before being approved for testing outdoors and in other less controlled environments. Even then, outdoor tests are first conducted on a small scale in remote places under careful observation. Under these conditions, the efficiency, safety, control, and environmental impact of new organisms are determined. If the new organism poses an unmanageable threat, it can be destroyed.

Customer resistance to new food products developed through biotechnology has been demonstrated since the first of these foods arrived at supermarkets. For example, some customers demanded that milk from cows treated with the biotech product BST (bovine somatotropin) should be labeled to distinguish it from other milk. Some believed that it should not be marketed at all. Despite assurances from the Food and Drug Administration (FDA) that no differences exist in comparisons of BST milk and milk from untreated cows, some customer resistance to BST persists. A few milk processors initially joined the resistance movement because they were afraid that their products might be boycotted by consumers. In some instances, these processors refused milk shipments from dairy farms that treated their cows with BST. Customer resistance to biotech food products appears to be diminishing.

Biotechnology is rapidly gaining the public's confidence and has become an important part of our daily lives. Many of its potential benefits have already been realized and most believe that we have only scratched the surface. With proper safeguards, we can look confidently to a bright future in this emerging field. Many other applications of biotechnology are cited throughout this text.

ETHICS IN BIOTECHNOLOGY

INTERNET KEY WORDS:

biotechnology ethics

Ethics is a system of moral principles that defines what is right and wrong in a society. The ability to manipulate the genetics of living organisms raises important ethical questions about how the technology should be used. For example, should humans be cloned from the strongest athletes or the smartest scholars? Would it be right to sell cloned embryos so that parents with a genetic defect might have children that are free of the defect? Is it morally right or wrong to mass produce clones of popular human beings in a laboratory? Imagine what Hitler might have done to create his "master race" if modern biotechnology methods had been known to his scientists.

It seems appropriate that a discussion of ethics should be part of the biotechnology revolution that is occurring. Such a discussion would help scientists and consumers to decide how ethical issues related to biotechnology should be handled. At the very least, we can expect new laws to be passed and courtroom decisions to be rendered on the basis of ethics in biotechnology.

Science Connection

"FINGERPRINTING" ORGANISMS

USDA microbiologists at the National Animal Disease Center in Ames, Iowa, were able to "crack" the mysteries of the three worst known cases of food poisoning in North America. DNA matching, or "fingerprinting," was used to link (1) the persons who became ill, (2) the contaminated food that caused the poisoning, (3) the place where the food had been contaminated, and (4) the material where the food poisoning organisms had grown. Food poisoning is a life-threatening condition resulting from eating food containing toxic material produced by certain bacteria under unsanitary conditions.

Medical authorities wondered if the meningitis of a child in California was caused by food poisoning, possibly from a certain cheese she had eaten. Could the culprit be *Listeria monocytogenes*—bacteria that had caused deaths in southern California in 1985, in New England in 1983, and in Canada in 1981?

DNA analysis makes "fingerprinting," or individual identification, possible in all organisms. *(Courtesy USDA/ARS #K-4767-1)*

Researchers were using restriction enzyme analysis to check the isolates involving *Listeria monocytogenes* from previous listeriosis outbreaks. By "fingerprinting" the isolates from each of the outbreaks, it was found that the isolates showed a characteristic DNA pattern for each particular episode.

The relatively simple method can be used by laboratory technicians and public health safety workers to track the spread of *Listeria* from the food-processing environment to the food product and on to the human patient. In analyzing *Listeria* in the California case, the researchers found that the isolates recovered from the patient, the suspect cheese, and the cheese factory all matched!

STUDENT ACTIVITIES

1. Write the Terms to Know and their meanings in your notebook.
2. Read an article in an encyclopedia or other reference on the process of genetic engineering.
3. Report your findings from Activity 2 to the class.
4. Make a collage depicting some important discoveries, inventions, and developments in biotechnology.
5. Form a discussion group to explore the benefits and hazards of biotechnology.
6. Arrange for a resource person to speak on the ethical and moral issues surrounding developments in biotechnology.
7. Organize a class debate on the ethical and moral issues regarding research and the use of new discoveries in biotechnology.

SELF EVALUATION

A. Multiple Choice

1. Bio means
   a. a study of.
   b. life.
   c. three.
   d. science.

2. An example of a fermented food is
   a. applesauce.
   b. bologna.
   c. cheese.
   d. coffee.

3. The earliest method of livestock improvement was probably by
   a. biotechnology.
   b. crossbreeding.
   c. gene splicing.
   d. selection.

4. The person providing the foundation for scientific study of heredity was
   a. Gregor Johann Mendel.
   b. George Washington Carver.
   c. Joseph Glidden.
   d. Thomas Jefferson.

5. The genetic code of life is
   a. clone.
   b. DNA.
   c. progeny.
   d. thymine.

6. Adenine, Guanine, Cytosine, and Thymine are all
   a. acids.
   b. bases.
   c. DNA.
   d. genes.

7. Recombinant DNA technology is also known as
   a. bovine somatotropin.
   b. gene splicing.
   c. porcine somatotropin.
   d. X-Gal.

8. Genetic engineering can be done to change
   a. animals.
   b. microorganisms.
   c. plants.
   d. all of the above.

9. An important contribution of biotechnology to waste management is
   a. bacteria that consume oil.
   b. disease-resistant bacteria.
   c. ice-minus bacteria.
   d. human bacteria.

10. Chemical pollutants in water that may be decomposed or deactivated by bacteria include
    a. chlorine.
    b. fluorides.
    c. iron.
    d. PCBs.

B. Matching

_____ 1. Fruits and grains
_____ 2. Yeast
_____ 3. Silage
_____ 4. Genetics
_____ 5. Genes
_____ 6. Tall pea plants
_____ 7. DNA
_____ 8. Fertilized cell
_____ 9. Insulin
_____ 10. X-Gal

a. Makes "marked" bacteria turn blue
b. Controls blood sugar levels
c. Used to make alcoholic beverages
d. Embryo
e. Result of dominant gene
f. Deoxyribonucleic acid
g. Causes bread to rise
h. Fermented grains or forage
i. Heredity
j. DNA and bases

You and the New Millennium

Agriscience Career Options

Park Rangers

Credit Manager

Crop Scouts

Loan Officer

Financial Manager

Agriscience Educator

Growers

Foresters

Financial Analyst

Marketing Specialist

Fish and Wildlife Managers

Agricultural Loan Officer

Food Scientists

Range Managers

Ranchers

Farmers

Game Managers

Nutritionists

Timber Producers

You and Agriscience

The "agricultural industry" can use your energy in science and engineering, business and financial management, production, renewable natural resources, communications, or other challenging and rewarding careers.

Futurists predict that during the next quarter century the most important discoveries in genetic engineering will be made in agriscience. Here you could help develop plants and animals that grow better and more efficiently, even in adverse situations. Nutritionists and food scientists study links among food, diet, and health. As one of them, you might work to ensure that new convenience foods are nutritious and healthy; or, you might design a new way of preserving, processing, or packaging food.

Agriscience educators teach in high schools, colleges, and universities. They also work for land-grant universities as extension educators. The job of agriscience education is to teach new and proven scientific practices to students and farm families.

As a financial manager you might work for a bank or a credit agency as an agricultural loan officer, or for a company that sells supplies to producers and growers. As credit manager, loan officer, financial analyst, or marketing specialist you could utilize your knowledge of business and finance, as well as production, processing, or distribution.

The farmers, ranchers, timber producers, and other growers are the foundation of the food and fiber system. To join their ranks is to play a most basic role in the future of our planet. For others who love the outdoors, a look at the work of foresters, range managers, game managers, fish and wildlife managers, park rangers, and crop scouts should be of interest.

Like all parts of our world touched by science and technology, the food, fiber, and renewable natural resources system is changing rapidly. Because it contributes to and is strongly influenced by trade and consumer lifestyles, the system must adjust continuously. If you get the appropriate training and experience, any of these careers can be yours.

Career Options in Agriscience

MATERIALS LIST

✓ paper

✓ pencil or pen

✓ bulletin board materials

✓ agriscience magazines and pictures

✓ *Occupational Outlook Handbook* or other agriscience career references

✓ Internet access

COMPETENCIES TO BE DEVELOPED

After studying this unit, you should be able to:

- define agriscience and its major divisions.
- describe the opportunities for careers in agriscience.
- compare the scope of job opportunities in farm and off-farm agriscience jobs.
- list activities in middle school, high school, and thereafter to help prepare for agriscience careers.
- identify resource people for obtaining career assistance in agriscience.

SUGGESTED CLASS ACTIVITIES

1. Conduct a career day for agriscience classes. Invite agricultural professionals to present workshops about the career fields they represent. Assign class members to take notes on two or three workshops in which they are interested. Collect the notes for a class assignment and have the students keep them in their notebooks for later reference.

2. Invite the recruiter from the College of Agriculture at a state land-grant university to talk with the class about career opportunities in agriculture. Ask him or her to talk about what high school preparation is needed to succeed at the university. Encourage the recruiter to provide literature about specific career opportunities. Invite parents to attend. Repeat this exercise by inviting representatives of technical colleges to meet with students and parents.

3. Make a list of 10 to 15 interests you currently have. As you read this unit, look for careers that may incorporate some of your interests or that appear to be interesting to you. Write those careers next to the interests on your list and consider your future in those professions.

TERMS TO KNOW

production agriculture	horticulture	agricultural supplies and	profession
agricultural processing,	forestry	services	agriscience professions
products, and distribution		agricultural mechanics	

How Much is One Trillion Dollars?

You can count $1 trillion ($1,000,000,000,000) by using the following procedure:

- One dollar bill every second
- Sixty bills per minute
- Thirty-six hundred bills per hour
- Eighty-six thousand per day
- Thirty-one million five hundred thirty-six thousand per year
- And continue counting for thirty-one thousand seven hundred and ten years!

FIGURE 4–1 The agricultural industry in the United States has assets of nearly $1 trillion ($1,000,000,000,000).

400,000 people are needed each year as replacements

100,000 trained people available

300,000 openings each year for additional people trained in agriscience

FIGURE 4–2 The employment outlook is good for people trained in agriscience.

Life is possible without many of our modern conveniences, but not without food. Basic to life is an adequate supply of suitable food and other products of the soil, air, and water. This includes food for nourishment, fiber for clothing, and trees for lumber. Less obvious are alcohols for fuel and solvents, oils for home and industry, and oxygen for life itself. The industry that provides these vital basic commodities is agriculture. American agriculture is the world's largest commercial industry, with assets of nearly $1 trillion (Figure 4–1).

DEFINITION

Agriscience is a term that includes all jobs relating in some way to plants, animals, and renewable natural resources. Such jobs occur indoors and outdoors. They include people in banking and finance; radio, television, and satellite communications; engineering and design; construction and maintenance; research and education; and environmental protection. All are in the field of agriscience if their products or services are related to plants, animals, and other renewable natural resources.

PLENTY OF OPPORTUNITIES

Approximately 21 million people are employed in agriscience careers. About 400,000 people are needed each year to fill positions in this field. Of those vacancies, only 100,000 are currently being filled by people trained in agriscience (Figure 4–2). That means there are many opportunities for you. You can use what you learn today in your current job, on your farm, or in agriscience classes to go directly to full-time employment. However, if you choose to pursue a college degree in agriscience, many additional career opportunities will be open to you.

About 20 percent of the careers in agriscience require college degrees. Many professional careers in agriculture require four-year college or university degrees. These careers are in the fields of education, marketing, communications, production, social services, finance, management, science, engineering, and many others (Figure 4–3). Additional career opportunities are available to students who graduate with technical college certificates and degrees. Among these opportunities are careers as mechanics, sales representatives, field representatives, agriscience laboratory technicians, insurance adjustors, and many others. Many of these careers offer financial opportunities that are equivalent to those available to university graduates.

Careers That Help Others

Helping others is an extra bonus with a career in agriscience. Of the jobs in processing, marketing, production, natural resources, mechanics, banking, education, writing, and other areas, many are people-oriented jobs.

FIGURE 4–3 A college education opens additional doors in agriscience.

This means you have the extra benefit of being a product or process specialist and receive the special appreciation of others. Considering the new medical procedures discovered through agriscience, more and more opportunities for helping others will develop.

Careers That Satisfy

You might ask, "How can an agriscience career benefit me? What's in it for me?" Of course, there's the money. In agriscience, salaries vary tremendously from job to job. Generally, the better qualified individuals will be able to earn more. Agriscience industries employ one fifth of all workers in the United States. And there are job openings for skilled individuals at various levels of expertise. You can be an agricultural mechanic right out of high school, or you can work for an advanced degree and be an agricultural engineer. You can do what you decide is best for you.

Before you invest in a technical college or university education, it is a good idea to assess your talents and interests. What kind of work do you think you will enjoy doing? What special skills and talents do you have? How can your interests, skills, and talents be matched to a career in which your skills and talents can be developed and expressed? These are important considerations to think about before you enter advanced education and training programs.

Some valuable resources are available to help you explore your career options. One of these may be as close as the counseling center in your high school. Many high school counselors are trained to use computer programs to help identify careers that match your interests and talents. If your school does not have a computer-assisted program such as Career Information System (CIS), you may be able to identify some career choices using the Internet on the World Wide Web.

THE WHEEL OF FORTUNE

Agriscience is like a wheel with a large hub. The hub of that wheel is production agriculture, or farming and ranching. The rest of the wheel consists of the nonfarm and nonranch activities in agriscience. Because so many opportunities for rewarding careers exist in that wheel, it may be called a wheel of fortune.

FIGURE 4–4 The average U.S. farmer produces enough food and fiber for approximately 140 people. *(Courtesy USDA/ARS)*

FIGURE 4–5 Agriscience careers may be illustrated by a wheel of fortune.

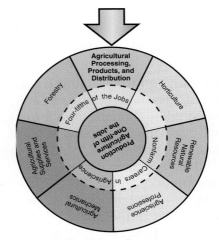

Ag Establishment Inspector
Butcher
Cattle Buyer
Christmas Tree Grader
Cotton Grader
Farm Stand Operator
Federal Grain Inspector
Food & Drug Inspector
Food Processing Supervisor
Fruit & Vegetable Grader
Fruit Distributor
Fruit Press Operator
Flower Grader
Grain Broker
Grain Buyer
Grain Elevator Operator
Hog Buyer
Livestock Commission Agent
Livestock Yard Supervisor
Meat Inspector
Meatcutter
Milk Plant Supervisor
Produce Buyer
Produce Commission Agent
Quality Control Supervisor
Tobacco Buyer
Weights & Measures Official
Winery Supervisor
Wood Buyer

FIGURE 4–6 Spin the wheel of fortune! What comes up for you? Agricultural processing, products, and distribution.

Production Agriculture

Production agriculture is farming and ranching. It involves the growing and marketing of field crops and livestock. Careers in this area account for one-fifth of all jobs in agriscience. Some estimates indicate the average U.S. farmer produces enough food and fiber for approximately 140 people (Figure 4–4). Large farm operators produce enough to feed more than 200 people.

Most other agriscience careers are involved with goods and services that flow toward or away from production agriculture. Workers in those careers permit U.S. farmers to supply goods so efficiently that U.S. consumers spend only 11 percent of their income on food. This is the lowest percentage in the world. Out of five workers in agriscience, four have jobs that are not on farms. The nonfarm agriscience jobs may be in rural, suburban, or urban settings. The agriscience wheel of fortune contains a hub with one fifth of the agriscience workers in production on farms and ranches. The rest of the wheel contains the four fifths of the agriscience jobs in nonproduction-type careers that are off the farm (Figure 4–5).

Agricultural Processing, Products, and Distribution

Spin the wheel of fortune! What comes up for you? Agricultural processing, products, and distribution (Figure 4–6)!

Agricultural processing, products, and distribution are those parts of the industry that haul, grade, process, package, and market commodities from production sources. Pick any item of food, clothing, or other commodity. Trace it back to its source. Except for metals and stone, most

INTERNET KEY WORDS:

food processing careers

FIGURE 4–7 The components of a deluxe cheeseburger may have come from several states or even different countries. *(Courtesy National FFA; FFA #18)*

INTERNET KEY WORDS:

horticulture careers
natural resource careers
forestry careers

objects can be traced back to a farm, ranch, forest, greenhouse, body of water, or other agricultural production facility. If you consider a deluxe hamburger, you can trace the beef, mayonnaise, tomato, lettuce, pickle, catsup, mustard, relish, bun, and sesame seeds back to farms where they were produced (Figure 4–7). The same is true of many ingredients in soda, coffee, chocolate, or any other beverage you choose.

Check the label in your coat. Is it made of cotton, leather, vinyl, rubber, or wool? Each can be traced to a farm, ranch, or plantation. Your search may take you to a Maryland farm, a California ranch, a Colombian rubber plantation, or a Utah mink farm.

People with careers and jobs in agricultural processing, products, and distribution make it all possible. From hauling to selling, processing to merchandising, inspection, and research—the commodity moves from its source to consumption. The U.S. Department of Agriculture (USDA) reports that the producer's share of the food dollar is as low as 15.4 cents for cereal and bakery products. The rest is for handling, processing, and distribution.

Horticulture

Spin the wheel of fortune! What comes up for you? Horticulture (Figure 4–8)!

Horticulture includes producing, processing, and marketing fruits, vegetables, and ornamental plants such as turfgrass, flowers, shrubs, and trees (Figure 4–9). Horticultural production could be thought of as farming. But, because it is generally done on small plots, the production of horticultural crops is classified with horticulture rather than farming. Horticultural commodities are high-labor and high-income commodities.

The landscape designer, golf course superintendent, greenhouse supplier, greenhouse manager, flower wholesaler, floral market analyst, florist, strawberry grower, vegetable retailer, and turf farmer (Figure 4–10) are all horticulturists. Recent data show more than 110,000 people employed by the floral industry alone.

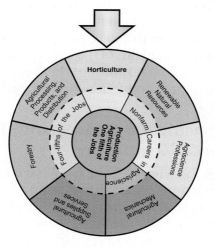

Floral Designer
Floral Shop Operator
Florist
Golf Course
   Superintendent
Greenhouse Manager
Greenskeeper
Horticulturist

Hydroponics Grower
Landscape Architect
Landscaper
Nursery Operator
Plant Breeder
Turf Farmer
Turf Manager

FIGURE 4–8 Spin the wheel of fortune! What comes up for you? Horticulture.

Hot Topics in Agriscience

A CAREER IN FOOD SCIENCE

A professional career in food science has become one of the hottest career fields available in agriculture. It combines the career fields of science and agriculture in the development of new food products. It also extends into the processing, packaging, distribution, and marketing of food products. This career requires a high degree of creativity in designing new products. Preparation for this career requires a college education with a strong emphasis in chemistry and the biological sciences. Salaries in this career field rank high in comparison with most other agricultural careers, and college graduates in this field are in high demand.

FIGURE 4–9 Horticulture provides many opportunities in urban as well as rural areas. *(Courtesy National FFA; FFA #242)*

FIGURE 4–10 Turf farming is the production of grass that is harvested, roots and all, and transplanted to provide "instant lawns."

Forestry

Spin the wheel of fortune! What comes up for you? Forestry (Figure 4–11)!

Forestry is the industry that grows, manages, and harvests trees for lumber, poles, posts, panels, pulpwood, and many other commodities. Americans have huge appetites for wood products.

Careers in forestry range from growing tree seedlings to marketing wood products. Many jobs in forestry are outdoors and require the use

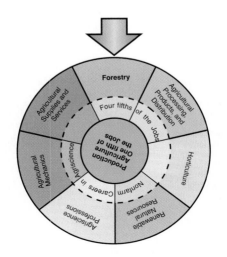

Forester
Forest Ranger
Heavy Equipment
  Operator
Log Grader
Logging Operations
  Inspector

Lumber Mill Operator
Nursery Operator
Park Ranger
Plant Breeder
Timber Manager
Tree Surgeon

FIGURE 4–11 Spin the wheel of fortune! What comes up for you? Forestry.

FIGURE 4-12 The forest industry provides many opportunities for outdoor work. *(Courtesy National FFA; FFA #187)*

of large machines to cut trees, drag logs, and load trucks (Figure 4–12). Other jobs are service oriented, such as the state or district forester whose job is to give advice and administer governmental programs. Many find enjoyable careers in forestry research, teaching, wood technology, and marketing.

Renewable Natural Resources

Spin the wheel of fortune! What comes up for you? Renewable natural resources (Figures 4–13 and 4–14)!

Renewable natural resources involve the management of wetlands, rangelands, water, fish, and wildlife. All fields require people with an appreciation for natural and scientific knowledge of plants and animals. This area of agriscience is attractive to those who enjoy working in parks, on game preserves, or with landowners to preserve and enhance natural habitat, plants, and wildlife. Water quality and soil conservation are state and regional concerns of high priority. New career opportunities in natural resource management are resulting from new efforts to save our oceans, lakes, wetlands, rivers, and bays.

Agricultural Supplies and Services

Spin the wheel of fortune! What comes up for you? Agricultural supplies and services (Figures 4–15 and 4–16)!

Agricultural supplies and services are businesses that sell supplies or provide services for people in the agricultural industry. Examples of supplies are seed, feed, fertilizer, lawn equipment, farm machinery, hardware, pesticides, and building supplies. These businesses are operated by owners, managers, mill operators, truck drivers, sales personnel, bookkeepers, field representatives, clerks, and others.

People in these jobs provide the supplies for the industry. However, there are many in agriscience who seldom handle the commodities themselves. Instead, they provide a service. Those who provide legal assistance, write agricultural publications, advise agriculturists on money matters, or provide advice on crops, livestock, pest control, or soil fertility are working in service occupations. Such jobs are for those who are more people oriented than commodity oriented.

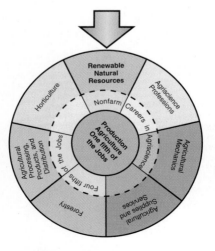

Animal Behaviorist	Ground Water Geologist
Animal Ecologist	Park Ranger
Animal Taxonomist	Range Conservationist
Environmental	Resource Manager
Conservation Officer	Soil Conservationist
Environmentalist	Trapper
Fire Warden	Water Resources Manager
Forest Fire Fighter/Warden	Wildlife Manager
Forest Ranger	
Game Farm Supervisor	
Game Warden	

FIGURE 4-14 Effective management of natural resources begins with education about resource management and care. *(Courtesy National FFA; FFA #226)*

FIGURE 4-13 Spin the wheel of fortune! What comes up for you? Renewable natural resources.

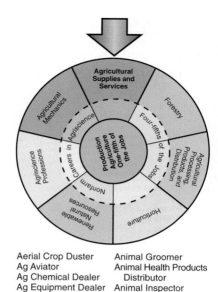

Animal Keeper
Animal Trainer
Artificial Breeding Distributor
Artificial Breeding Technician
Artificial Inseminator
Biostatician
Chemical Applicator
Chemical Distributor
Computer Analyst
Computer Operator
Computer Programmer
Computer Salesperson
Custom Operator
Dairy Management Specialist
Dog Groomer
Farm Appraiser
Farm Auctioneer
Farrier
Feed Mill Operator
Feed Ration Developer & Analyst
Fertilizer Plant Supervisor
Fiber Technologist
Field Inspector

Field Sales Representative, Agricultural Equipment
Field Sales Representative, Animal Health Products
Field Sales Representative, Crop Chemicals, Machinery
Harness Maker
Harvest Contractor
Horse Trainer
Insect & Disease Inspector
Kennel Operator
Lab Technician
Meteorological Analyst
Pest Control Technician
Pet Shop Operator
Poultry Field Service Technician
Poultry Hatchery Manager
Poultry Inseminator
Sales Manager
Salesperson
Service Technician
Sheep Shearer

Aerial Crop Duster
Ag Aviator
Ag Chemical Dealer
Ag Equipment Dealer
Animal Groomer
Animal Health Products Distributor
Animal Inspector

FIGURE 4–15 Spin the wheel of fortune! What comes up for you? Agricultural supplies and services.

FIGURE 4–16 Agricultural supplies and services provide the vital materials and services to keep a trillion dollar industry moving. *(Courtesy National FFA; FFA #87)*

INTERNET KEY WORDS:

agricultural sales career

Agricultural Mechanics

Spin the wheel of fortune! What comes up for you? Agricultural mechanics (Figures 4–17 and 4–18)!

Are you fascinated by tools and equipment? Are you challenged by something that does not work? Are you creative and like to build things? If so, a career in agricultural mechanics may be for you. **Agricultural mechanics** is the design, operation, maintenance, service, selling, and use of power units, machinery, equipment, structures, and utilities in agriculture.

Agricultural mechanics includes the use of hand and power tools, woodworking, metalworking, welding, electricity, plumbing, tractor and machinery mechanics, hydraulics, terracing, drainage, painting, and construction. Choose your level—indoors or outdoors. Choose your role—employee, employer, or professional.

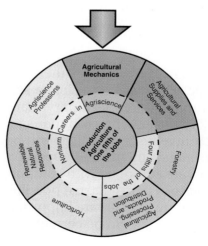

Ag Construction Engineer
Ag Electrician
Ag Equipment Designer
Ag Plumber
Ag Safety Engineer
Diesel Mechanic
Equipment Operator

Hydraulic Engineer
Irrigation Engineer
Land Surveyor
Machinist
Parts Manager
Research Engineer
Safety Inspector
Soil Engineer
Welder

FIGURE 4–17 Spin the wheel of fortune! What comes up for you? Agricultural mechanics.

FIGURE 4–18 Careers in agricultural mechanics are varied and challenging. *(Courtesy National FFA; FFA #290)*

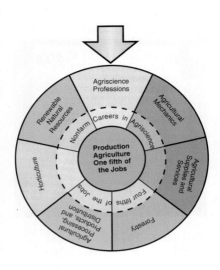

Ag Accountant	Avian Veterinarian	Ichthyologist	Plant Nutritionist
Ag Advertising	Bacteriologist	Information Director	Plant Pathologist
Executive	Biochemist	International Specialist	Plant Taxonomist
Ag Association	Bioengineer	Invertebrate Zoologist	Pomologist
Executive	Biophysicist	Land Bank Branch	Poultry Scientist
Ag Consultant	Botanist	Manager	Public Relations
Ag Corporation	Computer Specialist	Limnologist	Manager
Executive	Credit Analyst	Magazine Writer	Publicist
Ag Economist	Dairy Nutrition Specialist	Mammalogist	Publisher
Ag Educator	Dendrologist	Marine Biologist	Reproductive
Ag Extension Educator	Electronic Editor	Marketing Analyst	Physiologist
Ag Extension Specialist	Embryologist	Media Buyer	Rural Sociologist
Ag Journalist	Environmental Educator	Microbiologist	Satellite Technician
Ag Lawyer	Entomologist	Mycobiologist	Scientific Artist
Ag Loan Officer	Equine Dentist	Nematologist	Scientific Writer
Ag Market Analyst	4-H Youth Assistant	Organic Chemist	Silviculturist
Ag Mechanics Teacher	Farm Appraiser	Ornithologist	Software Reviewer
Ag News Director	Farm Broadcaster	Ova Transplant	Soil Scientist
Agriculture Attaché	Farm Investment	Specialist	Vertebrate Zoologist
Agronomist	Manager	Paleobiologist	Veterinarian
Animal Cytologist	Food Chemist	Parasitologist	Veterinary Pathologist
Animal Geneticist	Foreign Affairs Official	Pharmaceutical Chemist	Virologist
Animal Nutritionist	Graphic Designer	Photographer	Viticulturist
Animal Physiologist	Herpetologist	Plant Cytologist	Vocational Agriculture
Animal Scientist	Horticulture Instructor	Plant Ecologist	Instructor/FFA
Agriculturist	Hydrologist	Plant Geneticist	Advisor

FIGURE 4–19 Spin the wheel of fortune! What comes up for you? Agriscience professions.

FIGURE 4–20 Professional workers are becoming more important than ever as agriscience enters the information age. *(Courtesy USDA/ARS K-3401-10)*

Agriscience Professions

Spin the wheel of fortune! What comes up for you? Agriscience professions (Figures 4–19 and 4–20)!

The word **profession** means an occupation requiring specialized education, especially in law, medicine, teaching, or the ministry. **Agriscience professions** are those professional jobs that deal with agriscience situations. They cut across all divisions in the wheel of fortune.

Consider the agriscience teacher and the Cooperative Extension educator. Both must have a bachelor's or master's degree to be qualified. They may teach or consult about subjects in several or all divisions of agriscience.

Consider the veterinarian, agricultural attorney, research scientist, geneticist, or engineer—all of these professions require advanced degrees and high levels of training. If you can meet the standard, then a career in an agriscience profession may be for you.

Computers in Agriscience

The use of computers is extensive in agriscience. This means there are many opportunities to combine computer skills with agriscience settings. Among the agricultural uses of computers are machinery management, farm financial records, livestock management, crop management, commodity marketing, farm/ranch inventory management, ag business management, taxes, irrigation management, and precision farming. In addition, tractors and combines are equipped with computer systems that are highly sophisticated. These systems make it possible to operate the machines more efficiently. Computer skills are important in nearly every agriscience career. Every student should be certain that he or she is proficient in the use of computers.

Agri-Profile

CAREER AREA:

Agriscience Technician or Professional?

An assessment of career opportunities in agriscience by the United States Department of Agriculture revealed a bright outlook for college graduates. Other studies indicate a need for workers trained in agriscience for technical-level jobs.

The agriscience technician may be broadly trained in plant and animal sciences and employable in many fields. For the sharp individual with good work habits and a broad background in plant and animal sciences, the choices are extensive. Add agriscience mechanics skills to the package and the individual has access to dozens of career options.

Education for agriscience careers should begin at the high-school level or earlier. If one plans to work at the technician level, an early start will be especially helpful. The technician is expected to have first-hand experience and detailed knowledge of procedures and techniques. Such knowledge comes with experience in shops, laboratories, farms, greenhouses, fisheries, and on-the-job training situations, as well as in the classroom. Technicians generally have some special training beyond the high school level, whereas professionals are required to obtain degrees at the bachelor's, master's, or doctorate levels.

Whether you prefer to be a technician or a professional, agriscience offers a broad array of career possibilities. *(Courtesy of USDA #036)*

PREPARING FOR AN AGRISCIENCE CAREER

Career education is an important part of public education today. Since the early 1970s, the career education movement has spread from occupational education programs to the general curriculum. Many school systems now emphasize career education from kindergarten through adulthood.

As you consider a career in agriscience, it is important to consider how to meet the requirements to get started in that career. Some young people have an early start on careers in agriscience. They may have grown up on a farm or ranch. Their parents may have worked in one of the other areas of agriscience such as horticulture, resource management, business, or teaching. Perhaps they have obtained jobs or worked for friends or neighbors who had agriscience businesses. Following are some suggestions for preparing for a career in agriscience.

Agriscience Career Portfolio

An agriscience career portfolio is a collection of your best work on agriscience projects and other career-related materials that you have personally developed. The portfolio will be used to sell your skills to a prospective employer. Only your best work should be included, because a mediocre portfolio is an indicator of mediocre or average job skills.

Some of the materials that you may want to include in your portfolio are:

- Resume or vita
- FFA Agriscience Scholarship application
- Articles and papers written by you (published and unpublished)
- Photographs and written reports of projects that you have completed
- Letters of recommendation
- Personal letters and citations for services you have performed
- Personal and career goals
- Action plan for accomplishing your goals
- Newspaper articles, video clips, and sound bytes of media coverage of your projects

Career Plan While in Middle School

- Plan and conduct science projects with plants, animals, soil, water, energy, ecology, conservation, and wildlife.
- Research and report on the above.
- Join 4-H or Scouts and choose agricultural projects and merit badges.
- Volunteer to work on lawn, garden, greenhouse, farm, or conservation projects.
- Enroll in agriscience or other career education programs.

Career Plan While in High School

- Enroll in agriscience classes, including plant science, animal science, agricultural mechanics, agribusiness, and farm management.
- Enroll in college-preparatory courses in English, math, and science.
- Join the FFA organization and participate in leadership, citizenship, and agriscience activities.
- Develop a broad, supervised, occupational agriscience experience program.
- Acquire hands-on, skill-development experiences.
- Conduct an agriscience research project and enter it in the FFA Agriscience Scholarship and Agriscience Fair programs.

Career Plan after High School

- Obtain an agricultural job and plan ways to get additional training while on the job.
- Enter a community college and take courses that will transfer to the college of agriculture or life sciences at your state university.
- Enter a two-year program in technical agriculture.
- Enter a college of agriculture or life sciences and obtain a bachelor degree (B.S.), master's degree (M.S.), and/or doctorate (Ph.D.).

Agri-Profile

WORKFORCE SKILLS:

What Skills Will You Bring to Your Career?

The SCANS report is a document that was developed by the U.S. Secretary of Labor. The report was written by the Secretary's Commission on Achieving Necessary Skills (SCANS). Commission members represented business owners, unions, workers, supervisors, and public employers. The commission developed the following five competencies that students need to master so they will be prepared for productive careers:

- **Resources: Identifies, organizes, plans, and allocates resources**
  A. Time—selects goal-relevant activities, ranks them, allocates time, and prepares and follows schedules
  B. Money—uses or prepares budgets, makes forecasts, keeps records, and makes adjustments to meet objectives
  C. Material and Facilities—acquires, stores, allocates, and uses materials or space efficiently
  D. Human Resources—assesses skills and distributes work accordingly, evaluates performance, and provides feedback

- **Interpersonal: Works with others**
  A. Participates as Member of a Team—contributes to group effort
  B. Teaches Others New Skills
  C. Serves Clients/Customers—works to satisfy customers' expectations
  D. Exercises Leadership—communicates ideas to justify position, persuades and convinces others, responsibly challenges existing procedures and policies
  E. Negotiates—works toward agreements involving exchange of resources, resolves divergent interests
  F. Works with Diversity—works well with men and women from diverse backgrounds

- **Information: Acquires and uses information**
  A. Acquires and Evaluates Information
  B. Organizes and Maintains Information
  C. Interprets and Communicates Information
  D. Uses Computers to Process Information

- **Systems: Understands complex interrelationships**
  A. Understands Systems—knows how social, organizational, and technological systems work and operates effectively with them
  B. Monitors and Corrects Performance—distinguishes trends, predicts impacts on system operations, diagnoses deviations in systems' performance, and corrects malfunctions
  C. Improves or Designs Systems—suggests modifications to existing systems and develops new or alternative systems to improve performance

- **Technology: Works with a variety of technologies**
  A. Selects Technology—chooses procedures, tools, or equipment, including computers and related technologies
  B. Applies Technology to Task—understands overall intent and proper procedures for setup and operation of equipment
  C. Maintains and Troubleshoots Equipment—prevents, identifies, or solves problems with equipment, including computers and other technologies

Science Connection

WHEN I'M TWENTY SOMETHING!

Close your eyes and relax for a minute—then think about yourself in the year 2015. How old will you be then? Twenty something? You have finished your formal schooling and are into your career now. What kind of job do you have? Are you married or single? Is life enjoyable and your work challenging? Did you stay in your hometown, or are you in some other state or nation? Are you happy with your career? If you could start again, would you make the same choices? Now read the following questions and make mental notes of the answers that are most likely to be correct for you in the year 2015.

A Day on the Job in the Year 2015

DIRECTIONS: SELECT THE RESPONSE TO EACH ITEM THAT YOU THINK WILL BEST DESCRIBE YOU AND YOUR SITUATION IN THE YEAR 2015

1. What time do I generally wake up?
   - ❑ Early morning ❑ Afternoon
   - ❑ Evening

2. What is my work environment?
   - ❑ Indoor ❑ Outdoor ❑ Both

3. Where am I working?
   - ❑ In an office
   - ❑ In a factory or shop
   - ❑ In the community
   - ❑ In my home

4. What type of clothes do I wear to work?
   - ❑ Dress clothes (coat and tie or suit)
   - ❑ Uniform
   - ❑ Casual (open collar)
   - ❑ Worn jeans

5. Am I married?
   - ❑ Yes
   - ❑ No

6. If married, how long?
   - ❑ Less than 3 years
   - ❑ Three years or more

7. Number of children:
   - ❑ Zero
   - ❑ One
   - ❑ Two
   - ❑ Three
   - ❑ More than three

8. I will retire when I am:
   - ❑ 54 years or younger
   - ❑ 55 to 65
   - ❑ Older than 65

You can obtain information on careers, schools, and colleges from many sources. The following suggestions may be helpful to you:

High School Agriscience Teachers

- Agricultural mechanics
- General agriscience
- Animal and plant sciences
- Horticulture
- Biotechnology

9. How much education do I have? My highest level of education is:

- ❑ High school
- ❑ 2-year technical school degree or certificate
- ❑ 4-year college
- ❑ Master's degree
- ❑ Doctorate degree
- ❑ Other

10. I like keeping up with technical trends and procedures related to my work?

- ❑ Yes
- ❑ No

11. Where do I live?

- ❑ Hometown
- ❑ Home state, but not hometown
- ❑ Not home state, but in the U.S.
- ❑ Outside the U.S.

12. I am working with or for:

- ❑ Government
- ❑ Education
- ❑ Self-employed
- ❑ A large company (more than 200 employees)
- ❑ A small company (200 or less employees)

13. My income level makes me:

- ❑ Wealthy
- ❑ Comfortable
- ❑ Struggling financially
- ❑ Generally deprived

14. The amount of paid vacation is:

- ❑ Less than 5 days
- ❑ 6 to 10 days
- ❑ 11 to 15 days
- ❑ 16 days or more

15. The most important thing in my life is:

- ❑ Family
- ❑ Money and belongings
- ❑ Prestige and status
- ❑ Time off from work
- ❑ The place where I live

16. My work can best be described as:

- ❑ Producing information
- ❑ Producing or selling a product
- ❑ Distributing information
- ❑ Serving people

How well do your current career preparation activities mesh with your perception of yourself in the year 2015? The material in this and other units should be helpful in determining your preferences and making plans and preparations for a successful and challenging career.

High School Counselors

- Career brochures, bulletins
- Career information system
- Aptitude tests
- College/university catalogs

Cooperative Extension Service

- Listed in your phone book under county or city government
- 4-H career bulletins

Hot Topics in Agriscience

AGRISCIENCE CAREERS

Food Technician/Scientist
Environmental Technician
Information Technology
   Specialist

Plant Technician/Scientist
Global Positioning
   System Technician
Biotechnology Engineer

Farm/Ranch Managers
Urban Forester
Soil Technician/Scientist
Genetic Engineer

State Department of Education

- Specialist in agriscience, agribusiness, and renewable natural resources: scholarship, internships
- FFA State Executive Secretary: FFA career materials

Community Colleges, Technical Colleges, and Other Postsecondary Institutions

INTERNET KEY WORDS:

community and technical colleges

- Occupational Dean, Community College: program information, scholarships
- Director or Dean, Institute or Technical School: career/program information
  Typical programs include agriscience business management, farm production and management, ornamental horticulture, water resources, and wildlife
- Dean, College of Agriculture or Life Sciences
  Typical programs include agricultural education, agricultural and resource economics, agronomy (crops and soils), animal sciences, food science, forestry, horticulture, natural resources management, and poultry science
- Agricultural Education Coordinator, Department of Agricultural and Extension Education, Land-Grant University: ag education career information

As new technologies and job opportunities emerge, so will the need for well-trained and educated new people. Agriscience is a diverse field with job opportunities available at all levels. Pick your area of interest, determine the level at which you wish to operate, obtain the appropriate education for the job, and follow through with a rewarding career!

STUDENT ACTIVITIES

1. Write the Terms to Know and their meanings in your notebook.
2. Using paper and pencil, calculate the amount of time needed to count the dollar value of assets in agricultural in the United States, as suggested in Figure 4–1.
3. Develop a bulletin board that illustrates the "wheel of fortune," with its listing of the broad categories of jobs or divisions in agriscience.
4. Develop a collage that illustrates the many careers in agriscience.

5. Write the names of the divisions of agriscience, such as "Production Agriculture," "Agricultural Processing, Products, and Distribution," "Horticulture," and so on, and list five careers under each that may interest you.

6. Choose a career from the lists you developed for Activity 5 and write a one-page description for that career. Include the following sections in your description:
   a. career title;
   b. education/training required to enter the career and advance in the field;
   c. working conditions;
   d. advantages/benefits;
   e. disadvantages of the career;
   f. salaries of beginning and advanced workers in the field; and
   g. aspects of the career that you like.

7. Using the career you researched for Activity 6, or another job or career area, list the things you should do during and after high school to prepare for a career in that area.

8. Determine the name and address of an appropriate official of a school or college and request information from that person about educational opportunities for you in his or her institution.

9. Develop a list of agriscience careers in which computers are used.

10. At fast food restaurants, a popular order is a cheeseburger, fries, and a soda pop. Make a list of possible agriscience jobs that were involved in making this order a reality. For example, a wheat farmer was responsible for growing the wheat that is in the bun.

11. Make an outline of the unit. Phrases and words that are bold or in color should be included along with a brief description of each.

SELF EVALUATION

A. Multiple Choice

1. The industry that provides commodities that are basic to life is
   a. aerospace.
   b. agriscience.
   c. biotechnology.
   d. transportation.

2. The number of workers in agriscience in the United States is approximately
   a. 21 million.
   b. 100 million.
   c. 100,000.
   d. 400,000.

3. The percentage of total jobs in agriscience that require a college education to enter the job is
   a. 10 percent.
   b. 20 percent.
   c. 41 percent.
   d. 60 percent.

4. The products and services that are provided in most of the areas in the agriscience wheel of fortune seem to flow to or originate from
   a. agriculture processing, products, and distribution.
   b. agriscience professions.
   c. horticulture.
   d. production agriculture.

5. The management of wetlands comes under the area of
   a. agricultural processing, products, and distribution.
   b. agriscience professions.
   c. horticulture.
   d. renewable natural resources.

6. Of all agriscience jobs, the percentage that is *not* on farms or ranches is
   a. 20 percent.
   b. 40 percent.
   c. 60 percent.
   d. 80 percent.

7. A producer's share of each dollar spent for bread and cereals in the United States is about
   a. 11 cents.
   b. 15 cents.
   c. 25 cents.
   d. 75 cents.

8. The number of floral industry workers in the United States is about
   a. 110,000.
   b. 220,000.
   c. 500,000.
   d. 1,000,000.

9. A student may join 4-H or Scout groups to learn agriscience concepts as early as
   a. college.
   b. high school.
   c. middle school.
   d. none of the above.

10. Agriscience classes in high school usually include extensive instruction in
    a. plants, animals, and agribusiness.
    b. plants, animals, and social sciences.
    c. plants, mechanics, and higher math.
    d. food, fiber, and physics.

B. Matching

_____	**1.** Production	a. Teacher or veterinarian
_____	**2.** Processing and distribution	b. Hydraulics
_____	**3.** Horticulture	c. Seed, feed, or lawn supply
_____	**4.** Forestry	d. Farming or ranching
_____	**5.** Natural resources	e. Lumber
_____	**6.** Supplies and services	f. Ornamentals
_____	**7.** Mechanics	g. Grading and packaging
_____	**8.** Professions	h. Wildlife

UNIT 5

Supervised Agricultural Experience

OBJECTIVE

• To learn the rationale for and plan a supervised agricultural experience program.

COMPETENCIES TO BE DEVELOPED

After studying this unit, you should be able to:

• define supervised agricultural experience program (SAEP) terms.
• determine the place and purposes of SAEPs in agriscience programs.
• determine the types of supervised agricultural experience activities.
• explore the opportunities for SAEPs.
• set personal goals for an SAEP.
• plan your personal SAEP.

SUGGESTED CLASS ACTIVITIES

1. Conduct a Saturday or summertime bus tour to allow students to see the supervised agricultural experience programs (SAEPs) of other students. Encourage students who show their projects to describe why they chose the project, what they liked about the project, and what they learned through the project.

2. Invite a representative from the federal Farm Service Agency (or an agricultural lender) to visit with the class about financial assistance that is available to young people who wish to become involved in farming or other agricultural businesses. Provide students with information about these lending programs before the presentation to allow them to ask appropriate questions. Invite parents to attend the presentations.

3. Spend a class period looking through agriscience magazines, old SAEP project books, and any other sources your teacher can provide. After learning what projects have been done in the past, create two new ideas or ways to update an old project for an SAEP you could do.

TERMS TO KNOW

simulate	program	career exploration	placement supervised
real-world experience	FFA	career	agricultural experience
on-the-job training	project	agriscience research project	internship
supervised agricultural	enterprise	scientific method of inquiry	improvement activities
experience program	exploratory supervised	entrepreneurship supervised	agriscience skills profile
(SAEP)	agricultural experience	agricultural experience	Resources Inventory
supervised	mentor	production enterprise	resumé
experience	agriscience literacy		

INTERNET KEY WORDS:

supervised agricultural
experience program (SAEP)
National FFA

Agriscience programs in various schools teach basic principles and practices in plant and animal sciences, resource management, business management, agriscience mechanics, landscape design, leadership, and personal development. Such programs emphasize reading and math skill development and the application of scientific principles.

Classrooms are excellent places to learn fundamentals and theory through reading, study, discussion, and planning. School laboratories, such as greenhouses, agriscience mechanics shops, school land demonstration plots, farms, and animal production facilities, help provide experiences that simulate real-world experiences. **Simulate** means to look or act like. **Real-world experience** means conducting the activity in the daily routine of our society. Simulation is an excellent way to learn. It imitates the real world and may provide an ideal setting for the activity.

Classroom experiences and simulation lack the thoroughness of real-world experiences. Also, the personal relationships, such as employer–employee, supervisor–subordinate, owner–worker, salesperson–customer, owner–government, owner–community, fellow workers, and employee competition, are rarely present in simulated activities. Therefore, a program is needed for the student to obtain real-world experiences and on-the-job training if agriscience education is to lead to a successful career. **On-the-job training** means experience obtained while working in an actual job setting. In agriscience, the method used for students to obtain real-world experiences is referred to as the **supervised agricultural experience program (SAEP)**.

SUPERVISED AGRICULTURAL EXPERIENCE PROGRAM (SAEP)

An SAEP consists of all the supervised agricultural experiences that are learned outside of the regularly scheduled classroom or laboratory. **Supervised** means to be looked after and directed. Agriscience in this phrase means business, employment, or trade in agriculture, agribusiness, or renewable natural resources. **Experience** means anything and everything that is observed, done, or lived through. **Program** means the total plans, activities, experiences, and records of the supervised agricultural activity.

Agri-Profile

CAREER AREA:

Supervised Agricultural Experience

The phrase *supervised agricultural experience* has three important elements. Experience suggests hands-on or real-life activities in the work place. Attitudes, knowledge, and skills gained here will generally be advantageous to students as they seek employment. The term *"agricultural"* means that the setting and skills will be in the area of plant or animal sciences and related natural resources or management, mechanics, or technologies. *Supervised* means that experienced persons such as your teacher, parents, and/or employer recognize your interest in learning and will help direct the experience.

Agricultural experience is obtained in many ways. Generally, the teacher helps the student develop an understanding of the process and find a suitable location for a productive experience with cooperation from the parents. Sometimes sympathetic parents and supportive school programs are not available. In such cases, students must develop an extra measure of determination and seek experiences on their own.

Experiences may be for pay, or simply for experience. Many schools now provide supervised experience programs in school laboratories. Regardless of whether the student is being paid, the experiences obtained are worth the effort and give the participant an edge in the job market.

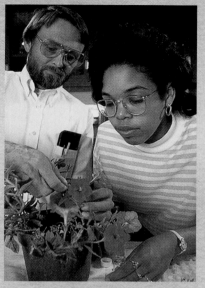

Learning by doing is generally accepted as the best way to become proficient in complex procedures and psychomotor skills. *(Courtesy USDA/ARS #K-3395-1)*

PURPOSE OF SAEPs

Supervised agricultural experience programs provide opportunities for learning by doing. They provide the means for you to learn with the help of your teacher, parents, employer, and other adults experienced in the area of your interest. Student SAEPs are an essential part of effective agriscience programs.

Some important purposes and benefits of SAEPs are to:

- provide opportunities to creatively explore a variety of agriscience careers;
- provide educational and practical experiences in a specialized area of agriscience;
- provide the opportunity to become established in an agriscience occupation;
- provide opportunities for earning while learning;
- create opportunities for earning after graduation;
- develop interests in additional areas of agriscience;
- develop valuable work skills such as
  - appreciating the importance of honest work;
  - improving personal habits;
  - developing superior work habits;

- establishing good relationships with others;
- keeping effective records;
- filling out useful reports;
- following instructions and regulations;
- contributing to your occupation;
- contributing to your family, community, and nation;
- permit individualization of instruction;
- permit recognition for individual achievement;
- become established in an agriscience business; and
- obtain experience as an entrepreneur.

SAEP AND THE TOTAL AGRISCIENCE PROGRAM

High school agriscience programs should be comprehensive in that they provide students with agricultural experience programs, leadership development programs, and laboratory experiences, together with classroom instruction. Leadership skills are developed through the FFA program, as discussed in Unit 6. These components are integrated so they complement each other (Figure 5–1). This integration should provide the most effective program for the student and make the best use of teacher time.

SAEP and Classroom Instruction

The SAEP is planned as part of the classroom instruction. Students use this instruction to learn how to plan an SAEP, identify possible SAEP choices, choose activities for their own SAEPs, and make appropriate arrangements with parents, teachers, and employers. The student conducts the SAEP under the supervision of the teacher, who provides instruction on the necessary topics. The teacher also arranges for small group and individual instruction. This permits the student to use the classroom and laboratory to solve problems encountered with the SAEP.

SAEP and the FFA

The SAEP overlaps with the FFA (an intracurricular youth organization for students enrolled in agriscience programs). The FFA has many activities that encourage students to do more in their SAEPs and provides competitive career development events that more fully develop useful skills for SAEPs. Furthermore, the FFA provides awards and other recognition for achievements in SAEPs. These frequently lead to trips and other travel experiences that greatly enrich the student's education.

FFA and Classroom Instruction

FFA is merged with classroom and laboratory instruction in a number of ways. The teacher has instructional units on the FFA in the classroom. Instruction is provided in public speaking, parliamentary procedure, and other leadership skills. The classroom setting can become the place

A COMPREHENSIVE AGRISCIENCE PROGRAM

FIGURE 5–1 A comprehensive agriscience program should provide students opportunities to learn through classroom/ laboratory instruction, supervised agricultural experiences, and personal growth through FFA.

INTERNET KEY WORDS:

FFA Proficiency Awards

where FFA teams polish their skills in preparation for upcoming contest or award activities. Therefore, the students who obtain the most benefit from their agriscience programs are those who take advantage of the opportunities provided through SAEPs and the FFA.

Proficiency Awards   The FFA has an extensive system of awards for individual members. These awards provide members with incentives and rewards for excellence in leadership and agriscience achievement. These awards change as the FFA expands to meet the changing nature of agriscience. Only those awards that have an industry sponsor are distributed in any given year. Examples of proficiency awards that may be available at the chapter, state, and national levels are:

Agricultural Communications	Floriculture
Agricultural Mechanical Technical Systems	Food Science and Technology
	Forage Production
Agricultural Processing	Forest Management and Products
Agricultural Sales and/or Service	Fruit and/or Vegetable Cluster
Aquaculture	Grain Production
Beef Production	Home and/or Community Development
Dairy Production	
Diversified Agricultural Production	Landscape Management
Diversified Crop Production	Nursery Operations
Diversified Horticulture	Outdoor Recreation
Diversified Livestock Production	Poultry Production
	Sheep Production
Emerging Agricultural Technology	Small Animal Production and Care
Environmental Science and Natural Resource Management	Specialty Animal Production
	Specialty Crop Production
	Swine Production
Equine Science	Turfgrass Management
Fiber and Oil Crop Production	Wildlife Production and Management

FFA members may apply for the proficiency awards that fit their own situations. Medals and certificates are awarded to winners at the chapter level, whereas plaques and financial awards are given to those at the state and national levels. More information on proficiency and other award programs is found in the FFA *Student Handbook.*

American Star Awards Program

The National FFA Organization recognizes strong supervised agriculture experience programs by selecting winners of the Star in Agriscience, Star in Agribusiness, Star in Agricultural Placement, and Star Farmer awards. These four awards are available at the chapter, district, state, and national levels. In addition, the National FFA Organization names local and national winners of Agricultural Entrepreneur awards. All of these awards are based on the quality of the students' SAEPs. Students receive cash awards, and they are recognized for their achievements at state and national conventions and in local chapter banquets.

FIGURE 5–2 Effective planning results in meaningful supervised agriscience experiences. *[Courtesy USDA/ARS #K-3405-11]*

INTERNET KEY WORDS:

National FFA Organization

TYPES OF SAE PROGRAMS

Supervised agricultural experiences grow out of planned programs (Figure 5–2). The word project is used to describe a series of activities related to a single objective or enterprise, such as raising rabbits, building a porch, or improving wildlife habitats. The term enterprise generally refers to a business that raises animals or plants. Examples are dairy, beef, or rabbits, and corn, hay, turf, or poinsettia. Students may choose from various types of SAEPs (Figure 5–3).

Exploratory SAE

An exploratory supervised agricultural experience program conducts supervised activities to explore a variety of subjects about agriscience and careers in agriscience. Such experiences help students gain understanding of and appreciation about agriscience to satisfy personal interests and needs.

Exploratory SAEs might include investigations and experiences in small animal health, biotechnology, water rights, agriscience journalism, aquaculture, hydroponics, air pollution, crop science, tissue culture, agriscience engineering, and many other areas. The student's exploratory SAEP is planned by the student under the direction of the teacher in cooperation with the parent/guardian, mentor, and others who will help the student obtain the exploratory experiences. A mentor is a person who you admire and who has skills that you would like to learn.

Exploratory SAEs are intended for students who wish to observe and experience a variety of areas in agriscience or to explore one or more areas not covered sufficiently in class to satisfy the student's interest. Such programs add an exciting dimension to courses at the elementary or middle school levels where agriscience literacy and career exploration are emphasized.

Agriscience literacy means education in or understanding about agriscience. Literacy does not require the student to become proficient in a given area, but it requires general knowledge about an area. Career exploration means learning about occupations and jobs that could possibly become one's future career or life's work.

Agriscience Research Project

Research is another possible area for students to obtain excellent SAEs. Research may be done in school laboratories, at home, on the job, or wherever suitable facilities may be found. Research is generally not regarded as a profit-making activity, but may be conducted to

TYPES OF SUPERVISED AGRICULTURAL EXPERIENCES				
Agriscience Research Project	**Exploratory SAEP**	**Entrepreneurship SAEP**	**Placement SAEP**	**Improvement Activities**
	Careers	Production Enterprise Agribusiness Enterprise	Farm/Ranch Placement (for wages and experience) Agribusiness Placement (for wages and experience) School Greenhouse or Farm	New Construction Repairs
Ag Problems/ Issues				

FIGURE 5–3 Supervised agriscience experience programs.

A production enterprise is a crop, livestock, or agribusiness venture. The student may own or be employed on a production enterprise.

Types of Crop Enterprises

- Corn Production
- Soybean Production
- Small Grain Production
- Greenhouse Production
- Nursery Production
- Hay Production
- Vegetable Production
- Fruit Production
- Forestry Production
- Christmas Tree Production

Types of Animal Enterprises

- Commercial Cow–Calf Production
- Registered Breeding Stock Production
- Market Beef Production
- Dairy Production
- Feeder Pig Production
- Market Swine Production
- Sheep Production
- Poultry Production
- Horse Production
- Rabbit Production

Types of Agribusiness Enterprises

- Lawn Service
- Custom Farm Work
- Trapping and Pelt Sales
- Hunting Guide Service
- Tree Service
- Farm and Garden Supply Service
- Artificial Insemination Business
- Animal Care and Boarding
- Winery
- Fishing and Crabbing for Sales
- Custom

FIGURE 5–4 Agriscience students have many production projects from which to choose.

answer questions that can be profitable if applied in production settings. Research projects may be part of community improvement activities or state research efforts such as stream monitoring, weather watch, forest fire watch, crop scouting, insect or weed monitoring, and crop reporting.

The **agriscience research project** is an original research project. It consists of identifying an agriscience problem, reviewing the scientific literature, applying the **scientific method of inquiry** to the problem, and reporting the results. A real problem should be identified, and some kind of experiment or testing procedure should be used to test the hypothesis that you have developed concerning the problem. You will usually be able to find a mentor to help you design an accurate experiment or testing procedure. The mentor can be a teacher in a local high school or college, or perhaps he or she is a science professional in a business or industry. In many cases, the mentor will help you find a laboratory to work in and equipment that can be used to conduct tests and measurements.

Research skills are valuable for solving problems and for working in interesting and good-paying careers. As a special incentive for teaching and learning research skills, the National FFA provides awards annually to outstanding agriscience students and teachers.

The final step is to report the results of your research to the public. This is done by writing a report. Sometimes a press release to local newspapers will attract the interest of a reporter who is willing to help you. The ultimate step is to enter the National FFA Agriscience Student Scholarship competition and/or submit the report to a scientific journal.

Enterpreneurship SAE

Entrepreneurship supervised agricultural experience refers to supervised activities conducted by students as owners or managers for profit (Figure 5–4). Emphasis is placed on developing the skills of the job or enterprise and working in a profitable and professional manner. Students may develop and own plant or animal enterprises where they grow commodities such as flowers, fruits, vegetables, field crops, turfgrass, Christmas trees, nursery stock, forestry, small animals, wildlife, beef, sheep, swine, honey bees, earth worms, and other commodities. A project that produces raw materials such as crops or livestock is a **production enterprise**. Agriculture students are encouraged to own or work with a production enterprise to learn about agriculture through hands-on experiences.

Entrepreneurship activities may also be conducted in agribusiness. An agribusiness entrepreneurship enterprise is one in which the student buys and sells an agricultural commodity for profit, rather than raising or growing the commodity. Some examples include a pet business, florist shop, livestock sales business, game dressing service, crop scouting service, crop spraying service, feed sales, seed sales, flower vendor, auctioneer, agriscience mechanic, and trucker.

Placement SAE

Placement supervised agricultural experience programs place the student with an employer in a production unit such as a farm, ranch, green house, nursery, and aquaculture facility to produce commodities

Science Connection

AGRISCIENCE RESEARCH PROJECT

Agriscience research projects benefit everyone. For example, a newspaper article motivated an agriscience student in Boise, Idaho, to conduct research on contamination of surface water in the delivery and drainage canals of a large irrigation project. The newspaper had reported that irrigation practices were causing the Snake River to become polluted with nitrates and phosphates that were dissolved in runoff water from the farm fields. Renee Burton gathered water samples throughout the irrigation season and tested them at a local university science laboratory with the help of the chemistry professor. Her research data showed that the algae problems in the river did not originate from the irrigated farm land. Her efforts in this project resulted in being named a national agriscience student finalist by the National FFA Organization.

INTERNET KEY WORDS:

FFA
Agriscience
Research

for wages. The student may also be placed with an employer or mentor in an agency or agribusiness where commodities are bought and sold or services are rendered. Some examples of agribusinesses are veterinary centers, kennels, feed or seed stores, pet shops, nursery outlets, florists, and garden centers. Some examples of agencies where students may be placed are Cooperative Extension Program, Farm Service Agency (FSA), Forest Service (FS), wildlife and environmental agencies, and school laboratories. The emphasis in placement programs is learning real skills and becoming a professional in a chosen agriculturally related career.

Agriscience Internship

Many agricultural businesses sponsor internships within their organizations. An **internship** may be a paid or unpaid work experience that allows a student to work in an industry. The student learns what a career in the industry is really like by experiencing a variety of jobs within the industry and gaining valuable career experiences to list on a resume. The sponsoring organization benefits by having the opportunity to evaluate the student as a potential employee after graduation.

Improvement Activities **Improvement activities** are activities that improve the appearance, convenience, efficiency, safety, or value of a home, farm, ranch, agribusiness, or other agriscience facility. For example, construction of a deck on a home is a challenging and worthwhile improvement activity (Figure 5–5). The student does not receive a wage or profit for conducting improvement activities. However, the student benefits by learning new skills and enjoying the benefits of the improvements. The owner of the facility should provide the materials and cover other expenses. Improvement activities provide the student with many opportunities to learn without the risks and commitment necessary for entrepreneurship activities (Figure 5–6).

FIGURE 5-5 Construction of a deck is an improvement activity.

AGRISCIENCE SKILLS PLAN AND PROFILE

Each student is encouraged to develop an agriscience skills profile to be placed in his or her portfolio. An agriscience skills profile is a record of skills developed and the level of competence in each. The better that competent people can document what they can do, the more likely they are to be hired in the jobs and careers they desire. Time and resources are limiting factors in preparing for a career, so every effort should be made to learn the most useful and interesting skills. Skills may be obtained in the classroom, laboratory, community, and through participating in SAE. Lists of appropriate skills for various areas in agriscience should be helpful for the student and teacher for planning a curriculum (Figure 5–7).

EXPLORING OPPORTUNITIES FOR SAEPS

Students should use great imagination when considering SAEPs. Some students may not have very many opportunities for meaningful SAEPs. Yet other students in similar circumstances find or create opportunities for effective programs. Seek the advice of your teacher for ideas. Also, observe what successful students have done in the agriscience program and in your community. Then develop an SAEP that provides the opportunity to learn and earn.

Personal Interest

Personal interest is an important factor in the success of an SAEP. Consider the kinds of activities you like to do and then build on those interests. A Student Interest Survey or Inventory should help you assess your natural interests and provide some guidance in developing an SAEP (Figure 5–8).

EXAMPLES OF IMPROVEMENT ACTIVITIES

Soil Improvement Programs

- Liming
- Fertilizing
- Drainage
- Erosion control
- Plow under green manure
- Introduce a cropping system
- Soil sampling

Building Improvement Programs

- Painting
- Window repair
- Roof repair
- Foundation repair
- Floor repair
- Siding repair
- Door repair
- Electric wiring
- Water systems installation
- Heating
- Lighting protection
- Feeding floor construction
- Remodeling
- Home sewage system installation

Fence Improvement

- Construction of new fence
- Fence replacement and repair
- Construction of flood gates
- Construction and repair of gates

Homestead Improvement Programs

- Plan and set out a windbreak
- Seed or reseed lawn
- Plant shrubs and trees
- Clean up homestead

Orchard, Small Fruits, and Vegetables Improvement Programs

- Plan and set out a fruit tree orchard
- Plan and set out a small fruit garden
- Plan and grow a home vegetable garden
- Renovate an old orchard

Weed Control Programs

- Spray major weed areas on farm
- Mow or spray weeds in fence rows
- Pull weeds in corn
- Clip weeds in permanent pastures

Insect and Pest Control

- Rat control
- Corn borer control
- Japanese beetle control
- Livestock parasite control

Farm Management Programs

- Keep farm accounts
- Plan farm safety program
- Inventory farm equipment
- Keep checking account records

Agricultural Shop Programs

Agricultural Machine Repair and Reconditioning Programs

Livestock Improvement Programs

Crop Improvement Programs

Landscaping Improvement Programs

General Clean-up Program
- Remove dead trees or shrubs
- Remove and discard dead branches
- Remove unsightly junk, trash, and woodpiles
- Provide a specific storage area for all lawn and horticulture equipment
- Repair or remove broken lawn furniture
- Improve the grade if necessary
- Remove, replace, or repair fences, sidewalks, step railings or porches
- Transplant trees, shrubs, or flowers
- Make or improve the driveway
- Pick up nails and broken glass
- Remove unsightly rocks

Grounds Maintenance Practices
- Mow the lawn
- Trim or prune trees and shrubs
- Fertilize the lawn, trees, shrubs, and flowers
- Apply herbicides
- Apply insecticides
- Water the lawn, trees, shrubs, and flowers
- Repair trees
- Brace trees
- Prevent sunburn of plants
- Prevent insect damage and disease
- Set up, adjust, and move a sprinkler
- Edge the lawn
- Stake trees
- Set up rain gauge
- Record precipitation from rain gauge
- Mulch trees, shrubs, and flowers
- Rake the lawn

Improvement and Beautification Activities
- Plant new trees and shrubs
- Draw a landscape plan
- Seed, plug, or sod lawn
- Plant flowers
- Relocate and replant trees and shrubs
- Renovate existing lawn
- Build a patio
- Make a window box
- Build a trellis
- Plant a windbreak

FIGURE 5–6 Improvement activities are available for agriscience students regardless of the home situation.

EXAMPLES OF AGRISCIENCE SKILLS

Agribusiness

- Operate cash register
- Display merchandise
- Keep inventory records
- Write sales tickets
- Deliver products
- Set up machinery
- Assemble equipment
- Greet customers
- Close sales
- Answer telephones
- Order merchandise
- Operate adding machine
- Wrap meat
- Cut carcass into wholesale cuts
- Cut wholesale cuts into retail cuts
- Compute sales tax

Agriculture Mechanics

- Arc weld metals
- Oxy-acetylene weld metals
- Cut details with oxy-acetylene
- Operate farm machinery
- Operate wood power tools
- Operate metal power and hand tools
- Recondition and sharpen tools
- Service air cleaner
- Service electric motor
- Store machinery
- Install rings or pistons
- Grind valves
- Wire electrical convenience outlet
- Change transmission fluid
- Fasten sheet metal with rivets
- Repair flat tire
- Calibrate field sprayer
- Pour concrete
- Lay reinforcement steel
- Use farm level

Corn

- Select seed
- Plant seed
- Prepare seedbed
- Calibrate corn planter
- Adjust planter for depth
- Apply dry fertilizer
- Apply anhydrous ammonia
- Conduct corn variety test
- Check harvest losses
- Apply herbicides and insecticides
- Operate combine
- Operate corn picker
- Identify weeds
- Identify insects
- Cultivate corn
- Dry corn artificially

Forages

- Innoculate legume seeds
- Rotate pastures
- Greenchop forages
- Bale hay
- Renovate permanent pastures
- Combine grasses and legumes
- Graze pastures properly
- Rotate pastures

Horticulture

- Plant vegetable garden
- Prepare garden plot
- Plant fruit trees
- Bud-graft
- Cleft-graft
- Whip-graft
- Prepare growing medium
- Sterilize soil
- Pot plants
- Water greenhouse plants
- Root cuttings
- Harvest crop
- Process vegetables
- Store produce
- Fertilize plants
- Determine plant diseases
- Treat deficiency symptoms
- Force blooming

Small Grains

- Calibrate grain drill
- Plant small grains
- Broadcast fertilizer
- Select adapted varieties
- Clean seeds
- Harvest small grains

Soil Management

- Test soil for lime requirements
- Lime soils
- Seed grass waterway
- Rotate crops
- Plant windbreak
- Test for fertilizer
- Fertilize soils
- Lay drainage tile
- Terrace fields
- Farm on contour
- Aerate soil
- Control erosion

Soybeans

- Test for germination
- Innoculate seed
- Take soil sample
- Control weeds
- Treat seed for storage
- Market beans
- Harvest soybeans
- Check harvest losses

Beef Cattle

- Assist cow in calving
- Dehorn
- Castrate bull calves
- Disinfect navel of baby calves
- Select herd sire
- Select replacement heifers
- Creep feed calves
- Cull poor producers
- Ear tag
- Tatoo
- Remove warts
- Drench
- Treat for bloat
- Treat for external parasites
- Select feeders
- Trim hooves
- Vaccinate for blackleg
- Vaccinate for infectious bovine rhinotracheitis (IBR)
- Vaccinate for brucellosis
- Formulate a balanced ration
- Ring bull
- Fit animal for show or fair
- Analyze production records
- Palpate to determine pregnancy

FIGURE 5–7 There are hundreds of agriscience skills that are useful for employment in agriscience.

EXAMPLES OF AGRISCIENCE SKILLS *(cont'd.)*

Dairy Cattle

- Select replacement heifers
- Dry cows
- Control external parasites
- Assist newborn calves in getting colostrum
- Vaccinate heifers for brucellosis
- Test cows for T.B.
- Cull low producers
- Dehorn
- Production test cows
- Participate in Dairy Herd Improvement Association (DHIA)
- Treat mastitis
- Operate milking machines
- Clean facilities after milking
- Artificially inseminate cows
- Prevent milk fever
- Formulate balanced ration
- Wash udder with disinfectant
- Detect abnormal milk with strip-cup
- Feed according to production
- Clean and sterilize utensils

Poultry

- Clean and disinfect brooder
- Cull poor producers
- Select pullets
- Sex baby chick or poults
- Prevent cannibalism
- Keep production records
- Dress broilers
- Prevent breast blisters
- Stimulate egg production
- Disinfect laying house
- Grade eggs by candling
- Size eggs
- Debeak chicks
- Vaccinate for fowl pox
- Treat for external parasites
- Worm poultry
- Provide sanitary water
- Feed balanced rations
- Provide litter

Sheep

- Select ram
- Flush ewes
- Treat for external parasites
- Assist ewes at lambing
- Creep feed lambs
- Dock lambs
- Castrate lambs
- Determine pregnancy in ewes
- Ear tag
- Shear sheep
- Tie fleece
- Cull farm flock
- Determine estrus in ewes
- Trim hooves
- Worm for internal parasites
- Keep production records

Swine

- Select herd boar
- Select replacement gilts
- Flush gilts
- Vaccinate sows for leptospirosis
- Provide farrowing stalls
- Clean facilities before farrowing
- Clean sow before farrowing
- Assist sow at farrowing
- Treat navels of baby pigs
- Clip needle teeth
- Creep feed pigs
- Ring hogs
- Treat for external parasites
- Inject iron in baby pigs
- Vaccinate for erysipelas
- Vaccinate for brucellosis
- Wean pigs at 4–6 weeks
- Weigh pigs at 56 days
- Formulate balanced ration
- Castrate boar pigs

FIGURE 5–7 (Continued)

STUDENT INTEREST SURVEY

Place an X in the blank by the tasks that you like to do or would like to learn how to do.

Tasks Typical of Agribusiness	Tasks Typical of Horticulture	Tasks Typical of Production
_____ Delivering merchandise	_____ Applying pesticides	_____ Applying pesticides
_____ Displaying merchandise	_____ Arranging flowers	_____ Baling hay
_____ Driving trucks	_____ Balling and burlapping trees	_____ Building fences and buildings
_____ Keeping records	_____ Building patios	_____ Castrating animals
_____ Mowing lawns	_____ Edging flower beds	_____ Cleaning animals
_____ Operating cash registers	_____ Identifying plants	_____ Feeding animals
_____ Operating equipment	_____ Lifting heavy materials	_____ Getting up early
_____ Pricing merchandise	_____ Making Christmas decorations	_____ Handling manure
_____ Processing meat, milk, grains	_____ Making cuttings	_____ Harvesting crops
_____ Repairing equipment	_____ Mowing lawns	_____ Helping parents
_____ Selling merchandise	_____ Mulching beds	_____ Keeping records
_____ Stocking shelves	_____ Operating power machinery	_____ Lifting heavy materials
_____ Taking customer orders	_____ Planting bulbs	_____ Milking cows
_____ Taking inventory	_____ Planting grass	_____ Operating machinery
_____ Taking telephone orders	_____ Planting seeds	_____ Painting buildings
_____ Unloading trucks	_____ Planting trees and shrubs	_____ Planting crops
_____ Working outside	_____ Protecting plants from weather	_____ Plowing fields
_____ Working with people	_____ Pruning plants	_____ Repairing buildings
_____ Working with plants	_____ Raking leaves	_____ Repairing machinery
	_____ Selling plants	_____ Shearing sheep
	_____ Watering plants	_____ Showing animals
	_____ Weeding by hand	_____ Taking soil samples
	_____ Working in various weather conditions	_____ Working in various weather conditions
	_____ Working with people	_____ Working with animals

FIGURE 5–8 A Student Interest Survey should be helpful to you in choosing SAEP activities.

Resources Inventory

A **Resources Inventory** is a summary of the resources that may be available for conducting SAEP activities. It includes information about your home, farm, work setting, and community that might be useful in considering your SAEP (Figure 5–9). Part of the inventory is a scale drawing of the property where you live or work. Making the scale drawing will help you realize what is available and help your teacher to suggest SAEP possibilities.

SELECTING AND IMPLEMENTING YOUR SAEP

After completing the Personal Interest Survey and the Resources Inventory, you should arrange a conference with your teacher. The conference should include discussions on your interests and a look at the possible production enterprises, improvement projects, and supplementary skills. After the conference you should record tentative plans for the SAEP (Figure 5–10). At this point you and your teacher should discuss the plan with your parents or guardians. If an employer is involved, he or she should become a partner in the planning process.

Securing a Job

INTERNET KEY WORDS:

prepare a resume

If placement on a farm or in an agribusiness is part of the SAEP plan, you will need a brief **resumé** (a one-page summary of information about a job applicant; Figure 5–11). The prospective employer will be interested in your educational and occupational background, which will help determine your qualifications and experiences. Be sure to ask a minimum of three adults who know your character and qualifications if you may list them as references. Your prospective employer will probably want to contact them.

When your resumé is complete, be sure your teacher has approved the final version. Then you are ready to approach employers for a job that will help achieve the objectives of your SAEP plan. Your approach is critical, because first impressions are lasting impressions. It is important to dress in a businesslike manner, be courteous and confident, and conduct yourself according to standard interview procedures (Figure 5–12).

Hot Topics in Agriscience

SAEP–INSIDE THE DAIRY BUSINESS

The Dairy Heifer Replacement Project is a program that is in place in many areas of the United States. It is designed to increase the knowledge and interest of young people in the dairy industry. The goal is to enhance life skills of its youth participants. The project begins when the participant purchases a heifer calf from a program-approved seller. For the next 18 months, the participant cares for the animal. Specific care must be provided, vaccinations must be given, and a magnet needs to be administered. Quality feed must be provided to ensure proper weight and good health. The participant is required to keep detailed records on the animal. In time, the heifer is bred according to set standards. The project culminates in the pregnant heifer's second year when she is presented for show and sale. The Dairy Replacement Project is a nice supervised agricultural experience project and allows each participant to gain an understanding of a crucial part of the dairy industry. Contact your local extension office for more information.

Resources available to the student for the Supervised Agricultural Experience Program (SAEP).

MAP OF HOME FARM AND/OR BUSINESS

Let each square represent any convenient acreage or square footage, i.e., 20, 40, 80, 100, etc.

One square = _____

LEGEND

————	Public Road
— — —	Private Drive
⌐	Farmstead
∿	Terrace
∿	Terrace Outlet
⇗	Natural Drain
∿	Gullies
—→	Stream
⬭	Pond
⌂	Trees
✳✳✳	Fence
▦▦▦	Railroad
= =	Crossing

Resources Inventory

1. Name ———————— Age ———— Class ————

2. Address ———————————— Phone ————————

3. Parent's Name ———————————— Occupation ————————

4. Number in my family ———— Boys ———— Girls ————

5. I live: on farm ———— in town ———— on an acreage ————.

6. Is land available for you to rent to produce crops? ———— yes ———— no

   a. If yes, how many acres? ————————

   b. Which crops? ————————

   c. Location of land? ————————

7. Are facilities available for you to rent to produce livestock or livestock products? If so,

   a. What type of livestock? ————————

   b. Number ————————

   c. Location of facilities ————————

8. Do you have available space for a garden? ———— yes ———— no

9. Do you have facilities for mechanical work? ———— yes ———— no

10. Do you have a greenhouse available for your use? ———— yes ———— no

11. Would you be interested in producing livestock or crops on the school farm? ————————

12. Do you have an agricultural job available to you?

   ———— yes ———— no   If so, what type? ————————

FIGURE 5–9 The Resources Inventory is especially helpful in planning production enterprises and improvement activities.

Personal Resumé

Name: Jamison Ledeoux

Address: 1234 Honeylocust Drive
Frederick, Maryland 21701

Telephone: (301)662-0000

Education: Junior at Frederick High School

Career Interest: Landscaping

Subjects Studied:
Horticulture I
Horticulture II
Landscaping I
Landscaping II
Typing I

Student Activities:
President: FFA
Editor of school yearbook
Tennis team and softball team

Special Skills:
Ball and burlap trees, operate cash register, water
plants, transplant plants, and operate a tractor.

Employment Experience:
Worked as a cashier and cook
Landscaped neighbor's yard

References:
Mr. Ralph Rece, Principal, Frederick County Vo-Tech,
Frederick, MD.
Mrs. Holly Deane, Instructor, Frederick County Vo-Tech,
Frederick, MD.
Mr. Bernard Rose, Manager, Hardees Fast Food, Frederick,
MD.

FIGURE 5–11 A personal resumé will help when seeking a job.

Selecting a Supervised Agriscience Experience Program for

_____
(Name of Student)

Instructions: Use this form to tentatively decide on a beginning agriscience
SAEP. This information will be helpful in agriscience classes
to develop detailed plans for obtaining experiences.

My interest areas in agriscience/horticulture are _____

Based upon my interest and opportunities available to me to get practical
experience in agriscience, I plan to include the following in my SAEP.

1. Production or Productive enterprises (examples: beef, dairy, nursery
production, Christmas trees)
_____
_____

2. Placement in an agribusiness (examples: supply store, florist shop, nursery,
golf course, landscape contracting)
_____
_____

3. Improvement activities (examples: landscape your home, fertilize your
lawn, plant trees)
_____
_____

4. Skills (examples: change spark plugs, weld, change the oil in small engines)
_____
_____

5. Other activities (example: projects on school facility)
_____
_____

FIGURE 5–10 Enterprises, improvement activities, and agriscience
skills should be selected and recorded early in the school year.

PREPARING FOR THE JOB INTERVIEW

What Employers Look for in an Employee

- **Attitude**—The prospective employee should have a positive attitude about the job. He/she should show enthusiasm and a willingness to learn and work. Employers stress this as being one of the most important qualities they look for in prospective employees.
- **Experience**—Previous experience of the prospective employee is important. However, employers are usually willing to train the person with a positive attitude.
- **Appearance**—The prospective employee should be neat and clean, have hair combed, and be well dressed. It is better to be overdressed than underdressed for an interview.
- **Posture**—It is important to stand and sit up straight. The employer will be observing the way you carry yourself and will make judgments accordingly.
- **Mannerisms**—Mannerisms are gestures that are made that may be annoying or could be welcomed. However, one should be aware of mannerisms. Do you:
  1. Yawn a lot? If so, others will think you're bored or worse—lazy.
  2. Fidget? Squirming may indicate lack of confidence or disinterest in the job.
  3. Daydream? Give your full attention to the interviewer.
- **Handshake**—Have a firm handshake; not bone crushing and not limp.

What Questions Should be Asked?

The following questions may be asked if the information is not provided by the interviewer.
- What type of jobs or tasks are to be done?
- What are the policies and procedures for workers?
- What are the working hours?
- What is the rate of pay?
- What arrangements are needed for time off?
- If you are uncertain about something that has been discussed in the interview, you should ask the employer to clarify or explain.
- IN SUMMARY, BE POLITE AND ATTENTIVE DURING YOUR JOB INTERVIEW.

FIGURE 5-12 Preparation for the job interview will permit you to relax and give your total attention to the interviewer.

SETTING GOALS FOR PRODUCTION IN THE SAEP

What goals should be set for the supervised agricultural experience program?

DEFINITION: A goal is the hoped-for end result of hard work and should be challenging and realistic.

- Goals should be challenging!
- Goals should be reachable!
- SAEP goals should focus on scope, learning opportunities, and production efficiency factors.
- Parents, employers, agriculture teachers, and other qualified adults should help develop SAEP goals.
- Goals should be recorded.
- Goals should be analyzed and evaluated periodically, and new goals should be developed.
- Goals provide direction and organization.
- Settings realistic goals should help increase profits.

What are efficiency factors?

DEFINITION: Efficiency factors are measures of production success and encourage enterprise improvement and profit. Examples of efficiency factors are as follows.

- **Size of Enterprise.** For animal weight or livestock products produced.
- **Rate of Gain and Production.**

Beef: Percent of calf crop $= \dfrac{\text{Calves born alive}}{\text{Cows bred}}$

Swine: Pigs farrowed per litter $= \dfrac{\text{Live pigs farrowed}}{\text{Sows bred}}$

Poultry: Percent of egg production $= \dfrac{\text{Average eggs per hen}}{\text{Number days in production}}$

Weight produced per litter $= \dfrac{\text{Total production lbs.}}{\text{Number of litters}}$

Sheep: Percent of lamb crop $= \dfrac{\text{Lambs born alive}}{\text{Ewes bred}}$

- **Returns and Feed Costs.** Round total income and value of feed fed to the nearest whole dollar.

Returns per \$100 feed fed $= \dfrac{\text{Total income}}{\text{Dollars worth of feed fed}} \times 100$

Average weight sold $= \dfrac{\text{Total sales weight}}{\text{Animals sold}}$

Returns per \$100 invested $= \dfrac{\text{Total income}}{\text{Total expenses}} \times 100$

Average price received $= \dfrac{\text{Total sales value}}{\text{Units sold}}$

Expense Per Cwt. of Production $= \dfrac{\text{Total expenses}}{\text{Total production}} \times 100$

- **Feeding Efficiency**

Note: Convert all corn to shelled corn basis (56 lb per bu.) before figuring efficiency factors.
Note: Poultry—1 unit is equal to 1 dozen eggs or 1.5 lb.
Note: Dairy—1 unit is equal to 1000 lb milk or 100 lb weight.

Feed cost per Cwt. or per unit

a. For swine or beef cattle $= \dfrac{\text{Total feed cost}}{\text{lb weight produced}} \times 100$

c. For dairy or poultry $= \dfrac{\text{Total feed cost}}{\text{Units of production}}$

b. For sheep $= \dfrac{\text{Total feed cost}}{\text{lb wool + lb weight}} \times 100$

Feed per Cwt. produced

a. For hogs, beef cattle, or sheep $= \dfrac{\text{lb feed fed}}{\text{lb weight produced}}$

b. For dairy or poultry $= \dfrac{\text{lb of each feed fed}}{\text{Units of production}}$

(For sheep, include wool with weight as in lb above)

- **Death Loss**

Percent death loss $= \dfrac{\text{Number of dead animals}}{\text{Total number of dead produced and purchased}}$

A low percentage of death loss means a high enterprise rating for this item.

FIGURE 5–13 Goals should state the number, size, time lines, and efficiency factors you hope to achieve.

			EXPERIENCE INVENTORY			

Directions: Complete the following information sheet by listing any experiences you have had or would like to gain in the field of agriculture.

Tasks or Jobs	Can perform without help	Can perform with help	Can help perform	Cannot or have not performed	Would like to learn how to perform	How or where to obtain experience
Examples						
• Drive tractor		X				Landscaping business
• Take cuttings					X	Agriscience Class
• Keep records				X		SAEP & Technical Skills
1._____						
2._____						
3._____						
4._____						
5._____						
6._____						

FIGURE 5–14 The Experience Inventory is a device to plan and record experiences you hope to gain. It is also a mechanism for recording how well you have learned new skills.

Refining the Plan

Plans should be worked out for each production enterprise. It is important to estimate the anticipated expenses and income, which will help you make financial arrangements to conduct the project. Also, some goals for the enterprise should be set, including the size of the project and some efficiency factor goals (Figure 5–13).

Employers are generally impressed with students who are eager to learn. In this regard, both the student and the employer can benefit from a carefully thought out statement of skills to be developed on the job. You should develop the list with the guidance of your teacher and prepare an Experience Inventory to record the completion of tasks or jobs (Figure 5–14). It should be frequently updated. The experience inventory helps the student, teacher, and employer track progress in achieving goals and in developing a skills profile.

The Placement Agreement document helps finalize the plans for placement on a job (Figure 5–15). Such agreements need the signature of the student, parents or guardians, employer, and teacher. Once all parties are in agreement regarding the student's placement experiences, the chances for success are great.

Another document that will help plan and conduct the SAEP is the Improvement Project Plan and Summary (Figure 5–16). This plan directs the student to describe the conditions found, plans for improvement,

PLACEMENT AGREEMENT

To provide a basis of understanding and to promote business-like relationships, this memorandum is established on _____, 20 ____. This work will start on _____, 20 ____, and will end on or about _____, 20 ____, unless the arrangement becomes unsatisfactory to either party before the ending date. Person (employer) responsible for training _____

The usual working hours will be as follows:

1. While attending school working hours shall be _____
   When not attending school working hours shall be _____
2. Provisions for overtime _____
3. Provisions for time off _____
4. Liability insurance coverage (type and amount) _____

Wages will be at the following rate(s): _____
Trial period _____
Remainder of the agreement period _____
And will be paid (when?) _____

A. IT IS UNDERSTOOD THAT THE EMPLOYER WILL (check the items that apply):

___ Provide the student with opportunities to learn how to do well as many jobs as possible, with particular reference to those contained in the planned program;
___ Coach the student in methods found desirable in implementing project activities and handling management problems;
___ Help the student and teacher make an honest appraisal of the student's performance;
___ Avoid subjecting the student to unnecessary hazards;
___ Notify the parents and the school immediately in case of accident or sickness and if any other serious problem arises;
___ Assign the student new responsibilities when he/she can handle them;
___ Cooperate with the teacher in arranging conferences with the student on supervisory visits; and/or
___ Other:

B. THE STUDENT AGREES TO (check the items that apply):

___ Do productive work, recognizing that the employer must profit from the student's labor in order to justify employment;
___ Keep the employer's interest in mind and be punctual, dependable, and loyal;
___ Follow instructions, avoid unsafe acts, and be alert to unsafe conditions;
___ Be courteous and considerate of the employer, the family, and others;
___ Keep such records of work experience and make such reports as the school may require;

PLACEMENT AGREEMENT (cont'd.)

___ Develop plans for management decisions with the employer and teacher; and/or
___ Other:

C. THE TEACHER, ON BEHALF OF THE SCHOOL, AGREES TO (check the items that apply):

___ Visit the student on the job at frequent intervals for the purpose of instruction and assurance that the student gets the most education out of the experience;
___ Show discretion in the time and circumstances of these visits, especially when the work is pressing; and/or
___ Provide appropriate job-related instruction at school; and/or
___ Other:

D. THE PARENTS AGREE TO (check the items that apply):

___ Assist in promoting the value of the student's experience by cooperating with the employer and the teacher of agriscience;
___ Satisfy themselves in regard to the living and working conditions made available to the student; and/or
___ Other:

E. ALL PARTIES AGREE TO:

___ An initial trial period of ____ working days to allow the student to adjust and prove himself/herself;
___ Discuss any issues concerning the job with the teacher before ending employment and/or
___ Other:

STUDENT's Signature _____   EMPLOYER's Signature _____
Address _____   Address _____

Telephone Number _____   Telephone Number _____
Social Security No. _____
PARENT's Signature _____   TEACHER's Signature _____
Address _____   School Address _____

Telephone Number _____   Telephone Number _____
                                  School Telephone Number _____

FIGURE 5–15 A Placement Agreement states what every party is expected to do. It promotes good planning and reduces misunderstandings and conflicts.

IMPROVEMENT ACTIVITY PLAN AND SUMMARY

Improvement Project No. _____

A. Conditions found:

_____

_____

_____

_____

B. Plans for improvement (including costs):

_____

_____

_____

_____

_____

C. Value of improvement when completed:

_____

_____

_____

_____

_____

AGRICULTURAL IMPROVEMENT PROJECT SUMMARY

Started: _____ , _____   Completed: _____ , _____

Date	Jobs Done	Hours of Labor	Cost of Materials & Equipment

FIGURE 5–16 Improvement activities should be planned and records kept on the jobs, hours, and costs involved.

and estimated value of the improvement when completed. The summary is filled out as the improvement project progresses and serves as the record when finished.

Finally, a skills plan should be developed. These skills are chosen from lists that apply to the student's community and are recorded on a Agriscience Skills Plan and Record (Figure 5–17). The date completed should be added when the skill is acquired.

SUMMARY

A carefully planned SAEP is both challenging and rewarding. The student learns by doing and earns while learning. Learning is at the highest level, because the laboratory is the real world. The teacher benefits from having highly motivated students, because the learning is interesting and rewarding. The employer benefits by having better-trained employees. The community benefits from the improvement projects and better-educated citizens.

Directions: Using the list of Supplementary Practices supplied by the instructor, complete the chart below by choosing skills you would like to include as part of your SAEP.

Skills, Practices, Job, or Experience	Place to obtain skill	Date planned to obtain skill	Date completed
Examples:			
• Operate Cash Register	On-Job	Sept. 16	Sept. 16
• Bud-Graft	School Farm	February	Feb. 20
1. _____			
2. _____			
3. _____			
4. _____			
5. _____			
6. _____			
7. _____			

FIGURE 5–17 Agriscience skills should be selected at the beginning of the year with plans for times and places to complete each item.

Science Connection

TRACKING THE DISAPPEARING GENE

The USDA Agriculture Research Service (ARS) is on the lookout for good students who are seeking experience in research. For more than 15 years, plant geneticist Thomas E. Devine of the USDA-ARS plant molecular biology laboratory in Beltsville, Maryland, has been looking at the genetic structures and characteristics of soybean plants. Surprisingly, there has been a succession of talented high school students working with him. These students do work of real scientific significance and are making original contributions to science. Together, they have conducted long and detailed studies of thousands of soybean plants to track down the genes responsible for disease resistance and nitrogen fixation.

Students can get valuable research skills in supervised agriscience experience programs.
(Courtesy USDA/ARS #K-4604-1)

Nikola Lockett, while a high school junior, started a two-summer experience program at the USDA Southern Regional Research Center. Later, while a student at Xavier University in New Orleans, she was able to continue her work with a plant physiologist as part of the Cotton Fiber Bioscience team at the Center. Similarly, a high school research apprenticeship program attracts students to the ARS Arthropod-Borne Animal Diseases Research Laboratory in Laramie, Wyoming. Students there assist with research projects involving insects and diseases of livestock and poultry.

The USDA has numerous plant, animal, disease, insect, food, fiber, nutrition, and other research laboratories throughout the United States. For information on research assistance opportunities for students, write to: USDA-ARS, Personnel Operations Branch, 6305 Ivy Lane, Greenbelt, MD 20770-1435.

STUDENT ACTIVITIES

1. Write the Terms to Know and their meanings in your notebook.
2. Describe the relationships between (1) classroom instruction and supervised agriscience experience and (2) the FFA program and supervised agriscience experience.
3. Construct a bulletin board showing the relationships presented in Figure 5–1.
4. Study Figure 5–4 and write five ideas for production projects or enterprises to discuss with your teacher.
5. Discuss Figure 5–6 with your parents/guardians and select two or three improvement activities that you would like to conduct.
6. Review the examples presented in Figure 5–7 and choose 20 agriscience skills from enterprises other than your production projects and improvement activities.
7. Examine the tasks in the Student Interest Survey (Figure 5–8). Determine whether your interests are more in agribusiness, horticulture, production, or other areas of agriscience.
8. Using Figure 5–9 as a guide, draw a map (to scale) of your home, farm, or business where you can conduct an SAEP. Make an inventory of the resources that may be available for you to use in conducting an SAEP.
9. Talk with your teacher and parents/guardians. Write the names of the projects, activities, and skills that you definitely plan to do during the current year (Figure 5–10).
10. Prepare a personal resumé.
11. Apply and interview for a part-time job.
12. Develop an Experience Inventory (Figure 5–14).
13. Work out a Placement Agreement with an employer (Figure 5–15).
14. Set goals for production projects (Figure 5–13).
15. Make detailed plans for improvement activities (Figure 5–16).
16. Select definite agriscience skills. Write down where and when you plan to accomplish the skills (Figure 5–17).

SELF EVALUATION

A. Multiple Choice

1. Conducting an activity in the daily routine of our society is said to be
   a. laboratory experience.
   b. real-world experience.
   c. simulation.
   d. supervised occupational experience.

2. Which is not a purpose or benefit of SAEPs?
   a. Become established in an agriscience occupation.
   b. Permit early graduation.
   c. Permit individualized instruction.
   d. Provide educational and practical experiences.

3. Which is not a major component of a comprehensive agriscience program?
   a. classroom/laboratory instruction
   b. FFA
   c. memorization and recitation
   d. supervised occupational experience

4. SAEPs should be planned
   a. at home with parents/guardians.
   b. in the classroom.
   c. on the job with employers.
   d. all of the above.

5. A student-drawn map of the home property is an important part of a

    a. job interview.

    b. Placement Agreement.

    c. Resources Inventory.

    d. resumé.

6. A production or productive project

    a. is the same as an improvement activity.

    b. is the same as a skill.

    c. may involve either ownership or placement for experience.

    d. must be done without pay or profit.

7. Development of agriscience skills is important because

    a. a skill inventory is part of a career portfolio.

    b. personal agriscience skills help identify career interests.

    c. skills lead to part-time jobs.

    d. all of the above.

8. Improvement activities

    a. are always connected with employment.

    b. focus primarily on leadership development.

    c. must improve some part of the instructional program.

    d. should be without pay.

B. Matching

_____ **1.** Enterprise

_____ **2.** Experience

_____ **3.** SAEP

_____ **4.** FFA

_____ **5.** Improvement Project

_____ **6.** Agriscience

_____ **7.** Production Enterprise

_____ **8.** Program

_____ **9.** Project

_____ **10.** Skill

a. A project or experience in agriculture under the direction of your teacher

b. A project conducted for wages or profit

c. Plans, activities, records, and experiences related to an agricultural enterprise

d. Activities related to a single enterprise

e. The application of science to agriculture

f. A national organization for agriscience students

g. Ability to do something well

h. A business raising animals or crops

i. Activities that improve the appearance, convenience, efficiency, safety, or value of a home or other facility

j. Anything that is observed, done, or lived through

Leadership Development in Agriscience

OBJECTIVE

To develop basic leadership skills.

MATERIALS LIST

✓ paper
✓ pencil or pen

✓ bulletin board materials

✓ Internet access

COMPETENCIES TO BE DEVELOPED

After studying this unit, you should be able to:

• define *leader* and *leadership*.

• explain why effective leadership is needed in agriscience.

• list some characteristics of good leaders.

• describe the opportunities for leadership development in FFA.

• demonstrate positive leadership skills.

SUGGESTED CLASS ACTIVITIES

1. Assign class members to attend a public meeting to observe how leadership skills are used to conduct the business of the community. Include several different kinds of meetings, assigning groups of students to each meeting. Provide a worksheet to each group to be sure that students observe and report on critical leadership activities or skills. For example, you may want to have students critique the use of parliamentary procedure skills, explain how the committee process was used, or describe how community leaders used other leadership techniques.

2. View a segment of C-Span on television while a congressional committee is meeting or while the Congress is in session debating a national issue. Identify and discuss the parliamentary rules used. How are they different from Robert's Rules of Order, which are used by the National FFA Organization? How are they the same?

3. In small groups, describe the characteristics of a good leader. Produce examples of people who were/are good leaders and explain why. Present your ideas to the rest of the class.

TERMS TO KNOW

leadership	enthusiasm	extemporaneous speaking	motion
lead	selflessness	parliamentary procedure	main motion
plan	loyalty	business meeting	amend
manage	Cooperative Extension	presiding officer	refer
citizenship	System	secretary	lay on the table
integrity	4-H	minutes	point of order
knowledge	Girl Scout	order of business	adjourn
courage	Boy Scout	executive meeting	
tact	extemporaneous	gavel	

What is leadership in agriscience? Agriscience has been described as a broad and diverse field. It is not just horticulture or supplies and services; not just professions or products, processing, and distribution; not just mechanics or forestry; and not just renewable natural resources or production. Agriscience is all of these. Then what is leadership in agriscience?

LEADERSHIP DEFINED

INTERNET KEY WORDS:

leadership defined

Leadership may be defined as the capacity or ability to set a direction. To **lead** is to show the way by going in advance or guiding the actions or opinions of others. To do this in agriscience, you must have knowledge of technical information and people. You must know how to organize and manage activities. Most jobs are too big for one person. We can only do part of what needs to be done. Therefore, we need to use the help of others.

A leader uses the knowledge and skills of others to achieve a common goal. For instance, a quarterback on a football team will use leadership skills to coordinate the team players to achieve a touchdown. Similarly, the wise batter in baseball will hit the ball in a place that not only gets the batter on first base, but permits other runners to advance around the bases. A properly placed hit supports the common goal of achieving runs. A single base hit, where everyone advances and no one gets out, may be the best for the team. Conversely, a line drive that doesn't quite make the fence may result in a third out, with no chance for other players to score.

WHY LEADERSHIP IN AGRISCIENCE?

Agriculture is a highly organized industry. It involves people and complex processes. Leadership skills are necessary whenever people are assembled. Those who teach in agriscience are part of a team of teachers, principals, supervisors, community advisory groups, and others. Those in agribusiness are typically part of teams consisting of the manager, office staff, sales representatives, field personnel, and board of directors (Figures 6–1 and 6–2). Those on farms may be the owner, manager, spouse, children, hired help, or neighbors who assist at times. The manager of a farm or business must **plan** (think through, determine

FIGURE 6–1 A Leader must learn to coordinate the work of all of the members of a team to avoid having a team member working on something that has already been completed. *(Courtesy National FFA)*

FIGURE 6–2 Agribusinesses rely on team efforts to achieve goals. *(Courtesy USDA)*

procedure, assemble materials, and train staff to do a job). Once a job is planned, it may be accomplished through management. To **manage** is to direct people, resources, and processes to reach a goal. A manager uses leadership skills continuously in working with others on a day-to-day basis (Figure 6–3).

A landscaper is someone who plans, plants, builds, or maintains outdoor ornamental plants and landscape structures. Sometimes working alone and sometimes working with others, the landscaper is a leader in many respects. When developing a plan for a customer, the landscaper exerts leadership. The landscaper knows the name, function, and performance of each plant. This information is used to develop an acceptable plan. Because the customer has personal ideas about

FIGURE 6–3 Managers are leaders. *(Courtesy National FFA)*

FIGURE 6–4 Participation in class discussions is the first step toward developing qualities of leadership. *(Courtesy National FFA; FFA #1)*

INTERNET KEY WORDS:

leadership traits
leadership skills

landscaping, a professional must consider these ideas in the plan. It may test the landscaper's leadership skills to lead the customer to an acceptable plan with plants that survive the climate and conform to acceptable landscape practices.

The landscaper and other agriscience personnel may be called on as officers or members of professional organizations to give testimony before legislators or other public officials on the need for laws, regulations, or other actions that affect their work. It may be quite possible to end the day presiding over a meeting or taking minutes at a professional or civic meeting.

For the agriscience student, the need for leadership skills is also apparent. Having confidence to participate fully in class is important (Figure 6–4). To meet prospective employers and conduct a supervised agricultural experience program requires leadership skills. Functioning in the community, participating in group meetings, and making friends readily all require acceptable leadership skills. Good leadership skills greatly improve individual marketability in the working world.

To be a good citizen, you must earn your way in life without infringing on the rights of others. Useful citizenship uses leadership to promote the common good in society. **Citizenship** means functioning in society in a positive way.

TRAITS OF GOOD LEADERS

Good leaders must have **integrity** (honesty). Without it, others cannot entrust an individual with the power to manage or control, even in minor things. A leader must have **knowledge**, which means familiarity, awareness, and understanding. Good leaders are dependable and have the **courage** (willingness to proceed under difficult conditions) and initiative to carry out personal and group decisions. To lead, one must demonstrate the initiative to carry out personal and group decisions. To lead, one must also communicate. This requires good speaking and listening skills.

Hot Topics in Agriscience

THE AGRICULTURAL LOBBY

Agriculture is affected in many ways by the laws that are approved by Congress and state legislatures. In addition, state and federal agencies write regulations to implement new laws. Farmers, ranchers, and agricultural processors are subject to both the laws and the regulations. In a time when politics affects the agricultural industry in so many ways, agricultural lobby efforts have become important.

Special leadership skills are required by those who lobby on behalf of agriculture. A lobbyist must be registered in most states before he or she can actively lobby a legislature or other political organization. Most lobbyists represent a particular group of farmers or processors, such as the United Dairymen or the Wheat Commission. The lobbyist is paid to negotiate, persuade, and be persistent in promoting legislation that is favorable to the segment of the industry that he or she represents.

INTERNET KEY WORDS:

National 4-H Clubs
National FFA Organization
Boy Scouts of America
Girl Scouts of the United States
of America

In working with others, tact, or the skill of encouraging others in positive ways, is useful. Similarly, a sense of justice to ensure the rights of others is important. Enthusiasm, or energy to do a job and inspiration to encourage others, is useful. Selflessness means placing the desires and welfare of others above yourself. It, too, is an important quality for good leadership. These traits encourage loyalty, which results in reliable support for an individual, group, or cause. These and other traits are achieved through effective leadership development.

LEADERSHIP DEVELOPMENT OPPORTUNITIES

Modern schools provide extensive opportunities for agriscience students to develop leadership skills. Students develop leadership in school organizations, athletics, and in classroom and shop situations. Some become leaders at home, on the job, or in community organizations.

4-H Clubs

The Cooperative Extension System is an educational agency of the U.S. Department of Agriculture (USDA) and an arm of your state university. It provides educational programs for both youth and adults. Its programs include personal, home and family, community, and agriscience resources development. The Cooperative Extension System also sponsors 4-H clubs. The 4-H network of clubs is directed by Cooperative Extension System personnel to enhance personal development and provide skill development in many areas, including agriscience (Figure 6–5). The four Hs in 4-H stand for head, heart, hands, and health. These provide the basis for the 4-H pledge, which is, "I pledge my head to clearer thinking, my heart to greater loyalty, my hands to larger services, and my health to better living for my Club, my community, my country, and my world."

FIGURE 6–5 Many agricultural youth experience their first leadership training as members of 4-H clubs. *(Courtesy Rensselaer County 4-H Urban Summer Program)*

Scout Organizations

Girl Scout and Boy Scout organizations provide opportunities for leadership development and skill development in agriscience and other areas. Scouts focus heavily on outdoor activities and provide excellent

Agri-Profile

CAREER AREA:

Leadership Development

Leadership development may be a career area or specialty for teachers, consultants, personnel managers, coaches, and others. However, many agriscience positions require good leadership capabilities as a tool for everyday use. Auctioneers, salespersons, managers, entrepreneurs, corporate executives, politicians, and anyone who directs others or routinely meets the public must have strong leadership skills.

Leadership involves good planning, goal setting, and the ability to inspire others to work toward a common goal. Such skills as committee interaction, parliamentary procedure, and self-expression are important leadership techniques that are developed through study and practice. Such skills are used in church, civic, and community organizations, as well as in work places. Group projects and club activities in agriscience provide excellent opportunities for leadership development.

The art of public speaking is one of the most powerful tools of a leader. *(Courtesy National FFA)*

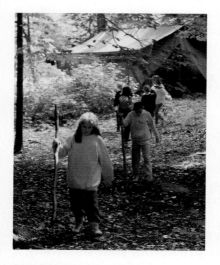

FIGURE 6–6 Boy Scouts of America and Girl Scouts of the United States of America provide opportunities for young people to develop skills in providing leadership to other members of their organizations. *(Courtesy Joni Conlon)*

leadership development and natural resources skills. They provide recognition through a system of merit badges, which are earned by learning skills and obtaining experiences in many areas, including agriscience (Figure 6–6).

FFA

The FFA is a youth organization that was developed specifically to expand the opportunities in leadership and agriscience skill development for students in public schools. Only students enrolled in agricultural courses are eligible for membership in FFA.

Aim and Purposes   The FFA is part of the agriscience curriculum in most schools where agriscience programs are offered. It is an important teaching tool. It serves as a laboratory for developing leadership and citizenship skills. These, in turn, are helpful in learning agriscience skills. The primary aim of the FFA is the development of agriscience leadership, cooperation, and citizenship. The specific purposes of the FFA may be paraphrased as follows:

- to develop competent and assertive agriscience knowledge and leadership;
- to develop awareness of the global importance of agriscience and its contribution to our well-being;
- to strengthen the confidence of agriscience students in themselves and their work;
- to promote the intelligent choice and establishment of an agriscience career;
- to stimulate development and encourage achievement in individual agriscience experience programs;
- to improve the economic, environmental, recreational, and human resources of the community;

- to develop competencies in communications, human relations, and social abilities;

- to develop character, train for useful citizenship, and foster patriotism;

- to build cooperative attitudes among agriscience students;

- to encourage wise use and management of resources;

- to encourage improvement in scholarship; and

- to provide organized recreational activities for agriscience students.

The Emblem The FFA emblem contains five major symbols that help demonstrate the structure of the organization (Figure 6–7). They are as follows:

Eagle—The emblem is topped by the eagle and other items of our national seal. The eagle was placed in the emblem to represent the national scope of the organization. It could also represent the natural resources in agriscience.

Corn—Corn is grown in every state in the United States. It reminds us of our common interest in agriscience, regardless of where we live.

Owl—The owl represents knowledge and wisdom. Use of this symbol in the emblem recognizes the fact that people in agriscience need a good education and that education must be tempered with experience to be of greatest usefulness.

Plow—The plow has been used to represent work-labor-effort. These qualities are needed to cause things to happen and to get results in agriscience.

Rising Sun—The rising sun is a symbol of the progressive nature of agriscience. It is symbolic of the need for workers in agriscience to cooperate and work toward common goals.

The FFA emblem may be constructed one symbol at a time. When assembled and dissembled in this manner, it is a good device to help others understand the FFA.

The Colors The official FFA colors are blue and gold. The shade of blue is national blue. The shade of gold is the yellow color of corn. Therefore, the colors are called national blue and corn gold.

Motto The FFA motto contains phrases that describe the philosophy of learning and development in agriscience. The motto is:

- Learning to Do

- Doing to Learn

- Earning to Live

- Living to Serve

"Learning to Do" emphasizes the practical reasons for study and experience in agriscience. It also suggests ambition and willingness to productively use the hands, as well as the mind. "Doing to Learn" describes procedures used in agriscience instruction at the doing level. Experiencing results from doing is the most permanent result of learning. "Earning to Live" suggests that FFA members intend to develop their skills and support themselves in life. And "Living to Serve"

FIGURE 6–7 The FFA emblem contains five important symbols. *(Courtesy National FFA; FFA #148)*

Agri-Profile

BE ALL YOU CAN BE

On the day after Christmas in 1960, a 12-year-old boy with polio was delivered to a ranch for needy boys, together with his two brothers. Their broken home was without heat, and food was scarce. After 5 years of growing up on the ranch, this individual enrolled in a high school agriculture program and joined an FFA chapter.

Although love of animals attracted him to the FFA, it was recognition for his successes on the parliamentary team that spurred him on. "It was the first time I had ever won anything," was his later observation. From the parliamentary team, he advanced to area FFA president and on to state president and, eventually, he was a delegate to the national FFA convention. It was there that he presented the historic motion to open FFA membership to female members.

After graduation from college, this individual followed in the footsteps of his FFA advisor, teaching agriculture and inspiring young people to be all they could be. As FFA advisor, he insisted that his students could prepare themselves for any career through leadership training, talking on their feet, developing responsibility, learning to manage money, experiencing team work, learning respect, setting goals, and planning ahead. These experiences help the individual win in life.

The experiences in FFA and teaching and advising helped develop the confidence and skills he needed for a career in public life. After teaching for a while, he served 8 years in the Texas Senate before going on to the United States House of Representatives. There his leadership skills were soon recognized, he was elected president of the Freshman Class of Congressmen, and eventually became a valued member of the House Agriculture Committee.

What advice would this highly successful statesman give to young people pursuing careers in agriscience? The Honorable Congressman Bill Sarpalius suggests:

1. Be in the right frame of mind. Don't dwell on your handicaps or lack of ability like in public speaking or running. Forget, "I can't."
2. Avoid negatives.
3. Stay physically and mentally sharp. Don't let yourself get lazy.
4. Develop a religious background.
5. Concentrate on doing for others—not yourself.

In summary, he asserts, "To achieve all that is possible—we must attempt the impossible. To be as much as we can be—we must dream of being more!"

indicates an intention to help others through personal and community service.

Salute The Pledge of Allegiance to the American flag is the official FFA salute. The words of the pledge are: "I pledge allegiance to the flag of the United States of America and to the Republic for which it stands, one nation under God, indivisible, with liberty and justice for all."

Degree Requirements The FFA has four degrees, which indicate the progress a member is making. These are Greenhand, Chapter, State, and American degrees. The Greenhand and Chapter FFA degrees are awarded by the FFA chapter in the agriscience department of the school. The State FFA degree is awarded by the State Association. The American FFA degree is awarded by the National FFA.

The Greenhand degree is so named to indicate the member has learned beginning skills in FFA and is studying agriscience. To receive the Greenhand degree, the member must meet the requirements spelled out in the current FFA *Official Manual.* In general, the requirements for the Greenhand degree are to:

- be enrolled in an agricultural education course;
- have satisfactory plans for a supervised agricultural experience program;
- recite the FFA creed, motto, and salute;
- describe the FFA emblem, colors, and symbols;
- explain the FFA Code of Ethics and proper use of the FFA jacket;
- have satisfactory knowledge of the history of the organization and of the Chapter Constitution and Program of Activities;
- know the duties and responsibilities of members;
- own or have access to a copy of the *Official Manual* and FFA *Student Handbook;* and
- submit a written application for the Greenhand degree.

INTERNET KEY WORDS:

FFA career development events

The requirements for the other three degrees help the member to learn and grow professionally from one level to the next in the organization. The *Official Manual* contains the exact requirements for all degrees, details of membership, and chapter operation for the FFA.

Career Development Events The FFA sponsors competitive career development events for a wide range of career interests. The first level is in the local chapter at the school. The second level is the district or regional level within the state. The third level is the state association level, and the fourth is the national level. Local FFA advisors determine which events are appropriate for students in their programs. The competitions that are conducted at each level should reflect the content of the instructional programs.

The purpose of FFA career development events is to encourage agriscience students to develop technical and leadership skills and to practice these skills in friendly competition with other FFA members. These events include:

Agricultural Communications	Floriculture
Agricultural Issues	Food Science and Technology
Agricultural Mechanics	Forestry
Agricultural Sales	Horse
Agronomy	Job Interview
Creed Speaking	Livestock
Dairy Cattle	Marketing Plan
Dairy Foods	Meats Evaluation
Environmental and Natural Resources	Nursery and Landscape
	Parliamentary Procedure
Extemporaneous Speaking	Poultry
Farm Business Management	Prepared Speaking

Some FFA career development events require contestants to know how to grade agricultural products, such as eggs, meat, poultry, fruits, and

FIGURE 6-8 FFA Career Development Events provide motivation to students to develop leadership skills through public speaking and by using parliamentary procedure to conduct the business of the organization. *(Courtesy National FFA)*

vegetables. Other events require students to evaluate live animals, such as beef and dairy cattle, horses, poultry, sheep, and swine. Some require mechanical abilities, such as welding, plumbing, electronics, irrigation, surveying, engine trouble-shooting and repair, painting, wood-working, and general tool use.

Some events, such as Floriculture and Horticulture, require knowledge of the art and science of floral arrangement. Some career development events include Forestry, Vegetables, and Nursery/Landscape competitions, which require students to identify plants, plant materials, insects, and diseases. Land Judging involves evaluating the soil and land and recommending appropriate management practices. Parliamentary procedure teams demonstrate their ability to conduct meetings using their knowledge of correct parliamentary practices. These procedures are used to open meetings, conduct business, close meetings, and write minutes according to acceptable practice in the real world. Various types of speaking contests help individuals to sharpen their speaking skills.

Each of these skill events is organized so the participants are competing as individuals and as teams of three or four participants (Figure 6–8). An example of a chapter contest is the Chapter Award. This competition involves most or all of the students in the agriscience program, and encourages them to make valuable improvements to the program, school, and community.

PUBLIC SPEAKING

INTERNET KEY WORDS:

how to give a speech

Oral communication skills are important for good leadership. Effective leaders must speak with individuals, committees, small groups, and in large forums. The ability to relax, speak clearly, and state what is pertinent to the subject at hand is useful. These skills are developed by applying basic principles of speech preparation and organization.

Speeches may be prepared or extemporaneous. An **extemporaneous** speech is a speech delivered with little or no preparation. Most speaking is extemporaneous. The ability to do effective extemporaneous speaking is enhanced by delivering prepared speeches. Therefore, the prepared speech is used to teach public speaking skills in FFA.

The National FFA Organization sponsors three different types of speech competitions: the Creed Speaking, Prepared Speaking, and Extemporaneous Speaking contests. Each of these events provides opportunities for students to stand before an audience and deliver their speech.

Creed Speaking

The Creed Speaking event is for students who are enrolled in an agriscience class for the first time. It requires the speaker to repeat the FFA Creed from memory. The emphasis is on accuracy and delivery. When the speaker is finished, a specific statement taken from the creed is cited. The speaker explains what the statement means to him or her. Each contestant responds to the same statement.

The Creed Speaking event gets students started in speaking without undue concern over the content of the speech. It is a good way to help students succeed before they are expected to prepare the written content of a speech. Once confidence is gained, students often become motivated to participate in other speech contests.

Prepared Speaking

The Prepared Speaking contest provides an opportunity for a student to research an agricultural topic and develop his or her own ideas. The content of the speech must be original, not copied from someone else. The length of the speech should be 6–8 minutes, and points are deducted if the speech is too long or too short.

Part of this speech competition is based on the quality of the written manuscript. The neatly typed manuscript (double spaced) is given to the judges ahead of time. Questions are developed from the manuscript. Each contestant responds to the questions for 5 minutes following the delivery of the speech. Contestants are judged on the quality, effectiveness, accuracy of the written manuscript, speech delivery, and response to the questions.

Extemporaneous Speaking

Extemporaneous speaking is a valuable skill for life. This skill is used every day by most agriscience professionals. Sales people use extemporaneous speaking skills to negotiate sales. Agricultural educators use these skills to teach classes or to teach individuals how to deal with problems and issues. Agricultural executives and administrators use extemporaneous speaking skills to convince their employees and stockholders to support their leadership and business plans. Nearly everyone can benefit from learning extemporaneous speaking skills.

Extemporaneous Speaking contests require students to gather original documents and materials in a notebook or file, but no written preparation of a manuscript may be done before the competition. Each competitor draws for a speech topic, and he or she is allowed to prepare for 30 minutes using only the materials that were assembled earlier. The speech length is 5–8 minutes, and the judges are allowed 5 minutes to ask questions after the speech has been delivered.

Planning the Speech

A speech should have at least three sections: the introduction, body, and conclusion. The plan should clearly identify the sections.

Introduction   The introduction indicates the need for and importance of the speech. It should be carefully planned and spoken with confidence. The introduction may be in the form of statements or questions. If the introduction is not to the point, does not fit the occasion, or is not delivered in a spirited manner, the audience may not listen to the rest of the speech. The introduction might be only a few lines long, but it must capture the attention of the audience.

Body   The body of the speech contains the majority of the information. It should consist of several major points that support one central theme or objective. Each major point is supported by additional information to explain, illustrate, or clarify the point. It is best to write the body of the speech in outline form (Figure 6–9).

SPEECH OUTLINE

Introduction

Honorable judges, instructors, and fellow students
1. Rabbits, cows, plants, and plows—WE STILL NEED THEM!
2. When the family farm goes under, a piece of America goes under with it

Body

The 2000 census revealed . . .
1. The midsize farm is likely to be the true family farm
   • Owned and operated by the family
   • Family receives benefit from their work
2. Some believe the family farm is a relic of the past!
   a. Press fascination with bankruptcy sales
   b. Farms not in view from interstate highways
   c. Small population on farms
   d. Animal rights groups and unionizing efforts
3. The family farm endures in the United States
   a. Better managed farm businesses buy the weaker farms
   b. Disease epidemics threaten specialized operations
   c. Farm retailing is on the rise
   d. Small farms are becoming legitimate

4. Farm bankruptcy must be minimized
   a. Farm failure affects general businesses
   b. We will miss fresh farm produce
   c. We will see more pollution
   d. We will depend more on imports

5. Are there remedies? Yes!
   a. Remove politics from marketing
   b. Recognize farmers as astute business people
   c. See the total industry of agriculture
   d. Tax farm land for farm use
   e. Increase agriculture land preservation programs

Conclusion

Indeed . . .
1. General George Washington nearly lost the continental army at Valley Forge from lack of food, shelter, and clothing.

2. It can happen to us if we don't maintain a healthy farm situation in the United States today

3. Don't give away our most valuable resource—the ability to feed, clothe, and shelter ourselves!

FIGURE 6–9 An outline of a winning speech on the importance of family farms.

FIGURE 6–10 Speaking in public can be a pleasant and exciting experience once you have learned the proper way to organize and present your thoughts. Speaking is a skill that is learned only through practice as you stand and speak in front of an audience.

An outline will help direct the thought and delivery of the speech. After outlining the speech, a carefully worded narrative may be written to help the speaker fully develop the content of the speech. However, it must be emphasized that a speech should be given from an outline to avoid the temptation to memorize the speech. Memorization of a speech has two serious pitfalls. First, there is the danger it will sound like someone else's words and lack authenticity. Second, if the line of thought is lost during delivery, the speaker may not be able to find the location in the narrative. This can greatly damage the quality of the speech.

Conclusion The conclusion should remind the audience of the major theme or central points of the speech and briefly restate them. The conclusion should leave the audience feeling like they want to take action to implement or adopt what you have said. Some speeches call for action, whereas others call for changes in attitude or perception. The more powerful speeches move people to action. The words needed to do such a big job must be carefully planned.

Giving the Speech

Giving the speech can be fun and provide much satisfaction (Figure 6–10). However, this fun and satisfaction does not come easily. The speaker must prepare the plan well and practice the speech extensively. Practicing the speech until the content becomes familiar helps speaking become nearly automatic. The speech should be given orally to yourself several times. Then practice in front of a mirror to observe facial expressions, posture, and gestures. Finally, give the speech in front of others, and invite them to make suggestions to improve the delivery (Figure 6–11).

Books have been written on techniques to enhance speech delivery. However, for the beginner, a few basic and time-tested procedures should be helpful for effective speaking. Some suggested procedures for giving speeches are:

- Have your teacher read and make suggestions on the content of your written speech.

- Learn the content thoroughly through repeated thought and practice.

- Record the speech on a tape recorder and observe the sound, speed, power, and effectiveness of your voice. Make corrections to improve the delivery.

- Practice the speech in front of a mirror to observe posture, hand gestures, and facial expressions. Your posture should be erect and natural, with hands at your sides or resting lightly on the edges of the podium. Your hands should be used occasionally for gestures that emphasize a point, show direction, or indicate count.

- Ask your teacher for a score sheet for judging speeches. Deliver the speech in front of a trusted person who can check your delivery against the score sheet and provide suggestions for improvements. This may be a friend, relative, or teacher.

- Deliver your speech in front of your class for experience and suggestions.

- Ask your teacher to critique your speech for final approval.

Explanation of Score Sheet Points

Part I–For Scoring Content and Composition

1. *Content of the manuscript* includes:
   Importance and appropriateness of the subject
   Suitability of the material used
   Accuracy of statements included
   Evidence of purpose
   Completeness and accuracy of bibliography

2. *Composition of the manuscript* includes:
   Organization of the content
   Unity of thought
   Logical development
   Language used
   Sentence structure
   Accomplishment of purpose-conclusions

Part II–For Scoring Delivery of Production

1. *Voice* includes:
   Quality
   Pitch
   Articulation
   Pronunciation
   Force

2. *Stage Presence* includes:
   Personal appearance
   Poise and body posture
   Attitude
   Confidence
   Personality
   Ease before audience

3. *Power of expression* includes:
   Fluency
   Emphasis
   Directness
   Sincerity
   Communicative ability
   Conveyance of thought and meaning

4. *Response to questions* includes:
   *Ability to answer satisfactorily the questions on the speech which are asked by the judges indicating originality, familiarity with subject and ability to think quickly.

5. *General effect* includes:
   Extent to which the speech was interesting, understandable, convincing, pleasing, and held attention.

*NOTE: Judges should meet prior to the contest to prepare and clarify the questions asked.

National Public Speaking Contest   **Judge's Score Sheet**

PART I. FOR SCORING CONTENT AND COMPOSITION

Items To Be Scored	Points Allowed	Points Awarded Contestant												
		1	2	3	4	5	6	7	8	9	10	11	12	13
Content of Manuscript	200													
Composition of Manuscript	100													
Score on Written Production	300													

PART II. FOR SCORING DELIVERY OF THE PRODUCTION

Items To Be Scored	Points Allowed	Points Awarded Contestant												
		1	2	3	4	5	6	7	8	9	10	11	12	13
Voice	100													
Stage Presence	100													
Power of Expression	200													
Response to Questions	200													
General Effect	100													
Score on Delivery	700													

PART III. FOR COMPUTING THE RESULTS OF THE CONTEST

Items To Be Scored	Points Allowed	Points Awarded Contestant												
		1	2	3	4	5	6	7	8	9	10	11	12	13
Score on Written Production	300													
Score on Delivery	700													
TOTALS	1000													
* Less Overtime Deductions, for each minute or major fraction thereof	20													
* Less Undertime Deductions, for each minute or major fraction thereof	20													
GRAND TOTALS														
Numerical or Final Placing														

* From the Timekeeper's record.

FIGURE 6–11 A score sheet for evaluating speeches. *(Courtesy National FFA)*

- Deliver your speech in front of civic groups and/or in FFA public-speaking contests.
- Videotape your speech for later critique and review.

PARLIAMENTARY PROCEDURE

INTERNET KEY WORDS:

how to use parliamentary procedure

What is parliamentary procedure? Why are so many people familiar with it? Why is it important? Why should agriscience students be interested in learning parliamentary procedure? Parliamentary procedure is a system of guidelines or rules for conducting meetings. Most Americans who are influential in their communities are familiar with parliamentary procedure.

Parliamentary procedure is used to guide the meetings conducted by school groups, church groups, and civic organizations, such as Lions, Rotary, and Ruitan clubs. Agriscience students should be interested in learning this procedure so they can have their opinions heard and influence decisions that affect their lives. Parliamentary procedure is important because it permits a group to:

- discuss one thing at a time;
- hear everyone's opinion in a relaxed, courteous atmosphere;
- protect the rights of minorities; and
- make decisions according to the wishes of the majority of the group.

Requirements for a Good Business Meeting

A good business meeting is a gathering of people working together to make wise decisions. Wrong decisions cause unhappiness, loss of income, inefficiency in business and social activities, injury, and other problems. The most common outcome of poorly run business meetings is waste of time and lack of results. Meetings run by groups of individuals who know and use parliamentary procedure are smooth, efficient, orderly, brief, and get much done (Figure 6–12). Some requirements for a good business meeting are as follows:

Effective Presiding Officer A presiding officer is a president, vice president, or chairperson who is designated to lead a business meeting. He or she should be committed to the goals of the organization and should want to lead the group in making good group decisions. A good presiding officer must know and use proper parliamentary procedure.

Competent Secretary A secretary is a person elected or appointed to take notes and prepare minutes of the meeting. Minutes is the name of the official written record of a business meeting. Minutes should include the date, time, place, presiding officer, attendance, and motions discussed at the meeting. They should be written clearly, include all actions taken by the group, and be kept in a permanent secretary's book.

Informed Members Informed members are members who are active in the organization and want to be part of the group. They give previous thought to issues to be discussed and gather useful information about the issues. They share these thoughts with others in the meeting. This permits everyone to have the benefit of the best thinking in the group and permits the best decisions to be made. Effective members know and use parliamentary procedure.

FIGURE 6–12 Membership in the FFA provides opportunities for students to lead class discussions and to learn to conduct the business of an organization by taking a turn as the presiding officer. *(Courtesy National FFA; FFA #141)*

A Comfortable Meeting Room The meeting place must be comfortable and free of distractions. A moderate temperature and good lighting are essential. Members should be seated so they can hear and see each other. For small groups, seating at a table or in a circle works well. Large groups must rely on a good sound system for the presiding officer to be heard, and members must speak clearly with good volume to be heard. Good public speaking skills and thorough knowledge of parliamentary procedure help the members to conduct effective meetings.

Conducting Meetings

The Order of Business The order of business refers to the items and sequence of activities conducted at a meeting. The order of business is usually made up by the secretary. This generally grows out of an executive meeting. An executive meeting is a meeting of the officers to conduct the business of the organization between regular meetings. They also consider what needs to be discussed by the total membership at the regular meeting. The essential items in an order of business are:

- call to order
- reading and approval of minutes of the previous meeting
- treasurer's report
- reports of other officers and committees
- old business
- new business
- adjournment

Other items or activities that are frequently included in orders of business are programs, speakers, or entertainment.

FIGURE 6–13 The presiding officer and each member should take time to learn how to use the gavel as a tool to conduct the business of an organization. *(Courtesy National FFA; FFA #138)*

Parliamentary Practices

Use of the Gavel The gavel is a small wooden hammer-like object used by the presiding officer to direct a meeting (Figure 6–13). It is used to call the meeting to order, announce the result of votes, and adjourn the meeting. It is also used to signal the members to stand, sit down, or reduce the noise level of the group. The gavel is a symbol of the authority of the office of president or chairperson, and it should be respected by all attending the meeting.

In some organizations, such as the FFA, a system of taps is used to signal the audience to do certain things. In FFA meetings, the gavel is used as follows:

- one tap—the outcome of or decision about the item under consideration has been announced by the presiding officer
- two taps—the meeting will come to order, members should sit down if standing, or members should be quiet except when recognized
- three taps—members should stand up

Obtaining Recognition and Permission to Speak For a meeting to be orderly, members must speak one at a time and in some logical and fair sequence. The presiding officer is regarded as the "traffic controller," and calls on members as they request to be recognized according to certain rules. To be recognized, the member should raise a hand to get the presiding officer's attention. The presiding officer should call the member by name; then the person should stand and address the presiding officer as Madame or Mr. Chairperson, or Madame or Mr. President. The individual should then proceed to speak. Both the presiding officer and the members should understand the correct classification of motions (Figures 6–14 and 6–15).

Presenting a Motion A motion is a proposal, presented in a meeting, to be acted upon by the group. To present a motion, the member raises a hand and is recognized by the presiding officer. Then the member states, "Madame/Mr. President, I move that . . ." (and continues with the rest of the motion). The words "I move" are important to say when beginning the motion. Otherwise, you will be regarded as incorrect and not using good parliamentary procedure. For a motion to be discussed by the group, at least one other member must be willing to have the motion discussed. That second individual expresses this willingness by saying, "Madame/Mr. President, I second the motion."

Some Useful Motions There are dozens of motions, but a few are basic ones that are generally known and widely used. These include:

- **main motion**—a basic motion used to present a proposal for the first time. The way to state it is to obtain recognition from the chairman and then say, "I move . . ."
- **amend**—a type of motion used to add to, subtract from, or strike out words in a main motion. The way to present an amendment is to say, "I move to amend the motion by . . ."
- **refer**—a motion used to refer to a committee or person for finding more information and/or taking action on the motion. The way to state a referral is to say, "I move to refer this motion to . . ."

SUMMARY OF MOTIONS

Motion	Debatable	Amendable	Vote Required	Second	Reconsider
PRIVILEGED					
Fix time which to adjourn	No	Yes	Majority	Yes	Yes
Adjourn	No	No	Majority	Yes	No
Recess	No	Yes	Majority	Yes	No
Question of privilege	No	No	None	None	Yes
Call for orders of the day	No	No	None/2/3	None	No
INCIDENTAL					
Appeal	Yes	No	Majority	Yes	Yes
Point of order	No	No	None	No	No
Parliamentary inquiry	No	No	None	No	No
Suspend the rules	No	No	2/3	Yes	No
Withdraw a motion	No	No	Usually none	No	No
Object consideration of question Negative vote only	No	No	2/3	No	Yes
Division of the question	No	Yes	Majority	Yes	No
Division of the assembly	No	No	No	No	No
SUBSIDIARY					
Lay on table	No	No	Majority	Yes	No
Previous question before vote	No	No	2/3	Yes	Yes
Limit debate	No	Yes	2/3	Yes	Yes
Postpone definitely	Yes	Yes	Majority	Yes	Yes
Refer to committee	Yes	Yes	Majority	Yes	Yes
Amend	Yes	Yes	Majority	Yes	Yes
Postpone indefinitely	Yes	No	Majority	Yes	Yes vote only
Main motion	Yes	Yes	Majority	Yes	Yes
UNCLASSIFIED					
Take from table	No	No	Majority	Yes	No
Reconsider	No	No	Majority	Yes	Negative vote only

For more details on parliamentary procedure, see a parliamentary procedure book such as *Robert's Rules of Order*.

FIGURE 6–14 Parliamentary skills are useful in FFA and other organizations. *(Courtesy National FFA)*

- **lay on the table**—a motion used to stop discussion on a motion until the next meeting. The way to table a motion is to say, "I move to table the motion."

- **point of order**—a procedure used to object to some item in or about the meeting that is not being presented properly. The procedure to use is to stand up and say, "Madame/Mr. President, I rise to a point of order!" The presiding officer should then recognize the member by saying, "State your point." The member then explains what has been done incorrectly.

- **adjourn**—a motion used to close a meeting. The procedure is to say, "I move to adjourn."

ORDER OF PRECEDENCE OF MOTIONS

In using the chart, a motion lower on the pyramid is out of order if a motion above it is being considered.

Whenever a motion has been decided, it loses its precedence, as it is no longer on the floor.

Key to symbols used:

- Amendable
- Debatable
- Not Debatable
- Requires 2/3 vote

ADJOURNMENT
RECESS
POINTS OF ORDER
LAY ON TABLE
PREVIOUS QUESTION
POSTPONE TO DEFINITE TIME
REFER TO COMMITTEE
AMENDMENT TO AMENDMENT
AMENDMENT TO MAIN MOTION
POSTPONE INDEFINITELY
THE MAIN MOTION

FIGURE 6-15 Correct order of precedence of motions. *(Courtesy National FFA)*

The FFA *Student Handbook* provides a listing of additional motions and explains how to use parliamentary procedure for more effective meetings.

Agriscience students are encouraged to develop effective leadership skills. The ability to work effectively as a member of a group is essential for all. The ability to function as a chairperson or officer creates more opportunities to serve and influence the communities in which we live. The development of self-confidence is essential. Self-confidence is a product of knowledge and skill. Therefore, all should strive to learn to speak well, to function in groups through parliamentary procedure, and to use the opportunities in FFA for personal development.

STUDENT ACTIVITIES

1. Write the Terms to Know and their meanings in your notebook.
2. Make a list of the many ways that you exercise leadership in your family, school, and community.
3. Develop a bulletin board showing the symbols of the FFA emblem.
4. Develop a bulletin board illustrating the purposes of the FFA. Include the FFA colors.
5. Write down five career development events and five proficiency awards for which you would like to try out. Discuss these with your classmates and teacher.
6. Prepare and present a 3-minute speech on an agriscience topic to your class.
7. Ask your teacher to let you use a gavel and direct a mock class or FFA meeting to learn parliamentary skills.
8. Form a parliamentary procedure team consisting of you and your classmates, and then demonstrate various parliamentary skills to the class.
9. Under the supervision of your teacher, develop an appropriate solution for a school or district issue. Nominate and appoint a small committee to present your idea, in the form of a main motion, to the principal or school board.
10. Consider running for a leadership position in your FFA chapter. Find which offices are available and make a list of the requirements.
11. Attend an FFA activity you have never participated in. This could be a local Chapter, District, State, Regional, or National activity or convention.

SELF EVALUATION

A. Multiple Choice

1. To lead is to
   a. manage.
   b. organize.
   c. show the way.
   d. all of the above.

2. Which is not a trait of a good leader?
   a. courage
   b. integrity
   c. selfishness
   d. tact

3. A leadership organization of the Cooperative Extension System is
   a. Boy Scouts.
   b. FFA.
   c. Girl Scouts.
   d. 4-H.

4. Membership in FFA is limited to youth who
   a. are in the country.
   b. enroll in an agriscience program in school.
   c. plan careers in agriscience.
   d. seek leadership training.

5. Which is *not* a purpose of FFA?
   a. develop leadership
   b. intelligent choice of agriscience occupations
   c. promote scholarship
   d. promote self above others

6. The symbol that signifies that the FFA is a national organization is the
   a. corn.
   b. eagle.
   c. owl.
   d. rising sun.

7. The first line of the FFA motto is
   a. doing to learn.
   b. earning to live.
   c. learning to do.
   d. living to serve.

8. One requirement for the Greenhand degree is
   a. prepare a plan for supervised agricultural experience.
   b. $70 earned from agriscience experience.
   c. school grades of C or above.
   d. unselfish attitude in FFA activities.

9. An FFA activity not generally organized as a career development event is
   a. dairy foods.
   b. forestry.
   c. land judging.
   d. agricultural sales.

10. One of the last items in an order of business is
    a. new business.
    b. officer reports.
    c. reading of the minutes.
    d. treasurer's report.

11. The largest part of a speech is the
    a. body.
    b. conclusion.
    c. introduction.
    d. summary.

12. The only acceptable way to start a motion is to say
    a. "I believe . . .".
    b. "I make a motion that . . .".
    c. "I move . . .".
    d. "I think . . .".

B. Matching

_____	**1.** Adjourn	a.	Present a new proposal
_____	**2.** Amend	b.	Leave it to a committee
_____	**3.** Lay on the table	c.	Correct some procedure
_____	**4.** Main motion	d.	Close the meeting
_____	**5.** Point of order	e.	Consider it at the next meeting
_____	**6.** Refer	f.	Change a motion
_____	**7.** Three taps of gavel	g.	Prepare minutes
_____	**8.** Secretary	h.	Members stand

Natural Resources Management

Stewards of the Land

Hikers, bikers, hunters, anglers, farmers, foresters, ranchers, caretakers, and occupants all must be good stewards of the land. All are dependent on the land and rely on the soil, air, water, wildlife, and other natural resources around us. Farmers, ranchers, and foresters rely on the land to grow the crops and animals of their businesses. Similarly, hunters and other recreational users of the land and water rely on the habitat to grow and sustain the plants and wildlife. All enjoy wildlife for sport and recreation.

Land for farming and ranching is typically owned by the families who occupy the land; and they have definite property rights to grant or deny access of others to their property for hunting, fishing, or other recreational pursuits. However, the good-citizen hunter or angler seeks permission of the owner to access private property and strives to protect or enhance the fish and wildlife population and habitat through legal and good stewardship practices. Both owners and good-citizen users share a love for animals and a respect for crops, pasture, and woodland. Both wish to conserve the quality of land, air, and water.

Farmers and ranchers can do much to encourage growth of food, cover, and habitat for wildlife. They interact with wildlife biologists, game specialists, game officers, and other public authorities to nurture game populations and enforce game laws. Hunters and anglers (including farmers and ranchers), wildlife specialists, and the general public all help keep wildlife populations in check by harvesting the excess game. Animals that are taken by humane and legal hunting and fishing practices are generally the excess mature males in the population that compete with the females and young for food and cover.

Game hunting limits are set by wildlife specialists. Scientific methods should be used in an effort to permit and encourage the removal of excess wild animals by hunting. Those animals that are not removed by hunting compete with young game animals for food and shelter. As a result, the young often perish by disease, predation, and starvation. This is nature's way of keeping animal and plant populations in balance.

Public lands are owned by federal, state, or local governments and require employees in many specialty occupations to care for them. Foresters, biologists, fish and game managers, game officers, park rangers, horticulturists, scientists, technicians, and others all contribute to the upkeep and improvement of public lands.

To be counted as good stewards of the land, the owners, managers, and users of both private and public lands must all cooperate to use and conserve soil, water, trees, crops, wild plants, livestock, wildlife, and wildlife habitats.

UNIT
7
Maintaining Air Quality

MATERIALS LIST

✓ pencil and paper

✓ several specimen aerosol cans

✓ Internet access

COMPETENCIES TO BE DEVELOPED

After studying this unit, you should be able to:

- define the term *air* and identify its major components.
- analyze the importance of air to humans and other living organisms.
- determine the characteristics of clean air.
- describe common threats to air quality.
- describe important relationships between plant life and air quality.
- discuss the greenhouse effect and global warming.
- list practices that lead to improved air quality.

SUGGESTED CLASS ACTIVITIES

1. Identify the pollutants that are most often responsible for reducing air quality. Create a bulletin board in the classroom that illustrates the sources of these pollutants.

2. Invite an official from a government agency such as the Environmental Protection Agency (EPA) or the Department of Environmental Quality (DEQ) to do a class presentation on air quality. After the presentation, conduct a visual inspection of your community to identify potential sources of air pollution. Be sure to note that visible emissions, such as steam, do not necessarily indicate serious pollution problems. Consider ways to control those pollutants that are identified.

3. In groups of three or four students, identify one source of air pollution. Make a suggestion for a new law, in the form of a main motion, that would improve the air quality in your community.

TERMS TO KNOW

air	nitrous oxides	ozone	greenhouse effect
water	tetraethyl lead	particulates	respiration
soil	carbon monoxide	pest	photosynthesis
habitat	radon	pesticide	
sulfur	radioactive material	toxic	
hydrocarbons	chlorofluorocarbons (CFCs)	asbestos	

Life as we know it on our planet requires a certain balance of unpolluted air, water, and soil. **Air** is a colorless, odorless, tasteless mixture of gases. It occurs in the atmosphere around the Earth and is composed of approximately 78 percent nitrogen, 21 percent oxygen, and 1 percent mixture of argon, carbon dioxide, neon, helium, and other gases (Figure 7–1). **Water** is a clear, colorless, tasteless, and nearly odorless liquid. Its chemical makeup is two parts hydrogen to one part oxygen. **Soil** is the top layer of the Earth's surface that is suitable for the growth of plant life.

AIR QUALITY

INTERNET KEY WORDS:

air quality standards

air pollutants

air pollution organizations

Without a reasonable balance of air, water, and soil, most organisms would perish. Slight changes in the composition of air or water may favor some organisms, whereas causing others to diminish in number or in health. Unfavorable soil conditions usually mean inadequate food, water, shelter, and other factors related to habitat. **Habitat** is the area or type of environment in which an organism or biological population normally lives (Figure 7–2).

Threats to Air Quality

The mixture of gases we call air is absolutely essential for life. The air we breathe should be healthful and life supporting. Air must contain approximately 21 percent oxygen for human survival. If a human stops breathing and no life-supporting equipment or procedures are used, the brain will die in approximately 4 to 6 minutes. Air may contain

COMPONENTS OF AIR

Argon, Helium, Neon, Carbon Dioxide, Other gases — 1%

Oxygen — 21%

Nitrogen — 78%

FIGURE 7–1 The atmosphere of the Earth is composed mostly of nitrogen and oxygen and small amounts of a few other gases.

FIGURE 7–2 Clean air, clean water, and productive soil are necessary for a good habitat for plants, animals, and humans. *(Courtesy Michael Dzaman)*

FIGURE 7–3 Automobiles, trucks, homes, and factories burn gasoline, oil, coal, and wood, which release products of combustion that pollute the air. *(Courtesy Michael Dzaman)*

poisonous materials or organisms that can decrease the body's efficiency, cause disease, or cause death through poisoning.

The circumference or distance around the Earth at its largest part is 24,000 miles. Yet abuse of the atmosphere in one area frequently damages the environment in distant parts of the world. Air currents flow in somewhat constant patterns, and air pollution will move according to those patterns. However, when warm and cold air meet, the exact air movement will be determined by the differences in temperature, the terrain, and other factors. Because we cannot control the wind, humans have an obligation to keep the air clean for their own benefit as well as that of society at large.

There are major worldwide threats to air quality. These threats include sulfur compounds, hydrocarbons, nitrous oxides, lead, carbon monoxide, radon gas, radioactive dust, chlorofluorocarbons (CFCs), pesticide spray materials, and others (Figure 7–3). Most of these products are poisonous to breathe, and they have other damaging effects.

Sulfur   Sulfur is a pale yellow element occurring widely in nature. It is present in coal and crude oil. It also combines with oxygen to form harmful gases, such as sulfur dioxide, when these and other fuels are burned. Most smoke and exhaust from homes, factories, or motor vehicles contains some harmful compounds unless special equipment is used to remove them. Once these invisible gases are in the air, they combine with moisture to form sulfuric acid, which falls as acid rain. Acid rain has been found to damage and kill trees and other plants. It also has a corrosive effect on metals.

Hydrocarbons   Hydrocarbons have become serious problems as factories and motor vehicles have become so prevalent during the 20th century. In the United States, hydrocarbon emissions (by-products of combustion or burning) are held in check by special emission-control equipment on automobiles and special equipment called stack scrubbers in large industrial plants. Hydrocarbon output is controlled on automobiles by crankcase ventilation, exhaust gas recirculation, air injection, and engine refinements, such as four valves per cylinder. Without this equipment, air pollution would be much more intense in our major cities and in heavy stop-and-go traffic on major highways.

Nitrous Oxides and Lead   Nitrous oxides are compounds containing nitrogen and oxygen. They make up about 5 percent of the pollutants in

Science Connection

NATURAL SELECTION AS A RESULT OF POLLUTION

Air pollution affects many aspects of life on earth. An example of this was observed in England during the industrial revolution. In the early 1800s, coal was largely used as fuel to power factories. The burning of coal polluted the air with dark soot. Some of this soot would settle out of the air and fall to the earth. Moths are small-winged insects that spend the daylight hours hiding on the bark of trees from birds. Their wings are camouflaged to match the trees they hide on. In time, the soot darkened the bark of the trees. The majority of the moths had lighter colored wings. They became easy targets for predators. Because the tree bark was now darker, the moths with slightly darker wings were able to survive and reproduce. The lighter winged moths were not. The change in the moths' wing color was influenced by the pollution. Later, in the 1900s, this pollution was reduced. The tree bark returned to its original color, as did the wing color of the moths. This phenomenon is called natural selection.

INTERNET KEY WORDS:

tetraethyl lead gasoline

INTERNET KEY WORDS:

pollution, radon
pollution, carbon monoxide
radioactive dust pollution
chlorofluorocarbon pollution
ozone pollution

automobile exhaust. Yet they are damaging to the atmosphere and must be removed from exhaust gases. This group of chemicals is the most difficult and perhaps the most costly pollutant to remove from automobile emissions. Scientists and engineers solved the problem partly by installing catalytic converters in automotive exhaust systems.

Hot exhaust gases from the engine flow through a honeycomb of platinum in the catalytic converter. The reaction converts the nitrous oxides into harmless gases. Prior to 1986, all gasoline contained **tetraethyl lead**, a colorless, poisonous, oily liquid. This lead improved the burning qualities of gasoline and improved the control of engine knocking. However, tetraethyl lead ruined catalytic converters. Therefore, catalytic converters could not be used until a substitute for tetraethyl lead was found in the 1970s.

Tetraethyl lead is a poisonous product that is still used in third world countries. Lead and nitrous oxides are still major pollutants of the atmosphere, but the level of pollution has been reduced as the use of lead in gasoline has declined. Today, only fuel for small airplanes contains tetraethyl lead.

Carbon Monoxide   **Carbon monoxide** is one of the automotive gases that cannot be removed with our present technology. It is the colorless, odorless, and poisonous gas that kills people in automobiles with leaking exhaust systems. A similar hazard exists when engines are run in closed areas without adequate ventilation. Victims fall asleep and die. Carbon monoxide emissions may be reduced by keeping engines in good repair and properly tuned.

Radon   Radon has become a hazard in homes in many parts of the United States. **Radon** is a colorless, radioactive gas formed by disintegration of radium. It moves up through the soil and flows into the atmosphere at low and usually harmless rates.

However, if a house or other building is constructed over an area where radon gas is being emitted, a hazardous condition can result. Radon gas can accumulate in buildings with cracks in the basement floor or walls, or it can enter through sump holes. The problem can be prevented by tightly sealing all cracks and/or providing continuous

Agri-Profile

CAREER AREA:

Air Quality Control

Careers in air quality are available with the weather services of local, state, and national agencies. Local and network radio and television stations all use weather reporters and meteorology forecasters. Of equal importance are those who monitor and help to improve air quality. Technicians collect and chemists analyze samples of air taken from various places in the atmosphere, buildings, and homes. Employees of environmental protection agencies and environmental advocacy groups are important links in our efforts to maintain a healthful environment.

Air quality specialists advise and assist industry in reducing harmful emissions from motor vehicles and industrial smokestacks. Not to be overlooked are the entomologists, who monitor the winds for signs of invading insects, and plant pathologists, who watch for airborne disease organisms. Career opportunities in air quality maintenance and improvement undoubtedly will increase in the future.

Weather scientists at work.
(Courtesy USDA/ARS #K-2228-11)

ventilation either below the basement floor or throughout the building (Figure 7–4).

Radioactive Dust and Materials Radioactive material is material that is emitting radiation. Dust resulting from an atomic explosion or other nuclear reaction, and materials contaminated by atomic accidents or

POTENTIAL RADON ENTRY ROUTES
A. Cracks in concrete slabs.
B. Spaces behind veneers supported on uncapped hollow walls.
C. Pores, cracks, and mortar joints in un-treated concrete blocks.

D. Floor/wall joints
E. Exposed soil.
F. Weep (drain) tile, if drained to an open sump.
G. Openings around pipe penetrations.
H. Open tops of hollow walls.
I. Building materials.
J. Water from private wells.

FIGURE 7–4 Ventilation systems must be correctly designed and carefully maintained to keep interior areas free of radon gas pollution.

wastes are of growing concern. The damage from radioactivity ranges from skin burns to sickness to hereditary damage to death. Controversy over the possibility of worldwide contamination and other hazards from serious nuclear accidents has caused a reduction in construction of atomic-powered electric generation plants.

Chlorofluorocarbons Chlorofluorocarbons (CFCs) are a group of compounds consisting of chlorine, fluorine, carbon, and hydrogen. They are used as aerosol propellants and refrigeration gas. These materials are very stable. Once released from an aerosol can or cooling system, they bounce around in the air and eventually float upward into the high atmosphere. It is believed that CFCs will survive in the upper atmosphere for about 100 years. Meanwhile, their chlorine atoms destroy ozone without themselves being destroyed. In newer equipment, CFCs are being replaced by less-polluting agents.

Ozone is a compound that exists in limited quantities about 15 miles above the Earth's surface. It filters out harmful ultraviolet rays from the sun. There is evidence that the ozone layer is being damaged. Increased problems with skin cancer and damage to the body's immune system are occurring. Most living organisms will be exposed to the damaging effects of ultraviolet rays.

In 1987, there was an historic international conference where at least 37 nations agreed to schedule cutbacks in the production of CFCs. However, given the possible damage by this pollutant, it may make sense to stop all production immediately.

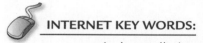

INTERNET KEY WORDS:

particulate pollution
global warming,
greenhouse effect,
climatic change

Particulates Small particles that become suspended in air are known as particulates. These tiny particles appear as smoke or dust clouds. Particulates eventually will settle out of the air because of gravity, but they are so light in weight that the slightest breeze keeps them suspended. They are especially harmful to people who suffer from respiratory diseases such as asthma or emphysema. Most particulate matter coming from industrial processes can now be removed from gas emissions by a process called *scrubbing*.

Pesticide Spray Materials A pest is a living organism that acts as a nuisance. Examples include house flies, cockroaches, fleas, and mosquitoes. A pesticide is a material used to control pests. Many pesticides are chemicals mixed with water so they can be sprayed on plants, animals, soil, or water to kill or otherwise control diseases, insects, weeds, rodents, and other pests. Spray materials are pollutants if they carry toxic (poisonous) materials or are harmful to more than the target organism. Spray materials used to control pests, such as insects, weeds, rodents, and others, are generally harmful to the air if they are not used exactly as specified by the government and the manufacturer. Poisons may be thinned out or diluted by air movement, but excessive toxic materials can overburden the ability of the atmosphere to cleanse itself. Abuse of chemicals to control pests is an area of growing concern in maintaining air quality.

Asbestos Asbestos is a heat- and friction-resistant material. It was used extensively in the past for brake and clutch linings of vehicles, shingles for house siding, steam and hot-water pipe insulation, ceiling panels, and other products.

Unfortunately, asbestos fibers are damaging to the lungs and cause disease and death. There are now state and federal laws and codes requiring the removal of asbestos from public buildings, industrial settings, and general use.

The Greenhouse Effect and Global Warming

It is well understood that one of the factors that makes it possible for life to exist on Earth is the proximity of Earth to the sun. If the distance between the sun and Earth was any less, the surface of the planet would become too hot to sustain life. If the distance was any greater, the temperature would be too cold to support many life forms. Sunlight also has other effects on living organisms. For example, excessive heat and

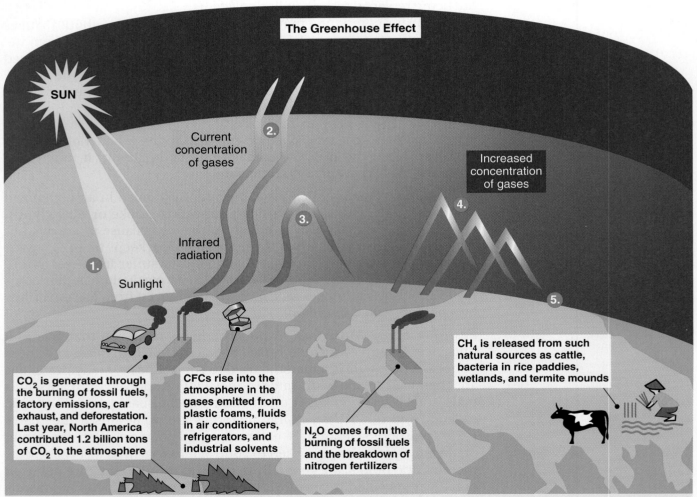

The Greenhouse Effect

SUN

Current concentration of gases

2.

Increased concentration of gases

4.

3.

Infrared radiation

1.

Sunlight

5.

CH$_4$ is released from such natural sources as cattle, bacteria in rice paddies, wetlands, and termite mounds

CO$_2$ is generated through the burning of fossil fuels, factory emissions, car exhaust, and deforestation. Last year, North America contributed 1.2 billion tons of CO$_2$ to the atmosphere

CFCs rise into the atmosphere in the gases emitted from plastic foams, fluids in air conditioners, refrigerators, and industrial solvents

N$_2$O comes from the burning of fossil fuels and the breakdown of nitrogen fertilizers

1. Sunlight travels through the atmosphere and warms the Earth's surface.

2. The surface radiates heat (infrared radiation) back into the atmosphere, where some of it escapes into space.

3. Some rays are absorbed by water vapor and other greenhouse gases in the atmosphere. These gases act like the panes of glass in a greenhouse by trapping and reflecting the heat back to Earth.

4. As greenhouse gases from factory emissions and other sources build in the atmosphere, more heat is trapped and reflected to Earth. The oceans warm, producing more water vapor, which traps still more heat.

5. The temperature of the planet rises to achieve a new balance.

FIGURE 7–5 The greenhouse effect on land, sea, and air. *(Adapted from material provided by Electric Power Research Institute)* drought of summer is frequently accompanied by parched and withered crops. Authorities also have found a definite link between extensive exposure to the Sun and the occurrence of the deadly melanoma skin cancer. Furthermore, there is some scientific evidence suggesting that skin-damaging and life-threatening ultraviolet rays from the sun now reach the Earth's surface with greater intensity than in the past.

The Greenhouse Effect   When sunlight passes through a clear object, such as a glass window, it heats the air on the opposite side of that object. If the warmed air is not cooled or flushed out, heat builds up under the glass. Examples of the phenomenon may be observed under skylights, auto windshields, and in greenhouses. The glass also absorbs some of the energy from the light and gives it off or radiates it as heat to the interior area. The result is a buildup of heat causing the interior to become warmer than the outside. This heat buildup from the rays passing through the clear object and the resulting heat being trapped inside is known as the greenhouse effect.

The sun's rays include many different colors of light and types of rays. Some of the more familiar rays are ultraviolet and infrared. The ultraviolet rays are known for their extensive skin damage and other life-threatening effects of overexposure to the sun. Infrared rays are emitted from any warm object, such as a hot stove, glass warmed by the passage of ultraviolet rays, or the presence of an open fire.

The gases in the Earth's atmosphere serve as a relatively clear object through which sunlight passes. The crust of the Earth absorbs and radiates, as well as reflects, heat back into the air above it. The atmosphere encircles the Earth and creates the greenhouse effect (Figure 7–5).

Scientists do not agree on the concept of global warming; but they do report that the buildup of heat within the global "greenhouse" is increasing and is creating changes in average temperatures. This increased warming trend is believed by many to be a serious threat to the environment and to life itself. If the warming trend continues, scientists contend that the greenhouse effect must be stabilized or reduced if we are to protect air quality (Figure 7–6).

Predicted Change	Global Average	Regional Average
Temperature	+4° to +9° F	−5 to + 18° F**
Sea Level	+4 to +40 in	
Precipitation	+7% to +15%	−20 to +20%
Direct Solar Radiation	−10 to +10%	−30 to +30%
Evaporation/Transpiration	+5 to +10%	−10 to +10%
Soil Moisture	?	−50 to +50%
Runoff	increase	−50 to +50%

Source: Schieder, S. (1990). Prudent planning for a warmer planet. *New Scientist,* 128(1743).
**To interpret, if the greenhouse effect produces the results that some scientists predict, these kinds of changes could occur. The average global temperature could increase between 4°and 9° F. Regional changes could range from a drop of 5° to an increase of 18° F in different parts of the world. Other categories of possible changes are given.

FIGURE 7–6 Global warming is believed to be caused by the greenhouse effect.

Hole in the Ozone Layer   The ozone layer over the South Pole has been found to be less dense, or thinner, than in the past. The thinner ozone layer lets more of the damaging ultraviolet rays through. This thinner area is referred to as a "hole in the ozone layer," which is enlarging at an alarming rate. Scientists have identified the buildup of CFCs as the cause of this hole. Of equal concern is the buildup of air pollutants that could contribute to global warming even where the ozone layer has been damaged less than over the South Pole. Scientists and governments of the world are trying to find ways to make continued improvements in living conditions without experiencing additional reductions in air quality and other environmental factors.

Global Warming—A Topic for Scientific Debate   Scientists around the world are divided on the issue of global warming. Some of them insist that a recent upward trend in the temperature of Earth's surface is evidence of dangerous warming of the global environment if current trends continue (Figure 7–7). They cite human activities as factors that contribute to the intensity of the greenhouse effect and to global warming.

Other scientists cite the lack of scientific data to support the global warming theory. They believe that the trend toward a rise in global temperatures may be partly because of long-term climatic cycles and weather patterns. Daily temperatures have been recorded for a little more than 100 years, and we do not have actual historical records from which to draw conclusions on climatic cycles and weather patterns of

Carbon Dioxide (CO₂) The relative contribution of the greenhouse gas CO_2 to the global warming trend is expected to be about 50 percent by 2020

Chlorofluorocarbons (CFCs) About 25 percent of greenhouse effect by 2020

Methane (CH₄) About 15 percent of greenhouse effect by 2020

Nitrous Oxide (N₂O) About 10 percent of greenhouse effect by 2020

FIGURE 7–7 Increases in the concentrations of four major pollutants and their predicted effects on global warming by 2020.

longer duration. An example of this line of thinking is evident in the following statement by Dr. Roy Spencer, Senior Scientist for Climate Studies, NASA/Marshall Space Flight Center:

> The adjusted satellite trends are still not near the expected value of global warming predicted by computer climate models. The Intergovernmental Panel on Climate Change's (IPCC) 1995 estimate of average global warming at the surface until the year 2100 is +0.18 deg. C/decade. Climate models suggest that the deep layer measured by the satellite and weather balloons should be warming about 30% faster than the surface (+0.23 deg. C/decade). None of the satellite or weather balloon estimates are near this value.

In contrast with this scientific view is a scientists' statement on global climatic disruption that was issued in June 1997. The text of the statement is as follows:

> We are scientists who are familiar with the causes and effects of climatic change as summarized recently by the Intergovernmental Panel on Climate Change (IPCC). We endorse those reports and observe that the further accumulation of greenhouse gases commits the earth irreversibly to further global climatic change and consequent ecological, economic, and social disruption. The risks associated with such changes justify preventive action through reductions in emissions of greenhouse gases. In ratifying the Framework Convention on Climate Change, the United States agreed in principle to reduce its emissions. It is time for the United States, as the largest emitter of greenhouse gases, to fulfill this commitment and demonstrate leadership in a global effort.
>
> Human-induced global climatic change is under way. The IPCC concluded that global mean surface air temperature has increased by between about 0.5 and 1.1 degrees Fahrenheit in the last 100 years and anticipates a further continuing rise of 1.8 to 6.3 degrees Fahrenheit during the next century. Sea-level has risen on average 4–10 inches during the past 100 years and is expected to rise another 6 inches to 3 feet by 2100. Global warming from the increase in heat-trapping gases in the atmosphere causes an amplified hydrological cycle resulting in increased precipitation and flooding in some regions and more severe aridity in other areas. The IPCC concluded that "The balance of evidence suggests a discernible human influence on global climate." The warming is expected to expand the geographical ranges of malaria and dengue fever and to open large new areas to other human diseases and plant and animal pests. Effects of the disruption of climate are sufficiently complicated that it is appropiate to assume there will be effects not now anticipated.
>
> Our familiarity with the scale, severity, and costs to human welfare of the disruptions that the climatic changes threaten leads us to introduce this note of urgency and to call for early domestic action to reduce U.S. emissions via the most cost-effective means. We encourage other nations to join in similar actions with the purpose of producing a substantial and progressive global reduction in greenhouse gas emissions beginning immediately. We call attention to the fact that there are financial as well as environmental advantages to reducing emissions. More than 2000 economists recently observed that there are many potential policies to reduce greenhouse-gas emissions for which total benefits outweigh the total costs. . . .

The United Nations Framework Convention on Climate Change met in Milan, Italy, in 2003. There were 188 parties to the convention, and the following statement is part of a press release that was issued:

INTERNET KEY WORDS:

Framework Convention on Climate Change

FIGURE 7–8 A lawn mower may create 50 times more pollution per horsepower than a modern truck engine. *(Courtesy National FFA; FFA #28)*

Ministers agreed that climate change remains the most important global challenge to humanity and that its adverse affects are already a reality in all parts of the world.

Precautions Against Global Warming   Reversing the perceived trend toward continued global warming and the problems it could bring may require real changes in the way we do some things. With good research, wise government, and environmentally sensitive business, we can slow down or stop the decline of our air quality brought about by human-caused pollution. Even small-engine lawn and garden equipment is now seen as serious contributors to pollution, and changes are needed to address this problem. For instance, leaf blowers create pollution through engine exhaust. Pollution is created by the generation of electricity consumed by the electric motor when expelling dust and dirt into the air. Experts now observe that a gasoline lawn mower can cause 50 times more pollution per horsepower than a modern truck engine (Figure 7–8). Similarly, a lawn mower running 1 hour may create as much pollution as an automobile traveling 240 miles. A chain saw running 2 hours emits as much pollution from hydrocarbons as a new car running from coast to coast across the United States (Figure 7–9).

Carbon Dioxide (CO$_2$)   Carbon dioxide is a major product of combustion. Our robust appetites for food, clothing, consumer goods, heated and air conditioned spaces, and transportation have led to a lifestyle that requires huge volumes of fuels to be burned to raise food, manufacture goods, and move vehicles. Our highways in and around large cities are clogged with automobiles. Interstate highways are loaded with trucks, and rivers and other waterways carry heavy shipping traffic. Homes and commercial buildings use extensive volumes of electricity for light and temperature control, and farms and factories have huge machines and heating devices—all consuming fuel that expels carbon dioxide and other pollutants into the air.

Green plants can use some carbon dioxide from the air and convert it into plant food, oxygen, and water. However, in most parts of the world, the mass of green plants is being reduced, whereas the expulsion of pollutants increases dramatically. The practice of slash and burn in the jungle regions of the world threatens to remove vast areas and volumes of plant growth with little hope of replacement. Increased carbon dioxide in the atmosphere is projected to account for 50 percent of the increase in global warming by 2020.

Chlorofluorocarbons (CFCs)   Since discovery of the damage that CFCs do to the ozone layer, other propellants have been used in pressurized spray containers, and CFCs are now recaptured from refrigeration units before they are discarded. Still, the escape of CFCs into the air and their effect on the ozone layer is projected to account for 25 percent of the increase in global warming by 2020.

Methane (CH$_4$)   Most methane gas comes from naturally decaying plant materials such as leaves and debris on the soil surfaces of forests and jungles. Some of it is a product of decaying organic matter, such as human and animal waste. Methane rises from piles, pits, and other accumulations of animal manure, peat bogs, sewage, and other decaying waste. Some large farms are now using carefully engineered systems to capture the gas from large manure holding areas. The methane gas is then used as fuel for engines driving generators to provide electricity for

FIGURE 7–9 A chain saw is powered by a two-cycle engine that emits large amounts of air pollutants. *(Courtesy National FFA; FFA #186)*

Science Connection

OVERCOMING THE EFFECTS OF AIR POLLUTION

U.S. Department of Agriculture (USDA) scientists are tackling what could be the toughest battle of the century—the battle to breathe. Whereas ozone depletion is a serious problem in the upper atmosphere, ozone as a product of combustion hovers just above the Earth's surface as a pollutant that decreases air quality.

It has been estimated that U.S. farmers experience at least 1 billion dollars per year in lost crop yields because of air pollution. Research indicates that cutting the level of ozone in the air by 40 percent would mean an extra $2.78 billion for agricultural producers.

USDA and University of Maryland scientists discovered that treatment of plants with the growth hormone Ethylenediura (EDU) can reduce damage by ozone. One soil drench was found to effectively protect some plants from damage and reduce the sensitivity of others to excessive ozone in the air. Similarly, injection of EDU in the stems of shade trees could protect them from damage if located in highly polluted areas. EDU alters enzyme and membrane activity within the leaf cells where photosynthesis takes place.

A biological aide uses a porometer to measure the impact of CO_2 enrichment of the air on the transpiration rate and stomata activity of a soybean leaf. *(Courtesy USDA/ARS #K-3204-1)*

In addition to ozone, other major air pollutants that are damaging to crops include peroxyacetetyl nitrate, oxides of nitrogen, sulfur dioxide, fluorides, agricultural chemicals, and ethylene. The task of reducing the amounts of pollutants in the air and finding ways to reduce the effects of pollutants on living organisms will continue to challenge future generations.

the farm and for sale. Methane is projected to account for 15 percent of the increase in global warming by 2020.

Nitrous Oxide (N₂O) Nitrous oxide has long been a troublesome pollutant from gasoline engines. The ever-increasing number of automobiles, trucks, tractors, heavy equipment, aircraft, chain saws, lawn motors, boats, and other engine-driven applications has off-set the tremendous improvements made in emission reduction from individual engines. Because nitrous oxide emissions are still increasing, it is projected that they will account for 10 percent of the increase in global warming by 2020.

Air and Living Organisms

Oxygen in the air is consumed by plants and animals during a process called respiration (Figure 7–10). Animals, as well as humans, use oxygen to convert food into energy and nutrients for the body. Animals breathe in or inhale to obtain oxygen. They exhale (breathe out) carbon dioxide gas. Plants give off oxygen during the day. They create oxygen through the process of photosynthesis (a process in which chlorophyll in green plants enables those plants to use light, carbon dioxide, and water to make food and release oxygen) (Figure 7–11).

FIGURE 7–10 Respiration is the process by which plant tissues are broken down to produce heat, water, and carbon dioxide.

Maintaining and Improving Air Quality

Air quality can be improved by reducing or avoiding the release of pollutants into the air and removing the pollutants that are there. Some specific practices to reduce air pollution are:

- Stop using aerosol products containing CFCs.
- Provide adequate ventilation in tightly constructed and heavily insulated buildings.
- Have buildings checked for the presence of radon gas.
- Use exhaust fans to remove cooking oils, odors, solvents, and sprays from interior areas.
- Clean and service furnaces, air conditioners, and ventilation systems regularly.

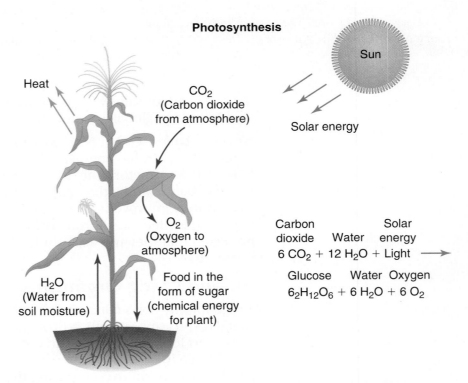

FIGURE 7–11 Photosynthesis is the process by which carbon dioxide is combined with water to store energy obtained from sunlight. Chlorophyll supports this reaction in which sugar and oxygen are produced.

Hot Topics in Agriscience

RESEARCH TO IMPROVE PHOTOSYNTHESIS

An area of emphasis for plant research is to improve the efficiency of the photosynthesis process. Scientists have identified an enzyme that plays an important role in photosynthesis. When it is present in large amounts, plant growth increases. Small amounts of the enzyme cause plants to grow slowly. In addition, the enzyme is known to react with oxygen during hours of darkness to reverse photosynthesis. This process is called respiration.

Scientists hope to be able to stimulate plant growth through a more efficient process of photosynthesis. They also hope to interrupt the respiration process during hours of darkness. Total plant yields will increase if science is successful in maintaining the gain in plant tissue that occurs during the day.

- Maintain all systems for the removal of sawdust, wood chips, paint spray, welding fumes, and dust to assure that they function at their most efficient levels.
- Keep gasoline and diesel engines properly tuned and serviced.
- Keep all emissions systems in place and properly serviced on motor vehicles.
- Observe all codes and laws regarding outdoor burning.
- Report any suspicious toxic materials or conditions to the police or other proper authorities.
- Reduce the use of pesticide sprays as much as possible.
- Use pesticide spray materials strictly according to label directions.

STUDENT ACTIVITIES

1. Write the Terms to Know and their meanings in your notebook.
2. Make a pie chart illustrating the components of air.
3. Stand in a safe location about 5 ft behind and to the side of a parked truck or bus with its gasoline engine running. Notice the smell of the exhaust. Stand in the same relative position from an automobile that is less than 3 years old or has traveled less than 40,000 miles. What differences do you observe in the truck and automobile exhausts? Why? Which do you think causes more pollution of the air?
4. Examine three different aerosol cans. Which products use chlorofluorocarbons for the propellant? Why is it unwise to use such products?
5. Talk with an automobile tune-up specialist about the effect of engine adjustments on the content of exhaust gases.
6. Ask your teacher to invite an air pollution specialist to your class to discuss the problems of air pollution in your town, county, or state.
7. Do a research project on the greenhouse effect and its relationship to global warming.
8. Obtain a radon test kit and perform a test for radon in your home.
9. Create four original drawings showing the four major pollutants. Under each of the drawings, write one sentence that explains where this pollution comes from and what can be done to improve air quality.

SELF EVALUATION

A. Multiple Choice

1. Air is
   a. 78 percent argon.
   b. 21 percent nitrogen.
   c. 21 percent oxygen.
   d. 10 percent carbon dioxide.

2. Pure water is
   a. a mixture of gases.
   b. metallic tasting.
   c. one part hydrogen to two parts oxygen.
   d. odorless.

3. Without proper air to breathe, a human can survive only about
   a. 6 minutes.
   b. 12 minutes.
   c. 2 hours.
   d. 12 hours.

4. Radon gas is a widespread threat to air quality
   a. on the highway.
   b. in factories.
   c. in homes.
   d. in wooded areas.

5. Radioactive dust is likely to be caused by
   a. improperly adjusted furnaces.
   b. cracks in basement floors.
   c. a damaged ozone layer.
   d. nuclear reactions.

6. Chemicals used to kill insects are called
   a. pests.
   b. pesticides.
   c. pollutants.
   d. toxic materials.

7. One ingredient *not* associated with photosynthesis is
   a. carbon dioxide.
   b. oxygen.
   c. radon.
   d. water.

8. Chlorofluorocarbons have been found to damage
   a. aerosol sprays.
   b. the ozone layer.
   c. refrigeration units.
   d. water pumps and equipment.

9. Poisonous gas we cannot remove from auto exhaust is
   a. carbon monoxide.
   b. hydrocarbons.
   c. nitrous oxides.
   d. radon.

10. The most reliable source of information on the use of a pesticide is
    a. experienced applicators.
    b. an extension service.
    c. personal experience.
    d. the product label.

11. The buildup of heat resulting from sunlight passing through glass and heating trapped air in the interior area is called
    a. the greenhouse effect.
    b. infrared energy.
    c. radiation.
    d. ultraviolet.

12. A product of decaying plant or animal matter is
    a. chlorofluorocarbons.
    b. methane.
    c. nitrous oxide.
    d. ozone.

B. Matching

_____	**1.** Carbon monoxide	a. Tetraethyl
_____	**2.** Chlorofluorcarbons	b. Pale yellow
_____	**3.** Hydrocarbons	c. 5 percent of auto exhaust
_____	**4.** Lead	d. Damages ozone layer
_____	**5.** Nitrous oxides	e. Filters ultraviolet rays
_____	**6.** Ozone	f. Diseases, insects, weeds
_____	**7.** Radon	g. Chlorophyll, light, carbon dioxide
_____	**8.** Pests	h. Causes death from auto exhaust
_____	**9.** Photosynthesis	i. Leaks into houses
_____	**10.** Sulfur	j. Pollutant from autos and factories

UNIT
8

Water and Soil Conservation

OBJECTIVE

To determine the relationships between water and soil in our environment and the recommended practices for conserving these resources.

MATERIALS LIST

✓ three growing plants in 4- to 6-inch pots for Student Activity 5

✓ kitchen scale or laboratory balance scale

✓ plant watering containers

✓ plant growing area

✓ Internet access

COMPETENCIES TO BE DEVELOPED

After studying this unit, you should be able to:

- define *water, soil,* and related terms.
- cite important relationships between land characteristics and water quality.
- discuss some major threats to water quality.
- describe types of soil water and their relationships to plant growth.
- cite examples of enormous erosion problems worldwide.
- describe key factors affecting soil erosion by wind and water.
- list important soil and water conservation practices.

SUGGESTED CLASS ACTIVITIES

1. Make a bulletin board display in the school or classroom about conservation of water and soil. Illustrate causes of soil and water losses, and illustrate ways to prevent losses. Include some statistics about the value of lost agricultural production caused by soil and water losses.

2. Prepare a class presentation and demonstration on soil and water conservation. Make arrangements with elementary school teachers to present the demonstration to elementary school students. Include a model demonstration on the effects of slope or soil type on soil loss caused by erosion by water. Assign teams of class members to give short presentations on conservation topics. Topics are easily identified on the World Wide Web using key words.

3. Add large amounts of three different pollutants to carefully labeled gallons of water. Leave another gallon of water unpolluted; this will be your control for comparison with the polluted treatments. Purchase four inexpensive houseplants. Apply one-half cup of a specific water treatment to each plant daily. Be sure to label each plant with the water pollutant that each has been given. Observe and discuss the short-term appearance and health of the plants after a few days, and then after a few weeks (long-term).

TERMS TO KNOW

potable	watershed	cover crop	crop rotation
fresh water	water table	erosion	organic matter
domestic	fertility	port	aggregate
tidewater	saturated	aquifer	lime
food chain	free water	sheet erosion	fertilizer
universal solvent	gravitational water	gully erosion	grass waterway
water cycle	capillary water	mulch	terrace
desert	hygroscopic water	conservation tillage	overgrazing
irrigation	purify	plant residue	conservation plan
precipitation	no-till	contour practice	
evaporation	contour	strip cropping	

INTERNET KEY WORDS:

conservation, soil, water

As you fly across the vast continents of the Americas, Europe, Asia, and Africa, there is a feeling that land is an endless resource (Figure 8–1). The great oceans of the world combine to provide an even larger area than that covered by the landmasses of Earth. Yet both land and water resources have become limited and there is real concern that we are rapidly depleting them. In developed countries, a safe and adequate water supply is generally threatened only by temporary shortages. Such shortages usually create only mild inconveniences, such as restrictions on the use of water for nonessential tasks such as washing cars and watering lawns. But in Third World countries, a safe water supply is a luxury. Although there seems to be a sufficient volume of water in most areas of the world, supplies of safe water are generally insufficient because of misuse, poor management, waste, and pollution.

Similarly, productive land is becoming a scarce commodity, and ownership is expensive. Good land is sought by individuals for homes, farms, and recreation. Land is also needed by businesses for banks, stores, warehouses, car lots, and other things. Industry needs land for factories and storage areas. Governments need land for roads, bridges, buildings, parks, recreation, and military facilities. Most parties look for the best land—land that is level, with deep and productive soil. Such land is in great demand because it provides a firm foundation for roads

FIGURE 8–1 It is difficult to view our land as a resource that is in short supply because the amount of farmland seems so vast. It is important to realize that we are already farming most of the land that is suited to producing crops. *(Courtesy USDA #K-5051-05)*

A

B

C

D

FIGURE 8–2 Good land is in high demand. (A) Housing developments, (B) Apartments, condominiums, business, and industry, (C) Roads and bridges, (D) Farm land. *(Courtesy of Elmer Cooper)*

INTERNET KEY WORDS:

American Farmland Trust

and buildings, grows crops, supports trees and shrubs, and is easy to modify according to the desired use (Figure 8–2).

THE NATURE OF WATER AND SOIL

Water

We live on the water planet! Most of the Earth's surface is covered with water (Figure 8–3). The oceans and lakes are vast. Most people in the United States live near an ocean, river, lake, or stream. Those who do not live near bodies of water have access to water from deep wells. Our bodies and the tissues of plants and animals are about 90 percent water. Therefore, we can survive only a few days if our supply of **potable** (drinkable—that is, free from harmful chemicals and organisms) water is cut off.

Water is an essential nutrient for all plant and animal life (Figure 8–4). Water transports nutrients to living cells and carries away waste products. Water cools the body. It serves so many useful functions that a sufficient supply of potable water is one of the first considerations for a healthy community.

FIGURE 8–3 Most of the Earth's surface is covered by water. *(Courtesy Chesapeake Bay Foundation)*

FIGURE 8–4 Clean, fresh water serves many life-supporting functions including nutrient carrier, waste transporter, coolant, home for aquatic plants and animals, oxygen for fish, and cleanser for humans and animals. *(Courtesy Wendy Troeger)*

INTERNET KEY WORDS:

food chain web

Fresh versus Salt Water   Most of the water on Earth is not fresh water and is not suitable for humans to drink or use except for transportation. Fresh water refers to water that flows from the land to oceans and contains little or no salt. The water in our oceans contains heavy concentrations of salt. Similarly, our bays and tidewater rivers contain too much salt for domestic or household use. Tidewater refers to the water that flows up the mouth of a river as the ocean tide rises or comes in. Salt water is not fit for animal consumption or for plant irrigation.

Aquatic Food Chains and Webs   A food chain is made up of a sequence of living organisms that eat and are eaten by other organisms living in the community (Figure 8-5). Each member of the chain feeds on lower ranking members of the chain. The general organization of an aquatic food chain moves from organisms known as producers (food plants) to herbivores (plant-eating water insects). Herbivores are eaten, in turn, by carnivores (meat-eating animals).

The Universal Solvent   Water has been described as the universal solvent (a material that dissolves or otherwise changes most other materials). Nearly every substance will rust, corrode, decompose, dissolve, or otherwise yield to the presence of water. Therefore, water is seldom seen in its pure form. It generally has something in it.

Some minerals in water are healthful and give the water a desirable flavor. However, water sometimes carries toxic or undesirable chemicals or minerals. Water may contain decayed plant or animal remains, disease-causing organisms, or poisons. Ocean water may be described as a thin soup. It is like our blood. It gathers and transports nutrients, and it is the habitat for microorganisms. It carries life-supporting oxygen, and it neutralizes and removes wastes.

Scientists tell us that the purity levels of our rivers, bays, and seas are in trouble. People are polluting air, water, and land faster than nature can cleanse and purify these resources (Figure 8–6). The sea contains all of the dissolved elements carried by the rivers through all time to the low places of the Earth's crust. The water itself may evaporate and be carried back to the land as moisture in the clouds. Eventually it may fall to the land again as rain or snow. However, the minerals remain in the

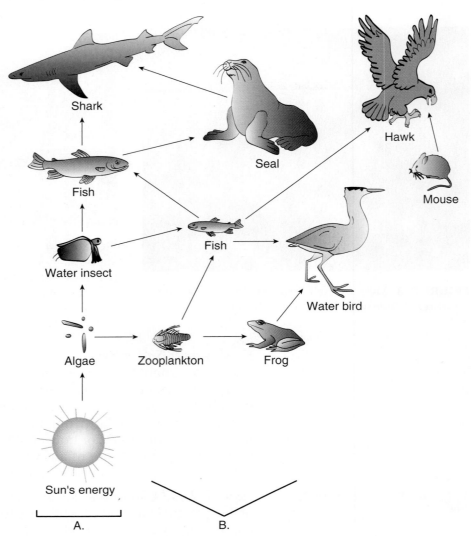

FIGURE 8–5 (A) A food chain model showing nutritional energy flowing in one direction. (B) A food web model showing the nutritional energy flow throughout the ecosystem.

Science Connection

FOOD CHAINS AND WEBS

Bays and oceans provide excellent support for growth of algae. Algae, in turn, become a major food source for water-dwelling insects. Insects provide nutritional energy for shellfish. Shellfish are eaten by larger fish that are finally consumed by sharks. In this example, energy is passed from one organism to another in the form of food. This situation is referred to as a food chain. A food chain is a model that describes the interdependence of plants and animals for nutrition. All food chains begin with the sun, which is the primary energy source for most living organisms.

Producers capture the energy of the sun and convert it into nutrients such as sugars and proteins. Plants and algae are good examples of producers. The next step in the food chain takes place when a consumer eats a producer. Animals that depend on producers or other animals for food are called consumers. In a food chain, energy is passed from the sun to the producers, and finally to consumers in one straight line. In most ecosystems, feeding relationships are much more complex than a single food chain. Food webs represent these situations. A food web is made up of many different overlapping food chains. They are the sum of all feeding relationships in an ecosystem. Using the example above, it is easy to see the involvement of other organisms. For example, birds will also eat the fish, as do the sharks. Water insects may eat different aquatic plants than algae. Foods chains and food webs explain the feeding relationships in given ecosystems. Both of these relationships can be observed on land and in water.

OUT OF A TOTAL AREA POPULATION OF MORE THAN 12 MILLION, 4 MILLION WERE ADDED IN THE LAST 30 YEARS.

AS FIN FISH AND SHELLFISH HABITAT IS LOST, WATERWORKERS LOSE THEIR LIVELIHOOD.

60 PERCENT OF BAY'S SUBMERGED AQUATIC VEGETATION HAS DISAPPEARED IN THE LAST TWO DECADES.

OSPREY AND EAGLE POPULATIONS HAVE RISEN SINCE DDT WAS BANNED. BUT PROBABLY HAVE NOT REACHED 1920 LEVELS.

WINTERING WATERFOWL GREATLY REDUCED. SPECIES INCLUDE: PINTAILS, PIGEONS, BLACK DUCKS, CANVASBACKS, AND REDHEADS.

BALD EAGLES ARE STILL THREATENED AS SHORELINE IS DEVELOPED AND OLD FORESTS ARE CUT.

SOFT-SHELLED CLAMS HAVE DECLINED EVEN MORE THAN OYSTERS.

ROCKFISH (STRIPED BASS) POPULATIONS SEVERELY DEPRESSED. COMMERCIAL CATCHES FELL FROM 6 MILLION POUNDS IN 1970 TO 600,000 POUNDS IN 1983.

SHAD CATCHES LOWEST EVER: 3 MILLION POUNDS IN 1984, DOWN FROM PEAK OF 50 MILLION IN 1890s.

OYSTER HARVESTS HAVE DROPPED BY TWO-THIRDS IN THE LAST 20 YEARS.

CRAB POPULATIONS FLUCTUATE GREATLY YEAR TO YEAR.

BAY FACTS

SIZE: 195 miles long, 3.5 to 30 miles wide. Average depth: 24 feet.
DRAINAGE BASIN: 64,000 square miles, 50 major tributaries feed it.
WATER: Fresh water from tributaries mixes with salt water from Atlantic. Bay acts like sink, trapping pollutants. Only 1 percent are flushed out to sea.
PLANTS, ANIMALS: More than 2,000 species in bay shoreline.
1 INDUSTRIAL WASTES: Thousands of commercial, industrial facilities discharge water containing toxic chemicals, metals, nitrogen, phosphorous into bay.

Also: cooling needs can lead to corrosion, chlorine contamination.
2 MUNICIPAL SEWAGE: Water discharged from treatment plant contains nitrogen, phosphorous, toxic chemicals.
3 RAIN RUNOFF: Sediment from farms, forests, urban areas carries fertilizers, pesticides, herbicides. Over past 30 years, farmers have doubled the amount of fertilizers they use. Since 1960, herbicide use has tripled.
4 NUTRIENTS: High nutrient levels are most severe on northern, middle bay areas and tributaries, leading to excessive algae growth and less light filtering down to allow grasses to grow below water surface. This

gives waterfowl less food. Algae dies quickly, using up oxygen in the water and making it unsuitable for aquatic life. This leads to fewer fish, shellfish, oysters.
5 TOXIC CHEMICALS: Rain washes sediment into bay. This often includes chemicals, herbicides, pesticides, toxic wastes from industry.
6 CHLORINE: It's used in bay for disinfecting drinking water, sewage, industrial processes. It's suspected of hindering spawning runs of migrating fish.
7 HEAVY METALS: Cadmium, mercury, copper from industrial wastewater, sewage treatment plants can be toxic. Lead from auto exhaust, iron and zinc from industrial discharge, shore erosion.

FIGURE 8–6 The water resources upon which the world depends are becoming stressed due to pollution effects and the inability of aquatic species to maintain stable populations. *(Adapted from, and used with permission from The Washington Post and "MD Magazine," Autumn 1988)*

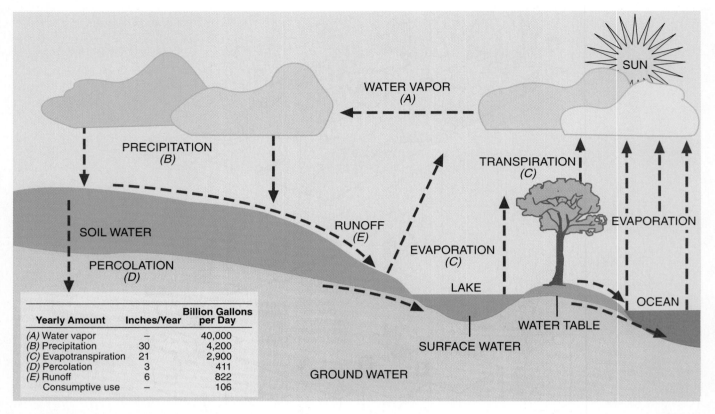

Yearly Amount	Inches/Year	Billion Gallons per Day
(A) Water vapor	–	40,000
(B) Precipitation	30	4,200
(C) Evapotranspiration	21	2,900
(D) Percolation	3	411
(E) Runoff	6	822
Consumptive use	–	106

FIGURE 8–7 The water cycle is a natural process by which water moves in a circular flow from oceans to land and back to the oceans.

INTERNET KEY WORDS:

properties of water
hydrologic cycle

ocean. Many of those substances may be toxic or otherwise threatening to the living organisms in the sea. The water that falls as rain or snow would be pure if it were not contaminated by pollutants as it falls through the air.

The Water Cycle Moisture evaporates from the Earth, plant leaves, freshwater sources, and the seas to form clouds. Clouds remain in the air until warm air masses meet cold air masses. This causes the water vapor to change to a liquid and fall as rain, sleet, or snow. The cycling of water among the water sources, atmosphere, and surface areas is called the water cycle. (Figure 8–7).

Land

Land provides us solid foundations for buildings, nutrition and support for plants, and space for work and play. It provides storage for water. It also serves as a heat and compression chamber that has converted organic material into coal and oil throughout Earth's existence.

Soil is an important component of land. To be productive, soil must be made up of the correct proportions of soil particles, have the correct balance of nutrients, contain some organic matter, and have adequate moisture. Much of the Earth's crust is too rocky or has an incorrect balance of nutrients for crop production. Similarly, much land is covered by only a thin layer of productive soil or is too steep to permit cultivation. Where there is some useful soil, however, trees may survive. Forest lands provide lumber, poles, paper, and other products. Here, too, humankind can benefit from the pleasures of wildlife and recreation.

FIGURE 8–8 Soil acts like a huge sponge that soaks up excess water in times when a surplus exists and releases water in a somewhat uniform flow from springs and wells throughout the year. *(Courtesy of PhotoDisc)*

INTERNET KEY WORDS:

irrigation systems science, watershed

INTERNET ADDRESSES:

http://www.zimmatic.com

http://www.epa.gov/owow/
watershed

http://water.usgs.gov/wsc/

http://soil.hostweb.org.uk/

Large areas of Earth have soil with a usable balance of soil particles and minerals for plant growth, but large areas have insufficient water. Areas with continuous, severe water shortages are called **deserts**. Fortunately, some desert areas have become productive through the use of modern irrigation practices. **Irrigation** is the addition of water to plants to supplement the water provided by rain or snow. Many nations have desert areas, but do not have the money, technical knowledge, or water to make their deserts productive. Most nations of the world have such limited land resources that all land must be used to its greatest capacity.

During the 1930s, a large area of western Oklahoma and neighboring states became known as the "dust bowl" when lack of rainfall and increased wind shifted the dry soil and ruined the productivity of the land. Later, the productivity returned as weather patterns changed and soil and water conservation practices were implemented.

Relationships of Land and Water

Precipitation   Land and water are related to each other in many ways. Land in cold regions or high altitudes holds moisture on its surface in the form of snow. This moisture is then released gradually to feed the streams and rivers after the **precipitation** (moisture from rain and snow) has stopped falling. Precipitation is caused by the change of water in the air from a gaseous state to a liquid state. It then falls to the land or bodies of water. Moisture-laden, warm-air clouds contact cold-air masses in the atmosphere and precipitation results. The clouds are formed by water changing from a liquid to a gas when it is evaporated by air movement over land and water. **Evaporation** means changing from a liquid to a vapor or gas.

A **watershed** is a large land area in which water is absorbed from rain or melting snow, and from which water drains as it emerges from springs and moves into the streams, rivers, ponds and lakes. It acts as a storage system by absorbing excess water and releasing it slowly throughout the year (Figure 8–8).

Land as a Reservoir   Land serves as a container or reservoir for water. Water soaks down into the soil and forms a **water table** (the level below which soil is saturated or filled with water). Water held below the water table may run out onto the surface at a lower elevation in the form of springs. Springs feed streams, which, in turn, form rivers and flow into bays and oceans. Since ancient times, people have known to dig wells below the water table, so water could be extracted for human needs.

Even above the water table, excess water is held in the soil to provide water for plant roots. From the roots, water travels throughout the plant; and much of it evaporates from the leaves to contribute to the moisture supply in the atmosphere. Water helps soil by improving its physical structure. It is essential for microorganisms that live in the soil and contribute to the soil's **fertility** (the amount and type of nutrients in the soil). Water also contributes to soil structure.

Types of Groundwater   Soil is **saturated** when water is added until all the spaces or pores are filled. Plants will die from lack of air around their roots if soil remains saturated for too long. The water that drains out of soil after it has been wetted is called **free water**, or **gravitational water**. Gravitational water feeds wells and springs. However, when gravitational water leaves the soil, some moisture remains. Plant roots can absorb

Hygroscopic Water — held tightly against soil particles and not available to plant roots

Capillary Water — held loosely against soil particles and can be absorbed by plant roots

Pores — spaces filled with air where roots penetrate and absorb water nutrients

Soil Particles

Droplet of **gravitational water** escaping from the soil

FIGURE 8–9 Plants use only capillary water. However, hygroscopic water contributes to soil structure, and gravitational water is held in reserve for future use of plants and animals.

or take up this moisture called capillary water. Water that is held too tightly for plant roots to absorb is called hygroscopic water (Figure 8–9).

Groundwater is easily polluted by chemicals and manure through abandoned wells and other depressions. Also, the constant use of irrigation can lead to increased salt concentration. These and other problems are being addressed by soil conservation and water management districts.

Benefits of Living Organisms Both soil and water benefit from living organisms. Plants break the fall of raindrops and reduce damage to the soil from the impact of water. When plants drop their leaves and plant materials accumulate, they provide a rain-absorbing layer on the surface. When plant materials decay, they improve the structure of the soil. Plants, therefore, contribute to water absorption and reduce soil erosion. Decayed plant and animal matter contributes substantially to the nutrient content of the soil. Plants assist in moving water around by taking it up through the roots and releasing it in the atmosphere. Worms, insects, bacteria, and other small and microscopic plants and animals contribute by decomposing plant and animal matter. They also contribute in other important ways.

CONSERVING WATER AND IMPROVING WATER QUALITY

How can we reduce water pollution? How can soil erosion be reduced? What is the most productive use of water and soil without polluting or losing these essential resources? These are important questions deserving correct answers now! Farmers have long appreciated the value of these resources and generally have used them wisely. However, economic conditions, governmental policies, production costs, farm income, personal knowledge, and other factors influence the use of conservation practices by farmers and other land users. Every citizen, business, agency, and industry impacts on air and water quality. Similarly, all have some influence on the use of our land and water resources (Figure 8–10).

THE COLORADO RIVER BASIN ... AND ITS PLUMBING

PLUMBING COMPILED
AND EDITED BY MARY MORAN
RENDERED BY LESTER DORÉ

FIGURE 8–10 Water sources and uses from the Green, Colorado, and other rivers to the Gulf of California. *(Courtesy Dinosaur National Monument, National Park Service)*

Agri-Profile

CAREER AREA:

Soil Conservation Hydrology

Some major challenges in soil and water conservation are to keep rain water on the land where it falls, keep soil in place and stabilized with plants, minimize pollution of fresh water, and wisely manage land and water to maintain ecological balance. For instance, soil and water scientists at the University of California are developing ways to reuse salt-laden irrigation water. It is estimated that California alone may be generating upward of 1.6 trillion gallons of sub-surface drainage water annually.

Soil and water conservation career options provide extensive opportunities for indoor and outdoor work. Work is available in offices of the Federal Farm Service Agency. One may work in the field served by such offices in most counties of the United States. Typical job titles are conservation technician, farm planner, soil scientist, and soil mapper.

The United States Department of Interior, state departments of natural resources, city and county governments, industry, and private agencies hire people with soil and water conservation expertise. Such workers manage water resources for recreation, conservation, and consumption. They may work as consultants, law enforcement officers, technicians, administrators, heavy equipment operators, among other emerging careers.

Salt-laden waste water flows from irrigated land in California's Imperial Valley into the Salton Sea. *(Courtesy USDA/ARS #K-1918-10)*

The improvement of water quality can be achieved by proper land management, careful water storage and handling, and appropriate use of water. Some practices that help reduce water pollution are as follows:

- Save clean water. Whenever we permit a faucet to drip, leave water running while we brush our teeth, or flush toilets excessively, we are wasting water. Taking long showers, leaving water running while we wash automobiles or livestock, and using excessive water for pesticide washup waste clean water. Once clean water is mixed with contaminants, it is not safe for use for other purposes until it is cleaned or repurified. **Purify** means to remove all foreign material.

- Dispose of household products carefully. Many products under the kitchen sink, in the basement, or in the garage are threats to clean water (Figure 8–11). Never pour paints, wood preservatives, brush cleaners, or solvents down the drain. They will eventually enter the water supply, rivers, or oceans. Use all products sparingly and completely. Put solvents into closed containers and permit the suspended materials to settle out. Then reuse the solvent. Stuff empty containers with newspaper to absorb all liquids before discarding. Then send the containers to approved disposal sites.

Science Connection

INDICATOR SPECIES

One way scientists can tell that water has been polluted is by studying "indicator species." Indicator species are organisms that are used to determine the health of an environment. The four categories of indicator species in water are fish, aquatic invertebrates (including worms, snails, and aquatic insects), algae, and aquatic plants. A pollutant in the water affects these organisms in different ways. Depending on the type of pollution, the organisms can die off, be reduced, or increase in number. An algae bloom is an example of how an indicator species is affected by pollution.

Runoff water from farm lands where fertilizer is too heavily applied may contain excess fertilizer. As this fertilizer enters a lake or river, algae can take it up. The fertilizer works on the algae as it does on crops; it encourages it to grow. The result is an overgrowth called an algal bloom. These conditions often are seen during the late growing season. They indicate to scientists that a fertilizer pollutant is in the water. Problems arise because of the bloom. Algae can produce toxins that are harmful to fish, animals, humans, and other organisms. When the algae die and decompose, oxygen levels in the water are used up. Other organisms that depend on the oxygen can die as a result. Clean and safe water is needed for drinking, watering crops, and maintaining a healthy living environment. Using indicator species can help scientists determine the safety of the water supply.

FIGURE 8–11 Many products found around the home are hazardous products that require correct disposal to avoid pollution of ground water. *(Courtesy of Chesapeake Bay Foundation)*

INTERNET KEY WORDS:

wetlands, water

• Care for lawns, gardens, and farmland carefully. Improve soil by adding organic matter. Mulch lawn and garden plants (Figure 8–12). Use proper amounts and types of lime and fertilizer. Only till soil that will not erode excessively. Cover exposed soil with a new crop immediately. Water soil only when it is excessively dry, and then water until the soil is soaked to a depth of 4 to 6 inches.

FIGURE 8–12 A mulching system allows plant vegetation from lawns and gardens to be added back to the soil as a nutritious plant mulch.

FIGURE 8–13 Construction of terraces on sloping land distributes water from precipitation along the terrace and helps to prevent soil erosion.

- Practice sensible pest control. Many insecticides kill all insects—both harmful and beneficial ones. Insecticides also pollute water. Therefore, use cultural practices, such as crop rotations and resistant varieties, instead of insecticides, whenever possible. Encourage beneficial insects and insect-eating birds by improving habitat around lawn, garden, and fields. Eliminate pools of stagnant water to eliminate mosquitoes. Keep untreated wood away from soil to avoid termite, other insect, and rot damage. Follow all pesticide label instructions exactly.

- Control water runoff from lawns, gardens, feedlots, and fields. Keep soil covered with plants. Construct livestock facilities so manure can be collected and spread on fields. Where it is feasible use no-till cropping. **No-till** means planting crops without plowing or disking the soil. Alternate strips of close-growing crops (such as small grains or hay) with row crops (such as corn or soybeans). Farm on the **contour** (following the level of the land around a hill). Plant cover crops where regular crops do not protect the soil. A **cover crop** is a close-growing crop planted to temporarily protect the soil surface.

- Control soil erosion. Reduce the volume of rainwater carried away by minimizing the amount of blacktop or concrete surface constructed. Use grass waterways in low areas of fields. Add manure and other organic matter to soil to increase water-holding capacity. Construct terraces on long or steep slopes (Figure 8–13). Leave steep areas in trees and sloping areas in close-growing crops.

- Avoid spillage or dumping of gasoline, fuel, or oil on the ground or in storm drains. Turn over used petroleum products to truck and automobile service centers for recycling.

- Keep chemical spills from running or seeping away. Do not flush chemicals away. Chemicals will damage lawns, trees, gardens, fields, and groundwater. Sprinkle spills with an absorbent material such as soil, kitty litter, or sawdust. Then place the chemical-laden absorbent material in a strong plastic bag and discard it according to local recommendations.

Hot Topics in Agriscience

CONSTRUCTED WETLANDS

Constructed wetlands are man-made marshes that can be used to treat polluted water. Natural marshlands contain plants that are tolerant of standing water. These marshlands also provide habitat for bacteria that are capable of breaking down materials that cause water pollution. As water passes slowly through the marsh, these bacteria and other microbes cleanse the water by consuming pollutants and changing them to harmless compounds. Some water plants also take up pollutants.

A wetland in nature performs the same function—it removes impurities from the water. Many of the natural wetlands have been drained to prepare the land for farming. Currently, we are beginning to recognize the value of the wetlands. Constructed wetlands usually consist of rectangular plots arranged in a series. They are filled with gravel or porous soil and lined to prevent pollutants from leaching into the groundwater. Plants are added to the plots to simulate a natural marshland. Constructed marshlands do a good job of mimicking the advantages of natural marshlands. An added benefit of constructed marshlands is that excellent wildlife habitat is created.

- Properly maintain your septic system (if your home has one). Avoid using excessive water. Do not flush inappropriate materials down the toilet. Do not plant trees where their roots might interfere with sewer lines, septic tanks, or field drains. Do not run tractors and other heavy equipment over field drains.

These recommendations are some practices that should help increase water supplies and decrease water usage and pollution.

LAND EROSION AND SOIL CONSERVATION

The Problem

Land Erosion—A Worldwide Problem Land erosion (wearing away) is a serious problem worldwide (Figure 8–14). Both wind and water are capable of wearing away soil. The food and fiber production capabilities of large nations are being compromised because of extensive damage from soil erosion. Numerous cases can be cited from history where soil erosion has caused enormous problems. Yet, as we continue in the 21st century, the people of the world are making the same mistakes in many areas that caused extensive losses in the past.

Consider, for example, an important colonial port (a town having a harbor for ships to take on cargo) along the eastern coast of what is now the United States. The colony was established in 1634. By 1706, the port had been replaced by another town as the central port for the colony. The harbor area of that original port was rapidly filling with soil that was washed in from the surrounding area. The water was no longer deep enough for the ships to move.

Consider the massive size of the Mississippi River Delta (land created by the soil that water deposits at the mouth of a river). The Mississippi River Delta is more than 15,000 square miles in area. For the Mississippi River Delta to be formed, the river had to deposit soil on the bottom of the Gulf of Mexico and continue until it reached the surface of the water.

FIGURE 8–14 Erosion of soil means loss of productive soil, damage to machinery, additional costs of production, and pollution of streams, rivers, lakes, bays, and oceans. *(Courtesy USDA/ARS #K-1021-15)*

INTERNET KEY WORDS:

soil conservation stabilization

INTERNET ADDRESS:

http://www.nhq.nrcs.usda.gov/WSR

The soil gets into the river from smaller rivers and streams receiving runoff water from fields in the heartland of the United States. Therefore, one must conclude that the Mississippi River Delta was built from the bottom to the surface of the Gulf of Mexico with the best topsoil in the United States. Yet, in these modern times, it is estimated that the amount of soil being dumped by the river into the Gulf everyday would fill a freight train 150 miles long. It has been estimated that the amount deposited at the mouth of the Mississippi River in 1 year is sufficient to cover the entire state of Connecticut with soil 1 inch deep, or more than 8 feet in 100 years! Similar deltas exist at the mouth of the Nile River and most other great rivers of the world.

During the late 1960s and early 1970s in the People's Republic of China, much of the forested land was cleared. It soon became apparent that much of that land depended on the trees to prevent soil erosion. Seeing the extent of damage from soil erosion, the country has responded with a national policy of rapid reforestation. There are more than 1 billion people in China and all are expected to contribute to the tree planting effort. However, it will take many years for newly planted trees to grow sufficiently to protect the soil again. Unfortunately, the soil washed away during the absence of trees can never be returned to the land. Soil scientists report that it takes 300 to 500 years for nature to develop 1 inch of topsoil from bedrock.

Another serious threat to large areas of the world is the farming system known as "slash and burn." This is used in the tropical rain forests of South America, Africa, and Indonesia. In these areas, impoverished farmers cut or slash the jungle growth and burn the plant residues. The extensive burning causes serious air pollution and destroys useful fuel for cooking. Then these farmers raise crops for several years until the soil is drained of its meager fertility and soil erosion takes its toll. They then move on to uncleared land and repeat the cycle. Sadly, once the land is cleared, it becomes difficult for plant growth to become re-established and the land is left permanently damaged. Reportedly, there are 50 acres of tropical rain forest lost every minute through this process. It is said that, in some regions, 80 percent of the rainfall is attributed to transpiration from trees. When the trees are removed, such areas may become subject to drought.

Soil Loss: 1 Acre of Highly Erodible Soil

You would have to load and carry:

1 30 pound bucket of soil
Every 30 seconds
Working 8 hours per day
For over 4 days

To replace the amount of soil lost from 1 acre in 1 year

FIGURE 8–15 Imagine how much work it would be to replace the soil that is lost each year from 1 acre of highly erodible land!

National Problems Each year, about 1.6 billion tons of soil are worn away from 417 million acres of U.S. farmland and deposited into lakes, rivers, and reservoirs. One ton equals 2,000 lbs. Although some soils are deep and can tolerate a certain amount of erosion, many fragile soils cannot. The U.S. Department of Agriculture's (USDA's) National Resources Inventory indicates 41 million acres (or about 10 percent) of our nation's cropland are highly erodible at rates of 50 or more tons per acre per year (Figure 8–15).

Of growing concern is the contamination of groundwater under large areas of the United States. Groundwater pollution emerged as a public issue in the late 1970s. The first reports documented sources of contamination associated with the disposal of manufacturing wastes. By the early 1980s, several incidences of groundwater contamination by pesticides used on field crops were confirmed. Groundwater contamination can threaten the health of large populations. For instance, in New York State, the vast **aquifer** (a water-bearing rock formation) that underlies Long Island represents the only supply of drinking water for more than 3 million people. In the United States, major aquifers underlie areas that contain thousands of square miles of land, which often include several states. Contamination of the aquifer in one place means contamination of the water for large areas.

Soil Conservation

Preventing and Reducing Soil Erosion Soil erosion can be reduced or stopped by good land management. Fortunately, management practices that reduce soil erosion increase water retention (Figure 8–16). This, in turn, helps stabilize the supply of fresh water available for crops, livestock, wildlife, and people. It also reduces dryness and dustiness, which are threats to good air quality. Most soil conservation methods are based on either (1) reducing raindrop impact, (2) reducing or slowing the speed of wind or water moving across the land, (3) securing the soil with plant roots, (4) increasing absorption of water, or (5) carrying runoff water safely away. If water is free to run down a hill, it increases in volume and speed, picks up and pushes soil particles, and carries these particles off the land. This results in sheet and gully erosion and deposits soil, nutrients, and chemicals into streams. **Sheet erosion** is removal of layers of soil from the land, whereas **gully erosion** is soil removal that leaves trenches.

A

B

C

FIGURE 8–16 Plant growth holds the key to soil and water conservation. (A) Surface mulch protects the soil around cotton seedlings, (B) Extensive root systems of soybeans, and (C) Revegetation of sandbars with willow trees. *(A: Courtesy USDA/ARS #K-3226-6; B: Courtesy USDA/ARS #K-3213-1; C: Courtesy USDA/ARS #K-5Z10-18)*

The following are some recommended practices to reduce or prevent wind and water erosion:

- Keep soil covered with growing plants. Plants reduce the destructive impact of raindrops, reduce the speed of wind and water movement across the land, and hold soil in place when threatened by wind or moving water.

- Cover the soil with a mulch. **Mulch** is a material placed on the soil surface to break the fall of raindrops, prevent weeds from growing, and/or improve the appearance of the area.

- Utilize conservation tillage methods. **Conservation tillage** means using techniques of soil preparation, planting, and cultivation that disturb the soil the least, whereas leaving the maximum amount of plant residue on the surface. **Plant residue** is the plant material left when a plant dies or is harvested.

- Use contour practices in farming, nursery production, and gardening. **Contour practice** means conducting all operations, such as plowing, disking, planting, cultivating, and harvesting, across the slope and on the level. This way, any ruts or ridges created by machinery go around the slope or hill. When water tries to run down the hill, it encounters the grooves and ridges. Because these are level, the water tends to soak into the soil. This holds the water for future use.

- Use strip cropping on hilly land. **Strip cropping** means alternating strips of row crops with strips of close-growing crops. Examples of close-growing crops are hay, pasture, and small grains, such as wheat, barley, oats, and rye. Examples of row crops are corn, soybeans, and most vegetables. The strips of close-growing crops capture runoff water from the row crops and prevent it from entering streams.

- Rotate crops. **Crop rotation** is the planting of different crops in a given field every year or every several years. Crop rotation permits close-growing crops to retain water and soil, and it tends to rebuild the soil countering losses incurred when row crops occupied the land.

- Increase organic matter in the soil. **Organic matter** is dead plant and animal tissue. Nonliving plant leaves, stalks, branches, bark, and roots decay and become organic matter. Similarly, animal manure and dead insects, worms, and animal carcasses decompose to make organic matter. The decomposed organic matter forms a gel-like substance that holds soil particles in absorbent granules called **aggregates**. An aggregated soil is a water-absorbing and nutrient-holding soil. Organic matter also releases nutrients that improve plant growth.

- Provide the correct balance of lime and fertilizer. **Lime** is a material that reduces the acid content of soil. It also supplies nutrients such as calcium and magnesium to improve plant growth. **Fertilizer** is any material that supplies nutrients for plants.

- Establish permanent grass waterways. A **grass waterway** is a strip of grass growing in an area of a field where water tends to flow, causing erosion.

- Construct terraces. A **terrace** is a soil or wall structure built across the slope to capture water and move it safely to areas where it will not cause erosion.

- Avoid overgrazing. **Overgrazing** results in damage to plants or soil due to animals eating too much of the plants at one time. This reduces the ability of the plant roots to hold soil or recover after grazing.

Science Connection

SAFE-GUARDING GROUNDWATER

Authorities estimate that nearly 330,000 tons of pesticides are applied yearly on U.S. crops. Added to this are the chemicals used to control fleas, flies, ticks, termites, roaches, and other pests of livestock and humans. Agricultural chemicals have enabled us to feed and clothe ourselves and many others in the world; but pollution from these same chemicals has become a problem and a threat to our air, water, and land.

Applying these pesticides where they are intended and keeping them there until they change or biodegrade into harmless products is a major concern in protecting our water supplies. To help safe-guard our surface water and groundwater from pesticide pollution, the USDA Agricultural Research Service has developed a specific strategy. Research priorities are shaped by this strategy.

The ultimate goals of the plan are to (1) provide the U.S. farmer with cost-effective best management practices to ensure ample supplies of food and fiber at a reasonable cost, whereas reducing pesticide movement into the groundwater; (2) identify the factors that accelerate or retard pesticide movement; and (3) provide computer models that will quickly and accurately predict contamination.

USDA water research is being concentrated in three areas to protect against contamination of groundwater. These areas are:

- agricultural watershed management

- irrigation and drainage management

- water quality protection and management

The computer models under development are decision-enhancing tools that use a database and computer program to aid the selection of management practices. Models integrate data from many sources that are frequently not available to the decision makers.

Computer models help identify pesticide processes related to movement in various soils, climates, and other environmental conditions. New technologies, such as advances in slow-release formulations, improved pesticide application scheduling, and selective placement, are under study. Today groundwater contamination by agricultural chemicals is considered a risk in many of the major crop and livestock production areas of the United States.

Areas of Potential Groundwater Contamination from Agricultural Chemicals

Nitrates Only
Pesticides Only
Nitrates and Pesticides

Groundwater contamination by agricultural chemicals is considered a risk in many areas. *(Courtesy USDA/ARS August '89 p. 2-USDA/ERS, AER 576. Neilson & Lee, "The Magnitude and Cost of Groundwater Contamination from Agricultural Chemicals: A National Perspective," 1987)*

- Use land according to a conservation plan. A **conservation plan** is a plan developed by soil and water conservation specialists to use land for its maximum production and water conservation without unacceptable damage to the land.

The USDA's Natural Resources Soil Conservation Service (NRCS) and Farm Service Agency (FSA) provide advice, technical assistance, and funds to assist land owners with soil and water conservation practices. The Environmental Protection Agency (EPA) monitors groundwater quality and enforces point and nonpoint pollution laws to help protect our land and water.

There are many careers in the field of water and soil conservation. The material in this unit hints at the problems created by humans as they use the natural resources around them to provide food, water, shelter, recreation, and other resources for living. The health, wealth, peace, and general welfare of humanity depends on our skill in conserving these most basic resources.

Constructed wetlands are a growing trend in agriculture. Farmers understand that water is priceless. Some have chosen to set aside a fraction of their property as a wetland to help purify water and ensure its safety.

STUDENT ACTIVITIES

1. Write the Terms to Know and their meanings in your notebook.

2. Collect a sample of drinking water from each of five sources as follows: a safe spring or well; bottled pure water; and faucets attached to (a) galvanized pipe, (b) plastic pipe, and (c) copper pipe. (Do not run off the water before obtaining the samples from the faucets.) Taste a small amount of each sample and describe the taste. Do the samples have different tastes? If so, what causes the differences?

3. Study the eating habits of one species of birds in your community. Describe the food chain that accounts for the survival of that species. What effect did the use of DDT as an insecticide have on that species of birds before DDT was banned from use?

4. Obtain a rain gauge and record the precipitation on a daily basis for several months. What variations did you observe from week to week? How do you explain the variations? What effects did these variations have on the agricultural activities of the community?

5. Conduct an experiment to determine the effect of soil water on plant growth. Obtain three inexpensive pots of healthy flowers or other plants. Each pot must be the same size and type, have the same amount and type of soil, and contain the same size and number of plants. Use the following procedure:

   a. Mark the pots "1," "2," and "3."

   b. Plug the holes in the bottoms of pots 2 and 3 so water cannot drain from the pots. Leave pot 1 unplugged so it has good drainage.

   c. Add water to pot 1 as needed to keep the plant healthy for several weeks. Use it as a comparison specimen (called the "control").

   d. Add water slowly to pot 2 until water has filled the soil and the water is just level with the surface of the soil. All pores of the soil are now filled and the soil is "saturated." Weigh the pot and record the weight as "A."

   e. Remove the plug from the bottom of pot 2 and permit the water to drain out. After water has stopped flowing from the drain hole, immediately weigh the pot again. Record the weight as "B." The water that flowed from pot 2 when the plug was removed is the free or gravitational water. Weight of the gravitational water, "D," should be calculated and recorded using the formula A – B = D. Do not add any more water to pot 2 for 2 weeks.

f. With the hole plugged in pot 3, add water slowly until the water is just level with the top of the soil. Do not remove the plug, and keep pot 3 filled with water to the saturation point for 2 weeks.

g. Keep all three pots in a good growing environment for 2 weeks and record all observations.

h. When the plant in pot 2 wilts badly due to lack of water, weigh the pot and record the weight as "C."

Make the following calculations:

Weight of the gravitational water (D) = A – B

Weight of the capillary water (E) = B – C

The weight of the hygroscopic water can only be determined by driving the remaining water from the soil by heating the soil in an oven.

6. Ask your teacher to help you design and conduct a project that demonstrates some or all of the following:

a. effect of the force of raindrops on soil

b. effect of soil aggregation on absorption

c. effect of slope on erosion

d. effect of living grass on erosion control

e. effect of plant residue on erosion control

7. Consider some feeding relationships in your area. Illustrate and label a food web or food chain that consists of five or more organisms. Write one paragraph to explain your drawing.

SELF EVALUATION

A. Multiple Choice

1. Most of the Earth's surface is covered with

a. crops. c. trees.

b. farms. d. water.

2. The bodies of plants, animals, and humans consist of about what percentage water?

a. 10 percent c. 70 percent

b. 40 percent d. 90 percent

3. The universal solvent is

a. gasoline. c. varsol.

b. paint thinner. d. water.

4. The content of ocean water may be likened to

a. fresh water. c. thin soup.

b. pure water. d. varsol.

5. The interdependence of plants and animals on each other for food is known as

a. domestic. c. symbiosis.

b. food chain. d. universal relationship.

6. The land serving as a heat and compression chamber gives us

a. building foundations. c. crops.

b. coal and oil. d. wildlife habitat.

7. Groundwater that is available for plant root absorption is called
   a. capillary.
   b. free.
   c. gravitational.
   d. hygroscopic.

8. Improvement of water quality can be achieved by
   a. appropriate use of water.
   b. careful water storage and handling.
   c. proper land management.
   d. all of the above.

9. The problem of land erosion is
   a. characteristic of developed countries only.
   b. found worldwide.
   c. mostly found in poor countries.
   d. not a substantial problem in view of food surpluses.

10. A conservation plan
    a. conserves soil and water.
    b. is prepared by professionals.
    c. maximizes productivity of land.
    d. all of the above.

B. Matching

_____	**1.** Contour	a. Amount and type of nutrients
_____	**2.** Delta	b. Without plowing or disking
_____	**3.** Evaporation	c. Town with a harbor
_____	**4.** Fertility	d. Removal of foreign material
_____	**5.** Irrigation	e. Soil deposited by water
_____	**6.** No-till	f. Top of saturated soil
_____	**7.** Port	g. Rain and snow
_____	**8.** Precipitation	h. Supplemental water
_____	**9.** Purify	i. On the level
_____	**10.** Water table	j. Change from liquid and gas

C. Completion

1. An _____ is a water-bearing rock formation.
2. The process of creating narrow and deep trenches eroded in soil is known as _____.
3. Material placed on soil to break the fall of raindrops is called _____.
4. Alternating row crops with close-growing crops is known as _____.
5. _____ is used to reduce acid in soil.
6. One inch of topsoil may be formed from bedrock in about _____ years.
7. _____ is dead plant and animal material in soil.
8. A _____ may be planted to temporarily protect soil from erosion.
9. In the United States, about _____ tons of soil are worn away from farmland each year.
10. About _____ acres, or 10 percent, of the U.S. cropland is highly erodible.

UNIT 9

Soils and Hydroponics Management

MATERIALS LIST

✓ writing materials
✓ newspapers
✓ topography map
✓ Internet access

COMPETENCIES TO BE DEVELOPED

After studying this unit, you should be able to:

- define terms in soils, hydroponics, and other plant-growing media management.
- identify types of plant-growing media.
- describe the origin and composition of soils.
- discuss the principles of soil classification.
- determine appropriate amendments for soil and hydroponics media.
- discuss fundamentals of fertilizing and liming materials.
- identify requirements for hydroponics plant production.
- describe types of hydroponics systems.

SUGGESTED CLASS ACTIVITIES

1. Assign class members to fill a 1-gallon container with soil and bring it to school. This will provide a wide range of soil types that can be used as teaching materials. Each student should measure his or her soil sample to determine the soil texture. Students should work in pairs, and each student should also test his or her partner's soil sample as a check for accuracy.

2. Conduct drainage tests on the soil samples. Begin by placing a filter paper at the bottom of a funnel. An acceptable funnel can be made by cutting off the top third of a plastic beverage bottle and inverting it. Next, place a measured amount of soil inside the funnel and on top of the filter paper. Make sure the oil is directly on the filter paper. Then carefully add a measured amount of water to the soil surface while using care not to disturb the placement of the soil. Observe the length of time that is required for the water to move through the soil to the container below the funnel. This is a good time to line up the soil samples from most permeable to least permeable. Have students check the texture of the soil with their fingers according to the chart that is provided in this unit.

SUGGESTED CLASS ACTIVITIES—*continued*

3. Invite a local expert to come and introduce hydroponics to the class. Students have the ability to do new solution culture experiments at little cost by using plants that have never been grown in solution culture before, such as most houseplants and bedding plants. Experiments could be done by students in the areas of nutrient deficiencies, toxicities, carbon dioxide and oxygen deficiencies, pH, fertilizer testing, growth regulators, nitrogen fixation, shoot-to-root ration, bulb forcing, and others. The possibilities for hydroponics projects are nearly endless.

TERMS TO KNOW

media	lacustrine deposit	C horizon	active ingredient
medium	loess deposit	bedrock	broadcasting
hydroponics	colluvial deposit	coarse-textured (sandy) soil	incorporated
decomposed	glacial deposit	medium-textured (loamy) soil	band application
leaf mold	percolation		side-dressing
compost	permeable	fine-textured (clay) soil	top-dressing
sphagnum	capability class	structure	starter solution
peat moss	capability subclass	crumb	foliar spray
bog	capability unit	amendment	knife application
waterlogged	O horizon	pH	nitrate
perlite	mineral matter	acidity	nitrogen fixation
vermiculite	A horizon	alkalinity	aggregate culture
leached	clay	neutral	water culture
microbe	silt	decomposer	solution culture
parent material	sand	petiole	nutriculture
horizon	tillable	gypsum	aeroponics
profile	topsoil	primary nutrient	continuous-flow system
residual soil	B horizon	complete fertilizer	inert
alluvial deposit	subsoil	fertilizer grade	

The roles of plants in our environment and their importance in our lives have been discussed in previous units. Plants are necessary to nourish the animals of the world and maintain the balance of oxygen in our atmosphere. However, they depend upon soil, water, and air as their media for support. **Media** is plural for medium. **Medium** is a surrounding environment in which something functions and thrives.

PLANT-GROWING MEDIA

For discussion in this unit, the word *media* is used to mean the material that provides plants with nourishment and support though their root systems.

Types of Media

Media comes in many forms. The oceans, rivers, land, and man-made mixtures of various materials are the principal types of media for plant growth. Seaweed, kelp, plankton, and many other plants depend on water for their nutrients and support. It is only recently that we have come to understand the tremendous amount of plant life in the sea. The plant life in oceans and rivers is important for feeding the animal life found in water.

It has long been known that water could be used to promote new root formation on the stems of certain green plants, and it can completely support plant growth for a short time. Recently, however, it has been found that food crops can be grown efficiently without soil. This is done using structures where plant roots are submerged in or sprayed with solutions of water and nutrients (Figure 9–1). These solutions feed the plants, whereas mechanical structures provide physical support. The practice of growing plants without soil is called **hydroponics**. Hydroponics has become an important commercial method of growing green plants.

Soil

Soil is defined as the top layer of the Earth's surface, which is suitable for the growth of plant life. It has long been the predominant medium for cultivated plants (Figure 9–2). In early years, humans accepted the soil as it existed. They planted seeds using primitive tools and did not know how to modify or enhance the soil to improve its plant-supporting performance.

Ancient civilizations discovered that plant-growing conditions were improved on some land where deposits were left after river waters flowed over the land during flood season. Similarly, other land was ruined by flood waters. Therefore, early efforts to improve plant-growing media were a matter of moving to better soil. Obviously, good soil was a valuable asset and, therefore, was the cause of intense personal disputes and wars among nations.

Other Media

In addition to water and soil, certain other materials will hold water and support plant growth. Fortunately, some of the best nonsoil and nonwater plant-growing media are partially **decomposed** (decayed) plant materials. Hence, plants tend to improve the environment where they grow. One common material available around most homes is leaf mold and compost. **Leaf mold** is partially decomposed plant leaves. **Compost** is a mixture of partially decayed organic matter such as leaves, manure, and household

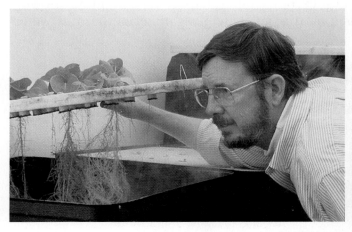

FIGURE 9–1 Plants can live and grow without any soil by spraying or submerging their roots in water in which plant nutrients are dissolved. *(Courtesy USDA/ARS)*

FIGURE 9–2 Soil is the most important plant medium for most crops.

Hot Topics in Agriscience

PRODUCING "HOT-HOUSE" VEGETABLES

The production of "hot-house" vegetables is done in a controlled greenhouse environment. In many cases, these vegetable production units are located near natural hot water springs to take advantage of the heat source during the winter months. The hot water is often piped into the greenhouse where it is an inexpensive source of supplemental heat. Most of these vegetable farms use some variation of hydroponics rather than using soil as a medium for plant growth. The plants are supported as they grow by tying them up using wires and strings. Tomatoes and other vegetable products are harvested frequently at their peak of quality, and then shipped immediately to markets. Most "hot-house" vegetables are produced during seasons when field production is limited or impossible because of climate restrictions. As a result, these vegetables command high prices in the markets.

plant wastes. Decaying plant matter should be mixed with lime and fertilizer in correct proportions to support plant growth (Figures 9–3 and 9–4).

There is a group of pale or ashy mosses called sphagnum. These are used extensively in horticulture as a medium for encouraging root growth and growing plants under certain conditions. Peat moss consists of partially decomposed mosses that have accumulated in waterlogged areas called bogs. Waterlogged means soaked or saturated with water. Both substances have excellent air- and water-holding qualities.

Many other sources of plant and animal residues may become plant-growing media. For instance, a fence post may rot on the top and hold moisture from rainfall. In time, a plant seed may be deposited in the rotted wood medium, germinate, root in the wood, and start to grow. Even horse manure mixed with straw is used extensively as a medium for growing mushrooms. In this instance, both animal residue (manure) and plant residue (straw) combine to make an effective medium.

Some mineral matter can also become plant-growing media. For instance, volcanic lava and ash eventually accumulates soil particles on the surface. Seeds settle into the cracks, and moisture causes the seeds to germinate. Roots then penetrate and break up the volcanic residue. As time passes, the area becomes covered with plant life.

Horticulturists use certain mineral materials in plant-growing areas, too. Perlite is a natural volcanic glass material that has water-

Layers should be turned and mixed together every several weeks as the materials decay into a fertile mass. Depending upon moisture, temperature, content, and frequency of turning, it should take 3 to 12 months to make compost.

Fertilizer, Lime and Small Amounts of Soil in Each Layer

ORGANIC MATERIAL

ORGANIC MATERIAL

ORGANIC MATERIAL (CROP RESIDUE)

ORGANIC MATERIAL (GRASS CLIPPINGS)

ORGANIC MATERIAL (MANURE)

ORGANIC MATERIAL (LEAVES)

Enclosure Made of Snow Fencing or Chicken Wire

Most Any Combination of Organic Materials

◄─── 4 Ft. to 10 Ft. in Diameter ───►

FIGURE 9–3 Compost is excellent organic matter. Most mineral soils can be improved by adding compost.

FIGURE 9–4 Plant materials from yards and gardens can be converted to valuable compost by placing the materials in an environment with moderate temperatures and where bacteria known as decomposers have adequate moisture and oxygen.

holding capabilities. Perlite is used extensively for starting new plants. Vermiculite, a mineral matter from a group of mica-type materials, is also used for starting plant seeds and cuttings.

ORIGIN AND COMPOSITION OF SOILS

Factors Affecting Soil Formation

Productive soils develop on the Earth's surface as the atmosphere, sunlight, water, and living things meet and interact with the mineral world. If soil is suitable for plant growth to a depth of 36 inches or more, the soil is regarded as "deep." Many soils of the Earth are much more shallow than this. Plants attach themselves to the soil by their roots where they grow, manufacture food, and give off oxygen. Plants and animals of various sizes live on and in the soil, using carbon dioxide, oxygen, water, mineral matter, and products of decomposition.

Soils vary in temperature, organic matter, and the amount of air and water they contain. The kinds of soils formed at a specific site are determined by the forces of climate, living organisms, parent soil material, topography, and time (Figure 9–5).

INTERNET KEY WORDS:

soil formation
soil formation and weathering,
topography,
climate,
bedrock materials

Factors Affecting Soil Formation	
Climate/location	Affects rate of weathering
Living organisms	Cause decay of organic material
Parent material	Influences fertility and texture
Topography	Affects distribution of soil particles and water
Temperature	Influences rate of weathering
Weathering	Causes soils to develop, mature, and age

FIGURE 9–5 Soil formation depends on a number of natural factors.

FIGURE 9–6 Living organisms like earthworms play important roles in breaking down organic matter such as leaves and grass. *(Courtesy of PhotoDisc)*

INTERNET KEY WORDS:

soil microorganisms

Climate and Location

Climatic factors, such as temperature and rainfall, greatly affect the rate of weathering. When temperature increases, the rate of chemical reactions increases, and growth of fungi, organisms (such as bacteria), and plants increases. The rate and amount of rainfall in a locality greatly affect the soil. In areas of high rainfall, the soils usually are leached and somewhat acidic. **Leached** means having certain contents removed by water. If the land is covered by trees, the action of high temperatures and moisture on leaf residues generally creates an acidic soil. Rainfall during cold weather has less effect on the soil than during warm or hot weather.

Slope and location of a field also affect erosion and drainage, thus influencing soil formation. Moreover, free water in the soil carries fine particles to the deeper layers and tends to produce "layering." Too much water prevents or retards microbial growth and may exclude air by water-logging. Water and temperature also have the effect of swelling and contracting soil particles.

Living Organisms

Living organisms, such as microbes, plants, insects, animals, and humans, exert considerable influence on the formation of soil. Certain types of soil bacteria and fungi aid in soil formation by causing decay or breakdown of the plant and animal residues in the soil. Carbon dioxide and other compounds essential to soil formation are released by microbic activity. **Microbes** are microscopic plants and animals. Without soil microbes, organic materials would not decay.

Numerous insects, worms, and animals contribute to the formation of soil by mixing the various soil materials (Figure 9–6). Earthworms consume and digest certain soil substances and discharge body wastes. This aids decomposition and soil mixing. All such dead organisms add to soil organic matter.

Human activity also influences soil formation. Activities such as cultivation, bulldozing, and construction projects disturb the surface layer. The clearing of land removes native plant life and greatly modifies soil-forming activities (Figure 9–7).

Parent Material

Parent material is the horizon of unconsolidated material from which a soil develops. **Horizon** means layer. Parent materials compose the C horizon of a typical soil profile. **Profile** means a cross-sectional view of soil (Figures 9–8 and 9–9).

Parent materials formed in place are called **residual soils**. Other soils are transported and deposited by water, wind, gravity, or ice. **Alluvial deposits** are transported by streams, and **lacustrine deposits** are left by lakes. **Loess deposits** are left by wind, **colluvial deposits** by gravity, and **glacial deposits** by ice.

The kind of parent material from which soil is formed influences the many characteristics of that soil. Natural fertility and texture are influenced greatly by the parent material of a profile.

Topography

Slope and drainage affect soil formation both directly and indirectly. On a steep slope, loose material is moved downward by runoff water,

FIGURE 9–7 Clearing natural vegetation from land usually contributes to soil loss at a much faster rate than it is formed. *(Courtesy National FFA)*

FIGURE 9–8 A soil profile consists of a cross-sectional view of the different layers of materials beginning at the surface and going down to bedrock.

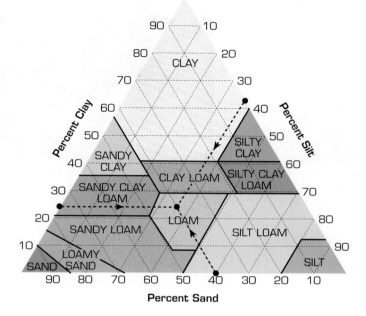

Example: Identify a soil that is 40% sand, 22% clay, and 38% silt.
1. Find 40 on the side for sand.
2. Draw a line in the direction of the arrow.
3. Do the same for clay (22%) and silt (38%).
4. The spot where the three come together is the soil texture. In this case, the soil is a loam.

A textural name may include a prefix naming the dominant sand size, as in "coarse sandy loam."

FIGURE 9–9 Soil texture is determined by the percentages of sand, silt, and clay found in a soil sample. The soil triangle is a tool that is used to determine texture. *(Courtesy USDA, Soil Conservation Service)*

gravity, and movement of humans and animals. This movement not only breaks up soil materials and adds them to the lower levels, but it exposes subsoil materials along the upper slopes. The movement of soil materials has a pulverizing effect on the material being moved, as well as on the material left behind.

Slope affects the distribution of water that falls on the Earth's surface. On level areas, the water soaks in and moves through the soil in a process called **percolation**. On nonlevel land, the water tends to run off and moves some surface soil with it. Soils that develop on level land at low elevations tend to be poorly drained, whereas soils on gentle slopes tend to be better drained and more productive (Figure 9–10).

Drainage (or lack of it) affects the water table in a particular field or area. The water table has a direct bearing on soil formation, especially if it is near the surface. When a soil is saturated with water, little or no air can penetrate it. The lack of air reduces the action of fungi, bacteria, and other soil-forming activities in the soil.

A wet soil, therefore, is a slow-forming soil and usually is low in productivity. Because of the lack of air, undecomposed organic matter will accumulate in a wet soil. This organic matter generally causes the soil to be a blackish color. Poor drainage, accompanied by free water in the soil, reduces or retards plant growth and affects soil formation.

FIGURE 9–10 Topography or slope of the land influences the formation of soil and how well excess water drains from the soil. *(Courtesy Michael Dzaman)*

Time

Soils are formed by the chemical and physical weathering of parent material over time, as affected by climate, living organisms, and topography. Therefore, time itself is regarded as a factor in soil formation. Chemical weathering is the result of the chemical reactions of water, oxygen, carbon dioxide, and other substances that act on the rocks, minerals, organic matter, and life that compose the soil. The leaching action of water hastens the weathering process by removing soluble materials, and chemicals react with each other to form new chemicals in the soil.

Weathering

Weathering refers to mechanical forces caused by temperature change such as heating, cooling, freezing, and thawing. As these processes occur, rocks, minerals, organic matter, and other soil-forming materials are broken into smaller and smaller particles until soil is formed. Soils at different stages of weathering will differ widely.

Weathering causes soils to develop, mature, and age much as people do. Soils develop rapidly, mature, and then develop certain characteristics of age. Plant nutrients are released quickly from the minerals, plant growth increases, and organic matter accumulates. Soils age more slowly during the later stages of weathering.

Eventually, nutrients in the soil are depleted. Water moving through the soil leaches away many soluble portions. At this stage, many soils are acidic because the limestone originally in them is gone. As the supply of nutrients in the soil decreases, the amount of plant growth is reduced to the point where the organic matter decomposes faster than it is produced. When soils become acidic and have lost their native fertility, they require expensive amendments to keep them productive.

In **permeable** soils that permit water movement, the fine clay particles tend to move downward from the surface soil into the subsoil during the weathering process. This movement, together with further breakdown of the rock material, accounts for the fact that many soil types have a greater percentage of clay in the subsoil than in the surface soil.

SOIL CLASSIFICATION

Soil scientists have developed a system for mapping soils according to the physical, chemical, and topographical aspects of the land. Such maps have lines showing the outline of soil types, and they provide numerical codes

Science Profile

THE ANATOMY OF A SINKHOLE

Soil and water scientists go underground to check water quality! Doug Boyer squeezes, winds, crawls, stoops, glides, jumps, splashes, plunges, soaks, wades, and slithers his way through the muddy, narrow passage. Beneath farms of West Virginia's Greenbrier County lies "The Hole." The Hole is part of a 6-square-mile hole basin. Here, underground water flowing through limestone bedrock has dissolved the limestone and left caverns, sinking streams, and sinkholes. Such sinkholes are found in many parts of the nation.

Sinkholes are funnel-shaped depressions in the Earth's surface formed when soil settled into underground cavities and caves. They can provide perfect collection basins for surface water and pollutants enroute to underground water reserves unless special management practices around each such sinkhole are followed.

The Hole is one of 37 drainage basins selected for study under the United States Clean Water Act. The purpose of these studies is to ensure that surface and underground water supplies are protected through the safe use of fertilizers, manure, and pesticides.

As part of the water quality study, easily accessible water samples are taken at least weekly, and cave samples are taken occasionally. Samples are analyzed for nitrates, selected bacteria, herbicides, and other indicators of pollution. Bill Balfour, a volunteer with the West Virginia Association for Cave Studies, has mapped much of The Hole and much of the 400 miles of other known caverns in Greenbrier County. He introduced Boyer and other soil and water scientists and technicians to the cave system and continues to push back the frontiers of knowledge in this field.

Water quality experts take water droplet samples and examine a well casing beneath a sinkhole to check for water pollution. A sinkhole is a depression in the surface of the Earth caused by surface soil caving into cavities below that are left after limestone or other rock materials have dissolved. *(Courtesy USDA/ARS: Agricultural Research, August 1993, p. 7)*

keyed to large amounts of information about the land. The experienced soil technician can obtain a wealth of information about the land by consulting a soils map and the accompanying material.

Land Capability Maps

INTERNET KEY WORDS:

land capability classes

Soil mapping and land classification have been a priority of the U.S. Department of Agriculture (USDA) throughout most of the 20th century. Much of the original work of mapping was completed by the Soil Conservation Service (SCS). However, in recent years, this agency has been incorporated into the Natural Resources Conservation Service (NRCS). Now the NRCS can provide maps and classification information for almost any area in the United States, including small areas on individual farms. Agency personnel work with local farmers to develop farm plans. These provide recommendations for land use, cropping systems, crop production practices, and livestock systems. These plans are designed to maximize production while controlling soil erosion and enhancing productivity.

Land capability maps indicate (1) capability class, (2) capability subclass, and (3) capability unit.

Capability Class Capability classes are the broadest groups on a soils map and are designated by Roman numerals I through VIII (Figure 9–11). The numerals indicate progressively greater limitations and narrower choices for practical use of land as follows.

Class I—soils have few limitations that restrict their use.

Class II—soils have moderate limitations that reduce the choice of plants or require moderate conservation practices.

FIGURE 9–11 The eight classes of land in the United States.

Class III—soils have severe limitations that reduce the choice of plants, require special conservation practices, or both.

Class IV—soils have severe limitations that reduce the choice of plants, require careful management, or both.

Class V—soils are not likely to erode but have other limitations that are impractical to remove and limit their use.

Class VI—soils have severe limitations that make them generally unsuitable for cultivation.

Class VII—soils have severe limitations that make them unsuitable for cultivation.

Class VIII—soils and geologic features have limitations that nearly always prevent their use for commercial plants.

Land capability maps are usually color-coded for ease in differentiating capability classes.

Capability Subclasses Capability subclasses are soil groups within one class. They are designated by adding a lowercase letter *e, w, s,* or *c* to the class numeral (for example, IIe). The letter *e* indicates the main limitation is risk for erosion unless close-growing plant cover is maintained; *w* indicates weather in or on the soil interferes with plant growth or cultivation; *s* indicates the soil is limited mainly because it is shallow, droughty, or stony; and *c,* used in only some parts of the United States, indicates the chief limitation is climate—too cold or too dry (Figure 9–12).

In Class I; there are no subclasses because the soils of this class have few limitations. Conversely, Class V contains only the subclasses

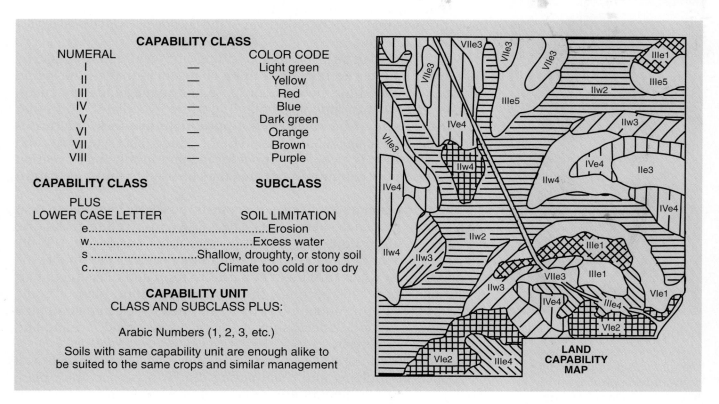

FIGURE 9–12 Land capability map. A land capability map is used to identify soil locations with similar characteristics. Similar land capability classes indicate that similar crops and management practices may be applied.

indicated by *w, s,* or *c,* because the soils in Class V are subject to little or no erosion. However, they have other limitations that restrict their use to pasture, range, woodland, wildlife habitat, or recreation.

Capability Units   The soils in one capability unit (soil groups within the subclasses) are enough alike to be suited to the same crops and pasture plants, and require similar management. They have similar productivity capabilities and other responses to management. The capability unit is a convenient grouping for making many statements about the management of soil.

Capability units are generally designated by adding numbers (0–9) to the subclass symbol (for example, IIIe4 or IIw2). Thus, in one symbol, the Roman numeral designates the capability class or degree of limitation; the lowercase letter indicates the subclass or kind of limitation; and the Arabic numeral specifically identifies the capability unit within each subclass. A map legend for each soil grouping is included with the soil and land capability map. This type of map is also an example of how symbols can be effectively used. They make it possible to place much information in a small space on the map by use of a code.

Use of Maps

When the landowner and the soil conservationist start planning for the most intensive use of a farm, they need a soil and land capability map. This will help them prepare a conservation plan or proposed land-use map. The soil conservationist and the landowner, through the use of the soil and land capability map, discuss the kinds of soil on the farm. The current and original land uses are also discussed and noted on the map. In developing a proposed land-use map, the soil conservationist must know the personal goals or objectives of the landowner and the plans for developing the land.

Many things can be involved in reaching land-use decisions, field by field. Field boundaries may need to be changed so that all the soil in each field is suited for the same purpose and management practices. The desired balance among cropland, pasture, woodland, and other land uses needs to be considered. Appropriate livestock enterprises should also be considered to match the land's characteristics and potential. If there is a good potential for income-producing recreation enterprises in the community, an area may be used for hunting, campsites, or fishing.

The landowner and soil conservationists must consider how to treat each field to get the desired results. The NRCS conservationist can give many good suggestions, but the landowner must decide what to do, when to do it, and how to do it. As planning decisions are made, the conservationist will record them in narrative form and make them part of the plan map. This becomes the farm conservation plan. It is the guide for the farming operation in the years ahead. A conservation plan is required to obtain federal support to encourage good farming and conservation practices.

Workers from the NRCS are also available to give on-site technical assistance in applying and maintaining the farm conservation plan. They provide management advice and services to nonfarm landowners, developers, strip-mining operations, and other activities where soil is involved (Figure 9–13).

INTERNET KEY WORDS:

land capability class map

INTERNET ADDRESS:

http://www.nhq.nrcs.usda.gov/
land/meta/m2771.html

FIGURE 9–13 Natural Resources personnel advise on the many uses of land. Here high-altitude photography helps document changing conditions. *(Courtesy USDA #K-5218-03)*

Horizon	Name	Colors	Structure	Processes Occurring
O	Organic	Black, dark brown	Loose, crumbly, well broken up	Decomposition
A	Topsoil	Dark brown to yellow	Generally loose, crumbly, well broken up	Zone of leaching
B	Subsoil	Brown, red, yellow, or gray	Generally larger chunks, may be dense or crumbly, can be cement-like	Zone of accumulation
C	Parent material (slightly weathered material)	Variable — depending on parent material	Loose to dense	Weathering, disintegration of parent material or rock

FIGURE 9–14 Characteristics of soil horizons.

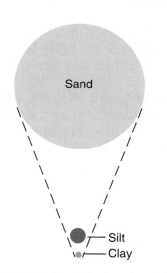

FIGURE 9–15 Relative sizes of sand, silt, and clay particles.

SIZES OF SOIL PARTICLES	
Name	Size, Diameter in Millimeters
Fine gravel	2 – 1
Coarse sand	1.00 – 0.50
Medium sand	0.50 – 0.25
Fine sand	0.25 – 0.10
Very fine sand	0.10 – 0.05
Silt	0.05 – 0.002
Clay	less than 0.002

FIGURE 9–16 Range of sizes of soil particles.

INTERNET KEY WORDS:

soil profiles, horizons

PHYSICAL, CHEMICAL, AND BIOLOGICAL CHARACTERISTICS

Soil Profile

Undisturbed soil will have four or more horizons in its profile. These are designated by the capital letters *O, A, B,* and *C* (Figure 9–14). The **O horizon** is on the surface and is composed of organic matter and a small amount of mineral matter. Organic matter originates from living sources such as plants, animals, insects, and microbes. **Mineral matter** is derived from nonliving sources such as rock materials.

The **A Horizon** is located near the surface and is mineral matter and organic matter. It contains desirable proportions of organic matter, fine mineral particles called **clay**, medium-sized mineral particles called **silt**, and larger mineral particles called **sand**. The appropriate proportion of these creates soil that is **tillable**, or workable with tools and equipment. With the presence of desirable plant nutrients, chemicals, and living organisms, the A horizon generally supports good plant growth. The A horizon is frequently called **topsoil**.

The **B horizon** is below the A horizon and is generally referred to as **subsoil**. The mineral content is similar to the A horizon, but the particle sizes and properties differ. Because organic matter comes from decayed plant and animal materials, the amount naturally decreases as distance from the surface increases.

The **C horizon** is below the B horizon and is composed mostly of parent material. C Horizon is important for storing and releasing water to the upper layers of soil, but it does not contribute much to plant nutrition. It is likely to contain larger soil particles and may have substantial amounts of gravel and large rocks. The area below the C horizon is called **bedrock**.

Texture

Texture refers to the proportion and size of soil particles (Figures 9–15 and 9–16). Texture can be determined accurately in the laboratory by mechanical analysis (Figure 9–17). The feel of the soil can also be used to analyze soil as follows:

1. Make a stiff mud ball.
2. Rub the mud ball between your thumb and forefinger.
3. Note the degree of coarseness and grittiness caused by the sand particles.
4. Squeeze the mud between your fingers, and then pull your thumb and fingers apart.
5. Note the degree of stickiness caused by the clay particles.
6. Make the soil slightly more moist and note that the clay leaves a "slick" surface on your thumb and fingers.

The outstanding physical characteristics of the important textural grades, as determined by the "feel" of the soil, are as follows (Figure 9–18).

Coarse-textured (sandy) soil is loose and single grained. The individual grains can be seen readily or felt. Squeezed in the hand when dry, it will fall apart when the pressure is released. Squeezed when moist, it will form a cast, but will crumble when touched.

Medium-textured (loamy) soil has a relatively even mixture of sand, silt, and clay. However, the clay content is less than 20 percent. (The characteristic properties of clay are more pronounced than those of sand.) A loam is mellow with a somewhat gritty feel, yet fairly smooth and highly plastic. Squeezed when moist, it will form a cast that can be handled quite freely without breaking.

Fine-textured (clay) soil usually forms hard lumps or clods when dry. It is usually very sticky when wet and is quite plastic. When the moist soil is pinched between the thumb and fingers, it will form a long, flexible "ribbon." A clay soil leaves a "slick" surface on the thumb and fingers

INTERNET KEY WORDS:

soil texture,
soil structure

DEMONSTRATION TO SHOW SAND, SILT, AND CLAY

MATERIALS: A quart jar with lid
1/2 teaspoon of Calgon (dispersing agent)
A pint of medium-textured soil

PROCEDURE: Fill the jar half-full of soil.
Add enough water to make the jar 3/4 full.
Add 1/2 teaspoon of Calgon.
Shake for 5 minutes.
Set jar aside and allow to settle undisturbed.
Note that the soil separates, settling into layers.

STEP 1: Add ingredients

STEP 2: Install lid and shake well

STEP 3: (24 hours after shaking)

FIGURE 9–17 Mechanical analysis of soil to accurately determine the soil texture.

when rubbed together with a long stroke and firm pressure. The clay tends to hold the thumb and fingers together because of its stickiness.

Structure

Soil **structure** refers to the tendency of soil particles to cluster together and function as soil units called aggregates. Aggregates, or **crumbs**, contain mostly clay, silt, and sand particles held together by a gel-type substance formed by organic matter.

Aggregates absorb and hold water better than individual particles. They also hold plant nutrients and influence chemical reactions in the soil.

Another major benefit of a well-aggregated soil, or a soil with good structure, is its resistance to damage by falling raindrops. When hit by falling rain, the aggregate stays together as a water-absorbing unit, rather than separating into individual particles. When aggregates on the surface of soil dry out, they remain in a crumbly form and permit good air movement. Dispersed soil particles run together when dry and form a crust on the soil surface. The crust prevents air exchange between the soil and the

SOIL TEXTURAL CLASSES

SAND

Dry: loose and single grained; feels gritty.
Moist: will form a ball that crumbles very easily.

LOAMY SAND

Dry: silt and clay may mask sand; feels loose, gritty.
Moist: feels gritty; forms a ball that crumbles easily; stains finger slightly.

SANDY LOAM

Dry: clods easily broken; sand can be seen and felt.
Moist: moderately gritty; forms a ball that can stand careful handling; definitely stains fingers.

LOAM

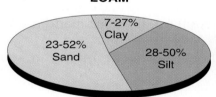

Dry: clods are moderately difficult to break; mellow, somewhat gritty.
Moist: neither very gritty nor very smooth; forms a firm ball; stains fingers.

SILT LOAM

Dry: clods are difficult to break; feels smooth, soft, and floury when pulverized; shows fingerprints
Moist: has smooth or slick "buttery" or "velvety" feel; stains fingers.

CLAY LOAM

Dry: clods very difficult to break with fingers.
Moist: has slightly gritty feel; stains fingers; ribbons fairly well.

SILTY CLAY LOAM

same as **CLAY LOAM** but very smooth.

SANDY CLAY LOAM

same as **CLAY LOAM.**

CLAY

Dry: clods cannot be broken with fingers without extreme pressure.
Moist: quite plastic and usually sticky when wet; stains fingers. (A silty clay feels smooth, a sandy clay feels gritty.)

FIGURE 9–18 Major soil textural classes.

atmosphere and decreases plant growth. The process and benefits of aggregation apply mostly to fine- and medium-textured soils.

Organic Matter

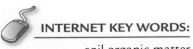

INTERNET KEY WORDS:

soil organic matter
soil, texture,
classification

As seen from the previous discussion, organic matter plays an important role in soil structure. Soil is a living medium with a great variety of living organisms. Some groups of organisms of the plant kingdom that are often found in soils are:

- roots of higher plants;
- algae: green, blue-green, and diatoms;
- fungi: mushroom fungi, yeasts, and molds; and
- actinomycetes of many kinds: aerobic, anaerobic, autotrophic, and heterotrophic.

Some examples of groups of organisms from the animal kingdom that are prevalent in soils are:

- those that subsist largely on plant material: small mammals, insects, millipedes, sow bugs (wood lice), mites, slugs, snails, earthworms;
- those that are largely predators: snakes, moles, insects, mites, centipedes, and spiders; and
- microanimals that are predatory, parasitic, and live on plant tissues: nematodes, protozoans, and rotifers.

Living Organisms

Living organisms excrete cell or body wastes that become part of the organic content of soil. The microbes of the soil and the remains of larger plants and animals decompose or decay into soil-building materials and nutrients.

People who grow indoor plants at home raise gardens, farm, produce greenhouse crops, or grow nursery stock generally find it useful to add organic materials to the soil. Popular sources of organic matter for soil amendments are peat moss, leaf mold, compost, livestock manure, and sawdust.

Some important benefits or functions of organic matter in soil include:

- making the soil porous;
- supplying nitrogen and other nutrients to the growing plant;
- holding water for future plant use;
- aiding in managing soil moisture content;
- furnishing food for soil organisms;
- serving as a store house for nutrients;
- minimizing leaching;
- serving as a source of nitrogen and growth-promoting substance; and
- stabilizing soil structure.

Other Properties of Soils

Soil scientists, managers, technicians, and operators must be aware of numerous other properties of soils in their work. Some of these are external

Science Connection

DIGGING LIFE

Soil is the most diverse ecosystem on Earth. The number of organisms per acre far outnumber that of any other place in the world. Most of us do not think about the abundant life under our feet as we cross a lawn. But just one square inch of this soil is teeming with busy creatures, most of which cannot be seen. Gardens and fields consist of fertile soil filled with living organisms. In 1 gram of this soil, there are:

3,000,000–500,000,000	Bacteria—A group of one-celled microscopic organisms.
1,000,000–20,000,000	Actinomycetes—Microscopic organisms that resemble both fungi and bacteria.
5,000–1,000,000	Fungi—Non-microscopic organisms that get their food from dead material. Mushrooms are an example of fungi.
1,000–100,000	Yeast—Single celled fungi. Many are used in producing food.
1,000–500,000	Protozoa—Small single celled organisms. An example is an amoeba.
1,000–500,000	Algae—One-celled organisms that contain chlorophyll.
1–500	Nematodes—Non-segmented round worms.

Large numbers of slime molds, viruses, insects, and earthworms are also present. Some of these organisms are harmful to plants and animals. Most are decomposers. A decomposer is an organism that breaks down material that was once living. They change things that are dead into the rich organic substances that add to the fertility of the soil.

The role these tiny creatures play in an ecosystem is irreplaceable. Imagine a world without decomposers. Leaves, dead animal carcass, and huge amounts of other dead organic matter would pile up. In a short period, dead and un-decomposed material would crowd out all life. Soil is a unique substance. It provides living space for billions of organisms. It is the medium that plants and other producers grow in. Soil is also a place where once-living things are broken down and changed into the fertile organic material of the soil.

FIGURE 9–19 The pH scale is neutral in the middle. It gets progressively more acidic from 7 down to 3 and progressively more alkaline from 7 up to 11.

factors such as land position, slope, and stoniness. Soil color, depth, drainage, permeability, and erosion are important considerations.

MAKING AMENDMENTS TO PLANT-GROWING MEDIA

The term amendment is used here to mean addition to or change in. Most soil amendments are made to add organic matter, add specific nutrients, or modify soil pH. The pH is a measure of the degree of acidity or alkalinity. Acidity is sometimes referred to as sourness, and alkalinity is referred to as sweetness. The pH scale ranges from 0 (maximum acidity) to 14 (maximum alkalinity). The midpoint of the scale is 7, which is neutral, meaning neither acidic nor alkaline (Figure 9–19).

Crops grow best in media with a narrow pH range unique to that plant species (Figure 9–20). Most plants require a pH somewhere between 5.0 and 7.5. Some crops, such as potatoes, prefer a soil pH around 5.5. Alfalfa responds best to a pH of 7.0 to 8.0.

Liming

Areas that were historically covered by trees, such as the northeastern, western, and northwestern parts of the United States, develop acidic soils.

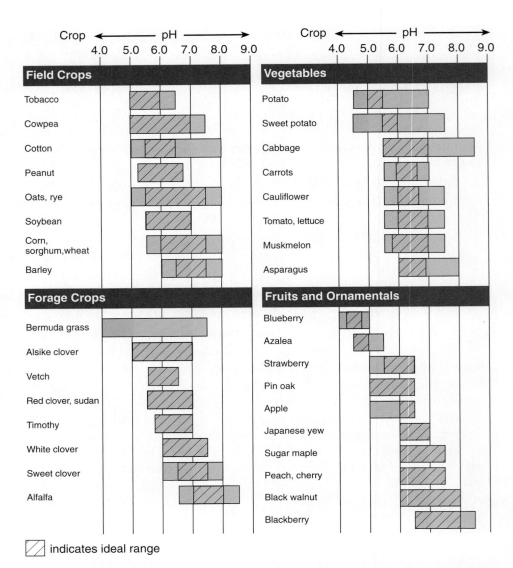

indicates ideal range

FIGURE 9–20 Plants grow in pH ranges from approximately 4 (very acidic) to 8.5 (very alkaline).

INTERNET KEY WORDS:

collecting a soil sample

When cleared of trees, such soils require additional lime to increase the pH for the efficient production of most farm crops.

A pH test can be performed using a pH test kit on-site, or soil samples can be sent to a university or commercial laboratory for analysis. Laboratory analyses generally include an analysis of phosphorus, potash, and magnesium, as well as pH. Liming and fertilizing recommendations may also be provided by testing laboratories and universities (Figure 9–21).

How to Take a Soil Sample in Your Lawn or Garden

1. Select an appropriate sampling tool (spade, auger, or soil tube) (Figure 9–22).

2. Make a sketch dividing the area into sampling areas—for example, front lawn, garden, flower bed, slope, and back lawn. Appropriately label each area (see Figure 9–21, view 1).

3. When taking samples, avoid wet or bare spots. Soils that are substantially different in plant growth or past treatment should be

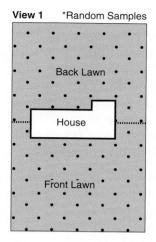

View 1 *Random Samples

Back Lawn

House

Front Lawn

View 2 *Random Samples

Wet Spot

Garden

Flower Bed

House

Slope

View 3

View 4

When using a spade, first make a V-shaped cut in the ground. Then, remove a 1-inch slice from one side of the cut (View 3). Take a 1-inch strip from the middle of the slice for the soil sample (View 4).

FIGURE 9–21 Procedure for taking a soil sample to conduct tests for pH and nutrient levels in soil. Such tests are necessary for matching lime and fertilizer applications to plant needs for optimum yields.

sampled separately, provided their size and nature make it feasible to fertilize or lime each area separately.

4. After removing surface litter, take a sample from the correct depth. This is 2 inches for established lawns, and about 6 inches for gardens, flower beds, farm crop land, and other areas to be tilled.

5. Submit a separate composite sample for each significantly different area—for example, front lawn, back lawn, and flower bed. Your composite sample for each area should include a small amount of soil taken from each of 10 to 20 randomly selected locations in the area represented by each sample.

6. When using a spade, first make a V-shaped cut. Then remove a 1-inch slice from one side of the cut (see Figure 9–21, view 3). Then take a 1-inch strip from the middle of this slice (see Figure 9–21, view 4). This represents the soil from one spot in the sample.

7. *Air-dry the soil; do not use heat.* Mix the soil from a composite in a clean bucket. Place about 1 pint of this mixture into the sample box. Use a separate box for each composite. Fill in the blanks on the box or information sheet for each box.

8. Send soil sample(s) and information sheet(s) to the soil test laboratory.

Soil samples are collected routinely to help determine appropriate management of soil amendments. The same soil sampling procedure that was described for lawns and gardens is used for farm fields, golf courses, and greenhouses. It is critical that adequate amounts of fertilizers and other soil amendments are provided to a crop without applying excessive amounts. Adding too much of a chemical to the soil is not only expensive, but it may leach into the groundwater and surface water.

Note: The Cooperative Extension System office in most counties and/or cities can arrange for soil testing.

FIGURE 9–22 A soil sample that is collected with a soil probe collects a sample that is uniform in volume at any given soil depth. *(Courtesy of James Strawser, The University of Georgia)*

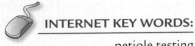

INTERNET KEY WORDS:

petiole testing
tissue testing

The Petiole Test Some crops are expensive to raise because they require large amounts of soil amendments. For example, potatoes require relatively high applications of nitrogen. When the nitrogen supply in the soil becomes depleted, the newest leaves on the plant will also have a nitrogen deficiency. By sampling and testing the petiole (young stem and leaves) of the potato plant, the deficiency can be identified and corrected before the crop yield is affected. The petiole test is effective when it is used to manage soil nutrients during the growing season.

The pH Test The amount of agricultural lime required to raise soil to the desired pH level is indicated by a pH test. The same pH value may require different amounts of agricultural lime. This is because soils contain varying amounts of organic matter, clay, silt, and sand. The greater the organic matter and clay content, the greater the amount of lime required to correct the acidity.

Even though all the lime required by your soil is applied, do not expect the pH to increase quickly. It generally will require 2 to 3 years for all of the lime to be used and the desired pH to be reached.

Mix Well The soil on 1 acre 6 inches in depth weighs about 2 million pounds. Therefore, it takes a lot of mixing to distribute a relatively small amount of lime with the soil. Liming recommendations are based on a specific plowing depth, such as 6 or 7 inches. Application rates may need to be adjusted for deeper or more shallow tilling.

Standard Ground Limestone Lime recommendations are based on standard ground limestone, which should contain a minimum of 50 percent lime oxides (calcium oxide plus magnesium oxide). About 98 percent should pass through a 20-mesh sieve, with a minimum of 40 percent passing through a 100-mesh sieve. The higher the percentage of limestone passing through a 100-mesh screen, the faster this limestone will correct soil acidity. The best limestone will have the greatest calcium and/or magnesium content and will be ground to a small particle size. It is more important to finely grind a high-magnesium stone than a high-calcium stone. A high-magnesium or dolomitic limestone should always be used when a magnesium deficiency is indicated by a soil test.

One ton [2,000 lb] standard ground limestone is approximately equivalent to 1,500 lb of hydrated lime or 1,100 lb ground burned limestone.

Correcting Excessive Alkalinity

A reduction of soil alkalinity is desirable where soils have a high pH and the alkaline condition causes unsatisfactory crop growth. In most cases, this condition will exist where heavy applications of lime are made at one time, or where lime has been applied to a soil with a high pH.

Gypsum is a soil amendment that is often used to reduce the alkalinity of agricultural soils. It is added when soil pH is too high for the intended crop to grow and thrive. In most instances, it should be added in the fall to allow time for it to be effective. For best results, it should be distributed in the upper layer of topsoil within the area where the roots of the crop are concentrated.

Another method that may be used to decrease the pH value of soil is to use sulfur or aluminum sulfate. Sulfur at 1.5 lb per 100 square feet or aluminum sulfate at 5 lb per 100 square feet will decrease the alkalinity, under most conditions, by 0.5 pH. Aluminum sulfate acts rapidly, producing acidification in 10 to 14 days. Sulfur requires 3 to 6 months.

INTERNET KEY WORDS:

organic, natural, fertilizer

Broadcast the material over the surface and thoroughly work it into the soil. For full benefit, sulfur should be applied in the fall, after garden crops are harvested. Aluminum sulfate may be applied in early spring.

Use chemicals cautiously to decrease alkalinity. Before these measures are taken, you should consider other factors that may be responsible for poor growth (such as drought, insect and disease injury, and fertilizer burning).

Fertilizers and Fertilizing

Essential plant nutrients are discussed in Unit 16, "Plant Physiology." However, it should be noted that nitrogen, phosphorus, and potassium are known as the three **primary nutrients**. These three ingredients must be present for a fertilizer to be called a **complete fertilizer**.

The proportions of nitrogen, phosphorus, and potassium are known as the **fertilizer grade**, expressed on a fertilizer container as percentages of the contents of the container by weight. Therefore, a 100-lb container of fertilizer with a grade of 10-10-10 contains 10 percent nitrogen, 10 percent phosphorus, and 10 percent potassium.

If the total amount of fertilizer (100 lb) is multiplied by the percentage of each ingredient, the pounds of each ingredient may be calculated. Therefore, 0.10 (percent of nitrogen) × 100 (total lb of fertilizer) = 10 lb nitrogen. The amount of phosphorus and potassium is also 10 lb each. The other 70 lbs consist of inert filler material.

Fertilizer is frequently shipped in 80-lb bags. Therefore, one bag of 10-10-10 fertilizer would have 8 lb nitrogen, 8 lb phosphorus, and 8 lb potassium for a total of 24 lb of **active ingredients** (components that achieve one or more purposes of the mixture).

Some popular grades of fertilizer are 5-10-5, 5-10-10, 10-10-10, 6-10-4, 0-15-30, 0-20-20, 8-16-8, and 8-24-8. These grades are formulated to meet the needs of a variety of crops on various soils. The amount and grade of fertilizer to apply are determined by: (1) the specific crop to be grown, (2) the desired yield or performance of the crop, (3) fertility of the soil, (4) physical properties of the soil, (5) previous crop, and (6) type and amount of manure applied. Therefore, decisions on rate of application must be made on a local basis. Soil tests, tissue tests, and plant observations are useful techniques for determining fertility needs (Figure 9–23).

Organic Fertilizers Organic fertilizers include animal manures and compost made with plant or animal products. Organic commercial fertilizers include dried and pulverized manures, bone meal, slaughterhouse tankage, blood meal, dried and ground sewage sludge, cottonseed meal, and soybean meal.

Organic fertilizers have certain definite characteristics. First, nitrogen is usually the predominating nutrient, with lesser quantities of phosphorus and potassium. One exception to this is bone meal, in which phosphorus predominates and nitrogen is a minor ingredient. Second, the nutrients are only made available to plants as the material decays in the soil, so they are slow acting and long lasting. Third, organic materials alone are not balanced sources of plant nutrients, and their analysis in terms of the three major nutrients is generally low.

Inorganic Fertilizers Various mineral salts, which contain plant nutrients in combination with other elements, are called inorganic fertilizers. Their characteristics are different from organic fertilizers. First, the

HEALTHY leaves shine with a rich, dark green color when adequately fed.

PHOSPHOROUS (phosphate) shortage marks leaves with reddish purple, particularly on young plants.

POTASSIUM (potash) deficiency appears as a firing or drying along the tips and edges of lowest leaves.

NITROGEN hunger sign is yellowing that starts at tip and moves along middle of leaf.

MAGNESIUM deficiency causes whitish stripes along the veins and often a purplish color on the underside of the lower leaves.

DROUGHT causes corn plants to have a grayish green color; leaves may roll up to the size of a pencil.

DISEASE helminthosporium blight, starts in small spots, gradually spreads across leaf.

CHEMICALS may sometimes burn tips, edges of leaves, and at other contacts. Tissue dies, leaf becomes whitecap.

Drawings: Maynard Reece

FIGURE 9–23 Corn plants are good indicators of nutrient deficiencies and other factors that impact plant health and productivity. *(Courtesy of Potash and Phosphate Institute, Norcross, GA)*

nutrients are in soluble form and are quickly available to plants. Second, the soluble nutrients make them caustic to growing plants and can cause injury. Care must be used in applying inorganic fertilizers to growing plants. They should not come in contact with the roots or remain on plant foliage for any length of time. Third, the analysis of chemical fertilizers is relatively high in terms of the nutrients they contain.

Fertilizers are needed to replenish mineral nutrients depleted from a soil by crop removal or by such natural means as leaching. Some soils with high fertility may need only nitrogen or manure. Use of fertilizers that also contain small amounts of copper, zinc, manganese, boron, and other minor elements is not considered necessary for most soils, but may be needed in certain soils and for certain crops.

PLANT NUTRIENTS IN FERTILIZER MATERIALS

Materials	Nitrogen N%	Phosphorus P_2O_5%	Potassium K_2O%	Calcium %	Magnesium %	Sulfur %
Ammonium Nitrate	33.5	—	—	—	—	—
Ammonium Sulfate	21	—	—	—	0.3	23.7
Urea	45	—	—	1.5	0.7	.02
Sodium Nitrate	16	—	0.2	0.1	.05	.07
Calcium Nitrate	15	—	—	19.4	1.5	.02
Calcium Cyanamide	21	—	—	38.5	.06	0.3
Anhydrous Ammonia	82	—	—	—	—	—
Superphosphate	—	20	0.2	20.4	0.2	11.9
Liquid Phosphoric Acid	—	58	—	—	—	—
Ammonium Phosphate	11	48	0.2	1.1	0.3	2.2
Potassium Chloride	—	—	62	—	0.1	0.1
Potassium Sulphate	—	—	53	0.5	0.7	17.6

FIGURE 9–24 Plant nutrients in common fertilizer materials.

There are also unmixed fertilizers that carry only one element (Figure 9–24). Most important of these unmixed materials are the nitrogen and phosphate carriers. Nitrogen carriers vary from 16 to 45 percent nitrogen. Be careful in using nitrogen materials. Too much may cause excessive and soft growth.

Superphosphate fertilizers carry only phosphorus. Phosphorus promotes flower, fruit, and seed development. It also firms up stem growth and stimulates root growth. Superphosphate may be added to manure to give a better balance of nutrients for plants (100 lb to each ton of horse or cow manure and 100 lb to each half-ton of sheep manure).

Fertilizer Applications There are many ways to apply fertilizers. For the home lawn, the most likely method is broadcasting (spreading evenly over the entire surface). In the case of gardens or fields, broadcasted fertilizer may be incorporated or mixed into the soil by spading, tilling, plowing, or disking.

Band application places fertilizers about 2 inches to one side of and slightly below the seed. This method is used extensively for row crops in gardens and fields.

Side-dressing is done by placing fertilizer in bands about 8 inches from the row of growing plants. This method is popular for row crops such as corn and soybeans.

Top-dressing is a procedure where fertilizer is broadcast lightly over close-growing plants. Top-dressing is used for adding nitrogen to small grain, hay, and turf crops after they are established.

Starter solutions are diluted mixtures of fertilizer used when plants are transplanted. Their purpose is to provide small amounts of nutrients that will not burn the tender roots of young plants.

Other methods of applying fertilizers include the application of foliar sprays directly onto the leaves of plants, and knife application of anhydrous ammonia gas into the soil. The practice of adding liquid fertilizer to irrigation water is also used extensively in the United States.

FIGURE 9–25 Animal manure is excellent fertilizer for most crops, and it also improves the soil by adding organic matter to the topsoil.

APPROXIMATE AMOUNTS OF PLANT NUTRIENTS AVAILABLE FROM ONE TON OF MANURE

	Cattle	Sheep	Swine	Poultry	Horse
Nitrogen (lb)	10	28	10	30	14
Phosphorus (lb)	5	10	5	20	5
Potassium (lb)	10	25	10	10	14

Using Manure

Animal manure is a valuable product when handled properly. Its content of plant nutrients makes it a valuable fertilizer material. In addition, the organic matter aids in developing and maintaining structure in soils (Figure 9–25).

To obtain the most nutrient value from manures, the following practices should be followed:

- Use adequate bedding to absorb all of the liquids.
- Balance the phosphorus in cow manure by adding superphosphate to the fields.
- Spread manure evenly over fields. An 8- to 12-ton application per acre is recommended. (Fifty bushels of manure and litter weigh about 1 ton.)
- When possible, incorporate manure into the soil immediately after spreading.
- Do not spread on steep slopes when the ground is frozen.
- When storing manure, keep it compact and under cover.
- Prevent liquid runoff from escaping from manure holding areas.
- Apply manure to crops that will give best response, such as corn, sorghums, potatoes, and tobacco.

When applying manure, the amount of commercial fertilizer should be reduced accordingly (Figure 9–26).

Legume Crops

Legumes are plants in which certain bacteria use nitrogen gas from the air and convert it to **nitrates** (the form of nitrogen used by plants). Some examples of legumes are beans, peas, clovers, and alfalfa. The process of converting nitrogen gas to nitrates by bacteria in the roots of legumes is called **nitrogen fixation**.

Nitrogen fixation reduces or eliminates the need for adding expensive nitrogen fertilizer to legume crops. When the roots of legume plants decay, large amounts of nitrates are left behind for the next crop. Crops such as corn, which requires large amounts of nitrogen, should follow alfalfa or clover in a field, because the legumes leave unused nitrogen behind.

Rotation Fertilization

Many crop rotations start with a small-grain crop. The preparation of the seedbed for small grains provides an excellent opportunity to put lime and fertilizer into the feeding zone for roots of the sod crop that

	N NITROGEN	**P** PHOSPHORUS	**K** POTASSIUM
Suppose the fertilizer recommendations <u>per acre</u> for a certain crop are:	150 lb	100 lb	100 lb
If well-managed cow manure is applied at the rate of 10 tons per acre, it can be determined from Figure 9–30 that the 10-ton application would provide:	100 lb	50 lb	100 lb
Therefore, the amount of nutrients that must be provided per acre through commercial fertilizer is:	50 lb	50 lb	0 lb
The shortfall of ingredients listed above could be made up by applying 500 lb of 10-10-0 commercial fertilizer per acre. NOTE: Caution must be exercised when estimating the nutrient values of manure due to variations in liquid and solid content captured from the animal, amount and type of bedding, and procedures used to handle and store the manure.			

FIGURE 9–26 When manure is applied to land, the amount of commercial fertilizer used should be adjusted to allow for the nutrients in the manure.

normally follows. Lime and phosphorus move slowly in the soil, and unless the roots contact these elements, they cannot provide the seedling's high requirements for phosphorus in the critical first year of growth.

A properly limed and fertilized sod crop is the backbone of a successful crop rotation. The growth of nutrient-enriched grass and legume roots throughout the soil provides a most favorable medium for natural soil-building processes. The addition of organic matter, the movement of fertilizer elements by the roots into the soil, and the production of root channels by sod roots produce soils that absorb water better, erode less, are easier to work, and grow better crops.

Phosphorus and potassium added to sod will benefit the sod and, in addition, will be placed in the best location and be in the best form to supply these elements to the long-season row crops that follow. Nitrogen produced by the legumes or added to grass will be present in organic form and will be released in the best possible form for the row crops and small-grain crops.

The composition of soils is complex and depends on many factors. Soil is dynamic and changing all the time. Nature has provided many cycles to help provide for soil renewal. The scientific management of soils makes them more productive and helps to ensure productive soils capable of efficient production for future generations.

HYDROPONICS

INTERNET KEY WORDS:

hydroponics

The term *hydroponics* refers to a number of types of systems used for growing plants without soil (Figure 9–27). Some major systems are:

- **aggregate culture:** in which a material such as sand, gravel, or marbles supports the plant roots;

- **water culture, solution culture,** or **nutriculture:** the plant roots are immersed in water containing dissolved nutrients;

- **aeroponics:** in which the plant roots hang in the air and are misted regularly with a nutrient solution; and

FIGURE 9–27 There are many types of hydroponics systems including (A) aggregate culture; (B) water culture; (C) aeroponics; and (D) continuous-flow systems. *(Adapted from material by Keith Staley, Middletown High School, Middletown, MD)*

- **continuous-flow systems:** in which the nutrient solution flows constantly over the plant roots. This system is the one most commonly used for commercial production.

Hydroponics is growing in importance as a means of producing vegetables and other high-income plants (Figure 9–28). In areas where soil is lacking or unsuitable for growth, hydroponics offers an alternative production system. Equally good crops can generally be produced in a greenhouse in conventional soil or bench systems.

When plants are grown hydroponically, their roots are either immersed in or coated with a carefully controlled nutrient solution. The nutrients and water are supplied by the solution alone and not by aggregates or other inert materials that support the roots. **Inert** means inactive. Without the presence of soil to absorb and release nutrients, nutrients must be carefully controlled in a hydroponic system.

Plant Growth Requirements

Hydroponically grown plants have the same general requirements for good growth as soil-grown plants. The major difference is the method by which the plants are supported and the source of nutrients that are supplied for growth and development.

FIGURE 9–28 High-value crops such as tomatoes are ideal for hydroponic production methods. *(Courtesy of PhotoDisc)*

INTERNET KEY WORDS:

plant nutrient, growth requirements

Water Providing plants with an adequate amount of water is not difficult in a water culture system. During the hot summer months, a large tomato plant may use one-half gallon of water per day. However, quality can be a problem. Water with excessive alkalinity or salt content can result in a nutrient imbalance and poor plant growth. Softened water may contain harmful amounts of sodium. Water that tests above 320 parts per million of salts is likely to cause an imbalance of nutrients.

Oxygen Plants require oxygen for respiration to carry out their functions. Under field and normal greenhouse conditions, oxygen is usually adequate as provided by the soil. When plant roots grow in water, however, the supply of dissolved oxygen is soon depleted, and damage or death soon follow unless supplemental oxygen is provided. Where supplemental oxygen is needed, a common method of supplying oxygen is to bubble air through the water. It is not usually necessary to provide supplementary oxygen in aeroponic or continuous-flow systems.

Mineral Nutrients Green plants must absorb certain minerals through their roots to survive. These minerals are supplied by soil, organic matter, or soil solutions. The elements needed in large quantities are nitrogen, phosphorus, potassium, calcium, magnesium, and sulfur. The nutrients needed in small amounts are iron, manganese, boron, zinc, copper, molybdenum, and chlorine. An oversupply of any nutrient is toxic or detrimental to plants. In hydroponic systems, all nutrients normally supplied by soil must be included in the water to form the solution or media.

Light All vegetable plants and many flowers require large amounts of sunlight. Hydroponically grown vegetables, such as those grown in a garden, need at least 8 to 10 hours of direct sunlight each day to produce well. Electrical lighting is a poor substitute for sunshine because most indoor lights do not provide enough intensity to produce a crop. Incandescent lamps supplemented with sunshine or special plant-growth lamps can be used to grow transplants, but they are not adequate to grow the crop to maturity. High-intensity lamps, such as high-pressure sodium lamps, can provide more than 1,000 foot-candles of light. They may be used successfully in small areas where sunlight is inadequate. However, these lights are too expensive for most commercial operations.

Spacing Adequate spacing between plants will permit each plant to receive sufficient light in the greenhouse. Tomato plants, pruned to a single stem, should be allowed 4 square feet per plant. European seedless cucumbers should be allowed 7 to 9 square feet, and seeded cucumbers need about 7 square feet. Leaf lettuce plants should be spaced 7 to 9 inches apart within the row and 9 inches between rows. Most other vegetables and flowers should be grown at the same spacing as recommended for a garden.

Greenhouse vegetables will not do as well during the winter as in the summer. Shorter days and cloudy weather reduce the light intensity, thus limiting production.

Temperature Plants grow well only within a limited temperature range. Temperatures that are too high or too low will result in abnormal development and reduced production. Warm-season vegetables and most flowers grow best between 60° and 80° F. Cool-season vegetables, such as lettuce and spinach, should be grown between 50° and 70° F.

Support In the garden or field, plants are supported by roots anchored in soil. A hydroponically grown plant must be artificially supported with string, stakes, or other means.

Hydroponics in the Classroom and Laboratory

Hydroponics has become an important teaching area in agriscience programs. The use of common plastic bottles, photographic film cans, and low-cost aquarium supplies permits students and teachers to set up and perform numerous research and demonstration projects in the classroom or laboratory. The material that follows was provided by Dr. David R. Hershey, Assistant Professor in Horticulture at the University of Maryland.

A solution culture system (Figure 9–29) may be constructed as follows:

- Fill two dark 2-liter plastic soda bottles with hot water to loosen the glued label and base. Remove the labels from both bottles and the base from one bottle; keep the base for future use.

FIGURE 9–29 A static solution culture system built from two plastic soda bottles. *(Adapted from material by Dr. David R. Hershey, University of Maryland)*

Aquarium tubing

Lid (constructed from bottom of a second 2-liter plastic bottle)

Plastic drinking straw

Reservoir

Air bubbles

Aquarium valve

Air Pump

Base of soda bottle (constructed from the bottom of a 2-liter plastic bottle)

Agri-Profile

CAREER AREAS:

Agronomy/Hydroponics

Electronic devices have become tools of the trade for soil and water research and management. Career options in soil and hydroponics management overlap those in soil and water conservation to a certain extent, and additional opportunities occur in management of farms and hydroponics greenhouses. The practice of hydroponics is not new, but hydroponics for commercial production has captured the imagination of the world. The recent popularity of hydroponics operations provides new career opportunities, especially in urban areas.

Soil and water management specialists are in demand on the global scene. Progressive Third World countries need help in policy development, education, and project management. They hope to leap from their primitive agriculture of the past to the agriscience of the present in a few short years.

Soil scientist Ronald Schnaber (left) and hydrologic technician Earl Jacoby study natural riparian zone processes that lessen the impact of upstream agriculture on water quality. *(Courtesy USDA/ARS #K-5050-3)*

- Using a fine-point felt-tip marker, draw a line around the bottle 9 inches up from the bottom.

- Using short-bladed scissors, cut on the line made in Step 2 and remove the upper portion of the bottle. This will be the reservoir.

- Using a cork-borer, drill, or scissors, make a hole in the center of the bottle base approximately 0.5 inches in diameter to accommodate a plant stem. Close all except one of the existing (prepunched) holes in the base with black vinyl electrician's tape to prevent light passage. Then place the base on the reservoir as a dome-shaped lid.

- Insert about 8 inches of 2-foot-long aquarium tubing into the open, prepunched hole. To keep the tubing in the reservoir rigid, place a plastic drinking straw over the tubing. Attach the free end of the aquarium tubing to an aquarium air pump with an air control valve.

Nutrient Solutions There are many nutrient-solution recipes, but Hoagland Solution No. 1 is used widely and can be modified to create nutrient deficiencies (Figure 9–30). Often, Hoagland solution is used at less than full strength. Nutrient stock solutions are prepared using a balance to weigh out the salts and a graduated cylinder or volumetric flask to measure the water. Plastic soda bottles can be used to store stock solutions. When stored in a dark place at room temperature, stock solutions should last for many years. To prepare the nutrient solution, add measured volumes of stock solutions to a measured volume of water.

Stock Solution		grams /liter	milliliters of stock solution per liter of nutrient solution[x]							
Formula	Name		Complete	–N	–P	–K	–Ca	–Mg	–S	–Fe
$Ca(NO_3)_2 \cdot 4H_2O$	Calcium nitrate, 4-hydrate	236	5	–	4	5	–	4	4	5
KNO_3	Potassium nitrate	101	5	–	6	–	5	6	6	5
KH_2PO_4	Monopotassium phosphate	136	1	–	–	–	1	1	1	1
$MgSO_4 \cdot 7H_2O$	Magnesium sulfate, 7-hydrate	246	2	2	2	2	2	–	–	2
FeNa EDTA[z]	Iron EDTA	18.4	1	1	1	1	1	1	1	–
Micronutrients[y]		–	1	1	1	1	1	1	1	1
K_2SO_4	Potassium sulfate	87	–	5	–	–	–	3	–	–
$CaCl_2 \cdot 2H_2O$	Calcium chloride, dihydrate	147	–	2	–	–	–	–	–	–
$Ca(H_2PO_4)_2 \cdot H_2O$	Calcium phosphate, monobasic	12.6	–	10	–	10	–	–	–	–
$Mg(NO_3)_2 \cdot 6H_2O$	Magnesium nitrate, 6-hydrate	256	–	–	–	–	–	–	2	–

[z]Ferric-sodium salt of ethylene-diamine tetraacetic acid. Differs from original Hoagland recipe, which used Fe tartrate.

[y]Contains the following in grams/liter:

2.86 H_3BO_3, (boric acid)   0.08 $CuSO_4 \cdot 5H_2O$, (copper sulfate, 5-hydrate)

1.81 $MnCl_2 \cdot 4H_2O$, (manganese chloride, 4-hydrate)   0.02 $H_2MoO_4 \cdot H_2O$, (85% molybdic acid).

0.22 $ZnSO_4 \cdot 7H_2O$, (zinc sulfate, 7-hydrate)

[x]For Mn-, B-, Cu-, Zn-, and Mo-deficient solutions, substitute micronutrient stock solutions missing one of the five salts in the regular micronutrient stock solution. For Cl-deficient micronutrient stock solution, substitute 1.55 $MnSO_4 \cdot H_2O$ (manganese sulfate, monohydrate) for 1.81 $MnCl_2 \cdot 2H_2O$.

FIGURE 9–30 Preparation of modified Hoagland nutrient solutions for nutrient deficiency system development. *(Courtesy D. R. Hoagland and D. I. Arnon, "The Water-Culture Method for Growing Plants Without Soil," California Agricultural Experiment Station Circular 347, revised 1950)*

Stock solutions can be dispensed using pipettes or graduated cylinders. Complete hydroponic salt mixtures can be purchased, and they greatly simplify nutrient solution preparation.

Aeration Solution cultures are typically aerated using an aquarium air pump and aquarium air tubing. In a soda bottle system, the aquarium air tubing is inserted into the reservoir through one of the prepunched holes, and the tubing in the reservoir is made rigid by slipping a plastic soda straw onto the tubing. To prevent clogging, a piece of cotton or aquarium filter floss is placed in the aquarium tubing as it exits the pump. The filter is necessary to either trap dirt from the air or pieces of the pump diaphragm. Adjust the aeration to a gentle rate of one to three bubbles per second. Some plants do not benefit from aeration but most do. An alternative to pump aeration is to let the top third of the root system remain uncovered by solution in the humid air in the top of the reservoir.

Maintenance Water must be added to static solution cultures every few days or so to replace water lost by transpiration and evaporation. Nutrient solutions are typically replaced with fresh solutions on a schedule, such as every 1 or 2 weeks. Frequency of solution replenishment depends on the rate of plant growth relative to the volume of nutrient solution. An electrical conductivity (EC) meter is useful for determining when a nutri-

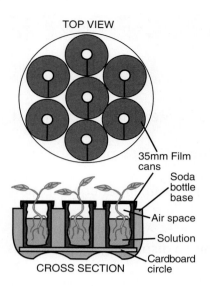

TOP VIEW

35mm Film cans

Soda bottle base

Air space

Solution

Cardboard circle

CROSS SECTION

FIGURE 9–31 Film cans with a quarter-inch hole punched in the center of each lid and a slit cut from the edge of each lid to the center hole make excellent static culture containers. Numerous film cans may be contained in the base taken from a 2-liter soda bottle. *(Adapted from material by Dr. David R. Hershey, University of Maryland)*

ent solution has been depleted. To prevent interruption of growth, replace the solution when the solution EC is about half the original value.

Plants Bean, corn, sunflower, and tomato are often used for teaching hydroponics because they grow rapidly from seed; but they are often difficult to handle because of their large size, high light requirements, and need for staking. Wisconsin Fast Plants, which are becoming a standard plant for teaching use, are excellent for solution culture in 35-mm film cans (Figure 9–31). Fast plants go from seed to flower in 2 weeks under a bank of six 4-foot, 40-watt fluorescent lamps and remain less than a foot in height.

Houseplants, such as piggyback (*Tolmiea menziesii*), wandering jew (*Tradescantia* species), evergreen euonymus (*Euonymus japonica*), coleus, common philodendron, and pothos (*Epipremnum aureum*), also are excellent in teaching hydroponics. They thrive with low-light levels, are readily available, and root quickly in solution. Radish and lettuce are easily grown from seed. A carrot placed in tap water in the 2-liter soda bottle system described earlier will sprout lateral roots, produce a shoot, and flower. Pineapple fruit tops easily root in solution.

Germination Seeds for hydroponics are conveniently germinated in containers of perlite, which is inert and is easily removed from roots before transferring the plants to solution. Small seeds, such as fast plants or coleus, can be germinated on a paper towel in a seed germinator built from a 2-liter soda bottle. The base is removed from a bottle and a 4-inch bottom section of the bottle is inverted over the base to complete the germinator. A wet paper towel is used to line the domed section, and small seeds are pressed into the towel. The seeds remain stuck until the seed germinates and the seedling is transplanted (Figure 9–32).

Future of Hydroponics

Hydroponics is increasing in use commercially and undoubtedly will become increasingly important in the future. Research is expanding, and new techniques are being developed. The use of nutrient solutions as media for growing plants will be an important part of agriscience in the future.

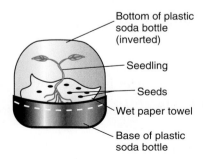

Bottom of plastic soda bottle (inverted)

Seedling

Seeds

Wet paper towel

Base of plastic soda bottle

FIGURE 9–32 A germination chamber may be constructed with the bottom cut from a 2-liter soda bottle inverted over a soda bottle base. The dome section is lined with a wet paper towel, and seeds are pushed against the towel to absorb moisture and germinate before transplanting. *(Adapted from material by Dr. David R. Hershey, University of Maryland)*

STUDENT ACTIVITIES

1. Write the Terms to Know and their meanings in your notebook.
2. Observe soils and other media used in flower pots, trays of vegetable seedlings, greenhouses, gardens, road banks, construction sites, and other places.
3. Examine the living organisms in a shovelful of soil taken from an outdoor area that is damp, moist, and high in organic matter.
4. Invite a Soil Conservation Service professional to the classroom to discuss soil mapping and land-use planning.
5. Obtain a land-use map from a Farm Service Agency (FSA) for your home, farm, or other area. Study the material and report your findings to the class.
6. Observe a profile at a cut in a road bank, stream bank, or hole. Identify the O, A, B, and C horizons.
7. Do a mechanical analysis of a sample of soil using a fruit jar, water, and a dispersing agent such as Calgon (see Figure 9–17).
8. Observe the feel of some of the soil used in Activity 7.
9. Obtain a pH test kit and test samples of soil from around your home.
10. Do research on models and procedures for a home or school hydroponics unit. Plan, build, and use the unit to experiment with hydroponic production of various plants.
11. Build and use a composting structure.
12. Add 1 teaspoon of lime to a quart of distilled water and mix well. Be sure to wear gloves and protective eye wear. Find the pH of this solution by using a pH test kit. Compare the pH of the sample to the pH of the soil taken in Activity 9. Write a short explanation describing the pH differences and what impact lime may have on soil.
13. Learn more about one type of crop grown without soil using the Internet or other agriscience resources. In one or two paragraphs, tell which crop you learned about, what nonsoil media it was grown in, and what procedure the grower used to produce crops.

SELF EVALUATION

A. Multiple Choice

1. An example of plant-growing media is
   a. soil.
   b. water.
   c. perlite.
   d. all of the above.

2. Which is *not* organic matter?
   a. leaf mold
   b. peat moss
   c. sphagnum moss
   d. vermiculite

3. Decay of organic matter is caused by
   a. large animals.
   b. microbes.
   c. rodents.
   d. water.

4. Which is *not* a factor affecting soil formation?
   a. hydroponics
   b. gravity
   c. ice
   d. water

5. Humans affect soil formation by
   a. landscaping.
   b. irrigating.
   c. bulldozing.
   d. weathering.

6. The land class with the fewest limitations is
    a. Class I.
    b. Class III.
    c. Class VI.
    d. Class VIII.

7. The land classes suitable for field crop production are
    a. I, II, IV, VI.
    b. I, II, III, IV.
    c. I, IV, V, VIII.
    d. I, II, VI, VII.

8. The horizon that is most supportive of plant growth is
    a. horizon O.
    b. horizon A.
    c. horizon B.
    d. horizon C.

9. The smallest soil particle is
    a. clay.
    b. gravel.
    c. sand.
    d. silt.

10. A term related to soil structure is
    a. aggregate.
    b. loam.
    c. silt.
    d. topsoil.

11. Soil pH is generally increased by adding
    a. sulfur.
    b. nitrogen.
    c. lime.
    d. complete fertilizer.

12. Hydroponics refers to
    a. aggregate culture.
    b. aeroponics.
    c. nutriculture.
    d. all of the above.

13. The importance and use of hydroponics is
    a. increasing.
    b. decreasing.
    c. about the same.
    d. a new field.

14. Hydroponically grown plants differ principally from soil-grown plants by
    a. light requirements.
    b. nutrient requirements.
    c. oxygen requirements.
    d. support mechanisms.

15. The best and most economical light source for growing plants hydroponically is
    a. incandescent light.
    b. fluorescent light.
    c. sodium.
    d. sunlight.

B. Matching (Group I)

_____ **1.** Colluvial
_____ **2.** Alluvial
_____ **3.** Glacial
_____ **4.** Lacustrine
_____ **5.** Loess
_____ **6.** Leaching
_____ **7.** Organic matter
_____ **8.** Residual
_____ **9.** IIe
_____ **10.** Profile

a. Deposited by ice
b. Deposited by lakes
c. Formed in place
d. Deposited by gravity
e. Class II land with erosion problem
f. Cross section of soil
g. Makes soil dark in color
h. Removal of soluble materials
i. Deposited by streams
j. Deposited by wind

B. Matching (Group II)

_____ **1.** Horizon A   a. Mostly organic matter

_____ **2.** Horizon B   b. Parent material

_____ **3.** Horizon C   c. Topsoil

_____ **4.** Horizon O   d. Subsoil

_____ **5.** Coarse   e. 5-10-5 problem

_____ **6.** Medium texture   f. Equal parts of nitrogen, phosphorus, and potassium

_____ **7.** Fine texture   g. Sulfur

_____ **8.** 10-10-10   h. Loamy soils

_____ **9.** Complete fertilizer   i. Clay soils

_____ **10.** Decreases pH   j. Sandy soils

C. Completion

1. Soil is defined as the top layer of the Earth's surface suitable for the growth of _____.

2. Three groups of plants found in soil are _____, _____, and _____.

3. Three groups or types of animals that live in the soil are _____, _____, and _____.

4. Five important benefits of organic matter in soil are _____, _____, _____, _____, and _____.

5. In solution culture, about one third of the roots must be exposed to air, or air must be bubbled through the solution to prevent damage or death from lack of the element _____.

6. The six mineral elements taken up by roots and needed in large quantities by plants are _____, _____, _____, _____, _____, and _____.

7. The seven mineral elements taken up by roots and needed in small quantities by plants are _____, _____, _____, _____, _____, _____, and _____.

Forest Management

MATERIALS LIST

✓ bulletin board materials

✓ samples of forest and wood products

✓ tree leaf collection

✓ forestry reference materials

✓ Internet access

COMPETENCIES TO BE DEVELOPED

After studying this unit, you should be able to:

- define forest terms.
- describe the forest regions of the United States.
- discuss important relationships among forests, wildlife, and water resources.
- identify important types and species of trees.
- describe how a tree grows.
- discuss important properties of wood.
- apply principles of good woodlot management.
- describe procedures for seasoning lumber.

SUGGESTED CLASS ACTIVITIES

1. Use a chainsaw to cut some 3- to 4-inch crosscut sections from a large log. The grain of the tree should be distinct. Assign class members to determine the age of the tree by counting the annual rings. Identify the dates of important historical events by pinning a label to the appropriate annual ring. Seek an opportunity with an elementary school class for students to present a mini-history lesson using the crosscut visual to identify important dates.

2. Invite a forestry professional to instruct the class on the kinds of forest management activities he or she encounters. The forestry professional might be employed by a federal or state agency, or self-employed in a privately owned forest operation. In addition, many towns and cities employ urban foresters to care for trees and shrubs in parks and along streets. Prepare the class for the visit by compiling a list of questions that the students might ask. Assign students to write a short report of the forester's presentation.

3. Take a walk around the schoolyard. Collect leaves or needles from the trees. Identify as many trees as you can using the illustrations contained in this unit.

195

TERMS TO KNOW

forest land	softwood	xylem layer	woodlot
timberland	deciduous	sapwood	silviculture
forest	hardwood	phloem	arboriculture
tree	pulpwood	inner bark	seedling
shrub	clear cut	heartwood	forester
lumber	plywood	hardness	virgin
board foot	veneer	shrinkage	
evergreen	cambium	warp	
conifer	annual ring	ease of working	

In the United States, there are 483 million acres of timberland and 248 million acres of other forest land for a total of 731 million acres. This represents about one third of the total land in the United States. **Forest land** is defined by the U.S. Department of Agriculture (USDA) as land at least 10 percent stocked by forest trees of any size. **Timberland** is defined as forest land that is capable of producing in excess of 20 cubic feet per acre per year of industrial wood, and that has not been withdrawn from timber utilization by statute or administrative regulation.

A **forest** is a complex association of trees, shrubs, and plants that all contribute to the life of the community (Figure 10–1). A **tree** is a woody perennial plant with a single stem that develops many branches. Trees vary greatly in size, but they usually grow to more than 10 feet in height. A **shrub** is a woody plant with a bushy growth pattern and multiple stems. A productive forest is one that is growing trees for lumber or other wood products on a continuous basis. **Lumber** consists of boards that are sawed from trees. It is bought and sold by the board foot. A **board foot** is a unit of measurement for lumber that is equal to $1 \times 12 \times 12$ inches (Figure 10–2).

Forest land may include parks, wilderness land, national monuments, game refuges, and other areas where harvesting of trees is not permitted. When you consider that there are 860 species of trees in the United States, it is evident that forestry is an important part of the econ-

FIGURE 10-1 A forest contains trees, shrubs, and other plants, as well as animal life. *(Courtesy of USDA)*

EXAMPLES OF ONE BOARD FOOT

The formula for calculating board feet is —
bd. ft. = number of pieces × thickness in inches × width in inches × length in feet ÷ 12.
Calculate the board feet in the following:

Problem 1. 5 pieces 2" × 4" × 8'.

Solution: $\dfrac{5 \times 2" \times 4" \times 8'}{12} = 26^{2}/_{3}$ bd. ft.

Problem 2. 6 pieces 1" × 8" × 10' *or*, the same pieces dressed would be 6 pieces $^{3}/_{4}$" × $7^{1}/_{2}$" × 10'. (Fractions from $^{1}/_{2}$" to 1" are considered as 1".)

Solution: $\dfrac{6 \times 1" \times 8" \times 10'}{12} = 40$ bd. ft.

Problem 3. 8 pieces 2" × 6" × 38".
(If the length is in inches, divide the product by 144 instead of 12. Why?)

Solution: $\dfrac{8 \times 2" \times 6" \times 38"}{144} = 25^{1}/_{3}$" bd. ft.

FIGURE 10–2 Lumber is bought and sold by the board foot. One board foot has a volume of 144 cubic inches. Any combination of dimensions that equals 144 cubic inches is 1 board foot.

omy. Forestry is the management of forests. An unbelievable variety of products comes from trees (Figure 10–3).

Trees in the forests of the United States are divided into two general classifications: evergreen and deciduous. Evergreen trees do not shed their leaves on a yearly basis. Evergreen trees of commercial importance are mostly conifers. Conifers are evergreen trees that produce seeds in cones, have needle-like leaves, and produce lumber called softwood. Deciduous trees shed their needles or leaves every year and produce lumber called hardwood.

FOREST REGIONS OF NORTH AMERICA

INTERNET ADDRESS:

http://www.wri.org/wri/ffi/
maps/namerica.htm

Eight major forest regions are generally recognized by forestry educators and other forestry professionals (Figure 10–4). They include the Northern Coniferous Forest, Northern Hardwoods Forest, Central Broad-leaved Forest, Southern Forest, Bottomland Hardwoods Forest, Pacific Coast Forest, Rocky Mountain Forest, and Tropical Forest. In addition, some experts make reference to the Wet Forest and the Dry Forest in Hawaii (Figure 10–5).

FOREST PRODUCTS FROM COMMERCIAL TREES

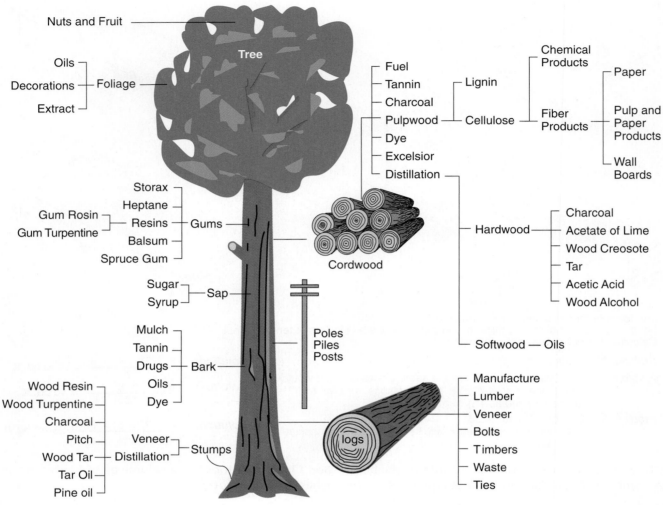

FIGURE 10–3 Many kinds of commercial products are obtained from trees and wood by-products.

Northern Coniferous Forest

The Northern Coniferous Forest region contains vast regions of softwoods. Some areas along the border between Canada and the United States contain a mixture of softwoods and hardwoods. The region is characterized by swamps, marshes, rivers, and lakes. The climate is cold. This forest is the largest region in North America, extending across Canada and Alaska. The most dominant type of tree is the evergreen, and large amounts of **pulpwood** are harvested in this region. The most important species include the white spruce, Sitka spruce, black spruce, jack pine, black pine, tamarack, and western hemlock.

Northern Hardwoods Forest

The Northern Hardwoods Forest region reaches from southeastern Canada through New England to the northern Appalachian Mountains. It blends with the Northern Coniferous Forest on the northern border and the Central Broad-leaved Forest on the south. The region extends westward

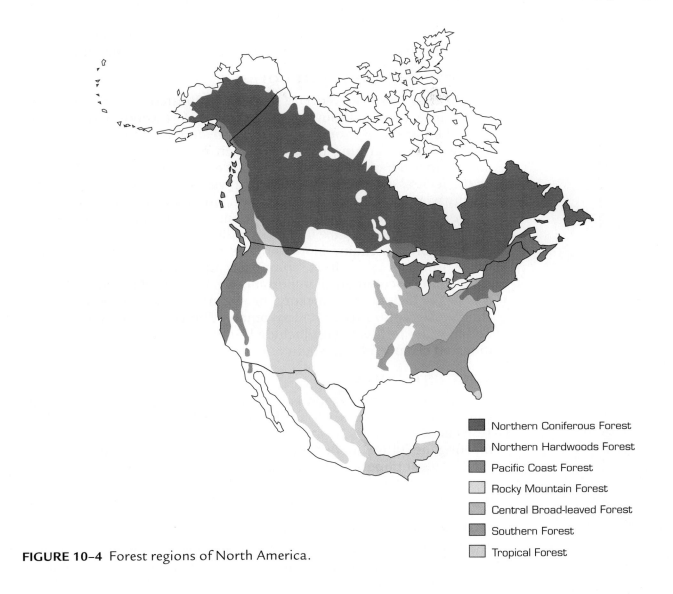

FIGURE 10–4 Forest regions of North America.

Northern Coniferous Forest

Northern Hardwoods Forest

Pacific Coast Forest

Rocky Mountain Forest

Central Broad-leaved Forest

Southern Forest

Tropical Forest

FIGURE 10–5 Forest regions of Hawaii. *(Adapted from material provided by USDA Forest Service)*

beyond the Great Lakes region, and it is populated by a number of important hardwood species including beech, maple, hemlock, and birch trees.

Central Broad-leaved Forest

The Central Broad-leaved Forest region is located mostly east of the Mississippi River and south of the Northern Hardwoods Forest. It consists of an arbitrary grouping of several distinctly different forest subgroups. It is a farming region in which most of the land has been cleared to produce cultivated crops. Little of the forested area is owned by the federal government, in contrast with some other regions. High-quality wood is produced in this region, and much of it is used to construct high-quality furniture. Hardwoods of lesser quality are used for construction and to make industrial pallets.

This forest contains more varieties and species of trees than any other forest region. It is composed mostly of Hardwood trees. Hardwoods of commercial importance in the Central Broad-leaved forest region include oak, hickory, beech, maple, poplar, gum, walnut, cherry, ash, cottonwood, and sycamore. The conifers that are of economic value in this region include Virginia pine, pitch pine, shortleaf pine, red cedar, and some hemlock.

Southern Forest

The Southern Forest region is located in the southeastern part of the United States. It extends south from Delaware to Florida, and west to Texas and Oklahoma. It is the forest region with the most potential for meeting the future lumber and pulpwood needs of the United States.

The most important trees in the Southern Forest are conifers. They include Virginia, longleaf, loblolly, shortleaf, and slash pines. Oak, poplar, maple, and walnut are hardwood trees of economic importance.

Bottomland Hardwoods Forest

The Bottomland Hardwoods forests occur mostly along the Mississippi River. They contain mostly hardwood trees and are often among the most productive of the U.S. forests. This is because of the high fertility of the soil in this area. Oak, gum, tupelo, and cypress are the major hardwood species found here.

Pacific Coast Forest

The Pacific Coast Forest region is located in northern California, Oregon, and Washington. It is the most productive of the forest regions in the United States and has some of the largest trees in the world. The approximately 48 million acres of Pacific Coast Forest provide more than 25 percent of the annual lumber production in the United States. About 19 percent of the pulpwood and 75 percent of the plywood produced in the United States comes from trees grown in the Pacific Coast Forest region.

Trees in the Pacific Coast forests include 300-foot-tall redwoods and giant Sequoias that may be as much as 30 feet in diameter. Douglas fir, ponderosa pine, hemlock, western red cedar, Sitka spruce, sugar pine, lodgepole pine, noble fir, and white fir are conifers that are important in this region. Important hardwood species include oak, cottonwood, maple, and alder.

Rocky Mountain Forest

The forests of the Rocky Mountain Forest region are much less productive than those of the Pacific Coast region. This region is divided up into many small areas and extends from Canada to Mexico. About 27 percent of the lumber produced in the United States comes from the 73 million acres in this forest region.

Most of the trees of commercial value in the Rocky Mountain forests are the western pines. They are western white pine, ponderosa pine, and lodgepole pine. Spruce, fir, larch, western red cedar, and hemlock also grow there in small quantities. Aspen is the only hardwood of commercial importance in the Rocky Mountain Forest region.

Tropical Forest

The tropical or subtropical forests of the continental United States are located in southern Florida and in southeastern Texas. These compose the smallest forest region in the United States. The major trees in this region are mahogany, mangrove, and bay, which are unimportant commercially. However, they are very important ecologically.

Hawaiian Forest

The Wet Forest region of Hawaii produces ohia, boa, tree fern, kukui, tropical ash, mamani, and eucalyptus. Most of these woods are used in the production of furniture and novelties.

The Dry Forest region of Hawaii produces koa, haole, algaroba, monkey pod, and wiliwili. None of these is of commercial value.

RELATIONSHIPS BETWEEN FORESTS AND OTHER NATIONAL RESOURCES

INTERNET KEY WORDS:

water cycle,
forest relationships

The relationships between forests and other natural resources, such as water and wildlife, are important to the overall well-being of the ecological system.

Forests play important roles in the water cycle. As water circulates between the oceans and the land areas, forests reduce the impact of falling rain on the soils and serve as storage areas for vast amounts of water. The stored water is released slowly from a forest watershed, allowing streams to flow throughout the year. In this manner, a forest regulates the flow of water, making it possible for fish and other aquatic plants and animals to survive.

Forests filter rain as it falls and help reduce erosion of the soil. They trap soil sediment and help maintain water quality. Trees and shrubs of the forest are also instrumental in removing many of the pollutant materials from the air and from water runoff. Similarly, the roots of trees filter excess nutrients from water runoff. They also help reduce the harmful effects of excess fertilizer nutrients that enter streams and underground water systems.

The relationships between forests and many types of wildlife are numerous. Algae, fungi, mosses, and numerous other plant and animal forms make their homes in forests. Forests provide food, shelter, protection, and nesting sites for many species of birds and animals. They also help maintain the quality of streams so that fish and other types of

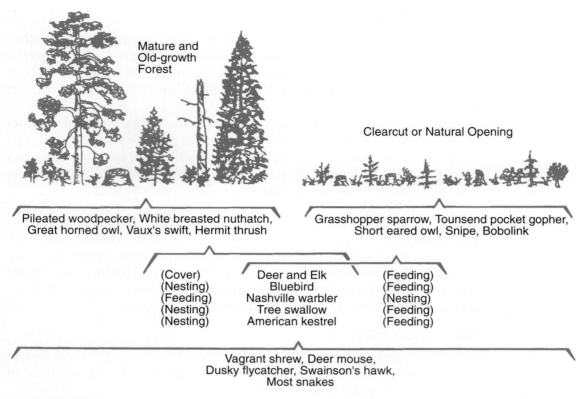

Mature and
Old-growth
Forest

Clearcut or Natural Opening

Pileated woodpecker, White breasted nuthatch,
Great horned owl, Vaux's swift, Hermit thrush

Grasshopper sparrow, Tounsend pocket gopher,
Short eared owl, Snipe, Bobolink

(Cover)
(Nesting)
(Feeding)
(Nesting)
(Nesting)

Deer and Elk
Bluebird
Nashville warbler
Tree swallow
American kestrel

(Feeding)
(Feeding)
(Nesting)
(Feeding)
(Feeding)

Vagrant shrew, Deer mouse,
Dusky flycatcher, Swainson's hawk,
Most snakes

FIGURE 10–6 Preference of 20 wildlife species for different forest habitats.

aquatic life can live and thrive. The shade provided by forests helps to maintain proper water temperatures for the growth and reproduction of aquatic life.

The wildlife found in the different types of forests varies considerably (Figure 10–6). Some species must have the open areas that occur naturally or that have been clear cut by people. A forest that has been **clear cut** has had all of the marketable trees removed from it. Other wild birds and animals depend to a greater extent on mature forests to provide their needs in life. As a forest changes naturally or as a result of harvesting, the types of wildlife that inhabit it also change.

SOME IMPORTANT TYPES AND SPECIES OF TREES IN THE UNITED STATES

INTERNET KEY WORDS:

tree species

INTERNET ADDRESS:

http://www.washington.edu/
home/treetour/

Trees may be described in terms of the lumber they produce. Characteristics such as hardness, weight, tendency to shrink and warp, nail- and paint-holding capacity, decay resistance, strength, and surface qualities are used to evaluate the usefulness of lumber.

Softwoods

The softwoods, or needle-type evergreens, that are important commercially in the United States include Douglas fir, balsam fir, hemlock, white pine, cedar, southern pine, ponderosa pine, and Sitka spruce (Figure 10–7).

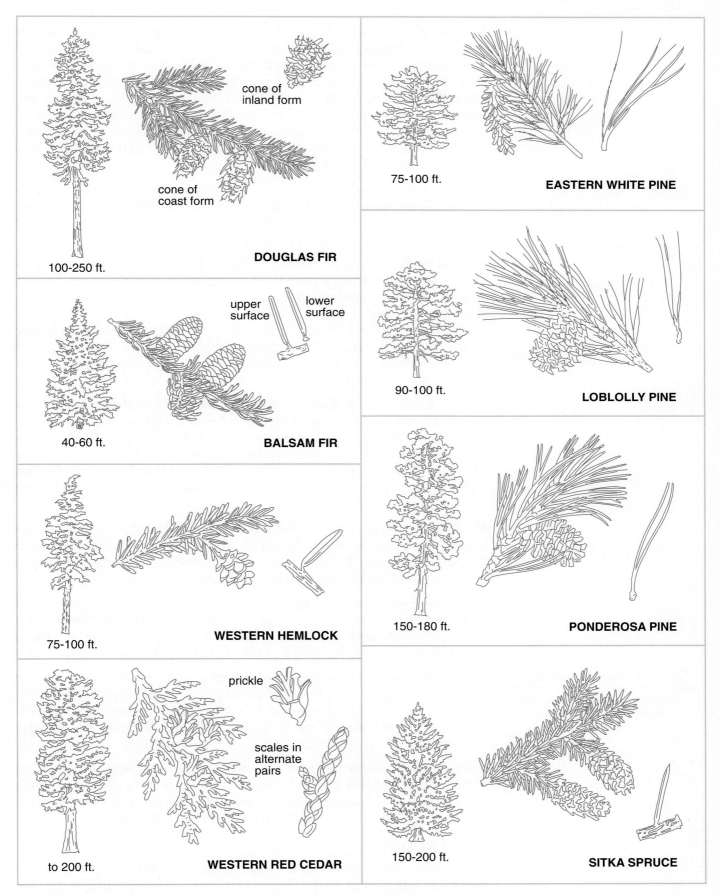

cone of
inland form

cone of
coast form

DOUGLAS FIR

100-250 ft.

75-100 ft.

EASTERN WHITE PINE

upper
surface

lower
surface

40-60 ft.

BALSAM FIR

90-100 ft.

LOBLOLLY PINE

75-100 ft.

WESTERN HEMLOCK

150-180 ft.

PONDEROSA PINE

prickle

scales in
alternate
pairs

to 200 ft.

WESTERN RED CEDAR

150-200 ft.

SITKA SPRUCE

FIGURE 10–7 Some species of softwood trees commercially important for wood products.

Douglas Fir Douglas fir is probably the most important species of tree in the United States today. It grows to a height of more than 300 feet and a diameter of more than 10 feet. About 20 percent of the timber harvested in the United States each year is Douglas fir. One-hundred-year-old stands of Douglas fir can produce 170,000 board feet of lumber per acre. This is five to six times the production of most other softwood species. Douglas fir is popular as construction lumber and for the manufacture of plywood. Plywood is a construction material made of thin layers of wood glued together.

Balsam Fir Found in the forests of the Northeast, the lumber from balsam fir is used mostly for framing buildings. Balsam fir trees have soft, dark green needles and a classic triangular shape when grown at low densities. They are often used as Christmas trees.

Hemlock: Eastern and Western Eastern hemlock is strong and is often used for building material. At times, it can be brittle and difficult to work. Eastern hemlock grows over most of the Northern Coniferous Forest range.

Western hemlock grows in the Pacific Coast Forest region, where yearly rainfall averages 70 inches. Western hemlock lumber is very strong and is one of the most important sources of construction-grade lumber. It is also important for pulpwood.

Cedar: Eastern Red, Eastern White, and Western Red Eastern red cedar is used for fence posts because it is resistant to decay. It is also used for the lining of chests and closets because its odor repels many insects. White cedar is a swamp tree with decay-resistant wood that is often used for shingles and log homes. Western red cedar resembles redwood in appearance. It is used where decay resistance, rather than strength, is important.

White Pine White pine lumber is soft, light, and straight-grained. It has less strength than spruce or hemlock, and it is more popular as a wood for cabinetmaking. Eastern white pine grows from Maine to Georgia. Western white pine is found in the Rocky Mountain Forest region.

Southern Pine Included in the category of southern pine is longleaf pine, shortleaf pine, loblolly pine, and slash pine. The southern pines grow in the southern and south Atlantic states. Lumber from southern pine is used for construction; it is also used for pulpwood and plywood.

INTERNET KEY WORDS:

http://www.washington.edu/
home/treetour/ppine.html

Ponderosa Pine The ponderosa pine is a large tree that grows up to 130 feet in height and 4 feet in diameter. It is widely distributed in the western United States. The wood is heavy, and it can be brittle; however, it is reasonably free of knots and other defects. Its most valuable use is for construction of wooden windowpanes and doors.

Sitka Spruce Growing from California to Alaska, Sitka spruce trees attain a height of 300 feet and a diameter of 18 feet. Lumber from Sitka spruce is of very high quality. It is strong, straight, and even-grained. Sitka spruce is also used in large quantities for pulpwood.

Hardwoods

Hardwoods come from deciduous trees. Commercially important species of hardwoods in the United States include birch, maple, poplar, sweetgum, oak, ash, beech, cherry, hickory, sycamore, walnut, and willow (Figure 10–8).

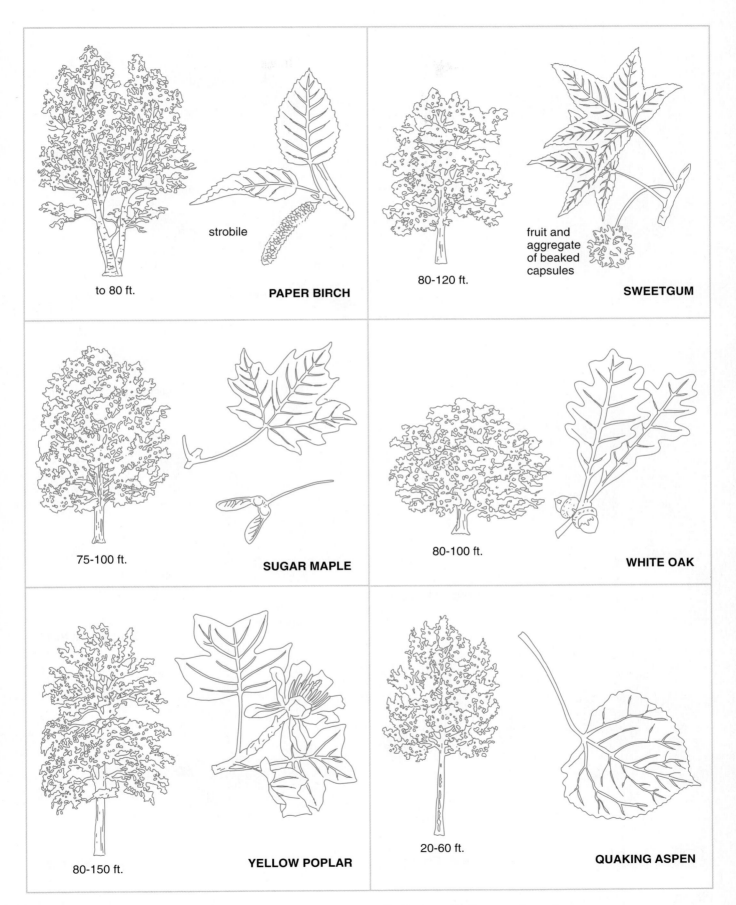

FIGURE 10–8 Some species of hardwood trees commercially important for wood products.

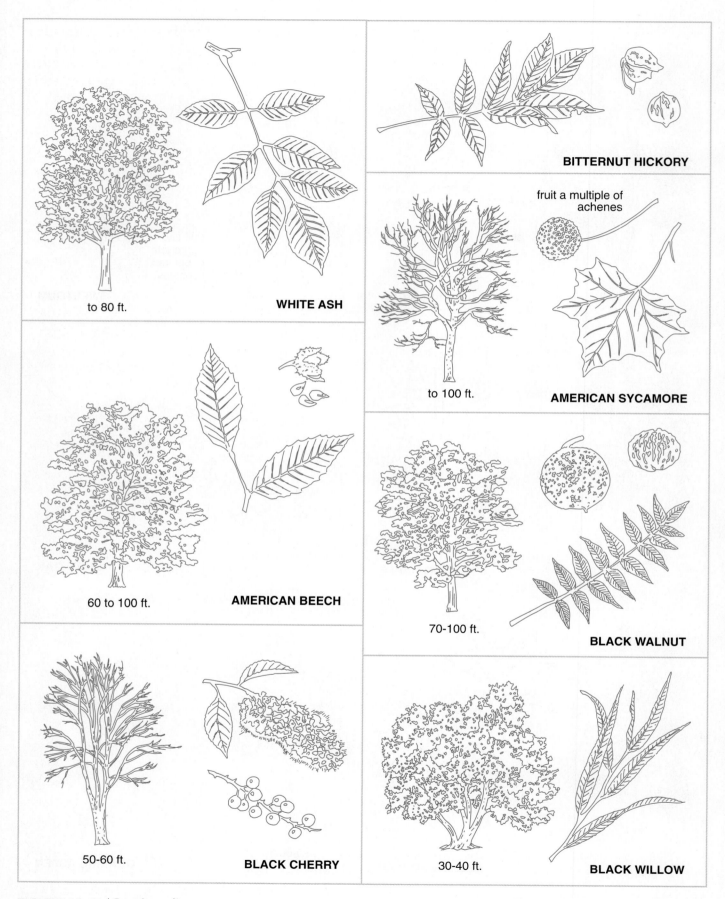

to 80 ft. **WHITE ASH**

BITTERNUT HICKORY

fruit a multiple of achenes

to 100 ft. **AMERICAN SYCAMORE**

60 to 100 ft. **AMERICAN BEECH**

70-100 ft. **BLACK WALNUT**

50-60 ft. **BLACK CHERRY**

30-40 ft. **BLACK WILLOW**

FIGURE 10–8 (Continued)

Birch Easily recognized by their white bark, birch trees grow in areas where summer temperatures seldom exceed 70° F. Birch lumber is dense and fine-textured. It is used for furniture, plywood, paneling, boxes, baskets, and veneer, as well as for many small novelty items. Veneer is a very thin sheet of wood glued to a cheaper species of wood that is used in paneling and furniture making.

Maple Maple lumber is classified as both hard and soft. Hard maple lumber is heavy, strong, and hard. It is used for butcher blocks, workbench tops, flooring, veneer, and furniture. Soft maple is only about 60 percent as strong as hard maple, and it is used in the same applications. Some species produce sweet sap that is made into maple syrup.

Poplar Poplar grows over most of the eastern United States. Poplar is classified as a hardwood because of its deciduous structure. However, lumber from poplar trees is soft, light, and usually knot-free. Poplar lumber may be white, yellow, green, or purplish and can be stained to resemble most of the fine hardwoods. Poplar is used for furniture, baskets, boxes, pallets, and building timbers.

Sweetgum The sweetgum tree is easily recognized by its star-shaped leaves and distinctive ball-shaped fruit. Sweetgum trees grow to as much as 120 feet in height and 3 to 5 feet in diameter. Lumber from sweetgum trees has interlocking grain and is used for house trim, furniture, pallets, railroad ties, boxes, and crates. The gum that comes from wounds in trees can be used as natural chewing gum or as a flavoring or perfume.

Oak: White and Red There are two general types of oak in the United States: white and red. White oak lumber is hard, heavy, and strong. Its pores are plugged with membranes that make it nearly waterproof. It is used for structural timbers, flooring, furniture, fencing, pallets, and other uses where strength of wood is a necessity.

Red oak is similar to white oak, except that it is very porous. It is not very resistant to decay and must be treated with wood preservatives when used outside. Chief uses of red oak include furniture, veneer, and flooring.

Aspen Aspen trees grow in the Northeast, Great Lake states, and the Rocky Mountains. Aspen is rapid-growing, making lumber that tends to be weaker than most construction-grade timber. It is also used for pulpwood.

Ash Ash lumber is heavy, hard, stiff, and has a high resistance to shock. It also has excellent bending qualities. It is popular for use in handles, baseball bats, boat oars, and furniture. It resembles oak in appearance.

Beech Grown in the eastern United States, beech is heavy and hard, and it is noted for its shock resistance. It is hard to work with and is prone to decay. Beech is used in veneer for plywood, and for flooring, handles, and containers.

Cherry Cherry can be found from southern Canada through the eastern United States. Cherry wood is dense and stable after drying. It is desirable and popular in the production of fine-quality furniture. It is expensive and in limited supply; therefore it is used mostly for veneer and paneling. It may, however, be used for other woodworking purposes.

Hickory   Hickory grows best in the eastern United States. Hickory lumber is hard, heavy, tough, and strong. It is somewhat stronger than Douglas fir when used as construction lumber. Other uses for hickory include handles, dowel rods, and poles. Hickory is also popular for use as firewood and for smoking meat.

Sycamore   Sycamore wood is used for flooring in barns, trucks, and wagons because of its strength and shock resistance. It grows from Maine to Florida, and west to Texas and Nebraska. Boxes, pallets, baskets, and paneling are other uses for sycamore.

Black Walnut   A premier wood for the manufacture of fine furniture, black walnut grows from Vermont to Texas. The wood has straight grain and is easily machined with woodworking tools. Because walnut is slow-growing and desirable, it is often made into veneer to get more use from its chocolate brown heartwood. It is also the source of black walnut nut meats.

Black Willow   Most of the black willow of commercial value is grown in the Mississippi River Valley. It is soft and light and has a uniform texture. Willow is used mostly in construction for subflooring, sheathing, and studs. Some willow is also used for pallets and for interior components of furniture. Black willow is sometimes a low-cost substitute for black walnut, because it has a similar brown appearance when finished.

There are many other domestic softwoods and hardwoods that grow in the United States. Many of these are important in local areas. The types discussed here are but a sampling, rather than a definitive list, of the commercially important trees in the United States.

Agri-Profile

CAREER AREAS:

Dendrology/Silviculture/Forestry

Forestry is known as a career area for rugged individuals who prefer the outdoors and like to work in relative isolation. However, many jobs in forestry are in urban areas and involve much indoor work. The United States Forest Service hires large numbers of forestry technicians and managers. Many forestry jobs do involve an extensive amount of outdoor work, but most jobs provide a desirable mix of outdoor and indoor work.

Forestry includes the work of the dendrologist engaging in the study of trees, the silviculturist specializing in the care of trees, the forestry consultant advising private forest land owners, lumber industry workers, government foresters, loggers, national and state forest rangers, and firefighters. A relatively new position is that of the urban forester, who is responsible for the health and well-being of the millions of trees found in parks, along streets, and in other areas of our cities.

The arborist is an urban forester whose work may include planting, transplanting, pruning, fertilizing, or tree removal.

There are many challenging careers in the forest industry.
(Courtesy FFA; FFA #186)

TREE GROWTH AND PHYSIOLOGY

Trees use carbon dioxide (carbon and oxygen) from the air and water (hydrogen and oxygen) from the soil to manufacture simple sugars in their leaves. The leaves then use additional carbon, hydrogen, and oxygen to convert simple sugars into complex sugars and starches. Nitrogen and minerals from the soil are then used to manufacture proteins, the building blocks for growth and reproduction.

A tree typically starts from a seed. For instance, oak trees grow from seeds called acorns, pine trees start from seeds in pine cones, and peach trees grow from peach seeds. Trees may also sprout and grow from stumps or other tree parts. When a seed germinates, a shoot grows upward to form the top growth, and roots grow downward and outward to form the root system.

Both roots and shoots extend themselves by growth at the tips through cell division and elongation. At the same time, tree roots, stems, and trunks grow in diameter by adding cell layers near their outer surfaces (Figure 10–9). This growth layer in a tree root, trunk, or limb is called the

INTERNET KEY WORDS:

tree, age, annual rings

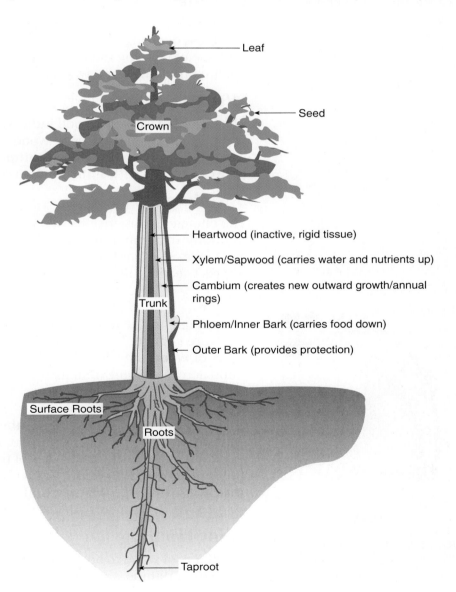

FIGURE 10–9 Major parts of a tree.

cambium. The outward growth of the cambium in 1 year creates an **annual ring** as seen in the cross section of a root, trunk, or limb.

Water and minerals are taken in by the roots and transported up to the leaves through a layer of cells called the **xylem layer**, or **sapwood**. The xylem is located just inside of the cambium layer. Just outside the cambium is another layer of cells called the **phloem**, or **inner bark**, which carries food manufactured in the leaves to the stems, trunk, and roots. Each year the tree grows new cambium, xylem, and phloem tissues, and the older sapwood becomes heartwood. **Heartwood** is the inactive core that gives a tree strength and rigidity.

PROPERTIES OF WOOD

The various properties of wood determine the uses for which it is best adapted. It should be noted that there are wide variations within a specific species of trees, and the properties discussed here are general in nature.

Hardness

The property of **hardness** refers to a wood's resistance to compression. Hardness determines how well it wears. It is also a factor in determining the ease of working the wood with tools. Splitting and difficulty in nailing are problems when wood is too hard.

Weight

The weight of wood is a good indication of its strength. In general, heavy wood is stronger than light wood. The moisture content affects the weight of wood; therefore, comparisons should always be made between woods with similar moisture contents.

Shrinkage

The change in dimensions of a piece of wood as it reacts to changes in humidity or temperature is referred to as **shrinkage**. The amount of expansion and contraction of wood greatly affects construction techniques used in building items from wood (Figure 10–10).

Warp

Warp refers to the tendency of wood to bend permanently because of moisture change. The tendency to warp is more of a problem in some types of wood than in others. Warping is caused by uneven drying of wood across its three dimensions.

Ease of Working

In wood, **ease of working** refers to the level of difficulty in cutting, shaping, nailing, and finishing the wood. It is influenced by the hardness of the wood and the characteristics of the grain. The grain of wood is hard and soft patterns caused by growth of the tree as it adds successive annual growth rings (Figures 10–11 and 10–12). In general, soft wood is easier to work than hard, dense wood.

Paint Holding

The ability of wood to hold paint may be determined by the type of paint, surface conditions, and methods of application, as well as characteristics

FIGURE 10–10 Dimensions of wood shrinkage.

Radial
Longitudinal
Tangential

Outer Bark
Phloem/Inner Bark
Cambium
Xylem/Sapwood
Ray
Heartwood
Pith

FIGURE 10–11 The structures of a tree illustrated in a cross section of a log or stump.

FIGURE 10–12 The annual rings in the cross section of a log or stump reflect the conditions of growth experienced by the tree on a year-to-year basis. These conditions influence the appearance of the grain.

of the wood itself. Moisture content, amount of pitch in the wood, and the presence of knots will all affect how well wood holds paint.

Nail Holding

Nail-holding capacity refers to the resistance of wood to the removal of nails. In general, the harder and denser the wood, the better it will hold nails.

Decay Resistance

The resistance of wood to microorganisms that cause decay is a chief factor in selecting wood. Natural resistance to insects that live in, tunnel through, or devour wood also affects the choice of species for a given application.

Bending Strength

The ability of wood to carry a load without breaking is a measure of bending strength. Bending strength is important when determining types and sizes of lumber to use for rafters, beams, joists, and other building applications.

Stiffness

The resistance of wood to bending under a load is referred to as stiffness. When wood used in construction is not sufficiently stiff, ceilings and walls may flex and buckle, and the wallboards will crack under the load.

Toughness

The ability of wood to withstand blows is called toughness. Hardwoods are much more resistant to shock and blows than softwoods. This characteristic is particularly important in woods used for tool handles.

Surface Characteristics

The appearance of the wood's surface is sometimes the determining factor in its use. The pattern of the grain greatly influences the staining characteristics and beauty of the finished product. The number and type of knots and pitch pockets are surface characteristics that need to be considered when selecting lumber for a project.

WOODLOT MANAGEMENT

INTERNET KEY WORDS:

woodlot management

The proper management of a wooded area or woodlot involves more than just the harvest of trees and the removal of unwanted species. A woodlot is a small, privately owned forest. The production of trees for harvest is a long-term investment, and mistakes in management take a long time to correct.

Some of the factors that need to be considered in the management of woodlots include soil, water, light, type of trees, condition of trees, markets available, methods of harvesting, and replanting. Using scientific methods in the management of forests is called silviculture. The care and management of trees for ornamental purposes is called arboriculture.

Restocking a Woodlot

The least expensive method of replacing trees harvested from a woodlot is natural seeding. Sources of seed for the desired species must be available in the forest, and conditions must be right for seed germination for natural seeding to take place. If seed from natural sources is not available, seeds from other sources may be planted on the forest site.

A surer method of restocking a woodlot is to plant trees of the desired species, rather than relying on seeds to do the reforestation. In most cases, seedlings (young trees started from seeds) are planted during late winter and early spring, before the new season's growth begins. Woodlots can be planted with one species of tree or a mixture of several compatible species (Figure 10–13).

Management of a Growing Woodlot

Management of a woodlot is much more involved than just sitting back and watching it grow. Proper care and management is important if the forestry enterprise is to be successful.

Trees that are of no commercial value should be removed as soon as possible to eliminate competition for light, moisture, and nutrients. Because these "weed" trees are removed when they are small, there is seldom a market for them, and they are left on the woodlot floor to decay. If weed trees are of sufficient size, however, they may be used for firewood.

When all the trees of a woodlot are nearly the same age (typically 15–30 years old), they often need to be thinned. Trees should be thinned any time that the crowns or branches occupy less than one third of the height of the trees. Usually, about one fourth to one third of the trees in a woodlot are removed during thinning.

Trees that are being grown for lumber are often pruned of side branches to produce a better quality log. Prompt removal of side branches helps keep logs free of knots. Only rapidly growing trees should be pruned. Branches should be pruned flush with the trunk of the tree. Pruning is usually done during the fall and winter, when the trees are dormant.

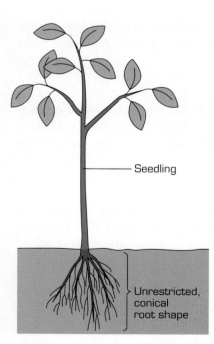

Seedling

Unrestricted, conical root shape

FIGURE 10–13 It is important to the survival of the seedling to maintain a conical root shape as the young tree is transplanted into the soil.

Science Connection

CONSERVING OUR BIODIVERSITY

Our biodiversity is at risk! Not far from the shores of Florida and Texas, the tropical rain forests of Central and South America, as well as parts of Africa and Southeast Asia, are disappearing from the face of the Earth. Commonly called a jungle, a tropical rain forest is a hot, wet, green place characterized by an enormous diversity of life and a huge mass of living matter. Here, biologists estimate that more than half of the plant and animal species of the world are found. Unfortunately, many of these species are not found anywhere except in the tropical rain forests. The rain forests occupy only 7 percent of the land area of the Earth, and their total area is diminishing at an alarming rate!

Rain forests are our richest source of plant and animal species and biodiversity.

Abundant moisture and warm temperatures encourage tremendous plant growth year-round. Plants and moisture then provide an excellent environment for animals and microorganisms to grow and develop. Rain forests are covered by a dense canopy formed by the crowns of trees up to 150 feet in height. Under the shade of this canopy are smaller trees and shrubs creating a second layer called the understory. The two layers are typically woven together by strong woody vines, which may exceed 700 feet in length.

The third layer is the forest floor. Palms, herbs, and ferns dot the forest floor. The jungle floor, with its shrubs and trees, is home to a myriad of insects and animals. Annual rainfall may reach 400 inches, and the daily rate exceeds the rate of evaporation. Life is plentiful. However, the soil beneath the jungle is poor, shallow, and erodible. As human populations expand into the rain forests, the foliage is cut and burned to make way for logging, grazing, and cropping.

Sadly, the land will generally support grazing or crop production for only a few years until it is worn out, depleted of minerals, and eroded. The occupants then typically move on to virgin jungle land to slash and burn a new tract and repeat the cycle. Once the jungle is destroyed, crops extract the scant nutrients, the soil is soon eroded, and the land often does not support the resurgence of jungle. Therefore, once cleared, the jungle area is often lost forever. It is estimated that one-half of the world's original tropical rain forests is gone. At the current rate of destruction, only a quarter will remain beyond 2000.

Because the rain forests are the storehouse for more than half of the genetic diversity in the entire world, species are being lost at an alarming rate. Many species are becoming extinct before they are discovered and recorded. In fact, it is estimated that scientists have discovered and named only a sixth of all the plant and animal species in rain forests. Earth's biodiversity is indeed threatened by human hands.

Planning a Harvest Cutting

A woodlot should never be harvested without a plan. A harvesting plan can maximize the income from the woodlot over many years. The use of a forester's services is usually wise in developing a harvesting plan. A forester is a person who studies and manages forests.

There are several systems of harvesting a woodlot. They include clear cutting, seed-tree cutting, shelterwood cutting, diameter limit cutting, and selection cutting.

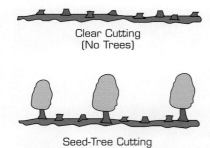

Clear Cutting
(No Trees)

Seed-Tree Cutting

FIGURE 10–14 Clear cutting is generally done in small patches, but seed-tree cutting is an economical way of harvesting and reseeding large areas.

INTERNET KEY WORDS:

trees, salvage, harvest

Clear Cutting   Clear cutting is a system of cutting timber where all of the trees in an area are removed. It was used extensively in cutting the **virgin** forests (those that have never been harvested) of the United States when there was little thought of reforestation of the cut area.

Clear cutting is usually done in small patches, ranging in size from one-half acre to 50 acres. Prompt reforestation with desirable species of trees on the clear cut areas is essential. This prevents erosion and the loss of fragile forest soil.

Seed-Tree Cutting   Harvesting trees by the seed-tree cutting method is very similar to clear cutting. The primary difference is that enough seed-bearing trees are left uncut to provide seeds for reforestation (Figure 10–14). The trees left to produce seeds should be representative of the desired species and free from insect, disease, or mechanical damage.

The removal of the seed trees usually takes place after the cut area is repopulated with desirable seedlings.

Shelterwood Cutting   In shelterwood cutting, enough trees are left standing after harvesting to provide for reseeding of the woodlot. They will also protect the area until the young trees are well established. After the young trees become established, the residual trees are harvested.

Diameter Limit Cutting   When harvesting trees by the diameter limit method, all trees above a certain diameter are cut. Slow-growing or diseased trees are often left standing when harvesting is done using this method. Rapid-growing trees of desirable species may be cut before they reach their production potential. Often this method leaves a woodlot with only slow-growing trees that may be undesirable. Diameter limit cutting is used advantageously to remove trees left from previous harvesting, which reduces competition with younger trees.

Selection Cutting   Selection cutting is the usual method of harvest used when the trees in a woodlot are of different ages. Selecting the trees of harvest can be a difficult task for the average person, and obtaining the services of an experienced forester is usually wise. This system of harvesting allows for fairly frequent income and maintains the woodlot's aesthetic value.

Salvage Harvesting   Natural disasters occur regularly in forests, and salvage harvesting often follows. Trees that are dying from insect damage and diseases can still be used for lumber products if harvested before they are completely dead. Physical damage sometimes occurs to trees. Examples include charred trunks from forest fires and broken trunks and limbs caused by high winds. Once damage has occurred to the trees, they need to be harvested as soon as possible. Waiting longer than 2 or 3 years to harvest damaged trees often will result in poor-quality lumber.

Protecting a Woodlot

A woodlot must be protected from fire, pests, and domestic animals if it is to yield up to its potential. It is estimated that more than 13 billion board feet of timber are destroyed each year by pests. This represents nearly 25 percent of the estimated net growth of forests and woodlots each year.

Hot Topics in Agriscience

INTEGRATED PEST MANAGEMENT

An insect management plan that uses a variety of control methods is the most acceptable form of pest control, because it does minimal damage to insect species for which the control method is not intended. Integrated pest management is a concept for controlling harmful insects or other pests, whereas providing some protection for useful organisms. It involves the use of some chemical pesticides in emergency situations, but it also relies on natural enemies and other biological control strategies to control harmful pests. Integrated pest management is a control program that does not attempt to kill all of the harmful insects because insect control of this kind also kills the natural enemies of the pest. To survive, natural insect enemies must have a small population of the harmful pest upon which they can depend for food.

FIGURE 10–15 Fire can cause major damage to timber resources when it gets out of control. It is important to plan for fire protection in forests that are highly valuable. *(Courtesy Boise National Forest)*

Fire Millions of dollars worth of timber are destroyed by fire each year (Figure 10–15). In addition to actually killing trees, fire slows the growth of others and damages some so that insects and diseases may destroy them. Fire also burns the organic matter on the forest floor, which takes nutrients away from the trees and exposes the soil to erosion.

To help prevent fires, debris should be removed from around trees. Weeds, brush, and other trash around the edges of a woodlot should be removed. Not allowing human use, such as hunting and fishing, during dry periods may also reduce the potential for fire. The construction of permanent firebreaks is useful for fire control.

A plan of how to deal with fire is important in minimizing damage should a fire occur. State and county foresters can assist in developing fire prevention and control plans. These strategies include planning for water storage, and setting procedures for notifying proper authorities and obtaining appropriate equipment and help.

Pests Insects and diseases cause more damage to existing forests than fires (Figure 10–16). Pests cause trees to be weak and deformed. They consume leaves, damage bark, and retard the growth of trees. They kill billions of trees each year. However, control of diseases and insects in forest lands is difficult and expensive.

Removal of dead, damaged, or weak trees may help reduce disease and insect problems. Not only are weak trees attacked, but healthy ones are sometimes favorite targets of pests. Prompt action in dealing with outbreaks of insects and diseases will do much to ensure a profit at harvest time.

Domestic Animals The grazing of woodlots by cattle and sheep usually results in the destruction of all small seedlings in the forest. It also eliminates most of the woodlot floor coverage. Livestock may also strip the bark from trees, causing them to die. Woodlots do not provide much food for grazing animals, so it is still wise to exclude livestock from them.

Harvesting a Woodlot

Before timber is harvested, a market for it must be found. Various methods of marketing the timber should be explored to determine the most

FIGURE 10–16 Insects and diseases become epidemic in a forest, killing many trees during periods of drought or stress.

profitable alternatives. There are many uses for forest products, and trees should be marketed to maximize their value.

The actual harvest of a woodlot can be done by the owner or by a contractor skilled in timber harvest. Another alternative is to sell the standing timber to a company that later harvests and markets it.

Hot Topics in Agriscience

TO FIGHT OR NOT TO FIGHT

Before humans began managing our forests, this was nature's job. Fire was a natural occurrence. When it erupted, it would burn underbrush, ground litter, and trees. Often, these fires would not kill the larger mature trees. It would kill the smaller trees that were in competition with these older trees for nutrients. The giant sequoia or ancient red woods, that stand today in some parts of our country, would not have grown as tall or as thick as they have if fire had not thinned the population of competing trees. Some types of trees actually need fire to reproduce. The seeds of the Lodge Pole Pine tree cannot germinate without first going through the intense heat of fire. In the early 1900s, people in the United States began fighting forest fires. For almost 100 years, few forest fires have been allowed to burn. The result is thicker, more dense forests than were present a century ago. There are more trees, and they are much closer together in today's forests. Now, when fires start, the damage is much more severe. Fires burn hotter, faster, and with more intensity than in the past.

Forest fires demolish forests, kill wildlife, destroy ecosystems, burn homes, and threaten human life. Although fire was once nature's way of cleaning and thinning forests, it now can wipe out all life in its path. The job of keeping our forest healthy now turns to the forest industry. Some types of logging practices thin trees; the wood is then sold for construction, papermaking, and many other uses.

Some people are strongly opposed to logging and fighting fire. They would prefer that whole forests burn and lives of forest creatures be lost than fight fires or thin by logging. Dr. Moore, co-founder and former president of Greenpeace, made the following statement to the Wall Street Journal. "The root of the problem is that when we protect our forests from wildfires, over time they become susceptible to disease and to catastrophic wildfires as fuel loads build up. The only way to prevent this is to actively remove dead trees and to thin the forest. The active management of these forests is necessary to protect human life and property, along with air, water and wildlife." Individuals who are truly concerned with the health of our forests understand that thinning our forests by logging and fighting dangerous forest fires is the best way to preserve this natural resource for generations to come.

Regardless of how the timber is harvested, it is extremely important that contracts covering all pertinent information be drawn up and signed by all parties involved.

SEASONING LUMBER

The proper seasoning of lumber is essential to protect it from damage. As soon as a tree is cut, it starts to lose moisture. If the wood dries too slowly, it may be subject to rot, stain, or insect damage. If it is allowed to dry too quickly, lumber may twist and warp or split. This may make it unsuitable for most uses.

Wood that is sawed for lumber should be stacked immediately after sawing to allow for even drying (Figure 10–17). The stacking should take place on a level, well-drained location. It should be off the ground and stacked so that air can circulate freely around each board. The stacked lumber should also be protected from weather. The amount of time required for lumber to dry depends on the thickness of the lumber and the species of tree from which it was cut. Times range from 30 to 200 days.

Lumber can be seasoned quickly by placing the stacked lumber in a heated kiln for a few days to dry it out. This procedure will keep the lumber from becoming warped as it dries. Under the controlled conditions that exist inside a heated kiln, most of the moisture is removed from the wood within a few hours.

The forestry industry in the United States produces about 16.3 billion cubic feet of wood products each year. The demand for wood products, such as lumber, pulp wood, posts, and pilings, is almost 19 billion cubic feet per year. Only by carefully managing our forest resources and replacing our trees can the United States fulfill its need for wood and wood products in the future.

½" air space between boards

1" × 3" strips

24" spacings

FIRST COURSE

Material for weight

Building paper or sheet metal for waterproofing.

6"

6"

COMPLETED STACK

FIGURE 10–17 Freshly sawed lumber must be properly stacked to prevent warping and end splitting as the lumber dries.

STUDENT ACTIVITIES

1. Write the Terms to Know and their meanings in your notebook.
2. Make a collection of forest products.
3. Make a collection of tree leaves, bark, or twigs that are of economic importance in your area.
4. Make a bulletin board showing the many uses of forests and forest products.
5. Write a report on a species of trees or a type of forest product of interest to you.
6. Visit a local forest and identify the types of trees growing there.
7. Have a forester visit the class and speak about forest management.
8. Visit a wood-processing plant operation to learn how trees are processed into other products.
9. Write a report on a career in the forest or wood products industry.
10. Study Figure 10–2 and learn how to calculate board feet (BF). Determine the number of board feet in the following:
    a. a board that is $1'' \times 12'' \times 12'$: BF = _____
    b. a plank that is $2'' \times 8'' \times 16'$: BF = _____
    c. a $4'' \times 4'' \times 8'$ board: BF = _____
    d. six $2'' \times 4'' \times 8'$ boards: BF = _____
    e. twenty $2'' \times 10'' \times 16'$ boards: BF = _____
11. What is the total board feet in Activity 10?
12. If the items in Activity 10 were rough-sawed lumber and were at $0.20 per board foot, what would be the total cost of the order? $_____
13. Look up "rain forests" on the Internet or using the school library. Find one organism that is used to benefit humans (besides building products). Write a one page report describing the location where the organism is found, its importance to humans, and its survival expectations if forests are destroyed.
14. Obtain a coring tool from your teacher and find a large tree. Core the tree by following the instructions of your teacher. Then, count the rings in the core to determine the age of the tree.

SELF EVALUATION

A. Multiple Choice

1. Trees that are used for making paper are called
   a. timber.
   b. lumber.
   c. pulpwood.
   d. veneer.

2. The scientific management of forests is
   a. silviculture.
   b. arboriculture.
   c. pomology.
   d. olericulture.

3. There are about _____ acres of productive forests in the continental United States.
   a. 105 million
   b. 235 million
   c. 500 million
   d. 735 million

4. The most important commercial species of trees in the United States is
   a. oak.
   b. Douglas fir.
   c. redwood.
   d. walnut.

5. About 75 percent of the wood for plywood is harvested from the _____ Forest region.
   a. Northern Coniferous
   b. Central Broad-leaved
   c. Hawaiian
   d. Pacific Coast

6. A forest that has never been harvested is called
   a. virgin.
   b. hardwood.
   c. clear cut.
   d. seedling.

7. The seed-tree method of harvesting
   a. cuts all trees over a certain diameter.
   b. cuts all trees under a certain diameter.
   c. cuts about one third of the trees in a woodlot.
   d. cuts all but a few trees left for seed.

8. Which of the following was not stated as a property of wood?
   a. nail-holding capacity
   b. bending strength
   c. color
   d. surface characteristics

9. The yearly demand for wood products in the United States is about _____ cubic feet.
   a. 19 billion
   b. 23 billion
   c. 190 billion
   d. 253 million

10. In a woodlot with even-aged trees, the lot usually needs to be thinned when the trees are _____ years old.
   a. 1 to 5
   b. 5 to 10
   c. 10 to 15
   d. 15 to 30

B. Matching

_____ **1.** Tree
_____ **2.** Shrub
_____ **3.** Woodlot
_____ **4.** Veneer
_____ **5.** Plywood
_____ **6.** Lumber
_____ **7.** Forestry
_____ **8.** Softwood
_____ **9.** Hardwood
_____ **10.** Conifer

a. Thin layers of wood glued together
b. Wood from conifers
c. Small, multi-stemmed plant
d. Management of wooded land
e. Wood from deciduous trees
f. Thin slices of wood
g. Small forest
h. Tree with needle-like leaves
i. Woody, single-stem plant
j. Boards sawed from trees

C. Completion

1. The air drying of lumber is referred to as _____.

2. Nearly _____ percent of the estimated net growth of forests is lost each year to fire and pests.

3. _____ is a type of harvesting where every tree over a certain size is cut.

4. Two general types of oak are _____ and _____ lumber.

5. Douglas fir is an example of an _____ or _____ tree.

6. Cutting all of the trees in an area is called _____.

7. The _____ Forest region is located along the Mississippi River.

8. The forest region with the most potential for meeting future needs for forest products is the _____ Forest region.

UNIT 11

Wildlife Management

MATERIALS LIST

✓ bulletin board materials

✓ reference materials on wildlife, wildlife management, and pollution

✓ Internet access

COMPETENCIES TO BE DEVELOPED

After studying this unit, you should be able to:

- define wildlife terms.
- identify characteristics of wildlife.
- describe relationships between types of wildlife.
- understand the relationships between wildlife and humans.
- describe classifications of wildlife management.
- identify approved practices in wildlife management.
- discuss the future of wildlife in the United States.

SUGGESTED CLASS ACTIVITIES

1. Conduct a wildlife survey of the area where you live. Create a master list in the classroom that is updated each day as students report sightings of different species of birds, mammals, reptiles, and amphibians. Conclude the activity by writing a press release that details the findings for a local newspaper.

2. Identify a local wildlife expert such as a professional who works with a fish or wildlife agency or a local bird watcher. Invite this person to tell the class about the local species of wildlife and where they may be viewed. Encourage the expert to bring photos or other materials to illustrate the presentation.

3. A week before beginning this unit, encourage the class to find and bring in newspaper or magazine clippings that focus on wildlife issues in your area. To introduce the topic, a few students may want to share some of the articles and direct a discussion pertaining to their articles.

TERMS TO KNOW

wildlife	prey	mutualism	competition
vertebrate	parasitism	predation	wetlands
predator	warm-blooded animals	commensalism	

Wildlife has been part of the life of humans since the beginning of time. **Wildlife** includes animals that are adapted to live in a natural environment without the help of humans. Early humans followed herds of wild animals and killed what they needed to live and survive. They observed what the animals did and what they ate to determine what was safe for human consumption. Early humans also used wildlife as models for their artwork and in many of their ceremonial rites.

As settlers came to the new world, wildlife often provided the bulk of the food available until food production systems could be developed. Supplies of wildlife seemed to be inexhaustible as the skies were blackened with the flight of millions of passenger pigeons, and herds of bison created dust storms as they migrated on the vast prairie (Figure 11–1).

Unfortunately, supplies of wildlife were not and are not unlimited. Human activities have damaged or destroyed wildlife habitat (the area where a plant or animal normally lives and grows). Humans have polluted the air and water supplies, killed wildlife in tremendous numbers, and in some instances, generally disregarded the needs of wildlife. As a result, many species of wildlife in the United States now require some degree of management.

Fortunately, humans also have the ability to restore wildlife habitat and manage many wildlife species successfully. The Great Plains may never sustain vast herds of roving bison, but the bison has been brought back from the brink of extinction. Populations of large-game animals such as deer, elk, moose, and black bear have expanded to fill the available habitat. Many smaller species such as eagles have also experienced population growth. There are a few species, such as the whooping crane, that have not responded well to management efforts. These species of wild creatures will require our best efforts to accomodate their survival needs.

CHARACTERISTICS OF WILDLIFE

All **vertebrate** animals (animals with backbones), except humans, are included in the classification of wildlife. They have many of the same characteristics as humans. Growth processes, laws of heredity, and general cell structure are common to both humans and animals. When populations become too dense, disease outbreaks occur, populations suffer from starvation, and disposal of waste becomes a problem.

The wildness of the animal itself is a characteristic that allows the animal to survive without interference or help from humans. The animal's wildness often contributes to the interest that humans have in wildlife. Characteristics identified as wildness are what attract hunters to hunting

FIGURE 11–1 Wildlife populations declined in the early 1900s, but management efforts have been successful in restoring the populations of many species. *(Courtesy of U.S. Fish and Wildlife Service)*

FIGURE 11–2 The nature of many wild animals is to avoid humans. It is human nature to be curious about elusive wild animals, but we must be careful to make sure that our interests do not intrude on the needs of wild animals. *(Courtesy Chesapeake Bay Foundation)*

INTERNET KEY WORDS:

wood tick, parasite, mutualism, positive relationships

and fishermen to fishing. Bird watching and wildlife photography would be far less fascinating if wildlife were less wild and wary of humans (Figure 11–2).

With few exceptions, wildlife species live in environments over which they have no control. Wildlife must be able to adapt to whatever they are presented in terms of food and environment, or they will perish. They must also possess natural senses that allow them to avoid predators and other dangers. A **predator** is an animal that feeds on other animals. The animal being eaten by the predator is the **prey.**

The ability to avoid overpopulation is a characteristic of many groups of wildlife. Establishing and defending territories is one way that wildlife may naturally avoid overpopulation. The stress of overpopulation causes some animals to slow or stop reproducing altogether.

WILDLIFE RELATIONSHIPS

Every type of wildlife is part of a community of plants and animals where all individuals are dependent on others. Any attempt to manage wildlife must take into account the relationships that exist naturally. This is because relationships within the wildlife community are constantly changing, and it is difficult to set standard procedures for their management.

The balance of nature is actually a myth, because wildlife communities are seldom in a state of equilibrium. The numbers of various species of wildlife are constantly increasing and decreasing in response to each other and to many external factors such as natural disasters. These include fires, droughts, and disease outbreaks. The interference of humans also often upsets sensitive relationships in nature.

Some of the natural relationships that exist in the wildlife community include parasitism, mutualism, predation, commensalism, and competition.

Parasitism

The relationship between two organisms, either plants or animals, in which one feeds on the other without killing it is called **parasitism.** Parasites may be either internal or external. An example of a parasitic relationship is the wood tick, which lives on almost any species of warm-blooded animal. **Warm-blooded animals** have the ability to regulate their body temperatures.

Mutualism

Mutualism refers to two types of animals that live together for mutual benefit. There are many examples of mutualism in the wildlife community. Tick pickers are birds that remove and eat ticks from many of the wild animals in Africa, to the mutual benefit of both. The wild animals have parasites removed from them, and the birds receive nourishment from the ticks. A moth that lives only on a certain plant is also the only pollinator of that plant in several relationships. Some plant seeds will germinate only after having passed through the digestive tract of a specific bird or animal.

Predation

When one animal eats another animal, the relationship is called **predation** (Figure 11–3). Predators are often very important in controlling popula-

FIGURE 11–3 Predatory animals play an important role in nature by keeping populations of rodents, birds, and other animals from expanding beyond the capacity of their environments to provide food and shelter for them. *(Courtesy U.S. Fish and Wildlife Service; photo by Pedro Ramirez, Jr.)*

FIGURE 11–4 Competition among wildlife species helps to keep the animal populations in balance. *(Courtesy U.S. Fish and Wildlife Service; photo by John D. Wendler)*

INTERNET KEY WORDS:

predatory animals

tions of wildlife. Foxes are necessary to keep populations of rodents and other small animals under control. Populations of predators and prey tend to fluctuate widely. When predators are in abundance, prey becomes scarce because of overfeeding. When prey becomes scarce, predators may starve or move to other areas. This permits the population of the prey species to increase again.

Commensalism

Commensalism refers to a plant or animal that lives in, on, or with another, sharing its food, but not helping or harming it. One species is helped, but the other is neither helped nor harmed. Vultures waiting to feed on the leftovers from a cougar's kill is an example of commensalism.

Competition

When different species of wildlife compete for the same food supply, cover, nesting sites, or breeding sites, competition exists. Competition may exist between two or more species that share the same resources. It also exists among members of the same species, especially when food or shelter is in short supply or during the mating season (Figure 11–4). When competition exists, one species may increase in number, whereas the other declines. Often, the numbers of both species decrease as a result of competition. For example, owls and foxes compete for the available supply of rodents and other small animals.

The various relationships that exist among species of wildlife make it necessary to consider more than just one species any time that management is contemplated. Understanding the relationships that exist in the entire wildlife community is essential if wildlife management programs are to be successful.

A   B   C

FIGURE 11–5 Ecology is the branch of biology that describes relationships between living organisms and their living environments. Shown in their natural habitats are the (A) blue heron *(Courtesy of Chesapeake Bay Foundation)*; (B) eastern cottontail *(Courtesy of U.S. Fish and Wildlife Service; photo by William Janus)*; and (C) arctic hare *(Courtesy of U.S. Fish and Wildlife Service)*

FIGURE 11–6 Hunting and fishing are popular recreational uses of wildlife. *(Courtesy of U.S. Fish and Wildlife Service; Photo by Richard Baldes)*

RELATIONSHIPS BETWEEN HUMANS AND WILDLIFE

Relationships between humans and wildlife may be biological, ecological, or economic. Biological relationships exist because humans are similar to wildlife in the biological processess that control life. Relationships may be ecological because humans are but one species among nearly 1 million species of creatures that inhabit Earth, sharing its resources and environments (Figure 11–5).

The economic relationship that exists between humans and wildlife is important. Originally, humans were dependent on wildlife for food, clothing, and shelter. Today, there are six positive values of wildlife relationships with humans—commercial, recreational, biological, aesthetic, scientific, and social.

The harvesting and sale of wildlife and/or wildlife products is an example of the commercial relationship between humans and wildlife. Raising wild animals for use in hunting, fishing, or other purposes also falls into this category.

Hunting and fishing, as well as watching and photographing wildlife, are included in the category of recreational relationship (Figure 11–6). Although it is estimated that more than $2 billion is spent each year on hunting and fishing and at least another $2 billion on other recreational uses of wildlife, many of the recreational values of wildlife are intangible.

The value of the biological relationship between humans and wildlife is difficult to measure. Examples of the biological relationship include pollination of crops, soil improvement, water conservation, and control of harmful diseases and parasites (Figure 11–7).

Aesthetic value refers to beauty. Watching a butterfly sipping nectar from a flower, a fawn grazing beside its mother, or a trout rising to a

Hot Topics in Agriscience

SUPPLEMENTAL FEEDING OF WILDLIFE

Every year as the snow starts to fall in big-game country, a major controversy begins. During especially difficult winters, it is common for elk, deer, and other big-game animals to die of starvation. The question of feeding the animals is debated again each time large numbers of animals are found dead and dying. In mild years when conditions are not as difficult, most big game animals survive to the next season. Wildlife managers are concerned with keeping animal population sizes constant. When a dramatic decline in the number of animals occurs during the winter, managers become concerned that the animals will not be able to recover on their own. Sometimes, the

decision is made to feed these animals during food shortages. Some individuals believe that when humans become involved in this way, the animals are no longer self-sufficient, and that they are therefore no longer wild. This is a concern of the Fish and Game Departments in some states. It becomes a balancing act to keep wildlife wild and to keep wildlife alive. For this reason, many factors must be considered before a decision is made to provide supplemental feed for big-game animals. Some of these factors include whether the animals constitute a threat to private property. Are excess animals a threat to public safety? Will excess deaths affect recovery of the animal population? Will saving the animals create a bigger problem later because of a limited or unavailable supply of food? Even though wildlife managers make every effort to limit supplemental feeding practices, the controversy continues. Will supplemental feeding take the "wild" out of "wildlife"?

FIGURE 11–7 Biological values of wildlife and human relationships include the pollination of crops by honeybees. *(Courtesy USDA/ARS #K-4716-1)*

hatch of mayflies are all examples of the aesthetic value of wildlife. Wildlife also provides the inspiration for much artwork. Even though the aesthetic value of wildlife is not measurable in economic terms, wildlife can contribute greatly to the mental well-being of the human race.

Using wildlife for scientific studies often benefits humans. The scientific relationship between humans and wildlife has existed from the beginning of time when early humans watched wild animals to determine which plants and berries were safe to eat.

The value of the social relationship between humans and wildlife is also difficult to measure. However, wildlife species have the ability to enhance the value of their surroundings simply by their presence (Figures 11–8 and 11–9). They provide humans the opportunity for variety in outdoor recreation, hobbies, and adventure. They also make leisure time much more enjoyable for humans.

CLASSIFICATIONS OF WILDLIFE MANAGEMENT

Wildlife management can be divided into several classifications for ease in developing management plans. Techniques for management vary

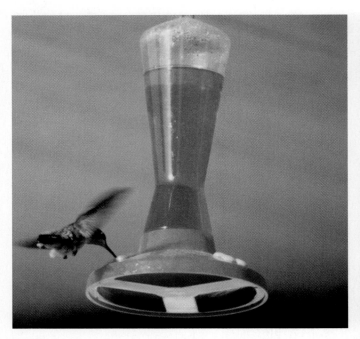

FIGURE 11–8 The presence of birds can be encouraged by providing them with feeding stations.

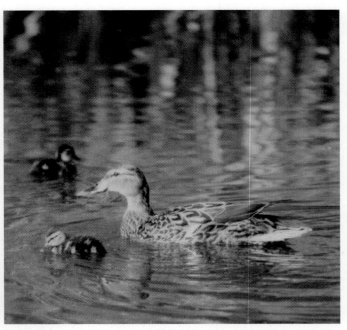

FIGURE 11–9 Wildlife sanctuaries are created to provide safe habitats for animals. They are also attractive to humans who go there to observe and enjoy the birds and other animals.

tremendously according to classification. Some wildlife management classifications are farm, forest, wetlands, stream, and lakes and ponds.

The management of farm wildlife is probably the most visible wildlife management classification. The development of fence rows,

Science Connection

POPULATION DYNAMICS

A population is a group of individuals of a species that live in a specific area. A wolf pack in Yellowstone National Park is an example of a population. Dynamics refers to the factors that cause changes in a population. Biologists are interested in population dynamics that affect animal behavior and the number of animals in a population. Wildlife biologists spend a lot of time observing, counting, and calculating the information they gather from studying a population. The goal of this work is to predict what changes may take place within a population in time. Factors that can affect a population's size include changes in the number of births or the number of deaths. The number of animals that move into and out of the group affect the population, as do such factors as predation, disease, and competition for resources.

Another factor that must be considered by scientists is the carrying capacity of the environment. Carrying capacity is the maximum number of individuals an ecosystem's resources can support. This is an important aspect to determine. If the carrying capacity of the area is not determined, then it would be difficult to see what effect the other factors may have. The Bighorn sheep population size in the Northwest has been on the decline in recent years. Wildlife biologists are trying to determine what is causing the sheep to die. Through observations, counting, blood sampling and using radio collars, they hope to identify what is causing the decline. Once these things are understood, they hope to be able to predict the future for the Bighorn sheep more accurately.

FIGURE 11–10 Management of wildlife in forest environments is often difficult, because many species of plants and animals occupy the living environment. Interactions among all of the organisms in the environment must be considered in developing the management plan. *(Courtesy National FFA)*

INTERNET KEY WORDS:

petroleum remediation, bioremediation

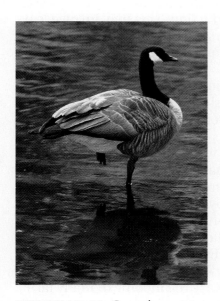

FIGURE 11–11 Canada goose populations have benefitted from man-made nesting platforms. The nesting geese are protected from predators and flood water when they use the platforms. *(Courtesy of PhotoDisc)*

minimum tillage practices, improvement of woodlots, and controlled hunting are all techniques that have long been used to manage farm wildlife. Rabbits, quail, pheasants, doves, and deer are the types of wildlife that are normally managed in this category.

Forest wildlife is more difficult to manage. Plans should be developed so that timber and wildlife can exist in populations large enough to be sustained and possibly harvested. Management of forest wildlife may include population controls to prevent destruction of habitat. Deer, grouse, squirrels, and rabbits are usually wildlife species that are included in forestry wildlife management programs (Figure 11–10).

The most productive wildlife management areas are wetlands. Wetlands include all areas between dry upland and open water. Marshes, swamps, and bogs are all wetland areas. Because these areas are sensitive to changes in environmental conditions, careful management of them for wildlife is essential. The wetlands are home to ducks, geese, beaver, muskrats, raccoons, deer, pheasants, grouse, woodcock, fish, frogs, and many other species of wildlife (Figure 11–11).

The management of running water or streams is often a difficult task. Water pollution and the need for clean water for a growing human population continue to increase at a rapid pace. Potential damage to the wildlife in streams from chemical pollution, the building of dams and roads, home construction, and the drainage of swampland are critical considerations for the stream wildlife manager.

Management of wildlife in lakes and ponds is normally somewhat easier than it is in streams, because water is standing rather than running. Population levels of pond wildlife, oxygen levels, pollutants, and the availability of food resources are all concerns of the pond and lakes wildlife manager.

FIGURE 11–12 Providing clean water for wildlife is an important wildlife management practice. *(Courtesy of Wendy Troeger)*

APPROVED PRACTICES IN WILDLIFE MANAGEMENT

Farm Wildlife

Management of wildlife on most farms is usually a by-product of farming or ranching. It is often given little attention by the farmer or rancher, except when wildlife causes crop damage and financial loss.

Much of the management of farm wildlife involves providing a suitable habitat for living, growth, and reproduction. This may involve leaving some unharvested areas in the corners of fields, planting fence rows with shrubs and grasses that provide winter feed and cover, or leaving brush piles when harvesting wood lots.

The timing of various farming operations is also important in a farm wildlife management program. Crop residues should be left standing over the winter to provide food and cover. Planting crops attractive to wildlife on areas that are less desirable as cropland is an excellent farm wildlife management practice. Providing water supplies for wildlife during dry periods is often necessary to maximize the numbers of farm wildlife on the area being managed (Figure 11–12).

Harvesting farm and ranch wildlife by hunting has been shown, by extensive research, to have little impact on spring breeding populations. Excess populations of farm and ranch wildlife that are not harvested by humans usually die during the winter. Even heavy hunting pressures seldom result in severe damage to wildlife populations. The sale of hunting rights to hunters is a way to increase the income of many farms and ranches. In addition, it often means the difference between profit and loss in the farming enterprise.

Management of wildlife on game preserves or farms set up specifically for hunting often differs drastically from other wildlife management programs. Species of animals and birds that are not native to the area are sometimes raised and released on the preserve. Native wildlife species may also be raised in pens and released to the farm or preserve expressly for harvest by hunters.

Hot Topics in Agriscience

ENVIRONMENTAL CLEANUP... THROUGH BIOTECHNOLOGY

One of the remarkable scientific developments in the last 20 years is bio-engineered bacteria. They have been developed for a variety of purposes. Bacteria that are capable of breaking down crude oil have been developed. Accidental spills of crude oil from ships have caused serious damage to marine animals and environments. Bio-engineered bacteria are important tools in reducing environmental damage caused by oil spills. The bacteria ingest the oil and convert it to a form that is more compatible with the environment.

Forest Wildlife

The types and numbers of forest wildlife in any specific woodland are dependent on many factors. These include type and age of the trees in the forest, density of the trees, natural forest openings, types of vegetation on the forest floor, and the presence of natural predators.

Management of forest wildlife is usually geared toward increasing numbers of desired species of wildlife. If desired populations of wildlife are present, the management goal is usually to maintain those populations. Sometimes, numbers of certain species of forest wildlife increase to the point where destruction of habitat occurs. When this happens, control measures may have to be instituted to restore proper balance.

The steps in developing a forest wildlife management plan should include taking an inventory of the types and numbers of wildlife living

Agri-Profile

CAREER AREAS:

Wildlife Biologist/Manager/Officer

Wildlife biologists work with fish and game species of land, freshwater streams and lakes, tidal marshes, bays, seas, and oceans. Wildlife biologists generally have Master or Doctorate level degrees in biology. They use the basic sciences in their work.

Wildlife managers typically have Associate or Bachelor level degrees. They work in government agencies, advising land owners and managing game populations on public lands. Their work frequently requires the use of helicopters, small planes, snowmobiles, all-terrain vehicles, horses, and land rovers, as well as time in the wild on foot or horseback.

Wildlife officers interact continuously with the hunting and fishing community. They advise governments in establishing fish and game laws and programs for habitat improvement. Wildlife officers have the backing of strict laws and stiff penalties for offenders. However, much of their time is spent on educating the public and obtaining private assistance in improving habitats and maintaining game populations.

Conducting game counts and doing habitat analyses provide information for wildlife managers. *(Courtesy USDA/ARS; USDA #K-5213-3)*

FIGURE 11–13 Habitat improvement is an important part of the wildlife manager's job. *(Courtesy of National FFA #255)*

FIGURE 11–14 Deer management is a key activity of fish and game agencies. Management units are studied carefully to determine the size of the deer population and to set hunting season regulations. *(Courtesy of PhotoDisc)*

in the forest area to be managed. Goals for the use of the forest and the wildlife living in it need to be developed. The third step in the development of a forest wildlife management plan is determining the types and populations of wildlife that the forest area can support and how best to manage the forest so that required habitat is provided.

The requirements for forest wildlife include food, water, and cover. These necessities must be readily available to the desired species of forest wildlife at all times. Management practices that meet these requirements include making clearings in the forest so that new growth will make twigs available for deer to feed on. Another practice is selective harvesting so that trees of various ages exist in the forest to make a more suitable habitat for squirrels and many other species of forest wildlife. Leaving piles of brush for food and cover is also a management practice that leads to increased production of forest wildlife. Care in managing harvests of forest products so that existing supplies of water are not contaminated is also important in good wildlife management.

Deer, grouse, squirrels, and rabbits are the forest wildlife species that are usually targeted for management because they are valuable for recreational purposes, especially hunting (Figures 11–13 and 11–14). They may also be managed to prevent the destruction of valuable forest trees and other products.

Notably, during times of overpopulation of forest animals, especially deer, it is seldom a good idea to provide supplemental food. Natural losses should nearly always be allowed to occur, including starvation of excess animals or allowance of heavier-than-normal hunting pressures. Artificial feeding of wildlife populations usually results in further population increases and an extension of the problem.

Wetlands Wildlife

No area of U.S. land is more important to wildlife than the wetlands. Wetlands include any land that is poorly drained—swamps, bogs, marshes, and even shallow areas of standing water (Figure 11–15). The wetlands are constantly changing as wet areas fill in with mud and decaying vegetation. They eventually become dry land that contains forests.

Wetlands provide food, nesting sites, and cover for many species of wildlife. Ducks and geese are probably the most economically important type of wildlife that depends on the wetlands for survival (Figure 11–16). Other types of wildlife found in the wetlands include woodcock, pheasants, deer, bears, mink, muskrats, raccoons, and many other lesser known species.

Management of wetlands for wildlife may include impounding water. Open water areas should occupy about one third of the wetlands for optimum use by wildlife. The depth of the standing water should not be more than about 18 inches.

The management of the plant life in the wetlands is also important. This may include cutting trees to open up the wetland area. Many species of wildlife require large, open areas in order to thrive. Care must be taken not to remove hollow trees that are used as nesting sites for some species of wildlife. Wetland areas can also be opened up by killing excess trees rather than cutting them. This provides resting areas for many types of wetlands wildlife.

FIGURE 11–15 No area of U.S. land is more important to wildlife than the wetlands. *(Courtesy of Chesapeake Bay Foundation)*

FIGURE 11–16 Wetlands are important nesting areas for millions of ducks and geese in the United States. *(Courtesy of U.S. Fish and Wildlife Service)*

INTERNET KEY WORDS:

wetlands, wildlife

Establishing open, grassy areas around wetlands and planting millet, wild rice, and other aquatic plants in the wetlands also helps to attract many types of wildlife to the area.

A serious hazard to wetlands wildlife is pollution. Pollution of water flowing into the wetlands area may come from agriculture, industry, or the disposal of domestic wastes. Because pollutants are trapped in the mud and silt of the wetlands, the effects of pollutants are often long-term.

In areas lacking natural nesting sites, populations of some wildlife species can be greatly increased by providing artificial nesting sites. Wood duck boxes, old tires, and islands surrounded by open water provide safe nesting sites for many species of wetlands wildlife (Figure 11–17).

Raising certain species of ducks in captivity and later releasing them in wetlands areas has helped in maintaining viable duck populations. This has been important as more and more natural duck nesting areas have been destroyed to meet the needs of people.

Stream Wildlife

Stream wildlife can be divided into two general categories: warm-water and cold-water wildlife. These categories are based on the water tem-

FIGURE 11–17 Artificial nesting sites are beneficial to many kinds of birds. Wood ducks would benefit from the artificial nesting site that is illustrated here. *(Courtesy of Cameron Waite)*

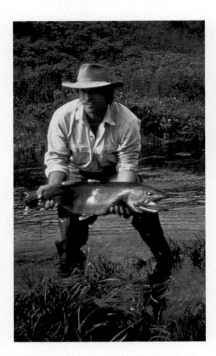

FIGURE 11–18 Cool, rapid-flowing brooks and mountain streams provide appropriate habitats for trout. *(Courtesy of U.S. Fish and Wildlife Service)*

FIGURE 11–19 Riparian areas, which include stream banks, can be protected by fencing livestock out. Such practices protect stream environments and the animal and plant life that is found there.

peratures at which the wildlife, primarily fish, can best grow and thrive. There is little or no difference in the management practices between the two categories of wildlife. In general, fish are the type of stream wildlife for which management plans are developed, although many other types of wildlife also depend on streams for their existence (Figure 11–18).

As land is developed, forests are harvested, and civilization is expanded, streams and their wildlife populations come under increasing pressure. Because we cannot build new streams, it is essential that existing ones be managed properly.

Management practices for streams include preventing stream banks from being overgrazed by livestock. Fencing the stream to limit access by livestock is also wise to reduce pollution and the destruction of stream banks (Figure 11–19).

Good erosion control practices on lands surrounding streams is important to help maintain clear, clean water. It is also important to prevent silt or chemical pollutants from entering streams (Figure 11–20).

Maintaining stream-side forestation is important in regulating stream temperatures during the warm summer months. Some species of fish stop feeding and may even die when stream temperatures become too high. The amount of dissolved oxygen in warm water is also much lower than it is in cold water. Without adequate oxygen, aquatic wildlife dies.

Anything that impedes the flow of a stream also serves to change it, to the detriment of many species of stream wildlife and to the benefit of other species. Trout must have swiftly moving, cool water in which to thrive, whereas catfish are adapted to sluggish streams (Figure 11–21).

FIGURE 11-20 Soil erosion causes siltation of streams, which destroys the habitat for desirable fish species. *(Courtesy of Chesapeake Bay Foundation)*

FIGURE 11-21 Some fish require cold, clear water in order to survive. Catfish are well adapted to slow-moving streams that are too warm for cold-water species such as trout. *(Courtesy of U.S. Fish and Wildlife Service)*

Care must be taken to maintain desirable species of wildlife in the stream. Introducing new species of wildlife in the stream may result in reducing native wildlife already in the stream.

The maintenance of population levels of stream wildlife that are in balance with the available food supply is important. Too many fish for the available food supply normally results in stunted fish that are of no value to fishermen. This situation does provide an increased food supply for some types of birds and animals that use streams for their food supply. Overfishing of predatory species of fish, such as bass or northern pike, may allow perch or sunfish to overpopulate the stream and become stunted. Often, the only way to restore streams to a desired mix of fish species is to remove the unwanted species. This is accomplished by netting, poisoning, or electric shocking. These techniques are legal only for authorized officials and should be done only by specially trained personnel.

The artificial rearing and stocking of desired species of stream wildlife is a management practice that is important in many streams. Typically, game species of fish are stocked in streams for fishermen to catch and remove. Often few or no fish survive to reproduce, and stocking must take place each year.

The regulation of sport fishing is often necessary to maintain desirable populations of game fish. This may include closed seasons, minimum size limits, creel limits, and restricted methods of catching fish.

Lakes and Ponds Wildlife

The management practices for wildlife in lakes and ponds are usually very similar to those for managing stream wildlife. Pollution must be controlled. Wildlife populations must be managed to maintain desired mixes of species. Harvest and use must also be controlled to ensure wildlife for the future.

However, there are some differences between management of wildlife in streams and management of wildlife in ponds and lakes. Because

Science Profile

ALASKAN ADVENTURE

Sierra Stoneberg of Hinsdale, Montana, had developed extensive knowledge and experience of the outdoors by the time she graduated from high school. As early as the seventh grade, she became interested in botany through the Montana Range Days program. She experienced first-hand elements of botany, biology, and other range sciences. Her summers during high school were filled with FFA activities and experiences with the Soil Conservation Service.

The summer after high school graduation provided her with an Alaskan adventure studying moose, mountain sheep, and other major wildlife species. As a volunteer at the Kenai Lake Work Center, she, together with others at the center, worked 40 hours a week for lodging, board, and plane fare.

Incredible, huge green trees and mosses created vivid memories of the high alpine country. Similarly, the tiny plants and lichens on the trees and rocks contributed a unique hue to the surroundings.

High school and college summers in the field provide valuable experiences for future wildlife biologists or range scientists. *(Courtesy National FFA)*

Scientific observations and data collection were important facets of her many treks into the wilderness. These included monitoring of fertilizing results on rangeland for sheep herds and range grass response to management practices. Scouting for eagle nests and moose habitat created an interesting and stimulating blend of mental and physical activities. Even a close encounter with a black bear provided valuable experience in becoming a National FFA Proficiency Award winner and for a future career in wildlife biology and range science.

FIGURE 11–22 Fish species such as sunfish and bluegill provide both sport fishing and food for predatory species such as bass. *(Courtesy of Chesapeake Bay Foundation)*

the water in lakes and ponds is normally standing, the amount of oxygen available for aquatic life sometimes becomes critical in the hot summer months. In small ponds, artificial means of incorporating oxygen into the water may be used to prevent fish deaths.

Water temperatures in lakes and ponds are more variable than they are in streams. This means that different species of fish are usually dominant in ponds and lakes. In many ponds and lakes, fish populations are predominantly largemouth bass and sunfish (Figure 11–22).

When it is necessary or desirable to rid a pond or lake of unwanted species of fish such as carp, it is often much easier to do so, because the water is contained. Sometimes it is possible to drain the body of water to remove the unwanted species of wildlife. More often, the body of water is simply poisoned so that all fish and other species of pond wildlife are killed. The pond is then restocked with desirable wildlife species.

The management of wildlife is an imprecise business. Often, specific species of wildlife are managed to the detriment of others. It is reasonably clear that any human interference in the wildlife community results in changes that are not always to the benefit of much of that community.

Science Profile

LEAD POISONING IN WATERFOWL

Nobody does more to promote wetlands and waterfowl populations than the hunters who go to the waterways to hunt ducks, geese, and other waterfowl. A few years ago, it was discovered that the lead shot that is used in shotgun shells was killing waterfowl long after the hunting season was over. Waterfowl do not know the difference between lead shot and the small rocks they must eat to grind the seeds and plant materials that make up their food. When lead shot is eaten, it is gradually absorbed into the body, causing lead poisoning to occur. Eventually it causes the death of the affected birds. Simply replacing lead shot with steel shot eliminates this problem. Steel shot can be eaten by waterfowl without poisoning them.

THE FUTURE OF WILDLIFE IN THE UNITED STATES

INTERNET KEY WORDS:

wildlife management

A bright future for all of the wildlife species is not ensured in the United States, because the needs of the human population continue to compete with those of the wildlife community. The outlook for wildlife in the future is not all bleak, however. Humans have recognized that it is possible to satisfy the needs of wildlife and the demands of humans if careful management practices are instituted (Figure 11–23).

Careful studies of how humans and wildlife can peacefully co-exist are continuing as wildlife management plans are refined to include new scientific findings. Sincere attempts are being made to reduce the pollution of the environment. Polluted areas are being cleaned up. More extensive testing of new chemicals and other pesticides and their environmental effects is being conducted. The effects of new construction on wildlife habitat is also studied before and during the construction process.

Establishing large acreages in national parks and wildlife refuges is also important for the future well-being of the wildlife in the United States. More emphasis is being placed on management of wildlife

FIGURE 11–23 Ponds and lakes can accommodate fish and wildlife as well as provide recreation for the family. *(Courtesy of Chesapeake Bay Foundation)*

resources, rather than simply exploitation. This also bodes well for the future of the country's wildlife. Fish and game laws are important elements in preserving and enhancing wildlife.

Realistically, some species of wildlife will decline and cease to exist in the future, whereas other species will proliferate. This has been the case in the past, and it seems likely to continue to be so in the future. The human population can choose to help the process or to interfere with it.

STUDENT ACTIVITIES

1. Write the Terms to Know and their meanings in your notebook.
2. Visit a wildlife area and identify the species of wildlife that you see there.
3. Construct a bulletin board showing the species of wildlife that are important to hunters and fishermen in your area.
4. Have a wildlife manager visit your class and explain management practices that are used in the area that he or she manages.
5. Write a report on the effects of pollution on wildlife in a particular area of the United States.
6. Participate in a New Year's Day bird count.
7. Plant a feed patch area for wildlife.
8. Participate in a stream or other wildlife area cleanup program.
9. Visit a zoo and list the species of wildlife housed there that are endangered.
10. Develop a list of endangered species of wildlife in your area and what is being done to prevent their extinction.
11. Write a report on a species of wildlife that interests you.
12. Have a local farmer or rancher speak to the class on what measures he or she is taking to enhance the environment for wildlife.
13. Invite a birdwatcher or wildlife photographer to discuss his or her given hobby or profession with the class.
14. Using the plans in Appendix B, construct and place birdhouses and nesting boxes.
15. Spend 1 hour outside of class observing an animal in the wild. Determine what activities the animal is involved in and how much time it spends on each activity. Write a one-page summary of your findings.

SELF EVALUATION

A. Multiple Choice

1. _____ is a species of fish adapted to cold, running water in streams.
   a. Carp
   b. Catfish
   c. Trout
   d. Sunfish

2. Forest wildlife generally survive best in forests that are
   a. of mixed-age trees.
   b. deciduous.
   c. evergreen.
   d. of even-age trees.

3. Trees growing along streams help to
   a. regulate water flow.
   b. provide food for aquatic wildlife.
   c. provide oxygen for fish.
   d. regulate stream temperatures.

4. The most important wildlife management area is the
  a. farm.
  b. stream.
  c. wetlands.
  d. forest.

5. When two species of wildlife live together for the benefit of both, the relationship is called
  a. mutualism.
  b. predation.
  c. commensalism.
  d. competition.

6. The raising and selling of wildlife for hunting is an example of what type of relationship between humans and wildlife?
  a. biological
  b. recreational
  c. social
  d. commercial

7. One species of fish that is often undesirable in lakes and ponds is
  a. bass.
  b. carp.
  c. sunfish.
  d. trout.

8. Rabbits, quail, pheasants, doves, and deer are wildlife often targeted for management in
  a. farm areas.
  b. forests.
  c. wetlands.
  d. none of the above.

9. Management of deer is a concern of _____ wildlife managers.
  a. forest
  b. farm
  c. wetlands
  d. all of the above

10. Wetlands should be made up of about _____ shallow, standing water for optimum wildlife use.
  a. one-fourth
  b. one-third
  c. one-half
  d. two-thirds

B. Matching

_____ **1.** Predation

_____ **2.** Parasitism

_____ **3.** Mutualism

_____ **4.** Commensalism

_____ **5.** Competition

  a. One type of wildlife living and feeding on another type
  b. Two types of wildlife eating the same food
  c. One type of wildlife eating another type of wildlife
  d. One type of wildlife living in, on, or with another type but without helping or harming it
  e. Two types of wildlife living together for the benefit of both

C. Completion

1. Inspiration for artwork is a/an _____ value of wildlife.

2. Fishing is a/an _____ value of wildlife.

3. Pollination of crops is a/an _____ value of wildlife.

4. The ability of wildlife species to increase the value of their surroundings simply by their presence is a/an _____ value of wildlife.

5. The observation of wildlife by early humans to determine what was safe to eat is a/an _____ value of wildlife.

UNIT
12 | Aquaculture

OBJECTIVE

To recognize the biological requirements necessary for the production of aquatic plants and animals.

MATERIALS LIST

✓ bulletin board materials

✓ small freshwater aquarium (optional)

✓ 30 g NaCl with 1,000-ml flask

✓ Internet access

COMPETENCIES TO BE DEVELOPED

After studying this unit, you should be able to:

- describe the food chain in a freshwater pond.
- discuss water quality and list eight measurable factors.
- identify three major aquaculture production systems.

SUGGESTED CLASS ACTIVITIES

1. Identify some species of fish or shellfish that are adapted to aquaculture production methods. Gather resource materials and assign teams of two or three students to prepare written and oral reports on the needs of a particular species. The written report should be the basis for the oral report that is presented to the class by each team.

2. Plan and conduct a field trip to an aquaculture facility. Have class members prepare some questions ahead of time about specific practices such as monitoring water quality, nutrition, disease management, waste treatment, and other practices.

3. A few days before introducing Unit 12, remove the shells from four eggs. This is done by soaking the eggs in vinegar. When the shells have dissolved, gently remove the eggs from the vinegar. Weigh the eggs and record their individual weight. Place two eggs in a jar with corn syrup or salt water. Place the other two eggs in distilled water. Predict what will happen to the eggs. After 2 days, observe and reweigh the eggs, and then discuss any changes.

TERMS TO KNOW

aquaculturist	brackish water	water quality	salmonid
natural fishery	fry	dissolved oxygen	spat
salinity	spawn	ppm	seining
gradient	osmosis	water hardness	rolls over
terrestrial	adaptation	buffer	ppt
amphibian	gills	turbidity	larvae
bay	shellfish	ammonia/nitrite/nitrate	
estuary	crustacean	TAN	
saltwater marsh	molt	toxin	

INTERNET KEY WORDS:

aquaculture

Water covers three quarters of the surface of the earth. This resource produces both plants and animals that are used to feed the world. Aquaculture is the management of this and other water environments to increase the harvest of usable plant and animal products. An **aquaculturist** is a person trained in aquaculture who must understand where and how organisms live, eat, grow, and reproduce in water. The manipulation of these factors determines the success of the aquaculture system. Aquaculture systems are becoming important to our food system because natural fisheries have reached their capacities to produce more fish (Figure 12–1).

These production systems are part of an integrated industry that requires specialized products and services. Aquaculturists include nutritionists, feed mill operators, pathologists, fish hatchery managers, processing managers, researchers, and growers. These services are used to produce fresh and processed seafoods, shellfish, and ornamental fish and plants.

Natural fisheries are fish production areas that occur in nature without human intervention. They include the oceans, continental shelves, reefs, bays, lakes, and rivers. These fisheries are currently fished by sophisticated fishing fleets that are so efficient that yearly catches (yields) have leveled off or are decreasing because of insufficient supply of fish. The population of

FIGURE 12–1 Natural fisheries have reached their capacities to produce more fish. *(Courtesy of Chesapeake Bay Foundation)*

FIGURE 12–2 The oceans represent our largest water resource.

the world continues to increase; therefore, aquaculture must be used to produce more aquatic plants and animals for food. Understanding the aquatic environment, the biology of the organisms, and how to control the production of aquatic plants and animals is essential.

THE AQUATIC ENVIRONMENT

The oceans represent the largest expanse of water resources in the world (Figure 12–2). They represent a reservoir of water containing soluble nutrients and materials washed from the land. Over time, the evaporation of water into the atmosphere has increased the concentration of these nutrients until the salinity, or mineral content, of the water is high. We call this seawater. The concentrations of these nutrients and salts are so high that land plants and animals are unable to survive if this water is used for irrigation or drinking.

Rain contains only small amounts of salts. Therefore, accumulation of water on land and its flow into the oceans generates a gradient, or measurable change over time or distance, in salinity. The water cycle is the means by which the increase in salinity has occurred in ocean water (Figure 12–3). This affects the types of organisms that can flourish.

The Salinity Gradient

Freshwater wetlands, such as marshes, ponds, and streams, are generated by rainwater (Figure 12–4). Here, natural rainfall accumulates and provides aquatic environments that represent the transition between aquatic and terrestrial (land) plants and animals. Amphibians, such as frogs, turtles, and reptiles, are animals that live part of their lives in these fresh waters and the remaining period on land. Several plant species, such as cattails (*Typha* spp.), watercress *(Nasturtium officinale)*, water spinach *(Ipomoea reptans)*, and rice, also require a transitional period of flooding and drainage to flourish.

As water flows from fields and urban areas into large streams and lakes, this runoff accumulates more soluble nutrients, and the salinity increases. The profile of plants and animals changes as other organisms that are more adapted to this changed environment displace the original

FIGURE 12–3 The natural water cycle removes water from the oceans in the form of precipitation that falls on the land masses. The water then gradually returns to the ocean through springs, streams, lakes, and rivers.

FIGURE 12-4 Freshwater marshes serve as transition areas where some species of animals and plants are able to adjust while they make the transition between the aquatic and terrestrial phases in their life cycles.

FIGURE 12-5 Osmosis is a process by which water moves across a membrane from an area of high concentration to one of lower concentration.

residents. Flows accumulate into rivers that empty into **bays, esturaries, and saltwater marshes.** Bays are open waters along coastlines where freshwater and saltwater mix. Estuaries are ecological systems influenced by brackish or salty water. The saltwater marshes are lowlands influenced by tidal waters. The freshwater mixes with the seawater to create unique growing conditions for various fish, shellfish, and aquatic plants. This **brackish water** is a mixture of freshwater and saltwater that fluctuates with the tide, flow of the rivers, and weather.

Several types of seaweeds are commercially produced in the Orient. These include types of red, green, and brown algae. Similar plant growth of algae must be maintained in intensive fish systems to feed newly hatched fish called **fry.**

The migration of saltwater salmon upstream to **spawn** (lay eggs) in freshwater streams illustrates the gradient effect on the life cycle of aquatic organisms.

Saltwater and Freshwater Fish

Osmosis is the process by which water moves from an area of high concentration to an area of low concentration through a selectively permeable membrane. A selectively permeable membrane will allow some molecules to pass through, whereas other molecules (for example, salts) cannot. Figure 12-5 illustrates water movement from side A that has a greater water concentration to side B that has a lower concentration. This happens because water tries to make both sides of the membrane equal in concentration.

The skin of a fish and the cells of its gills can act as selectively permeable membranes that control water movement. In a marine environment, water would naturally move from the inside of the fish to the outside of the fish in an attempt to dilute the saltwater outside of the body. If this took place without the membranes of the fish reacting to controlled water loss, the fish would become dehydrated and die. Conversely, in a freshwater environment where there is more salt in a fish's body than outside of it, water would move inside in an attempt to dilute the salt concentration in the body. This would result in a waterlogged fish, which would also cause severe problems and death.

Adaptation occurs when heritable traits that increase the chances of an organism for survival are passed from one generation to the next. One adaptation that fish have made over time is the ability to live in freshwater, saltwater (marine), or both environments. Marine fish drink water to make up for water loss. They also have specialized "chloride" cells that rid the body of salt. Freshwater fish, in contrast, have no need to drink water. They also have chloride cells in their gills. In freshwater fish, these cells work conversely to those in their marine relatives. Their chloride cells absorb salt from the water. They also excrete a large amount of very diluted urine. Some fish, such as salmon, have adapted even further and can change the function of their chloride cells to live in either freshwater or marine environments.

The Aquatic Food Chain

INTERNET KEY WORDS:

aquatic food chain

The aquatic environment constantly changes to maintain a balance of organisms that function in the food chain or system. A simple illustration is the makeup of a freshwater pond. The food chain is fueled by sunlight. Green plants and algae use this energy to grow and use nutrients they absorb from the water. These plants are eaten by fish and other animals, which are preyed on by larger animals (fish, reptiles, and others) and sometimes caught by humans. These large animals return nutrients to the water as waste, or carrion, that is reabsorbed by plants for growth. The maintenance of this food chain supports life (Figure 12–6).

The aquaculturist must understand the effect of any management activity on this cycle and make adjustments using technology or design in an aquaculture production system.

General Biology

The biology of aquatic plants and animals is similar to that of terrestrial plants and animals. Both assimilate nutrients, grow, reproduce, and interact with the environment. Green plants harvest energy from the sun through photosynthesis and absorb nutrients from water to manufacture carbohydrates, proteins, fats, and cellulose. They serve as the waste recyclers in the aquatic environment by constantly absorbing waste products (nutrients) and contributing to the food chain. Like any land plant, they respond to fertilization, shading, competition, insects, disease, and weather. Aquatic plants are composed of many parts, as discussed in Unit 15, "Plant Structures and Taxonomy." Certain parts help them compete within the aquatic environment. Green plants absorb carbon dioxide from water and release oxygen during photosynthesis. During the night, plants reabsorb a smaller amount of oxygen and release carbon dioxide.

Aquatic animals, particularly fish, complement the relationship with plants by generating carbon dioxide during respiration and releasing soluble nutrients through waste products and decay. Figure 12–7 illustrates the anatomy of a fish and shows the specialized gills, which exchange gases by absorbing oxygen from water and releasing carbon dioxide into the water.

FIGURE 12–6 The aquatic food chain in a freshwater pond.

The Anatomy of a Fish

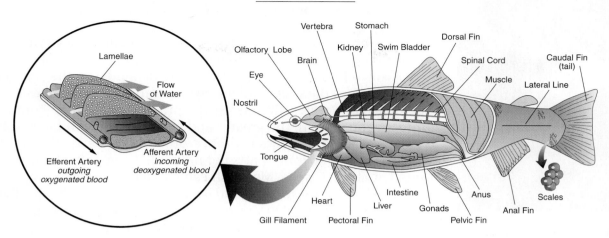

FIGURE 12–7 Fish are equipped with specialized organs such as gills and swim bladders that make it possible for them to survive in a water environment.

In this competitive environment of predator and prey, plants provide the shelter and food that are essential to the life cycles of these animals.

Shellfish are aquatic animals with a shell or shell-like extensions. Some adult shellfish are nonmotile—that is, they cannot move about. They include clams, mussels, oysters, and others, and they occupy a unique niche in the aquatic environment. Located on the bottom of a body of water, these organisms have developed an efficient pumping mechanism that filters great quantities of water. It filters out edible microscopic plants and animals.

Crustaceans are a group of aquatic organisms with exoskeletons. These organisms **molt,** or replace their outer shells as they grow. Saltwater lobsters (*Homarous* spp.), crawfish (*Procambarus* spp.), and the various crabs, shrimps, and prawns are important crustaceans. These mobile organisms are characterized by hard exoskeletons that must soften and split as the animals molt and secrete larger shells.

Aquaculture Production

The aquaculturist, in an attempt to increase the production of any aquatic organism, must monitor and maintain the optimum **water quality.** Water quality has several chemical and physical characteristics that

Science Profile

THE GRADIENT EFFECT

Several species of fish migrate from freshwater to saltwater, where they grow to maturity. During their outward migration, their bodies adapt to thrive in a saltwater environment. Eventually they migrate back to freshwater to spawn. This is necessary because the young fry must hatch in low-salinity water. For most fish that cross the saltwater gradient, the spawning migration is a suicide mission because their bodies are no longer adapted to survive in freshwater. Among the fish that live on both sides of the saltwater gradient are salmon, steelhead trout, and some species of bass.

FIGURE 12–8 Water quality must be monitored closely to maintain an environment in which aquatic plants and animals can live. Water quality is measured using electronic instruments and chemical tests. *(Courtesy of USDA/ARS #K-4246-12)*

interact within the water and must be measured and maintained within a narrow range to promote growth and development of aquatic plants and animals (Figure 12–8).

There are several different types of water-quality test kits available to test the characteristics discussed in the following paragraphs.

Dissolved oxygen (oxygen in water) depends on the temperature and pressure of the water and the concentration of atmospheric oxygen. The cooler the water, the higher the pressure. The more contact with atmospheric oxygen, the greater and faster oxygen can be dissolved in water. Measured by oxygen probes or chemical tests, the results are reported as 0 to 10 **ppm** (parts per million). Water at 85° F (30° C) is saturated at about 8 ppm. Most fish can survive at levels as low as 3 ppm but quickly become stressed and succumb to other problems. Rainbow trout must have excellent, or high, levels of dissolved oxygen and can only be cultured in oxygen-saturated water (Figure 12–9 and 12–10).

The measurement of acidity or alkalinity in water is the pH. This factor affects the toxicity of soluble nutrients in the water. This measurement is recorded, using a pH meter or litmus paper tape, as a number from 1 to 14. Readings less than 7 are acidic solutions, 7 is neutral, and numbers greater than 7 are alkaline. Most aquatic plants and animals grow well in water with a pH between 7 and 8.

Water hardness is measured by chemical analysis and is expressed as ppm calcium. This element is essential in the development of the exoskeletons of shellfish and crustaceans. It also serves as a chemical **buffer** that stabilizes rapid shifts in pH.

Turbidity in water is caused by the presence of suspended matter. High turbidity limits photosynthesis and visibility. A simple method for estimating pond turbidity uses a white Secchi disc that is lowered into the water. When visibility is impaired, the depth is recorded. The greater the depth, the less turbid the water (Figure 12–11).

FIGURE 12–9 Oxygen content of water is increased by churning and mixing water through air.

FIGURE 12–10 Trout require high levels of dissolved oxygen and a constant environment of cold water to survive and grow in an aquaculture setting. The water environment is controlled carefully to provide adequate oxygen and to eliminate waste.

FIGURE 12–11 The turbidity of water is measured with a Secchi disc. It is lowered in the water until visibility is impaired. The depth at which this occurs is used to calculate the turbidity of the water.

INTERNET KEY WORDS:

http://www.newmex.com/ platinum/data/index2.html

Temperature limits the adaptive range of almost all aquatic organisms. Sunlight warms the upper surface of open water but does not penetrate it. Deep waters, cool-region currents, and melting winter covers can affect water temperature.

Ammonia/nitrite/nitrate compose a group of nitrogen compounds generated by aquatic animals, first as urea and ammonia. These waste products are converted first to nitrite by microscopic organisms in the water, and then to nitrate. They are ultimately converted to nitrogen gas or absorbed by plants. The accumulation of both ammonia and nitrite is toxic to fish and often limits commercial production. Total ammonial nitrogen, or TAN, is recorded by chemical assay in ppm. This does not reflect the toxicity of the measured amount, because the toxicity of ammonia is dependent on the pH. Generally, levels of TAN are maintained at less than 1 ppm.

Toxins represent a host of materials that act as poisons, adversely affecting the growth and development of aquatic plants and animals. These include agricultural chemicals, pesticides, municipal wastes, and industrial sludges. Chemical analyses are difficult and often inconclusive.

Selection of Aquaculture Crops

The actual selection of aquatic crops that may be grown is dependent on the resources and experience of the aquaculturist. Like terrestrial crops, each species of aquatic crop has a particular set of water-quality standards to ensure survival and reproduction. A discussion of a few of the well-known aquatic crops should indicate the diversity of this commercial industry. Characteristics will also vary between species. Ideal conditions should be requested from your local county extension office and local aquaculturists.

Trout and salmonids are high-quality fish products in high demand (Figure 12–12). The trout flourishes in high-quality water. Dissolved oxygen must be kept at greater than 5 ppm. Salmonids also require low salinity, cool temperature (60° F/15° C), and low turbidity. The TAN must be maintained at less than 0.1 ppm.

Catfish (*Ictalurus* spp.) represent one of the fastest-growing aquaculture industries in the United States. Current figures project that more than 130,000 acres of ponds are producing catfish annually. Catfish thrive at 75° to 79° F (24–26° C), a pH between 6.6 and 7.5, a water hardness of 10 ppm, and dissolved oxygen greater than 4 ppm.

Crawfish (*Procambarus* spp.) thrive in freshwater lakes and streams. Good growth occurs at 70° to 84° F (21–29° C), a pH close to neutral, a water hardness of 50 to 200 ppm, salinity up to 6 ppt, and dissolved oxygen greater than 3 ppm.

Clams, crabs, and oysters are cultivated in bays and estuaries that are subject to tidal flows (Figure 12–13). Good growing conditions vary greatly with species, but include approximately 6 to 20 ppt salinity, 15° to 30° C, dissolved oxygen greater than 1 ppm, and adequate amounts of microscopic organisms (low turbidity). Production should improve on cultivated sites when improved spat, or young oysters, are developed for stocking.

Shrimp and prawn are being cultured in brackish-water ponds and estuaries. Conditions for good growth are temperatures greater than 25° C, a salinity of 20 percent, high levels of microorganisms, and dissolved

FIGURE 12–12 Trout are high-quality fish that are produced on fish farms in large numbers. A high-quality source of cool water is required to raise trout successfully.

oxygen greater than 4 ppm. Hatchery techniques are complex and involve several distinct growing stages.

Production Systems

The cultivation of any aquatic organism by the aquaculturist integrates the necessary cultural requirements with existing resources. This blend has developed three general production programs.

Open ponds, rivers, and bays are stocked with natural or cultured young and are maintained with densities that are balanced with the existing ecosystem. Competing species are controlled, and natural recycling techniques are encouraged. This form of aquaculture can use both natural and constructed ponds (Figure 12–14). Some care must be taken to prevent any peaks in TAN. Overfertilization stimulates rapid algae blooms, with high levels of dissolved oxygen during photosynthesis but very low levels during early morning hours. Stress leads to high levels of mortality. Even lower dissolved-oxygen levels and fish deaths can occur. The aquaculturist must be careful to monitor incoming water for toxins and other suspended materials. Harvesting must involve draining or **seining** the entire production area. Seining is the removal of fish with nets.

Caged Culture

Caged culture represents a more capital-intensive program. Aquatic animals or plants are contained in a small area, and waste products are removed by the flushing action of the natural waters (Figure 12–15). Confining growing fish in floating pen cages or shellfish on suspended float tables is a technique for managing increasing densities of aquatic organisms. As is true on a cattle feedlot, the aquaculturist confines the animals in a limited area and provides the necessary feed and cultural management. The natural ebb and flow of the water removes waste products and replenishes dissolved oxygen.

Cage culture can be designed for both natural waters and newly constructed ponds. Aquaculturists have a better idea of growth rates and can adjust feeding ratios more economically. Some growers have reported problems when a pond **rolls over** (that is, changes water quality suddenly during certain weather conditions and brings the less-oxygenated water to the surface). Fish in cages are unable to move and can be stressed or killed. In this intensive production system, the aquaculturist must ensure adequate nutrition, disease control, predatory control, and physical maintenance like any terrestrial animal producer.

FIGURE 12–13 The highly prized blue crab of the Chesapeake Bay. *(Courtesy of USDA/ARS #K-3692-3)*

FIGURE 12–14 Constructed ponds made of concrete are used by aquaculturists to produce large numbers of fish and other freshwater species in controlled water environments. *(Courtesy of Rick Parker)*

Science Profile

SALINITY

Salinity is the measurement of total mineral solids in water. It is measured by either electrical conductivity (ohms/cm) or is calculated against known standards and converted to **ppt** (parts per thousand). The higher the concentration of salts, the greater the conductivity reading. A seawater standard can be made with artificial sea salts or by dissolving 29.674 g NaCl in 1 liter H_2O. Measurements of salinity usually range from 0 to 32 ppt (freshwater to saltwater).

INTERNET KEY WORDS:

cage fish culture
production

The young stock must be caught from natural waters or produced in controlled hatcheries. Successful operations include the production of Atlantic salmon off the coast of Norway, Nova Scotia, and Maine. Hybrids of striped bass have been cultured in cages in Maryland and California. Trout have been cultured in net pens suspended in mountain streams and ponds.

Shellfish growers in Japan and the United States have demonstrated the production of oysters, clams, and mussels on suspended float tables or ropes. Production is increased by higher water quality and ease of harvest.

FIGURE 12–15 Fish can be raised in pens or cages that sit on the bottom or are suspended in water.

Agri-Profile

CAREER AREA:

Aquaculture Researcher

Catfish farming and other aquaculture enterprises are big business in many nations of the world. Similarly, catfish farming is one of the fastest growing food production enterprises in the United States. Cultured seafood production is rapidly approaching the volume of that taken from natural waters.

The rapid growth in aquaculture has spurred research and development activities. These, in turn, have stimulated career opportunities in animal science, nutrition, genetics, physiology, aquaculture construction, facility maintenance, pollution control, fish management, harvesting, marketing, and other areas.

As productive land becomes more scarce among the global resources, aquaculture will play an increasing role in providing high-quality food at affordable prices.

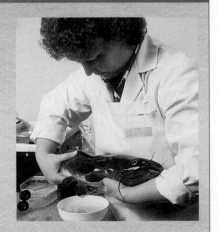

Animal physiologist Cheryl Goudie of the Catfish Genetic Research Unit at Stoneville, Mississippi applies gentle pressure to cause a female catfish to release her eggs for artificial fertilization. *(Courtesy of USDA/ARS #K-5325)*

Science Connection

WHIRLING DISEASE

A major problem for trout and salmon is a small parasite called Myxobolus cerebralis. This harmful organism is native to Europe and Asia, but it was accidentally transported to the United States in 1955. It was not until the early 1990s that scientists discovered that a major epidemic was occurring in the waters of many states. Both wild fish and hatchery fish were affected.

The parasite is virtually indestructible. The spores can live in water systems for as many as 30 years. Eventually, a common aquatic worm ingests these spores. Inside the worm, the second life stage develops. This is the free-floating stage that moves through the water until it comes in contact with a young fish called a fry. The parasite then burrows into the head and spinal cartilage of the fry. Here, it begins to multiply very quickly. This puts pressure on the fry's brain, causing the fish to swim in circles. The fish appear to be whirling. The fry that are affected have a hard time eating and cannot avoid predators well. Whirling disease, as it is now called, is a concern of many sportsman, aquaculturists, fish farmers, and scientists. In the future, whirling disease will be the focus of much aquaculture research as an attempt is made to control it.

INTERNET KEY WORDS:

aquaculture, cage culture
aquaculture, hatchery

FIGURE 12–16 New technology is making fish production profitable in tanks located in sheds, warehouses, greenhouses, and other enclosures. *(Courtesy of National FFA)*

Recirculating Tanks

Many areas of the world lack sufficient water resources to maintain a viable aquaculture industry. Recirculating systems must circulate the waste water through a biological purifier and return it to the growing tank (Figure 12–16). This complicated process is similar to that in a city waste-treatment plant. The system must remove the solid fish wastes, soluble ammonia/nitrite/nitrates, and carbon dioxide and replace depleted oxygen. The pH must be maintained and integrated into the biological needs of the bacteria that populate the biological filters. As in any pond, if any single parameter is ignored, the organisms become stressed and production is decreased.

Extensive research in aquaculture has been done at the University of Wisconsin with yellow perch; the University of Maryland with striped bass, trout, and tilapia; the Walt Disney World EPCOT Center with numerous species; Mississippi State University with catfish; and many other institutions in the United States.

Hatcheries

The development of more intensive aquaculture systems will depend on a constant supply of high-quality, young organisms. The natural fisheries are threatened by overfishing, pollution, and habitat destruction. Hatcheries are investigating the parameters that affect fish breeding habits and attempts to induce spawning, fry, or larvae production. **Larvae** are mobile aquatic organisms that grow into nonmobile adults.

These advances in fish and shellfish management allow the industry to develop improved breeds and hybrids to support improved production. Improved genetic lines of trout, catfish, and salmon are already commercially available.

Bio-Tech Connection

AQUACULTURE RESEARCH IS BIG TIME!

The Aquaculture Research Center in Baltimore, Maryland, was established to provide combined research facilities for the University of Maryland Center of Marine Biotechnology, the $160 million Columbus Center for Marine Sciences, and the National Aquarium in Baltimore. The center was established by converting a harborside shipbuilding structure into a state-of-the-art aquaculture research facility. The Columbus Center eventually will occupy two large piers in the Inner Harbor area to complete the aquaculture research complex.

Scientists are growing high densities of valuable fish, such as striped bass, in closed systems using biological filters for cleaning the water to make the facility environmentally friendly. Tools of biotechnology are being used to acquire knowledge about native species and to develop novel solutions to problems

Acidity (left) and solids (right) tests are among the many tests needed periodically to keep fish growing in this 7,400-gallon recycled water aquaculture system. *(Courtesy USDA/ARS #K-4248-2)*

associated with commercial finfish and shellfish culture. Studies in the environmental, hormonal, and molecular regulation of spawning will enable growers to have access to seedstocks whenever they need them, rather than being dependent on the normal spawning cycles.

The study of fish genes and hormones that control fish growth and the study of fish nutrition will decrease the time and cost of raising fish to market weight. Of primary importance is the study of devastating diseases that have nearly eliminated the commercial oyster population and certain other species in many natural waters. The worldwide demand for seafood is projected to increase by 70 percent by 2025. Yet the world's natural fisheries are already stressed beyond sustainable limits. To meet the increased demand for seafood with decreasing harvests from natural waters, it is believed that aquaculture will have to increase its output by seven times the 1993 levels.

With its working participatory science laboratory and public spaces for observation, the Aquaculture Research Center is an international focal point for scientists. It includes a major research center and an educational center.

INTERNET KEY WORDS:

Fish hatchery
practices culture

Aquaculture and Resource Management

The demand for aquaculture products will continue to increase. This demand will stimulate a tremendous growth in commercial aquaculture. During the expansion of the commercial aquaculture industry, sources of clean, pure water will have to be located or developed. Conflicts for scarce natural resources and clean water, the impact on recreational areas, and the potential pollution effects will need to be resolved by trained specialists.

Science Profile

AQUACULTURE—A SOURCE OF FISH AND SEAFOOD

It is well known that the harvest of fish and other seafood from the oceans of the world is not likely to increase greater than current production levels. At the same time, the demand for seafood such as fish and shellfish is continuing to increase. These two trends have created ideal economic conditions for expanding the aquaculture industry. The key of success will be to produce and market high-quality fish and other seafood products economically to take advantage of this opportunity. Obtaining access to enough pure water to support aquacultural production is the first challenge in establishing this kind of farming enterprise. The second big challenge is to return the water to streams and rivers without polluting them. Aquaculture appears destined to be a growth industry to support the demand for fish and other seafood products.

STUDENT ACTIVITIES

1. Write the Terms to Know and their meanings in your notebook.
2. Visit a local supermarket or seafood market and list the seafood products. Classify them as fish, shellfish, or crustaceans; freshwater or saltwater products; or imported products.
3. Describe why some of the products in Activity 2 might not be produced in your area.
4. Locate three local aquaculturists and discuss their production systems. How do they maintain water quality?
5. Visit a local pet store. Describe the various parts of a freshwater aquarium. How is the water quality maintained in this recirculating system?
6. Make a bulletin board illustrating the food chain of a freshwater pond.
7. Set up a class aquarium and discuss the balance between plants and animals in the system.
8. Make saltwater and place a fresh cucumber in the water overnight. Observe any changes. In a few sentences, explain what changes you observed.
9. Research three aquaculture careers that interest you. In a short paragraph, describe the jobs and the training they require.
10. In your own words, describe the eight factors that affect water quality.

SELF EVALUATION

A. Multiple Choice

1. The aquaculturist must understand how aquatic organisms
   a. eat.
   b. reproduce.
   c. live.
   d. all of the above.

2. The yearly catch of fish from natural waters is
   a. increasing.
   b. holding constant or decreasing.
   c. difficult to determine.
   d. mostly catfish.

3. The highest salinity level is measured in
   a. pond water.
   b. irrigation water.
   c. creeks.
   d. ocean water.

4. The accumulation of salts in water occurs most often when
   a. water runs across agricultural land.
   b. water settles in a pond.
   c. water collects in a drainage ditch.
   d. water is lost through evaporation.

5. Brackish water is
   a. colored black.
   b. located in tidal areas.
   c. collected from small creeks and branches.
   d. mostly high in salinity (20–34 ppt).

6. Water quality is least affected by which of the following factors?
   a. fish density
   b. weather
   c. chemical runoff
   d. fish species

7. The greater the density of fish in a system, the
   a. smaller the tank.
   b. larger the fish.
   c. the more difficult the management.
   d. the greater the temperature.

8. A fish death can occur when a pond "rolls over"
   a. because of the temperature shock.
   b. because the cages sink to the bottom.
   c. because of low levels of dissolved oxygen.
   d. because the fish turn upside down.

B. Matching

_____ 1. Salinity
_____ 2. Dissolved oxygen
_____ 3. Turbidity
_____ 4. Temperature
_____ 5. Ammonia
_____ 6. Hardness

a. Measured in ppm calcium
b. Recorded as degrees F or C
c. Measured as ppm or percent
d. Depth of visual penetration
e. TAN
f. Conductivity or ppt

C. Completion

1. Both plants and animals are part of the food _____.
2. The _____ are where freshwater and seawater mix.
3. The _____ of a fish absorb oxygen from the water.
4. Between 32 and 24 ppt is the salinity of _____.
5. Trout need a _____ dissolved oxygen concentration than clams to grow and mature.
6. Crawfish must _____, or break out of their exoskeletons to grow.
7. Salmon must return to _____ water to spawn and complete their life cycles.
8. Pond culture relies mostly on _____ recycling of fish waste products.
9. The recirculating production systems must treat the fish wastes with _____ filters.

Integrated Pest Management

A Balancing Act!

Honeybees are a major pillar in the food, fiber, flower, and ornamentals production arena in the United States. We rely on them as the only method of pollinating certain plants and count on them to do some of the pollination of nearly all species of plants. Bees enter flowers of plants to gather nectar and pollen for their own food and nourishment of their young. Their service to humans and animals in pollinating plants, and thereby producing seeds and fruit, is a service we cannot do without. Most plants could not reproduce and survive without producing seeds. Their safety from pesticides used to control harmful insects is always at the top of the agenda for entomologists.

Honeybees are numbered among the many insects that are beneficial to humans. Though they can sting if threatened, honeybees in the United States are of the European type and are predictable. Except for the inexperienced person approaching a beehive, honeybee stings are generally a single sting by a single bee. Except for the relatively few individuals who have life-threatening allergic reactions to bee stings, honeybees pose little threat, because they act individually when they are aggravated enough to sting.

Enter the Africanized honeybee—a hybrid cousin of the domestic pollinator, famous for its aggressiveness and dubbed "killer bee"!

Africanized honeybees resulted in 1957 when bees were imported from Africa to Brazil by a Brazilian scientist. The plan was to experiment by crossbreeding them with the domestic European bees prevalent in the Americas to develop a better strain of honeybees for the tropics. Unfortunately, some African bees were inadvertently released in the countryside and promptly interbred with the domestic bees. The new hybrids and their descendents are known as Africanized honeybees. They have migrated as far south as Argentina and as far north as the United States.

On October 15, 1990, the first Africanized honeybee swarm to migrate naturally to the United States was identified by entomologists near Brownsville, Texas. The swarm was promptly destroyed, according to standard procedure.

Unfortunately, Africanized honeybees have different dispositions than the domestic bees of the United States. They tend to defend their colonies more vigorously, stinging in greater numbers and with less provocation. One bee is likely to inflict many stings. Therefore, there is greater danger in an encounter with the Africanized bees. U.S. agriculture and the beekeeping industry fear that domestic bees interbred with Africanized bees may become harder to manage as pollinators of crops and may not be as efficient as honey producers.

The challenge of observing, detecting, and stopping the northward migration of Africanized honeybees will be a top priority of government inspectors, entomologists, beekeepers, farmers, and citizens at large. At the same time, animal behaviorists will study the bees' habits and will look for ways to manage them. Geneticists will study the bees' genes and will look for ways to genetically engineer future bees so as to decrease their objectionable habits and enhance their abilities as pollinators and honey makers.

Honeybees are friends we cannot do without. However, the Africanized hybrids can be dangerous and threaten to decrease bee productivity in the United States. *(Courtesy USDA/ARS #K-3653-12)*

Biological, Cultural, and Chemical Control of Pests

OBJECTIVE

To develop an understanding of the major pest groups and some elements of effective pest-management programs.

MATERIALS LIST

✓ insect net
✓ killing jar

✓ insect mounting pins
✓ insect specimen labels

✓ pictures of pest
✓ Internet access

COMPETENCIES TO BE DEVELOPED

After studying this unit, you should be able to:

- define *pest, disease, insect, weed, biological, cultural, chemical,* and other terms associated with integrated pest management.
- know how the major pest groups adversely affect agriscience activities.
- describe weeds based on their life cycles.
- describe both the beneficial and detrimental roles that insects play.
- recognize the major components and the causal agents of disease.
- understand and explain the concept of integrated pest management.

SUGGESTED CLASS ACTIVITIES

1. Contact the agricultural extension office that serves your county or parish to obtain literature about pests that are problems in the area near your school. Ask the agent to talk with your class about the best way to control the pests that have been identified. Make a chart with the following elements: pest name, kind of damage, and methods of control.

2. Design and construct a public display that tells the story of integrated pest management. Prepare teams of students to explain this concept for controlling pests and help them set appointments to present what they have learned to civic clubs, community groups, and other interested parties.

3. Have each student bring five weeds, including the roots, to class. Spend a class period identifying the weeds using field guides. Discuss the type of damage each of the weeds is known to inflict on the environments of other plants, animals, and humans.

TERMS TO KNOW

disease	stolon	fungi	quarantine
vector	meristematic tissue	hyphae	targeted pest
insect	noxious weed	mycelium	eradication
arachnid	exoskeleton	bacteria	pheromone
defoliate	entomophagous	virus	cultivar
weed	metamorphosis	symptom	biological control
pathogen	instars	mosaic	cultural control
annual weed	plant disease	nematode	clean culture
biennial weed	causal agent	key pest	trap crop
perennial weed	disease triangle	pest population equilibrium	chemical control
rhizome	abiotic (nonliving) disease	economic threshold level	pesticide resistance
node	biotic disease	monitoring	pest resurgence

FIGURE 13–1 A sliced-open cotton boll showing a pink bollworm and the damage it has done. *(Courtesy of USDA/ARS #K-2886-13)*

FIGURE 13–2 Pear fruits that yellowed and shrivelled when the fire blight disease cut off the flow of nutrients from the tree to the fruit. *(Courtesy of USDA/ARS #K-5300-1)*

The ability to control pests by chemical, cultural, or natural control methods has afforded people in the United States an unprecedented standard of living. We often take for granted an unlimited food supply, good health, a stable economy, and an aesthetically pleasing environment. Without effective pest control strategies, our standard of living would decrease.

Good pest management practices have resulted in dramatic yield increases for every major crop. A single U.S. farmer in 1850 could only support himself and four people; but, currently, a farmer can provide food and fiber for approximately 140 people. The ability to control plant and animal **diseases** or disorders vectored by insects and arachnids has reduced the incidence of malaria, typhus, and Rocky Mountain spotted fever. A **vector** is a living organism that transmits or carries a disease organism. An **insect** is a six-legged animal with three body segments. An **arachnid** is an eight-legged animal, such as a spider or a mite.

The impact of pest management in maintaining a stable economy can be seen on a regional and national basis. The regional economy suffered shortly after the cotton boll weevil's introduction into the United States in 1892. The weevil devastated much of the cotton crop in the early 1900s (Figure 13–1). Similarly, the potato blight disease in Ireland caused famine and mass migration of Irish people to other parts of the world in 1845. Today many blights are still serious threats to our crops (Figure 13–2).

TYPES OF PESTS

The word pest is a general name for any organism that may adversely affect human activities. We may think of an agricultural pest as one that competes with crops for nutrients and water, tends to **defoliate** plants (strips a plant of its leaves), or transmits plant or animal diseases. The major agricultural pests are weeds, insects, and plant diseases. However,

Crabgrass
(Summer Annual)

Johnson Grass
(Perennial)

Henbit
(Winter Annual)

Mullein
(Biennial)

FIGURE 13-3 Different types of plants considered to be weeds. *(Courtesy of Maryland Cooperative Extension Service)*

INTERNET KEY WORDS:

weeds
annual weed
biennial weed
perennial weed
noxious weed

other types of pests exist. Some examples, and the classes of pesticides or chemicals used for killing them, are listed below.

Type of Pest	Class of Pesticide
mites, ticks	acaricide
birds	avicide
fungi	fungicide
weeds	herbicide
insects	insecticide
nematodes	nematacide
rodents	rodenticide

Damage by pests to agricultural crops in the United States has been estimated to be one third of the total crop production potential. Therefore, an understanding of the major pest groups and their biology is required to ensure success in reducing crop losses caused by pests.

Weeds

Weeds are plants that are considered to be growing out of place (Figure 13–3). Such plants are undesirable because they interfere with plants grown for crops. The definition of a weed is therefore a relative term. Corn plants growing in a soybean field or white clover growing in a field of turfgrass are examples of weeds, just as crabgrass is considered to be a weed when it grows in a yard or garden.

Weeds can be considered undesirable for any of the following reasons:

- They compete for water, nutrients, light, and space, resulting in reduced crop yields.
- They decrease crop quality.
- They reduce aesthetic value.
- They interfere with maintenance along rights-of-way.
- They harbor insects and disease pathogens (organisms that cause disease).

Weeds can be divided into three categories—annual, biennial, and perennial—based on their life spans and their periods of vegetative and reproductive growth.

Annual Weeds An annual weed is a plant that completes its life cycle within one year (Figure 13–4). Two types of annual weeds occur, depending upon the time of year in which they germinate. A winter annual germinates in the fall and actively grows until late spring. It then will produce seed and die during periods of heat and drought stress. Examples of winter annuals are chickweed, henbit, and yellow rocket.

A summer annual germinates in the late spring, with vigorous growth during the summer months. Seeds are produced by late summer and the plant will die during periods of low temperatures and frost. Examples of summer annuals are crabgrass, spotted spurge, and fall panicum.

Biennial Weeds A biennial weed is a plant that will live for 2 years (Figure 13–5). In the first year, the plant produces only vegetative growth, such as leaf, stem, and root tissue. By the end of the second year, the plant will produce flowers and seeds. This is referred to as reproductive growth. After the seed is produced, the plant will die. There are only a few plants that are considered biennials. Some examples are bull thistle, burdock, and wild carrot.

FIGURE 13–4 An annual weed is a plant that completes its life cycle within 1 year. Yellow mustard is a common example of this kind of weed.

FIGURE 13–5 A biennial weed is a plant that completes its life cycle in 2 years. The bull thistle is a widespread example of this type of weed

Perennial Weeds A **perennial weed** can live for more than 2 years and may reproduce by seed and/or vegetative growth (Figure 13–6). By producing rhizomes, stolons, and an extensive rootstock, perennial plants reproduce vegetatively. A **rhizome** is a stem that runs underground and gives rise to new plants at each joint, or **node.** A **stolon** is a stem that runs on the surface and gives rise to new plants at each node. These plant parts have **meristematic tissue** (tissue capable of starting new

Science Connection

A THORN IN YOUR SOCK

Bromus tectorum is a noxious weed that was introduced to the United States from Europe in the 1850s. It has subsequently invaded every state in the country. Early pioneers nicknamed the weed cheatgrass, because they believed the weed was cheating them out of greater wheat yields. Other common names include downy brome, downy chess, bronco grass, cheat, and six-weeks grass. In early spring, range animals take advantage of it as a nutritious source of food. Once the plant dies in late spring, its prickly seeds become a potential danger to range animals by irritating the mouth and throat, resulting in painful sores.

B. *tectorum* is a successful weed. A single plant can produce up to 5,000 seeds under favorable growing conditions. The wind, animal fur, and human clothing easily transport the seeds. Most people who spend time outdoors have collected these seeds in their shoes and socks. Cheatgrass germinates earlier than most plants and quickly uses most of the available water in the soil. When other plants begin to germinate, they often die because of lack of water. This is why cheatgrass is so harmful to crops. Cheatgrass and other noxious weeds must be dealt with aggressively using proven integrated pest management techniques.

plant growth). Examples of perennial weeds are dandelion, Bermuda grass, Canada thistle, and nutsedge.

Noxious Weeds A noxious weed is a plant that causes great harm to other organisms by weakening those around it. Most states have developed lists of noxious weeds, and great effort is made to control or eradicate them. Most noxious weeds are difficult to control, and they require extended periods of treatment followed by close monitoring. Noxious weeds should be handled carefully to avoid spreading seeds to unaffected areas. Noxious weeds are often spread when seeds become airborne, fall into flowing streams, become attached to the hair of an animal or to human clothing, or are eaten and distributed by birds.

Insects

Insects have successfully adapted to nearly every environment on the earth. There are more species of insects than any other class of organism. Part of their success is because of the large numbers of offspring they are capable of producing and the short time they require to reach physical maturity. The human race is dependent on insects in many ways, and they perform great service to us. Some of the most beneficial insects are the ladybug, preying mantis, parasitic wasps, and honeybees. Insects also cause great losses to crops, livestock, and people by damaging or injuring them.

Insect Pests When compared with the total number of insect species, there are relatively few species that cause economic loss. However, it is estimated that crop losses caused by such insects total more than $4 billion annually.

Insects can cause economic loss by feeding on forests, cultivated crops, and stored products (Figure 13–7). They can also vector plant and animal diseases, inflict painful stings or bites, or act as nuisance pests.

Insect Anatomy Insects are considered to be one of the most successful groups of animals present on Earth. Their success in numbers and species is attributed to several characteristics, including their anatomy, reproductive potential, and developmental diversity.

Insects are in the class Insecta and are characterized by the following similarities (Figure 13–8):

- Each insect has an exoskeleton, which is the body wall of the insect. It provides protection and support for the insect.

- The exoskeleton is divided into three regions: head, thorax, and abdomen.

- There are segmented appendages on the head called antennae, which act as sensory organs.

- Three pairs of legs are attached to the thorax of the body.

- Wings are present (one or two pairs) in the majority of species. This permits mobility and greater use of habitat.

Feeding Damage Insects have either chewing or sucking mouthparts. Damage symptoms caused by chewing insects are leaf defoliation, leaf mining, stem boring, and root feeding. Insects with sucking mouthparts produce distorted plant growth, leaf stippling, and leaf burn (Figure 13–9).

FIGURE 13–6 A perennial weed can live for more than two years and may reproduce by seed and/or vegetative growth. An example of a perennial weed is hoary cress, also known as white top.

INTERNET KEY WORDS:

cotton boll weevil
insects, beneficial, useful
insects, harmful, pests,
destructive

FIGURE 13–7 Some insects injure or kill trees by feeding on the leaves and needles. Such insects are called defoliators. *(Courtesy of Boise National Forest)*

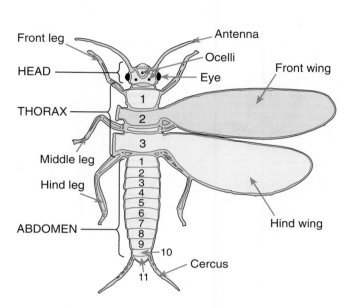

FIGURE 13–8 Diagram of an adult insect.

FIGURE 13–9 The different types of insect damage. *(Courtesy of Maryland Cooperative Extension Service)*

Science Profile

BENEFICIAL INSECTS

Scientists estimate that there are more than 1 million species of insects that inhabit the Earth. A majority of them are beneficial, or helpful, to humans. For example, insects are necessary for plant pollination. In the United States it is estimated that bees pollinate more than $1 billion worth of fruit, vegetable, and legume crops per year. Honey, beeswax, shellac, silk, and dyes are just a few of the commercial products produced by insects. Many insects are entomophagous and help in the natural control of their insect species. Entomophagous insects feed on other insects. Insects that inhabit the soil, act as scavengers, or feed on undesirable plants all play important roles. These insects increase soil tilth, contribute to nutrient recycling, and act as biological weed control agents. Insects are at the lower levels of the food chain. Thus, they support higher life forms, such as fish, birds, animals, and humans.

Honeybees' service to humans and animals in pollinating plants is a service we cannot do without.

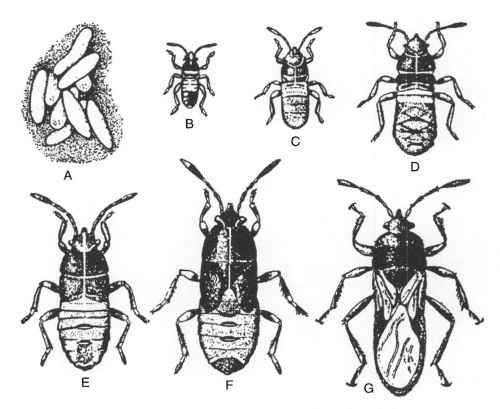

FIGURE 13–10 Gradual metamorphosis of the chinch bug (A) egg, (B–F) first to fifth instars, and (G) adult.

INTERNET KEY WORDS:

insect metamorphosis

INTERNET ADDRESS:

http://www.aphis.usda.gov/
ppq/weeds/

Development As an insect grows from an egg to an adult, it passes through several growth stages. This growth process is known as **metamorphosis.** The two types of metamorphosis are gradual and complete.

Gradual metamorphosis consists of three life stages: egg, nymph, and adult (Figure 13–10). As a nymph, the insect will grow and pass through several **instars** (the stage of the insect between molts). Each time the insect sheds its exoskeleton, or molts, it passes into the next instar. For example, chinch bugs have five instars before they reach adult form, but will vary in size, color, wing formation, and reproductive ability. When the insect reaches the adult stage, no further growth will occur.

Complete metamorphosis consists of four life stages: egg, larva, pupa, and adult (Figure 13–11). The larval stage is the period when the insect grows. As larvae molt, they pass to the next larval instar. A Japanese beetle will have three larval instars before developing to the pupa

Egg Young larva Full-grown larva Pupa Adult

FIGURE 13–11 Complete metamorphosis of the June beetle.

Agri-Profile

CAREER AREA:

Entomology/Plant Pathology

The work of entomologists and plant pathologists is never done. They do battle in the laboratory and in the field against insects and diseases that consume or ruin much of what we produce. The advice and service of these specialists are sought to control diseases and damaging insects, as well as to encourage beneficial ones.

Entomologists and plant pathologists attempt to control or reduce the buildup of damaging insect and disease populations. Such work may include assessing damage; attracting, trapping, counting, and observing insects; and advising, directing, and assisting those who attempt to control insects and plant diseases. The mysteries of some insects are so

Plant pathologists observe tree damaged by fire blight disease (foreground) and healthy, fire blight-resistant trees (background). *(Courtesy USDA/ARS #K-5310-1)*

great that scientists must specialize on just a few insects to be truly knowledgeable about them. Chemicals can no longer be our only means of controlling insects and diseases. Rather, we now use a variety of techniques collectively known as integrated pest management.

Career opportunities exist for field and laboratory technicians, as well as for degree-holding specialists. Neighborhood jobs may include termite control, scouting, spraying, crop dusting, inspecting, monitoring, selling, and managing field research projects. Honeybee specialists may manage hives to pollinate crops for improved seed and fruit production.

stage. The pupa is a resting period. It is also a transitional stage of dramatic morphological change from larva to adult.

Plant Diseases A **plant disease** is any abnormal plant growth. The occurrence and severity of plant disease is based on the following three factors:

1. A susceptible plant or host must be present.
2. The pathogenic organism, or causal agent, must be present. A **causal agent** is an organism that produces a disease.
3. Environmental conditions conducive to support of the causal agent must occur.

The relationship of these three factors is known as the **disease triangle** (Figure 13–12). Disease-control programs are designed to affect each or all of these three factors. For example, if crop irrigation is decreased, a less favorable environment will exist for the disease organism. Breeding programs have introduced disease resistance into new plant lines for many different crops. Pesticides may also be used to suppress and control the disease organism.

Causal Agents for Plant Disease

Diseases may be incited by either abiotic factors or biotic agents. **Abiotic (nonliving) diseases** are caused by environmental or man-made stress. Examples of abiotic diseases are nutrient deficiencies, salt damage, air pollution, chemical damage, and temperature and moisture extremes.

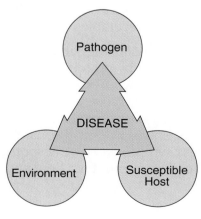

FIGURE 13–12 Components of the disease triangle.

FIGURE 13–13 Biotic diseases occur when living organisms such as fungi, bacteria, viruses, nematodes, and others cause damage to growing plants.

Biotic means living. **Biotic diseases** are caused by living organisms (Figure 13–13). Examples of causal agents or organisms are fungi, bacteria, viruses, nematodes, and parasitic seed plants. Organisms are parasites if they derive their nutrients from other living organisms. Examples and a discussion of causal agents for plant diseases follow.

Fungi   Fungi (plural for fungus) are the principal causes of plant disease. **Fungi** are plants that lack chlorophyll. Their bodies consist of threadlike vegetative structures known as **hyphae** (Figure 13–14). When hyphae are grouped together, they are called **mycelium.** Fungi can reproduce and cause disease by producing spores or mycelia. Spores can be produced asexually or sexually by the fungus. For example, a mushroom produces millions of sexual spores under its cap. These spores can be dispersed by wind, water, insects, and humans.

Bacteria   **Bacteria** are one-celled or unicellular microscopic plants. Relatively few bacteria are considered plant pathogens. Being unicellular, bacteria are among the smallest living organisms. Bacteria can enter a plant only through wounds or natural openings. Bacteria can be scattered in ways similar to fungi. Some important bacterial diseases are fire blight of apples and pears and bacterial soft rot of vegetables.

Viruses   Plant viruses are pathogenic, or disease-causing, organisms. **Viruses** are composed of nucleic acids surrounded by protein sheaths. They are capable of altering a plant's metabolism by affecting protein synthesis. Plant viruses are transmitted by seeds, insects, nematodes, fungi, grafting, and mechanical means, including sap contact. Viral diseases produce several well-known symptoms. A **symptom** is the visible change to the host caused by a disease. These symptoms are ring spots, stunting, malformations, and mosaics. A **mosaic** symptom is a light and dark green leaf pattern.

Nematodes   **Nematodes** are roundworms that may live in the soil or water, within insects, or as parasites of plants or animals. Plant parasitic nematodes are quite small, often less than a quarter inch (4 mm),

FIGURE 13–14 Powdery mildew of rose. (A) Conidia (a sexual spore), (B) Mycelium of fungus, (C) Fungal haustoria.

FIGURE 13–15 A nematode feeding on a plant cell. The stylet is a tiny needle-like feeding structure.

and produce damage to plants by feeding on stem or leaf tissue (Figure 13–15). The main symptom of nematode damage is poor plant growth, which results from nematodes feeding on the roots. The major plant parasitic nematodes are included in one of three groups: root-knot, stunt, or root-lesion.

INTEGRATED PEST MANAGEMENT

History

Integrated pest management (IPM) is a pest-control strategy that relies on multiple control practices. It establishes the amount of damage that will be tolerated before control actions are taken. The concept of integrated control is not new. Entomologists had developed an array of cultural and natural controls for the boll weevil and other insect pests by the early 1900s.

However, our approach to pest management during the period from 1940 to 1972 moved to a major reliance on chemical pesticides. Alternate control strategies were deemphasized, because chemical control gave excellent results at a low cost.

It was not until 1972 that a major change in policy occurred in the United States to encourage other pest-control strategies. Natural, biological, and cultural control programs began to be introduced as alternatives to chemical pest control.

The more recent trend toward reduced use of chemicals for pest control was triggered by a book published in 1962 entitled *Silent Spring,* by biologist Rachel Carson. After 1962, heavy reliance on chemical pest management began to be questioned. Carson's book created a public awareness of the environmental pollution that results from the overuse of pesticides. Adverse effects from pesticide misuse and/or overuse were beginning to occur as well. These effects included pest resurgence, resistance to pesticides, and concern over human health from exposure to pesticides.

Since the 1970s, great strides have been made in the development and implementation of IPM programs. The end result has been to reduce dependency on chemical use, while still achieving acceptable pest control.

Principles and Concepts of Integrated Pest Management

INTERNET KEY WORDS:

integrated pest management

The following concepts or principles are important in understanding how IPM programs should operate.

Key Pests A **key pest** is one that occurs on a regular basis for a given crop (Figure 13–16). It is important to be able to identify key pests and to know their biological characteristics (Figures 13–17 and 13–18). The weak link in each pest's biology must be found if management of the pest is to be successful.

Damage to crops:	weevils	aphids
	beetles	leafhoppers
	grasshoppers	crickets
	moths	locusts
	bugs	cockroaches
	nematodes	

PERCENTAGE OF TOTAL ARTHROPOD PESTS

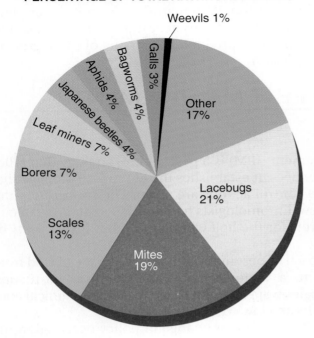

FIGURE 13–16 Arthropod pests and their percentages in six Maryland communities. *(Adapted from material from Entomology Department, University of Maryland)*

FIGURE 13–17 Grasshoppers, locusts, and crickets are capable of crop destruction to such an extent that famines have afflicted people throughout recorded history.

FIGURE 13–18 Vast forests are afflicted with severe insect damage throughout North America. Insects account for the death of many trees every year, and they contribute to the devastating fires that occur in unhealthy forests.

Damage to trees:	sawflies	cone beetles
	cone maggots	seed bugs
	cone worms	cone borers
	leafminers	needleminers
	moths	mites
	aphids	psyllids
	carpenter worms	termites
Damage to animals and people:	flies	mosquitoes
	ticks	keds
	lice	mites
	fleas	screwworms

Crop and Biology Ecosystem The integrated pest manager must learn the biology of the crop and its ecosystem. The ecosystem of the crop consists of the biotic and abiotic influences in the living environment of the crop. The biotic components of the ecosystem are the living organisms, such as plants and animals. The abiotic components are nonliving factors, such as soil and water. Examples of human-managed ecosystems are a field of soybeans, a turfgrass area, or a poultry production operation.

Ecosystem Manipulation With IPM, an attempt is made to understand the influence of ecosystem manipulation on reducing pest populations (Figure 13–19). To illustrate this concept, the manager must ask, "What would happen to the pest population equilibrium if a disease-resistant plant were introduced?" **Pest population equilibrium** occurs when the number of pests stabilizes or remains steady. The introduction of disease-resistant plants should decrease the pest population to less than the economic threshold level. The **economic threshold level** is the

FIGURE 13–19 The effect of lowering the equilibrium position of a pest. *(Adapted from material from University of Maryland, Entomology Department)*

point where pest damage is great enough to justify the cost of additional pest-control measures. Until the pest population increases to a high enough level that the cost of controlling the pest is less than the cost of the losses that the pest causes, no control actions will be taken.

Threshold Levels The level of a pest population is important. For instance, the mere presence of a pest may not warrant any control measures. But, at some point, the damage created by insects may be great enough to warrant control measures. Various threshold levels are developed to determine if and when a control measure should be implemented. This prevents excessive economic loss of plants to pest damage (Figure 13–20).

Economic Injury Threshold for Alfalfa Weevil; Number of Larvae From 30-Stem Sample

How to use table below:

1. Use plant height category that fits the field.

2. Estimate the value of crop in dollars per ton of hay equivalent and the cost to spray an acre.

3. From monitoring the field, find the number of alfalfa weevil larvae from a sample of 30 stems.

4. The number in each small box indicates the number of larvae per 30-stem sample that is required before a spray application would be profitable under these conditions.

EXAMPLE:

Plants in the field are 20 inches high (use Category II), hay is valued at $80 per ton, cost to spray is $8.00 per acre, and you collected 40 larvae from the sample of 30 stems. The number in the box common to $80 and $8 is 75. This means that under these conditions, 75 larvae are needed before a spray would be profitable. Since you collected only 40 larvae, a spray at this time will not be profitable.

Value of hay per ton	Category I plant height 12 to 18 inches						Category II plant height 18 to 24 inches						Category III plant height 24 to 30 inches					
$ 60	91	114	137	160	183	225	99	124	149	174	199	240	104	130	156	182	209	260
$ 80	68	85	102	119	136	171	75	94	113	131	150	186	78	97	117	137	157	195
$100	54	68	81	95	108	137	62	75	90	105	120	149	63	78	94	110	126	156
$120	45	57	68	79	91	114	50	62	75	87	100	124	52	65	78	91	105	130
$140	39	49	59	68	77	99	43	54	64	75	86	107	45	56	67	78	90	112
$160	34	43	51	60	68	86	37	47	56	65	75	93	39	49	58	68	79	98
	$8	10	12	14	16	20	$8	10	12	14	16	20	$8	10	12	14	16	20

Cost of insecticide application per acre

FIGURE 13–20 Chart to determine economic threshold level for the alfalfa weevil. *(Adapted from material from Pennsylvania State University)*

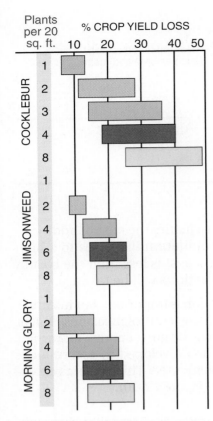

FIGURE 13–21 The pest-damage index for several weeds in soy beans.

INTERNET KEY WORDS:

pests, biological control

Economic threshold levels have been determined by first developing a pest-damage index (Figure 13–21). It is crucial in the decision-making process to know the level of pest infestation that will cause a given yield reduction. Pest populations are measured in several different ways. They can be counted in number of pests per plant or plant part, number of pests per crop row, or number of pests per sweep with a net above the crop.

Monitoring For IPM to be successful, a **monitoring** (checking) or scouting procedure must be performed. Different sampling procedures have been developed for various crops and pest problems (Figure 13–22). The presence or absence of the pest, amount of damage, and stage of development of the pest are several visual estimates a scout must make. The method used must be speedy and accurate.

Scouts are people who monitor fields to determine pest activity. They must be well trained in entomology, pathology, agronomy, and horticulture.

PEST-CONTROL STRATEGIES

Pest-control programs can be grouped into several broad categories. These include regulatory, biological, cultural, physical and mechanical, and chemical.

Regulatory Control

Federal and state governments have created laws that prevent the entry or spread of known pests into uninfested areas. Regulatory agencies also attempt to contain or eradicate certain types of pest infestations. The Plant Quarantine Act of 1912 provides for inspection at ports of entry. Plant or animal quarantines are implemented if shipments are infested with targeted pests. A **quarantine** is the isolation of pest-infested material. A **targeted pest** is a pest that, if introduced, poses a major economic threat.

If a targeted pest becomes established, an eradication program will be started. **Eradication** means total removal or destruction of a pest. This type of pest control is extremely difficult and expensive to administer. In California, the Mediterranean fruit fly was eradicated at a cost of $100 million in 1982. It has recurred since then, and the cost is high each time it is eradicated. This program relies on chemical spraying, sanitation, sterile male releases, and pheromone traps to ensure complete eradication. A **pheromone** is a chemical secreted by an organism to cause a specific reaction by another organism of the same species.

Hot Topics in Agriscience

INSECT CONTROL USING STERILE MALE INSECTS

Imagine a laboratory that raises millions of insects that are harmful to agricultural crops. Once the insects are mature, they are irradiated. This eliminates their ability to reproduce. The treated insects are released in huge numbers into areas where the same species of insect pest is found. Because the treated insects vastly outnumber the untreated insects, the odds of treated insects mating with untreated insects is high, but no offspring are produced from these matings. By repeating the releases of sterile insects over a 2- or 3-year period, the original population of harmful insects may be completely eradicated.

FIGURE 13–22 Random sampling of a plant stem and leaf to determine pest populations and damage. *(Courtesy of Maryland Cooperative Extension Service)*

INTERNET KEY WORDS:

insects, irradiation

Host Resistance

The development of plants having pest resistance is an extremely effective control practice. The advantages of resistant varieties are as follows:

- low cost
- no adverse effect to the environment
- a significant reduction in pest damage
- ability to fit into any IPM program

Breeding programs attempt to identify and select plants with pest resistance. Currently, new plant cultivars with improved resistance to pests are released annually. A **cultivar** is a plant developed by humans, as distinguished from a natural variety.

Biological Control

Biological control means control by natural agents. Such agents may be predators, parasites, and pathogens. A predator is an animal that feeds on a smaller or weaker organism. An example of a predator is the lady beetle. Aphids are the lady beetle's principal prey. Parasites are organisms that live in or on another organism. The Braconid wasp is parasitic on the caterpillars of many moths and butterflies. Pathogens are organisms that will produce disease within their hosts. For example, the bacterium *Bacillus popilliae* is a pathogen, because it causes the milky spore disease in Japanese beetle grubs.

Successful biological control programs reduce pest populations to less than economic thresholds and keep the pests in check. Such programs require a thorough understanding of the biology and ecology of the beneficial organism, as well as of the pest. Careful research can even match desirable plant pathogens against undesirable weeds (Figure 13–23).

Cultural Control

Cultural control is the attempt to alter the crop environment to prevent or reduce pest damage. It may include such agricultural practices as soil tillage, crop rotation, adjustment of harvest or planting dates, and irrigation

FIGURE 13–23 Plant pathologist Rick Bennent examines fungi that may be used for biological control of weeds. *(Courtesy of USDA/ARS#K-4652-1)*

schemes. Other practices that are considered cultural control are clean culture and trap crops. Clean culture refers to any practice that removes breeding or overwintering sites of a pest. This may include removal of crop leaves and stems, destruction of alternate hosts, or pruning of infested parts. A trap crop is a susceptible crop planted to attract a pest to a localized area. The trap crop is then either destroyed or treated with a pesticide.

Bio-Tech Connection

BIO-ENGINEERING A SAFER MOSQUITO!

The common mosquito is a scourge from the Alaskan tundra to the equatorial marshes. The biological need of the female mosquito to obtain a blood meal to provide protein before laying her eggs causes the mosquito to be an annoyance and a threat to humans and animals alike. The small amount of blood extracted during a "bite" is not the problem. First, there is discomfort of the bite. However, the biggest problem lies in the life-threatening yellow fever or malaria-causing microorganisms that certain mosquitos transfer to humans during their blood meals. The search for methods to rid the environment of mosquitos is continuous.

Is there any such thing as a "good" mosquito? Yes, if the mosquito is a nonmalaria mosquito that replaces a malaria carrier. Geneticist Andrew Cockburn has developed a genetic engineering technique that may be useful in creating a malaria-resistant mosquito. Such mosquitos could theoretically thin out or replace their protozoan-carrying cousins and reduce or eliminate the malaria disease in humans.

A promising technique is the use of tiny needles, new genes, and insect eggs spun in an ordinary laboratory vortex in saline solution. The needles gently pierce the eggs and allow new material to enter. Cockburn has transferred either a test gene or dye into eggs of house flies, fruit flies, and stable flies. So far, the mosquito eggs have proven to be too tough for penetration. However, Cockburn believes he eventually will find a combination of needle number and vortex speed that will permit him to slip material into mosquito eggs.

When the technique for egg penetration is worked out, perhaps a gene for malaria resistance will be found and a malaria-resistant population of mosquitos grown. The mosquito population in a given area could then be killed off using modern insecticides and the malaria-resistant mosquitos released to repopulate the area. You would still have biting mosquitos, but you would not get malaria!

Scientists have tried for decades to control the dreaded mosquito. (A) Mosquito in blood-sucking position *(Courtesy USDA/ARS #K-4158-7)*. (B) An arm treated with a modern mosquito repellent *(Courtesy of USDA/ARS #K-4158-19)*

Physical and Mechanical Control

Physical and mechanical control programs use direct measures to destroy pests. Examples of such practices are insect-proof containers, steam sterilization, hand removal, cold storage, and light traps. Implementation of these control practices is costly and provides varying pest-control results.

Chemical Control

INTERNET KEY WORDS:

pesticide resistance

Chemical control is the use of pesticides to reduce pest populations. Chemical-control programs have been very cost effective. However, various problems occur if this practice is misused or overused. Problems that can develop are environmental pollution, pesticide resistance, and pest resurgence. **Pesticide resistance** is the ability of an organism to tolerate a lethal level of a pesticide. **Pest resurgence** refers to a pest's ability to repopulate after control measures have been eliminated or reduced.

Integrated pest management seems to be our best defense against pests. Biological and cultural controls are favored when they are effective. However, we cannot control certain pests without the use of chemical pesticides. Under such circumstances, it is important to use chemical pesticides safely.

STUDENT ACTIVITIES

1. Write the Terms to Know and their meanings in your notebook.
2. Use reference materials to find and list 10 examples of beneficial insects and 10 examples of insect pests.
3. Name five insects that are beneficial some of the time but are pests at other times.
4. Make an insect collection, with the insects properly identified and named.
5. Make a drawing of an insect and label the various body parts, appendages, and mouthparts.
6. Research a major crop in your community and discuss the key pests and measures recommended to control those key pests.
7. Develop a collage showing pests of plants in your community.
8. Make a weed collection and identify each sample. Your instructor may have a weed identification key for this purpose.
9. Using the Internet, library, magazines, or other sources, prepare a 2-minute class presentation on a local pest. Be sure to include the type of harm the pest is responsible for causing and the integrated pest management techniques used to control it.
10. Make an outline of the unit. Phrases and words that are bold should be included, together with a brief description of key concepts.

SELF EVALUATION

A. Multiple Choice

1. A biennial weed will live for
   - a. 1 year.
   - b. 2 years.
   - c. 3 years.
   - d. more than 3 years.

2. The major causal agent of plant disease is
   - a. nematodes.
   - b. bacteria.
   - c. viruses.
   - d. fungi.

3. The number of insect species in the world is estimated to be
   - a. 100,000.
   - b. 500,000.
   - c. 1 million.
   - d. none of the above.

4. The term *instar* refers to the development stage of
   - a. plants.
   - b. fungi.
   - c. bacteria.
   - d. insects.

5. Plant diseases are vectored by
   - a. wind.
   - b. rain.
   - c. insects.
   - d. none of the above.

6. A nematode is a type of
   - a. fungus.
   - b. roundworm.
   - c. annual plant.
   - d. insect.

7. A type of regulatory control is
   - a. plant quarantine.
   - b. sanitation.
   - c. crop rotation.
   - d. soil tillage.

8. The control practice that relies on the introduction of parasites and predators is
   - a. cultural.
   - b. chemical.
   - c. biological.
   - d. host resistance.

9. A threshold level is also known as the
   - a. pesticide residues.
   - b. control program.
   - c. degree of pest control.
   - d. pest concentration.

10. A pesticide used to control diseases is a/an
    - a. fungicide.
    - b. nematacide.
    - c. insecticide.
    - d. acaricide.

B. Matching

_____	**1.**	Biotic
_____	**2.**	Pest
_____	**3.**	Summer annual
_____	**4.**	Acaricide
_____	**5.**	Instar
_____	**6.**	Appressorium
_____	**7.**	Abiotic
_____	**8.**	Entomophagous
_____	**9.**	Symptom
_____	**10.**	IPM

a. The insect stage between molts

b. A chemical used to control mites and spiders

c. Integrated pest management

d. A visible change to the host caused by pests

e. Adversely affects human activities

f. Diseases caused by living organisms

g. A plant that germinates in the summer and lives for only 1 year

h. Structure by which a fungus attaches itself to a plant

i. Nonliving factors

j. Insects on which other insects feed

C. Completion

1. Soil tillage is an example of _____ control.

2. An _____ will control weeds.

3. A _____ lives in or on another organism.

4. Insects have _____ pairs of legs.

5. Complete metamorphosis consists of the following stages: egg, _____, _____, and adult.

6. _____ resistance is when a plant has developed its own defensive response to pests.

7. _____ control will use quarantine practices.

8. Monitoring is essential for _____ programs to work successfully.

9. _____ pests occur on a regular basis for a given crop.

10. Causal agents of disease are _____, _____, _____, and _____.

Safe Use of Pesticides

OBJECTIVE

To determine the nature of chemicals used to control pests, to know important terms regarding chemical safety, and to practice the safe use of pesticides.

MATERIALS LIST

✓ writing materials

✓ encyclopedias

✓ assorted pesticide labels

✓ Internet access

COMPETENCIES TO BE DEVELOPED

After studying this unit, you should be able to:

- describe the previous and current trends of pesticide use in the United States.
- recognize some popular classes of chemicals used for pest management and their roles in pest control.
- read and interpret information on pesticide labels.
- state the components of protective clothing for individuals handling pesticides.
- describe the environmental and health concerns relating to pesticide use.

SUGGESTED CLASS ACTIVITIES

1. Divide the class into as many work teams as you have computer work stations with access to the Internet. Assign the students to search for Web sites that (1) identify new pesticides, and (2) describe licensing requirements of different states for pesticide consultants and applicators. Have the teams report their findings to the class.

2. Conduct a class debate on pesticide safety. Have students work in teams of two on an assignment to debate food safety. Each team will debate in both the affirmative and the negative positions on the issue of using pesticides on foods. Identify a champion debate team, and give the students appropriate recognition.

3. Invite a local organic farmer to come and discuss the advantages and disadvantages of organic farming. Invite the guest speaker to discuss what alternate methods he or she uses to keep yields high and pest infestations at bay.

TERMS TO KNOW

element	phloem tissue	general-use pesticide	LC_{50}
compound	preemergence herbicide	restricted-use pesticide	misuse statement
inorganic compound	postemergence herbicide	trade name	toxicity
organic compound	mode of action	formulation	acute toxicity
selective herbicide	photodecomposition	common name	chronic toxicity
nonselective herbicide	volatilization	net contents	carcinogen
contact herbicide	insecticide	signal word	drift
systemic herbicide	protectant fungicide	symbol	vapor drift
xylem tissue	eradicant fungicide	LD_{50}	

INTERNET KEY WORDS:

inorganic pesticide
organic pesticide

The development and use of pesticides has provided many benefits for both the producer of agricultural commodities and the consumer. However, there are also risks associated with pesticide use. These risks are reduced with proper pesticide application, storage, and disposal. Improper use of pesticides will increase the risk of environmental contamination and any adverse effects on human health.

Pesticide use currently is a controversial issue in the United States. It is important to objectively balance the benefits and the risks associated with pesticide use. The Environmental Protection Agency (EPA) conducts benefit-risk assessments on each pesticide that is registered. When a pesticide is registered for use and is applied according to label directions, the benefits greatly outweigh the risks.

Some benefits of pesticide use are summarized as follows:

- increased yields of food and fiber;
- reduced loss of stored products;
- increased crop quality;
- economic stability;
- better health; and
- environmental protection and conservation.

The increase in crop and animal yields and improved quality of crops and livestock products resulting from the use of pesticides have been adequately documented (Figure 14–1). It has been estimated that the average total income spent on food in the United States would increase from the current 11 percent to 30 percent without the protection offered by pesticides. Benefits to human health are best illustrated where insecticides are used to control insects that carry and spread malaria and typhus diseases.

Minimum-tillage or no-tillage practices have reduced soil erosion. These practices would not be possible without the use of pesticides. Many other substantial benefits to the quality of life resulting from the proper use of pesticides could be cited.

HISTORY OF PESTICIDE USE

The use of chemicals to control pests is not new. Elements such as sulfur and arsenic were among the first chemicals used for this purpose. An **element** is a uniform substance that cannot be further decomposed by

Crop	Estimated Percent Increase	Major Pests Controlled
Corn	25	Weeds, rootworms, corn borers, seedling blights
Cotton	100	Pink bollworms, boll weevils, nematodes, boll rots
Alfalfa seed	160	Weeds, alfalfa weevils
Potatoes	35	Tuber rots, blackleg, soft rots, blights
Onions	140	Botrytis blights, neck rot, smut, onion maggots

FIGURE 14–1 Yield increases from pesticide applications. *(Adapted from material from "The New Pesticide User's Guide")*

Pesticide Type	Targeted Pests
acaricide	mites, ticks
algaecide	algae
attractant	insects, birds, other vertebrates
avicide	birds
bactericide	bacteria
defoliant	unwanted plant leaves
desiccant	unwanted plant tops
fungicide	fungi
growth regulator	insect and plant growth
herbicide	weeds
insecticide	insects
miticide	mites
molluscicide	snails, slugs
nematicide	nematodes
piscicide	fish
predacide	vertebrates
repellents	insects, birds, other vertebrates
rodenticide	rodents
silvicide	trees and woody vegetation
slimicide	slime molds
sterilants	insects, vertebrates

FIGURE 14–2 Classification of pesticides based on the target pests. *(Adapted from material from "The New Pesticide User's Guide")*

ordinary chemical means. Homer, in about 1000 BC, wrote that sulfur had pest-control ability. The Chinese, in about AD 900, discovered the insecticidal properties of arsenic sulfide, a chemical **compound** or substance that is composed of more than one element.

Until the late 1930s, pest-control chemicals or pesticides were mainly limited to **inorganic compounds** (any compounds that do not contain carbon). Examples of other inorganic pesticides are mercury and Bordeaux mixture. Bordeaux mixture is a combination of copper sulfate and lime. It is used for plant disease control.

A majority of currently used pesticides are synthetically produced **organic compounds** (compounds that contain carbon). The organic chemistry involved in pesticide production is often complex and extremely diverse. However, a classification system for pesticides that is based on the type of pest being controlled is useful (Figure 14–2). The major pesticide groups are herbicides, insecticides, and fungicides.

In the United States, pesticide use totaled 500,000 tons of 600 different pesticides at a cost of $7.6 billion in 1999. Currently, EPA has registered more than 600 chemicals that are formulated into some 30,000 products for pest control. Of the three major pesticide categories, the largest volume was for herbicides, followed by insecticides and fungicides.

INTERNET KEY WORDS:

insecticide
fungicide
pesticide use

HERBICIDES

Herbicides can be grouped into several major categories based on application method, type of control, and chemical structure. The terminology for herbicide use, type of control, and chemical family follows.

Hot Topics in Agriscience

REDUCING NATURAL TOXINS USING PESTICIDES

One of the natural phenomena that occurs among all living things is the production of toxic substances. All of these substances are naturally formed chemicals. Dr. William Robertson is Medical Director of the Washington D.C. Poison Center. He suggests that diseased or stressed food plants are often infected with natural toxins: "Just like medicines can prevent harmful diseases from destroying the human body, appropriate pesticides can control or eliminate naturally occurring toxins that can contaminate our food supply and sometimes cause harm to humans. These toxins are produced by fungi, such as aflatoxins, by bacteria, in the case of shiga toxin, or by other microorganisms in our environment." He also suggests that the pesticides we use today pose much less serious risks than the arsenic, thallium, and strychnine that were used in the early 1900s.

Selective Herbicides

A **selective herbicide** kills or affects a certain type or group of plants. The selectivity of a herbicide can be caused by many different factors. Some of these factors include:

- differences in herbicide chemistry, formulation, and concentration;
- differences in plant age, morphology, growth rate, and plant physiology; and
- environmental differences such as temperature, rainfall, and soil type.

Nonselective Herbicides

INTERNET KEY WORDS:

selective herbicide, nonselective herbicide

A **nonselective herbicide** controls or kills all plants. These herbicides are used for many different purposes. Examples of their use are for railroad rights-of-way, industrial areas, fence rows, irrigation and drainage ditch banks, and renovation programs.

Contact Herbicides

Contact herbicides will not move or translocate within the plant. They only affect that part of the plant with which they come in contact. In addition, they are often used for controlling annual weeds.

Systemic Herbicides

A **systemic herbicide** is absorbed by the plant and is then translocated in either xylem or phloem tissue to other parts of the plant. **Xylem tissue** is where water and minerals are transported within the plant. **Phloem tissue** is responsible for transporting carbohydrates within the plant.

Preemergence and Postemergence Herbicides

A **preemergence herbicide** is applied before weed or crop seed germination. A **postemergence herbicide** is applied after the weed or crop is growing.

Chemical Families of Herbicides

Herbicides are chemicals that are used to control weeds. These compounds can affect plant growth in many different ways. Currently, more than 23 chemical families of herbicides have been developed. Each one

Bio-Tech Connection

PEST RESISTANCE USING GENETICALLY ENGINEERED PLANTS

Scientists have learned to identify the genes that make some plants resistant to a pest. Once such a gene is located on the chromosome of a pest-resistant plant, it can be isolated and transferred to other plants. This process is used to develop new plant varieties that are resistant to a particular kind of insect. For example, new potato varieties have been developed that have a genetically engineered resistance to the Colorado potato beetle. This beetle is capable of destroying entire fields of potatoes by eating all of the leaves and part of the stems of potato plants. It is a devastating pest to potato plants. The beetle-resistant potato plants have a new gene that produces a natural pesticide in the juices of the plant. Potato beetles that eat the foliage of these plants are poisoned without causing risk to the human population who eats the potato tubers.

has a unique chemical structure and a different site or mode of action. **Mode of action** is a term used to describe the way in which herbicides adversely affect plant growth. Several of the more important chemical herbicide families and their characteristics are described in this section.

Acetanilides   Acetanilides interfere with cell division and protein synthesis. They are applied as either preemergence or preplant applications for control of annual grasses and some annual broadleaf weeds. Two popular herbicides within this category are Lasso and Dual. They are used extensively for weed control in corn and soybeans.

Dinitroanilines   Dinitroanilines act on root tissue, preventing root development in seedling plants. They are preemergence herbicides applied to prevent weed germination and should be incorporated into the soil. The dinitroanilines are deactivated quickly by **photodecomposition** (chemical breakdown caused by exposure to light), **volatilization** (changing to gases), and chemical processes. Examples of these herbicides are Balan and Prowl. They are used for control of annual grasses and broadleaf weeds in many different crops.

Phenoxys   Phenoxy herbicides affect plants by causing overstimulation of growth. They perform best when applied as postemergence foliar sprays. They are selective herbicides that affect broadleaf weeds in grass crops. The herbicide 2, 4-D was first used in 1942, and is still widely used.

Triazines   Triazines are photosynthetic inhibitors that interfere with the process of photosynthesis. They are preemergence herbicides used to control both annual and broadleaf weeds in grass crops. Aatrex and Princep are two examples of these herbicides. They are primarily used in corn weed–control programs.

INTERNET KEY WORDS:

acetanilides;
Lasso, Dual

dinitroanilines;
Balan, Prowl

phenoxys;
2, 4-D

triazines;
Aatrex, Princep

Agri-Profile

CAREER AREAS:

Pesticide Applicator/Pesticide Specialist/Chemist

Tens of thousands of chemical formulations have been developed in recent years to control insects, diseases, weeds, nematodes, rodents, and other pests. We must use chemicals to help control pests. However, these materials are likely to be hazardous to the operator, plants, animals, or the environment if not properly used.

Pesticide applicators are used by farmers and ranchers, lawn service companies, farm and garden supply firms, termite control companies, highway departments, and railroads, as well as self-employed persons. Special training and licensing is required for handling most pesticides.

State agencies such as state departments of agriculture are responsible for developing the regulations for pesticide consultants and applicators. In most states, the people who apply for licenses must attend classes and seminars to learn how to handle pesticides properly. The training usually includes pesticide storage, mixing, application, and disposal procedures. Those who attend these classes and seminars may be required to show evidence of attendance before they are allowed to take the licensing tests that are given. Many states use these tests to establish that the candidates are qualified to prescribe and use pesticides.

Pesticide applicators may work in homes, buildings, fields, forests, and even the holds of ships. They may dust, spray, bait, or fumigate, depending on the setting and the pest. Pesticide applicators must always use protective clothing and safety devices to protect against accidental poisoning. Their tools may be as simple as aerosol cans or as complex as orchard spray rigs or specially equipped helicopters.

Airplane crews are important members of the teams that apply chemical and biological pesticides to large crop, range, and forest areas. *(Courtesy USDA/ARS #K-3663-15)*

Many companies throughout the world are continually researching new and existing chemicals to determine what characteristics they may exhibit, such as pesticide qualities, toxicity levels, persistence in the environment, risk to crops and animal life, and effectiveness in controlling pests. A tremendous amount of research is required to develop and gain government approval for a single pesticide. The cost is high, but the reward for developing an effective pesticide product is also high. A single product that is safe and effective has the potential to earn huge profits for the manufacturer. For this reason, chemical companies are willing to pay for the research to develop and gain government approval for a new product.

INSECTICIDES

Insecticides are chemicals used to control insects. They can affect the insect in many different ways. The classification system of insecticides may be based on chemical structure and/or mode of action, as shown in Figure 14–3.

The botanical, inorganic, and oil insecticides are some of the original chemicals used for insect control. Sulfur, for example, may be used to control mites. Its effectiveness as an insecticide was discovered thousands of years ago.

Chemical Group	Mode of Action	Common/Trade Names
Inorganics	Protoplasmic Poisons Physical Poisons	Sulfur Silica Aerogel
Oils	Physical Poisons	Superior Oils
Botanicals	Metabolic Inhibitors	Pyrethrum Rotenone
Synthetic Organics Chlorinated Hydrocarbons	Nerve Poison	Lindane Endosulfan-Thiodan
Organophosphates	Nerve Poison	Diazinon-Spectracide Parathion-Thiophos
Carbamates	Nerve Poison	Carbofuran-Furadan Carbaryl-Sevin
Pyrethroids	Nerve Poison	Permethrin-Ambush Fluvalinate-Mavrik
Insect Growth Regulators (IGRs)	Alter Insect Growth	Methoprene-Altosid Kinoprene-Enstar
Biorational/Microbial Insecticides	Wide Range of Activity	*Bacillus thuringiensis*

FIGURE 14–3 Insecticide classification system.

Rotenone is a chemical present in the roots of certain species of legume plants. It affects insects by inhibiting respiratory metabolism and nerve transmission. Rotenone, first used in 1848, is a botanical insecticide still used today. It is a contact and stomach-poison insecticide. Rotenone is principally used in vegetables for controlling fleas, beetles, loopers, Japanese beetles, and many other insects (Figure 14–4).

Superior oils are highly refined oils and are applied to ornamentals and citrus crops to control scale insects, mites, and other soft-bodied insects. Oils act by excluding oxygen from the insect, thus causing suffocation. It is also believed that oils may destroy cell membrane function. Oils were first used in the early 1900s to control San Jose scale in fruit orchards (Figure 14–5).

The synthetic organic group of insecticides was principally developed after 1940. This group includes the chlorinated hydrocarbons, organophos-

FIGURE 14–4 Rotenone is a chemical insecticide that is used to control beetles and other insects.

FIGURE 14–5 Superior oils are insecticides that are used to control insects such as mites, scale, and soft-bodied insects by blocking their supply of oxygen.

phates, carbamates, pyrethroids, insect growth regulators, and other minor classes. A majority of these insecticides adversely affect nerve transmission. The chlorinated hydrocarbons were the first to be synthesized (man made). Released in 1940, DDT (dicloro-diphenyl-trichloroethane) was the first and most popular chlorinated hydrocarbon. This chemical had excellent insecticidal properties and was used extensively. However, environmental and health problems were linked to DDT and other insecticides within this group. Because of these risks, many of the chlorinated hydrocarbons (such as DDT, aldrin, dieldrin, and chlordane) are banned from use in the United States.

The organophosphate and carbamate insecticides are the principal insecticides currently used to control insects. Approximately 60 percent of all insecticides produced in the United States are from these two groups. They control insects by affecting the nervous system. The potential for pesticide poisoning of people and livestock is high for these insecticides. They will affect the nervous systems of humans and animals in a manner similar to the way they affect the insects they are designed to kill. The dose, or the amount, of insecticide applied is the discriminating factor with respect to insect control or human poisoning.

FUNGICIDES

INTERNET KEY WORDS:

fungicide

Fungicides are chemicals used to control plant diseases caused by fungi. Chemical control of plant diseases is more difficult than it is for weed and insect control. Fungi are plants without chlorophyll. They are parasites of other plants. The fungicide must be selective enough to control the fungus but not adversely affect the host. Also, fungi have many generations each growing season. Therefore, reapplication of the fungicide is required to provide effective control.

Protectant Fungicide A **protectant fungicide** is applied before disease infection. This will provide a chemical barrier between the host and the germinating spores. However, this barrier will be broken down by environmental weathering of the fungicide. Rainfall, sunlight, temperature, and plant growth are the major causes of this breakdown (Figure 14–6). The fungicide will have to be reapplied if adequate disease control is expected.

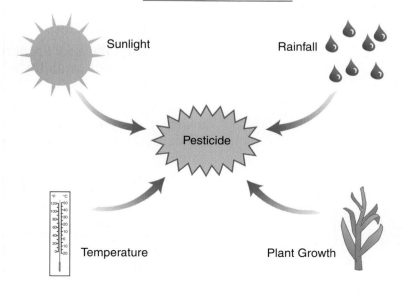

FIGURE 14–6 Pesticides are broken down by environmental weathering, which is caused by sunlight, rainfall, temperature, and plant growth.

Chemical Family	Fungicide Activity	Common/Trade Name
Benzimidazoles	Eradicant	Benomyl-Tersan 1991
Dicarboximides	Protectant	Captan-Captane
Dithiocarbamates	Protectant	Mancozeb-Dithane M45
Oxathiins	Eradicant	Carboxin-Vitavax

FIGURE 14-7 Fungicide classification system.

Eradicant Fungicide An **eradicant fungicide** can be applied after disease infection has occurred. These fungicides act systemically and are translocated by the plant to the site of infection. They offer a longer control period than do protectant fungicides, because they are not so prone to environmental weathering.

Chemical Structure

Fungicides, like insecticides and herbicides, can also be classified into different chemical families. Figure 14–7 lists some of the major groups of fungicides.

Inorganic Fungicides The elements sulfur, copper, mercury, and cadmium, or mixtures of them, are some of the oldest pesticides used by humans. They protect many ornamentals and turfgrasses from diseases and are examples of inorganic pesticides.

Organic Fungicides Organic fungicides are a newer group of fungicides. These are used both as protectants and as eradicants.

PESTICIDE LABELS

The information on a pesticide container label instructs the user on the correct procedures for application, storage, and disposal of the pesticide. The pesticide label is a summary of information gathered by the pesticide manufacturer and is required for product registration. The registration process is estimated to take 8 to 10 years and costs up to $20 million per pesticide. Several of the studies required to meet federal standards are summarized as follows:

- Chemical and Physical Properties: water solubility, volatility, movement in soils, stability to heat and light, and other factors affecting environmental stability

- Toxicology studies: determine acute oral, dermal, and inhalation toxicity to various animals; evaluate chronic toxicity, including any effect on reproduction or the tendency of the pesticide to be a carcinogen

- Residue analysis: determine the amount of pesticide residues at the time of harvest; develop safe tolerance levels for any pesticide residue

- Metabolism studies: determine application exposure, determine consumer exposure to pesticide residues, and establish safety practices that minimize exposure

The label is a legal document indicating proper and safe use of the product. Pesticide use that differs from that specified on the label is a misuse. Such use is illegal, and the offender can be charged with civil or criminal penalties. When improperly applied, a pesticide can pose danger to the applicator, the environment, animals, and other people. Therefore, it is important to read, understand, and follow the information on the pesticide label.

The pesticide label and other labeling information must meet federal standards. The label consists of a front panel and a back panel on the product (Figures 14–8 and 14–9). If there is insufficient room on the

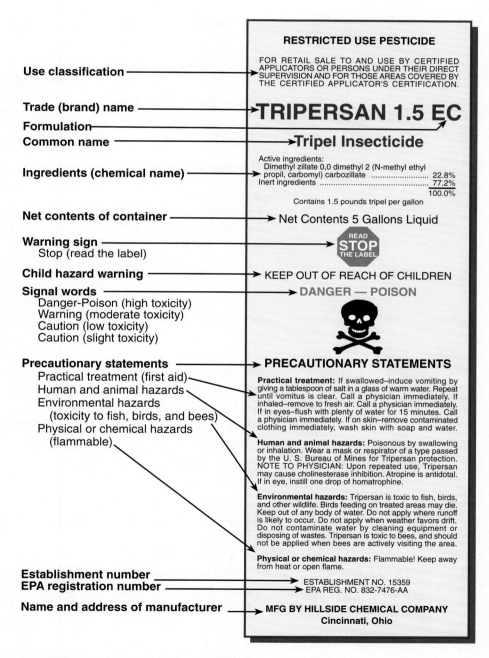

FIGURE 14–8 Front panel of a sample pesticide label. *(Adapted from material from Vo-Ag Services, University of Illinois)*

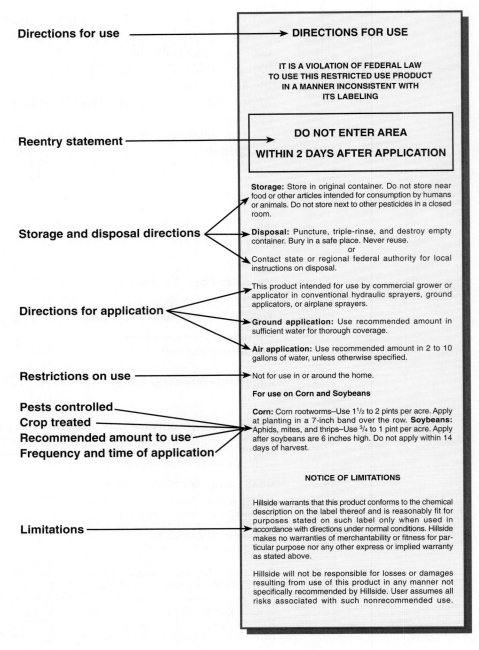

FIGURE 14–9 Back panel of a sample pesticide label. *(Adapted from material from Vo-Ag Services, University of Illinois)*

panels, additional labeling information will be attached, in booklet form, to the product. An outline of a label follows.

Use Classification

Pesticides are classified as either general use or restricted use. A **general-use pesticide** poses minimal risk when applied according to label directions.

Restricted-use pesticides pose a greater risk to humans and the environment. Therefore, anyone applying a restricted-use pesticide must be properly trained and certified. Applicator certification is administered by each state. An applicator must meet a minimum set of standards and is often evaluated by tests to be certified.

INTERNET ADDRESS:

http://ianrpubs.unl.edu/
pesticides/ec2500.pdf

Trade Name

A trade name is the manufacturer's name for its product. It appears on the label as the most conspicuous item. The manufacturer will use the name in all of its promotional campaigns. The same chemical may have several different trade names, depending on the type of formulation and patent rights.

Formulation

Formulation refers to the physical properties of the pesticide. The pesticide chemical or active ingredient will often have to be modified to allow for field use. These modifications may include adding inert ingredients, such as solvents, wetting agents, powders, or granules, to the pesticide. This will result in different formulations. Some examples of the different types of formulations and their abbreviated label names are as follows:

Granules—G

Solutions—S

Wettable Powders—WP or W

Soluble Powders—SP

Dry Flowables—DF

Emulsifiable Concentrates—EC or E

Dusts—D

Common Name

The common name is given to a pesticide by a recognized authority on pesticide nomenclature. A pesticide is identified by a trade name, common name, and chemical name. It may have several trade names but will have only one common and one chemical name. The common name, or generic name, identifies the active pesticide ingredient and can be used for comparison shopping.

Ingredients

The percentages by weight of both the active and inert ingredients are stated on the label. The active ingredient is identified by its chemical name. This is the name of the chemical structure of that pesticide. The inert, or inactive, ingredients do not have to be listed by their chemical names. The label must state only the total percentage of inert material.

Net Contents

The label will give the amount of product in the container. This is referred to as net contents. This quantity can be expressed in gallons, quarts, pints, and pounds.

Signal Words and Symbols

The signal words and symbols describe acute toxicity of the pesticide. The different categories are based on LD_{50} and LC_{50} values and on skin and eye irritation. The signal words used on pesticide labels are (1) danger—poison, (2) warning, and (3) caution. These words are used to alert the person handling or using the pesticide to the poisoning effect of contact with the chemical.

Pesticides with high toxicity—only a few drops to 1 teaspoon will kill a 150-lb person—are labeled "DANGER—POISON." Pesticides with moderate toxicity—1 to 2 tablespoons will kill a 150-lb person—

INTERNET ADDRESS:

http://ohioline.ag.ohio-state.edu/b512/b512_4.html

Bio-Tech Connection

SUBSTITUTING BIOLOGICAL FOR CHEMICAL PESTICIDES

The sound of a roaring cyclone sprayer or the sight of an aircraft fogging a field or treetops creates an image of chemical pesticide application in the minds of most people. However, more and more such sprays are the new, user-friendly, environmentally safe, biological-control pesticides. These materials generally contain bacteria, fungi, viruses, or other microbes that attack the targeted host but are harmless to other living organisms and the environment.

The newer, safer pesticides increase the chances of eradicating or permanently subduing some of our most troubling and costly pests. For instance, the efforts made to control cotton bollworms and tobacco budworms on the Mississippi Delta cost an estimated $50 million per year using chemical pesticides. Marion Bell, an Agriculture Research Service entomologist at Stoneville, Mississippi believes he has a better way—the *Heliothis* nuclear polyhedrosis virus. *Heliothis* has been found to be very specific to the bollworm and tobacco budworm, because it works only on insect larvae that have an alkaline midgut.

Bell specializes in microbial control of insects. He and other appropriate authorities have determined that treating the entire 4.7 million acres of the Mississippi Delta with *Heliothis* would cost about $7 million. However, the team is aware of the natural fears of the populace with such a widespread, general spraying program. Farmers are concerned about the effect of any general spraying program on their crops, aquaculturalists are concerned that such material could contaminate their catfish ponds, and the general public is concerned about any spray that may be damaging to the environment.

Therefore, they opted for an extensive educational program coupled with a spray program on small areas until the system is proven and more widely accepted by the public. The educational program included informational releases and face-to-face contacts with farmers, congressional representatives, extension personnel, environmental and health officials, physicians, civic clubs, and private interest groups. The message was, "The approach is solid and the virus is harmless to every living thing except the target pests."

Before a test site was sprayed, brochures were given to the people affected, and catfish farmers were contacted to get permission to spray the virus around their ponds. The first year, spray planes were used to apply the virus mixture at the rate of 2 ounces of virus mix in 2 gallons of water per acre. Spray trucks were used to spray certain areas not accessible to aircraft. The researchers found the control does work, and they will fine-tune the procedure for use in future years.

are labeled "WARNING." Those restricted-use pesticides requiring more than 2 tablespoons of the chemical to kill a 150-lb person are labeled "CAUTION." Obviously, even the pesticides with "CAUTION" as the signal word must be handled with extreme care.

Acute toxicity is measured by determining LD (lethal dose)$_{50}$ values when the pesticide is absorbed through the skin or is ingested orally. These values are determined by inhalation studies. LD_{50} is the amount or dose of the pesticide that is required to kill 50 percent of test populations. It is expressed in milligrams (mg) of pesticide per kilogram (kg) of body weight. The lower the LD_{50} value, the more toxic a pesticide.

LC_{50} is the lethal concentration of the pesticide in the air that is required to kill 50 percent of test populations. It is expressed in micrograms per liter (μg/l) or in parts per million (ppm). The lower the LC_{50} value, the more toxic the pesticide.

All labels must bear the statement, "KEEP OUT OF REACH OF CHILDREN," regardless of pesticide toxicity. Figure 14–10 shows the toxicity ratings for the various signal words.

Toxicity Rating	Label Signal Words	Lethal Oral Dose, 150 lb Person	Oral LD$_{50}$ (mg/kg)	Dermal LD$_{50}$ (mg/kg)	Inhalation LC$_{50}$ (μg/l or ppm)
High	Danger–Poison	few drops to 1 teaspoon	0-50	0-200	0-2,000
Moderate	Warning	1 teaspoon to 1 ounce (2 tablespoons)	50-500	200-2,000	2,000-20,000
Low	Caution	1 ounce to 1 pint+ or 2 pounds	500-5,000	2,000-20,000	200,000+
Very low	Caution	1 pint+ or 2 pounds+	5,000+	20,000+	—

FIGURE 14-10 Toxicity ratings and signal words for pesticides. *(Adapted from material from College of Agriculture, University of Illinois)*

Precautionary Statements

Precautionary statements on the label will list any known hazards to humans, animals, and the environment. They will advise the user how to minimize any adverse effect that the pesticide may have. The categories that are normally listed are as follows:

- Hazards to Humans and Domestic Animals
- Statement of Practical Treatment
- Environmental Hazards
- Physical and Chemical Hazards

Establishment and EPA Registration Numbers

The establishment number establishes the manufacturer. The EPA registration number indicates that the product has passed the review process imposed by the EPA.

Name and Address of Manufacturer

All pesticide labels must contain the name and address of the company that manufactures and distributes the pesticide.

Directions for Use

The correct amount, timing, and mixing of the pesticide is given under the Directions for Use section of the label. The label will also list the different pests that are controlled, the application technique, and any other specific directions for optimum control.

Misuse Statement

The misuse statement appears on the label to remind the user to apply the pesticide according to label directions. Problems associated with pesticides, whether they involve environmental pollution or human poisoning, often occur because of pesticide misuse.

Reentry Information

Specific directions on reentering a treated area will appear under this heading. Only a few pesticides require reentry times of more than a day after application. However, even if the pesticide label does not contain specific restrictions, no one should ever be allowed to enter a treated area until the pesticide has dried.

FIGURE 14-11 Leftover chemicals should be disposed of by taking the material to a hazardous waste station.

Storage and Disposal Directions

This section will describe the proper storage and disposal of the pesticide. It is recommended that you purchase only the amount of pesticides needed for the current season. Stockpiling them will only increase storage risks and, ultimately, the problem of pesticide disposal, if they can no longer be used.

Many states have enacted laws and established regulations that control the disposal of unused chemicals and chemical containers. It is never wise to bury these kinds of materials because they can ultimately pollute our groundwater and soil. It may also be illegal. The best choice is to take the containers and chemicals to designated sites where hazardous materials are collected so that they can be disposed of properly (Figure 14–11). Chemicals that are left over should be applied to another area or crop where they will not pollute the environment.

Notice of Limitations

The manufacturer guarantees that the product will perform as the label states. The company also conveys inherent risks to the user if the pesticide is applied in a manner inconsistent with the label. The manufacturer will limit its liability in case of lawsuits stemming from misuse of the pesticide.

RISK ASSESSMENT AND MANAGEMENT

Risk Measurement

An experienced pesticide applicator who understands the hazards of pesticides will take steps to reduce risks. The risk, or hazard, of pesticide applications has been expressed as:

Risk (Hazard) = Toxicity × Exposure

INTERNET KEY WORDS:

chemical, pesticide disposal

The hazard, or risk, is the relationship between the toxicity of the pesticide and the exposure or use of the pesticide. Toxicity is a measure of how poisonous a chemical is. These data may be expressed in several ways. Acute toxicity describes the immediate effects (within 24 hours) of a single exposure to a chemical. Acute toxicity data based on dermal (skin), oral (by mouth), and inhalation (breathing) exposure routes have been determined. Signal words on pesticide labels indicate acute toxicity values. Chronic toxicity measures the effect of a chemical over a long period. To determine this information, the chemical is administered at low levels, with repeated exposures to the test animals. The effect of the chemical on reproduction or as a potential carcinogen and any other adverse effects are evaluated.

Limiting Exposure

INTERNET KEY WORDS:

chemical, pesticide exposure

A pesticide's toxicity cannot be changed, but risk can be managed by addressing the exposure component of the formula. Many things can be done to reduce exposure. Examples of practices that can reduce exposure include the following:

- Select a pesticide formulation with a lower exposure potential. For example, granule formulations have a much lower exposure potential than do emulsifiable concentrates.

- Use protective clothing and other safety equipment during the time of pesticide mixing, application, and disposal.

- Apply pesticides during weather conditions that will not cause pesticide drift and that provide for the most effective control.

- Check all application equipment for proper working condition before applying pesticides.

- Store pesticides and application equipment properly.

The pesticide label will provide guidance concerning acceptable protective clothing or gear for the application of the pesticide (Figure 14–12). Recommended protective clothing and gear will vary according to the toxicity of the pesticide. Even if no special gear is required, it is best to minimize your exposure to all pesticides by selecting appropriate clothing.

Appropriate clothing includes long pants, a long-sleeved shirt, nonabsorbent shoes, and socks. Avoid the use of any leather clothing, particularly shoes, because leather absorbs pesticides. The use of heavy denim clothing provides good repellency to any pesticide and can be washed to remove any pesticide residue.

Special Gear

Gloves and Boots   Unlined rubber or neoprene gloves and boots significantly reduce pesticide exposure. Any type of cloth-lined or leather boot or glove will only increase exposure if the lining becomes contaminated. Gloves should be tucked inside sleeves if you are working below the waist. They should be left outside your sleeves if you are working above the waist. Pants legs should be placed over the boots. By following these rules, you can prevent material from entering the inside of protective clothing or gear.

Hat and Coveralls   Absorption of pesticides through the skin and into the body is greatest in the scrotal area, ear canal, forehead, and scalp (Figure 14–13). The use of an appropriate hat and coveralls minimizes

FIGURE 14–12 Proper clothing items and appropriate safety equipment are essential in the safe use of pesticides. *(Courtesy of USDA)*

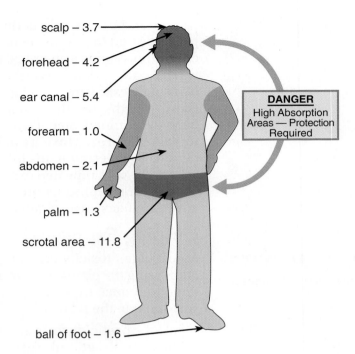

scalp – 3.7

forehead – 4.2

ear canal – 5.4

forearm – 1.0

abdomen – 2.1

DANGER
High Absorption
Areas — Protection
Required

palm – 1.3

scrotal area – 11.8

ball of foot – 1.6

FIGURE 14–13 Dermal absorption sites and rates of pesticide absorption into the body. Comparisons are based on forearm absorption rates.

exposure to pesticides in these areas. Lightweight, one-piece, repellent coveralls with hoods are available, and they provide excellent protection.

Apron During the mixing and loading operations, the applicator is exposed to pesticide concentrates. The use of a rubber or neoprene apron at this time will prevent pesticide concentrates from splashing onto the chest, waist, and legs of the applicator.

Goggles and Face Shield Eyes are extremely sensitive to many pesticides. The use of goggles and face shields is recommended when mixing pesticide concentrates or working in a spray, dust, or fog.

Respirators Respirators reduce the inhalation of pesticide fumes and/or dust. A recommended respirator for pesticide applications is a cartridge respirator that will absorb toxic fumes and vapors and filter any dust particles in the air. These respirators are often used during the mixing and application of a pesticide.

Personal Hygiene

Dermal exposure is the principal method of pesticide entry for the applicator. Personal hygiene can drastically reduce this type of exposure. Washing or showering at the end of the work day will remove any pesticide residue on the body. In case of a pesticide spill or splash at the work site, water can be used to immediately remove the material from the skin. After pesticide use, washing your hands before eating or using the restroom will further decrease pesticide exposure. Cleaning protective gear and clothing should also be done to prevent any residual exposure to pesticides on these objects.

INTERNET KEY WORDS:

pesticide storage

PESTICIDE STORAGE

Improper storage of pesticides can pose as much danger to the applicator, other people, animals, and the environment as misapplication of

FIGURE 14-14 Pesticides must be stored in approved storage units.

pesticides. Some important considerations in selecting a storage facility are the site location and building specifications.

Ideally, the site should be separate from other equipment or material storage facilities. This will reduce risk by decreasing exposure of individuals not involved in pesticide applications. The building should not be located on a floodplain, where flooding will introduce pesticides into surface water. It should be built to prevent any runoff or drainage from the site onto sensitive areas. Spill and drainage containment for large storage facilities is highly recommended. Containment systems would trap the pesticides and aid in emergency situations to minimize any environmental damage if the pesticides were to move from the site.

A well-planned storage building should be well ventilated, have a source of heat and water, be fireproof, have a secure locking system, and have sufficient storage area. The storage area should be well marked with placards indicating the presence of pesticides (Figure 14-14). The arrangement of the pesticides within the storage area should allow for ease in handling and safety. Tips to provide good storage conditions are:

• Separate each pesticide class for storage on its own shelf.

• Keep products off the floor.

• Store containers so the labels remain in good condition and the containers remain orderly.

• Practice good housekeeping.

HEALTH AND ENVIRONMENTAL CONCERNS

The current use pattern of pesticides has caused a heightened awareness of their risks. Human health and environmental quality are the major issues in assessing the hazards of pesticide use. The Environmental Protection Agency must conduct a benefit–risk assessment for each pesticide that is registered or re-registered for use. This is

Hot Topics in Agriscience

ORGANIC FOOD

There is a consumer group that has demonstrated a preference for organic food in contrast to foods produced traditionally. Some farmers are converting their farms into organic farms to take advantage of this market niche. The aim of organic farming is to produce safe, high-quality fruits and vegetables without the use of synthetic fertilizers and pesticides. Biological controls are used to keep insects at acceptable and manageable levels. To boost production, animal manures, legumes, and green manures are used as a natural alternative to chemical fertilizers. The U.S. Department of Agriculture has set strict standards for the production, handling, and labeling of "organic food." In a grocery store, you may see food labels with four different organic claims. "100% Organic"-labeled foods contain 100% organically produced ingredients. Labels with an "Organic" claim must contain at least 95% organic ingredients. Food "Made with Organic Ingredients," has to be composed of at least 70% organic ingredients; if a product claims that it "Has Some Organic Ingredients," it may have less than 70% organic ingredients. Whether a person prefers foods grown organically or traditionally, it is important to offer consumers a choice in the kind of food they buy.

INTERNET ADDRESS:

http://www.ams.usda.gov/
nop/indexIE.htm

an extremely controversial issue. The EPA presently defines acceptable risk to the public at one death per million caused by pesticide exposure.

Human Health

The number of lethal pesticide-related poisonings in the United States has decreased over the years. In one year, the total number of accidental deaths from all causes in the United States was 92,000, whereas deaths attributed to pesticide poisoning numbered only 27. However, more than 100,000 nonfatal pesticide poisoning cases per year have been estimated. A majority of the reported deaths were children involved in accidental ingestion of pesticides in the home. The causes have been traced to improper handling and storage practices by homeowners and other consumers.

The residues of a few pesticides used on food crops can pose potential health problems as carcinogens. A **carcinogen** is a material capable of producing a cancerous tumor. The National Academy of Science estimated that certain types of pesticide residues in food crops "may" cause up to 20,000 cancer cases per year. This estimate was based on a "worst-case" scenario. It assumes that exposure to the pesticide would be continuous over 70 years, with the maximum allowable pesticide residue on the food when eaten. Obviously, careful handling and preparation of food eliminates most of this risk.

Environmental Concerns

After a pesticide is applied, not all of the pesticide reaches or remains in the target area (Figure 14–15). When this happens, the pesticide is often considered an environmental pollutant.

The movement of a pesticide from the designated area may occur in several ways. Drift, soil leaching, runoff, improper disposal and storage, and improper application are some of the major causes of a pesticide becoming an environmental pollutant. Natural resources that can be contaminated are groundwater, surface water, soil, air, fish, and wildlife.

Surveys have shown that more than 50 percent of the counties in the United States have potential groundwater contamination from agrichemicals. The three main factors affecting groundwater contamination by agrichemicals are:

- soil type and other geological characteristics;
- the pesticide's persistence and mobility within the soil; and
- the production and application methods of pesticide users.

Pesticide drift is a major cause of soil and air contamination. **Drift** is the movement of a pesticide through the air to nontarget sites. It will occur at the time of pesticide application when small spray particles are moved by air currents to nontargeted areas. Also, vapor drift of a pesticide may occur after an application. **Vapor drift** is movement of pesticide vapors because of chemical volatilization of the product.

The adverse effects of pesticides on fish and wildlife may directly result in animal mortality. Pesticides may also indirectly influence animal feeding or reproduction. Pesticide labeling will indicate any potential harm to wildlife, and this information should be heeded to minimize risk. Fish, birds, bees, and other animals will be affected when pesticides reach them or their habitats.

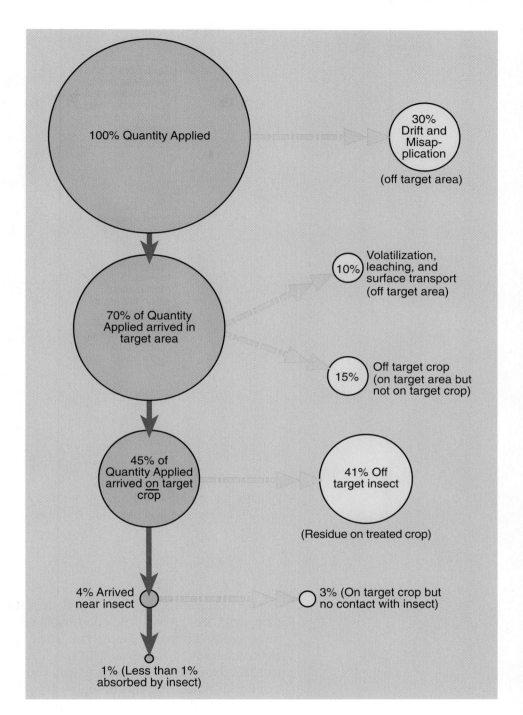

FIGURE 14–15 The movement of a pesticide after discharge from an aerial spray plane. *(Adapted with permission of Flint, M. L. and Van den Bosch, R. From* Introduction to Integrated Pest Management, *1981. Plenum Press, New York. 240 pp.)*

Within the figure:

100% Quantity Applied

30% Drift and Misapplication (off target area)

70% of Quantity Applied arrived in target area

10% Volatilization, leaching, and surface transport (off target area)

15% Off target crop (on target area but not on target crop)

45% of Quantity Applied arrived <u>on</u> target crop

41% Off target insect (Residue on treated crop)

4% Arrived near insect

3% (On target crop but no contact with insect)

1% (Less than 1% absorbed by insect)

Environmental contamination by agrichemicals can be decreased through several management practices. The use of integrated pest management (IPM) programs is reducing pesticide use. If pesticides are used, then proper mixing, application, storage, and disposal must be performed. These practices will decrease any adverse effect on the environment. An attempt must also be made to minimize any effect that temperature, soil type, rainfall, and wind patterns may have on a pesticide becoming an environmental pollutant.

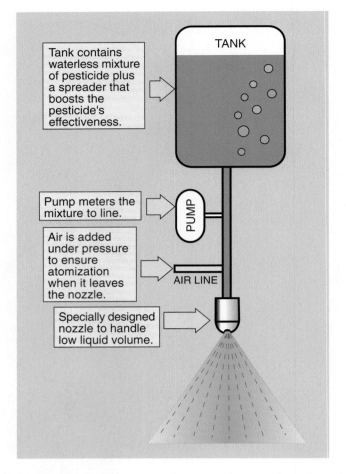

FIGURE 14–16 New technology permits pesticide application equipment to use less water and less pesticide to achieve equal or better control of pests.

Several strategies have been developed to reduce the amount of chemical that is required to control pests. One of these methods is to inject high-pressure air at the spray nozzle, causing a low-volume chemical mix to separate into tiny droplets. The air pressure also carries the droplets to the area where the chemical is intended to go. This application method has been proven effective in reducing the amount of chemical that is needed (Figure 14–16). Another method for reducing the amount of chemical that is needed is to induce opposite electrical charges in the chemical and on the plants to which it is being applied. This causes the chemical droplets to be attracted to the plants to which they are being applied. The net result is that a higher percentage of the chemical reaches the targeted area than it does when traditional application methods are used.

Chemical pesticides are an important part of our food- and fiber-production capability. They are necessary to maintain our current standard of living in the United States and the world. However, they do create risks if used by improperly trained or careless individuals. Therefore, it is important that those using pesticides be properly informed and follow label instructions (Figure 14–17).

Pesticide Safety
Choose the right pesticide for the job.
Carefully read the label and apply the chemical according to the directions for its use.
Mix the chemical properly to assure that the correct amount is applied.
Make sure that the applicator wears appropriate protective clothing.
Mix only the amount of chemical that is needed, and apply chemicals that are left over to a crop where it will not affect the environment in negative ways.
To avoid chemical drift, consider wind and weather conditions before making the decision to apply a chemical.
Clean the sprayer and carefully dispose of the wash water to avoid contamination to the environment.
Avoid inhaling or ingesting the chemicals and wash dangerous chemicals off the skin immediately.
Store chemicals in locked areas in the original containers away from people, animals, and feeds.
Dispose of chemical waste and empty containers according to the rules for handling hazardous waste materials.

FIGURE 14–17 Pesticide safety.

It is important that consumers wash and handle food products to minimize the intake of pesticide residues on food (Figure 14–18). Although the government does extensive research to arrive at safe pesticide-residue levels, it is in the best interest of the consumer to wash fruits, vegetables, and all plant parts that may contain pesticide residues. This precaution will further reduce the hazard of pesticides.

Similarly, the wise person will become familiar with the uses of pesticides. Such uses include pesticides to control roaches, termites, flies, insects, and diseases of lawn and garden plants; insect repellants and insecticides used for outdoor camping and recreation; and other everyday common practices. The benefits of pesticides far outweigh the risks.

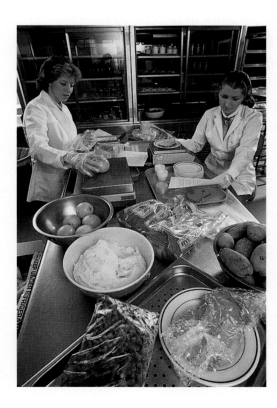

FIGURE 14–18 The government does extensive research to arrive at safe pesticide residue levels. *(Courtesy USDA/ARS #K-1992-9)*

STUDENT ACTIVITIES

1. Write the Terms To Know and their meanings in your notebook.

2. Ask your instructor to show examples of an approved pesticide applicator's respirator, goggles, gloves, boots, clothing, and other protective items.

3. Prepare and present a class demonstration on the proper use of one or more items of protective clothing and devices for safe handling of pesticides.

4. Do research and prepare to defend the position that pesticide residue on food in the United States is or is not a serious problem. Arrange for at least three classmates to prepare for a debate presenting various issues regarding the use of pesticides.

5. Ask a professional pesticide applicator with a special interest in pesticide safety to demonstrate safe pesticide-application principles to the class.

6. Collect newspaper and magazine articles on accidents involving pesticides. Study the articles and determine how each accident could have been avoided. Report your conclusions to the class. *Note:* Many pesticide accidents occur in or on the home, lawn, and garden—do not overlook such cases.

7. Conduct a survey of pesticide storage-and-use practices in your home, farm, and place of employment. Correct all safety violations.

8. List the classes of chemicals used for pest management. Give a brief description of each, and explain how they work to prevent or control pests.

SELF EVALUATION

A. Multiple Choice

1. An example of an inorganic pesticide is
   a. pyrethrum.
   b. rotenone.
   c. Bordeaux mixture.
   d. organophosphate.

2. The total amount of pesticides used annually in the United States is
   a. 820 million pounds.
   b. 420 million pounds.
   c. 620 million pounds.
   d. 1 billion pounds.

3. A preemergence herbicide is applied
   a. after the weed or crop is present.
   b. before the weed or crop is present.
   c. at the time of planting.
   d. none of the above.

4. Which herbicide family inhibits photosynthesis?
   a. acetanilines
   b. dinitroanilines
   c. phenoxys
   d. triazines

5. An example of botanical insecticide is
   a. 2, 4-D.
   b. rotenone.
   c. diazinon.
   d. sulfur.

6. Pesticide registration often takes
   a. between 1 and 2 years.
   b. between 3 and 4 years.
   c. between 8 and 10 years.
   d. 15 years.

7. Pesticide risk can be decreased by
   a. proper pesticide exposure.
   b. reading the label.
   c. minimizing pesticide exposure.
   d. all of the above.

8. The signal word(s) for a highly toxic pesticide is

    a. CAUTION.
    c. DANGER—POISON.

    b. WARNING.
    d. none of the above.

9. An example of protective clothing or gear that will minimize inhalation of a pesticide is

    a. a respirator.
    c. gloves.

    b. boots.
    d. coveralls.

10. The number of lethal pesticide poisoning cases per year is

    a. less than 20.
    c. between 40 and 60.

    b. between 20 and 30.
    d. more than 100.

B. Matching

_____	**1.** Carcinogen	a. The effect of a single exposure to a pesticide
_____	**2.** Chronic toxicity	b. Movement of a pesticide through the air to nontarget sites
_____	**3.** Water	c. Repeated exposures to low doses of a pesticide
_____	**4.** DF	d. A soluble powder pesticide formulation
_____	**5.** Signal word	e. A material capable of producing a tumor
_____	**6.** Acute toxicity	f. A fungicide used after disease infection
_____	**7.** SP	g. Used to remove pesticides from the body
_____	**8.** Drift	h. A dry, flowable pesticide formulation
_____	**9.** Protectant	i. Describes the acute toxicity of a pesticide
_____	**10.** Eradicant	j. A fungicide used before disease infection

C. Completion

1. An _____ will control insects.

2. A _____ herbicide will control all types of plants.

3. A systemic herbicide will be translocated in the _____ and _____ tissue.

4. The amount of money spent on pesticides in 1999 was _____ billion dollars.

5. A pesticide will have only one _____ and one _____ name.

6. To reduce exposure to pesticides, the use of _____ is recommended for clothing and boots.

7. Environmental and health hazards of a pesticide are listed under the _____ of the label.

8. The _____ number identifies where the pesticide was manufactured.

9. LD refers to the _____ of a pesticide.

10. The _____ name is the manufacturer's name for its product.

Plant Sciences

The Quest for More and Better Plants!

Scientists, growers, propagators, breeders—everyone gets into the act. How can we grow plants that are different, plants to provide materials for new medicines and industrial products; make our flowers and ornamentals more beautiful; make our present plants more disease, insect, drought, and cold or heat resistant? Where can we find trees that grow and produce wood faster? These questions all drive our quests in plant science.

During World War II, the USDA sent botanist Richard Schultes to South America in search of rubber trees resistant to diseases. Such trees were discovered, which opened the way for rubber plantations in the Western Hemisphere. These and other plants, which USDA explorers collected, changed agriculture around the world. For example:

- Curare vines provide a muscle relaxant used in surgery.
- Ucu uba trees in Colombia have bark that provides a highly desired suppressant and skin medicine.
- Rootstock from China strengthens peach trees.
- Navel oranges from Brazil have created a new California industry.
- Durum wheats from Russia set the standard for United States varieties for years.
- A peanut from Peru has genes that are resistant to two major diseases that impacted the U.S. peanut industry.
- Wild oats have resulted in one of the most disease-resistant oat varieties ever developed.
- California's avocado industry was started with germplasm from Mexico.
- Sorghum, dates, tung oil, and numerous forage grasses from around the world are now grown in the United States.

There are more than 1,600 species of plants used for medicinal purposes by people of the Colombian Amazon. However, only a few of these have been studied by scientists. The USDA Chief of the Agriculture Research Service coordinates about 10 trips a year in search of new plants and plant materials. The need for plant collecting, cataloging, and preserving is more urgent each year. The genetic base for many of the crops we take for granted is narrow. Furthermore, the encroachment of modern civilization on plants in remote places of the world is wiping out some important sources of new genes. Such plants could carry the genetic material needed to help important species survive. They may also be the source of genetic material that is needed to develop and introduce new plant species with new uses.

Small samples of some of the diversity of plant materials are brought back to the United States. The following guidelines are used by modern plant hunters and collectors when exploring for or making new discoveries:

- Trips are to be organized as collaborations between the United States and the host country.
- All collected material is divided at least equally with the host country.
- All germplasm collected with the support of the USDA is deposited in the National Plant Germ-plasm System and is available to all valid users.
- All collection must "be done with a conservation ethic in mind."
- Collection must not endanger natural plant populations, with enough left behind so that the plant population can regenerate naturally.

Scientists believe that genetic diversity can be better preserved when new plants are cataloged, and most of them are left and protected where they are found.

Centers of origin for some major fruits, vegetables, grains, and oil crops. (Courtesy USDA/ARS)

Plant Structures and Taxonomy

MATERIALS LIST

✓ plant collection materials

✓ bulletin board materials

✓ Internet access

COMPETENCIES TO BE DEVELOPED

After studying this unit, you should be able to:

- draw and label the major parts of plants.
- describe the major functions of roots, stems, fruits, and leaves.
- draw and label the parts of a typical root, stem, flower, fruit, and leaf.
- explain some of the variations found in the structures of root systems, stems, flowers, fruits, and leaves.
- describe the relationship of plant parts to fruits, nuts, vegetables, and crops.

SUGGESTED CLASS ACTIVITIES

1. Gather plant materials that can be used to illustrate as many of the Terms to Know as possible. Assign students to work in teams, and have them label as many of the plant parts as they can identify (reference books are encouraged). Follow up with a review to reinforce correct labels and to correct mistakes.

2. Collect, prepare, and laminate samples of plants that are found in your community. Accomplish this during the summer months before the students are in class. Add additional samples to the collection as students bring them in. Have class members memorize the common names of each of the samples.

3. As a class, brainstorm ways to remember the taxonomic order: kingdom, phylum, class, order, family, genus, and species.

TERMS TO KNOW

adventitious root
taproot
ornamental plant
fibrous root
root cap
area of cell division
area of cell elongation
xylem
area of cell maturation
root hair
stem
woody
herbaceous
bulb
corm
tuber
vascular bundle

internode
axillary bud
axil
lenticel
terminal bud
vegetative bud
flowering bud
leaf
phototropism
margin
simple leaf
compound leaf
leaf blade
petiole
cuticle
epidermis
palisade cell

spongy layer
chloroplast
mesophyll
stoma
guard cell
flower
bract
stamen
filament
anther
pollen
pistil
stigma
style
ovary
ovule
perfect flower

imperfect flower
pollination
petal
corolla
sepal
caylx
fruit
vegetable
nut
taxonomy
genus
species
binomial
variety

Plants are a basic part of the food chain. Without plants, the web of life cannot exist, and most animals and humans would die. A knowledge of plant growth is essential. To have a better understanding of plants, it is necessary to identify the parts that make up plants. The casual observer sees stems, branches, leaves, and possibly flowers and some nuts or fruits. The agriscience technician, however, will see a series of interconnected tissues and organs that depend on each other to function. The technician knows that all of the organs do not need to be present at one time, but is aware that each cell or organ (composed of cells) has an important role in the successful growth of the plant. Plant technicians and scientists are concerned with the efficient growth of plants. A plant may become stressed if one or more parts are absent or not functioning properly.

The basic industry of agriculture is dependent on the proper functioning of plants. The animal grower needs many kinds of plants to feed livestock and poultry. The plant industry also needs superior plants for feed and seed production. The horticulture industry needs plants for seeds and cuttings and for food, such as fruits, vegetables, and nuts. Plants are also needed for landscaping the inside and outside of homes and office buildings.

To be successful with plants, one must have a knowledge of plant parts and how they function. Such knowledge is essential, whether you are growing, selling, or using plants.

THE PLANT

Plants are composed of many parts. Each part is important in the overall life and function of a plant. The root system is normally under the ground and is responsible for anchoring the plant and supplying water and nutrients. The stem, or trunk, is normally above the ground and functions as a support system for the rest of the aboveground parts. Leaves constitute the food-manufacturing parts of plants. Flowers come in many sizes, colors, and shapes, and function as the seed-producing

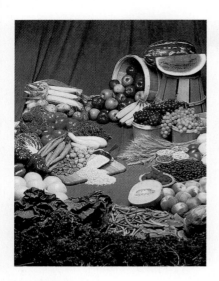

FIGURE 15–1 Seeds, nuts, and fruits are plant parts commonly used for food. *(Courtesy USDA/ARS #K-3839)*

INTERNET KEY WORDS:

plant root systems

INTERNET ADDRESS:

http://metalab.unc.edu/botnet/
glossary/a_ii.html

parts of the plant. Healthy plants produce seeds, nuts, fruits, and vegetables. These parts are popular foods for animals and humans. They are also used for reproduction of the plant (Figure 15–1).

Basic Necessities of Plant Life

For a plant to survive, its basic needs must be met. These needs include light, water, air, and minerals. Animals must capture their food to have sufficient energy. Plants also must capture an energy source. Some plants can thrive in quite shady areas, whereas others need direct sunlight. The sun's light is required by plants to perform photosynthesis. In this process, plants convert the sun's energy into food. This process is discussed in Unit 16.

Another requirement is water. The amount of water plants need varies quite a bit. A desert cactus needs far less water than a tropical fern, but no matter the amount, all plants need it. Most water is taken in through the roots, however many plants can bring small quantities in through leaves. Plants also need air. Oxygen is used during plant respiration, whereas carbon dioxide is required for photosynthesis. Minerals are necessary to supply nutrients for plants. Many of these minerals are found free in the soil, but others must be supplemented by providing fertilizer.

ROOTS

Root Systems

The largest part of the plant is often the root system. Roots take up more space in the soil than does the top part of the plant seen in the air above the ground. In fact, some roots will go down into the soil 6, 8, or even 10 feet. Some plants, such as squash, have massive root systems.

Although many roots are in the soil, there are other types seen above the ground that may not be considered roots. Some plants, such as poison ivy *(Rhus radicans)* and English ivy *(Hedera helix),* have roots that help them climb trees, walls, and sides of buildings. These are called adventitious roots. **Adventitious roots** appear where roots are not normally expected (Figure 15–2). Adventitious roots also prop up plants such as corn, strengthening them against the wind.

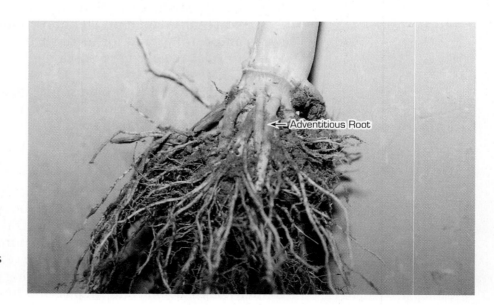

Adventitious Root

FIGURE 15–2 Adventitious roots develop above the ground, such as the prop roots located at the root of the corn plant.

FIGURE 15–3 Mistletoe is a parasitic plant with roots that grow through the bark and into the tissue of the tree branch. *(Courtesy Boise National Forest)*

FIGURE 15–4 Corn plants showing extensive fibrous root systems at an early age. *(Courtesy USDA/ARS #K-5169-5)*

The mistletoe, a popular Christmas plant, has roots that penetrate the bark of trees in the upper branches, or crown (Figure 15–3). These roots grow into the xylem and phloem tissues of the host plant and extract nutrients that originate in the soil. The dodder *(Cuscuta campestris)* has soil roots that die off as the plant gains a foothold in a plant. The dodder plant forms rootlike attachments that penetrate the stem of the host plant and extract nutrients from that plant.

Root systems are generally either taproot or fibrous roots (Figure 15–4). Knowledge of these two types of root systems can be of value in caring for and handling plants. The **taproot** is the main root of

Agri-Profile

CAREER AREAS:

Botany/Biology/Taxonomy

Everyone relies heavily on our system of plant classification. Without it, we could not order, buy, sell, or instruct others about a plant unless the plant was in the presence of both parties at the same time. Taxonomists devote their careers to identifying, classifying, and teaching others about plants. An important and exciting part of their work is that of discovering, studying, and naming new plants in their appropriate places in the classification system. Consider the excitement of discovering plants in faraway lands or even at home that have not been observed or recognized even by the most knowledgeable specialists to date.

Botany is the study of plants, and biology is the study of both plants and animals. A knowledge of plant structures is critical to both. Similarly, consumers of plants for food, ornamentation, medicine, shade, wood, and other uses should have some knowledge of plant structures. Plant structures affect plant nutrition, functions, disease susceptibility, adaptation, and use.

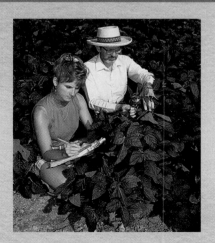

Plant scientists have studied individual plant specimens and devised a system of classification of plants based on characteristics of plant parts. *(Courtesy USDA/ARS)*

FIGURE 15–5 Taproots of carrots, beets, and radishes are excellent sources of food.

a plant and generally grows straight down from the stem. It is a heavy, thick root that does not have many side, or lateral, branches. Taproots are often used for human and livestock consumption, because they are food-storage organs. Carrots *(Daucus carota)* and sugar beets *(Beta vulgaris)* are examples (Figure 15–5). Some plants with taproots are used for ornamental purposes. An ornamental plant is used to improve the appearance of a structure or area.

Plants that have taproot systems have the ability to survive periods of drought. Because they grow deep into the soil and have few fine secondary roots, taproots do not stabilize the soil well. Fibrous roots are generally thin, somewhat hairlike, and numerous. The fibrous root system is normally shallow. Grasses, corn, and many ornamentals, such as *Begonia semperflorens,* are good examples of plants with fibrous root systems. There are many small, thin-branched roots in this type of system. The result is that they are able to hold soil much better than taproot systems. However, fibrous root systems dry out more easily. Therefore, they cannot tolerate drought conditions.

Root Tissues

Although there are different systems, all roots look similar when they are examined on the inside (Figure 15–6). The parts of the root have very specific functions in the plant. A knowledge of these parts is helpful in diagnosing diseases and other dysfunctions of plants.

The Root Cap   The root cap is the outermost part of the root. It protects the tender growing tip as the root penetrates the soil. The root cap is a tough set of cells that is able to withstand the coarse conditions that the root encounters as it pushes its way through soil with rock and small sand particles. As the root cap wears away, the cells are replaced by more cells that develop at the root tip. This portion of the root is known as the area of cell division.

Area of Cell Division and Elongation   The area of cell division provides new cells that allow the root to grow longer. The cells in this area multiply in two directions. The small and tougher cells are produced on the front edge of this region. They replace cells of the root cap that are worn off or destroyed as the tip pushes its way through the soil. Small and more tender cells are produced on the back of this area. They are used as the root tip grows longer. The area of cell division is actually quite thin, maybe as thin as a strand of hair.

The next area, as you move toward the base of the plant, is the area of cell elongation. In this area, the cells start to become longer and specialized. They also begin to look like the older cells and will start to do their specific jobs.

Xylem and Phloem   There are many types of cells in the root. Perhaps the most important ones are the xylem and the phloem cells. The xylem cells are responsible for carrying the water and nutrients that are in the soil to the upper portion of the plant. The phloem cells function as the pipeline to carry the manufactured food down from the leaves to other plant parts, including the roots, where it is used or stored. There are other cells in the root, and some are discussed later in this unit.

Area of Cell Maturation   The area of cell maturation is where cells become fully developed. This is also where the root hairs emerge. Root hairs are small microscopic roots. They will rise from existing cells

ROOT STRUCTURE

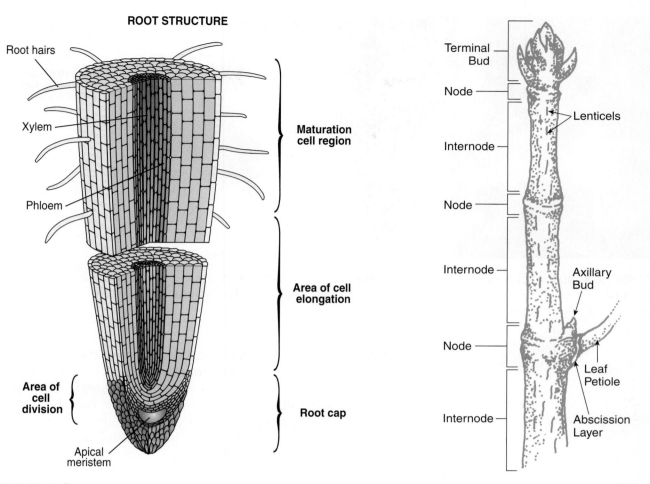

FIGURE 15–6 Root structure.

FIGURE 15–7 Important parts of stems.

INTERNET KEY WORDS:

plant parts functions

located on the surface of the root. It is the job of root hairs to take in water and nutrients. Water and nutrients move into the root hairs, enter the xylem, and move to the upper portions of the plant. Root hairs are small and tender. They will break off very easily. This means plants must be handled very carefully when transplanting. Once the root hairs are broken off, they cannot regrow or be replaced.

Although roots are normally hidden from view, they are very important parts of the plant. Roots need the same care and consideration as the other parts for plants to grow well.

STEMS

Stems are among the first things seen by the casual observer when looking at plants (Figure 15–7). Stems and branches are noticeable in the winter when the leaves are gone. They are easily seen as the plant grows. **Stems** support the leaves, flowers, and fruit.

Types of Stems

Two types of stems grow above the ground: woody and herbaceous. **Woody** stems are tough and winter hardy. They often have bark around them. **Herbaceous** stems are succulent, often green, and will not survive winter in cold climates.

FIGURE 15–8 Bulbs, such as Easter lilies and onions, are shortened stems surrounded by modified leaves called scales. *(Courtesy USDA/ARS)*

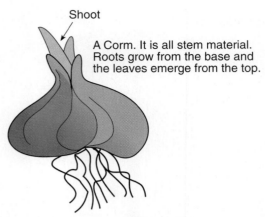

Shoot

A Corm. It is all stem material. Roots grow from the base and the leaves emerge from the top.

EXAMPLE: GLADIOLUS

FIGURE 15–9 Corms, such as the gladiola, are stems that are thick, compact, and fleshy.

Not all stems are erect, aboveground structures. Some grow along the ground or even underground. Some stems have specialized jobs to perform. Such stems are referred to as modified stems. Examples of modified stems are bulbs, corms, rhizomes, and tubers. **Bulbs** are shortened stems that are surrounded by modified leaves called scales. Some examples of bulbs are Easter lilies *(Lillium longiflorum)* and onions *(Allium* sp.; Figure 15–8). **Corms** are thickened, compact, fleshy stems. An example of a corm is the gladiola *(Gladiola* sp.; Figure 15–9). Rhizomes are thick stems that run below the ground. Johnson grass and the iris *(Iris germanica)* are examples of plants with rhizomes (Figure 15–10). **Tubers** are thickened, underground stems that store carbohydrates. We often eat an example of this type of stem, the Irish potato *(Solanum tuberosum)* (Figure 15–11).

Parts of Stems

INTERNET KEY WORDS:

rhizomes, tubers, bulbs, corms

Stems have some of the same internal parts as roots. The xylem and the phloem continue to run the length of the stem and into all the branches of the plant. In a subclass of plants called dicotyledons, the xylem and phloem occur together in tissues called **vascular bundles.** In another important subclass called monocotyledons, the xylem and phloem occur in separate areas (Figure 15–12).

Some important external parts of plant stems include the node, internode, axillary bud, lenticels, and terminal bud. The node is the portion of the stem that is swollen or slightly enlarged where buds and leaves originate. The **internode** is the area between the nodes. The **axillary bud** grows out of the axil. The **axil** is the angle above a leaf stem or flower stem and the stalk. The function of the axillary bud is to develop into a leaf or branch. The **lenticels** are pores in the stem that allow the passage of gases in and out of the plant. The **terminal bud** is located on the tip or top of the stem or its branches. It may be either a vegetative or flowering bud. The **vegetative bud** produces the stem and leaf growth of the plant. The **flowering bud** produces flowers.

Bio-Tech Connection

INSIDE A CELL: PLANT BIOTECHNOLOGY AT WORK

This electron micrograph shows some of the components in a soybean seed cell. For better visibility, computer processing has colored seed storage proteins purple and stored oil yellow. Red areas are cellular compartments where synthesis of the stored protein and oil occurs. The electron micrograph was taken by Eliot Herman of the Agriculture Research Service, and the color enhancement was performed by Terry Yoo of the Science and Technology Center for Computer Graphics and Scientific Visualization at the University of North Carolina.

A computer-enhanced view of the interior of a soybean cell. *(Courtesy USDA/ARS)*

Through the use of standard and electron microscopes, together with other modern technology, scientists can peer into the most minute parts of plant structures. They can improve the images seen through various instruments via computer enhancement. The fields of molecular biology, cellular biology, and genetics search the mysteries of the cell and its components, and seek to manipulate its biology to bring about desired changes. Larger parts of plants are observed with hand glasses or the naked eye.

Plant structures permit the plant to function as a whole. If one part becomes diseased, it will limit the performance of the rest of the plant and may lead to plant death. If the roots cannot obtain sufficient water and nutrients, the leaves cannot manufacture food for the plant. If the stem grows too fast without developing strength, the plant may topple and die. If seeds do not form and mature properly, the plant may not be able to reproduce and perpetuate itself. The whole is dependent on the health of its parts.

FIGURE 15–10 Rhizomes are thick stems that grow underground near the surface and give rise to new plants at each node.

FIGURE 15–11 The potato is really a specialized stem called a tuber. *(Courtesy USDA/ARS #K-4016-5)*

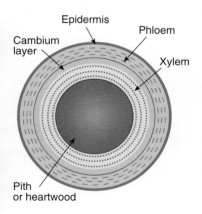

FIGURE 15–12 This cross section of a stem shows the xylem and phloem cells that make up the vascular system of a plant.

Hot Topics in Agriscience

SHORTER STRAW—A STRONGER WHEAT PLANT

A significant improvement has occurred in some varieties of wheat because of the efforts of scientists. They have discovered ways to improve the wheat plants by reducing the length of the stem and increasing the thickness to make it stronger. This change in the structure of the wheat plant results in fewer problems caused by lodging. Lodging occurs when wind and moist conditions combine to cause the wheat plants to fall over and remain flat on the surface of the soil. This condition makes it difficult to harvest the wheat, and some of the grain is usually left in the field because the grain combine cannot pick it up. The short, thick straw of some of the new wheat varieties makes them much less susceptible to this problem.

LEAVES

The leaf of a plant has an important function. It manufactures food for the plant by using light energy.

A plant leaf is capable of adjusting its angle of exposure to the sun. The leaves of some plants turn to allow full sunlight to shine on the leaf surfaces as the position of the sun changes during the day. The process by which this occurs is called phototropism. Without this important plant reaction to sunlight, plant growth would be less.

Leaf Margins

INTERNET KEY WORDS:

plants, phototropism

Plants may be identified by the edges, shape, and arrangement of the leaves. The leaf edges are known as margins. Leaf margins are named or described according to the toothed pattern on each leaf edge (Figure 15–13).

Leaf Shape and Form

Leaves vary in shape and form according to their species. Therefore, knowledge of the name given to each leaf shape and form is useful in identifying the plant (Figures 15–14, 15–15).

Types of Leaves

Leaf types vary according to the species. Therefore, leaf type is also used to identify plant species. A single leaf arising from a stem is called a simple leaf. Two or more leaves arising from a common point on the stem are referred to as compound leaves (Figure 15–16).

Leaf Parts

INTERNET KEY WORDS:

internal leaf structure

A leaf consists of a petiole and blade. These are the most familiar parts of a leaf. The leaf blade is the wide portion. It may be of many shapes and sizes. The petiole is the stem of the leaf. It may be almost absent or may be very long.

Internal Structure

The leaf is the food-manufacturing unit for the plant. The process of manufacturing food is called photosynthesis. The food that is created through photosynthesis in leaves enables the plant to grow. The process of photosynthesis is illustrated in Figure 15–17 and is discussed in greater detail in Unit 16.

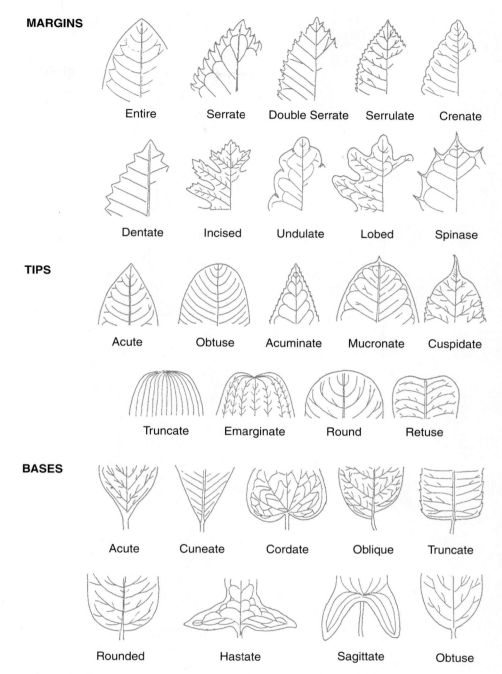

FIGURE 15–13 Leaf margins are helpful in identifying plants.

The **cuticle** is the topmost layer of the leaf. It is waxy and functions as a protective covering for the rest of the leaf. The **epidermis** is the surface layer on the lower and upper sides of the leaf. The epidermis protects the inner leaf in many ways. The elongated, vertical **palisade cells** give the leaf strength and are the sites for the food-manufacturing process. These cells, as well as the lower **spongy layer,** contain chloroplasts. **Chloroplasts** are the parts of the cells that contain chlorophyll. They are necessary for photosynthesis to be carried out. The lower layer is irregular and allows the veins, or vascular bundle, to extend into the leaf. The layer of palisades and spongy tissue is often referred to as the **mesophyll.**

LEAF SHAPES

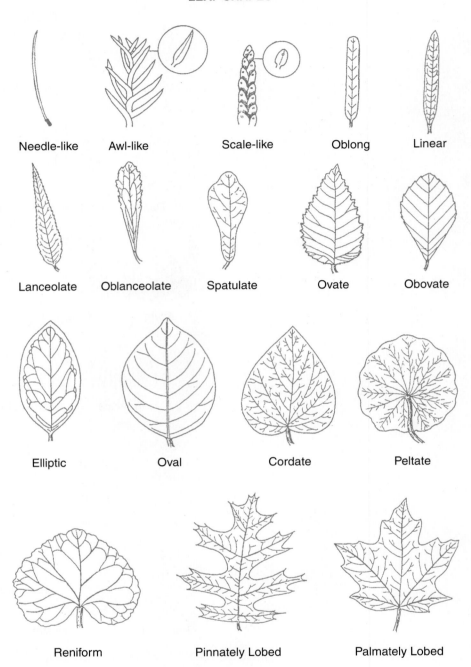

FIGURE 15–14 Names given to various leaf shapes.

The vascular bundles contain the xylem and the phloem. These are extensions of the same tissues that are located in the root. They extend through the stem to the leaves. The xylem brings the water and minerals from the root. The phloem carries the manufactured food from the leaf to the various parts of the plant to nourish plant tissue or be stored.

The lower epidermis contains some special cells called **stoma.** They are openings that allow for the exchange of carbon dioxide and oxygen, as well as some water. The stoma are surrounded by **guard cells** that open and close the stoma. If the plant is stressed by the lack of water or by a low light

INTERNET KEY WORDS:

leaf shapes
leaf parts plant

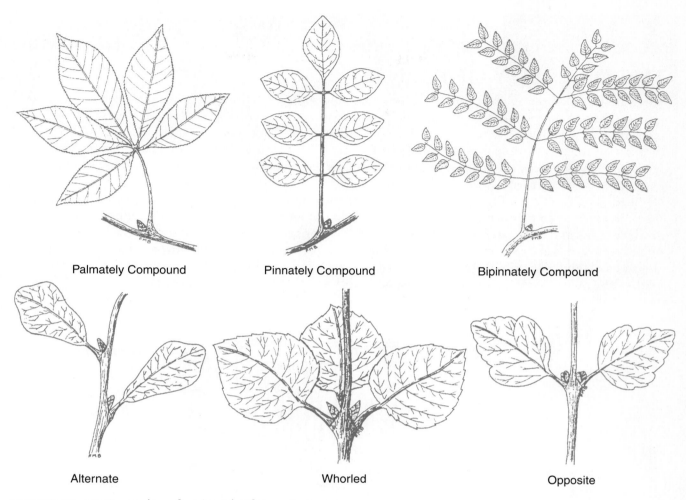

FIGURE 15–15 Examples of various leaf arrangements.

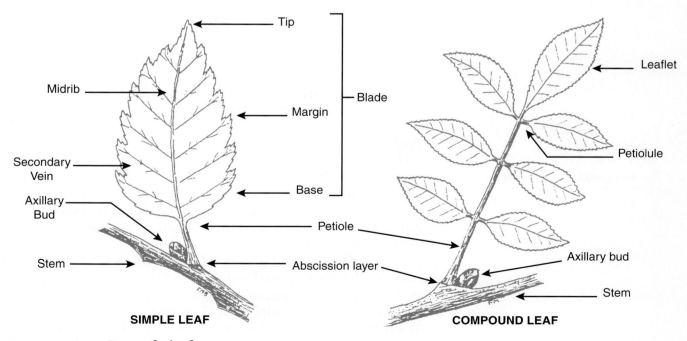

FIGURE 15–16 Parts of a leaf.

Science Profile

THE DIVERSITY OF PLANTS

There are four different plant types in the Kingdom Planta. Bryophytes are plants that lack a vascular system. They do not produce seeds, and they must depend on water for reproduction. A vascular system is a system of vessels in a plant, each of which carries water and nutrients independently from each other throughout the plant. Moss is an example of a Bryophyte. Seedless vascular plants have distinct vascular systems. They also need water to reproduce, but unlike the Bryophytes, they have true roots, leaves, veins, and stems. Ferns are seedless vascular plants.

Seed plants include both the Gymnosperms and the Angiosperms. Both groups of plants produce seeds. In contrast to plants mentioned earlier, they do not require water to reproduce. Both plant types produce seeds and have extensive vascular systems. Gymnosperms bear their seeds in cones, whereas Angiosperm seeds are protected in a thick tissue called fruit. Examples of gymnosperms are pine trees, ginkgo trees, and all conifers. Angiosperms, which are called flowering plants, are the most abundant of all of the plants. Grasses, corn, petunias, and most crop plants are angiosperms.

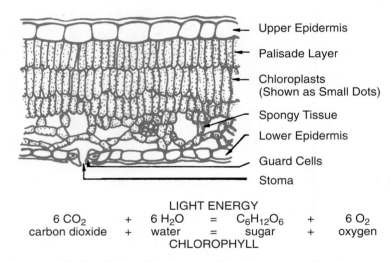

- Upper Epidermis
- Palisade Layer
- Chloroplasts (Shown as Small Dots)
- Spongy Tissue
- Lower Epidermis
- Guard Cells
- Stoma

LIGHT ENERGY

$$6 \, CO_2 + 6 \, H_2O = C_6H_{12}O_6 + 6 \, O_2$$

carbon dioxide + water = sugar + oxygen

CHLOROPHYLL

FIGURE 15–17 The leaf is the food-manufacturing part of a plant.

level, the guard cells will close the stoma. The result is that the plant cannot manufacture food because it will not have all the necessary ingredients.

FLOWERS

INTERNET KEY WORDS:

flower structure

Many people see plants mostly for the beauty of the flower. Others see only a fruit to eat. Fruit production is only part of the job of the flower. The **flower** has, as its primary function, the production of seeds needed to continue the species. It is with this structure that the plant scientist will work to produce new and different varieties.

Not all of the beauty that is seen as flowers are actually flowers. The poinsettia *(Euphorbia pulcherrima)* and the flowering dogwood *(Cornus florida),* for example, have modified leaves called bracts. A **bract** is a modified leaf that is often brightly colored and showy. People often see the red or white bracts and call them flowers. Their function is to protect the flower parts, as well as attract insects for pollination.

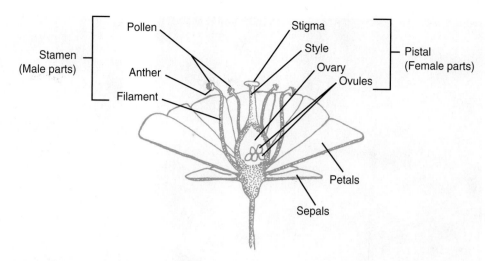

FIGURE 15–18 Major parts of a flower.

Flower Structure

Flowers are composed of many parts, including the filament, anther, pollen, stigma, style, ovary, petals, and sepals (Figure 15–18). These are the most important parts of the flowers.

The male part of the flower is the **stamen.** It consists of the filament, anther, and pollen. The **filament** supports the anther. The **anther** manufactures the pollen. The **pollen** is the male sexual reproductive cell. The female part of the flower, the **pistil,** is made up of the stigma, style, and ovary. The **stigma** receives the pollen. The pollen travels down the **style** and into the **ovary.** The ovary contains **ovules.** These are the eggs, which are the female reproductive cells. When the eggs are fertilized by the pollen, they will ripen into seeds.

If a flower contains all of the parts just mentioned, it is a **perfect flower.** If one or more of the parts are missing, it is considered an **imperfect flower.** In some plants, particularly the flowering plants used in horticulture, it is desirable to remove the anther sacs before the pollen ripens. This prevents pollination and stains from the pollen on the petals of the flower. **Pollination** means the union of the pollen with the stigma. In some cases, orchids for example, the unfertilized flower will last for many months.

In plant breeding, the anther sac is removed from the plant to prevent natural pollination. It may be destroyed, or it may be used to pollinate another flower to create a new variety. Many hybrids are created in this way.

The colored **petals** attract insects or other natural pollinators. The flower petals are collectively called the **corolla.** The **sepals** together function as a protective device for the developing flower. Collectively, sepals are called the **caylx.**

Fruits, Nuts, and Vegetables

After fertilization, the ripening seed develops in the pistil. The pistil then enlarges and becomes the **fruit** (Figure 15–19). The fruit may be of many different shapes and sizes (Figure 15–20). The true fruit consists of the seeds that carry the male and female genetic characteristics of the plant. However, the fleshy material surrounding the mature seed, and the seed itself, is commonly called the fruit of the plant. The purpose of the fleshy

CROSS SECTION OF AN APPLE

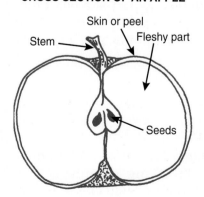

FIGURE 15–19 The ovary of a flower matures into a fruit that surrounds the seeds. When the fruits are eaten, the seeds are scattered to new locations.

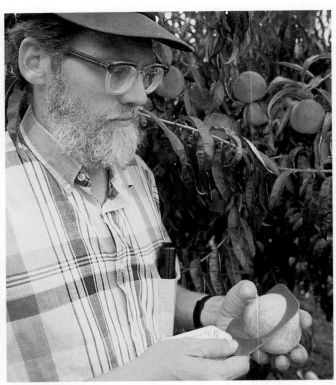

A

B

FIGURE 15–20 Fruits from different plants vary in size, shape, and taste. (A) Pineapple growing in Hawaii and (B) peaches in Georgia. *(Courtesy (A) USDA/ARS #K-4281-1 (B) USDA/ARS #K-4964-13)*

part of the fruit is to attract animals and humans to the seed to help spread it over wide areas. This helps in the reproduction of the plant. Entire fruit or just the seed may be moved by wind, water, animals, and humans. Often the fruit is eaten by animals and humans as a source of food. Then the seeds may be discarded, where they can take root and grow into new plants. People assist greatly in spreading seeds and starting new plants when they plant seeds for crops.

There are many kinds of fruits and vegetables. The two terms are sometimes used incorrectly. A **vegetable** can be any part of a plant that is grown for its edible parts. This can be a root, stem, leaf, or ripened flower. However, fruit, which is a ripened or mature ovary, is a specific plant part. A **nut** is also a type of fruit.

PLANT TAXONOMY

Importance of Classifying Plants

Taxonomy is the science, laws, and principles of classification. In biology, taxonomy provides the means for classifying organisms into established categories according to characteristics. Such classification makes it easier to understand and remember plants and animals by the similarities and differences found in their structures and parts. Living organisms are given Latin names to help scientists and techni-

	Yellow Corn	Petunia
Common Name:	Corn	Petunia
Kingdom:	Plant	Plant
Phylum:	Spermatophyta (seed plants)	Embryophyta
Subphylum:	Angiosperm (seed in fruit)	Angiosperm
Class:	Monocotyledonae (single leaf seed)	Dicotyledonae (two-seed leaf)
Order:	Graminales (grasslike families)	Tubiflorea
Family:	Gramineae (grass family)	Solanaceae
Genus:	*Zea* (the corns)	*Petunia*
Species:	*Mays* (dent corns)	*Hybridea*
Variety:	Reid's yellow dent	Blue Moon

FIGURE 15–21 Example of a field crop (yellow corn) and an ornamental plant (Petunia) showing their complete botanical classifications.

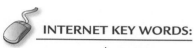

INTERNET KEY WORDS:

plant, taxonomy, binomial system

cians around the world communicate better. Latin is regarded as the universal language for those in professions dealing with the biological sciences. Agriscience has its origin in the biological sciences.

There are about 300,000 species of plants that have been identified and classified. The plant names are based on Latin descriptions and must be approved by a special committee of plant scientists. Carl Linnaeus, a Swedish botanist, developed the current system of plant classification in 1753.

Without the botanical classification system to identify and classify plants, many different species would carry one common name. For example, all clovers would be identified as clover, even though crimson clover is a winter annual, sweet clover is a biennial, and white clover is a perennial.

Complete classifications of field corn and petunia are shown in Figure 15–21.

Binomial System Used in Classifying Plants

When identifying a plant by its scientific name, it is not necessary to give its entire classification. Rather, a specific plant can be identified by using the genus and species only, because the genus and species name is not used in combination for any other plant or animal. For example, grain sorghum is *Sorghum* (genus) *vulgare* (species). **Genus** is the taxonomic category between family and species. It is customarily capitalized when written together with a species name. **Species** is the subgroup under genus. A species name is generally not capitalized when written in combination with its genus. It is appropriate for both the genus and the species names to be printed in italics. To compare the name of a plant or animal with that of a human, the species corresponds to the person's first name and the genus to the person's last name. The system of using genus and species in combination is referred to as a binomial system of classification. **Binomial** means consisting of two names.

Some species are broken down into varieties. A **variety** is a subgroup of plants developed by people, as opposed to species that originate in the wild. Variety is a rank within a species. When writing a plant variety name, it is generally capitalized. An example is Triumph wheat.

STUDENT ACTIVITIES

1. Write the Terms to Know and their meanings in your notebook.
2. Observe the plants that are commonly grown in your area. Classify them by the type of root system that they have.
3. Make a chart of the plants listed in Activity 2. Indicate their common and scientific names, and discuss their responses to drought or their water maintenance requirements.
4. Make a collection of different kinds of leaves. Classify each according to its shape and type of margin.
5. Sketch the parts of roots, stems, and leaves. Label all items.
6. Make a bulletin board showing the major parts of plants.
7. Ask your teacher to provide a microscope and slides of plant tissue. Diagram the plant parts and label the cells and other structures that you see.
8. Record the common and scientific names of plants listed in this unit and learn the correct spelling of each.
9. List the kingdom, phylum, class, order, family, genus, and species for three plants using the Internet, library, and other resources. Be sure to use proper punctuation.

SELF EVALUATION

A. Multiple Choice

1. The primary function of the root is to
   a. make sure that the plant will grow.
   b. anchor the plant and supply water and nutrients.
   c. ensure that the plant can be propagated.
   d. hold up the stem of the plant and provide propagation material.

2. The portion of the root that takes in the water and plant nutrients is the
   a. root cap.
   b. area of root division.
   c. root hair.
   d. area of cell maturation.

3. The major types of root systems are
   a. area of cell division and fibrous.
   b. fibrous and root cap.
   c. cuttings and root hairs.
   d. fibrous and taproot.

4. The area of cell division is
   a. responsible for the production of new cells on the tip of the root.
   b. where the cells will start to specialize.
   c. located in the area where the root hairs start to erupt from the wall of the epidermal cell.
   d. where the roots drop off on special plants such as the dodder.

5. The phloem
   a. is the pipeline that carries the water and nutrients from the soil to the leaves.
   b. is the part of the stem that gives support to the node.
   c. is the part of the leaf that holds it to the stem.
   d. carries the manufactured food from the leaves to the roots.

6. Herbaceous stems
  a. are tough and have bark around them.
  b. come from herbs.
  c. are green and are not winter hardy.
  d. are part of the bulb.

7. The node
  a. is the part of the stem that supports the flower.
  b. is the part of the stem where the leaf is attached.
  c. is the part of the stem that carries the nutrients.
  d. will become detached when dry weather sets in.

B. Matching

_____ **1.** Root cap
_____ **2.** Terminal bud
_____ **3.** Leaves
_____ **4.** Cuticle
_____ **5.** Blade
_____ **6.** Guard cells

  a. Located on the tip of the stem
  b. The wide portion of the leaf
  c. Protects the root tip as it moves in soil
  d. Manufacture food for the plant
  e. The topmost layer on the leaf
  f. Surround the stoma

C. Completion

1. The roots are responsible for _____ the plant.

2. An _____ plant is a plant used to improve the appearance of an area.

3. The area of _____ _____ is where the cells start to become specialized.

4. _____ are thick stems that run below the ground.

5. _____ are pores in the stem that allow the gases to pass through the stem.

6. _____ _____ are cells that give the leaf strength.

7. The _____ is the male part of the flower.

8. When a flower contains the stamen, pistil, petals, and sepals, it is considered a _____ flower.

9. The _____ is an enlargement that results after fertilization.

10. A _____ can be any part of a plant that is grown for its edible parts.

D. True or False

_____ **1.** Plants with many thin, hairlike roots have fibrous root systems.

_____ **2.** Plants with taproot systems are more likely to survive in a dry period.

_____ **3.** The root cap protects the young growing tip of the root.

_____ **4.** The xylem carries the nutrients and water down the stem of the plant.

_____ **5.** Herbaceous stems are tough and winter hardy.

_____ **6.** Bulbs are stems that are thick and compact.

_____ **7.** The vascular bundle is only in the leaf of the plant.

_____ **8.** The leaf consists of the petiole and the blade.

_____ **9.** The stomates allow for the passage of gases through only the leaf surface.

_____ **10.** The bract is the colored petal located in the flower.

UNIT
16

Plant Physiology

MATERIALS LIST

✓ writing materials
✓ encyclopedias
✓ corn or bean plant
✓ Internet access

COMPETENCIES TO BE DEVELOPED

After studying this unit, you should be able to:

- explain how plants make food.
- describe the roles of air, water, light, and media in relation to plant growth.
- trace the movement of minerals, water, and nutrients in plants.
- describe the ways that various plants store food for future use.
- compare the activity in a plant during exposure to light with periods of darkness.
- explain how plants protect themselves from disease, insects, and predators.

SUGGESTED CLASS ACTIVITIES

1. Meet as a class in the science laboratory or obtain sufficient microscopes to conduct a plant science laboratory class in another location. Stain some of the thin tissue from an onion bulb with iodine and observe the plant cells under the microscope. Observe under both low and high power. Identify the parts of a plant cell as they appear under the microscope.

2. Obtain two small potted plants. For one of the plants, fashion a covering from aluminium foil that completely blocks out sunlight to the plant. Place both plants in a window or other sunny location and wait 1 week. At the end of the week, remove the covering from the plant and examine the two plants to observe differences. Discuss the differences with the class.

3. Examine pondweed under a microscope on high magnification. Observe the plant cells. Make a drawing and label all visible structures.

TERMS TO KNOW

physiology	transpiration	macronutrient	acid
chlorophyll	osmosis	micronutrient	alkaline
glucose	semipermeable membrane	ion	chlorosis
light intensity	pore	anion	marginal burn
respiration	plant nutrition	cation	scorch
turgor	plant fertilization	precipitate	

INTERNET ADDRESSES:

<http://plantcell.lu.se/cells.html>

<http://www.cellsalive.com>

The life of a plant from its beginning to its maturity is a complex process. Many factors influence and directly control how a plant grows and what it produces. Growth, as in all living organisms, occurs by the division of cells and their enlargement as the plant increases in size (Figure 16–1). As the plant grows to maturity, the cells are produced, divide, grow, and become specialized organs. These specialized organs are stems, leaves, roots, flowers, fruits, and seeds (Figure 16–2). The study of how these organs function and the complex chemical processes that permit the plant to live, grow, and reproduce is **physiology.** An understanding of the processes of germination, photosynthesis, respiration, absorption of water and nutrients, translocation, and transpiration will enable the agriscience technician to maximize production of plants. The technician who works with interior and other ornamental plants must especially understand the environment of the plant, because it is not in its native habitat.

PHOTOSYNTHESIS

INTERNET KEY WORDS:

photosynthesis, plants

The most important life-sustaining process is photosynthesis. Without this chemical process, maintenance of life on this planet would not exist. Plants need carbon dioxide to manufacture food. Animals need oxygen to live. The complex chemical process of photosynthesis permits both to live and support each other.

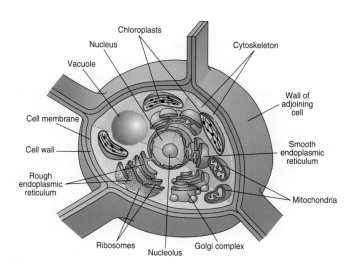

FIGURE 16–1 Major parts of a plant cell.

FIGURE 16–2 Major parts of a typical plant.

Science Profile

WHY TREES CHANGE COLOR

In the fall of every year, trees put on a display of magnificent beauty. Leaves turn from green to orange, red, yellow, brown, crimson, purple, and scarlet. The change is a beautiful reminder that the cold temperatures of winter will soon arrive. Although the change may appear to be art, science is responsible. During the spring and summer months, the leaves of trees appear green. Chlorophyll, the green light capturing pigment, also contains pigments called carotenoids. They are responsible for the yellow, brown, and orange colors found in bananas, rutabagas, and carrots. Anthocyanins are the red pigments that are found in the fluid within the cytoplasm of a plant cell. During the fall, anthocyanins are much more plentiful in response to sugars being trapped in the leaves. Chlorophyll is much more abundant than the other two pigments in the spring and summer and masks their colors until the fall when the day length gets shorter. As the days become shorter, plants slowly stop producing chlorophyll. This allows the other colors to appear and to produce a colorful show before winter arrives.

Photosynthesis is a series of processes in which light energy is converted to a simple sugar. Chlorophyll and chloroplasts are also essential in this process. **Chlorophyll** is the green material inside the leaves and stems of the plant. It is the substance that gives the green color to plant leaves. Chloroplasts are small, membrane-bound bodies inside cells that contain the green chlorophyll pigments. The chloroplasts are located in the mesophyll of the leaf. They are the sites of the actual conversion of solar energy (light) into stored energy (simple sugars).

Photosynthesis is the conversion of carbon dioxide and water in the presence of light and chlorophyll into glucose, oxygen, and water. **Glucose** is a simple sugar and contains the building blocks for other nutrients. A simple chemical definition of photosynthesis is:

$$6CO_2 + 12H_2O \xrightarrow[\text{chlorophyll}]{\text{light energy}} C_6H_{12}O_6 + 6O_2 + 6H_2O$$

carbon dioxide   water   sugar   oxygen   water

The rate at which the food-making process occurs depends on the light intensity, temperature, and concentration of carbon dioxide in the atmosphere. **Light intensity** is also known as the quality of light, or the brightness of light. Light must be present with sufficient brightness for the process to be successful. Some plants are able to adapt to various levels of light brightness. A knowledge of the level of light required for plants to grow well is essential, particularly for indoor plant production.

Temperature is also an important factor in the process of food manufacturing in the leaf. Photosynthesis occurs best in a temperature range of 65° to 85° F (18–27° C). Extremes of temperatures slow down or completely stop the process of photosynthesis. A lack of carbon dioxide also will affect photosynthesis. Carbon dioxide is especially important in the beginning of the process. Under normal outdoor conditions, its availability is not a problem. However, in enclosed conditions such as those found in a greenhouse, a carbon dioxide shortage could be a limiting factor. To correct this problem, a carbon dioxide generator might be used (Figure 16–3).

FIGURE 16–3 A technician measures the effect of carbon dioxide enrichment of the atmosphere on the transpiration rate and level of activity of the stomata that occur in the plant leaves.
(Courtesy USDA/ARS #K-3750-7)

RESPIRATION

INTERNET KEY WORDS:

plants, respiration
transpiration, plants

All living cells carry on the process of respiration. **Respiration** is a process by which living cells (plant or animal) take in oxygen and give off carbon dioxide. Unlike photosynthesis, which occurs only in the light, respiration occurs both day and night. It is not easily measured during the day, because the presence of photosynthesis will mask or obscure the occurrence of respiration. Respiration is a breaking-down process. It uses the sugars and starches produced by photosynthesis and converts them into energy. The chemical equation for respiration is:

$$\underset{\text{sugar}}{C_6H_{12}O_6} + \underset{\text{oxygen}}{6O_2} \longrightarrow \underset{\substack{\text{carbon} \\ \text{dioxide}}}{6CO_2} + \underset{\text{water}}{6H_2O} + \text{heat (energy)}$$

A comparison of the activities that occur during photosynthesis and respiration may be helpful in understanding the two processes (Figure 16–4).

TRANSPIRATION

Water saturates all of the spaces between the cells throughout the plant. About 10 percent of the water that enters from the roots is used in chemical processes and in the plant tissues. Functions of this water include transporting minerals throughout the plant, cooling the plant, moving sugars and plant chemicals, and maintaining turgor pressure. **Turgor** is a swollen or stiffened condition as a result of the plant cells being filled with liquid. When the plant does not have enough water, turgor pressure is lost, and the plant becomes wilted (Figure 16–5).

The exchange of gases is important to the plant, because air is needed for photosynthesis to occur, and water vapor must exit the plant to draw more dissolved nutrients into the plant. Both of these important functions occur through the tiny openings in the leaf called stoma. The stoma are surrounded by specialized cells called guard cells. The guard cells control the size of the opening in the surface of the leaf, depending

Photosynthesis	Respiration
1. Food is produced.	1. Food is used for plant energy.
2. Energy is stored.	2. Energy is released.
3. It occurs in cells that contain chloroplasts.	3. It occurs in all cells.
4. Oxygen is released.	4. Oxygen is used.
5. Water is used.	5. Water is produced.
6. Carbon dioxide is used.	6. Carbon dioxide is produced.
7. It occurs in sunlight.	7. It occurs in dark as well as light.

FIGURE 16–4 Comparison of the activities that occur during photosynthesis and respiration.

FIGURE 16–5 When a plant is unable to obtain enough moisture from the soil to replace moisture that is lost to the atmosphere, it loses turgor pressure and becomes wilted.

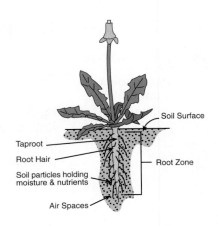

FIGURE 16–6 The process of transpiration occurs as water vapor and air are exchanged through the stoma.

FIGURE 16–8 Root hairs extend from the main root into the pore spaces in the soil from which they absorb water and dissolved nutrients.

FIGURE 16–7 Factors that influence transpiration.

on the amount of water available to the plant and other conditions in the environment surrounding the plant.

Transpiration is the process by which a plant gives up water vapor to the atmosphere (Figure 16–6). Transpiration takes place primarily through the stoma, which open to allow water vapor and air to be exchanged by the leaf. Most plants transpire about 90 percent of the water that enters through the roots.

Transpiration is greatly influenced by humidity, wind and other air movement, and temperature (Figure 16–7). As humidity in the air around the plant increases, the rate of transpiration decreases. Conversely, as humidity decreases, the rate of transpiration increases. Increased air movement around the plant increases the rate of transpiration, due to the evaporation caused by air movement. Similarly, as temperature increases, the rate of transpiration increases.

Often during dry weather or when plants are not watered, transpiration causes the plant to lose water faster than it can be replaced by the root system. When this occurs, the guard cells will close the stoma in the leaves, thus slowing down the rate of transpiration. This mechanism enables the plant to preserve the water it contains. If there is water in the soil, the plant may wilt slightly, but it will recover. However, if there is insufficient moisture in the soil, the plant may not be able to recover.

SOIL

Productive soil provides a natural environment for the root zone. It provides air, water, and nutrients for the plant. Root hairs penetrate the pore spaces in the soil and absorb plant nutrients (Figure 16–8). A process called osmosis is used to get water and nutrients into root cells so they can be transported to the remainder of the plant. **Osmosis** is the process by which water moves from an area of high concentration to an area of low concentration through a semipermeable membrane that is separating two solutions (Figure 16–9). A **semipermeable membrane** will allow certain things to pass through, whereas other things cannot. The epidermis or outside cell layer of a root hair is a semipermeable membrane. The root allows things such as water, minerals, and nutrients to come inside the plant. The movement of water and dissolved minerals tends to concentrate minerals and nutrients inside the plant at

Osmosis

FIGURE 16–9 During osmosis, water moves from areas of high concentration to low concentration. (A) The water outside the balloon has a greater concentration at 100 percent than the water inside the balloon at 80 percent. (B) Over time, some of the water from outside the balloon has moved inside, crossing the selectively permeable membrane, in an attempt to balance the concentrations.

INTERNET KEY WORDS:

osmosis,
semipermeable membrane

greater levels than in the soil. Soil moisture containing a lower concentration of dissolved nutrients and a greater concentration of water is able to move into the root hairs. Once inside the root-hair cell, nutrients can be transported to other parts of the plant as needed.

To allow the root hairs to move through the soil, the soil must have spaces between the particles of sand, silt, and clay. Such spaces are called **pores.** Their role is to store air, water, and nutrients and to permit root penetration.

AIR

INTERNET KEY WORDS:

plants, carbon dioxide
plants, water

The air or atmosphere that surrounds the portion of the plant that is above the ground must supply carbon dioxide, as well as oxygen. Generally, this is not a problem when the plants grow outdoors in fields. When plants are transplanted into artificial or unnatural environments, consideration must be given to the quality of the air surrounding the plant. In a greenhouse or other enclosed system, the quality of the atmosphere needs to be monitored. The presence and levels of carbon dioxide and pollutants must be understood for maximum production to occur. With certain crops in greenhouses, such as carnations and roses, the addition of some carbon dioxide might be desirable to increase crop production. In areas where crops are growing near industrial plants or cities, or along major highways, the technician must be aware of the many types of pollutants that might reduce production or severely damage the plants.

WATER

A consistent supply of pure water is absolutely necessary for plant and animal growth. The most essential ingredient for all living things is water. Nutrients in the soil must first be dissolved in water before they

Agri-Profile

CAREER AREA:

Plant Physiology

Physiology refers to the many functions that occur inside of plants. These include familiar activities such as osmosis, nutrient uptake, translocation, respiration, photosynthesis, food movement, and food storage. Other complex functions are known only to specialists in the field.

Plant physiologists work closely with technicians and scientists in other fields of plant science. They may be consultants to or collaborators with specialists in agronomy and horticulture. The work of plant physiologists is typically done by college or university faculty, employees of state or national research institutes, or specialists with agriscience corporations developing and selling seeds and plant materials.

Plant physiologists Marcia Holden and Douglas Luster inspect tomato plants for iron deficiency. *(Courtesy USDA/ARS #K-4191-5)*

can be absorbed through the plant roots. When inside the plant, water carries the nutrients to the leaves. These nutrients chemically combine with water in the process of photosynthesis. Sugars and other plant foods manufactured in the leaves are then transported throughout the plant by water. Water helps to control the temperatures in and around plants through transpiration. Water gives the plant support by maintaining rigidity in the cells. It is important, therefore, that water used for plant production be of good quality and in adequate supply.

PLANT NUTRITION

INTERNET KEY WORDS:

plant nutrient deficiency symptoms

Plant nutrition is often confused with plant fertilization. There is a difference. **Plant nutrition** refers to availability and type of basic chemical elements in the plant. **Plant fertilization** is the process of adding nutrients to the soil or leaves so these chemicals are added to the growing environment of the plant. Before chemicals that are supplied as fertilizer can be taken up and used by plants, they generally undergo various changes.

Essential Nutrients

INTERNET KEY WORDS:

plant micronutrients
plant macronutrients

There are 16 elements that are essential for normal plant growth. They are required in various amounts by plants and must be available in the relative proportions needed if the plants are to produce well. Three are used in huge amounts and are obtained from the atmosphere and water around the plant. They are carbon (C), hydrogen (H), and oxygen (O). There are six elements that are used in relatively large amounts. They are called **macronutrients.** The macronutrients are nitrogen (N), phosphorus (P), potassium (K), calcium (Ca), magnesium (Mg), and sulfur (S). They are all obtained from the soil.

An additional seven elements are used in small quantities. They are called **micronutrients** (trace elements). The micronutrients are also

FIGURE 16-10 Soil and nutrient management specialists do cooperative research to determine the best rates of fertilizer to optimize corn growth. *(Courtesy USDA/ARS #K-3694-5)*

obtained from the soil. They are boron (B), copper (Cu), chlorine (Cl), iron (Fe), manganese (Mn), molybdenum (Mo), and zinc (Zn). For a plant to grow at maximum efficiency, it must have all essential plant nutrients (Figure 16–10). The absence of any one of these nutrients will cause the plant to grow poorly or show some signs of poor health.

Remembering the 16 Plant Nutrients

Various schemes have been devised to help you remember the names of the 16 plant nutrients. One technique is to first learn the chemical symbols. Use the symbols to make a logical string of words that are easy to remember. One such string of words that uses the symbols of most of the nutrients is "C. Hopkin's cafe, mighty good." By remembering this phrase, you can recall the symbols of 10 of the 16 nutrients as follows: C HOPKNS CaFe Mg (carbon, hydrogen, oxygen, phosphorus, potassium, nitrogen, sulfur, calcium, iron, and magnesium). The remaining ones are boron, copper, chlorine, manganese, molybdenum, and zinc. Can you devise a string of words to help you remember the symbols of these micronutrients?

Ions

Plant nutrients are absorbed from the soil-water solution that surrounds the root hairs of the plant. In fact, 98 percent of the nutrients obtained from the soil is absorbed in solution, whereas the other 2 percent is extracted by the root directly from soil particles. Most of the nutrients are absorbed as charged ions. An **ion** is an atom that has an electrical charge.

Ions that are negatively charged are called **anions.** Positively charged ions are called **cations.** The electrical charges in the soil are paired so that the overall effect in the soil is not changed. These ions compete and interact with each other according to their relative charges. For example, nitrogen, in its nitrate form, has a negative charge and chemical formula NO_3^-. Therefore, nitrates are anions with negative charges.

Conversely, potassium has a positive charge and is an example of a cation (K^+). Potassium nitrate, $K^+NO_3^-$, is a combination of potassium and nitrate consisting of one nitrate ion and one potassium ion.

1. **HEALTHY** *stalk has normal size. Cut-away stalk section below ear shows healthy, white pith area.*

2. **POTASSIUM** *is needed when the cut-away shows dark brown discoloration at the nodes.*

3. **PHOSPHORUS** *shortage leads to weak, spindly stalks, often barren of ears. Note purple coloration of lower leaves.*

4. **SUCKERS** *may form when corn gets too much nitrogen early in the season. Cut-away section also shows corn borer damage.*

5. **DISEASE** *symptoms found in stalks include the black bundles in the upper cut-away, and the darker pith in the lower cut-away. Stalk rot works inside first, causing early dying and breakage of the stalk, and shrivelled ears with chaffy, low-test-weight grain.*

FIGURE 16–11 Nutrient deficiencies decrease plant health, vigor, and growth. Deficiency of many nutrients can be determined by observing the plant. *(Courtesy of Potash and Phosphate Institute, Norcross, GA)*

Calcium nitrate, $Ca^{++}(NO_3^-)_2$, has two nitrate ions and one calcium ion. The reason is that the calcium cation has two positive charges. As you might guess, this could be confusing, but it illustrates the need to understand chemistry to manipulate plant fertility if conditions are not ideal in the natural environment.

The balance of ions is important and needs to be carefully monitored for good plant growth. Opposite charges attract each other, but ions with similar charges compete for chemical reactions and interactions in the soil-water environment. Some ions are more active than others and might be able to compete better in the soil. Further study would be needed to thoroughly understand why soil tests may indicate the presence of a certain element in sufficient amounts for plant growth, yet the plants may show deficiency symptoms (Figure 16–11).

Deficiency means a shortage of a given nutrient that is available for plant use. A good example of a nutrient deficiency symptom is blossom-end rot of tomato. This is common in gardens and occurs when there is not enough water to dissolve and carry calcium to the plant in sufficient quantities. The end opposite the stem is called the blossom end. The calcium deficiency produces a tomato that looks good from the top, but, when picked, the bottom end is rotten.

Soil Acidity and Alkalinity

INTERNET KEY WORDS:

soil, anion, cation, ion plants, "soil pH"

INTERNET ADDRESS:

<http://www.greenair.com>

The chemistry of plant elements in the soil can be affected by pH. Soil pH is a measurement of acidity (sourness) and alkalinity (sweetness) (Figure 16–12). Many of the nutrients in soil form complex combinations and are capable of precipitating out of solution, where they are unavailable to the plant. Precipitate occurs when a solid is dropped out of solution. If the soil pH is acid, or extremely low, some micronutrients become too soluble and occur in concentrations great enough to harm the plants (Figure 16–13).

In contrast, if soil pH is high, in the alkaline range, many of the nutrients can be precipitated out and not be available to the plants. The pH of soils can be determined with low-cost test kits. Fortunately, soil pH can be corrected by adding lime if the pH needs to be increased, and sulfur or gypsum if it needs to be decreased. Such practices are common, because soil pH is seldom perfect for the crop being grown (Figure 16–14).

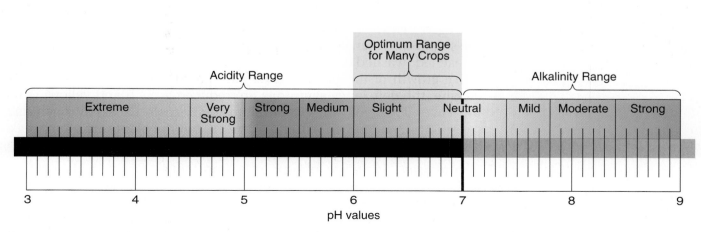

FIGURE 16–12 The pH scale measures the balance of positive and negative ions in a solution or in the soil.

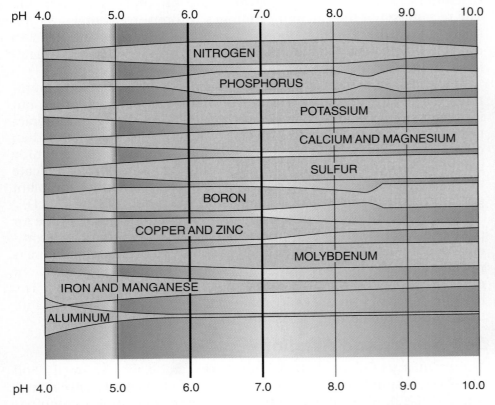

pH 4.0 5.0 6.0 7.0 8.0 9.0 10.0

NITROGEN

PHOSPHORUS

POTASSIUM

CALCIUM AND MAGNESIUM

SULFUR

BORON

COPPER AND ZINC

MOLYBDENUM

IRON AND MANGANESE

ALUMINUM

pH 4.0 5.0 6.0 7.0 8.0 9.0 10.0

FIGURE 16–13 This chart illustrates the effect of soil pH on the availability of plant nutrients. A wide section indicates high availability of the nutrient, and a narrow section indicates that the nutrient is not available to a plant.

FIGURE 16–14 Soil scientist Charles Foy compares barley plants grown in soils having different pH levels. *(Courtesy USDA/ARS #K-3212-1)*

Plant Nutrient Functions

The importance of carbon, hydrogen, and oxygen has already been discussed under the topic of photosynthesis. The other nutrients have very specific functions and must be available in the appropriate amounts and form. The effects of plant nutrients may be likened to a chain—the weakest link will determine how much the chain will pull. Similarly, the nutrient in shortest supply will determine the maximum growth that can be achieved by the plant.

Nitrogen   Nitrogen is present in the atmosphere as a gas. It is added to the soil in some fertilizers. Because it exists in nature as a gas, it is easily leached (washed out of the soil). Nitrogen is responsible for the vegetative growth of the plant and its dark green color. When nitrogen is lacking, the deficiency symptons are reduced growth and yellowing of the leaves. This yellowing is referred to as **chlorosis.** Excess nitrogen can cause succulent growth that is dark green, but the plants are often weak and spindly.

Phosphorus   In nature, phosphorus is present as a rock and is not easily leached out of the soil. It is important in the growth of seedlings and young plants, and it helps the plants develop good root systems. Some symptoms of a deficiency of phosphorus are reduced growth, poor root systems, and reduced flowering. Thin stems and browning or purpling of the foliage are also signs of poor phorphorus availability.

Nutrient	Excess	Deficiency
Iron	Rare	Interveinal chlorosis, especially on young growth
Zinc	Might appear as an iron deficiency	Interveinal chlorosis, reduction in leaf size, short internodes
Molybdenum	Not known	Interveinal chlorosis on older leaves; may also affect leaves in the middle of the plant
Boron	A blackening or death of tissue between veins	Failure to set seed; death of tip buds
Copper	Might occur in low pH; will appear as an iron deficiency	New growth small and misshapen, wilted
Manganese	Brown spotting on leaves; reduced growth	Interveinal chlorosis of the leaves and brown spotting; checkered effect possible

FIGURE 16–15 An excess or deficiency of most nutrients will cause predictable symptoms in plants.

INTERNET KEY WORDS:

chlorosis, leaves

Potassium Potassium is mined as a rock and made into fertilizer, but it can be leached from the soil. If too much potassium is present, it can cause a nitrogen deficiency. A lack of potassium will appear as reduced growth or shortened internodes, and sometimes as **marginal burn** or **scorch** (brown leaf edges). Dead spots in the leaf and plants that wilt easily are also indications of a potassium deficiency.

Calcium Calcium is often supplied by adding lime to the soil. It can be leached out and does not move easily throughout the plant. Too much calcium can cause a high pH and reduce the availability of some elements to the plant. A lack of this element can stop bud growth and result in death of root tips, cupping of mature leaves, and blossom-end rot of many fruits. Pits on root vegetables are also signs of calcium deficiency.

Magnesium Magnesium can be added by using high-magnesium lime. It can also be leached from soil. If magnesium is lacking, some reduction of growth and marginal chlorosis can be noticed. In some plants, even interveinal chlorosis can be seen. Cupped leaves and a reduction in seed production can also be symptoms of magnesium deficiency. Foliage plants commonly lack this nutrient.

Sulfur Present in the atmosphere as a result of combustion, sulfur is often an impurity in fertilizers carrying other nutrients. As a result, it is rarely deficient. However, if sulfur is deficient, a yellowing of the entire plant may result.

Deficiencies and exesses of nutrients cause typical symptoms in plants (Figure 16–15).

FOOD STORAGE

When the plant makes its food through photosynthesis, it often manufactures more than it needs to maintain itself. This excess is often stored in the plant for future use. Such food may be stored in roots, stems, seeds, or fruits.

Roots

The most common type of root that serves as a storage organ is the tap-root. Some common examples of plants with extensive storage capacity are sugar beets, carrots, radishes, and turnips. The sugars and carbohydrates are transported down the phloem and into the root cells. They are held there as the root enlarges. Most of the time, this type of plant is a short-term crop that does not take long to mature. This type of root system is easy to dig or harvest.

Stems

The stems of plants usually contain cells that are necessary for plant support. Some specialized stems, however, are excellent food storage organs (Figure 16–16). Some are used for propagation and some for food. The most common specialized stem used for food is the tuber. It is an enlarged portion of a stem containing all of the parts of a normal stem. Nodes, internodes, and buds can be identified in tubers. The Irish potato is an example of a tuber.

Corms and bulbs are other examples of specialized stems that contain large amounts of food made by photosynthesis. A rhizome is yet another example. Often, these are used for propagation purposes.

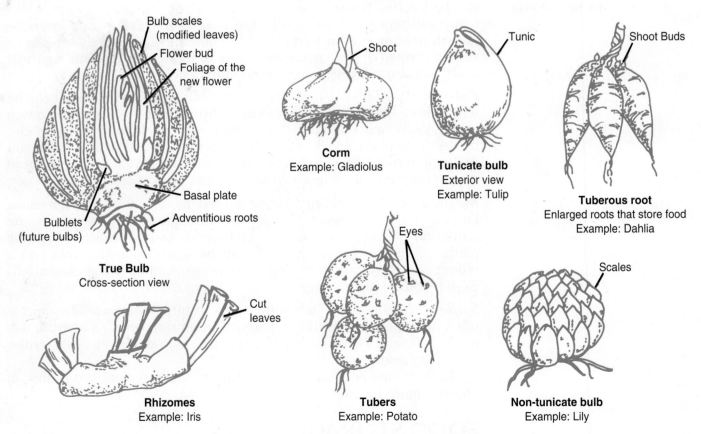

FIGURE 16–16 Examples of stems that are major food storage organs for the plant.

Science Connection

DIAGNOSING AN EMINENT KILLER

A subtle and mysterious killer, citrus blight, has eluded plant pathologists and other scientists for more than a century. Recognized by citrus growers and the federal government as a deadly disease of citrus trees, the U.S. Horticultural Research Laboratory was established in Eustis, Florida in 1892 to research this problem. Citrus blight is a mysterious disorder that renders a tree worthless for fruit production. It is most likely to attack young trees that are just beginning to produce fruit, but it also kills trees 20, 30, or even 50 years old. No cure or prevention is known for the citrus blight. First diagnosed in Florida in 1874, it causes physiological changes in the tree and leads first to a yellowing of leaves and eventually to leaf wilt. With no cure and no chance of recovery, diseased trees are destroyed as soon as the disorder is observed. Authorities estimate that a half million trees valued at $60 million are lost each year to the blight.

U.S. Department of Agriculture (USDA) plant pathologist Michael Bausher believes he has found a way that may lead to early diagnosis, and thus reduce losses from the disorder. At the U.S.

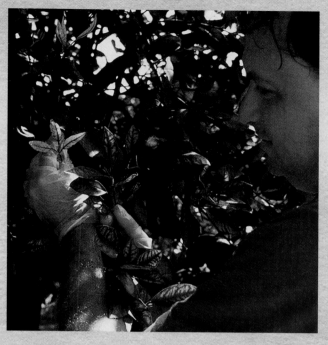

Pale green, chloritic leaves and reduced leaf size typical of citrus blight are displayed by plant pathologist Michael Bausher. *(Courtesy USDA/ARS #K-3323-3)*

Horticultural Research Lab in Orlando, Bausher discovered unique proteins in the leaves of diseased trees that are not present in healthy trees or in trees with other diseases or stress problems. Bausher prepared, freeze-dried, and partially purified an antigen derived from ground-up leaves of blighted trees. When the antigen was injected into rabbits, he discovered the rabbits produced a unique antisera. The rabbit-produced antisera was then found to react positively with the unique proteins from the blighted trees. However, the antisera reacted negatively when exposed to proteins derived from tissue of healthy trees and trees with other disorders or diseases.

Growers usually remove trees at the first indication of citrus blight to cut the losses from the disease. Unfortunately, other disorders may cause yellowing or mottling of leaves, which are the first observable symptoms of the blight. Therefore, it is hard to tell how many trees with correctable disorders are sacrificed. It is hoped that a process can be developed wherein proteins can be used as biological markers to help identify trees with citrus blight before the visual symptoms appear. This early detection would permit growers to remove only trees that really have the citrus blight and spare those that have other correctable disorders. Biological markers may help researchers develop trees that are resistant to the blight.

Seeds

As the ovule of a plant matures, it stores food for the young embryo to start its growth when it germinates. Both humans and animals use seeds as major food sources. They help the plant by spreading the seed to new locations, increasing the ability of the plant to survive.

Hot Topics in Agriscience

NEW PLANT VARIETIES—KEY TO AN ABUNDANT FOOD SUPPLY

Agricultural crop production is becoming more efficient all the time. Yields and quality are both increasing. These improvements are partially because of the development of better plants. Scientists are continually seeking ways to improve our food crops. Most of our highly productive crop plants are resistant to one or more diseases or pests. Some of them are able to withstand drought conditions or other environmental conditions that are known to reduce crop production.

Some new plants have been modified to produce food with different nutrient levels than those of the parent varieties. For example, some new wheat varieties contain less protein in the grain than the same quantity of grain produced by the parent stock. This is important to the industries that make pasta and crackers. The quality of these products is improved by using low-protein wheat. New plant varieties will always be in demand because the world will always be seeking greater crop yields and higher quality.

INTERNET KEY WORDS:

plants, new varieties
biology, plants, seeds

SUMMARY

This unit has covered some basic principles of plant physiology. Physiology is complex and must be studied in great depth to gain understanding. Plant physiologists typically have Master's or Doctorate degrees. However, most technicians and scientists in the plant sciences will have some training in plant physiology. A basic knowledge of soils and how plants use nutrients and function in general will help greatly in the successful production and management of plants. Farmers, ranchers, and growers generally have access to sophisticated technology (Figure 16–17).

FIGURE 16–17 Missouri farmer Bill Holmes and specialists from the Space Remote Sensing Center examine soil fertility variations in Holmes's cropland. *(Courtesy USDA/ARS #K-4914-2)*

STUDENT ACTIVITIES

1. Write the Terms to Know and their meanings in your notebook.

2. Make a bulletin board showing the cross section of a leaf. Label the various cells and leaf parts. Include the formula for photosynthesis.

3. Write an article for a newspaper or magazine that explains the importance of photosynthesis.

4. Collect plants or pictures to make a display that could be used by others to help identify nutrient deficiencies.

5. Select a crop that interests you and conduct research to determine the optimum nutrient requirements.

6. Set up a demonstration to explain osmosis.

7. Make a list of all the chemical symbols of plant nutrients. Write a sentence or story to help you remember them.

8. In your own words, explain the process of photosynthesis. Be sure to include all of the steps in the process.

9. In your own words, explain respiration. Be sure to include all of the conditions that are necessary for the steps to occur.

SELF EVALUATION

A. Multiple Choice

1. The study of functions and the complex chemical processes that allow plants to grow is known as
   a. plant taxonomy.
   b. plant physiology.
   c. plant nutrition.
   d. photosynthesis.

2. Chlorophyll is important in plants because it
   a. creates an atmosphere where it can determine the osmotic pressure.
   b. allows the plant to make good xylem tissue.
   c. makes it possible for plants to grow.
   d. is also known as the chloroplasts.

3. The rate at which photosynthesis is carried out depends on
   a. the amount of fertilizer in the water.
   b. the amount of oxygen in the atmosphere.
   c. the amount of respiration carried on during the daylight hours.
   d. the light intensity, temperature, and concentration of carbon dioxide.

4. Photosynthesis will work best in which temperature range?
   a. 50° to 60° F
   b. 60° to 70° F
   c. 65° to 85° F
   d. 85° to 95° F

5. Respiration
   a. uses food for plant energy.
   b. stores energy.
   c. occurs in cells that contain chlorophyll.
   d. uses carbon dioxide.

6. Plant nutrition is

a. plant food added to the plant pot.

b. use of basic chemical elements in the plant.

c. chemical processes providing plants with elements for growth.

d. the measurement of acidity (sourness) and alkalinity (sweetness).

B. Matching

_____ 1. pH

_____ 2. Osmosis

_____ 3. Corm

_____ 4. Root

_____ 5. Sulfur

_____ 6. Leaves

_____ 7. Fertilization

a. Movement through a semipermeable membrane

b. Addition of nutrients to the plant-growing environment

c. Storage organ for excess plant food

d. Site of photosynthesis

e. Measurement of acidity and alkalinity

f. Macronutrient

g. Specialized stem

C. Completion

1. Extremes of temperature will slow down or completely stop _____.

2. Respiration will occur only in the _____.

3. When the temperature increases, the rate of transpiration _____.

4. Soil provides a natural environment for the _____ _____.

5. The spaces in between the soil particles, where the soil water is found, are called the _____.

D. True or False

_____ 1. Respiration is a building process that uses sunlight to work.

_____ 2. The process by which a plant loses water is perspiration.

_____ 3. All plant elements perform the same function in the plant.

_____ 4. Humidity refers to the amount of water in the atmosphere.

UNIT 17

Plant Reproduction

OBJECTIVE

To determine the methods used by plants to reproduce themselves and to explore new propagation technology.

MATERIALS LIST

- ✓ seed catalog
- ✓ grafting knife
- ✓ grafting rubber
- ✓ grafting wax
- ✓ cutting knife
- ✓ rooting hormone

- ✓ stock plants
- ✓ rooting media
- ✓ tissue-culture tubes
- ✓ tissue-culture media
- ✓ scalpel

- ✓ razor blade (single edge)
- ✓ tweezers
- ✓ 50 percent alcohol solution
- ✓ sanitary work area
- ✓ Internet access

COMPETENCIES TO BE DEVELOPED

After studying this unit, you should be able to:

- distinguish between sexual and asexual reproduction.
- explain the relationship between reproduction and plant improvement.
- draw and label the reproductive parts of flowers and seeds.
- state the primary methods of asexual reproduction and give examples of plants typically propagated by each method.
- explain the procedures used to propagate plants via tissue culture.

SUGGESTED CLASS ACTIVITIES

1. Assign each class member to conduct an experiment to determine how long it takes for several kinds of seeds to germinate. Place three to five seeds of a single species of plant in damp tissue paper. Place the tissue in a section of an egg carton. Keep the tissue damp without excess water. Label the seeds by writing on the carton. Check the seeds daily and add water as necessary. Keep a record of the time that is required for germination to occur.

2. Conduct a cloning exercise using African violets or potatoes as parent stock. The procedure to be followed is described later in this unit. Have students work alone or in pairs. A written report should always be part of a classroom experiment. It should describe the process that was followed, supplies and equipment that were used, and results that were achieved.

3. Spend the first half of a class period collecting pollen from different plants. This can be done outdoors during the appropriate seasons. If outdoor collection is not possible, check with the local flower shops for discarded flowers. Next, set up a number of microscopes and examine the pollen. As a class, discuss the similarities and differences that can be seen.

333

TERMS TO KNOW

propagation	viable	double-eye cutting	grafting
reproduction	germination rate	leaf cutting	scion
sexual reproduction	cuttings	leaf petiole cutting	rootstock
asexual reproduction	fungicide	leaf section cutting	stock
vegetative	rooting hormone	split-vein cutting	graft union
hybrid vigor	stem tip cutting	root cutting	bud grafting
germinate	stem section cutting	layering	T-budding
dormant	cane cutting	simple layering	budding rubber
imbibition	heel cutting	tip layering	tissue culture
scarify	single-eye cutting	air layering	agar

INTERNET KEY WORDS:

pollen, sperm, ovule, egg

Plant **propagation,** or **reproduction,** is simply the process of increasing the numbers of a species, or perpetuating a species. The two types of plant propagation are sexual and asexual. **Sexual reproduction** is the union of an egg (ovule) and sperm (pollen), resulting in a seed. Two parents creating a third individual is referred to as sexual propagation. In plants, it involves the floral parts. It may involve one or two plants. **Asexual reproduction** uses a part or parts of only one parent plant. The purpose is to cause the parent plant to make a duplicate of itself. The new plant is a clone (exact duplication) of its parent. Because this type of reproduction uses the vegetative parts of the plant, namely the stems, roots, or leaf, it is often referred to as **vegetative** propagation.

Sexual propagation has some distinct advantages. It is often less expensive and quicker than some other methods. It is the only way to obtain new varieties and also capture hybrid vigor. A hybrid is a plant obtained by crossbreeding. **Hybrid vigor** refers to the tendency of hybrid plants to be stronger and survive better than plants of a pure variety. Sexual propagation is a good way to avoid passing on some diseases. In some plants, sexual propagation is the only way they can reproduce.

Asexual propagation has many advantages as well (Figure 17–1). In some cases, it is easier and less expensive to obtain plants this way. In some species or cultivars, it is the only way they can be propagated.

PLANT PROPAGATION	
Sexual Propagation Advantages	**Asexual Propagation Advantages**
• Less expensive • Many plants can be produced quickly • Crosses result in hybrid vigor • Avoids passing on some diseases	• Less time is required to produce a salable plant • Plants are genetically identical • The only way to reproduce some plant varieties

FIGURE 17–1 Two methods of plant reproduction, sexual and asexual propagation, are in common use in the plant industry. Each method has distict advantages over the other.

Science Connection

PLANT REPRODUCTION

It takes two chromatids to make up one chromosome. One chromatid comes from the male and one from the female parent. A haploid cell has half the normal amount of genetic material because only a single chromatid from one of the parents is present. A diploid cell has two chromatids, one from the female and one from the male parent. These two chromatids, when paired together, make one chromosome. The number of chromosomes in an organism depends on its species. Genes from both parents are represented in a diploid cell.

Plants sexually reproduce by a process called alternation of generation. In their life cycles, plants change back and forth between producing haploid cells and producing diploid cells. During the haploid generation, the plant produces sperm, eggs, and, in some cases, both. In flowering plants, the sperm is called pollen, and the egg is called an ovum. Once the sperm and the egg are fused, their chromatids become paired. This starts the diploid generation. After fertilization, the membrane around the ovum hardens and becomes a seed with a developing embryo inside. Under the right growing conditions, the seed will develop into a plant. At this point, the alternation of generation is complete. Once the plant matures, the cycle begins all over again.

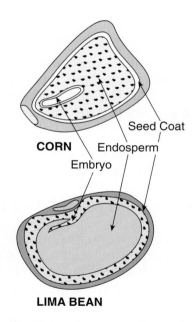

CORN
Seed Coat
Endosperm
Embryo

LIMA BEAN

FIGURE 17–2 Parts of a seed.

SEXUAL PROPAGATION

A seed is made up of the seed coat, endosperm, and embryo (Figure 17–2). The seed coat functions as a protector for the seed. Sometimes it is thin and soft, or it may be hard and impervious to water or moisture. The endosperm functions as a food reserve. It will supply the new plant with nourishment for the first few days of life. The embryo is the young plant itself. When a seed is fertilized and matures, it will be dormant. When it is given favorable growth conditions, it will **germinate,** which means that the seed will sprout and begin to grow.

Seed propagation starts with quality seed. Crop production by sexual reproduction allows consideration to be given to the type of plant that is needed or the variety that is best adapted to a particular area or purpose. Hybrid plants are developed by cross-pollinating two different varieties. Many varieties on the market are the result of hybridization or crossbreeding. Seeds of hybrid plants cost more than open-pollinated varieties. However, the increased quality of the plants generally offsets the increased cost of the seed. New varieties are being developed for disease and insect resistance. It is natural to expect the seeds of such improved varieties to cost more than standard or regular seeds. Some varieties have unusual cultural or product characteristics (Figure 17–3).

Seeds collected from plants used for commercial production generally will not save money in the long run. Seeds from such plants are often small. They are often poorly managed and improperly handled and stored (Figure 17–4). It is recommended that only certified seed be used. Seed saved from season to season should be stored in a sealed jar at 40° F (4.4° C) and at low humidity.

FIGURE 17-3 At the National Seed Storage Laboratory in Fort Collins, Colorado, seeds are packaged in flexible, moisture-proof bags. *(Courtesy USDA/ARS #K-1657-16)*

FIGURE 17-4 Hybrid seed is produced by crossbreeding two different varieties of the same species. This usually results in greater production and greater plant vigor.

INTERNET KEY WORDS:

seed, germination

FIGURE 17-5 Seeds require water, air, light, and favorable temperature to germinate and begin to grow.

Germination

When seed is harvested, or collected, it is normally mature and in a **dormant,** or resting, state. To germinate and start to grow, it must be placed in certain favorable conditions (Figure 17–5). The four environmental factors that must be right for effective germination are water, air, light, and temperature.

Water **Imbibition** is the absorption of water. It is the first step in the germination process. The seed, in its dormant stage, contains little water. The imbibition process allows the seed to fill all its cells with water. If other conditions are favorable, the seed then breaks its dormant stage and germinates.

A good germination medium is important. The medium must not be too wet or too dry. An adequate and continuous supply of water must be available. This is often difficult to control with crops directly seeded in the field. It is much easier to control in crops planted for transplanting. A dry period during the germination process will result in the death of the young embryo. Too much water will result in the young seed rotting. In some species, the seed coat is very hard and water cannot penetrate to the endosperm. In these cases, it is necessary to scarify the seed.

A common way to **scarify** seed is to nick the seed coat with a knife or a file. Another method is to soak the seeds in concentrated sulfuric acid. This requires special care and experience, because sulfuric acid is a dangerous material. Another technique is to place seeds in hot water that is 180° to 212° F (82.2–100° C) and allow them to soak as the water cools. This process takes 12 to 24 hours. A warm, moist scarification process may be used by simply placing the seeds in warm, damp containers and letting the seed coat decay over time.

Air Respiration takes place in all viable seed. Viable seed is alive and capable of germinating. Oxygen is required. Even in nongerminating seeds, a small amount of oxygen is required even though respiration is low. As germination starts, the respiration rate increases. It is important that the seed be placed in good soil or media that is loose and well drained. If the oxygen supply is limited or reduced during the germination process, germination will be reduced or inhibited.

Light Some seeds are stimulated to grow by light. Some are inhibited by the presence of light. It will be necessary to have some knowledge of the presence or absence of special light requirements. Many of the agronomic crops do not require light for germination. In fact, light will inhibit germination. Ornamental bedding plants are more likely to require light for germination. Some crops requiring light for germination are ageratum, begonia, impatiens, and petunia. Lettuce also requires light for successful germination. Seeds of these plants are often deposited on the surface of the soil by nature, and the grower should follow the same procedure for successful germination.

Temperature Heat is another important requirement for germination. The germination rate, or percentage of seed that germinates, is affected by the availability of heat. Some seeds will germinate over a wide range of temperatures, whereas others have more narrow limits. In the agronomic crops that are directly seeded in the field, the only way to control heat is to plant when the ground is warm. In horticultural crops, particularly bedding plants and perennials, knowledge of a plant's specific heat requirements for germination should result in more efficient production. The germination requirements are often listed in the seed catalogs (Figure 17–6).

Plant	Time to Seed Before Last Frost (Weeks)	Germination Time (Days)	Germination Temperature Requirements (Degrees F)	Germination Light Requirements Light (L) Dark (D)
Ageratum	8	5-10	70	L
Aster	6	5-10	70	–
Begonia	12+	10-15	70	L
Coleus	8	5-10	65	–
Cucumber	4	5-10	85	–
Eggplant	8	5-10	80	–
Marigold	6	5-10	70	–
Pepper	8	5-10	80	–
Portulaca	10	5-10	70	D
Snapdragon	10	5-10	65	L
Tomato	6	5-10	80	–
Watermelon	4	5-10	85	–
Zinnia	6	5-10	70	–

FIGURE 17–6 Time, temperature, and light requirements for germination of some common flower and vegetable plants.

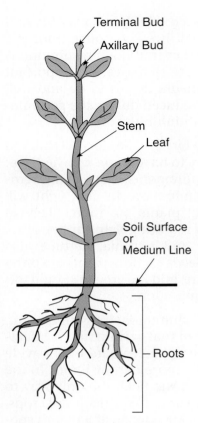

Terminal Bud

Axillary Bud

Stem

Leaf

Soil Surface or Medium Line

Roots

FIGURE 17-7 Vegetative parts of plants.

Stem cutting with unwanted portions (leaves, seed heads, and flowers) removed

FIGURE 17-8 Stem tip cuttings include the terminal bud and are 2 to 4 inches in length.

ASEXUAL PROPAGATION

As stated previously, asexual propagation is using the vegetative parts of the plant to increase the number of plants (Figure 17–7). Its primary advantages are economy, time, and plants that are identical to the parents. The primary methods of asexual propagation are cuttings, layering, divison, grafting, and tissue culture.

Stem Cuttings

Herbaceous and woody plants are often propagated by **cuttings** (vegetative parts that the parent plant used to regenerate itself). Types of cuttings are named for the parts of the plant from which they come. There are stem tip cuttings, stem cuttings, cane cuttings, leaf cuttings, leaf petiole cuttings, and root cuttings.

The procedure for taking cuttings is relatively simple. The equipment needed is a sharp knife or a single-edge razor blade. Sharp equipment will make the job easier and will reduce injury to the parent plant. To prevent the possibility of diseases spreading, it is best to dip the cutting tool in bleach water made with one part bleach to nine parts water. The tool can also be dipped in rubbing alcohol.

The flowers and flower buds should be removed from all cuttings. This allows the cutting to use its energy and food storage for root formation instead of flower and fruit development. A rooting hormone containing a fungicide is used to stimulate root development. A **fungicide** is a pesticide that helps prevent diseases.

Rooting hormone is a chemical that will react with the newly formed cells and encourage the plant to develop roots faster. The proper way to use a rooting hormone is to put a small amount in a separate container and work from that container. This procedure will ensure that the rooting hormone does not become contaminated with disease organisms. Do not put the unused hormone back in the original container.

Cuttings are normally placed in a medium consisting of coarse sand, perlite, soil, a mixture of peat and perlite, or vermiculite. It is best to use the correct medium for a specific plant to obtain the most efficient production in the shortest possible time. The rooting medium should always be sterile and well drained, with moisture retention ability to prevent the medium from drying out. The medium should be moistened before inserting the cuttings. It should then be kept continuously and evenly moist while the cuttings are forming roots and new shoots.

Stem and leaf cuttings do best in bright, but indirect, light. However, root cuttings are often kept in the dark until new shoots are formed and start to grow. For most plants, the most popular method of making cuttings is by stem cutting. On herbaceous plants, stem cuttings may be made almost any time of the year. However, stem cuttings of many woody plants are normally taken in the fall or the dormant season or both.

Stem Tip Cuttings **Stem tip cuttings** normally include the terminal bud. They are taken from the end of the stem or branch. A piece of stem between 2 and 4 inches long is selected, and the cut is made just below the node. The lower leaves that would be in contact with the medium are removed. The stem is dipped in the rooting hormone and is gently tapped to remove the excess rooting hormone. The cutting is then inserted into the rooting medium (Figure 17–8). The cutting should be

Agri-Profile

CAREER AREAS:

Plant Breeding/Plant Propagation/ Crop Improvement/Tissue Culture

Plant breeders' objectives might include developing plants that are faster growing, disease resistant, drought tolerant, insect resistant, wind resistant, frost tolerant, more beautiful, or better flavored, depending on the uses of the plants. Much plant breeding occurs in greenhouses. Some plants have small flowers that require the use of magnifying glasses and tweezers to transfer pollen or to remove reproductive parts. Generally, such research is followed by field trials and seed production, which gives the plant breeder some variety in the settings where work is done.

Asexual reproduction involves rooting, budding, grafting, layering, and other procedures in addition to pollination. Tissue culture, a procedure developed in biotechnology, permits the production of thousands of new plants identical to a single superior parent plant. The procedure is relatively cheap and easy, and it is used extensively to reproduce ornamental plants. Many jobs are available in the area of plant reproduction.

The plant breeder or plant geneticist searches for plants, transfers genes, or crossbreeds plants from various sources in an effort to develop new varieties with desired characteristics. *(Courtesy USDA/ARS #K-5146-16)*

FIGURE 17–9 A cane cutting is made from the stem of a plant with a canelike growth structure. One to two adjacent nodes are selected, and the cut is made to include the node or nodes that are desired.

inserted deep enough so the plant material will support itself. It is important that at least one node be below the surface of the medium because new roots will grow from it.

Stem Section Cuttings Stem section cuttings are prepared by selecting a section of the stem located in the middle or behind the tip cutting. This type of cutting is often used after the tip cuttings are removed from the plant. The cutting should be between 2 and 4 inches long, and the lower leaves should be removed. The cutting should be made just above a node on both ends. It is then handled as a tip cutting. Make sure that the cutting is positioned with the right end up. The axial buds are always on the tops of the leaves.

Cane Cuttings Some plants, such as the dumbcane (*Diffenbachia* sp.), have canelike stems. These stems are cut into sections that have one or two eyes, or nodes, to make cane cuttings. The ends are dusted with activated charcoal or a fungicide. It is best to allow the cane to dry in the open air for 1 or 2 hours. The cutting is then placed in a horizontal position with half of the cane above the surface of the medium. The eyes, or nodes, should be facing upward. This type of cutting is usually potted when the roots and new shoots appear (Figure 17–9).

Heel Cuttings Heel cuttings are used with woody-stem plants. A shield-shaped cut is made about halfway through the wood around the leaf and the axial bud. Rooting hormone may be used in the same manner as in the other types of cuttings. The cutting is inserted horizontally into the medium (Figure 17–10).

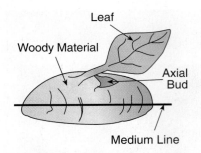

FIGURE 17–10 To make a heel cutting, a shield-shaped cut is made about halfway through the wood around a leaf and axial bud. The shield is placed horizontally into the growth medium with the leaf and axial bud above the medium line.

INTERNET KEY WORDS:

propagation, heel cutting
propagation, single-eye cutting
propagation, double-eye cutting
propagation, leaf cutting
propagation, petiole cutting
propagation, leaf section cutting
propagation, split-vein cutting

Single-eye Cuttings When the plant has alternate leaves, single-eye cuttings are used. The eye refers to the node. The stem is cut about a half inch above and below the same node (Figure 17–11). The cutting may be dipped in rooting hormone, and then placed either vertically or horizontally in the medium.

Double-eye Cuttings When plants have opposite leaves, double-eye cutting is the preferred type of cutting. It is often used when the stock material is limited. A single node is selected, and the stem is cut a half inch above and below the node with a sharp tool. The cutting should be inserted vertically in the soil medium (Figure 17–12).

Leaf-type Cuttings

For many of the indoor herbaceous plants, a leaf-type cutting will produce plants quickly and efficiently. This type of cutting will not normally work for woody plants, however.

Leaf Cuttings A cutting made from a leaf with a petiole cut to less than a half inch is referred to as a leaf cutting. To prepare a leaf cutting, detach the leaf from the plant with a clean cut and dip the leaf into the rooting hormone. Place the leaf cutting vertically into the medium. New plants will form at the base of the leaf and may be removed when they have formed their own roots (Figure 17–13).

Leaf Petiole Cuttings For leaf petiole cuttings, a leaf with a petiole about 0.5 to 1.5 inches long is detached from the plant. The lower end of the petiole is dipped into the rooting medium and is then placed into the medium. Several plants will form at the base of the petiole (Figure 17–14). These plants may be removed when they have developed their own roots. The cutting may be left in the medium to form new plants.

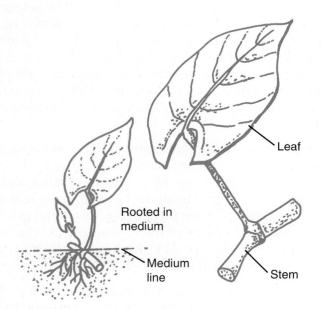

FIGURE 17–11 A single-eye cutting is made by cutting a section of stem above and below a single node and placing the cutting either vertically or horizontally into the growth medium.

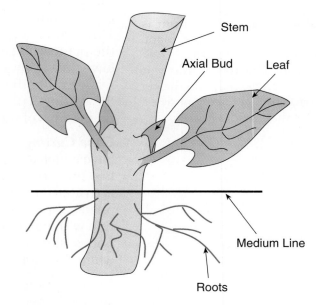

FIGURE 17–12 A double-eye cutting is used for propagating plants with an opposite leaf pattern. The stem is cut on either side of a single node, and it is inserted vertically into the growth medium.

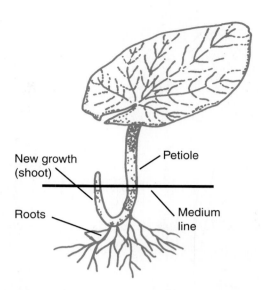

FIGURE 17–13 A leaf cutting is prepared by cutting a single leaf from a plant, dipping it in rooting hormone, and planting it vertically in the growth medium.

FIGURE 17–14 A leaf petiole cutting uses a leaf with an attached petiole that is 0.5 to 1.5 inches in length. The lower end of the petiole is dipped in the rooting compound and planted in the growth medium.

Leaf Section Cuttings Fibrous-rooted begonias are frequently propagated using **leaf section cuttings.** The begonia leaves are cut into wedges, each containing at least one vein (Figure 17–15). The sections are then placed into the medium.

New plants will form at the vein that is in contact with the medium. A section-type leaf cutting is made with the snake plant (*Sanseveria* sp.). The leaf is cut into sections 2 to 3 inches long. It is a good practice to make the bottom of the cutting on a slant and the top straight. This is done so you can tell the top from the bottom (Figure 17–16). The sections are placed in the medium vertically. Roots will form reasonably soon, and new plants will start to appear. These are to be cut off from the cutting as they develop root systems. The original cutting may be left in the medium for more plants to develop.

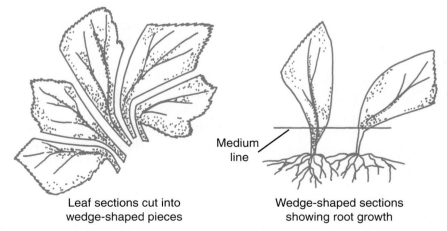

Leaf sections cut into
wedge-shaped pieces

Wedge-shaped sections
showing root growth

FIGURE 17–15 A leaf section cutting is obtained by cutting a single leaf into strips. The lower end of each leaf section is dipped in the rooting compound, followed by planting in the growth medium.

FIGURE 17–16 A snake plant leaf is cut into sections 2 to 3 inches in length, with the bottom cut at a slant and the top straight. Each section is placed vertically in the growth medium.

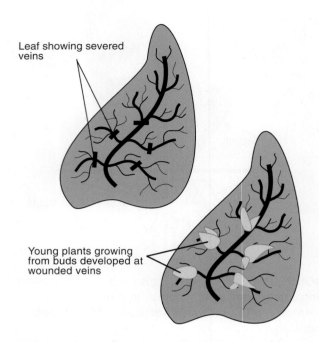

FIGURE 17–17 A split-vein cutting is prepared by making slits across the veins on the underside of the leaf. The cutting is then secured on the growth medium with the lower side down.

Split-vein Cuttings Split-vein cuttings are often used with large leaf types, such as begonias and other large-leaf plants. With split-vein cuttings, the leaf is removed from the stock plant, and the veins are slit on the lower surface of the leaf (Figure 17–17). The cutting is then placed on the rooting medium with the lower side down. It might be necessary to secure the leaf to make it lie flat on the surface. A good method is to use small pieces of wire, bending them like hair pins and pushing them through the leaf to hold it in place. The new plants will form at each slit in the leaf.

Root Cuttings

It is best to use plants that are at least 2 to 3 years old for making root cuttings. The cuttings should be made in the dormant season when the roots have a large supply of carbohydrates in reserve. In some species, the root cuttings will develop new shoots, which, in turn, will develop root systems. In others, the root system will be produced before new shoots develop.

If the plant has large roots, the root section should be 4 to 6 inches long. To distinguish the top from the bottom of the root, make the top cutting a straight cut and the bottom one a slanted cut. This type of cutting should be stored for 2 to 3 weeks in moist peat moss or sand at a temperature of about 40° F (4.4° C). When removed from the storage area, the cutting is inserted into the medium in a vertical position. The slanted cut should be down, and the top straight cut just level with the top of the medium. If the plant typically has small roots, a section 1 to 2 inches long is used. The cutting is placed horizontally a half inch below the surface of the medium.

Pegs hold the layer upright and in the ground   **Rooted layer cut from parent plant**

FIGURE 17–18 Layering is a method of promoting root growth by placing an attached stem or a cut stem beneath the soil partway along its length, with the last 6 to 10 inches of the stem exposed to sunlight.

INTERNET KEY WORDS:

propagation, plants, layering

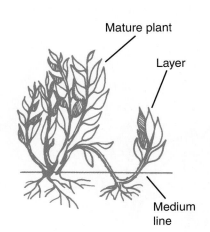

FIGURE 17–19 Tip layering is accomplished by making a hole in the medium next to a growing plant and burying the tip of a plant shoot in it to promote the growth of new roots and shoots.

LAYERING

In many plants, stems will develop roots in any area that is in contact with the media while still attached to the parent plant. After roots form, shoots develop at the same point. An advantage of this type of vegetative propagation is that the plant does not experience water stress, and sufficient carbohydrates are supplied to the new plant that is forming. The following sections discuss some of the more common methods of layering.

Simple Layering

Simple layering is an easy method that can be used on azaleas, rhododendrons, and other plants. A stem is bent to the ground and is covered with medium. It is advantageous to wound the lower side of the stem to the cambium layer. The last 6 to 10 inches of the stem is left exposed (Figure 17–18).

Tip Layering

Raspberries and blackberries are propagated using tip layering. With this method, a hole is made in the medium, and the tip of a shoot is placed in the hole and covered. The tip will start to grow downward and will then turn to grow upward. Roots will form at the bend. When the new tip appears above the medium, a new plant is ready to be transplanted. It will be necessary to separate the new plant from the parent by cutting the stem just before it enters the medium (Figure 17–19).

Air Layering

Many foliage plants are propagated using air layering. Some ornamental trees, such as dogwood, can also be reproduced by this type of layering. The stem is girdled with two cuts about 1 inch apart. The bark is

1. Prepare stem.

2. Soak sphagnum moss and squeeze excess water.

3. Pack damp moss over girdled area and tie.

4. Wrap with plastic and tape ends tightly.

FIGURE 17–20 Air layering.

removed. The wound is dusted with a rooting hormone and is surrounded with damp sphagnum moss (Figure 17–20). Plastic is wrapped around the moss-packed wound and tied at both ends. In a few weeks, depending on the plant, roots will appear throughout the moss. The stem is cut just below the newly formed root ball, and the ball is planted into a well-drained potting medium.

Division

Some plants are easily propagated by dividing or separating the main part into smaller parts. If the plant has rooted crowns, these crowns are separated by cutting or pulling them apart. The resulting clumps are planted separately. If the stems are not attached to each other, they are pulled apart. If the crowns are joined together by horizontal stems, they are cut apart with a knife (Figure 17–21). It is a good practice to dust the divided plants with a fungicide.

Some plants that grow from bulbs or corms form little bulblets or cormels at their bases. To produce more plants from this type, simply separate the newly formed plant part and place it in a good medium (Figure 17–22).

Hosta root clump before division

Hosta root divisions

FIGURE 17–21 Some plants are propagated by separating the rooted crowns of the mature plant.

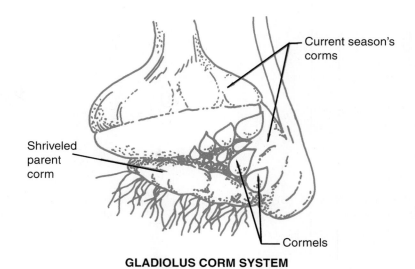

Current season's
corms

Shriveled
parent
corm

Cormels

GLADIOLUS CORM SYSTEM

FIGURE 17–22 Plants that grow from bulbs or corms may be separated by pulling the small bulblets or cormels from the base of the mature bulb or corm.

INTERNET KEY WORDS:

reproduction, plants, grafting

Grafting

Grafting is a procedure for joining two plant parts together so they grow as one. This method of asexual propagation is used when plants do not root well as cuttings, or when the root system is inadequate to support the plant for good growth. Grafting will allow the production of some unusual combinations of plants. For instance, several varieties of apples can be grown on one tree. Some nut trees can be made to grow varieties other than their own. Some unusual foliage plants can also be made by grafting. Finally, dwarf fruit trees are created by grafting regular varieties on dwarfing root stock obtained from related trees with similar, yet different genetics.

The top part of the plant that is to be propagated is called the **scion.** The **rootstock,** or **stock,** will be the new plant's root system and will supply the nutrients and water. The **graft union** is where the two parts meet.

To ensure successful grafting, the following conditions are necessary: (1) the scion and the rootstock must be compatible, (2) each must be at the right stage of growth, (3) the cambium layer of each section must meet, and (4) the graft union must be protected from drying out until the wound has healed (Figure 17–23).

There are many types of grafts. Some common grafts are the whip, or tongue, graft; bark graft; cleft graft; bridge graft; and bud graft. Each type is used for a special purpose. The most commonly used and the easiest to perform is the bud graft.

Bud Grafting

INTERNET KEY WORDS:

plants, rootstock, scion

The union of a small piece of bark with a bud and a rootstock is called **bud grafting.** It is most useful when the scion material is in short supply. This type of grafting is faster, and it will make a stronger union than other types of grafting.

T-Budding

T-budding is a popular type of bud graft in which a vertical cut about a quarter inch long is made on the rootstock. A horizontal cut is made at the top of the vertical cut. The result is a T-shape. The bark is loosened

(A) The scion before any cuts are made.

(B) The first cut is made in the scion.

(C) The second cut is made in the scion.

(D) The root, before any cuts are made.

(E) The first cut is made in the root.

(F) The second cut is made in the root.

(G) The scion and root are positioned for joining.

(H) The scion and root are pushed together. (Cambium must match on at least one side.)

(I) The two pieces are tied together.

(J) A covering of grafting wax may be necessary to prevent drying. This is especially important if the scion and the rootstock are not the same.

FIGURE 17–23 The process for performing a whip graft.

by twisting the point of a knife at the top of the T. A small, shield-shaped piece of the scion, including a bud, bark, and a thin section of the wood, is prepared. The bud is pushed under the loosened bark of the stock plant. The union is wrapped with a piece of rubber band called a **budding rubber.** The bud is exposed. After the bud starts to grow, the remainder of the rootstock plant is cut off above the bud graft (Figure 17–24).

TISSUE CULTURE

INTERNET KEY WORDS:

plants, tissue culture

A relatively new method of plant propagation is micropropagation, or **tissue culture.** Instead of using a large part of the plant as in other types of vegetative, or asexual, propagation, a small and actively growing part of the plant is used (Figure 17–25). The result is that many new plantlets may be obtained from a section of a leaf. The process must be done in a very clean atmosphere, and it is not successful in the greenhouse or other traditional propagation areas. Tissue culture requires the use of sanitary conditions. There are many commercial tissue-culture laboratories currently producing a large variety of plants.

SCION
The bud shield
is cut and removed.

STOCK
A T-cut is made
through the bark.

The shield is
inserted until tops
of the shield and
T-cut are even.

The bud is left
exposed while
remaining surfaces
are wrapped
tightly with
rubber strips.

FIGURE 17–24 The process of performing a T-bud graft.

Advantages over Traditional Methods

The greatest advantage of tissue-culture propagation is that numerous plants can be propagated from a single disease-free plant. Plants can be propagated more efficiently and economically than with traditional methods of asexual reproduction. The main disadvantage is that the work area must be very clean. All of the equipment must be sterile. In commercial production by tissue culture, there is more expense in equipment and facilities than there is with traditional methods of propagation.

The materials necessary for tissue culture are: (1) a clean, sterile area in which to work; (2) clean plant tissue; (3) a multiplication medium; (4) a transplanting medium; (5) sterile glassware; (6) sterile tools; (7) a scalpel, razor, or an X-acto® knife; and (8) tweezers.

Preparing Sterile Media

The first step in preparing for tissue culture is to prepare the medium in which the tissue will grow. The medium is called agar, which is available commercially from many of the scientific supply houses. The Virginia Cooperative Extension Service offers the following formula and procedure for preparing media for experimentation in tissue culture on a small basis.

1. Use a quart jar to mix the following materials:
   - ⅛ cup of sugar
   - 1 tsp of soluble, all-purpose fertilizer. The label will indicate that all of the major and minor elements are present. It is especially important that the soluble fertilizer contain ammonium nitrate. If it is lacking, add ⅓ tsp of 35–0–0 soluble fertilizer.
   - 1 tablet (100 mg) of inositol (myo-inositol). This can be obtained from most health-food stores.
   - ¼ of a pulverized tablet containing 1 to 2 mg of thiamine
   - 4 tbsp of coconut milk, the source of cytokinin. This is obtained from a fresh coconut. Freeze the remainder for later use.
   - 3 to 4 grains of a rooting hormone containing 0.1 active ingredient Indole-3-butyric acid (IBA)

FIGURE 17–25 Only small amounts of tissue are needed to produce an exact clone of a plant when propagating through tissue culture.
(Courtesy USDA/ARS #K-4825-1)

Hot Topics in Agriscience

CLONING A BETTER POTATO

Much of the seed stock that is used to produce potatoes is replaced with better cultivars over a period of a few years. Farmers who produce "seed potatoes" are constantly seeking plants that are resistant to diseases and pests. Individual potato plants that are identified in the field as being superior to other potato plants are often selected as parent stock. From the plant materials obtained from these plants, many new potato plants are cloned. These valuable young plants are initially raised in a greenhouse environment, and the supply of "seed potatoes" is expanded in the field until an adequate supply is available for commercial plantings. The "seed potatoes" are then cut into pieces for planting. A field of potatoes is a perfect example of massive cloning efforts that have been advanced to a commercial scale of operation.

2. Fill the jar with purified, distilled, or deionized water.

3. Shake the jar to dissolve all materials.

After the medium is dissolved, prepare the culture tubes using either test tubes with lids or other suitable glass containers. Fill the culture tubes one quarter of the way with sterile cotton balls. Use one or two per tube. They do not need to be packed tightly. Pour the prepared medium into the culture tubes to just below the top levels of the cotton. Place the lids on loosely.

After all medium is placed in culture tubes, it is ready to be sterilized. Sterilization may be done in two ways: (1) heat in a pressure cooker for 30 minutes, or (2) heat in an oven for 4.5 hours at 320° F. After they are removed, place the culture tubes in a clean area and allow them to cool (Figure 17–26). If several days will go by before using all of the tubes, wrap them in small groups in plastic wrap or foil before sterilizing.

Sterilizing Equipment and Work Areas

The tools and equipment used for tissue culture must also be sterilized. This can be done as the medium is sterilized by placing the tweezers, razor blade scalpel, or knife in the pressure cooker or oven. After the initial sterilization, they may be cleaned by dipping them in alcohol before and after each use.

1. Prepare culture medium

2. Place cotton balls and medium in culture tubes

3. Sterilize culture tubes and medium in oven or pressure cooker

4. Sterilized medium in culture tubes

FIGURE 17–26 Procedures for sterilizing tissue-culturing medium.

FIGURE 17–27 Plant material in culture tube.

The work area must be thoroughly cleaned and sterilized. Wash the area with a disinfectant. Keep a mist bottle filled with a mixture of 50 percent alcohol and sterilized water to spray work areas and tools, as well as the hands and arms of the propagator.

Preparing Plant Tissue and Placing in the Culture Tube

After the growing medium is properly prepared and cooled and the work area properly cleaned, the next step is to prepare the plant tissue. Various parts of the growing plant may be tissue cultured. For the production of vigorous plantlets, use only actively growing portions. With some species of plants, only a small, quarter-inch-square section of the leaf is used, whereas for others, a half inch of the shoot tip is used. With ferns, a quarter inch of the tip of the rhizome is used.

Remove the part of the plant to be used and discard the excess plant material. Submerge the plant part in a solution of one part commercial bleach and nine parts water for about 10 minutes. This will disinfect the plant tissue. Remove the tissue with sterile tweezers and rinse the material in sterile water. When the plant part has been disinfected in the bleach solution, it can be handled only with sterile tweezers and must not touch any nonsterile surface.

When the plant material has been disinfected and rinsed, remove any damaged tissue with a sterilized scalpel or razor blade. Remove the lid from a properly prepared culture tube or jar and place the plant material on the agar. Take care that the plant material is not completely submerged (Figure 17–27). Recap quickly to avoid contamination from the air. It is best if the material is placed in front of the propagator so that work is not done over the uncapped culture tube.

FIGURE 17–28 A young peach tree after being moved from the original culture tube and allowed to grow on sterile medium in a sterile container. *(Courtesy USDA/ARS #K-3279-11)*

> Reminder!
> *IT IS IMPORTANT THAT ALL WORK AND THE TRANSFERRING OF MATERIALS BE DONE QUICKLY AND IN A CLEAN ENVIRONMENT. Scrub all areas with disinfectant and clean all tools with a disinfectant solution. Any contamination may lead to unsuccessful work. Bacteria and fungus will grow in unclean culture tubes and will overtake the new plant growth.*

Storing Tissue Cultures

After all plant material has been cultured, put the cultures in an evenly warmed (70–75° F/21.1–23.9° C) and well-lighted area. The plant tissue will NOT do well in direct sunlight. If any contamination has occurred, it will be evident in 48 to 96 hours as mold or rotting on the medium. If contamination occurs, remove the contaminated tubes and wash them for reuse.

When the plantlets have grown to a satisfactory size, take them out of the culture tubes and transplant them into a good growing medium. Remember, the plantlets are fragile, so handle them carefully. As each plant is removed from its culture tube, wash the plant thoroughly and transplant it into its own culture container (Figure 17–28).

FIGURE 17–29 African violets can be readily propagated through tissue culture. *(Courtesy Michael Dzaman)*

When the plant is well established in its own culture container with viable roots and top growth, transplant the plant again into a suitable potting mix. Place the plant inside a protected area with high humidity. The plants are coming out of a well-protected environment with plenty of humidity and light. After they have adapted to the pot and are growing well, they may be treated like any other growing plant. This process will take about 3 to 6 weeks from the beginning to a successfully growing plant.

Laboratory: Cloning of African Violets

Plant cloning by tissue culture is one of the most widely used biotechnologies. Most potatoes and many houseplants are propagated by cloning. Cloning generates multiple, genetically identical offspring from the nonsexual tissues of a parent plant.

In theory, cloning is simple: Cut a leaf off a plant, disinfect it, cut it into fragments, then plant the fragments in nutrient agar. This may take 30 minutes.

In practice, contaminants from the air, hands, and tools quickly take over. Instead of healthy clones, you get colorful molds and bacteria. You can minimize contamination by using a simple hood and aseptic handling techniques.

INTERNET KEY WORDS:

cloning African violets

Cloning African violets in the classroom is a long-term project, but it can be done within a few class periods (Figure 17–29). The first stage takes about 30 minutes. The violets can be transplanted to a mini-greenhouse 6 to 8 weeks later. In another few weeks, the plant will be ready for repotting.

Aseptic Technique

Aseptic handling is critical for successful plant tissue culture. The culture vessel is a battleground between rapidly growing microbes and slowly regenerating plant fragments. Five simple techniques of aseptic handling will minimize transfer of contaminants, and thus growth of molds and bacteria. These techniques are:

1. Wash hands thoroughly and scrub nails using regular soap and paper towels. Do not touch your face or other objects or put your hands in your pockets. Such practices put contaminants back on your hands and can recontaminate your plants, equipment, work area, or medium.
2. Keep your hands from passing over open vessels.
3. Touch vessels far from the rim, neck, and similar areas. Keep the caps on when not in use.
4. Grasp tools as far from the working ends as possible.
5. Use sterile materials.

Practice Activities

1. Make thumbprints on bacteriologic nutrient agar plates before and after handwashing. Incubate overnight in a warm (not greater than 99° F/37° C) place. Bacterial colonies will appear on both plates. These bacteria are normal for us but will hinder cloning.
2. Study violet leaf fragments using a hand lens or dissecting microscope. Note the hairs protruding from the leaf's upper surface. Now look at your fingers. Note the many ridges. Where do contaminants hide in each case? How does washing the hand and leaf change how each looks?
3. Practice aseptic handling of tools. How do you pass scissors at home? How would you do this aseptically? Why might a scalpel be better for cloning work than a single-edge razor when both cut just fine?
4. Conduct a dry run of the cloning procedure using a spinach leaf. A spinach leaf bruises as easily as a violet leaf, so it readily shows how gently it has been handled.

A Hood for Cloning

Materials:

60- by 80-inch sheet of 2-, 3-, or 4-mil clear plastic (painter's tarp)
bulldog clips (2- to 3-inches in length)
support frame for hanging file folders

Assembly:

1. Place a file folder frame on the table with its arms facing you.
2. Fold the plastic sheet so it is two layers thick.
3. Drape the folded sheet with the foldline in front, so it overhangs the arms by about 2 inches.
4. Clamp the sheeting to the top of the arms to form a flat roof.
5. Straighten the sheeting to minimize creases.
6. Spray the inside of the hood and the work surface with 70 percent ethanol. Dry only the work surface, not the plastic.

The hood will look like a lean-to with a short curtain valance in front. This fits well on a student's desk. The overhanging plastic on the sides and back creates a larger work area than just the frame.

When working, stand over the hood. Do not breathe onto the cloning materials or work area.

Materials Preparation

To sterilize materials, autoclave for 15 minutes at 15 to 21 lbs. of pressure. Open only in a hood that has been surface sterilized with ethanol.

1. Mix and sterilize plant nutrient agar. One type is Murashige African violet/gloxinia multiplication medium, available from Carolina Biological Supply Co. One pack makes 30 to 40 plates. Stir one pack into 1 liter distilled water. Add 30 g sucrose and 15 g agar. The agar will not dissolve until it is heated. Loosely cap the flask with aluminum foil and autoclave or otherwise heat for 15 minutes. Cool in hot tap water of about 122° F (50° C). Pour 25 to 30 ml agar solution into each plastic petri dish (20 × 100 mm size). Gel at room temperature, then store in a refrigerator.

2. Sterilize a 100-ml beaker to hold disinfected leaves. Cap with a 4-inch square of heavy-duty aluminum foil.

3. Sterilize 150 ml tap water in a foil-capped, 250-ml Erlenmeyer flask.

4. Wrap the glass petri plate (cover and bottom assembled) in a double layer of white T-shirt rag, then autoclave. The inside of the cover will be used as the cutting surface, and the rim of the bottom will be used to support the tools as they drain.

5. Assemble but do not sterilize:
   - 500-ml beaker for waste liquid
   - 20 × 150-mm test tube filled to the brim with 70 percent ethanol (support in a 250-ml Erlenmeyer flask)
   - curved 8-inch forceps and a 6-inch scalpel handle with #11 blade
   - 200-ml beaker containing 100 ml of 70 percent ethanol to dip the leaf
   - single-edge razor to cut the leaf from the plant
   - 100-ml beaker to hold the leaf during disinfection

6. Prepare the disinfectant. Mix a 20 percent solution of liquid chlorine bleach and add one drop of Joy detergent per 500 ml. Swirl gently to mix; too many bubbles will inhibit wetting of the leaf surface. (Remember those little hairs.)

7. Soak the forceps and scalpel in the ethanol tube for at least 5 minutes before use. Do not store in the ethanol because the blade will rust.

8. Place in the hood:
   - plant nutrient agar plate
   - sterile H_2O flask
   - sterile beaker
   - sterile glass petri dish
   - 500-ml waste beaker
   - test tube of ethanol
   - forceps and scalpel
   - 200-ml beaker with ethanol

INTERNET KEY WORDS:

plant, nutrient, agar, medium

1.

Petiole

Leaf

2.

Beaker

Disinfectant

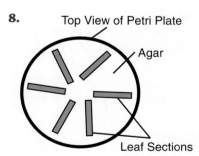

7.

Midrib Cut

Cross Cuts

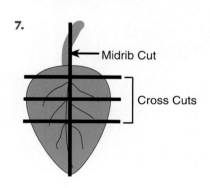

8. Top View of Petri Plate

Agar

Leaf Sections

9.
Petri Plate with Cover (side view)

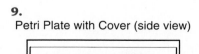

Cloning Procedure

1. Cut a young leaf so the petiole (stem) remains attached.

2. Put the leaf and petiole in the nonsterile beaker with disinfectant to remove dirt, mites, or other vermin. Leave it there for 10 minutes, but swirl it occasionally.

3. Working under the plastic cover, use the alcohol-soaked forceps to transfer the leaf, by the petiole, to the small sterile beaker. Pour the disinfectant into the waste beaker and remove the nonsterile beaker.

4. Rinse the leaf with 50 ml sterile water. Swirl and pour the water into the waste beaker. Gently hold the leaf in the small beaker with the forceps.

5. Wash again.

6. Disinfect the leaf by dipping the leaf in the ethanol leaf soak, count to 10, and remove the leaf. Meanwhile, open the glass petri plate. Use the bottom plate to drain the forceps and blade. (Resoak the forceps before cutting.)

7. Cut the leaf into fragments. Use the lid of the glass petri plate as the operating table.
    a. Cut off the petiole. Do not plant it.
    b. Cut down the midrib firmly.
    c. Cut across each half into four pieces, being sure to cut through a branching vein.

8. Plant each fragment in the plant nutrient agar so each piece is in, not just on, the agar. Plant in a spoke arrangement to minimize spread of contaminants.

9. Put the lid back on and cover the plastic petri plate with plastic cling wrap to keep moisture in.

10. Store in the dark for about 1 week, then place where the fragments have a daily light cycle and 68° to 77° F (20–25° C) temperatures.

11. Observe weekly. If part of a plate becomes contaminated, transfer healthy fragments to a fresh plate. Even with a commercial hood with sterile air, only about 50 percent of the plates remain completely free of contaminants.

12. Plantlets will appear on some fragments after about 8 weeks. When plantlet leaves grow to about 0.5 cm, aseptically remove the fragment. Separate its plantlets and return the plate to incubate. Gently cut the plantlets, being sure to have some root and some shoot. Place the individual plantlets in growing medium to develop under aseptic conditions.

13. After about a month, the plants should be large enough to transplant into a pot in sterile potting medium.

Plants should be planted with at least two per pot and fertilized with African violet food.

Adapted from *Science of Food and Agriculture*, January 1989, Council for Agricultural Science and Technology (CAST).

A mini-greenhouse for maintaining high humidity around fragile plants can be made by setting the plants in a plastic sandwich bag and adding enough water to maintain some moisture on the inside of the bag.

Science Connection

THE SEARCH FOR PERFECT PLANTS

The plants we use today for food, clothing, fiber, and ornamental use are quite different from those found in the wild. Domestic plants or plants grown for a specific use have generally been selected or bred to survive better, grow faster, look different, or in some way perform differently from their ancestors in the wild. However, it is becoming increasingly apparent to plant breeders that we must have wild plants that are not closely related to our favorite domestic species to inject new characteristics into our favored domestic plants.

Pest resistance is an area that requires a continuous reserve of foreign genetic sources. This is to be expected because the very pests that we breed plants to resist are constantly adapting to our plants through survival of the fittest among their kind. Insects and disease-causing pathogens have an amazing capacity to adapt to and eventually break crop resistance. Resistant varieties usually become obsolete in 3 to 10 years. It generally takes 8 to 11 years to breed a new variety to resist the changing individuals of a given pest. Therefore, plant breeding programs must function on a continuous basis to maintain our current capability to feed, clothe, and otherwise supply our population.

Plant geneticist Keith Schertz examines grain sorghum bred for tropical climates. Bags prevent the sorghum flowers from cross-pollinating, so the plant breeder can control which plants provide the male pollen to fertilize the female part of any given plant. *(Courtesy USDA/ARS K-2492-13)*

U.S. farmers grow more than 200 varieties of wheat, 85 varieties of cotton, 200 varieties of soybeans, and many varieties of fruit, vegetables, and ornamental crops. Disease and insect resistance must be bred into each of these varieties. Many of them thrive only in specific, limited growing areas, such as one part of one state. Keeping up with known and persistent pests is a relatively manageable process, as long as our state and federal experiment stations are reasonably well funded. However, the sudden appearance or introduction of a pest without resistant varieties or natural biological enemies can be devastating.

For instance, the Russian wheat aphid first appeared in the United States in 1986 and has cost wheat growers hundreds of millions of dollars since its arrival. U.S. wheat varieties have little resistance to the Russian wheat aphid. Therefore, chemical insecticides have to be used until either resistance can be bred into our domestic wheats or biocontrol agents can be found or developed. Entomologist Robert Burton of the Plant Science Laboratory in Stillwater, Oklahoma, believes the answer will be found by introducing selected genes from wheat varieties from Southwest Asia. It is estimated that it will take 5 to 10 years to breed in strong Russian aphid resistance.

STUDENT ACTIVITIES

1. Write the Terms to Know and their meanings in your notebook.

2. List the crops that are grown commercially in your area. Visit a site or sites where plants are propagated and ask the grower to discuss the propagation methods used.

3. List the prevalent crops that are grown from seed in your area and the popular varieties of each. Study a seed catalog and determine the requirements for germination of each variety and why the varieties are popular in your community.

4. Plant a seed in a jar filled with medium. Place the seed near the edge of the jar so that you can see what is happening. Keep a journal of daily observations as the seed or plant changes. You may want to do this with several seeds in different jars and vary the amount of water, light, air, or temperature in each jar. Note your observations and the conditions regarding each seed. Write your conclusions about what is best for maximum germination results.

5. Experiment with different kinds of plants and various kinds of cuttings. Keep a journal to determine the best type of cutting for specific plants. Keep notes on different media, temperatures, light, and rooting hormones.

6. Practice making each type of graft discussed in this unit under the supervision of your instructor.

7. Research additional grafting methods.

8. Conduct an experiment with the tissue-culture method of propagation. Keep notes on the different kinds of plants used, the time required for root formation, and observations regarding the benefits of propagation by tissue culture.

9. Prepare a statement about the propagation methods best suited to your purpose.

SELF EVALUATION

A. Multiple Choice

1. Propagation is defined as
   a. the union of an egg and sperm.
   b. the process of increasing the numbers of a species.
   c. a cheaper method of propagation than with seeds.
   d. the only way to reproduce some species and cultivars.

2. A seed consists of
   a. a root, stem, and flower.
   b. a root, seed coat, and endosperm.
   c. a seed coat, endosperm, and embryo.
   d. an embryo, cotyledons, and new plant.

3. A type of stem cutting used where stock material is limited and has alternate leaves is a
   a. stem tip cutting.
   b. cane cutting.
   c. simple layering.
   d. single-eye cutting.

4. A cutting that is usually made from a large-leaf plant with the veins split is
   a. split-vein cutting.
   b. leaf petiole cutting.
   c. terminal tip cutting.
   d. tissue propagation.

5. Grafting is
   a. a type of sexual propagation.
   b. a type of hybridization.
   c. a method by which two plants are propagated.
   d. a method of joining two parts of two different plants.

6. The most common type of grafting is
   a. T-budding.
   b. simple layering.
   c. stem-tip propagation.
   d. scion cut out of the stock plant.

7. Tissue culture may be used for
   a. cloning.
   b. disinfecting.
   c. sexual reproduction.
   d. sterilization.

8. With tissue culture, contamination may be a problem from
   a. air.
   b. hands.
   c. tools.
   d. all of the above.

9. The first stage of cloning African violets takes approximately
   a. 1 to 3 weeks.
   b. 4 to 5 weeks.
   c. 6 to 8 weeks.
   d. 10 to 12 weeks.

10. To make a very inexpensive hood for tissue culturing, the recommended covering is
    a. aluminum.
    b. plastic.
    c. sheet steel.
    d. wood.

B. Matching

_____ **1.** Outer seed coat

_____ **2.** Germinate

_____ **3.** Imbibition

_____ **4.** Tip cutting

_____ **5.** Root cutting

_____ **6.** Division

a. A cutting from an end of a branch containing a terminal bud

b. Best taken when the plant is dormant

c. The absorption of water into the young seed

d. Functions as a protector for the seed

e. When plants are separated and then replanted

f. When a seed starts to sprout

C. Completion

1. _____ propagation uses a part or parts of one parent plant.

2. The _____ will supply food to the young seedling until it is able to make its own.

3. A _____ might contain fungicide and is used to help plants produce roots more quickly.

4. _____ cuttings are normally made of a section containing one or two nodes.

5. Tissue culture is also known as _____.

6. A major aspect of tissue culture is that the area to be worked in must be _____ and _____.

D. True or False

_____ **1.** A clone is almost like the parents.

_____ **2.** The embryo is actually the young plant.

_____ **3.** All seeds need light to germinate.

_____ **4.** An advantage of tissue culture is that only one plant can be made from a disease-free plant.

_____ **5.** Double-eye cuttings are often used when plants have opposite leaves.

_____ **6.** Herbaceous plants are propagated by cuttings.

Crop Science

Crops in Space!

As humans push back the frontiers of outer space and consider the like-lihood of space travel taking years instead of days, we face the age-old dilemma—how do we feed the crew? Early manned space voyages were accommodated by freeze-dried food and food in tubes. Now scientists are gearing up for food production in space. Not only is there a need to produce food, but there is a need to generate oxygen, eliminate carbon dioxide, and dispose of organic waste. The solution is a controlled eco-logical life support system (CELSS). A controlled ecological life sup-port system is a system that will provide the basic components to sus-tain life without requiring inputs from external sources.

Consider a crew in space. For basic survival, they need food, water, and oxygen. The food could come from fruits, vegetables, grains, meat, milk, or eggs. Without a source of oxygen to replenish what is breathed in, the crew would soon die. In addition, if carbon dioxide were per-mitted to build up in the air, the crew would soon be poisoned. Similarly, if human waste could not be decomposed, its accumulation would become unbearable.

It has become clear that an effective CELSS is needed. Through photosynthesis, plants can take up nutrients from soil or water and use carbon dioxide from the air to manufacture food for themselves, as well as for animals. They give off oxygen as a by-product of this process. Fish and poultry are excellent sources of protein for humans and are excep-tionally efficient converters of plant material into essential proteins. They are small, grow fast, and would be excellent candidates for a space farm. Like humans, they give off carbon dioxide for plants to use. Also, like humans, they produce organic wastes that can be decomposed to replace the nutrients in the soil and water, where plants are growing. Microorganisms, worms, insects, and other forms of life would round out the system. Heat and combustible gases would be energy sources generated in the process.

Could it work as a closed system? Scientists believe it will. However, many questions must be answered, and problems must be overcome first. Steven Britz, a plant physiologist in the Plant Photobiology Laboratory in Beltsville, Maryland, has been trying to ask the right questions for the scientific community to answer to create a CELSS in space. Some questions raised so far are as follows:

1. What is critical to producing a harvest when you must supply every need in a closed environment?

2. What kind of and how much light is needed, and how can it be supplied?

3. How much room is needed for plant roots, and what are the effects of root area restriction?

4. What type of medium is needed—soil, water, or another medium?

5. What compounds will plants release in a closed environment, and are these useful or toxic?

6. What trace metals are picked up and transported in recirculated water?

7. What are the effects of little or no gravity on plant and animal growth and reproduction?

8. What light/dark cycles are best to optimize plant performance?

There are many more questions that need to be answered before we grow plants and animals in space. Can you suggest some? Answers to the questions about functioning in a space environment are helping us raise better plants today. A delegation to the United States from the Chinese Agricultural Ministry was interested in the details of raising max-

imum yields in extremely confined spaces. In China and other parts of the world, there is a growing industry raising hydroponic vegetables for the hotel trade. Using controlled environments that are not natural to plants allows us to discover the optimum conditions for the functioning of a particular plant.

Controlled environment facilities are gaining acceptance as a viable means of commercially growing high-value crops under scheduled production management in an environment free from harmful insects and pollution.

At Kennedy Space Center in Florida, scientists use a pressure chamber from a previous space craft to develop a controlled ecological life support system, where plants recycle air, water, and waste to produce food. *(Courtesy of NASA #89-HC-130)*

UNIT 18 | Home Gardening

OBJECTIVE

To plan, plant, and manage a home garden.

MATERIALS LIST

✓ grid-type paper
✓ pencil and eraser
✓ seed catalogs
✓ Internet access

COMPETENCIES TO BE DEVELOPED

After studying this unit, you should be able to:

- analyze family needs for homegrown fruits, vegetables, and flowers.
- determine the best location for a garden.
- plan a garden to meet family needs.
- establish perennial garden crops.
- prepare soil and plant annual garden crops.
- list recommended cultural practices for selected garden crops.
- protect the garden from excessive damage caused by drought and pests.
- harvest and store garden produce.
- describe the use of cold frames, hotbeds, and greenhouses for home production.

SUGGESTED CLASS ACTIVITIES

1. Obtain some potting soil and plant containers. Instruct the class on proper procedures for planting vegetable seeds such as tomatoes, cabbage, melons, flowers, and others, to transplant into home gardens. Have each student plant some vegetable and flower seeds and care for the young plants as the seeds sprout and grow. Send the plants home with the students when it is time to transplant them outside in the garden.

2. Visit a commercial greenhouse to see what kinds of plants are available for sale. Note the kind of potting soil that is used, and make a list of the varieties of plants that are offered. Ask the manager to explain why these varieties have been chosen. Inquire about the methods of controlling insects and diseases and the sources of the greenhouse's trees, shrubs, and ornamental plants. Have class members prepare written reports on the field trip covering these topics and other points that come up during the visit.

3. Provide students with tasting samples of produce that can be grown in a home garden. Include some foods that have been properly preserved by canning or dehydrating. As a class, discuss the advantages and disadvantages of growing a garden.

TERMS TO KNOW

seasonal	clod	biennial	herbicide
square foot	furrow	perennial	cold frame
successive	climate	cultivation	hotbed
loamy	annual		

FIGURE 18–1 Some people seem to have a natural talent for gardening; however, the growth of plants and the procedures for obtaining best results are based on scientific principles. *(Courtesy of USDA/ARS #K-5134-04)*

Gardening is an activity that can be enjoyed by all members of the family. It provides fresh fruits, vegetables, and flowers for immediate use or to sell for profit. Gardening is both an art and a science. It is demanding of the gardener, both in skill and creativity. A garden is alive and changing everyday, presenting new challenges to the gardener (Figure 18–1).

ANALYZING A FAMILY'S GARDENING NEEDS

First, one must decide what vegetables and flowers the family likes. It would not make sense to plant sweet potatoes, green beans, and marigolds if no one in the family cared for these vegetables and flowers. The home gardener should provide a seasonal and continuous variety of vegetables and flowers. **Seasonal** means pertaining to a certain season of the year. Fresh vegetables are important to everyone's diet. Plan to have plenty available during the growing season and to store some for future use. Choose mostly those vegetables that will produce high yields (Figure 18–2).

When it has been decided what vegetables and flowers the family likes, it is time to figure out how much ground will be needed. It is better to have a smaller garden that is well cared for than to have a garden that goes to waste because it becomes too much to handle. A good rule of thumb for four grown people is to start with a plot 10 feet wide and 26 feet long, or 260 square feet. A **square foot** is an area equal to 12×12 inches.

FIGURE 18–2 The garden plan should provide for high-yielding fruits, vegetables, and flowers that meet the needs and preferences of the family. *(Courtesy of USDA/ARS #K-2282-1)*

Agri-Profile

COMMUNITY GARDENS

A popular idea that has emerged in urban areas is to establish community gardens. A plot of land is prepared by measuring it into moderate-sized plots with clearly marked boundaries. Each plot is then rented to an interested person who enjoys gardening but does not have access to a garden plot. The plot owner establishes the rules that each gardener must follow. They might include weeding, observing established work times, taking water turns, respecting the privacy of other garden plots, among others.

Some community gardens are planted and tended together with no personal ownership of the produce. At harvest time, all of the produce is divided among the shareholders who paid the membership fee and helped do their share of the work. Whatever the arrangement, community gardens are appreciated by those who enjoy working with others and eating fresh produce.

THE GARDEN PLAN

INTERNET KEY WORDS:

garden plan

A prospective gardener should make a sketch on paper detailing the amount and placement of the various crops. At this stage, it is important to consider **successive** plantings. These crops follow each other in the season so that the ground is occupied throughout the growing season. Fall crops can follow spring and summer crops in many areas. This can be done easily by planting perennial crops and different varieties of specific annual crops. Variety is a category within a species of plant. Allow adequate space between the rows for cultivation.

LOCATING THE GARDEN

Depending on where you live—in the city, suburbs, or a rural area—the location of the garden is an important consideration. The garden should be convenient to the house. It should also be accessible to a water supply; on loamy, well-drained soil; in a sunny area; and visible from the home, if possible. **Loamy** soil is a granular soil with a balance of sand, silt, and clay particles.

One must visualize where to locate the garden. What happens to the proposed spot when there is a heavy rain? Or, what happens when it is very dry? From the chosen spot, look up to determine whether trees or branches will cause problems by excessively shading the garden. When planting flower beds around the house, remember that along the south and west sides, the heat will be reflected onto these beds, and they may require extra water. Select the best site you can for both vegetable and flower gardens.

PREPARING THE SOIL

Conditioning the Soil

Garden soil should be loose and well drained. The ideal soil type should be granular—like coffee grounds—so that water will soak in rapidly. Few soils are originally found this way. Soil-building practices can improve

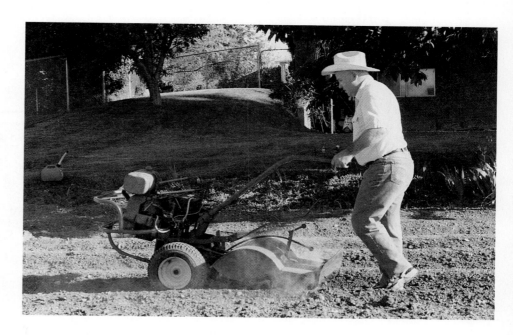

FIGURE 18–3 Soil preparation is one of the most important aspects of gardening. Soil is tilled to kill weeds, to mix materials into the soil, and to break up large chunks of soil into a granular texture.

the soil over time, and the gardener must prepare the soil to achieve the best results.

If the proposed site has not been previously planted, it is a good idea to add organic matter and plow, spade, or rototill it into the soil. The decayed organic matter and the freezing and thawing of the soil during the winter months will help improve the soil's physical condition. The application of fine organic matter in the spring can also improve soil condition and fertility.

Materials such as composted leaves and grasses, peat moss, composted sawdust, and sterilized weed-free manure are good soil conditioners. Peat moss is a soil conditioner made from spagnum moss. A good rate of application for organic matter is 1 lb dry material per square foot of surface area.

Preparing to Plant

If the site selected has never been planted, it is best to remove the sod before tilling the soil; then spread the organic matter over the soil. When moisture conditions are favorable, turn the soil with a shovel, spade, plow, or rototiller (Figure 18–3). When turning the soil, it is important to break up all clods. A clod is a lump or mass of soil.

After the spading (turning of the soil) is completed, make the planting beds. A good procedure is to heap the soil to make raised rows with grooves, or furrows, between them. Another method is to prepare raised beds that are 4 to 8 feet wide with sunken walks between them (Figure 18–4). The furrows or sunken walkways can drain off excess rain.

Next, level the raised beds with a rake, but do not push the soil back into the furrows. The next step is to walk over the beds or tamp them with the head of a garden rake to firm them and to eliminate air pockets (Figure 18–5). Finally, use the hoe and back of the garden rake to push and pull the soil to level it and to break any remaining clods into fine particles (Figure 18–6). The soil is now ready for planting.

A. Layout of a rectangular garden area— 10' x 26' = 260 sq. ft.

B. Layout of a square garden area with 2 beds.

FIGURE 18–4 Layouts for rectangular and square gardens for a family of four.

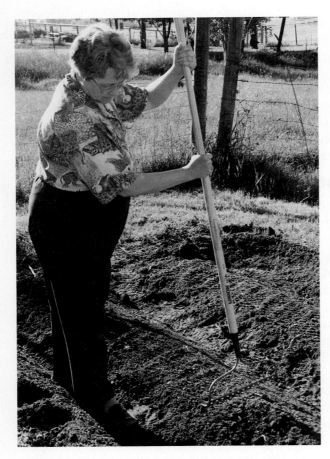

FIGURE 18–5 Raised beds in a garden allow the soil to warm up earlier in the spring; they also drain away excess moisture.

FIGURE 18–6 The garden rake and hoe are used to break up clods and loosen, smooth, and level the soil.

COMMON GARDEN CROPS AND VARIETIES

Consider the Climate

When choosing varieties of vegetables and flowers to plant in the home garden, consider the climate. **Climate** refers to the weather conditions for a specific region. Do not plant crops outdoors until all danger of frost is gone. Except for a few mountainous areas and the northern tier of states, almost all areas of the United States are frost free from June through August (Figure 18–7).

Plants grow rapidly in frost-free weather. In some areas, there is not enough time for long-season vegetables, such as eggplant, cantaloupe, and watermelon, to mature reliably. Much of the country has an intermediate growing season, with 5 to 6 months of frost-free weather. Long-season crops can be grown in these areas. Growing seasons of 7 to 10 months are common across the mid-South, South, and low-elevation regions of the Southwest and West. In these regions, gardeners can grow both spring and fall crops in many varieties. However, these same regions are susceptible to summers that are so hot or dry that only a few crop varieties can survive the intense summer heat.

Consider the Variety

INTERNET KEY WORDS:

<http://www.gurneys.com>

There are both warm-season and cool-season crops. Certain flowers and vegetables must have continuous cool weather to do well. Heat will quickly make the plants dry up or go to seed. Where the growing season is 5 months or longer, outdoor seeding in the late summer usually results in an excellent harvest during the cool fall season. Most of the cool-season plants can withstand light frosts with little damage.

There are two types of warm-season flowers and vegetables: those that mature quickly, and those that require 4 months or more from planting to harvest. The quick-maturing types are almost always started in the garden when the soil is warm. The later maturing types are usually started indoors and are transplanted to the garden after all danger of frost is past.

Most packets of home garden seeds are labeled to indicate the number of days that are required for the plants to mature. For example, the Blue Lake pole bean variety is labeled to mature in 60 to 65 days (Figure 18–8). Seed packet labeling is the most accurate information we have with regard to how long a plant will require to reach maturity. There are exceptions, however, such as the extreme northern latitudes where the sun goes down only briefly during the summer and the number of hours of daylight are much greater than in other areas of the country.

Annuals, Biennials, and Perennials

Flowers and vegetables are classified as annuals, biennials, or perennials (Figure 18–9). An **annual** is a plant whose life cycle is completed in one growing season. Growth is rapid. Practically all vegetables, except asparagus, rhubarb, and parsley, are annuals.

A **biennial** is a plant that takes two growing seasons or 2 years from seed to complete its life cycle. Some biennials bloom very little or not at all the first year, but they come into full bloom the second year, and then go to seed.

Vegetable Planting Guide

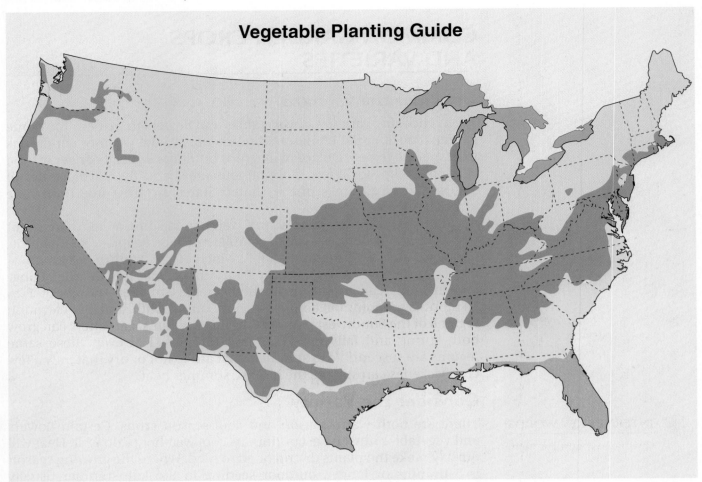

This map and accompanying chart are based largely on U.S. Department of Agriculture records showing the average dates of the last killing frosts in various parts of the country. Use this information to plan the planting and harvesting periods for your garden, but remember that these are averages, and individual conditions may vary.

	SUBTROPIC	WARM	MILD	COOL	CALIF.	PLANTING DEPTH	ROW SPACING	FIRST HARVEST	HARVEST LASTS
Beans	Apr.-Aug.	Apr.-June	May-June	May-June	Mar.-Aug.	1"	2-3 ft.	8 weeks	Until frost
Beets	Jan.-Dec.	Feb.-Oct.	Mar.-July	Apr.-July	Jan.-Dec.	1/2"	15-18"	50-78 days	6 weeks†
Broccoli	July-Oct.	Feb.-Mar.	Mar.-Apr.	Mar.-Apr.*	Sep.-Feb.	1/4"	3 ft.	65-70 days	To frost
Brussels sprouts	Feb.-May	Feb.-Apr.	Mar.-Apr.*	Mar.-Apr.*	Sep.-Feb.	1/2"	3 ft.	14-20 wks.	Past frost
Carrots	Jan.-Dec.	Jan.-Mar.	Mar.-June	Apr.-June	Sep.-May	1/4"	1 1/2-2 ft.	8 weeks	8 weeks†
Chard	Jan.-Dec.	Feb.-Sep.	Mar.-Aug.	Apr.-July	Jan.-Dec.	3/4"	1 1/2-2 ft.	8 weeks	To frost
Chives	Feb.-May	Mar.-May	Mar.-May	Apr.-June	Mar.-May	1/4"	1 1/2 ft.	8 weeks	To frost
Corn	Apr.-June	Mar.-June	May-July	May-June	Mar.-Aug.	1 1/2"	3 ft.	9-12 weeks	10 days
Cucumbers	Apr.-June	Apr.-June	Apr.-June	May-June	Mar.-Aug.	1/2"	6 ft.	7 weeks	5 weeks
Eggplant	Feb.-Mar.	Feb.-Apr.	Mar.-May*	Apr.-May*	Mar.-May	1/4"	3-4 ft.	10-14 wks.	To frost
Endive	July-Sep.	Aug.-Sep.	Mar.-May	Apr.-June	Jan.-Dec.	1/4"	2-3 ft.	10-12 weeks	7 weeks
Lettuce	Jan.-Dec.	Aug.-May	Mar.-June	Apr.-June	Sep.-May	1/4"	15"	6 weeks	6 weeks†
Mustard	Feb.-May	Feb.-May	Mar.-June	May-July	Mar.-Aug.	1/4"	1 1/2-2 ft.	40-50 days	2 weeks
Okra	Apr.-June	Apr.-June	Apr.-June	May-June*	Mar.-May	1/2-3/4"	3 ft.	55-60 days	To frost
Onions	Dec.-Mar.	Dec.-Apr.	Feb.-May	Mar.-June	Sep.-May	1/2"	18"	Variable	—
Parsley	Jan.-Dec.	Jan.-June	Feb.-June	Mar.-June	Jan.-Dec.	1/4"	1 1/2-2 ft.	10 weeks	To frost
Parsnips	Mar.-June	Feb.-June	Apr.-June	May-June	Dec.-May	1/2"	1 1/2-2 ft.	14-17 wks.	Past frost
Peas	Jan.-May	Jan.-Apr.	Feb.-May	Mar.-June	Sep.-May	1 1/2"	2-3 ft.	9 weeks	2 weeks†
Peppers	Feb.-Mar.	Feb.-Apr.	Mar.-May*	Mar.-May*	Mar.-May*	1/4"	3 ft.	9 weeks	To frost
Potatoes	Jan.-Dec.	Feb.-Oct.	Mar.-Apr.	Apr.-May	Jan.-Dec.	5"	2 1/2-3 ft.	100 days	To frost
Pumpkin	Apr.-June	Apr.-June	Apr.-June	May-June	Mar.-May	1"	8-10 ft.	14-17 wks.	To frost
Radishes	Jan.-Dec.	Feb.-Oct.	Mar.-Aug.	Apr.-July	Sep.-May	1/4"	1-2 ft.	3-6 weeks	1-2 weeks†
Squash, Summer	Apr.-June	Apr.-June	Apr.-June	May-June	Mar.-Aug.	1"	4-6 ft.	60 days	To frost
Squash, Winter	"	"	"	"	"	1"	6-8 ft.	100 days	"
Tomatoes	Jan.-Mar.	Feb.-Mar.	Mar.-May*	Mar.-May*	Mar.-May*	1/4"	3-4 ft.	9-12 weeks	To frost
Turnips	Feb.-Mar.	Jan.-Mar.	Feb.-Apr.	Mar.-May	Jan.-Dec.	1/2"	1-2 ft.	6-10 weeks	3 weeks†

*Transplants recommended.
†Following harvest, space may be used for late planting of carrots, beets, or bush beans.

FIGURE 18–7 Planting times, planting depths, row spacing, and harvest times for common garden vegetables.

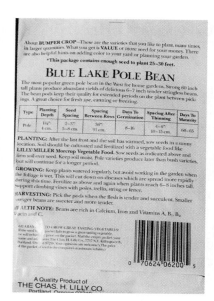

FIGURE 18–8 The length of the growing season varies according to the area of country in which you live. Choose those varieties of garden seeds that are capable of maturing before the growing season ends.

A **perennial** is a plant that lives on from year to year. A gardener commonly treats some plants that are true perennials as either annuals or biennials. This is done because, when planted each year from seed, perennials produce higher quality blooms than do older plants that remain from year to year. Perennial crops, such as asparagus, artichokes, rhubarb, and some herbs and flowers, should be planted in one section of the garden, separate from the annuals. By such separation, the perennials do not interfere with the cultivation of the annuals.

Vegetables or Flowers

The kinds of vegetables and flowers to plant depend on the individual tastes of the members of your family. There are certain top-ranking vegetables that are popular everywhere. Certain ones are popular only within certain regions of the country. The most popular vegetables are tomatoes, snap beans, onions, cucumbers, peppers, radishes, lettuce, carrots, corn, beets, cabbage, squash, and peas. Favorites of specific regions are artichokes in Louisiana and California, southern peas in warm southern climates, and melons in warm summer regions.

The gardener has a wide selection of flowers from which to choose. Generally, flower varieties are chosen because of their characteristics. Various varieties may survive well in poor soil, have a wonderful fragrance, or grow well in the shade. A few of the flowers that do well in poor soils are alyssum, cactus, cosmos, marigolds, petunias, and phlox. Flowers that are particularly fragrant are alyssum, carnations, petunias, sweet peas, and sweet william. Some that do well in partial shade are ageratums, begonias, coleus, impatiens, and pansies.

FIGURE 18–9 Characteristics of annual, biennial, and perennial plants.

Annual	Biennial	Perennial
Lives only one year	Lives two years	Lives more than two years
Germinates, matures, and reproduces in one growing season	Germinates and grows roots, a short stem, and a cluster of leaves in the first year	Germinates, matures, and reproduces in the first year
Produces flowers, fruits and seeds in same growing season	Produces flowers, fruits and seeds in second growing season	Produces new shoot systems, flowers, fruits, and seeds each growing season

Hot Topics in Agriscience

MASTER GARDENER PROGRAMS

The Cooperative Extension System associated with land-grant universities has developed a program for "Master Gardeners." Under this program, people who are expert gardeners are available as consultants to answer questions about home gardening problems. This service is not available in all counties in the United States, but information about many of these programs is available on the Internet using the key words "Master Gardener." People who wish to participate by becoming Master Gardeners must fill out applications, enroll in training programs, and commit to work as a volunteer to consult with people in the community on gardening problems.

CULTURAL PRACTICES FOR GARDENS

Cultivating

Never put off cultivation. **Cultivation** is the act of preparing and working the soil. It is easier to do a little each day than to let weeds get ahead or the soil get too hard. Once this happens, it is difficult to get the garden back into good condition. The main purpose of cultivation is to control weeds and to loosen and aerate the soil. A sharp hoe is a necessary tool for any gardener. Buy one that can fit between plants for easy cultivation. Begin cultivation as soon as weeds or grass break through the soil. Do not wait until the weeds and grasses are tightly rooted and threaten to take over the garden. When cultivating, take care not to damage the roots of the vegetables and flowers. Use short, shallow scraping motions instead of chopping deeply into the soil. Deep hoeing or cultivation will destroy desirable plant roots.

Weeding

Begin weeding as soon as the weeds appear. When weeds grow near the plants, their roots can become intertwined with the crop. Therefore, take care to avoid pulling out both the weed and the plant. When weeding, select a time of day when the soil and plants are fairly dry, so the weeds wither when pulled. Be careful to shake the soil from weed roots so the weeds will die quickly. Rake persistent weeds such as purslane, crabgrasses, and Bermuda grasses out of the garden to prevent them

Agri-Profile

CAREER AREAS:

Gardener/Caretaker/Homeowner/Market Gardener

Fruit and vegetable gardening has long been an important enterprise for providing fresh and wholesome food for rural and suburban families. Similarly, flower gardening provides a rewarding pastime and greatly increases the beauty of homes in urban, as well as rural and suburban, settings. Career opportunities exist for gardeners on estates, institutions, colonial farms, truck farms, and residential neighborhoods.

Nationally, there is a resurgence of small farms and truck gardens where roadside stands and farmers' markets have created new interest in farm-fresh fruits and vegetables. Pick-your-own operations provide opportunities for people to harvest their own field-fresh produce and save the producer labor costs.

People have a new awareness of the benefits of fresh food and pay premium prices when freshness is ensured. This has created new markets for garden produce and has opened new career opportunities in production, processing, and marketing.

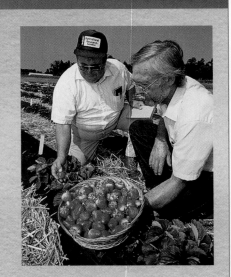

Home gardens, pick-your-own farms, roadside markets, and farmers' markets have all developed because people like to eat farm-fresh fruits and vegetables. *(Courtesy of USDA/ARS #K-4173-2)*

from taking root again. Mulches of various types are excellent for controlling weeds (Figure 18–10.)

Weed-control chemicals called **herbicides** can be used in the home garden, but always should be handled with care. However, the use of herbicides is best suited to large gardens. Chemicals are specific in their actions in certain types of plants, so they must be used with caution to prevent injury to the vegetables and flowers. Before using any herbicides, obtain detailed information from an extension agent or an informed garden center salesperson. Always read the label on the container and use the chemical strictly in accordance with the label instructions.

Watering

During a growing season, there will probably be some days or weeks of dry weather when the garden must be watered. Most soils require at least 1 inch of water per week, either through rain or irrigation. If water is needed, make it a practice to soak the soil to a depth of 6 inches. Frequent, light waterings tend to promote shallow root development and should be avoided.

When watering, run the water on the soil near the plants. A good sprinkler head, soaker hose, or water breaker is needed to ensure even distribution of water. Let the garden hose run for 15 to 30 minutes to water 100 square feet at a depth of several inches. You should water whenever the soil lacks moisture 1 to 2 inches below the surface.

Protection from Pests

Gardens can be damaged from a variety of insects and diseases. To help prevent severe damage, the following practices should be helpful:

- Rotate crops so that the same or a related crop does not occupy the same area every year. This helps control soilborne diseases.

- Watch closely for insects. Pick them off by hand or knock them off with a hard stream of water. Introduce predatory insects (Figure 18–11).

- When using sprinklers, water early in the day, so the foliage can dry before nightfall. Diseases of leaves and fruit prosper in damp conditions.

- Keep weeds out of the garden. Weeds may harbor diseases or insects and will interfere with spraying and dusting of crops.

- Use enough fertilizer and lime to promote vigorous growth.

HARVEST AND STORAGE OF GARDEN PRODUCE

Harvesting

Harvest the vegetables and flowers at the peak of quality or at the stage of maturity preferred by the users (Figure 18–12). Some vegetables will lose their quality within a day or two, whereas others hold it for a week or more.

For the best results for flowers, pick them in the early morning or late afternoon. Immediately place the cut flowers in lukewarm water for a short time. Display the flowers in a cool place.

FIGURE 18–10 Plant materials, newspaper, and plastic films all make excellent mulches that help control weeds and conserve moisture. *(Courtesy of USDA/ARS #K-4175-13)*

FIGURE 18–11 Lady beetles feed on pea aphids and help keep the damaging insect population under control. *(Courtesy of USDA/ARS #K-4179-19)*

Vegetable	Maturity Level	Vegetable	Maturity Level
Asparagus	From seeds—3rd year. From roots—after 1st year. Cut when spears are 6" to 10" high, in spring.	Lettuce	Leaf—When tender and desired size. Head—When round and firm.
Beans, Pole	When pods are nearly full size.	Muskmelons	When stem separates easily from fruit.
Beans, Bush	When pods are nearly full size.	Mustard Greens	When large leaves are still tender.
Beans, Bush Lima	When tender, pods are nearly full size.	Okra	Cut pods when about 2" or 3" long.
Beets	When bulbs are 1¼" to 2" in diameter.	Onions	For fresh use—When ¼" to 1 ½" in diameter. For storage—When tops shrivel at the bulb and fall over.
Swiss Chard	Outer leaves can be harvested anytime.	Parsley	Any time the outer leaves are desired size.
Broccoli	Before green clusters begin to open.	Parsnips	After hard frost. Can leave in ground all winter for spring use.
Brussels Sprouts	When sprouts are firm, pick from bottom up on stalks.	Peas	When pods are well-filled, before seeds are largest.
Cabbage	When heads are solid, before splitting.	Peppers	When solid and nearly full size.
Cauliflower	After blanching, when firm curds are 2" to 3" in diameter.	Pumpkins	When skin is hard and not easily punctured. Cut with some stem on.
Carrots	When top of root is 1" to 1½" in diameter.	Radishes	When desired size.
Celery	When about ⅔ mature, harvest as needed.	Rhubarb	When stems are of desired size, twist off near base of plant and discard leaves. Do not pick more than ⅓ of plant during a season.
Collards	When leaves are large but tender. Can harvest up to winter.		
Corn	When kernels are filled out and milky. The silk at the tip of the ear should be dry and brown.	Spinach	When outer leaves are large enough.
Cucumbers	When slender and dark green.	Squash	Summer—When skin is soft. Winter—When skin is hard, cut with part of stem on, before frost.
Eggplant	When half grown, glossy, and bright.		
Endive	When leaves are tender and desired size. About 15" in diameter.	Tomatoes	When uniformly red and firm.
		Turnips	When 2" to 3" in diameter.
Kale	When young and tender.	Watermelons	When underside is yellow and thumping produces a muffled sound.
Kohlrabi	When 2" to 3" in diameter.		

FIGURE 18–12 Levels of maturity for harvesting vegetables for top quality.

INTERNET ADDRESS:

<http://www.postharvest.com.au/storage.htm>

Storage

Many vegetables that are not canned or frozen for later use can be stored successfully if proper temperature and moisture conditions are met. Only vegetables that are of good quality and at the proper stage of maturity should be stored.

Warm Storage Vegetables that tolerate greater storage temperatures are squash, pumpkins, and sweet potatoes. These vegetables may be

Science Connection

Preserving Food

During the growing season a gardener spends hours weeding, watering, fertilizing, and finally harvesting. With the amount of time and energy that is put into a home garden, one would not want to see any of the produce go to waste. That is why many gardeners consider food preservation the final step in home gardening.

Within the fertile soil of a home garden, millions of microorganisms thrive. Many of these creatures are beneficial to the plants as they grow. Once inside humans, however, they can cause disease. During the food preservation process, care must be taken to ensure that food is free from living microorganisms.

Common organisms that are found in improperly preserved foods include fungi and bacteria. If left in food, the food will be unsafe to eat. Fungi include molds and yeasts. Bacteria are plentiful on freshly harvested foods. Many bacteria that are commonly found in the soil can cause serious illness if ingested.

Local agricultural extension offices can provide material, classes, and advice on safe and germ-free food preservation. Canning, freezing, and dehydrating are all methods that can keep garden produce safe and delicious for long periods. Molds and some bacteria are destroyed between 190° and 212° F. The boiling point of water at sea level is 212° F. By boiling canned food and blanching food that will be dehydrated or frozen, the chances that any microorganisms will remain in the food are slight. However, it takes temperatures of 240° F to kill some species of bacteria. Natural acids found in some foods can kill most bacteria. In produce with low acid levels, the food must be heated to 240° F using a pressure cooker. This will kill any remaining organisms. Although a lot of care must be taken to preserve the foods grown in a home garden, the benefits can last all year long as the food is enjoyed.

stored on shelves in an upstairs storage area. Damp basement areas are not recommended. Squash and pumpkins should be kept in a heated room, with a temperature between 75° and 85° F (23.9° and 29.4° C) for 2 weeks to harden the shells. Long-term storage temperature should then be reduced to 50° to 55° F (10–13° C).

Cool Storage Most vegetables require cool temperatures and relatively high humidity for successful storage. The storage area should be cool, dark, and ventilated. The room should be protected from frost, heat from a furnace, or high outdoor temperatures.

The suggested temperature and relative humidity for storage of crops are shown in Figure 18–13. This list will provide you with a general idea of the type of storage area that each vegetable requires.

COLD FRAMES, HOTBEDS, AND GREENHOUSES FOR HOME PRODUCTION

Cold Frames

Cold frames are an extremely convenient aid to the gardener. A cold frame is a bottomless wooden box with a sloping glass roof (Figure 18–14). It can be constructed by using plywood or common lumber and a window sash. The size of the sash should determine the dimensions of the box. The length and width should be 2 inches smaller than the overall dimensions of the cover sash.

Crop	Temperature °F	Relative Humidity Percent
Asparagus	32	85-90
Beans, snap	45-50	85-90
Beans, lima	32	85-90
Beets	32	90-95
Broccoli	32	90-95
Brussels sprouts	32	90-95
Cabbage	32	90-95
Carrots	32	90-95
Cauliflower	32	85-90
Corn	31-32	85-90
Cucumbers	45-50	85-95
Eggplants	45-50	85-90
Lettuce	32	90-95
Cantaloupes	40-45	85-90
Onions	32	70-75
Parsnips	32	90-95
Peas, green	32	85-90
Peppers, sweet	45-50	85-90
Potatoes	38-40	85-90
Pumpkins	50-55	70-75
Rhubarb	32	90-95
Rutabagas	32	90-95
Spinach	32	90-95
Squash, summer	32-40	85-95
Squash, winter	50-55	70-75
Sweet potatoes	55-60	85-90
Tomatoes, ripe	50	85-90
Tomatoes, mature green	55-70	85-90
Turnips	32	90-95

FIGURE 18–13 Recommended temperatures and relative humidity for storage of fresh vegetables. *(Courtesy of University of Maryland Cooperative Extension Service)*

The following procedures should be helpful in constructing a cold frame.

1. Make the front of the box approximately 8 inches high and the back of the box about 12 inches high.

2. Cut the two sides on an angle, from 12 inches at the back to 8 inches in the front.

3. Nail the four sides together.

4. Try the sash or top for size.

5. Hinge the sash to the back of the frame for easy use. The hinge allows you to prop the cover open slightly in warm weather for ventilation.

6. Place the cold frame in a southern exposure, with good protection from the wind and proximity to a water supply.

When the cold frame is built, it can serve three purposes:

1. It can be used as a protective home for seedlings that have been started indoors. The seedlings can continue to grow inside the cold frame and become "hardened off" before they are transplanted to the garden.

2. You can begin plants in the cold frames. Seeds can be planted directly into the soil in the cold frame. After they are 4 to 6 inches tall, they can be transplanted straight into the garden.

3. Vegetables, such as lettuce and endive, grow well into the fall. Sow seeds in early autumn. Cover the cold frame with a blanket or tarp during extremely cold nights.

Hotbeds

A **hotbed** is simply a cold frame with a heat source. In many areas, a cold frame is usually adequate. However, in colder areas, some type of artificial heat is necessary. The hotbed should be located on well-drained land.

Electricity is a convenient means of heating hotbeds. Use either lead- or plastic-coated electric heating cables or 25-watt frosted light bulbs. Temperature can be controlled with a thermostat. When using light bulbs, attach the electrical fixtures to strips of lumber and suspend the bulbs 10 to 12 inches above the soil surface. Allow one 25-watt bulb per 2 square feet of space. When using a heating cable, a standard 60-foot length will heat a 6 × 6 or a 6 × 8 foot bed. Lay cable loops about 8 inches apart for uniform heating. Lay the cable directly on the floor of the bed, unless drainage material is needed.

FIGURE 18–14 Cold frames are structures with glass or plastic covers used for starting garden plants.

Science Connection

FINDING AND PROMOTING THE GOOD ONES

Across the road from the giant Chrysler automobile assembly plant in Newark, Delaware, is a different facility specializing in foreign imports—insects! The imports are flies, wasps, beetles, and other insects that parasitize, or feed on, the insects that devour our crops, devastate our gardens, cause our lawns to turn brown, and infest our houseplants. These insects have been found doing their "good deeds" in countries all over the world. Most of our recent imports have come from Argentina, Brazil, Canada, Chile, China, France, Germany, India, Indonesia, New Zealand, South Korea, and Russia and other previously Soviet nations.

The facility is called the U.S. Department of Agriculture (USDA) Beneficial Insects Introduction Research Laboratory. Its job is to see that insects brought into the United States are, indeed, beneficial, and that the eggs, larvae, pupae, or adults of harmful insect species do not hitchhike along with the beneficial ones. The work of the entomologist at the Newark Laboratory includes picking up incoming shipments of beneficial insects at the Philadelphia airport; transporting them to the nearby Newark laboratory; taking them through a special receiving process; disinfecting all shipping materials that arrive with the insects; isolating the insects in special refrigeration units; exercising them daily; examining each insect under magnification to confirm identity; separating eggs, larvae, pupae, and adults for rearing purposes; increasing their numbers through reproduction; and shipping them under carefully controlled conditions to USDA and state research facilities throughout the United States and cooperating foreign nations.

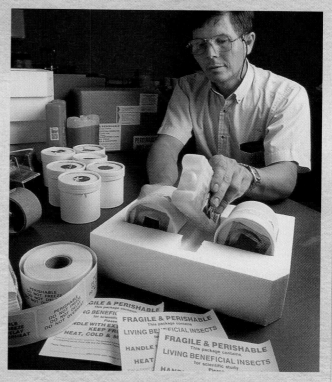

Entomologist and quarantine officer Larry Ertle at the Beneficial Insects Introduction Research Laboratory in Newark, Delaware, gently places beneficial insects in custom-designed packaging for shipment. *(Courtesy of USDA/ARS #K-4183-12)*

Three sets of heavy steel doors lead into the quarantine's main work area. There, a large sink and autoclave are used to clean and sterilize materials used in the laboratory. Sterilization is essential before discarding materials to avoid accidental release of both beneficial insects and "hitchhiking intruders." Interestingly, insects must receive regular exercise to remain healthy, just like humans! Once a day, even on weekends and holidays, batches of insects are removed from cool storage and released in restricted, warm, and lighted areas to warm up, stretch, soak up light, and move around in the greater space.

Many innovations in packaging and handling have evolved and have become standardized over the years. One of the more unusual ones is the parasite pill, or Trichocap. The parasite pill is a shipping capsule made of soft, gray cardboard and is about the size of an oyster cracker. It is used to ship insect eggs. One parasite pill holds about 500 eggs of the Mediterranean flour moth. Inside each egg, there is one future parasitic wasp used as a biocontrol against the devastating European corn borer. The discovery and introduction of parasitic wasps have helped control some of our most damaging insects, whereas reducing our reliance on chemical sprays.

FIGURE 18–15 A home greenhouse can be attached to the house for convenience and efficiency.

Greenhouses

Growing flowers and vegetables in a greenhouse can be enjoyable, as well as profitable. The conventional greenhouse is designed primarily to capture light and control temperature. It can be freestanding, but most often it is attached to a building to provide convenient access, simplified construction, and a potential source of supplemental heat if needed (Figure 18–15). The greenhouse can provide an environment for starting plants, hardening them off, or completely growing the plants.

In summary, gardening is both an art and a science. You can keep a garden for fun, reduce the cost of food for the family, generate income, or improve the beauty of the surroundings. If gardening is to be profitable, the operator must read extensively, get help occasionally with disease and insect problems, and perform garden chores in a timely fashion. For many people, the garden is a real source of satisfaction and creates a feeling of self-sufficiency and achievement.

STUDENT ACTIVITIES

1. Write the Terms to Know and their meanings in your notebook.
2. Discuss with family members the types of vegetables and flowers that they want in the home garden.
3. Select vegetable and flower varieties for the home garden from seed catalogs.
4. Sketch a garden plot containing both flowers and vegetables that are to be grown.
5. Make a calendar indicating the dates to plant and harvest the various crops in the garden.
6. Start and manage your own garden area. Use the flowers and vegetables at home or sell them for profit.
7. Learn to calculate area in square feet. A square foot is an area that is 1 foot long and 1 foot wide. An area that is 1 foot long and 2 feet wide contains 2 square feet. Area (A) in square feet is calculated by multiplying length (L) in feet times width (W) in feet. Therefore, $A = L \times W$, or $A = LW$.
   a. How many square feet are in a garden that is 10 ft long × 5 ft wide?
      A = _____ sq ft
   b. How many square feet are in a lawn that is 40 × 70 ft?
      A = _____ sq ft
   c. What is the area of a building lot that is 109 × 150 ft?
      A = _____ sq ft

    d. Suppose the lot in item c (above) is covered with lawn except for the house, which is 28 × 42 ft. What is the area of the house?

A = _____ sq ft

What is the area of the lawn?

A = _____ sq ft

8. Using the produce you grew during Activity 6, dehydrate, freeze, and/or can some of it for future use. Be sure to follow approved techniques. Contact your local extension education office for more information.

SELF EVALUATION

A. Multiple Choice

1. Gardening is both a
   a. science and art.
   b. chore and hard work.
   c. science and hobby.
   d. hobby and job.

2. A good rule of thumb for planning a garden for four people is to start with a plot that is
   a. 40 × 60 feet.
   b. 15 × 25 feet.
   c. 10 × 26 feet.
   d. 3 × 7 feet.

3. The south and west sides of a house may not be the best locations for a flower garden because they
   a. are not warm enough.
   b. reflect heat on the planting.
   c. have poor exposure to the sun.
   d. collect rainwater.

4. The furrow in a garden is a
   a. sunken walkway.
   b. hole.
   c. place to plant seeds.
   d. pile of weeds.

5. An annual is a plant whose life cycle is completed in
   a. two growing seasons.
   b. one growing season.
   c. four growing seasons.
   d. three growing seasons.

6. When cultivating, a gardener should use a _____ for best results.
   a. rake
   b. rototiller
   c. shovel
   d. hoe

7. The technical name for weed-control chemicals is
   a. pesticides.
   b. fungicides.
   c. herbicides.
   d. weed killers.

8. After cutting flowers from the garden, it is best to immediately put them in
   a. hot water.
   b. cold water.
   c. salty water.
   d. warm water.

9. To heat a hotbed, frosted light bulbs or _____ are commonly used for heat sources.
   a. a furnace
   b. lead- or plastic-coated heating cables
   c. a fire
   d. a small stove

10. The conventional greenhouse is designed to
   a. add more living space to the home.
   b. keep plants warm.
   c. capture light and control temperature.
   d. protect plants from pests and diseases.

B. Matching

_____	**1.** Perennial	a. 12×12 inches
_____	**2.** Peat moss	b. Fragrant flower
_____	**3.** Square foot	c. More than two growing seasons
_____	**4.** Loamy	d. Long-season vegetable
_____	**5.** Watermelon	e. Persistent weed
_____	**6.** Sweet pea	f. Soil type
_____	**7.** Bermuda grass	g. Form of organic matter

C. Completion

1. A cold frame is a bottomless wooden box with a sloping _____ _____.

2. You should store only the vegetables that are of good quality and at the proper stage of _____.

3. You should plant only the vegetables and flowers that are liked by _____ _____.

4. Fertilizer and lime should be used to promote _____ _____.

5. Watering should be done to a depth of _____.

Vegetable Production

MATERIALS LIST

✓ seed catalogs
✓ scissors, index cards, and glue
✓ pen, paper, eraser, and ruler

✓ lima bean seeds, containers, and soil mix
✓ hand trowels

✓ tomato seeds, soil mix, and flats
✓ Internet access

COMPETENCIES TO BE DEVELOPED

After studying this unit, you should be able to:

• determine the benefits of vegetable production as a personal enterprise or career opportunity.

• identify vegetable crops.

• plan a vegetable production enterprise and prepare a site for planting.

• describe how to plant vegetable crops and use appropriate cultural practices.

• list appropriate procedures for harvesting and storing at least one commercial vegetable crop.

SUGGESTED CLASS ACTIVITIES

1. Invite a local fruit and vegetable wholesaler to visit the class, or take the class to the distribution center of the wholesale business. Discuss the different fruits and vegetables with respect to local demand, where they were produced, length of time they can be stored, proper storage conditions to maintain quality, and what precautions are taken to ensure they are free of dangerous pesticides. Discuss the amount of produce that the business markets in a week, a month, and a year.

2. Assign members of the class to choose a vegetable crop on which they will prepare an oral report. They should seek information about the vegetable from vegetable growers' organizations in major vegetable-producing states, such as the potato commission in Idaho, vegetable seed companies, state university extension office publications, and Internet sites. Help each class member to organize his or her material in a predetermined format. Have students report their findings to the class.

SUGGESTED CLASS ACTIVITIES—*continued*

3. As a class, create a business plan for a virtual vegetable production enterprise. The plan should include all estimated costs, such as for fuel, seed, hired help, and so on. The size of the property and the type of crop to be planted should be given. Assume that the business will have a fair yield and the crop will sell at the average market value. Find the estimated profit by subtracting the income from the costs. As a class, discuss the business plan.

TERMS TO KNOW

olericulture	olericulturalist	aeration	semiarid
home gardening	angiosperm	green manure	precooling
market gardening	monocotyledon (monocot)	transplant	hydrocooling
truck cropping	dicotyledon (dicot)	arid	

INTERNET KEY WORDS:

olericulture

The study of vegetable production is **olericulture.** A vegetable is the edible portion of an herbaceous plant (Figure 19–1). Herbaceous describes a plant that has a stem that withers away at the end at each growing season. The production of vegetables can be classified into three categories: home garden, market garden, or truck crop. **Home gardening** usually refers to the vegetable production for one family and does not involve any major selling of the crops. **Market gardening** refers to growing a wide variety of vegetables for local or roadside markets. **Truck cropping** refers to large-scale production of a few selected vegetable crops for wholesale markets and shipping.

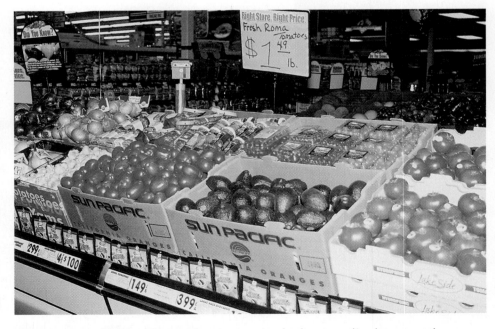

FIGURE 19-1 Vegetables are important in the human diet because they supply basic nutrients. They are good to eat and are good for you.

VEGETABLE PRODUCTION FOR HOME PROFIT OR AS A CAREER

Home Enterprise

Growing vegetables in a home garden is enjoyed by millions of people in the United States. It has become a part of the lifestyle of most families with access to a little bit of ground. The vegetables produced can be used for fresh table consumption or can be stored for later use. Vegetable gardening not only produces nutritious food but also provides outdoor exercise from spring until fall.

The gardener who enjoys this type of activity can plant enough to provide for the family and harvest vegetables for sale on a small scale for extra income. Fresh, homegrown vegetables are usually superior in quality to those found in the supermarkets. In addition, the gardener can grow those vegetables that may be too expensive to buy.

Career Opportunities

The vegetable industry is a large and complex portion of the horticultural industry today. Even though there are millions of homeowners who garden, the majority of vegetables consumed by the public are grown commercially. The commercial vegetable industry is fast moving, intensive, and competitive. It is a business that is continually changing as the demands for certain vegetables fluctuate with the tastes of the U.S. public.

There are numerous and varied career opportunities in the vegetable industry. These include being the owner of a small market gardening business, a member of a larger truck-crop business, a vegetable wholesaler or retailer, or a worker in a vegetable processing plant. With adequate education and training, a person can become an olericulturist—that is, someone who develops pest-resistant strains and new varieties of vegetables, and does other specialized work. The opportunities are endless in the vegetable industry.

IDENTIFYING VEGETABLE CROPS

Vegetables can be identified in various ways: by their botanical classifications, according to their edible parts, or by the kind of growing season that is required by the plant.

Botanical Classification

All vegetables belong to the division of plants known as angiosperms. These are plants with ovules and an ovary. From this division, the vegetables can be grouped into either Class I, monocotyledons (monocots) (having only one seed leaf), or Class II, dicotyledons (dicots) (having two seed leaves) (Figure 19–2). A vegetable can be further grouped into a family, a genus, a species, and sometimes a variety. One of the more popular vegetable families is Cruciferae—the mustard family, which contains Brussels sprouts, cabbage, cauliflower, collards, cress, kale, turnips, mustard, watercress, and radish. Other families include Leguminosae—the pea family, which contains bush beans, lima beans, cow peas, kidney beans, peas, peanuts, soybeans, and scarlet runner

INTERNET KEY WORDS:

monocot, monocotyledon
dicot, diocotyledon

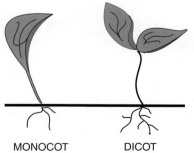

MONOCOT   DICOT

FIGURE 19–2 Monocots have one seed leaf, whereas dicots have two.

Plants of which the Fruits or Seeds are Eaten		Plants of which the Leaves, Flower Parts, or Stems are Eaten		Plants of which the Underground Parts are Eaten	
Family	Vegetable	Family	Vegetable	Family	Vegetable
Grass, Gramineae	Sweet corn, Zea mays	Lily, Liliaceae	Asparagus, Asparagus officinalis var. altilis Chives, Allium schoenoprasum	Lily, Liliaceae	Garlic, Allium satvium Leek, Allium porrum Onion, Allium cepa Shallot, Allium ascalonicum Welsh onion, Allium fistulosum
Mallow, Malvaceae	Okra (gumbo), Hibiscus esculentus	Goosefoot, Chenopo-diaceae	Beet, Beta vulgaris Chard, Beta vulgaris var. cicla		
Pea, Leguminosae	Asparagus or Yardling bean, Vigna sequipedalis Broad bean, Vicia faba Bush bean, Phaseolus vulgaris Bush Lima bean, Phaseolus limensis Cowpea, Vigna sinensis Edible podded pea, Pisum sativum var. macrocarpon Kidney bean, Phaseolus vulgaris Lima bean, Phaseolus limensis Pea (English pea), Pisum sativum Peanut (underground fruits), Arachis hypogaea Scarlet runner bean, Phaseolus coccineus Sieva bean, Phaseolus lunatus Soybean, Glycine max White Dutch runner bean, Phaseolus coccineus	Orach, Atriplex hortensis Parsley, Umbelliferae Sunflower, Compositae Mustard, Cruciferae	Spinach, Spinacia oleracea Celery, Apium graveolens Chervil, Anthriscus cerefolium Fennel, Foeniculum vulgare Parsley, Petroselinum crispum Artichoke, Cynara scolymus Cardoon, Cynara cardunculus Chicory, witloof, Chichorium intybus Dandelion, Taraxacum officinale Endive, Chichorium endivia Lettuce, Lactuca sativa Brussels sprouts, Brassica oleracea var. gemmifera Cabbage, Brassica oleracea var. capitata Cauliflower, Brassica oleracea var. botrytis Collard, Brassica oleracea var. viridis Cress, Lepidium sativum Kale, Borecole, Brassica oleracea var. viridis Kholrabi, Brassica oleraceae var. gongylodes Mustard leaf, Brassic juncea Mustard, Southern Curled, Brassica juncea Pak-Choe, Chinese Cabbage, Brassica chinensis var. crispifolia Pe-tsai, Chinese cabbage, Brassica pekinensis Seakale, Crambe maritima Sprouting Broccoli, Bra-sica oleracea var. italica Turnip, Seven Top, Brassica rapa Upland cress, Barbarea verna Watercress, Rorippa nasturtium-aquaticum	Yam, Dioscoreaceae Goosefoot, Chenopo-diaceae Mustard, Cruciferae Morning Glory, Convolvulaceae Parsley, Umbelliferae Nightshade, Solanaceae Sunflower, Compositae	Yam (true), Dioscorea batatas Beet, Beta vulguris Horseradish, Armoracia rusticana Radish, Raphanus sativus Rutabaga, Brassica campestris var. napobrassica Turnip, Brassica rapa Sweet Potato, Ipomoea batatas Carrot, caucus carota var. sativa Celeriac, Apium graveolens var. rapaceum Hamburg parsley, Petro-selinum crispum var. radicosum Parsnip, Pastinaca sativa Potato, Solanum tuberosum Black salsify, Scorzonera hispanica Chicory, Chicorium intybus Jerusalem artichoke, Helianthus tuberosus Salsify, Tragopogon porrifolius Spanish salsify, Scolymus hispanicus
Parsley, Umbelliferae	Caraway, Carum carvi Dill, Anethum graveolens				
Martynia, Martyniaceae	Martynia, Proboscidea louisiana				
Nightshade, Solanaceae	Eggplant, Solanum melongena Groundcherry (husk tomato), Physalis pubescens Pepper (bell or sweet), Capsicum frutescens var. grossum Tomato, Lycopersicon esculentum				
Gourd or Melon, Cucurbitaceae	Chayote, Sechium edule Cucumber, Cucumis sativus Cushaw, Cucurbita moschata Gherkin, Cucumis anguria Cantalope (Muskmelon), Cucumis melo Pumpkin, Cucurbita pepo Summer squash (bush pumpkin), Cucurbita pepo Squash, Cucurbita maxima Watermelon, Citrullus lunatus Winter melon, Cucumis melo var. inodorus				

FIGURE 19-3 Classification of vegetable crops by botanical family and crop use.

beans; Cucurbitaceae—the gourd or melon family—contains pumpkins, cucumbers, cantaloupe, casaba melons, and watermelons; Solanaceae—the nightshade family, contains eggplants, ground cherries, peppers, and tomatoes.

Edible Parts

Vegetables can also be classified by the part of the vegetable that is eaten. There are three groupings of vegetables according to their uses: (1) vegetables of which leaves, flower parts, or stems are used; (2) vegetables of which the underground parts are used; and (3) vegetables of which the fruits or seeds are used. (See Figure 19–3 for a list of these categories and the vegetables in each one.)

Growing Seasons

There are basically two growing seasons: warm and cool. Cool-season vegetable crops grow best in cool air and can withstand a frost or two. Some of these crops, such as asparagus and rhubarb, can even endure winter freezing. This group of crops is planted early in the spring and late in the season for fall and winter harvest. Cool-season crops include mostly leaf and root crops (Figure 19–4).

The warm-season, or warm-weather, crops are those that cannot withstand cold temperatures, especially frosts. These vegetables and fruits require soil warmth to germinate and long days to grow to maturity. They must have very warm temperatures to produce their edible parts. The edible portions of these crops are basically what can be picked off the standing plant or the fruit. A list of warm-season crops can be found in Figure 19–5.

INTERNET KEY WORDS:

cool-season vegetable crops
warm-season vegetable crops

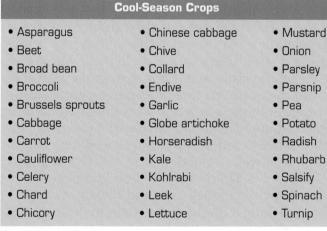

Cool-Season Crops		
• Asparagus	• Chinese cabbage	• Mustard
• Beet	• Chive	• Onion
• Broad bean	• Collard	• Parsley
• Broccoli	• Endive	• Parsnip
• Brussels sprouts	• Garlic	• Pea
• Cabbage	• Globe artichoke	• Potato
• Carrot	• Horseradish	• Radish
• Cauliflower	• Kale	• Rhubarb
• Celery	• Kohlrabi	• Salsify
• Chard	• Leek	• Spinach
• Chicory	• Lettuce	• Turnip

A

B

FIGURE 19–4 (A) Some cool-season crops. (B) Cool-season vegetable crops grow best in cool temperatures, and they are even able to tolerate occasional freezing temperatures.

Warm-Season Crops	
• Cowpea	• Pumpkin
• Cucumber	• Snap bean
• Eggplant	• Soybean
• Lima bean	• Squash
• Muskmelon	• Sweet corn
• New Zealand spinach	• Sweet potato
• Okra	• Tomato
• Pepper, hot	• Watermelon
• Pepper, sweet	

A

B

FIGURE 19–5 (A) Some warm season crops. **(B)** Warm-season vegetable crops respond best in summer conditions. They have little tolerance for temperatures that are near freezing.

PLANNING A VEGETABLE-PRODUCTION ENTERPRISE

Before planting a vegetable garden or truck crop, the operator needs to plan. The gardener must decide where to plant, as well as what to plant. Without some advance planning, the vegetable garden or truck farm is not likely to be successful.

Selecting the Site

Choose a site that is convenient to a water supply. The site should also be exposed to the sun a minimum of 50 percent during the day. A minimum of 8 to 10 hours of direct sunlight is needed. Also consider the type of trees that are around the proposed site. Trees can provide excessive shade. They can also take away nutrients from the soil that are needed by the vegetable plants. Some kinds of trees can also produce toxins that are harmful to certain vegetables. For example, the walnut tree is toxic to the tomato. Buildings and structures cast shade that can slow down or prevent growth of a vegetable crop. Avoid trying to grow vegetables closer than 6 to 8 feet from the northern side of a one-story structure—farther for higher structures. Both the south and west sides of a building have good light and often radiate heat late in the day (Figure 19–6).

The type of soil in the area selected is important as well. The majority of vegetables grow best in a well-drained, loamy soil. Avoid ground that develops puddles after it rains. This is a sign of poor drainage. To test the drainage of a site, dig a trench 12 inches wide and 18 inches deep. Fill it with water and observe how long it takes for the water to drain away. If it takes 1 hour or less, the soil can be considered well drained. If the area selected has supported vegetation before, even if it has only been weeds, it will probably support a vegetable crop. If the site selected needs some alterations to its structure, adding organic matter can help. The organic matter can improve the drainage and allow air to move readily through the pores of the soil. If possible, the garden soil should be about 25 percent organic matter. To accomplish this, put a layer of organic matter

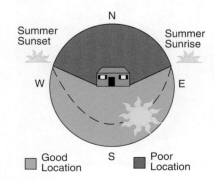

FIGURE 19–6 Avoid situating the garden in the shadow of a building or tree. The location of a building's shadow changes throughout the day and throughout the year.

Agri-Profile

CAREER AREAS:

Vegetable Producer/Processor/ Distributor/Produce Manager

Career opportunities in vegetable production are similar to those in fruit production. Vegetables and fruits are frequently grown, processed, and marketed by the same people. In addition to the careers mentioned in Unit 20, many people find rewarding careers that pay well in the area of distributing and marketing produce. In this area, truckers, wholesalers, and retailers move the product from producer to consumer.

Excellent jobs are available in supermarkets for produce stockers and produce department managers. Products include fruits, vegetables, and nuts. Tasks may include inventorying, ordering, handling, stocking, and displaying produce to keep it fresh and attractive. In large cities, as well as small towns, street vendors are frequently seen selling premium fruit and, occasionally, vegetables. Roadside stands provide opportunities for younger members of families to develop business skills while earning money for present and future needs.

Good french fries require the cooperative work of plant breeders, potato growers food specialists, and vegetable processors, among others.
(Courtesy of USDA/ARS #K-4018-12)

Amounts of Organic Matter Needed to Cover 100 Square Feet at Various Depths	
Depth (inches)	To Cover 100 sq. ft.
6	2 cubic yards
4	35 cubic feet
3	1 cubic yard
2	18 cubic feet
1	9 cubic feet
1 cubic yard = 27 cubic feet	

FIGURE 19–7 Amounts of organic matter to apply per 100 square feet of garden soil.

2 inches thick over the soil and work it in to a depth of at least 4 inches. If necessary, repeat this procedure until the final mix contains approximately 25 percent organic matter (Figure 19–7).

Deciding on the Size

The size of the vegetable growing area depends largely on the amount of ground that is available, the number of people served by the garden, and the use of the produce. A large garden 50 × 100 feet will produce enough vegetables for the annual needs of a large family. Figure 19–8 shows a suggested planting plan for a 50 × 100 foot home garden.

If this much land is not available, the fresh vegetable needs for a medium-sized family may be met with a 40 × 50 foot plot. A planting guide for such a garden can be found in Figure 19–9. When deciding on the size, it would be wise to remember that the larger the vegetable production, the more care it will take. Vegetable production will be more efficient with better crops if the area is a smaller and better managed garden than if you take on a bigger area that is neglected and full of weeds.

Deciding What to Plant

There are several items that a potential vegetable gardener or truck cropper must consider when deciding which vegetables to plant. If for family use only, the first thing to determine is which vegetables the family likes and how often each vegetable will be served. Also, consider which vegetables will be stored. Select vegetable varieties that the family enjoys and concentrate on the kinds that are definitely better when they are freshly harvested. Examples are tomatoes, carrots, corn, snap beans, and cantaloupe. Second, consider the maintenance that some vegetables require. The easier they are to grow and maintain, the better.

NORTH

Row	Crop					Feet between rows
1	Sweet corn		Sweet corn			3'
2	Sweet corn	1st planting	Sweet corn	2nd planting		3'
3	Sweet corn		Sweet corn			3'
4	Sweet corn		Sweet corn			3'
5	Sweet corn			3rd planting		3'
6	Sweet corn					3'
7	Tomatoes (staked)	Plant pole beans near tomato stakes in early July without disturbing the tomato plants				4'
8	Tomatoes (staked)					4'
9	Tomatoes (staked)					4'
10	Early Potatoes					3'
11	Early Potatoes					3'
12	Pepper	Eggplant	Chard (Swiss)			3'
13	Lima bean (bush)					3'
14	Lima bean (bush)					3'
15	Lima bean (bush)					3'
16	Snapbeans (bush)					3'
17	Snapbeans (bush)					3'
18	Broccoli					3'
19	Early cabbage					3'
20	Onion sets	This entire area may be replanted after harvest with such crops as: endive, cauliflower, Brussels sprouts, spinach, kale, beets, cabbage, broccoli, turnips, lettuce, carrots, and late potatoes.				3'
21	Onion sets					2'
22	Carrots					2'
23	Carrots					2'
24	Beets					2'
25	Beets					2'
26	Kale					2'
27	Spinach					2'
28	Peas					2'
29	Peas					2'
30	Lettuce	1st planting	Lettuce	2nd planting	Lettuce	3rd planting	2'
31	Radish		Radish		Radish		2'
32	Strawberries					3'
33	Strawberries					3'
34	Asparagus		Rhubarb			3'
35	Asparagus					3'
						3'

WEST — 100' — EAST

— 50' —

SOUTH

FIGURE 19–8 Example of a planting plan for a large home garden 50 × 100 feet.

A third factor to consider is the length of the harvest time for each vegetable. If the object is fresh consumption, then vegetables that can be harvested over a long period are best. For instance, carrots can be harvested for months, whereas cabbage should be picked at maturity. The last item to consider is the use for which the vegetables are planted.

Feet between Rows

Rows	Crop					1½'
1	Early Cabbage		(Turnips)			2'
2	Early Potatoes		(Late Cabbage)			2'
3	Peas	(Beans)		Peas	(Kale)	2'
4	Kale	(Carrots)		Turnips	(Beans)	2'
5	Parsley			Parsnips		2'
6	Onions and Radishes (same row)			Carrots	(Spinach)	2'
7	Swiss Chard	(Spinach)		Spinach	(Beets)	2'
8	Lettuce	(Beans)		Beets	(Kale)	2'
9	Beans	(Spinach)		Beans	(Lettuce)	2'
10	Peppers			Eggplant		2'
11	Lima Beans, Bush					2'
12	Tomatoes, Late					2'
13	Tomatoes, Early					2½'
14	Squash			Cucumbers		2½'
15						2½'
16	Corn, Sweet interplanted with Pole Lima Beans					2½'
17	Corn, Sweet interplanted with Winter Squash					2½'
18						2'

40'

←————————————— 50' —————————————→

NOTE: Items in parentheses are succession crops planted after the first crop is harvested or planted between the rows of mature plants to permit germination and early growth before the first crop is removed.

FIGURE 19–9 Example of a plan for a medium size garden 40 × 50 ft.

If the vegetables are going to be for fresh use, plant only for that purpose. Surplus vegetables can be stored in a root cellar or freezer, but plantings should match the estimated needs.

The truck cropper has a different set of questions to consider. First, which vegetables are in high demand in the area? Which ones will produce the greatest return for the cost and labor involved? The truck-cropping enterprise is a business and must make a reasonable profit to be considered a good business. Second, what vegetables can be raised in the soil and climate? Truck cropping should be a high-income business, so high quality and quantity of produce is important. Third, what combination of vegetables can be grown to keep the land occupied and provide cash flow from sales throughout the season?

Other questions to be explored are: (1) can the diseases and insect problems be managed? (2) will continuous use of the land for the same crops create pest buildup? (3) will the soil tolerate continuous cropping? and (4) what perennials may be included to reduce the labor involved?

Science Connection

ENCOURAGING THE MASTER TILLERS

Earthworms are nature's master tillers. Always a favorite for fishing bait, the lowly earthworm is now regarded as an ally of homeowners, gardeners, farmers, landscapers, and conservationists. They enhance soil tilth and crop growth by consuming and digesting organic matter and mixing it with the mineral content of the soil. The new mixtures left behind by feeding worms are called *worm casts* and are rich in nutrients for feeding plants. Worm holes in the soil start at the surface and go as deep as the worm needs to go to escape winter cold and summer heat, perhaps 3 feet or more in the central area of the United States. This network of holes and tunnels collects rainwater as it falls, and the enriched soil absorbs and holds water for future plant use.

The North Appalachian Experimental Watershed at Coshocton, Ohio, has been the scene of no-till experiments for more than 30 years. In a major experiment there, night crawlers *(Lumbricus terrestris)*, usually 4 to 8 inches in length and ⅜ inch in diameter, and other species were studied. It was found that earthworms may be encouraged by leaving crop residues on the surface or by incorporating them into the soil. However, if crop residues are removed from the soil, the earthworms do not have an adequate food source, and their populations do not expand as much.

Entomologist Edwin Berry of the National Soil Tilth Laboratory at Ames, Iowa, identified seven other species of earthworms. These all range in size from 2 to 6 inches in length and are about half the diameter of night crawlers. Most of their activity occurs in the first foot of topsoil. They make temporary burrows as they pass through the soil, consuming, digesting, and mixing soil particles and organic matter, and excreting their rich casts.

Worm farming has become an important industry, and earthworms may be purchased for release for soil improvement. Worms grow and multiply the fastest where plenty of organic matter and moisture are available. Unfortunately, worms do not function as well in coarse, sandy soils as they do in soils where medium and fine particles are available. Gardeners, horiticulturists, and farmers should consider practices that will encourage the work of the master tillers.

A

B

(A) Ken Ford at the Soil Tilth Research Lab in Ames, Iowa, uses a video camera adapted for microscopic work to study earthworm cocoons (top of monitor), newly hatched earthworm larvae, and (B) adult earthworms—nature's master tillers. *(Courtesy of (A) USDA/ARS #K-4024-1 and (B) USDA/ARS #K-3237-1)*

FIGURE 19–10 Soil moisture in clay or loamy soils may be judged by pressing a handful of soil into a ball. If the ball crumbles easily, the soil is ready to be worked.

INTERNET KEY WORDS:

preparing site
vegetable garden

INTERNET KEY WORDS:

vegetable garden fertility

PREPARING A SITE FOR PLANTING

Preparing the Soil

Before planting vegetables, the soil must be properly prepared. This will generally mean adding organic matter, lime, and fertilizer. The land should be plowed if it is in sod, and then left for 4 to 6 weeks for the sod to decay.

Plowing Land should be plowed or spaded to a depth of 6 to 8 inches. The soil can be plowed either in the spring or the fall. Fall plowing has advantages. First, the freezing and thawing of the winter months can improve the physical condition of the soil. Second, the exposure of the soil to the weather can result in a reduced population of insects. Spring plowing should be done only a few weeks before planting. Take care not to plow the soil when it is too wet. Doing so can destroy the physical structure of the soil, which results in hard clumps. The moisture content of loam or clay soil may be judged by pressing a handful of soil into a ball (Figure 19–10). If the ball crumbles easily, the soil is ready to plow. If the soil sticks together, it is too wet to plow.

Maintaining Organic Matter Organic matter increases the water-holding and absorption capacity of the soil. Furthermore, it helps prevent erosion and promotes aeration in the soil. Aeration refers to the movement of air through soil. If it is available, and can be secured cheaply, animal manure is the best material for maintaining the organic content of the soil. It is also a good source of nutrients. Manure from animals other than sheep and poultry can be turned under at the rate of 15 to 20 tons per acre, or 20 to 30 bushels per 1,000 square feet. Sheep and poultry manures can be used, but at half this rate. One note of caution—animal manures are low in phosphorus. Therefore, it may be necessary to use high-phosphorus fertilizer to balance the nutrients. For best results, apply fresh manure in the fall and well-rotted manure in the spring.

Another type of organic matter to use is green manure in the form of a cover crop, especially if it is a legume. Green manure is an active, growing crop that is then turned under to help build the soil. Green vegetation incorporated into the soil rots more quickly than does dry material. A cover crop should be planted at the end of the growing season. A cover crop is a close-growing crop planted to prevent erosion.

Liming The need for lime and fertilizer should be determined through the use of soil tests. A pH greater than 7.0 indicates alkalinity, a pH of 7.0 is neutral, and a pH less than 7.0 indicates acidity.

If the test indicates a pH of 6.0 or less, lime should be applied. The lime should be mixed into the top 3 to 4 inches of the soil. The amount of lime to add depends on the type of soil, the desired pH, and the form of lime used. Some plants, such as blueberries, require acidic soils. For such plants, special materials are available to make the soil acidic if the pH is too high.

Vegetable crops grow best at specific pH levels. Figure 19–11 lists vegetables and their optimum pH ranges for maximum growth.

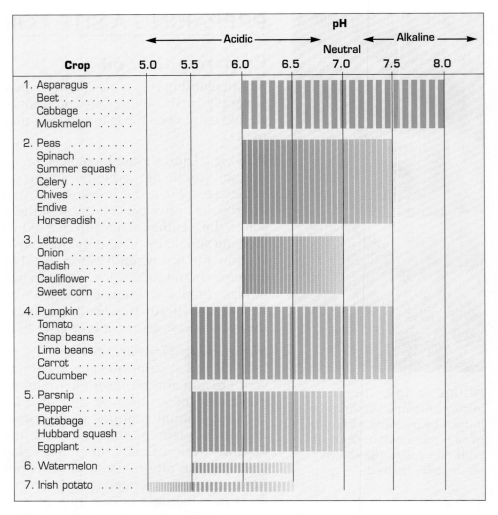

FIGURE 19–11 Optimum pH ranges for vegetable crops.

FIGURE 19–12 The analysis or ratio on a fertilizer bag indicates the percentages of nitrogen, phosphoric acid, and potassium in the fertilizer.

Fertilizing The amount and type of fertilizer to add can best be determined by a soil test. The test will help prevent applying too much or too little fertilizer and will determine the best timing. The application of commercial fertilizer is done to increase the amounts of nutrients available to the plants. Soils without additives are not likely to provide the best combination of conditions for all vegetables.

Commercial fertilizers usually contain nitrogen, phosphoric acid, and potassium. Each vegetable crop needs varying amounts of each of these three elements. In general, fertilizer ratios used for home gardens are 5-10-10, 10-10-10, and 5-10-5 (Figure 19–12). The first number in a fertilizer ratio or grade represents the percentage of nitrogen, the second shows the percentage of phosphorus, and the third indicates the percentage of potassium in the fertilizer. A rough guide for home gardeners is that most vegetables will need about 3 to 4 lb fertilizer per 100 square feet. Consult a county extension educator for exact fertilizer recommendations based on the soil types in the area. Extension educators are specialists on local and regional conditions that affect crops and gardens.

FIGURE 19–13 Some garden plants are planted in containers for later seeding. This allows the gardener to extend the length of the growing season by starting frost-sensitive plants in the greenhouse or hotbed.

PLANTING VEGETABLE CROPS

Vegetable crops are initially grown from seeds. Some seeds can be sown where the plants will grow. Others will need to be grown into plants, and the seedlings then transplanted to where the crop will grow. Transplants are plants grown from seeds in a special environment, such as a cold frame, hotbed, or greenhouse (Figure 19–13). The method used depends on the climatic requirements of the plant and the germinating characteristics of the seed.

Science Profile

WHAT IS GOOD ABOUT FUNGI?

Fungi are plentiful organisms that are found in the soil. Despite some species of fungi that are known to be harmful to crops, there are fungi that are useful to crops. Mycorrhizae are fungi that growers welcome in their fields. These microorganisms do not have chlorophyll, and they cannot use the process of photosynthesis to make their own food. To survive, they attach to the roots of plants in search of carbohydrate nutrients. Despite that fungi cover plant roots, they are considered beneficial.

Phosphorus is a nutrient that plants have a difficult time taking out of the soil. Mycorrhizae fungi produce an enzyme that frees nutrients from the soil so the plants can take them in. They also help the plant take in water and other nutrients. Both the plants and the fungi benefit from this symbiotic relationship. Scientists have estimated that 80% of all plants form this kind of relationship with fungi. Fields with adequate levels of these fungi require less water and less fertilizer. Mycorrhizae fungi are commercially available, and their benefits far outweigh their costs.

Planting Seed

Depending on the size and scope of the garden and the quantity of seed to be planted, vegetable seeds are planted (1) by hand in hills or rows, (2) through broadcasting by hand or machine, (3) with one-row hand seeders, and (4) with single- or multiple-row, tractor-drawn seeders.

Broadcasting is a planting method where seeds are scattered around on the soil surface. Regardless of the type of planting method used, the seeds should be planted at the proper depth. The soil should also be left smooth and firm over the seed. For most vegetables, seed must be fresh. Seed more than 1 year old will probably not germinate at the rate necessary to produce good yields.

Most vegetable seeds should be planted at a moderately shallow depth. It is best to plant in loamy-textured soil, with adequate moisture and no danger of frost. The seed will germinate best when it is planted at a depth no more than four times the diameter of the seed. Germinate means to sprout, grow, and produce a plant from a seed. Regardless of how deep the seeds are planted, the soil surface should be level and firmly pressed after the seed is planted. This will help prevent the seed from washing away or water from puddling above the seed after a rain.

Transplanting Vegetable Seedlings

INTERNET KEY WORDS:

seedling transplant procedure

There are three methods used in transplanting, depending on the number of plants that are being transplanted. These methods are: (1) hand setting, (2) hand-machine setting, and (3) riding-machine setting. In home gardening situations, hand setting is most appropriate.

Hand setting involves six steps: (1) dig a hole slightly bigger and deeper than the root ball of the plant being transplanted; (2) add some fresh soil to the hole; (3) place the plant in the soil a little deeper than it grew in its original container, trying to keep as much of the original soil as possible around the roots; (4) pull soil in and around the plant and firm it slightly; (5) add about a half pint of water and let it soak into the soil; and (6) pull in some dry soil to level off and cover the wet area (which helps to prevent loss of moisture and baking of the soil).

The vegetable transplants should be set out on a cloudy day or late in the afternoon. It is best to set plants just before or just after a rain.

CULTURAL PRACTICES

Cultivating

Cultivation, or intertillage of crops, is an old agricultural practice. The benefits of cultivation are: (1) weed control, (2) conservation of moisture, and (3) increased aeration. Cultivation increases the yield of most vegetable crops, mainly because the weeds are controlled.

All types of equipment are used to cultivate crops, from small hand tools to large tractors. The cultivation should be shallow and done only when there are weeds to be killed. Excessive cultivation is not beneficial and may even be damaging.

Controlling Weeds

Weeds are easier to control when they are small, using shallow cultivation. Weeds that are older and more established are more difficult to

Hot Topics in Agriscience

ORGANIC VEGETABLE FARMING

A farmer who raises "organic" vegetables produces them without using the chemical pesticides and fertilizers on which most vegetable farmers depend. This does not mean that the vegetable crop is produced without insect controls or fertilizers. Fertilizers are obtained from animal manure and/or "green manure" crops such as legumes that are plowed under when they are 6 to 12 inches in height. It simply means that an organic farmer first tries to maintain a healthy environment for the plants. This is because a healthy plant located in an environment that supports its needs is less susceptible to pests and diseases than a plant that is not well adapted to its environment. The organic farmer also introduces the natural enemies of pests to the environment to control their populations. In addition, pepper and garlic sprays are common methods for discouraging insect pests. As a final resort, the organic farmer may apply sprays consisting of horticultural oils, isopropyl alcohol, soaps, ammonia, baking soda, and even bug juice (a spray made with water and insects that have been pulverized in a blender). Some chemicals such as rotenone and sulfur are acceptable to organic farmers because they break down easily and are degraded quickly to materials that are no longer harmful. Organic vegetables that are of high quality command a premium price among some consumers in comparison with the price of produce that was produced using standard farming methods that rely on chemical fertilizers and pesticides.

FIGURE 19–14 Evaluating the weed-control effectiveness in plants passed over by the spyder (front), torsion weeder, side knife, and spring hoe attachments. *(Courtesy of USDA/ARS #K-5226-18)*

control. For them, deeper cultivation is required to rid the area of weeds. This can result in injury to the roots of vegetable crops (Figure 19–14).

Using herbicides is recommended for large home gardens and commercial vegetable production plots. The use of any herbicide depends on the registration of the herbicide by federal and state environmental protection agencies (EPAs). Do not use any herbicide unless it is clearly stated on the label that it is intended for a particular crop. Always seek the latest information for use of herbicides on vegetables.

Irrigation

Nearly all commercial vegetable operations irrigate their crops. Irrigation is especially important in California and other **arid** and **semiarid,** or dry, regions of the West. Irrigation is important because rainfall is rarely uniform and adequate for high yields.

Water for irrigation can be obtained from a stream, lake, well, spring, or stored storm water. Definite laws and regulations guide the practice of irrigation in each state. There are several types of irrigation systems, including sprinkler; drip surface, or furrow; and subirrigation.

Sprinkler systems are versatile (Figure 19–15). They can be used in almost any situation. They can deliver shallow and light irrigation to promote seed germination and can be used for the application of fertilizers.

Drip irrigation uses a system of pipes and tubes to deliver the water to individual plants rather than wetting all of the soil (Figure 19–16). It is gaining in popularity because of its water-conserving benefits.

Surface irrigation has the advantage of requiring a relatively low investment. The topography of the land is important for surface irrigation. The land must be gently sloping and uniform. This method is well

FIGURE 19–15 Sprinkler irrigation is a proven method for providing water to vegetable crops. The water can be applied uniformly in the proper amounts.

FIGURE 19–16 Drip irrigation delivers water to individual plants through small tubes. It is the most efficient method of irrigation for vegetable crops.

FIGURE 19–17 Mulching is the practice of adding material such as straw, leaves, or plastic film to the surface of the soil.

FIGURE 19–18 Insects such as the Colorado potato beetle and the Mexican bean beetle (shown above) can strip the leaves from vegetable plants in a matter of hours. *(Courtesy of USDA/ARS #K-1291-2)*

suited for irrigating vast areas of land, but it generally requires land leveling to work well.

Subirrigation requires large amounts of water. With this system, the water is added to the soil so that it permeates the soil from below. This method is expensive and is suited only to commercial enterprises.

Trickle irrigation is a form of subirrigation in which water is delivered under low pressure near the plants. The water seeps through porous hose under the rows. This system is expensive, but gives excellent results, with increased yields and less water consumed than with other methods.

Mulching

A mulch is created whenever the surface of the soil is modified artificially by using straw, leaves, paper, polyethylene film, or a special layer of soil (Figure 19–17). Mulching helps control weeds, regulates soil temperatures, conserves soil moisture, and provides clean vegetables. The mulching material should be spread around the plants and between the rows.

Plastic film is frequently used with cantaloupes, cucumbers, eggplants, summer squash, and tomatoes. It is important that the plastic be laid over moist soil that has been freshly prepared. It is advisable to lay plastic mulch 1 to 2 weeks before planting the vegetables. This allows the soil to warm up underneath the mulch.

Pest Control

Good pest-control practices are essential for quality vegetables. A combination of cultural, mechanical, biological, and chemical methods can be used. Insecticides and fungicides can be effectively used by both home and commercial vegetable growers.

Insect and disease infestations bring about losses that can be devastating to vegetable crops (Figure 19–18). Types of losses include: (1) reduced yields, (2) decreased quality of produce, (3) increased costs

A

B

C

FIGURE 19–19 (A) New disease-resistant varieties of sweet potatoes. (B) Spraying with nontoxic Bt to spread insect-fighting bacteria. (C) A spined soldier beetle making a meal of a leaf-devouring Mexican bean beetle larva. *(Courtesy of (A) USDA/ARS #K-3461-14; (B) USDA/ARS #K-2785-2; and (C) USDA/ARS #K-1461-3)*

FIGURE 19–20 The use of hydroponic methods to produce vegetables has become a big business in recent years. Water and nutrients may be delivered through a channel in the floor or by drip irrigation tubes.

of production and harvest, and (4) increased expenditures for materials and equipment. The use of chemicals is limited because of toxic and environmental considerations, so good biological controls are in great demand (Figure 19–19). There are many measures that can help the vegetable grower control insects and diseases. They are:

- Dispose of old crop residues that serve to shelter pests, and thus enable them to overwinter or survive from one crop to another.
- Rotate the location of individual crops.
- Choose insect- and disease-resistant vegetable varieties.
- Treat seeds with fungicides and insecticides.
- Control weeds.
- Use adequate amounts of fertilizer.
- Follow recommended planting dates.

Hydroponics

The ability to produce vegetables on a commercial scale without soil has contributed to the supply of tomatoes, cucumbers, and other fresh vegetable crops during the winter season (Figure 19–20). The plant most commonly produced using hydroponic methods is the tomato (see Unit 9). Although it is really a fruit, it is used as a vegetable, and it is included in this unit for that reason. Most commercial greenhouses that produce tomatoes plan to deliver most of the crop during the winter when tomato prices are highest. The plants may be rooted in inert filler material through which water and nutrients flow, or they may be placed in containers to which the nutrients and water are delivered using drip irrigation methods. Large hydroponic operations harvest daily or every other day during the production cycle. (See Unit 9 for more information on hydroponics.)

HARVESTING, MARKETING, AND STORING VEGETABLES

INTERNET KEY WORDS:

hydroponics,
irrigation systems

INTERNET ADDRESS:

<http://www.postharvest.com.au/
storage.htm>

Harvesting

For the best possible quality, vegetables should be harvested at the peak of maturity. This is possible when the vegetables will be used for home consumption, sold at a local market, or processed. The best harvesting time varies with each vegetable crop. Some will hold their quality for only a few days, whereas others can maintain quality over a period of several weeks. Frequent and timely harvest of crops is needed if the vegetable grower wishes to supply the market or the kitchen table with high-quality produce over time. Specifications for harvesting each kind of vegetable crop should be consulted to achieve the best results.

Harvesting is done by hand and through the use of machines (Figure 19–21). Mechanical harvesting is especially useful for the commercial grower. When harvesting a crop, care must be taken to prevent injury to the crop by machinery and handling.

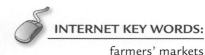

INTERNET KEY WORDS:

farmers' markets

INTERNET ADDRESS:

<http://www.ams.usda.gov/
farmersmarkets>

Marketing

The most important aspect of marketing fresh vegetables is to assure that the quality is high. This requires the vegetable plants to be protected from the stresses of disease and insects. Vegetables must be harvested when they are of optimum size and maturity. At this point in the process, it is important to market the produce as soon as possible.

Farmers' markets are ideal outlets for marketing fresh vegetables to local consumers. These markets are intended to bring several farmers together in a central location with vegetables and fruits of sufficient quality and quantities to attract local citizens to the site. In some instances, these markets are established on private property. In other instances, they may be located on federal or state property as long as permission is obtained from the proper government authorities. The U.S. Department of Agriculture has established rules under which these

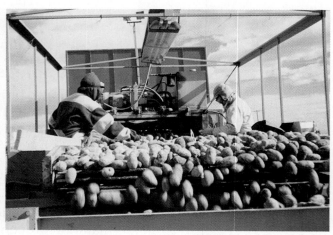

A   B

FIGURE 19–21 (A) Mechanical harvesting of vegetable crops such as potatoes requires special machinery that is designed to separate the crop from leaves, stems, and soil. (B) Metal parts are often rubber-coated to avoid bruising the tubers.

markets are legally established in local communities. It is important for a manager to assume the duties of running the business. Good business practices and high-quality fruit and vegetable products are keys to operating a successful farmers' market.

Storing

When storing vegetables, proper stage of maturity is essential. There are ideal temperatures and humidity levels for storage of vegetable crops (Figure 19–22). Most crops need 90 to 95 percent humidity. Most homeowners find it difficult to reproduce the exact temperatures and humidity levels, but commercial growers can accomplish storage through specialized facilities (Figure 19–23).

Vegetables differ in the amount of time that they can be stored. Some may be stored for several months, others for only a few days. For the vegetables that do store well, storage is essential to prolong the marketing period.

Fresh vegetables used for storage should be free of skin breaks, bruises, decay, and disease. Such damage or disease will decrease the life of the vegetable in storage.

Refrigeration storage is recommended (Figure 19–24). This type of storage reduces respiration and other metabolic activities. It slows ripening; moisture loss and wilting; spoilage from bacteria, fungus, or yeast; and undesirable growths, such as potato sprouts.

One activity that can increase the successful storage of vegetables is precooling. Precooling is the process of rapid removal of the heat from the crop before storage or shipment. One method of precooling is termed hydrocooling. This is a method wherein vegetables are immersed in cold water and stay under water long enough to reduce the temperature to the desired level.

FIGURE 19–22 Ideal temperature and humidity levels for storing common vegetables.

Vegetable Storage		
Crop	Temperature °F	Relative Humidity Percent
Asparagus	32	85-90
Beans, snap	45-50	85-90
Beans, lima	32	85-90
Beets	32	90-95
Broccoli	32	90-95
Brussels sprouts	32	90-95
Cabbage	32	90-95
Carrots	32	90-95
Cauliflower	32	85-90
Corn	31-32	85-90
Cucumbers	45-50	85-95
Eggplants	45-50	85-90
Lettuce	32	90-95
Cantaloupes	40-45	85-90
Onions	32	70-75
Parsnips	32	90-95
Peas, green	32	85-90
Potatoes	38-40	85-90

		Storage Conditions						
Vegetable	Temperature (°F)	Relative Humidity (%)	Storage Life	Vegetable	Temperature (°F)	Relative Humidity (%)	Storage Life	
Pea, English	32	95-98	1-2 weeks	Squash, winter	50	50-70	___[4]	
Pea, southern	40-41	95	6-8 days	Strawberry	32	90-95	5-7 days	
Pepper, chili (dry)	32-50	60-70	6 months	Sweet corn	32	95-98	5-8 days	
Pepper, sweet	45-55	90-95	2-3 weeks	Sweet potato	55-60[3]	85-90	4-7 months	
Potato, early	___[1]	90-95	___[1]	Tamarillo	37-40	85-95	10 weeks	
Potato, late	___[2]	90-95	5-10 months	Taro	45-50	85-90	4-5 months	
Pumpkin	50-55	50-70	2-3 months	Tomato, mature green	55-70	90-95	1-3 weeks	
Radish, spring	32	95-100	3-4 weeks	Tomato, firm ripe	46-50	90-95	4-7 days	
Radish, winter	32	95-100	2-4 months	Turnip	32	95	4-5 months	
Rhubarb	32	95-100	2-4 weeks	Turnip greens	32	95-100	10-14 days	
Rutabaga	32	98-100	4-6 months	Water chestnut	32-36	98-100	1-2 months	
Salsify	32	95-98	2-4 months	Watercress	32	95-100	2-3 weeks	
Spinach	32	95-100	10-14 days	Yam	61	70-80	6-7 months	
Squash, summer	41-50	95	1-2 weeks					

Adapted from R. E. Hardenburg, A. E. Watada, and C. Y. Wang, *The Commercial Storage of Fruits, Vegetables, and Florist and Nursery Stocks,* USDA Agriculture Handbook 66 (1986).

[1]Spring- or summer-harvested potatoes are usually not stored. However, they can be held 4-5 months at 40°F if cured 4 or more days at 60-70°F before storage. Potatoes for chips should be held at 70°F or conditioned for best chip quality.

[2]Fall-harvested potatoes should be cured at 50-60°F and high relative humidity for 10-14 days. Storage temperatures for table stock or seed should be lowered gradually to 38-40°F. Potatoes intended for processing should be stored at 50-55°F; those stored at lower temperatures or with a high reducing sugar content should be conditioned at 70°F for 1-4 weeks, or until cooking tests are satisfactory.

[3]Sweet potatoes should be cured immediately after harvest by holding at 85°F and 90-95% relative humidity for 4-7 days.

[4]Winter squash varieties differ in storage life.

FIGURE 19–23 Ideal temperature, humidity level, and storage life for common vegetables.

FIGURE 19–24 Climate-controlled vegetable storage units control humidity, temperature, and sometimes the level of oxygen in the atmosphere. This makes it possible to extend the storage life of vegetables and fruits.

STUDENT ACTIVITIES

1. Write the Terms to Know and their meanings in your notebook.
2. Using outdated seed catalogs, prepare flash cards of warm- and cool-season vegetables.
3. Bring in an unusual vegetable to show to the class. Be prepared to tell the class which part is edible.
4. Select a site for planting a vegetable crop and prepare a planting plan to scale.
5. Take a soil sample from the site selected in Activity 4. Submit the sample to the local county extension service for analysis and lime and fertilizer recommendations for the vegetables planned.
6. Plant lima bean seeds in containers to observe the growth of the stem and leaves. Determine whether the plant is a monocotyledon or a dicotyledon.
7. Plant tomato seeds in a greenhouse for later transplanting.
8. Develop a chart with pictures illustrating the various insects that attack vegetable plants. Find pictures of insects in vegetable catalogs or extension service bulletins.
9. Store vegetables under different conditions (light/dark, dry/humid, warm/cool) in the classroom or at home. Record what happens under each set of conditions. Note which conditions are best for each vegetable.
10. During a grocery store visit, count the number of vegetables offered to consumers. Notice that many species of one single variety are present. For example, how many varieties of lettuce are there? In your opinion, what are the advantages of producing more than one crop? What are the disadvantages?

SELF EVALUATION

A. Multiple Choice

1. The part of horticulture that has to do with the production of vegetables is called
   a. viticulture.
   b. olericulture.
   c. floriculture.
   d. pomology.

2. Vegetable gardening produces fresh, nutritious food and provides the gardener with
   a. fresh produce.
   b. weeds.
   c. pests.
   d. diseases.

3. The majority of vegetables grow best in
   a. well-drained, clay soil.
   b. well-drained, sandy soil.
   c. well-drained, silty soil.
   d. well-drained, loamy soil.

4. For vegetable gardening, the soil should be plowed to a depth of
   a. 1 to 2 inches.
   b. 3 to 5 inches.
   c. 6 to 8 inches.
   d. 10 to 12 inches.

5. The first number in the analysis of a fertilizer represents the percentage of _____ in the bag.
   a. nitrogen
   b. calcium
   c. phosphoric acid
   d. potash

6. The best time to transplant vegetable plants is on
   a. a bright, sunny day.
   b. a rainy day.
   c. a cloudy day.
   d. a snowy day.

B. Matching

_____ **1.** Cultivation   a. Weed-control chemicals

_____ **2.** Herbicides   b. Rapid removal of heat from vegetables

_____ **3.** Polyethylene film   c. A person involved in the study of vegetable production

_____ **4.** Olericulturalist   d. Intertillage of crops

_____ **5.** Monocotyledon   e. Having only one seed leaf

_____ **6.** Precooling   f. A type of mulching material

C. Completion

1. The scientific name for the mustard family of vegetables is _____.

2. A site for the vegetable garden should be one that is exposed to the sun a minimum of _____ percent of the day.

3. To correct the acidity level in the soil, _____ is commonly added.

4. The type of irrigation that allows water to permeate the soil from below is called _____ irrigation.

5. Whenever the surface of the soil is artificially modified through the use of straw, leaves, plastic film, or paper, a _____ is created.

6. Tomatoes for use in the local market are best harvested in the _____ stage of maturity.

Fruit and Nut Production

MATERIALS LIST

✓ paper, pencil, and eraser

✓ old seed catalogs

✓ pruning shears

✓ scissors, glue, and index cards

✓ Internet access

COMPETENCIES TO BE DEVELOPED

After studying this unit, you should be able to:

- determine the benefits of fruit and/or nut production as a personal enterprise or career opportunity.
- identify fruit and nut crops.
- plan a fruit or nut production enterprise and prepare a site for planting.
- describe how to plant fruit and nut trees and use appropriate cultural practices in fruit and nut production.
- list appropriate procedures for harvesting and storing at least one commercial fruit or nut crop.

SUGGESTED CLASS ACTIVITIES

1. Identify a farm in your area that produces fruit or nuts. Make arrangements with the manager for the class to visit the farm or for the farm manager to visit the class. Prepare as a class to ask appropriate questions that will stimulate a discussion of the management practices that are involved in fruit or nut production. Focus some questions on how new varieties are chosen; how the crop is processed, packed, and marketed; and the storage requirements for the crop. Serve a piece of fresh fruit at the end of the activity.

2. Obtain grafting supplies, rootstock, and scions from apple trees to demonstrate grafting techniques. Demonstrate the most common grafts and have students practice each of the techniques. Provide an opportunity for each student to graft a scion to a rootstock, and then plant it in a suitable container. Take care of the grafted trees and determine how many of the grafts were successful. Discuss possible problems that may have contributed to unsuccessful grafting attempts.

3. Invite a local fruit or nut grower to come and speak to the class. Have him or her discuss the activities that must be performed at different times throughout the year to achieve the best possible harvest.

TERMS TO KNOW

pomologist	dwarf	drupe	terminal
semidwarf	pome	nursery	cane

ome fruit operations can provide high-quality and bountiful varieties of fruit and nuts. The fruits grown at home can be enjoyed at the peak of ripeness. This quality is difficult to obtain from supermarket fruits. Home fruit and nut operations can be enjoyed as hobbies or turned into profitable commercial enterprises.

CAREER OPPORTUNITIES IN FRUIT AND NUT PRODUCTION

Like other agricultural enterprises today, the production of fruit and nut crops is more a business than a way of life. The orchardist, or small-fruit grower, has to be a good financial manager, as well as a scientist and a farmer. Producing food, however, can be an enjoyable and rewarding profession.

Career Descriptions

INTERNET KEY WORDS:

pomologist career

The production, harvest, and marketing of fruits is a large industry. The United States accounts for a sizable part of the combined world crops of apples, pears, peaches, plums, prunes, oranges, grapefruit, limes, lemons, and other citrus fruits. There is currently a shortage of trained pomologists in the fruit industry. A **pomologist** is a fruit grower or fruit scientist.

Career opportunities exist at all levels of the industry. One can be an owner/grower, orchardist, plant manager, foreman, or technician. People who work in this field must be able to propagate fruit and nut trees and vines, as well as plant and transplant, prune, thin, train, and fertilize them. They must also be able to provide protection and control of diseases and insects.

IDENTIFICATION OF FRUITS AND NUTS

Types of Fruits

There are three broad categories of fruit. Some fruits grow on trees, some grow on canes and small bushes, and others grow on vines. The shape and form of a plant, whether it is a tree, a bush, a cane, or a vine, is known as the *growth habit* of the plant. Fruit is produced on woody twigs and stems from plants that exhibit each of these growth habits.

Tree Fruits Tree fruits are popular with home gardeners, as well as large producers. One drawback, however, is that some types of trees can take several years before producing the first harvest. There are three types of fruit trees based on growth habit: standard, semidwarf, and dwarf. These three terms refer to the size of the tree and the length of time it takes from planting until the first crop is produced (Figure 20–1).

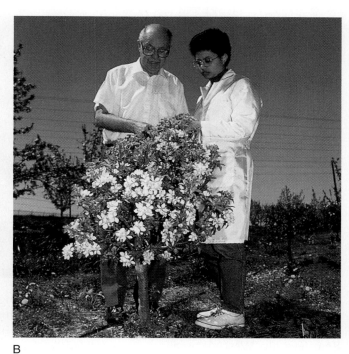

A   B

FIGURE 20–1 (A) Seven-year-old semidwarf and (B) twelve-year-old dwarf apple trees. *(Courtesy of (A) USDA/ARS #K-4048-17 and (B) USDA/ARS #K-4048-14)*

A standard tree is one that has its original rootstock and reaches normal size. A standard apple tree can grow to be 30 feet high. However, semidwarf and dwarf trees are smaller. They are standard varieties of trees, but they are grafted onto dwarfing rootstocks. Such rootstocks cause the tree to produce less annual growth. Thus, it remains smaller throughout its life. The semidwarf tree averages 10 to 15 feet in height, whereas dwarf trees average 4 to 10 feet. Dwarf trees are better suited to the home garden because they bear earlier and take up less space than semidwarf or standard trees (Figure 20–2).

FIGURE 20–2 The different types of rootstocks can be depended on to produce trees of a certain size. Each of the codes refers to a particular rootstock with known "dwarfing" characteristics.

FIGURE 20–3 Peach-growing research at the Appalachian Fruit Research Station in Kearneysville, West Virginia. *(Courtesy of USDA/ARS #K-3275-8)*

FIGURE 20–4 Trying to mechanize the harvest, agricultural engineer David Peterson watches his spiked-drum shaker as it traverses a row of blackberry bushes. *(Courtesy of USDA/ARS #K-2949-6)*

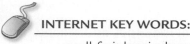

INTERNET KEY WORDS:

small-fruit horticulture

bush vine fruits

grape culture

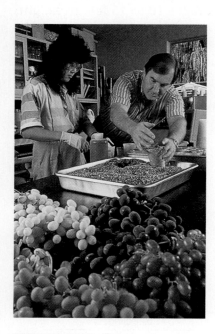

FIGURE 20–5 Grapes are popular as fresh fruit, dried fruit (raisins), juice, and wine. *(Courtesy of USDA/ARS #K-3681-7)*

Common fruits include apples, cherries, grapefruit, oranges, peaches, pears, plums, and quince. There are two types of tree fruits: pome and drupe. The common pome fruits are apples and pears. A **pome** is a fruit with a core and embedded seeds. A **drupe** is a fruit with a large, hard seed called a *stone*. Stone fruits include peaches, plums, quince, and cherries.

When growing tree fruits, particular attention must be paid to the climate of the area. Peaches, plums, and cherries are especially sensitive to climatic conditions (Figure 20–3).

Small-Bush Fruits Small-bush fruits are excellent selections for home gardens and small farms. The best choices are strawberries, blueberries, raspberries, and thornless blackberries. Small-bush fruits grow either low to the ground or only 3 to 4 feet high. They require less maintenance than do tree fruits or vine fruits. They tend to bear quickly—usually between 9 months and 1 year after planting.

Small-bush and cane fruits that are popular with gardeners and commercial operations are strawberries, blueberries, red raspberries, currants, gooseberries, and thornless blackberries (Figure 20–4). A factor to consider in growing bush fruits is to buy plants from a reliable **nursery**. Some of these fruits will bear a harvest in the summer and again in the fall.

Vine Fruits The best known vine fruit is the grape. Grapes will occupy the land for many years, so site selection is important. The vines need to be trained and pruned for adequate production and yield. Grapes require a growing season that has at least 140 frost-free days.

Grapes are also the most prevalent vine fruit. They are easily grown and have a wide range of flavors. The disadvantage is the 3- to 4-year wait for the vines to reach maturity before they produce fruit (Figure 20–5).

FIGURE 20–6 Nut trees work well in landscaping. They provide shade, food, and wood.

FIGURE 20–7 The black walnut tree can be used for shade in the landscape. It also produces a good-quality, edible nut.

TYPES OF NUTS

Nut trees are planted for numerous purposes. They can be useful for shade, landscaping, food, and wood. Wherever trees grow, a nut tree can also be grown. Popular nut trees grown in the United States are pecan, black walnut, filbert, hickory, almond, English walnut, chestnut, and macadamia (Figures 20–6 and 20–7).

PLANNING FRUIT AND NUT ENTERPRISES

Selecting Fruit or Nuts to Plant

Climatic Considerations Certain varieties and cultivars of fruits or nuts are more suited to one type of climate than another. For example, citrus fruits can be grown only in southern-type climates, such as in Florida, Texas, and California. This information is a key factor when deciding which fruit to grow. Even though you may absolutely love nectarines, they may not grow successfully in your area. Recommendations for each fruit crop for a given area can be obtained from your local county agricultural extension educator.

Rootstock Selection The variety of the root or rootstock of a grafted tree is important, because many fruit crops are propagated by grafting. Select a rootstock that is true to its variety and is hardy enough to support the tree. Most nurseries will readily give this information about the rootstock. The rootstock will determine whether the tree is a standard, semidwarf, or dwarf tree.

Frost Susceptibility Locate crops in areas as protected from frost as possible. Also, select varieties that are late bloomers if your area is susceptible to late frosts. Planting areas on hillsides, near ponds, or near cities will remain warmer on clear, frosty spring nights. Such locations decrease the

Agri-Profile

CAREER AREAS:

Pomology/Fruit Grower/Nut Grower/Packer

Pomology is the scientific study and cultivation of fruit, including nuts. Careers in pomology include work on farms, and in nurseries, orchards, and groves. Fruit and nut specialists may be plant breeders, propagators, educators, food scientists, or consultants. Technical-level jobs are available for graders, packers, supervisors, and managers.

In the past, fruit production meant hard labor because of hand picking. However, the need for mechanical harvesting has opened new opportunities for agricultural engineers, electronics technicians, plant breeders, systems managers, food scientists, and inventors. Mechanical harvesters range from hydraulically controlled buckets, to tree shakers, to sensor-directed robots.

Development of new pecan tree varieties is monitored continuously during the growing season.
(Courtesy of USDA/ARS #K-3318-8)

amount of damage caused by frosts. Apricots and sweet cherries are especially susceptible to late frost damage. The home gardener can protect some fruit crops, such as strawberries, from frost by remulching or irrigating and by placing cloth or plastic sheets over trees. Commercial growers use irrigation sprinklers, heaters, and fans to help reduce frost damage.

Fertility The production of high-quality fruit is dependent on the fertility of the soil. On average, soil should have a pH of 5.5 to 6.0. However, for growing blueberries, the soil should be more acidic—a pH of 5.2 or less. Fruit crops should be fertilized annually after they are planted. However, too much fertilizer can damage roots and create other problems. One pound of 10-10-10 fertilizer per 100 square feet of surface soil should give good results.

INTERNET KEY WORDS:

pollination fruits

Planning for Pollination Most fruit crops require pollination. Pollination is defined as the transfer of pollen from the male stamen to the stigma, or female part of the plant (Figure 20–8). Pollination is necessary for the fruit to set, or form. The gardener should choose crops that are known for having high-quality and abundant pollen. The pollen is distributed by wind or insects. Sometimes it is necessary to have several varieties that bloom at the same time for adequate cross-pollination. Similarly, it may be necessary to plant a tree just for its pollen quality and availability. Many fruit growers keep honeybees to help ensure good pollination (Figure 20–9).

When favorable conditions exist, just two varieties of a particular fruit crop are necessary for cross-pollination. Choose varieties that bloom at the same time or, at least, overlap for a period. Varieties of pears, plums, and cherries all tend to bloom at approximately the same time. However, apple varieties differ greatly in their times of bloom. Careful selection of apple varieties is important to ensure good cross-pollination.

FIGURE 20–8 The honeybee is one of our most important insects because it is such a good pollinator. It transfers pollen from one plant to another as it gathers nectar.

FIGURE 20–9 Honeybees are essential for good pollination of some fruit trees. The sale of honey can provide another source of income for the fruit grower. *(Courtesy of USDA/ARS #K-3438-1)*

Growing and Fruiting Habits The various types of fruit and nut crops have different kinds of growth patterns and fruiting dates. This fact is especially true of fruit trees. Some varieties of apple tree require as long as 8 years before they bear fruit. Newly planted peach and sour cherry trees can bear fruit in as little as 3 years. It pays to understand what to expect from the variety you have chosen.

Most fruit trees go through two stages of growth: a vegetative period and a productive period. The vegetative period consists of rapid, vigorous growth of the tree. In younger trees, 4 to 5 feet of growth may occur in a single season. When the tree enters the productive period of fruit bud formation, a change is noticeable. Fruit buds are larger and more plump. On apple, pear, cherry, and plum trees, short, spurlike growths occur (Figure 20–10). On the peach tree, the peach buds

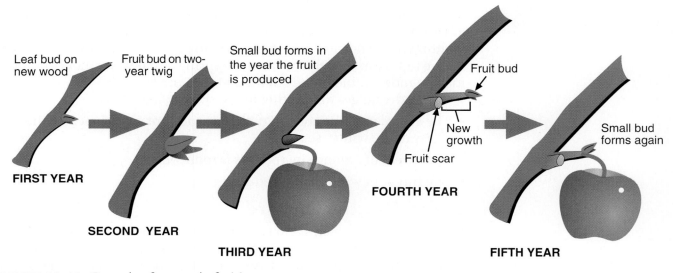

Leaf bud on new wood

Fruit bud on two-year twig

Small bud forms in the year the fruit is produced

Fruit bud

New growth

Fruit scar

Small bud forms again

FIRST YEAR

SECOND YEAR

THIRD YEAR

FOURTH YEAR

FIFTH YEAR

FIGURE 20–10 Growth of an apple fruiting spur.

become larger and occur in clusters, with two large ones surrounding a center one. The vigorous terminal growth slows down to only 18 to 20 inches in a season.

SOIL AND SITE PREPARATION

Selecting the Site

Because many fruits require full sunlight for maximum production and yield, choose a site that allows for maximum exposure to the sun. A southern exposure is best. To help prevent frost damage, select land with a slight slope and good air circulation. The soil for fruit crops should be well drained and of medium texture.

Soil Preparation

INTERNET KEY WORDS:

site selection fruit trees

The soil for fruit and nut trees should be limed, fertilized, and deeply plowed. These root systems grow deep into the soil. Deep plowing can be accomplished the year before the trees are planted. This permits the freezing and thawing action of the winter months to improve the texture of the soil. A soil sample should be taken a year in advance. Because many fruit and nut crops have specific pH and fertilizer requirements, a soil sample tested by a reputable testing laboratory will provide important information.

Any additional amendments, such as organic matter, lime, or fertilizer, should be added and disced into the soil before planting. Care should be taken when planting small-bush and cane fruits in soils that have been planted with vegetable crops, such as peppers, tomatoes, and potatoes. These vegetables may leave residues in the soil that contain disease organisms to which the fruit plants are susceptible.

PLANTING ORCHARDS OR SMALL-FRUIT AND NUT GARDENS

When to Plant

Tree stock from reputable nurseries can be planted either in the spring or fall. There are advantages to planting in each season. Fall planting is advisable if the soil will be less than desirable or wet in the spring. Planting in the fall also allows the crop to get its root system established before rapid top growth occurs in the spring. Finally, weather conditions may be more stable in the fall than in the spring.

The advantages of planting in the spring are:

- The crop can make considerable top and root growth before the stress of the winter season.
- The threat of damage by rodents is reduced.
- The threat of winter kill is reduced.

Laying Out the Orchard

Each fruit or nut species and type will have its own space requirements. It is important to follow the specific spacing recommendations for each variety. When planting small trees, the distance between trees seems enormous. However, the trees will grow and fill the spaces in time.

Science Connection

SAVING THE FRUIT

Each year, fruit growers lose money when their crops become frozen. A number of techniques have been tried to prevent this from happening. One successful method that is used often by fruit and vegetable growers appears to defy science. When air temperatures reach 32° F, the irrigation sprinklers are turned on. Water falls onto the fruit trees and becomes frozen. At this point, all a nervous farmer can do is wait to see if his or her efforts will save the harvest.

At first thought, this method makes no sense. Why would a person add water to trees that will surely become frozen? The explanation seems to be a trick of science. As the water is misted onto the fruit blossoms, the water begins to freeze. As this happens, heat that is trapped inside the blossom is released. What results is that a very thin layer of water vapor becomes trapped in between the blossom and the newly formed ice crystals. The water vapor acts as insulation protecting the potential fruit from freezing.

For aesthetic reasons, plant the trees in a straight line. A crooked row of trees will be obvious. Orchards are often organized in a grid pattern so that tractors and equipment may travel between rows and between the trees within the rows.

Planting the Fruit or Nut Crop

INTERNET KEY WORDS:

planting fruit and nut trees

Each fruit or nut variety has its own specifications for planting. To plant a tree, make the hole just wide enough to accommodate the roots without crowding them. The hole should be deep enough to allow the tree to be set 2 to 3 inches deeper than it was in the nursery. The two sides of the hole should be parallel, and the bottom should be as wide as the top. By making the hole to these specifications, the soil can be packed uniformly around the roots to avoid leaving air spaces. Air spaces permit the roots to dry out.

When planting trees, add some prepared soil to the hole. Then spread the roots, adding more soil until all roots are covered. Move the tree up and down several times, while firming the soil around the roots. This procedure allows all roots to be in contact with the soil and avoids air pockets.

Before the hole is completely filled, add about 2 gallons of water. After the water soaks in, add the final soil. Most trees will need to be staked for good support. A metal pipe may be driven 8 inches from the tree and the tree tied to the pipe by using wire strung through a piece of rubber hose (Figure 20–11). After planting, keep a 12-inch circular weed-free and grass-free area around the tree.

CULTURAL PRACTICES IN FRUIT AND NUT PRODUCTION

Fertilization

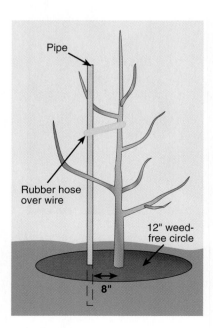

Pipe

Rubber hose over wire

12" weed-free circle

8"

FIGURE 20–11 Provide support for the newly planted tree.

A complete fertilizer, such as a 5-10-10 or a 10-10-10, should be used annually on fruit and nut trees. After filling the hole of a newly planted tree, apply 1 lb of the 5-10-10 or a half lb of 10-10-10 fertilizer around each tree. Do not place the fertilizer closer than 8 to 10 inches from the trunk. Excessive fertilization can result in damage or death to

Science Connection

THE RIPENING GENE—CAN IT BE CONTROLLED?

Inside the fruits and vegetables growing in gardens, vineyards, orchards, groves, farms, and in the wild, microscopic cells manufacture ethylene gas. This colorless and odorless gas is known to affect the ripening process. The more ethylene the plant produces, the faster the ripening process will be. The work of this compound is essential and good, and it transforms inedible, green produce to ripe and delicious food. However, there is a downside—that of changing delicious fruit into soft and mushy trash!

Molecular biologist Athanasios Theologis at Agricultural Research Service/University of California Plant Gene Expression Center in Albany, California, believes that much produce spoilage in and out of the field is the work of ethylene gas. He believes that it does not have to be that way. But rather, perhaps we can find the way to manage the production of ethylene in plants. If so, we can control when the fruit of a given crop will ripen. The dream is to modify plants through biotechnology to ripen at our convenience.

Theologis and his colleague, Takahide Sato at Chiba University in Japan, brought molecular biologists a step closer to the dream when they cloned a gene that plants apparently must have to manufacture ethylene gas. It is hoped that scientists will eventually find ways to turn down or turn off the effect of the gene, and thereby gain more control of the ripening process. Such a feat would enable produce to arrive at the supermarkets in fresher, better-tasting condition. Such produce would be better suited for export as well.

One possible new destination for U.S. fresh produce could be Japan's Mom-and-Pop-style produce stands. For this purpose, produce must be fresh, ripe, and edible on the spot. Without better control of the ripening process, we do not have the technology to move large volumes of produce from field to consumer fast enough to serve this market. Ethylene gas also causes plant parts to wither. The control of ethylene in the plant would slow down the wilting and withering processes in leafy vegetables and cut flowers. Imagine, perhaps some day, these commodities may be stored or shipped without spoiling and with little or no refrigeration!

A

B

(A) Jean Murphy at the Agricultural Research Service/University of California Plant Gene Expression Center examines the different stages of ripening of tomatoes, whereas (B) molecular biologist Athanasios Theologis uses electrophoresis to separate and match nucleotide sequences of DNA fragments. *(Courtesy of (A) USDA/ARS #K-5335-01; (B) USDA/ARS #K-3489-6)* young trees. After the first year, apply fertilizer at the rate of 1 lb times the age of the tree. After the tree is 6 to 8 years old, do not increase the rate of application. As the tree becomes larger, fertilize all of the ground area under the limb spread.

These rates are useful for all tree fruits except pears. Pears should not be fertilized, because increased susceptibility to fire blight more than offsets the benefit from fertilizing.

Fertilization requirements for the various bush and cane fruits vary considerably. Therefore, it is advisable to consult a grower's guide for each individual variety.

INTERNET KEY WORDS:

fruit tree pruning

Pruning

The purpose of pruning and training young fruit trees is to establish a strong framework of branches that will support the fruit. This framework will resemble the shape best suited for the tree and for the fruit it will bear. Each fruit tree has a specific way that it should be pruned. For example, peach trees are pruned for an open-center, V-shape (Figure 20–12). However, apple trees should have a strong central leader and be pruned into a Christmas-tree scaffold (Figure 20-13).

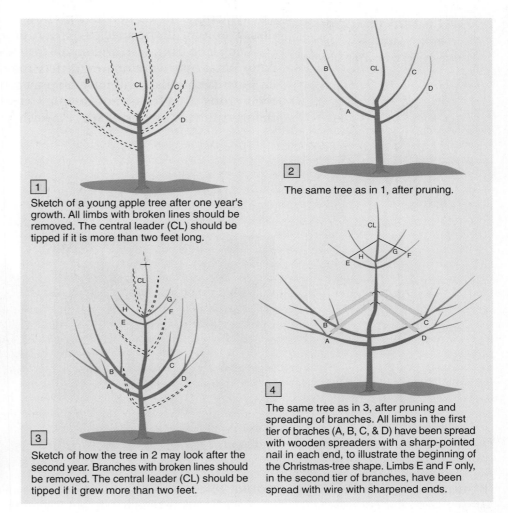

1 Sketch of a young apple tree after one year's growth. All limbs with broken lines should be removed. The central leader (CL) should be tipped if it is more than two feet long.

2 The same tree as in 1, after pruning.

3 Sketch of how the tree in 2 may look after the second year. Branches with broken lines should be removed. The central leader (CL) should be tipped if it grew more than two feet.

4 The same tree as in 3, after pruning and spreading of branches. All limbs in the first tier of braches (A, B, C, & D) have been spread with wooden spreaders with a sharp-pointed nail in each end, to illustrate the beginning of the Christmas-tree shape. Limbs E and F only, in the second tier of branches, have been spread with wire with sharpened ends.

FIGURE 20–12 A peach tree should be pruned to open up a V-shaped pattern in the center of the tree.

FIGURE 20–13 A young apple tree should be pruned to maintain a strong, central leader with strong, wide-angled main branches to support the weight of the fruit.

Pruning a grapevine depends on its fruiting habit. Next year's grape clusters are on shoots produced during the current growing season. Therefore, this new growth should be carefully pruned. When the shoots have fruited and the leaves have fallen, the shoots are called **canes.** The canes will produce the new shoots and next year's crop.

A common method of training grapevines is the Four-Arm Kniffin System. In this system, the vine is pruned to form a double T on a two-wire trellis system. Two arms originating at the two wires are trained to grow in opposite directions from the trunk. The vines are pruned to remove all canes except those that are saved for fruiting and renewal growth (Figure 20–14).

Pruning and thinning small-bush and cane fruits should be accomplished according to the specific standards for each fruit. Pruning, thinning, and training these smaller fruits are essential practices for high-quality yields. These practices also help prevent pests and diseases.

Disease and Pest Control

A great deal of the time and expense in fruit and nut production is spent controlling diseases, insects, and other pests. Control of diseases and insects is a major component of fruit and nut production (Figure 20–15). The type of pesticide needed and how often it is applied varies tremendously among all the species of fruits and nuts.

Each tree fruit, nut, and small-fruit variety must be treated individually. What may work on one variety may not be appropriate for another. Most orchardists will have a spray schedule worked out for each of their crops. Tree fruits require more pesticide than do any other fruits. Before applying any pesticide, however, read the label thoroughly.

INTERNET KEY WORDS:

diseases, fruit trees, fruits, nuts

INTERNET ADDRESS:

<http://www.caf.wvu.edu/kearneysville/wvufarm11.html>

A

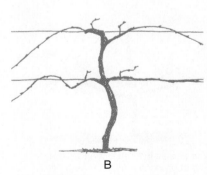

B

FIGURE 20–14 A grape vine trained to the Four-Arm Kniffin System (A) before and (B) after pruning. *(Courtesy of Maryland Cooperative Extension Service)*

A

B

FIGURE 20–15 (A) A Caribbean fruit fly laying eggs in the peel of a grapefruit. (B) Fruit fly larvae then hatch and devour the pulp. *(Courtesy of (A) USDA/ARS #K-898-8; (B) USDA/ARS #K-1674-7)*

HARVEST AND STORAGE

Harvesting

Harvesting fruits can be a daily practice. The advantage of homegrown fruit is that it can be picked at its peak of ripeness. However, fruits for commercial use must be picked several days ahead so they will withstand shipping and handling.

For the best quality of fruit, harvest apples, pears, and quince when they begin to drop, soften, and become fully colored. Some varieties will ripen over a 2-week period, requiring picking each day. Other varieties will ripen all at once.

Peaches, plums, apricots, and cherries should be harvested when the green disappears from the surface skin of the fruit. A yellow undercolor should be developed by this time. The fruit should be soft when pressed lightly in a cupped hand (Figure 20–16).

The nut varieties ripen from August to November. Harvest nuts immediately after they fall from the tree. Most of the nuts that do not fall can be knocked off the tree with poles or mechanical harvesters.

The nut varieties of pecan, hickory, chestnut, and Persian walnut will lose their husks when ripe. However, the husks of black walnuts and other types of walnuts will need to be removed.

Grapes should be harvested only when fully ripe. The best way to judge when the grape is at full maturity is to sample an occasional grape.

Small-bush and cane fruits, such as strawberries and raspberries, should be harvested when they are fully ripe. The best indicator is when the fruit is fully colored. When these fruits and berries reach maturity, they should be picked every day. When harvesting, pick when the fruit is dry to avoid mildew and mold damage.

Storage

Fruit storing is limited to those varieties that mature late in the fall or those that can be purchased at the market during the winter months. The length of time fruits can be stored depends on the variety, stage of maturity, and soundness of the fruit at harvest (Figure 20–17).

FIGURE 20–16 Fruit should be harvested for home use when it acquires the color of ripe fruit and when it begins to be soft to the touch. *(Courtesy of PhotoDisc)*

Commodity	Freezing Point	Storage Conditions		Length of Storage Period
		Temperature	Humidity	
	°F	°F		
Fruits:				
Apples	29°	32°	Moderate Moisture	Fall/Winter
Grapefruit	29.8°	32°	Moderate Moisture	4 to 6 weeks
Grapes	28.1°	32°	Moderate Moisture	1 to 2 months
Oranges	30.5°	32°	Moderate Moisture	4 to 6 weeks
Pears	29.2°	32°	Moderate Moisture	4 to 6 weeks

FIGURE 20–17 Fruit storage temperatures as adapted from the U.S. Department of Agriculture.

FIGURE 20–18 Red Delicious apples ready for harvest in an orchard near Wenatchee, Washington. *(Courtesy of USDA/ARS #K-3853-5)*

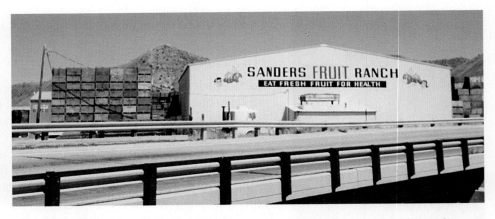

FIGURE 20–19 A fruit storage facility in which temperature and humidity are controlled.

For long-term storage of apples, the temperature should be as close to 32° F as possible (Figure 20–18). Apples can be stored in many ways, but they should be protected from freezing. A cellar or other area below ground level that is cooled by night air is a good place for apple storage (Figure 20–19). There should be moderate humidity in the storage area. Pears have the same storage requirements as apples.

Nuts should be air dried before they are stored. Nuts, especially pecans, keep longer if they are left in the shell and refrigerated at 35° F. They can also be frozen if they are kept in their shells. Chestnuts have special requirements. They should be stored at 35° to 40° F and high humidity, shortly after harvest.

Small-bush and cane fruits, including grapes, are not suitable for storage. These fruits perish too quickly to keep very long and are best used as soon as they are harvested.

STUDENT ACTIVITIES

1. Write the Terms to Know and their meanings in your notebook.
2. Develop a bulletin board display explaining opportunities in fruit- and nut-related occupations. Include those that can be found on the local, state, and national levels.
3. Bring a specimen of some type of unusual fruit or nut to class. Discuss its origin and the techniques for growing the fruit or nut.
4. Prepare fruit and nut flash cards using pictures from seed catalogs.
5. Prepare a planting plan for fruits and nuts, drawn to scale. Indicate the types, quantity, and location of plants to be grown.
6. Many fruits and nuts consumed in our country are imported from other countries. How does the government ensure that the produce is safe? Research the process that is followed before imported crops are allowed into the United States. Why is this process necessary?

SELF EVALUATION

A. Multiple Choice

1. Most fruits and nuts require _____ to grow properly.
   a. full sunlight
   b. partial sunlight
   c. partial shade
   d. no sunlight

2. The stock for a fruit or nut crop can be planted in the
   a. fall or spring.
   b. spring only.
   c. summer only.
   d. fall only.

3. When planting a fruit tree, you should make the hole deep enough to allow the tree to be set _____ deeper than in the nursery.
   a. 3 to 5 inches.
   b. 2 to 3 inches.
   c. 1 to 2 inches.
   d. 6 to 8 inches.

4. Nuts can be stored in the refrigerator at the optimum temperature of
   a. 40° F.
   b. 32° F.
   c. 38° F.
   d. 35° F.

5. The pear is an example of a _____ fruit.
   a. drupe
   b. pome
   c. bramble
   d. aggregate

B. True or False

_____ 1. A standard tree is one that has its original rootstock and grows to normal size.

_____ 2. A semidwarf tree averages a mature height of 30 to 40 feet.

_____ 3. Small-bush and cane fruits tend to bear quickly, usually between 9 months and 1 year after planting.

_____ 4. Rootstock selection is important to fruit and nut crops, because most of these crops are not propagated by seed.

_____ 5. Frost susceptibility is not a factor to fruit crops.

_____ 6. Too much fertilizer can be a hazard to fruit trees.

_____ 7. Control of diseases and insects is a major component of fruit and nut production.

_____ 8. An apple is an example of a drupe fruit.

Grain, Oil, and Specialty Field-Crop Production

OBJECTIVE

To determine the nature of and approved practices recommended for grain, oil, and specialty field-crop production.

MATERIALS LIST

✓ sample of field crops and products made from field crops

✓ bulletin board materials

✓ crop magazines

✓ reference materials on crops

✓ Internet access

COMPETENCIES TO BE DEVELOPED

After studying this unit, you should be able to:

- define important terms used in crop production.
- identify major crops grown for grain, oil, and special purposes.
- classify field crops according to use and thermo requirements.
- describe how to select field crops, varieties, and seed.
- prepare proper seedbeds for grain, oil, and specialty crops.
- plant field crops.
- describe current irrigation practices for field crops to meet their water needs.
- control pests in field crops.
- harvest and store field crops.

SUGGESTED CLASS ACTIVITIES

1. Most states have commissions to promote their major agricultural crops and products. For example, a Wheat Commission promotes and develops markets for wheat throughout the world. Obtain promotion and marketing materials from one of the agricultural commodity commissions within your state that is responsible for promoting and developing markets. Discuss the importance of marketing and promotion for the major crops that are produced in your state. Use the materials to demonstrate how this is done by the commodity commission that created the materials.

2. Obtain a variety of products such as cereals, vegetable oil, and pasta, and use them as object lessons as you discuss the products that are produced in your area. Obtain statistics from your state Department of Agriculture about the ranking of your state in the production of the various kinds of grains, oils, and specialty crops. Assign the students to make posters that show the state rankings in the production of these crops. Display the posters in the classroom. The rankings of states are available at the following Internet address: <http://www.usda.gov/nass/pubs/ranking/croprank/htm>

3. Bring field crop samples into class, including the roots and seeds. If fresh samples are not available, pressed samples or pictures may be used. Help students learn to identify each crop. Point out individual characteristics that are unique to each plant. This will help students with the identification process.

TERMS TO KNOW

field crop	ginning	sugar crop	broadcast planter
grain crop	seed piece	tuber crop	sprinkler irrigation
malting	cash crop	stimulant crop	surface irrigation
forage	thermo	conventional tillage	drip irrigation
oilseed crop	cereal crop	minimum tillage	mechanical pest control
linen	seed legume crop	row crop planter	genetic control
linseed oil	root crop	drill planter	threshing

INTERNET KEY WORDS:

field crops

The cultivation of the land and the growing of crops began about 10,000 years ago. The need to produce food for the animals that humans had captured and begun to domesticate caused early humans to change from hunters to farmers. There were no guidelines to follow in selecting plants. Early agriculturists had to rely on observing what the animals were eating to decide which plants to grow. Trial and error and thousands of years of selection have led to the crops that are grown today. New types, varieties, and uses of plants continue to be developed in response to current needs and in anticipation of future demands for food and plant fiber by an ever-increasing world population.

In the United States, the production of grain, oil, and specialty crops occupies more than 450 million acres. These crops are normally called **field crops**. This acreage represents nearly 20 percent of the land mass in the United States. U.S. agriculturists are among the most efficient in the world, producing enough food for themselves and people in many other countries. As a result, less than 2 percent of U.S. workers are engaged in the production of food and fiber. The efficiency of the U.S. agriculturist allows the U.S. population to spend only about 10.9 percent of its income on food. It also allows for sizable exports of food crops all over the world. This has helped maintain a favorable balance of trade for the United States in recent years. Today, we have so many products from agriculture that we no longer are aware of the plant and animal origins of common nonfood items (Figure 21–1).

FIGURE 21–1 New products from plants and animals through agriscience: (top row) plastics from vegetable oils and industrial uses for corn starch; (second row) cocoa butter substitute, improved cotton processing; (third row) tallow-based soap, detergents with improved surfactants, more uses for corn starch; and (fourth row) starch-based rubber products, plastics from vegetable oil, starch-based rubber. (Courtesy of USDA/ARS #K-4796-20)

MAJOR FIELD CROPS IN THE UNITED STATES

INTERNET KEY WORDS:

corn production

Grain Crops

There are seven major grain crops in the United States. Grain crops are grasses that are grown for their edible seeds. These crops are corn, wheat, barley, oats, rye, rice, and grain sorghum.

Corn Corn is the most important of the field crops grown in the United States. Corn is well adapted, high yielding, and is an important crop in most states. About 35 to 40 percent of the corn produced annually in this country is grown in the midwestern states, commonly called the Corn Belt. These states include Iowa, Illinois, Nebraska, Minnesota, Indiana, and Ohio, in descending order of production. The United States accounts for nearly 50 percent of the corn produced in the world (Figure 21–2).

Originating in Central America, corn served as a major part of the diet of the Indians of the area before the settlement of the New World by Europeans (Figure 21–3). Corn was unknown to the rest of the world until explorers found American Indians growing corn.

Less than 10 percent of the corn grown in the United States is for human consumption. The rest is used for livestock feed, alcohol production, and hundreds of other products.

The major types or classifications of corn are dent corn; flint corn; popcorn; sweet corn; flour, or soft corn; and pod corn. Most of the corn grown in the United States is dent corn.

FIGURE 21–2 Corn is a plant that produces high-quality grain for humans and animals, and forage (silage) for livestock. *(Courtesy of Michael Dzaman)*

FIGURE 21–3 Corn is an American Indian plant known as maize, or "Indian corn."

Hot Topics in Agriscience

CORN—AN AMERICAN INDIAN CROP

Long before the birth of the United States, and before corn became the number one grain crop produced here, it was the most important grain crop for American Indians. Corn was cultivated by American Indian tribes as their principal food source for many years before Columbus sailed to the New World. Plant breeders have modified the maize plant, also known as "Indian corn," as they have developed the plant for specific uses. For example, we now have popcorn, sweet corn, grain corn, and silage corn in many varieties. All of these corn plants trace back to the American Indian maize plants.

Corn is now used in many ways besides providing food for people and animals. It is used to make products that we use everyday, such as cooking oil, biodegradable plastic bags, and ethanol (a component of gasohol). Large amounts of corn are exported throughout the world, which helps sustain our balance of trade. Corn also remains the United States's most important feed grain for livestock.

INTERNET KEY WORDS:

Native maize

A general discussion of cultural practices for crops appears later in this unit. However, the following approved practices are useful for growing corn:

- Select a hybrid that will mature during the growing season.
- Obtain a current soil test.
- Lime and fertilize according to soil test and desired production.
- Select a field with deep, rich, well-drained soil.
- Select the proper tillage method.
- Prepare a firm seedbed or use a proper no-till planter.
- Plant corn when soil temperature is 50° F or warmer.
- Calibrate the planter for proper plant population.
- Match the plant population to soil-yield potential.
- Adjust the planter for proper depth of seed placement.
- Calibrate the sprayer for proper application of materials.
- Use selected herbicides for controlling problem weeds.
- Cultivate to help control weeds.
- Use insecticides as necessary for proper insect control.
- Apply fungicides when needed.
- Make a yield check.
- Check moisture content during the harvest season.
- Harvest the crop as soon as moisture permits, to reduce losses.
- Take a forage analysis on ensiled corn.
- Store ensiled or high-moisture corn in sound structures.
- Track market trends and habits.
- Market the crop at an optimum time.
- Keep accurate enterprise records.
- Summarize and analyze records.

INTERNET KEY WORDS:

cultural practices,
barley, oats, wheat, rye

Wheat Wheat is one of the most important grain crops in the world (Figure 21–4). In the United States, wheat ranks second to corn in bushels produced. Leading states in the production of wheat are Kansas, Oklahoma, Washington, Texas, and Montana.

Wheat is used primarily for human consumption. The wheat is ground into flour, which is then made into products such as bread, cakes, cereal, crackers, macaroni, and noodles. Other uses of wheat include the manufacture of alcohol and livestock feed.

Types of wheat grown in this country include common, durum, club, Poulard, Polish, Emmer, and spelt. Most of this wheat is of the common type. Classes of common wheat include hard red spring, hard red winter, soft red winter, and white.

Barley The leading states in the production of barley in the United States include North Dakota, Montana, Minnesota, Idaho, South Dakota, and Washington (Figure 21–5). Barley ranks fifth among the grain crops produced in the United States.

Most barley is used for livestock feed. It has slightly less food value than corn. The production of barley for malting is also important. Malting is the process of preparing grain for the production of beer and other alcoholic beverages.

The cultural practices for growing wheat, barley, oats, and rye are similar. The following list of approved practices for growing small grains should be useful for growing any of these crops:

- Select a variety adapted to your conditions.
- Buy certified seed.
- Obtain a current soil test.
- Lime and fertilize according to soil test and desired production.
- Select a field suitable for your grain choice.
- Select the proper tillage method.
- Prepare a firm and smooth seedbed.
- Calibrate the drill for proper plant population.
- Match plant population to soil-yield potential.

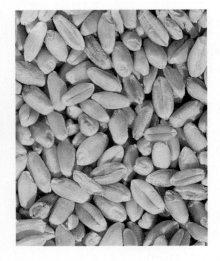

FIGURE 21–4 Worldwide, wheat ranks second to rice as the most important grain crop. *(Courtesy USDA/ARS #K-3597-18)*

A

B

FIGURE 21–5 (A) Most of the barley grown in the United States is used for livestock feed. (B) Barley is important as a feed grain, and it is also used in the production of malt.

A   B

FIGURE 21–6 (A) Oats provide bulk and protein to the diets of animals, as well as food for humans. (B) Oats contain greater fiber and protein than other grain crops.

- Adjust the drill for proper depth of seed placement.
- Calibrate the sprayer for proper application of materials.
- Use selected herbicides for controlling problem weeds.
- Use insecticides as necessary for adequate control of insects.
- Apply fungicides when needed.
- Make a yield check.
- Adjust combine for proper harvesting.
- Check moisture content during the harvest season.
- Harvest the crop as soon as the moisture level permits.
- Store grains in disinfected grain storage tanks.
- Bale straw as soon as moisture levels allow.
- Track market trends and habits.
- Market the crop at an optimum time.
- Keep accurate enterprise records.
- Summarize and analyze records.
- Make business decisions based on records.

Oats Oats as a grain crop are fourth in acres produced in the United States (Figure 21–6). Major oat-producing states are South Dakota, Minnesota, Wisconsin, Iowa, and North Dakota.

The value of oats in adding bulk and protein to the diets of livestock is well documented. However, about 5 percent of the oats produced in the United States are made into oatmeal and cookies. Oats are also used in the production of plastics, pesticides, and preservatives. In addition, they are important in the paper and brewing industries.

Rye Although rye is grown in nearly every state in the United States, it is the least economically important grain crop. It does, however, have many uses (Figure 21–7). Most rye is grown in South Dakota, Georgia, Minnesota, North Dakota, and Nebraska.

FIGURE 21–7 Rye is used as a livestock feed and for the production of flour. Economically, it is the least important grain crop produced in the United States.

FIGURE 21–8 Rice is the most important grain crop in the world. It is the only commercially grown grain crop that will grow in standing water. *(Courtesy of USDA/ARS #K-3444-1)*

INTERNET KEY WORDS:

rice food staple
rice production culture
space oil crops
soybean production

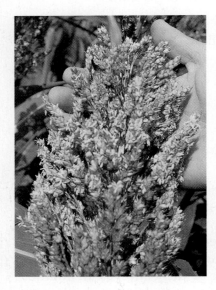

FIGURE 21–9 Sorghum is a grain crop that is produced in a few Midwestern states. It is mostly used for livestock feed, either as grain or forage.

About 25 to 35 percent of rye acreage is used for grain. The rest is used for forage, as a cover crop, or as a green manure crop. A forage is hay or grass grown for animal feed. Cover crops are planted to protect the soil from erosion. Green manure crops are grown to be plowed under to add nutrients and organic matter to the soil. The rye grown for grain is used for livestock feed, flour, whiskey, and alcohol production.

Rice Rice is the most important grain crop grown for human food in the world. Most of the rice in the United States is grown in Arkansas, California, Louisiana, Texas, and Mississippi. Rice is the only commercially grown grain crop that can grow and thrive in standing water (Figure 21–8). The types of rice grown in the United States are short grain, medium grain, and long grain.

The majority of the rice grown in this country is used for human consumption. The excess that is produced is exported to other countries of the world.

Sorghum Sorghum is grown in the United States primarily for livestock feed. It is about equal to corn in food value. Other uses of sorghum include forage, the manufacture of syrup or sugar, and the making of brooms (Figure 21–9). Leading states in the production of sorghum include Kansas, Texas, Nebraska, South Dakota, and Oklahoma. Based on acres harvested, sorghum is the third most important grain crop in the United States.

The five types of sorghum are grouped according to use. They are grain, forage, syrup, grass, and broomcorn.

Oilseed Crops

Crops that are grown for the production of oil from their seeds are called oilseed crops. These crops are increasing in importance each year, as people in the United States rely more and more heavily on vegetable oils and less on animal fats in their diets. Important crops grown for the oil contained in their seeds are soybeans, peanuts, corn, cottonseed, canola, safflower, flax, and sunflowers (Figure 21–10). Corn and cottonseed are discussed in other sections of this unit.

Soybeans There are approximately 60 million acres of soybeans grown in the United States each year. With an average yield of about 34 bushels per acre, the production of soybeans grosses more than $11 billion each year (Figures 21–11 and 21–12).

Oil and grain products are the major uses of soybeans. The meal resulting from the extraction of oil from soybeans is an important source of protein in livestock feeds. Soybeans are also used for hay, pasture, and other forage crops. Research has led to the development of hundreds of other uses for soybeans.

Large centers of production include the midwestern states of the Corn Belt. Major soybean-producing states include Illinois, Iowa, Minnesota, Indiana, and Missouri. Internationally, Brazil has become a major producer of soybeans. The approved practices for raising soybeans are similar to those for raising corn, except that soybeans and other legume seeds should be treated with the proper inocculant bacteria to ensure good nitrogen fixation.

Peanuts The peanut is actually a pea rather than a nut, despite its nutlike taste and shell. It is grown primarily in the South, where warm

FIGURE 21–10 The rape, or canola, plant is an old plant that is getting new attention as an oil-producing crop. This plant is part of the mustard family, as evidenced by the bright yellow flowers.

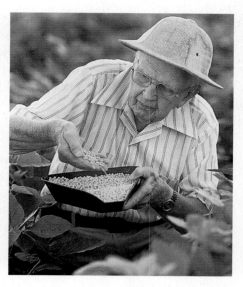

FIGURE 21–11 Sometimes called the Soybean Doctor, agronomist Edgar E. Hartwig has devoted half a century to soybean research. His focus has been on developing productive plants with built-in resistance to insects, nematodes, and diseases. *(Courtesy of USDA/ARS #K-5272-1)*

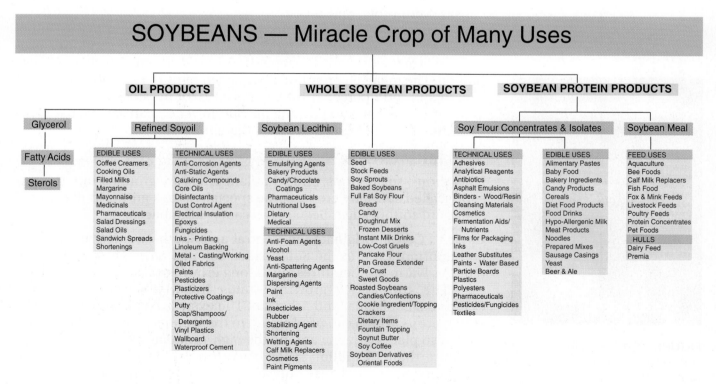

SOYBEANS — Miracle Crop of Many Uses

OIL PRODUCTS

Glycerol

Fatty Acids

Sterols

Refined Soyoil

EDIBLE USES
Coffee Creamers
Cooking Oils
Filled Milks
Margarine
Mayonnaise
Medicinals
Pharmaceuticals
Salad Dressings
Salad Oils
Sandwich Spreads
Shortenings

TECHNICAL USES
Anti-Corrosion Agents
Anti-Static Agents
Caulking Compounds
Core Oils
Disinfectants
Dust Control Agent
Electrical Insulation
Epoxys
Fungicides
Inks - Printing
Linoleum Backing
Metal - Casting/Working
Oiled Fabrics
Paints
Pesticides
Plasticizers
Protective Coatings
Putty
Soap/Shampoos/
  Detergents
Vinyl Plastics
Wallboard
Waterproof Cement

Soybean Lecithin

EDIBLE USES
Emulsifying Agents
Bakery Products
Candy/Chocolate
  Coatings
Pharmaceuticals
Nutritional Uses
Dietary
Medical

TECHNICAL USES
Anti-Foam Agents
Alcohol
Yeast
Anti-Spattering Agents
Margarine
Dispersing Agents
Paint
Ink
Insecticides
Rubber
Stabilizing Agent
Shortening
Wetting Agents
Calf Milk Replacers
Cosmetics
Paint Pigments

WHOLE SOYBEAN PRODUCTS

EDIBLE USES
Seed
Stock Feeds
Soy Sprouts
Baked Soybeans
Full Fat Soy Flour
  Bread
  Candy
  Doughnut Mix
  Frozen Desserts
  Instant Milk Drinks
  Low-Cost Gruels
  Pancake Flour
  Pan Grease Extender
  Pie Crust
  Sweet Goods
Roasted Soybeans
  Candies/Confections
  Cookie Ingredient/Topping
  Crackers
  Dietary Items
  Fountain Topping
  Soynut Butter
  Soy Coffee
Soybean Derivatives
  Oriental Foods

SOYBEAN PROTEIN PRODUCTS

Soy Flour Concentrates & Isolates

TECHNICAL USES
Adhesives
Analytical Reagents
Antibiotics
Asphalt Emulsions
Binders - Wood/Resin
Cleansing Materials
Cosmetics
Fermentation Aids/
  Nutrients
Films for Packaging
Inks
Leather Substitutes
Paints - Water Based
Particle Boards
Plastics
Polyesters
Pharmaceuticals
Pesticides/Fungicides
Textiles

EDIBLE USES
Alimentary Pastes
Baby Food
Bakery Ingredients
Candy Products
Cereals
Diet Food Products
Food Drinks
Hypo-Allergenic Milk
Meat Products
Noodles
Prepared Mixes
Sausage Casings
Yeast
Beer & Ale

Soybean Meal

FEED USES
Aquaculture
Bee Foods
Calf Milk Replacers
Fish Food
Fox & Mink Feeds
Livestock Feeds
Poultry Feeds
Protein Concentrates
Pet Foods
HULLS
Dairy Feed
Premia

FIGURE 21–12 The soybean has become famous for its many products and is growing in importance in world markets. *(Courtesy of American Soybean Association)*

FIGURE 21-13 George Washington Carver, the famous agricultural research pioneer, developed more than 300 products from peanuts. *(Courtesy of USDA/ARS #K-4297-18)*

INTERNET KEY WORDS:

sunflower production culture

FIGURE 21-14 A field of sunflowers in bloom is a sight to behold before the development of its oil-rich seed. *(Courtesy of USDA)*

temperatures and a long growing season are keys to success. Leading peanut-producing states are Georgia, Texas, Alabama, North Carolina, Oklahoma, and Florida.

One ton of peanuts in the shell will yield about 500 lb peanut oil and 800 lb peanut-oil meal. The remaining 700 lb is mostly shells. The oil meal is used for livestock feed and as a good protein source in human diets. Other food products produced from peanuts include peanut butter and dry roasted peanuts (Figure 21–13).

Safflower The production of safflower for oil occurs mainly in California. Safflower plants grow 2 to 5 feet in height and have flower heads that resemble Canadian thistles. The oil comes from the wedge-shaped seeds that the plant produces. The seeds contain from 20 to 35 percent oil.

Safflower oil is used in the production of paint and other industrial products. It is also used for cooking oil and for low-cholesterol diets.

Flax Originally, the production of flax was mostly for fiber. Flax fibers from the stems of plants are used to produce **linen** (a fabric made from flax fibers).

The oil produced from the seed of the flax plant is called **linseed oil**. It is an important part of many types of paint and has hundreds of uses in industry. The linseed-oil meal that is left after the extraction of the oil is an excellent source of protein for animal feeds.

Most flax is grown in North Dakota, South Dakota, Minnesota, and Wisconsin.

Sunflowers The production of oil-type sunflower seed has been important in the United States in recent years (Figure 21–14). Most of the sunflower production is located in North Dakota, South Dakota, Kansas, Minnesota, and Texas.

There are two types of sunflowers grown commercially in the United States: oil-type and non–oil-type. About 90 percent of sunflower production is of the oil-type. Oil-type sunflower seeds contain 49 to 53 percent oil. The meal left after the oil has been removed is high in bulk and contains 14 to 19 percent protein. It is used for livestock feed. The oil is used for margarine and cooking oil. Sunflower oil can also be used as a substitute for diesel fuel in tractors (Figure 21–15).

Specialty Crops

Fiber crops, sugar crops, and stimulant crops are grouped into the category called specialty crops. Some specialty crops grown in the United States are cotton, sugar beets, sugarcane, and tobacco.

Cotton Cotton originated in Central and South America. It has been an important crop in the South since colonial days. The cotton plant requires warm temperatures and a long growing season to reach maturity. The top cotton-growing states include Texas, Mississippi, California, Louisiana, and Arkansas. The extensive irrigated acreages of cotton crops, which can produce up to three crops per year on the same land, also places Arizona among the high-producing states.

More than 15 million bales of cotton are harvested per year in the United States. Because only about 9 million are needed by the textile industry, the rest is exported to other countries of the world (Figure 21–16).

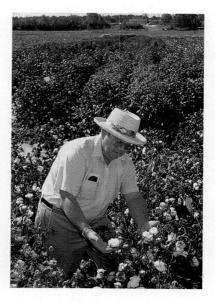

FIGURE 21–15 Some vegetable oils, such as rapeseed oil and sunflower oil, can be used to produce fuels for diesel tractors. A fuel of this kind is known as bio-diesel.

FIGURE 21–16 Cotton has been an important crop since colonial times, but care must be taken to protect cotton plants from insects and diseases.

The seed must be removed from cotton after it is harvested. This process is called **ginning.** The cotton seed is then processed to remove its oil, which is a major contributor to the vegetable-oil needs of the United States. The seed, after the extraction of the oil, is ground into a high-protein animal feed.

Sugar Beets The production of sugar beets for sugar accounts for about 35 percent of the refined sugar produced in the United States (Figure 21–17). This crop is grown for its thick, fleshy storage root in which sugars are accumulated. Centers of production of sugar beets are the western states and the upper Midwest.

INTERNET KEY WORDS:

sugar beets cane
sugar production

INTERNET ADDRESS:

<http://www.biodiesel.org>

Hot Topics in Agriscience

BIO-DIESEL—A RENEWABLE RESOURCE

One of the great concerns about world dependency on fossil fuels has been that these fuels are nonrenewable resources. Once the oil reserves are used up, this important resource will be lost to the human race. Bio-diesel is different because this product is produced by plants. It is obtained from oil seeds such as rapeseed and sunflower seeds.

Plant scientists and agricultural engineers have pooled their talents to develop ways that plant oils can be used to replace diesel fuel. Vegetable oils have shown great promise as fuels in diesel engines, but they are expensive in comparison with the cost of diesel fuel. One approach to reducing the cost of bio-diesel is to produce it from recycled vegetable cooking oils and from animal fat. These oils and fats form chemical compounds called *esters* when they are combined with methyl alcohol under alkaline conditions. This type of bio-diesel is an improvement over earlier biodiesels because this fuel product works better in cold weather than did the early bio-diesel products.

FIGURE 21–17 Sugar cane and sugar beets are the sources of nearly all of the domestic sugar supply. Large factories and refineries process the raw product into syrup, and then into the familiar sugar crystals we use to sweeten foods.

Sugar Cane Sugar cane production in the United States is concentrated in subtropical areas of Florida, the Gulf Coast states, and Hawaii. Sugar cane accounts for about 65 percent of the sugar refined in the United States.

This crop, which is a grass, is grown from sections of stalk called **seed pieces**, rather than from seed. It takes about 2 years for sugar cane to reach the harvesting stage in Hawaii. Compared with production in Hawaii, in the southern states sugar cane is harvested about 7 months after planting, with a corresponding loss of yield. The same field of sugar cane can be harvested several times before needing to be replanted (Figure 21–18).

Tobacco Tobacco is an original North American product that was used by the American Indians in religious rites. It is produced in the southeastern states, predominantly in North Carolina, Kentucky, Virginia, South Carolina, Tennessee, and Georgia, as a cash crop. A **cash crop** is grown strictly to sell for cash. Production of tobacco declined in the

A   B

FIGURE 21–18 (A) Sixty-five percent of U.S. sugar production comes from sugar cane produced in the southern states and in Hawaii. (B) Sugar beet production accounts for about 35 percent of the domestic sugar supply.

United States in the 1980s as cigarette smoking and other uses of tobacco declined, but it increased again in the 1990s as export markets expanded.

Tobacco production requires large amounts of labor and is therefore best adapted to small farming operations. Warm temperatures and plenty of rainfall are required for optimum production of high-quality tobacco.

Seed Crops

INTERNET KEY WORDS:

vegetable small-seed
production

The production of good-quality field crops depends on good seeds. An entire seed industry has emerged to supply the need for high-quality seeds for many varieties of field crops. Seeds must come from healthy plants that are free of diseases. Many of the vegetable crops that we raise are hybrid plants (see Unit 17) that come from different strains of parent stock. Seed crops are usually raised in areas that are isolated from related commercial crops. This helps in maintaining the purity of the genetic makeup of the seeds because stray pollen from undesirable sources is minimized.

Seeds for most field crops are produced in similar climates and growing conditions that will be encountered when the seeds are used to produce crops. This assures that they are adapted to the conditions under which they are expected to grow and produce. Care must be taken, however, to ensure that the seed crops do not become infected by diseases and destructive organisms such as fungi. Such an infection of a seed crop could spread the infection. In all cases, seed crops must be subjected to inspection and testing to ensure that they are pure and that the quality of the seed is high.

Most vegetable seeds are produced in isolation from commercial crops. An arid climate is desirable in the production of most vegetable seed crops because such a climate is not favorable to fungi, bacteria, and many other organisms that cause damage to crops. As a result, most vegetable seeds are produced on irrigated farms at isolated locations in the western desert region of the United States (Figure 21–19). Seeds are

FIGURE 21–19 Most vegetable seed crops are raised in arid climates at isolated locations to avoid contamination from the pollen of field crops and to minimize the exposure to fungal and bacterial diseases that thrive in damp environments.

high-value crops that must be protected from the stresses that are often found in the environment. The seeds must have favorable conditions that allow them to mature completely before they are harvested.

After the harvest, seeds are placed in storage under carefully controlled conditions to preserve their ability to germinate. Seed samples are routinely submitted to seed laboratories for certification. The seed laboratories test the germination rates of different seed samples, and they carefully inspect the seed samples for contamination with weed seeds. Seed crops must meet strict standards to be labeled as certified seed.

Classification of Field Crops

INTERNET KEY WORDS:

certified foundation seed

INTERNET ADDRESS:

<http://www.aosca.org/>

There are a number of ways of classifying field crops. Three ways are by use, thermo requirements, and life span. Thermo refers to the heat requirements or length and characteristics of the growing season that is required.

The classification of field crops according to use is as follows:

- **Cereal crops** are grasses grown for their edible seeds. They include corn, wheat, barley, oats, rice, rye, and sorghum.

- **Seed legume crops** are nitrogen-fixing crops that produce edible seeds. Included in this class are soybeans, peanuts, field peas, field beans, and cowpeas.

- **Root crops** are grown for their thick, fleshy storage roots. Beets, turnips, sweet potatoes, and rutabagas are root crops.

Agri-Profile

CAREER AREAS:

Broker/Elevator Manager/ Grain Handler/Agronomist/Producer

Corn, wheat and other small grains, sugar cane, soybeans, sugar beets, and other specialty crops are grown on large acreages in the United States, Canada, and many nations of the world. In the United States, grain, oil, and specialty crops account for large amounts of exports and do much to help maintain our balance of payments in foreign trade. Grain brokers, futures brokers, market reporters, grain elevator operators, crop forecasters, farmers, and others owe their jobs to these crop enterprises.

Custom combine operators, truck drivers, maintenance crews, cooks, and other service workers follow the grain harvest from Mexico to Canada. At season's end, the crews return south and prepare for the next season, when the cycle is repeated. At the same time, smaller farm owners and operators grow, harvest, and market millions of acres of field crops as part, or all, of the farm operations.

Along with crop enterprises are jobs in building and storage construction, systems engineering, machinery sales and service, welding, irrigation, custom spraying, hardware sales, agricultural finance, chemical sales, and seed distribution.

Plant technicians, as well as research specialists, are necessary to plan, research, and grow and manage crops, such as this pearl millet, to achieve crop improvement. *(Courtesy of USDA/ARS #K-3882-2)*

- Forage crops are grown for hay, silage, or pastures for livestock feed. Examples of forage crops include alfalfa, clover, timothy, orchard grass, and many other crops used for their stems and leaves.
- Sugar crops are grown for their ability to store sugars in their stems or roots. They include sugar cane, sugar beets, and sorghum. Corn is also used to produce sugar.
- Oil crops are produced for the oil content of their seeds. Examples are soybeans, peanuts, cottonseed, flax, rapeseed, sunflower seed, and castor beans.
- Tuber crops are grown for their thickened, underground storage stems. Potatoes and Jerusalem artichokes are examples of tuber crops.
- Stimulant crops are grown for their ability to stimulate the senses of the user. Examples of stimulant crops are tobacco, coffee, and tea.

Crops can also be classified according to their thermo requirements. The two major thermo groups are warm season and cool season.

Warm-season crops must have warm temperatures to live and grow. They are adapted best to areas where freezing or frost seldom occur. They also normally require longer growing seasons than do cool-season crops. Examples of warm-season crops are cotton, tobacco, and citrus.

Cool-season crops are normally grown in the northern half of the United States, where temperatures below freezing are normal. These crops often need a period of cool weather to attain maximum production. Most of the grains, tubers, and apples are cool-season crops.

Crops can also be classified according to their life spans. They may be annual, biennial, or perennial. An example of an annual crop is corn; red clover is a biennial crop; and alfalfa is a perennial crop.

Selection of Field Crops

There are a number of factors to consider when selecting which field crops to grow. Some of these factors are:

- Select crops that will grow and produce the desired yields under the type of climate available. Be sure to consider length of growing season, average yearly rainfall, average temperatures, humidity, and prevailing winds.
- Crops must be adapted to the type of soil available. Consider soil pH, soil type, soil depth, and soil response to fertilizers.
- Consider demand and availability of markets for the crop to be produced.
- Assess labor requirements and availability of labor for the crop.
- Identify machinery and equipment that will be needed to grow the crop.
- Consider the availability of enough land to justify production of the crop.
- Identify pest-control problems.
- Estimate expected yields.
- Anticipate production costs. Can a reasonable profit be expected?

FIGURE 21–20 A good seedbed consists of crumbly, mellow soil that is free of clods, rocks, and debris.

INTERNET KEY WORDS:

minimum tillage practices
seedbed preparation

Seedbed Preparation for Field Crops

The purpose of seedbed preparation for field crops is to provide conditions that are favorable for the germination and growth of the seed to be planted. Not only does the seedbed need to be prepared for seed germination, but the area under the seedbed must also be prepared for the root growth of the crop (Figure 21–20).

Eliminating competition from weeds and crop residues is a consideration when preparing a seedbed for planting. Proper seedbed preparation can also increase the availability of soil nutrients to plants.

Seedbeds should not be overworked. The texture of the soil should be porous and allow for free movement of air and water. Small seed requires a seedbed with a finer texture than larger seeds require. The seedbed should contain enough fertility to encourage germination and growth until additional fertilizer can be applied.

There are several methods of properly preparing seedbeds for field crops. They can be divided into three general categories: conventional tillage; reduced, or minimum, tillage; and no-tillage.

In **conventional tillage**, the land is plowed with a moldboard plow turning under all of the residue from the previous crop. The soil is then disced or harrowed to smooth and further pulverize the soil for the seedbed (Figure 21–21).

Reduced or **minimum tillage** is a system of seedbed preparation that works the soil only enough so that the seed can make contact with the soil and germinate. A chisel plow is frequently used (Figure 21–22). Minimum-tillage systems usually combine several operations into one pass across the field. This method reduces the amount of soil compaction, conserves soil moisture, and usually provides less opportunity for soil erosion (Figure 21–23).

No-till preparation of seedbeds involves planting seeds directly into the residue of the previous crop, without exposing the soil. Seed is usually planted in a narrow track opened by the seed planter. It is extremely important that good management practices be used when using the no-till method of seedbed preparation. Such practices will ensure control of insects and diseases and eliminate competition from previous crop residues (Figure 21–24).

FIGURE 21–21 With conventional tillage, the soil is turned via a moldboard plow so that all crop residues, livestock manure, lime, and fertilizer are mixed through the plow layer. *(Courtesy of Elmer Cooper)*

FIGURE 21–22 A chisel plow loosens the soil but leaves the crop residue on the surface.

Science Connection

DESIGNER FOODS

In the 1940s, food products were being fortified with vitamins and minerals. Before this time, many people experienced illness that was a result of vitamin and mineral deficiencies. Once the cause of such sicknesses was discovered, foods such as breads and cereals were supplemented with the needed nutrients. Currently, the back of any breakfast cereal box shows that fortification is still practiced. Biotechnologists are certain that there is another way in which the food we eat can improve our health. For example, a variety of bio-engineered rice has just been approved in the United States and Canada. Golden rice has been genetically altered to produce vitamin A.

Crops such as golden rice are created using a process called recombinant DNA technology. A simple explanation of this process begins when a desired gene is found and cut out of the DNA of the donor cells. This is done using proteins called restriction enzymes. These enzymes cut DNA at a specific place. Next, the donor gene is put into a plasmid. A plasmid is a circular piece of DNA that can be found in bacteria. The bacteria are then placed in a petri dish, where they begin to replicate. Many copies of the donor gene are made. Through bacterial action, the desired gene can be put into plant cells. The plant cells reproduce themselves with the new donor gene in them. Plant cells that contain the desired gene can be collected and put through another process that will result in the growth of a new plant that has the donor gene in it.

In many underdeveloped countries, many children have childhood blindness. This condition is caused by vitamin A deficiencies in young children. Rice is often the only food source available to poor families in these countries. When golden rice is made available to the citizens of these nations, the number of children affected by childhood blindness will decrease dramatically. Bio-engineered foods have the potential to do for this generation what food fortification has done for generations in the recent past.

INTERNET KEY WORDS:

planting field crops

Planting Field Crops

The invention of the seed planter was one of the most important events for U.S. agriculture. There are three general types of planters used in planting field crops today. They are row crop planters, drill planters, and broadcast planters.

Row crop planters plant seeds in precise rows and with even spacing within the rows. Three types of row crop planters are drill planters,

FIGURE 21–23 A tillage tandem disc/chisel-point tiller cuts trash and loosens surface soil via the discs, loosens the deep soil via chisel points, and leaves a fine seedbed on the surface.

FIGURE 21–24 No-till planting leaves crop residues on the surface for better erosion control. *(Courtesy of Dr. Allen Hammer)*

FIGURE 21–25 Three types of broadcast seeders are knapsack; spinner, or cyclone; and airplane.

hill-drop planters, and checkrow planters. Drill planters drop seeds individually in a row at set distances apart. Hill-drop planters drop two or three seeds together in rows. Checkrow planters plant several seeds together in a checkered pattern in the field to permit cross-cultivation. Row crop planters are used to plant corn, beans, sugar beets, soybeans, sorghum, and cotton.

Drill planters plant seeds in narrow rows at high population rates. Drills are available in many row spacings and planter widths. Seeding rates are less accurate than with row crop planters. Some field crops that are planted with drill planters are wheat, oats, barley, rye, and many grasses. Fertilizer and pesticides may be applied at the same time the seed is planted with drill planters.

Broadcast planters scatter the seed in a random pattern on top of the seedbed. The accuracy of seeding using this type of planter is the poorest of the planting methods. Broadcast seeders cover wide areas and usually plant seeds much faster than other methods. They are sometimes used when weather conditions make it difficult to get machinery into the fields. Airplanes are often used as broadcast planters. Knapsack seeders and spinners are other types of broadcast seeders (Figure 21–25).

One disadvantage of using broadcast seeders is that some seed needs to be covered to germinate and to protect it from loss. This means that a second trip often must be made over the field to cover the seed. Small grains, grasses, and legumes are sometimes planted by broadcasting.

There are other considerations in planting field crops. These include the date to plant, germination rate of seeds, uniformity of seed, weather conditions, and insect- and disease-control problems.

Meeting Water Needs of Crops

The soil that plants grow in acts as a storage vat for the water needed by the plant. Under ideal soil conditions, approximately half of the pore space is filled with water. About half of this water is available for use by plants. Unfortunately, this ideal condition seldom exists; often, much less water is available for plant use than is needed. Factors that affect the water available for crops include the type of soil, natural rainfall, water-table levels, and prevailing winds.

When conditions are such that sufficient water is not available for the crop being grown, irrigation may be the answer to obtaining profitable yields. Irrigation is the artificial supplying of water to crops.

The irrigation of crops has been practiced for more than 5,000 years. The Nile River was used by the Egyptians to irrigate crops grown in the fertile deserts of the area. The Chinese diverted the water from many rivers to irrigate their rice fields. Even the American Indians of the American West used irrigation to allow for the production of corn in arid areas.

The major methods of supplying irrigation water to crops are sprinklers, surface irrigation, and drip irrigation.

Sprinkler irrigation is one of the most efficient uses of supplemental water, because it places the water where it is needed at the time when it is needed. The amount of water that is applied is easily controlled by the length of time that the irrigation system is allowed to operate on the same setting. Unlike flood irrigation, which tends to overirrigate part of

FIGURE 21-26 Modern sprinkler irrigation systems are capable of delivering a uniform supply of water throughout the field. *(Courtesy of USDA/ARS #K-3231-17)*

FIGURE 21-27 Cucumbers located in a greenhouse are irrigated by a drip irrigation system that distributes water and nutrients.

INTERNET KEY WORDS:

pest control, field crops

FIGURE 21-28 A few Lesser Grain Borers left unchecked in stored grain soon reproduce to become a devastating horde of insects that is capable of ruining an entire crop of grain. *(Courtesy of USDA/ARS #K-3965-17)*

the field whereas underirrigating other areas, sprinkler irrigation delivers a uniform supply of water to all areas of the field (Figure 21–26).

In **surface irrigation**, water gets to the crop by gravity, flowing over the surface of the soil or in ditches or furrows. It is an inexpensive means of providing water to crops. However, the cropland may need to be leveled before it can be irrigated. Containment borders may also interfere with other farming operations.

Drip irrigation supplies water to the roots of crops in a uniform manner. Use of tubes located either above or under the ground to deliver the water to the crops makes a drip irrigation system expensive to set up. It has the advantage of low operating costs when in operation, however, and it permits the most efficient use of water (Figure 21–27).

Pest Control in Field Crops

The control of pests in field crops is often the factor that determines whether a profit is made. Pests of field crops may include diseases, weeds, insects, and animals. They may destroy the seed before it germinates, attack the growing crop, or render the harvested crop to be unsalable, unusable, and unfit for humans to eat. Economic losses from plant pests total billions of dollars each year (Figure 21–28).

There are three main categories of losses from plant pests. They are reduced yields, reduced quality, and spoilage.

Reduced yields occur when weeds germinate and grow faster than the crop being grown. Weeds compete successfully with the crop for moisture and nutrients, often causing the crop plant to be unhealthy. Parasitic plants actually feed on the host crop and may cause it to break off and die.

Damage from insects also causes the crop to yield less than expected. The damage may occur during the insect's feeding or while it lays eggs in and on the crop. Reduction of yield may also occur as insects spread disease from plant to plant.

Diseases can reduce yields by interfering with the ability of the plant to manufacture food. They can also cause other plant processes to be changed, affecting the health of the plant.

FIGURE 21–29 The use of this ridge tillage machine for cultivation reduces or eliminates preplant tillage. *(Courtesy of USDA/ARS #K-3239-1)*

FIGURE 21–30 Parasitic wasps lay eggs in the insect larvae of specific insect pests, which reduces the damage from and terminates the life cycle of those pests. *(Courtesy of USDA/ARS #K-5035-20)*

FIGURE 21–31 Plant scientists have learned methods by which genetic resistance to pests can be transferred from one plant to another.

The reduction of quality may result from such things as weed seeds or rodent hairs and droppings intermixed with the crop. Foreign materials may cause flavors that are objectionable to users of the crop. Most foreign materials must be removed before the crop can be used.

Damage from insects and diseases can make the crop less desirable in appearance and can increase processing costs tremendously. Food crops may be deemed unsuitable for human consumption and a total loss if they are too severely infested with insects and diseases.

Spoilage of crops sometimes results when weeds hinder the drying process. Insects may also cause stored crops to overheat and mold.

Methods of controlling pests in field crops include mechanical control, cultural control, biological control, genetic control, and chemical control.

Mechanical pest control refers to anything that affects the environment of the pest or the pest itself. Cultivation is the normal mechanical control of weeds (Figure 21–29). Cultivation of the soil may also expose insects and soilborne diseases to the air. The effect of the suddenly increased or decreased temperatures as a result of exposure to the sun and air often proves fatal to many pests. Other types of mechanical control of pests include the pulling or mowing of weeds and the use of screens, barriers, traps, and electricity.

Cultural control refers to adapting farming practices to control pests. Some cultural controls include timing farming operations to eliminate pests, rotating crops, planting resistant varieties, and planting trap crops that are more attractive to insects than is the primary crop.

Biological control of plant pests involves the use of predators or diseases as the control mechanisms. The release of sterile male insects and the use of baits and repellents are also examples of this type of pest control. When using insects or diseases to control crop pests, it is important that the control be specific to the intended pest (Figure 21–30).

The development of varieties of crops that are resistant to pests is called **genetic control** (Figure 21–31). This may involve making the crop less attractive to the pest because of taste, shape, or blooming time.

FIGURE 21–32 Environmentally friendly, ultra-low-volume herbicide application methods can significantly reduce the amount of agricultural chemicals used. *(Courtesy of USDA/ARS #K-5287-4)*

FIGURE 21–33 Late corn harvest using a combine near College Park, Maryland.

INTERNET KEY WORDS:

approved practices,
grain storage
post-harvest crop losses

Developing more rapidly growing crops that crowd out weeds is also an example of genetic pest control. Crops with resistance to diseases also fall into this category.

Chemical control of plant pests involves the use of pesticides to control pests of field crops. Excellent management practices must be exercised when using chemicals to control pests. Care should be taken to correctly identify the pest to be controlled and the chemical to be used. Dosage, runoff, and pesticide residues need to be carefully monitored (Figure 21–32).

Harvesting and Storing Field Crops

Harvesting field crops at the proper stage of maturity is a key to maximizing profits. The harvest of the crop is the culmination of a growing season of work and anticipation of the rewards of a job well done.

The development of mechanical harvesting equipment allowed field-crop producers to harvest thousands of bushels of grain daily and with less labor than previously required. This allowed for tremendous increases in the amount of food available for people and animals and a greatly improved standard of living.

The primary harvesting machine for field crops is the combine (Figure 21–33). It performs the tasks of cutting the crop, threshing it, separating it from debris, and cleaning it. Threshing refers to the separation of grain from the rest of the plant materials. There are many types of combines adapted to the harvest of specific crops.

FIGURE 21–34 Weather- and rodent-proof grain storage bins near Savannah, Georgia, are being used to find optimum storage conditions to minimize insect and mold damage to stored grain. *(Courtesy of USDA/ARS #K-4289-1)*

The proper storage of crops after harvesting is also important. The threats to the quality of stored crops include heat, moisture, fungi, insects, and rodents (Figure 21–34).

Drying grain to reduce moisture and heat is important for successful long-term storage. Much grain is harvested with a moisture content that is much too high. If stored without drying, the grain may heat up and encourage the growth of fungi, causing the grain to spoil. Foreign materials, such as weed seeds, may also cause stored crops to spoil.

Science Connection

NEW CROPS FOR NEW USES

As the medical profession scrambles to find new drugs and treatments for new and old medical problems, agriscience is becoming an increasingly important tool. At the Southern Weed Science Laboratory in Stoneville, Mississippi, scientists are developing a weed-control system to ensure the supply of a medicine for patients with malaria. Annual wormwood has been used for 2,000 years in China to treat malaria. Now a new, refined drug derived from the plant is in demand worldwide. The World Health Organization is seeking ways to increase the production of wormwood sufficiently to meet the need.

Annual wormwood grows throughout the United States as a weed. The woody-stemmed, cone-shaped plant reaches a height of 4 to 8 feet and could become a cash crop for farmers in the future. Wormwood is harvested with a sickle-type machine similar to that used to cut kenaf and sugar cane. Cut plants must be protected from prolonged exposure to the sun, because ultraviolet light breaks down artemisinin—the antimalarial substance.

Artemisinin is found in the leaves of growing tips, and it takes 2.2 lb of the leaves to process 1 gram of artemisinin. Authorities indicate that it would take thousands of acres of wormwood to meet the world demand for the drug. However, before commercial production of wormwood could occur, farming methods, including weed control, must be developed. The World Health Organization estimates that 110 million people throughout the world contract malaria each year.

Another newly emerging crop is lesquerella. Lesquerella has been developed from *Lesquerella fendleri*, a wild plant native to Arizona, New Mexico, Texas, and Oklahoma. The University of Arizona and two commercial firms joined with the U.S. Department of Agriculture to develop the crop. Machinery used for other grain and oil seed crops is suitable for lesquerella.

Cosmetic manufacturers and other companies are looking for the oils extracted from lesquerella seed. The oils can be used in resins, waxes, nylons, plastics, high-performance lubricants, corrosion inhibitors, coatings, and cosmetics such as hand soap and lipstick. The oil meal left after the oils are extracted is suitable as a high-protein livestock feed.

Checking lesquerella for seed set. New crops, such as lesquerella and wormwood, promise new advances in fabrics, plastics, lubricants, corrosion inhibitors, cosmetics, and medicines. *(Courtesy of USDA/ARS #K-4692-2)*

Stored crops must also be protected from insects and rodents if quality is to be maintained. Rodent droppings, hair, and urine, as well as insect parts, may render crops unfit for human consumption. Reduction of food value and spoilage are other hazards of stored crops when insects and rodents are not controlled.

The production of field crops generates more income for U.S. agriculturists than does any other production enterprise. With more than 20 percent of the land in the United States currently being used for growing crops, and with the excess production contributing to a more favorable balance of trade, crop production is likely to remain important in the future.

STUDENT ACTIVITIES

1. Write the Terms to Know and their meanings in your notebook.
2. Compile a list of the field crops grown in your area.
3. Write a report on a field crop of interest to you.
4. Select a field crop and determine as many uses for it as possible.
5. Visit a local crop farm and talk to the operator about the advantages and disadvantages of growing a particular crop.
6. Prepare an advertisement to promote a field crop.
7. Construct a bulletin board about field crops and products made from them.
8. Visit a farm-machinery dealer. Make a list of all the equipment sold there that is used in the production and harvesting of field crops.
9. Make a collection of as many different field crops as you can.
10. Do a germination test on the seeds of several types of field crops. Observe the number of days required for germination to take place and the percentage of the seeds that germinate. Also conduct germination tests under a variety of environmental conditions, such as warm and cool, wet and dry, or with and without light. Compare the results to determine optimum conditions for the germination of crop seeds.
11. Make sketches of each of the field crops (and their seeds) provided by your teacher for identification. Label each drawing with its name. Write any notes that will help you to identify the plant in the future. Review this activity from time to time to assure that you remember how to identify each field crop.

SELF EVALUATION

A. Multiple Choice

1. Flax is an example of a/an _____ crop.
   a. oilseed
   b. grain
   c. sugar
   d. fiber

2. The most important grain crop in the world is
   a. corn.
   b. wheat.
   c. rice.
   d. barley.

3. The moldboard plow is the primary tillage machine for the _____ tillage system.
   a. no-till
   b. minimum
   c. conventional
   d. none of the above

4. About 65 percent of the refined sugar produced in the United States comes from
   a. sugar cane.
   b. sugar beets.
   c. corn.
   d. sorghum.

5. The use of airplanes is an example of _____ seeding.
   a. row crop
   b. drill
   c. aerial
   d. broadcast

6. Grasses grown for their edible seeds are _____ crops.
   a. grain
   b. legume
   c. oil
   d. sugar

7. Ginning is the process used to
   a. prepare a seedbed for planting.
   b. prepare crops for storage.
   c. remove seeds from cotton.
   d. prepare grain for use as alcohol.

8. An example of genetic control of pests is
   a. planting a crop when insects are not present.
   b. releasing sterile male insects.
   c. genetically altering a plant so that insects that eat it will be killed.
   d. planting a variety of a crop that grows more rapidly than do weeds.

9. Tobacco falls into the category of crops called
   a. fiber.
   b. biennial.
   c. thermo.
   d. stimulant.

10. Soybeans are grown for
   a. oil.
   b. hay.
   c. rain.
   d. all of the above.

B. Matching

_____ **1.** Most important U.S. crop

_____ **2.** Provides raw material for breads and pasta

_____ **3.** Irrigation

_____ **4.** Preparing barley for alcohol production

_____ **5.** Primary harvesting machine

_____ **6.** Most important world grain crop

_____ **7.** Oil crop

_____ **8.** Cloth made from flax

_____ **9.** Warm-season crop

_____ **10.** Separation of grain from plant

a. Malting
b. Linen
c. Combine
d. Sugar cane
e. Corn
f. Sprinkler
g. Peanut
h. Rice
i. Threshing
j. Wheat

C. Completion

1. Heat, moisture, fungi, rodents, and insects are all problems to be dealt with in the _____ of crops.

2. _____ crops protect the soil from erosion.

3. The primary use for wheat is for _____ _____.

4. Cotton is grown for fiber and _____.

5. Mowing is a means of _____ control for weeds.

UNIT 22

Forage and Pasture Management

MATERIALS LIST

✓ several specimens of forage and pasture crops

✓ quart jars for Student Activity

✓ bulletin board materials

✓ samples of lactic and butytric acids

✓ agriscience magazines

✓ Internet access

COMPETENCIES TO BE DEVELOPED

After studying this unit, you should be able to:

- define important terms used in forage and pasture production and management.
- identify major crops grown for forage and pasture.
- select varieties for forage and pasture.
- prepare proper seedbeds for forage and pasture crops.
- plant forage crops and renovate pastures.
- control pests in forage crops and pastures.
- harvest and store forage crops.

SUGGESTED CLASS ACTIVITIES

1. Invite a producer of forage crops to discuss the production of high-quality forages with the class. Prepare the class with specific questions to ask. Have each class member prepare written answers to each of these questions as a follow-up to the presentation.

2. Have class members identify each of the important forage crops by studying laminated specimens of each of the crops. Prepare these materials in advance of the class.

3. Invite a representative from a local or state cattleman's or sheep producer's organization to come and discuss how range animals (ruminants) obtain adequate nutrition from grazing without being fed exact amounts of proteins, fats, and carbohydrates. Have the representative discuss the advantages and disadvantages of grazing animals on rangelands.

orage and pasture crops are the most important category of crops grown in the United States, regarding acres in production. There are more than 475 million acres of pasture and rangeland. Another 61 million acres are used to produce hay. The importance of forages is further emphasized when you consider that half of the pasture and rangeland in the United States is not suited for the production of cultivated field crops. Forages are crop plants that are produced for their vegetative growth.

Forage production can be divided into three general categories: hay, silage, and pasture. **Hay** is forage that has been cut and dried until it contains a low level of moisture. **Silage** is green, chopped forage that has been allowed to ferment in the absence of air. **Pasture** is forage that is harvested by livestock as they graze. Forages are generally planted or maintained with one of these uses in mind.

FORAGE AND PASTURE CROPS

Legumes

INTERNET KEY WORDS:

rangeland grazing, legume, forage crop

INTERNET ADDRESS:

<http://www.rangelands.org/srm.shtml>

Alfalfa The most important forage crop in the United States is alfalfa (Figure 22–1). It is often called the "queen of the forages." Alfalfa is a legume that adds nitrogen to the soil. It is high in protein and other nutrients and is productive in fertile soils (Figure 22–2). It is also one of the oldest cultivated forage crops; it was mentioned in the Bible and in other early writings.

When properly harvested and stored, alfalfa is the most economical source of nutrients for ruminant animals. With production of alfalfa and alfalfa mixtures accounting for nearly 40 percent of the hay produced in the United States, annual alfalfa production ranks third to corn and soybeans in dollar value.

FIGURE 22–1 Alfalfa is a forage crop that is high in protein and other nutrients that are needed by livestock. It is the most important forage crop in the United States.

FIGURE 22-2 Bacteria in the nodules on alfalfa roots are capable of converting nitrogen gas from the atmosphere into nitrates, a required plant nutrient. *(Courtesy USDA/ARS)*

FIGURE 22-3 Clover is an important hay and pasture crop in the United States. *(Courtesy of Sharon Rounds)*

FIGURE 22-4 Bird's-foot trefoil is a legume containing tannin, which reduces the threat of life-threatening bloat to grazing cattle. *(Courtesy of USDA/ARS #K-2610-10)*

The North Central states account for about two thirds of the yearly alfalfa production in the United States. However, alfalfa can be grown in nearly every state. Wisconsin ranks number one in alfalfa production, followed by South Dakota, Iowa, Minnesota, Nebraska, North Dakota, and Montana.

True Clovers True clovers include about 300 species; however, only about 25 have agricultural importance. Clovers of economic importance in the United States include red clover, white clover, crimson clover, ladino clover, and alsike clover (Figure 22–3).

Clover is the second largest group of the hays grown in this country. It is a legume with the ability to add nitrogen to the soil. Clover has the disadvantage of smaller yields than alfalfa under similar growing conditions. It is popular as hay, pasture, and silage and grows well in combination with many forage grasses. The Northeast and North Central states produce most of the clover grown in the United States.

Sweet Clover Sweet clover is used most often in areas that are hot, drought stricken, or both, where its ability to survive and produce a crop is unsurpassed. It is used as hay, pasture, and sometimes as a green manure crop. Sweet clover is also used in Texas in rotation with cotton to help control a cotton-root disease. It is also an excellent source of nectar, which honeybees make into honey.

There are three species of sweet clover grown in the United States: biennial yellow, biennial white, and annual. Biennial white sweet clover yields more than the other species, although biennial yellow sweet clover is usually of greater quality. Sweet clover will grow in all areas of the United States. Areas of large production include the Dakotas, Minnesota, Wisconsin, Michigan, and the Central states to Texas.

Bird's-foot Trefoil Bird's-foot trefoil is a comparatively new crop in the United States. It originated in Europe, where it has been a forage crop for 300 years. There are approximately 2 million acres in production in the United States. Bird's-foot trefoil is used as pasture in most cases. Some is also grown for hay (Figure 22–4).

The food value of bird's-foot trefoil is about equal to that of alfalfa. Because of smaller yields, bird's-foot trefoil is unlikely to seriously challenge alfalfa in importance. It does have the advantage over true clovers in being much longer lived. States that report large acreages of bird's-foot trefoil include California, Ohio, Iowa, New York, and Pennsylvania.

Lespedeza This legume is grown primarily in the South, where 1 to 2 million acres are harvested as pasture and hay each year. It can grow and thrive in soils that are low in fertility. Because lespedeza has a lower nutritive value than true clovers and alfalfa, it is recommended for feed for beef cattle but not for dairy cattle.

Types of lespedeza include annual and perennial. Most of the lespedeza grown in the United States is of the annual type (Figure 22–5).

Peanut Hay Peanut hay is a by-product of peanut production that is used in the southeastern region of the United States. Although it lacks the protein content of forage legumes, the addition of nitrogen compounds such as urea improves its quality. The extra nitrogen in the hay can be converted to protein by ruminant animals such as cattle and sheep.

Grasses

Bromegrass Bromegrass is an important forage grass throughout the northern half of the United States. It is extremely hardy and grows to a height of 2 to 3 feet (Figure 22–6). Because bromegrass produces many rhizomes, thin stands rapidly thicken with age. Rhizomes are horizontal, underground stems from which new plants arise. Fertile soil is required for the production of bromegrass.

FIGURE 22–5 Annual lespedeza may be grown as forage for beef cattle, whereas Korean lespedeza is used as a soil-stabilizing plant.

FIGURE 22–6 Bromegrass is grown throughout the northern half of the United States.

FIGURE 22–7 Orchard grass is known for its rapid, early-spring growth.

INTERNET KEY WORDS:

timothy grass
orchard grass
reed canary grass
Kentucky bluegrass
fescue grass

Orchard Grass   Orchard grass is known for its rapid germination and early-spring growth (Figure 22–7). It also recovers quickly after being harvested. Timing of the harvest of orchard grass is important because quality decreases rapidly after the plant reaches maturity. Orchard grass makes excellent pasture and is often used for hay in combination with legumes.

Timothy   Timothy is a cool-season grass that grows best when temperatures are between 65° and 72° F (Figure 22–8). The use of timothy as a forage grass has declined in recent years, because bromegrass and orchard grass outyield it in those areas where all three species are adapted. Most timothy is grown in the northeastern part of the United States.

Reed Canary grass   In areas that are wet or poorly drained, reed canary grass is often the only answer to producing a forage crop (Figure 22–9). It can produce more than 4 tons of forage per acre in the cool, damp areas where it thrives. This grass grows as tall as 7 feet and can produce high-quality feed if harvested before reaching maturity. Most reed canary grass is grown in Washington, Oregon, California, Iowa, and Minnesota.

Kentucky Bluegrass   Kentucky bluegrass is grown over much of the United States, even though it is best adapted to the Northeast. It is the major grass of many pastures, lawns, and even golf courses in areas where summers remain fairly cool. Kentucky bluegrass is a good native, permanent pasture grass, but it is outyielded by other grasses. It is not practical to use for hay because of low production. It also goes dormant during hot weather, making it necessary to use bluegrass in combination with other grasses such as ryegrass to have forages available all summer (Figure 22–10).

FIGURE 22–8 Timothy grows best when temperatures are between 65° and 72° F.

FIGURE 22–9 Reed canary grass grows in areas that are too wet for the production of other grasses.

FIGURE 22–10 Kentucky bluegrass is a popular cool-season grass that is widely used in lawns, golf courses, and pastures.

Hot Topics in Agriscience

RANGELAND FORAGE

In many parts of the country, ranchers use rangelands to graze their animals. Some of these lands are privately owned, whereas the government owns others. Some people are opposed to grazing on public land for fear that the animals will destroy the environment.

On rangelands, the forage is provided by nature, in the form of grasslands or edible wild forage. In native areas, forage management is much different as compared with forage production in pasture settings. Animals must be carefully watched to avoid overgrazing. Overgrazing occurs when animals are allowed to feed in an area for too long. The result is damage to the soil and loss of valuable native plants. When this occurs, the area may not produce enough forage in the subsequent years. Ranchers who rely on range-lands as feed for their animals must do all that they can to maintain the health of rangelands.

Range forage is a valued resource to owners of cattle and sheep. These ruminant animals are able to convert the forage to valuable meat for human consumption.

Some of the common techniques that are used to prevent overgrazing are fencing, rest rotation, deferred rotation, sup-plementation, and herding. Fencing off areas of the rangeland helps keep the animals where they are supposed to be. Fencing also plays a part in both rest rotation and deferred rotation. Rangelands can be divided into pastures; for example, a rancher may divide his or her range into four different pastures. Rest rotation is when one of these pastures is not grazed for a given period. This gives plants a chance to complete at least one life cycle. Ideally, this should be done every other year. Deferred rotation is when the normal rotation between pastures is changed. Nutrient supplements such as salt are also used to entice animals to stay where they are wanted.

Herders are people who spend a lot of time caring for range animals. They move them from place to place to make sure that the animals have sufficient forage, and that the land is preserved. When ranchers properly apply these methods, both the animals and the environment benefit.

INTERNET KEY WORDS:

Bermuda grass
Dallis grass
Johnson grass
corn silage

Fescue   Tall and meadow fescues are perennial grasses that are used for pasture and hay, usually in combination with other grasses and legumes. Fescue is best adapted to the Southeast, where 10 to 20 million acres are grown each year. The quality and palatability of the fescues are less than that of most other forage grasses.

Bermuda Grass   Bermuda grass is a warm-season grass that is dor-mant during cool weather. It is adapted to pastures and lawns because it grows only 6 to 12 inches tall. Common Bermuda grass is considered to be a weed in some areas, because it tends to crowd out other grasses. However, improved varieties of Bermuda grass make good pastures in the Southeast (Figure 22–11).

Dallis Grass   Dallis grass grows 2 to 4 feet tall in the warm areas of the South. It cannot stand continuous use as pasture, because it needs to be able to recover from close grazing. However, it is productive earlier in the spring than other warm-season grasses.

Johnson Grass   Johnson grass is so aggressive that it has been declared a noxious weed in many states of the country. A noxious weed is a plant

FIGURE 22-11 Improved varieties of Bermuda grass make good pastures in the Southeast.

that is prohibited by state law. It is coarse and grows to a height of 6 feet or more. It spreads quickly by rhizomes and often crowds out most other plant species.

Corn and Oats as Forages

Alfalfa and clovers are often planted with a nurse crop of oats. A **nurse crop** consists of plants that are planted with other plants that require protection during early stages of development. The oats protect the tiny legume plants during the early stages of growth. They are then cut for silage while the leaves are still nutritious and the grain is immature. Oats make excellent forage when they are harvested at this stage of maturity. The new alfalfa or clover crop responds to the reduction in competition, often producing a hay crop in the later part of the growing season.

Corn is produced as a forage crop on many farms where it is used for making silage. Corn silage is a highly nutritious feed for beef and dairy cattle, and high yields are possible when adequate water and fertilizer are provided. The corn plant is chopped and ensiled when the grain is in the full dent stage of maturity. The acids that are produced as the silage cures are highly digestible, and the nutritional value of the forage is maintained at a high level. It is a high-moisture feed; therefore, it is not recommended as a large part of the diet for young cattle, because they are unable to eat enough silage to satisfy their nutritional needs. Dairy cows also require dry hay in combination with silage to maintain high levels of milk production.

Selection of Forages

There are many considerations to be made when selecting forages for hay, silage, and pasture. Some of these are:

- intended use of the forage;
- expected or desired yield;
- nutrient value of the crop;
- climatic conditions under which the forage will be grown—warm season versus cool season, summer and winter temperatures, humidity, soil type, anticipated rainfall, nutrient level, length of growing season;
- pest-control measures;
- methods of establishment;
- compatibility with other forages when grown in mixtures;
- expected and desired life of the crop;
- care and maintenance; and
- equipment and labor necessary for growing, harvesting, and storing the crop.

Forage crops that are adapted to the production of hay include alfalfa, clover, bromegrass, orchard grass, timothy, and fescue. Forages used for pastures are clover, lespedeza, Kentucky bluegrass, and Bermuda grass. Almost any legume or grass crop can be used for silage. In addition, corn and most small-grain crops make excellent silage when harvested when the grain is nearing maturity and before the leaves and stalks lose their nutritional value.

Agri-Profile

CAREER AREAS:

Agronomist/Plant Physiologist/
Forage Manager/Range Manager

Forage and pasture provide a variety of career options, ranging from farm and ranch to plant breeding and physiology. Grains, such as corn, cut for silage provide enormous volumes of feed for dairy and beef cattle. Crops cut for silage, as well as those cut for hay, are numbered among the forages. Similarly, pasture and range plants are forages. Those specializing in the science of forage growth and use are called *agronomists*.

Forage specialists are hired by universities and agricultural research centers, as well as by large farms and businesses. In certain parts of the United States, hay businesses are on the increase, as more people do part-time farming, especially those with pleasure horses. In addition, racetracks need hay and straw, which creates a strong market for these forages in many areas.

Hay business managers, truckers, and dealers also conduct thriving businesses hauling straw and manure from horse barns at racetracks to mushroom farms, where the materials are used as the media for growing mushrooms. Large concentrations of dairy cattle and pleasure horses have created markets for cross-country shipping of hay, a practice once regarded as too expensive to be worthwhile. Large amounts of hay are also shipped during periods of drought.

Summer employee Aimee Crago stains an alfalfa cotyledon for microscopic examination. *(Courtesy of USDA/ARS #K-4264-8)*

Seedbed Preparation

In general, preparing a seedbed for forages is much the same as it is for the production of field crops. Residues from previous crops must be handled. The soil must be prepared for the planting of seed or the vegetative pieces used in the propagation of some grasses. Necessary seedbed pest-control measures must be practiced. The soil must be amended so that its pH and fertility are suitable for the germination and growth of the intended forage crop. And with some crops, the moisture level of the seedbed may need to be regulated to ensure germination.

One difference between preparing a seedbed for most forages and one for field crops is that the texture of the soil in the seedbed must be finer and the seedbed somewhat firmer than is necessary for most field crops. This is because most forages have very small seeds that have difficulty making firm contact with the soil in seedbeds that are not finely textured.

Conventional tillage traditionally has been used when preparing seedbeds for starting grass and many legume crops for forage. The residues from previous crops are plowed down or shredded. The soil is then pulverized and leveled by discing, harrowing, or both. For very small seeds, further preparation may be necessary. Final preparation of the seedbed should occur immediately before planting the seed. New seeding equipment has permitted grass seeding into established sod. This greatly reduces soil losses and provides protective vegetation while the seedlings become established (Figure 22–12).

FIGURE 22–12 The no-till grassland drill places seed, fertilizer, and pesticides in one pass across the field.

INTERNET KEY WORDS:

pasture renovation

Planting Forages and Renovating Pastures

Forage crops are usually planted by drilling or by broadcasting. The same grain drills that are used for planting small grains may be used to plant forages. There may be a separate seedbox on the drill for this purpose. The forage seed is dropped on the seedbed at the same time that the grain is planted. Many grasses are planted in this way. The grain crop germinates faster than the forage and acts as a nurse crop until the forage crop can become established. A nurse crop is used to protect another crop until it can get established.

Many forage crops are also planted alone, using either drills or broadcast planters. With many forages, the seed must be covered after it is broadcast onto the soil by broadcast planters. This is done to protect the seed from pests and the drying effects of sun and wind. Covering the seed may be accomplished by a light discing or by the use of a cultipacker, which has corrugated wheels to press the seed into the seedbed (Figure 22–13). Some types of forage seeds are broadcast into growing crops. They must be able to germinate fairly easily, because contact with the seedbed is often minimal. Red clover is often overseeded into stands

FIGURE 22–13 Fine grass and legume seeds must be slightly covered and left in a compact seedbed. The packing wheels create a firm seedbed, bringing the seeds in close contact with the soil particles.

of forage grasses to make mixed hay. **Overseeding** is the practice of seeding a second crop into one that is already growing. This is usually done during late winter and early spring, when freezing and thawing of the soil helps provide contact between seedbed and seed.

Some forages are being planted with no-till planters in live or killed crop residues. The no-till planter opens a narrow furrow in which the seed is planted. Advantages include fewer trips across the field with heavy tractors and tillage equipment. This means that there is less compaction of the soil. There is also little exposure of the seedbed to erosion. Care must be taken to control pests in no-till planting. Another concern is to minimize competition for water and nutrients from other plants growing in the seedbed.

There are several general methods of renovating pastures in the United States. The existing pasture can be killed with a herbicide, the soil disced heavily to prepare a fine seedbed, and the pasture then reseeded with the desired types of grasses or legumes. The pasture is usually treated for insects and diseases at this time if necessary. It is also an ideal time to fertilize and lime the pasture.

Another method of renovating a pasture involves using selective herbicides to kill unwanted species of plants. The pasture is then disced to break up the existing sod. This allows easier entry of moisture and nutrients into the soil. The pasture may or may not be overseeded with desirable species to improve yield. It is also fertilized and limed at this time, according to soil test results.

No-till grassland seeders may be used to place new seed in existing sod, which permits the thickening of existing seedings. These seeders can also introduce new, improved, and aggressive species or varieties of forage crops.

INTERNET KEY WORDS:

forage pest control

Pest Control in Forage Crops

The control of weeds, insects, diseases, and rodents in forage crops helps result in optimum yields. Because many forage crops grow for more than one growing season, pest control is usually an ongoing part of forage crop management (Figure 22–14).

The proper identification of the pest or pests affecting the crop is important. To that end, personal experience and the use of trained professionals are often necessary.

The actual methods of pest control are many, and they have been discussed in previous units. Chemicals, cultural practices, biological control, genetic control, and the timing of crops are all tools to be used in controlling the pests of forage crops. Not to be overlooked in pest-control management is the control of pests in stored forages. When chemicals are used to control pests, care must be taken to ensure that the pesticide is properly applied in the recommended dosages. Timing the application of the pesticide for maximum effect is also essential. Care must also be taken to ensure that overspray and runoff of pesticide materials do not adversely affect organisms other than the intended pest. Concern that the residue of pesticides does not end up in food sources is of utmost importance.

FIGURE 22–14 Grasshoppers are serious pests in forage crops because they eat the leaves and growing tips of the plants. This grasshopper represents only one of more than 600 species of grasshoppers in the United States. *(Courtesy USDA/ARS #K-1066-2)*

Harvesting and Storage of Forage Crops

The proper harvesting and storage of forage crops is extremely important in the management of the forage enterprise for maximum profits.

Hot Topics in Agriscience

PRODUCING HIGH-PROTEIN ALFALFA HAY

The alfalfa plant is a great forage plant because it is a succulent with a lot of leaves. Just as it begins to bloom, it reaches the peak of quality. If it is harvested any sooner, valuable tonnage is lost. Later harvesting results in a smaller percentage of protein. At just about the same time, young tillers, or shoots, begin to grow at the base of the plant. These tillers are the new stems for the next hay crop. If the harvest is delayed beyond this stage of maturity, the plant goes into seed production. The stems become thickened, and the proportion of stems to leaves will increase. This reduces the percentage of protein in the hay.

High-quality alfalfa hay can contain more than 20 percent protein when it is harvested at the proper time. Hay of high quality commands a premium price as dairy hay, and it is worth the extra cost. Cows that are fed top-quality, high-protein hay will increase their milk production. Only the best-quality alfalfa hay should be fed to dairy cows. Another key element in producing high-quality hay is to process the hay immediately and move it into a covered stack to protect it against the rain. Hay that is rained on between cutting and baling will never be high-quality hay. This is because nutrients and color are leached out, and leaves are usually lost during the extra days that are required to get it dry enough to bale.

Type of Hay	When to Harvest
Alfalfa	Pre-bud to $1/_{10}$ bloom stage
Clover	$1/_4$ to $1/_2$ bloom stage
Bird's-foot trefoil	$1/_4$ bloom stage
Sweet clover	Start of the bloom stage
Brome grass	Medium head stage
Timothy	Boot stage to early bloom stage
Lespedeza	Early bloom stage
Orchard grass	Full head but before blooming
Reed canary grass	When the first head appears

FIGURE 22–15 Maturity levels at which selected forages should be harvested for hay.

Most of the forages grown on U.S. farms and ranches are fed to livestock grown on the same farm. Forages are usually the least expensive sources of nutrients available for cattle and sheep.

Pastures The harvesting of pastures involves several factors. One of the most important is that the carrying capacity of the pasture must be determined to figure out how many animals can be fed by the pasture that is available. The carrying capacity of a pasture refers to the number of animals for which it will provide feed.

Pastures also need to have time to recover from the ravages of the animals that crop them. The rotation of the animals using the pastures is important if the pastures are to meet their potential yields. Pasture rotation also helps to control parasites of the animals grazing on them by breaking the pests' life cycles.

Hay Harvesting hay involves several operations that must be timed fairly accurately if high-quality hay is to be produced. The hay must be cut at the optimum time, with an eye on the weather forecast for the next several days. Hay is normally dried in the field by the sun. Nothing ruins hay faster than rain on the crop after it starts to dry and cure. The maturity of the forage being harvested for hay is also critical (Figure 22–15). The more mature the forage becomes, the lower the quality and food value of the hay.

Terms that describe forage crop maturity include *boot stage* (immature seed development), *pre-bud* (early stage of flower formation), and *1/10th bloom* (1 in 10 flower buds have opened).

Most hay is cut with either a sickle-bar mower or a rotary mower. The sickle-bar type mower cuts with the same action as a pair of scissors. These types of mowers are best adapted for forages that are standing up straight.

The rotary mower cuts with blades that move in a circular motion parallel to the ground. This type of mower will cut any type of forage, although greater amounts of horsepower are required when the stands of forage are extremely thick and heavy.

A swather, also known as a mower conditioner, is the most commonly used machine for cutting hay. It performs two processes in a single pass through the field. It cuts the forage plant off just above the ground with a sickle-bar mower, and it places the cut forage in a windrow for convenience in the baling or chopping operations.

Crushing and mashing of the stems is also performed by conditioners that are built into most swathers to reduce the drying time. In areas where rain is frequent, the hay may have to be cut with a mowing machine and allowed to lie in the swath where it falls. This spreads the hay sufficiently for the sun to dry it quickly. In areas of high humidity, chemicals to speed the drying of the forage are sometimes sprayed on the forage as it is cut.

The hay is often raked at least once before it is baled to allow it to dry more evenly and more rapidly. Legume-type hay should be raked during the part of the day when humidity is highest, to keep the loss of leaves to a minimum. Too much raking reduces hay quality because the high-protein leaves fall off. Raking hay also puts it into windrows so that it can be baled.

When hay has dried to the desired moisture level, it is baled or chopped and removed from the field. In some cases, it may be cubed or pelleted, particularly if it is being exported or transported for long distances. Most hay is baled and hauled with automated hay wagons that pick up the bales from the fields, transport them to the storage area, and stack them in place.

Most hay is harvested by forming it into square or rectangular bales or into round bales of various sizes (Figure 22–16). The baler gathers the dried hay from the winrow, compresses it into a bale, and ties twine or wire around the bale to hold it together and to allow for handling. The bale is then expelled from the baler. It is allowed to drop on the ground for later removal to storage. A recent trend has been to bale hay with balers that make large bales of approximately 1 ton each. These bales are preferred for shipping long distances (Figure 22–17).

There are a number of machines that are designed to move baled hay from the field to the storage site. These include trucks, wagons, bale loaders, bale handlers, and bale stackers (Figure 22–18).

Hay may also be harvested with machines called stack wagons. These machines gather forages from the field and form dense stacks or loaves of weather-resistant hay. With special handling equipment, stacks can be moved with few problems.

Hay cubes were developed to allow for the full mechanization of the feeding of hay. The cuber gathers the dry forage and compresses it into

INTERNET KEY WORDS:

alfalfa hay harvest

FIGURE 22–16 Hay is typically cut, conditioned, air dried, and then baled. *(Courtesy of USDA)*

FIGURE 22–17 Large, rectangular bales are preferred for shipping hay by truck because they can be loaded by machine, they are easier to secure on the truck, and the load is not easily shifted.

FIGURE 22–18 Bale stackers have done much to replace manual labor with machine labor. One person, operating a bale wagon, can haul more hay in a day than a crew of four using flatbed wagons. *(Courtesy USDA)*

FIGURE 22–19 Protective covers prevent baled hay from spoiling in the stack due to moisture entering the baled hay.

INTERNET KEY WORDS:

forage silage

cubes about $1.25 \times 1.25 \times 2$ to 3 inches. Normally, only legumes are cubed because of difficulty in getting grass cubes to bind together. Artificial binders are sometimes used.

Hay may be stored in buildings or in the open field. Regardless of the method of storage, care must be taken to ensure that quality is maintained. Hay that is stored in buildings must be kept free from pests, particularly rodents. Care must also be taken to ensure that the moisture level of stored hay is low enough that spoilage does not occur.

When hay is stored outdoors, it should be placed on land that is well drained and sloped to drain water away from the hay. The stacks or bales of hay stored outside must be dense enough so that little moisture can penetrate them and cause spoilage. Hay stored outside is often covered with plastic or other materials to protect it from adverse weather conditions (Figure 22–19).

Silage   Harvesting forages for silage is far more simple than making hay. Less equipment and labor is required, and the entire harvesting operation is usually completed in one pass over the field. A forage harvester cuts the green forage, chops it into small pieces, and deposits it into a wagon or truck to be hauled to the storage facility. Cutter heads on the forage harvester may be of either the cutter type or flail type.

Flail-type forage harvesters are used to harvest low-moisture silage called **haylage**. Forages used for haylage are cut and allowed to dry to a moisture content of 40 to 55 percent before they are placed in a silo. A **silo** is an airtight storage facility for silage or haylage. Haylage has the advantage of being lower in moisture than silage. This allows it to be fed to young cattle, and it requires less storage space than silage.

Silage may be stored in silos or in trenches or bunkers. It is also sometimes stored in piles on top of the ground and sealed with plastic or other materials to make it airtight. The production of silage is dependent on storage in the absence of air or oxygen. The fermentation that occurs in the absence of air preserves the silage. Spoilage occurs when green forages are stored in the presence of air (Figure 22–20). The chopped forage must be tightly packed to remove the air.

FIGURE 22–20 Silos are air-tight structures used to preserve and store silage. *(Courtesy of USDA)*

Science Connection

FIBER GIANT AND SUPER FORAGE?

The object of extensive research for several decades, the bamboo-like kenaf plant originated in Africa, but promises to be an important cash crop in the United States. The U.S. Department of Agriculture has conducted extensive research on kenaf to identify possible uses. Many products have been identified. These include fibers for making newsprint; bast for carpet backing or mats for seeding yards and slopes; absorbent fiber for cat litter; and core fibers for cleaning up oil spills, with oil and fibers usable as fuel. Another promising product is furniture made from a lightweight fiber board. However, that's not all kenaf can do!

The same tropical plant that may provide backing for your dining room rug and newsprint for your morning paper could also help grow your dinner steak, scientists say. Kenaf, or *Hibiscus cannabinus,* is considered an important contender against Southeast Asian jute

Growing 12 to 15 feet tall, kenaf promises to be the super crop of the future for fiber and possibly for forage. *(Courtesy of USDA/ARS #K-2975-9)*

for fiber and rope. It is also being researched for its usefulness as a roughage and protein source for sheep and cattle. Research at the Agricultural Research Service Forage and Livestock unit at El Reno, Oklahoma, indicates that animals will eat the leaves and stems, and the digestibility of certain parts of the plant is about the same as that of alfalfa. Kenaf leaves contain 30 to 35 percent crude protein. This combination of high fiber and high protein has attracted interest in the plants as cubes or pellets for cattle. The leaves are of growing interest as a small-animal feed.

Kenaf grows rapidly and could produce two crops in a season. When cut, two or more shoots will spring from each stalk to produce new plants. The potential is to use the stalks for fiber and the tops for forage, or to grow two forage crops in one season. Being an annual, the crop is easy to establish and has few or no insect and disease pests. Production in the United States is increasing as research continues on kenaf as a forage crop.

INTERNET KEY WORDS:

forage kenaf

The Fermentation Process   Silage-making is a fermentation process wherein bacteria breakdown sugars and starches to produce acids, which preserve the forage. Silage-making and silage-keeping require that several conditions be met and maintained. First, the crop must be chopped into short pieces and have sufficient moisture to pack tightly enough to exclude most of the air. Second, the crop must contain a sufficient amount of sugars and starches for the fermentation process to occur. Third, the silo or silage container must be airtight, so air cannot reenter the silage after it has fermented.

What happens in the silo to change the taste, texture, and chemical content of the forage? If the forage being ensiled contains grain, such as corn, sorghum, wheat, barley, oats, rye, or other seeds, it will ferment when packed in an airtight container or silo with about 60 percent or more moisture. If the mixture does not have sufficient sugars and starches, or if the moisture content is not within the acceptable range, then grains or preservatives must be added to the forage for good silage.

If the moisture content is less than 60 percent and the silo is sufficiently tight, a low-moisture silage known as haylage may be made with the addition of certain preservatives.

Immediately after suitable forage has been chopped and packed in an airtight space, bacteria that live in the presence of oxygen in the air that fills the pores of the forage start to multiply. They feed on the sugars and starches in the forage and consume the oxygen in the process. They feed and multiply until the oxygen is consumed, so long as there are sufficient sugars and starches in the forage to keep them feeding. The bacteria make lactic and acetic acids, which smell good and give the forage a pleasant taste for livestock.

The frantic burst of bacterial activity in freshly packed forage causes the temperature of the forage to increase within hours after the process starts. This continues until the oxygen is consumed and the bacteria die. The resulting level of acid in the forage, now called silage, prevents any other bacterial action so long as oxygen-laden air does not reenter the silage. Without bacterial action, the silage cools down and is preserved until air enters the silage again. In practice, air may get into silage and cause spoilage around poorly fitted silo doors, holes left in the plastic film used to seal silos, or movement through poorly packed or excessively dry silage. Spoiled silage can be detected by the presence of mold or the foul odor of butyric acid, which forms when undesirable fermentation takes place.

Approved Practices for Forage Crops

Many crops are usable as forages. Forages may be used in the form of grazing of young plants by livestock, cut and dried as hay, or harvested and stored in silos as medium- or high-moisture haylage or silage. Some approved practices for the production of forages are as follows:

- Select a perennial species for long-term stands.
- Select a perennial or annual species for short-term stands.
- Buy certified seed of the selected variety.
- Obtain a current soil test.
- Lime and fertilize according to the soil test and desired production.
- Select a field with soil suitable for your forage choice.
- Select the proper tillage method to prepare a firm and smooth seedbed.
- Calibrate the drill for a proper plant population.
- Match the plant population to the soil-yield potential.
- Adjust the drill for proper depth of seed placement.
- Decide if forage may best be grown with a companion crop.
- Calibrate the sprayer for proper application of materials.
- Use selected herbicides for controlling problem weeds.
- Use insecticides as necessary.
- Decide the stage of maturity to harvest the crop.
- Make a yield check.
- Keep harvesting equipment in good repair to save downtime.
- Check the moisture content of harvested forage.
- Harvest the crop as soon as the moisture level permits.

- Store hay in a clean, dry structure.
- Store silage in an airtight structure.
- Obtain a forage analysis.
- Track market trends and habits.
- Market the crop at the optimum time.
- Keep accurate enterprise records.
- Summarize and analyze records.

The production of forages in the United States accounts for about half of all the land in agricultural use. Much of this land is unsuitable for production of other crops. With the production of hay ranking third only to corn and soybeans, forages are extremely important to U.S. agriculture. Because forages are the least expensive sources of nutrients for cattle, sheep, and horses, they are likely to remain essential for years to come.

STUDENT ACTIVITIES

1. Write the Terms to Know and their meanings in your notebook.
2. Put fresh-cut forage in each of two quart jars. Pack the forage tightly into the jars. Seal one jar and leave the other open to the air. Compare the contents of the two jars after 10 days.
3. Make a collection of as many different forage seeds as you can.
4. Write a report on a forage of economic importance in your area.
5. Make a collection of local forages.
6. Visit a machinery dealership to learn about the types of seedbed preparation, planting, and harvesting equipment for forages.
7. Make a bulletin board about forages.
8. Obtain samples of hay and silage and compare them for firmness of leaves and stems, odor, and color.
9. Ask your instructor to obtain samples of lactic acid (found in good silage) and butyric acid (found in spoiled silage) from the science laboratory and compare the odors produced by these acids. *Do not smell* the vapors of these acids directly from the container.
10. A cow eats about 90 lb feed each day and drinks 35 to 50 gallons of water. Hay makes up about 30 lb of its total daily feed intake. Determine how many pounds of hay a cow will eat in 1 year in a confined feedlot with no supplemental grazing.
11. Determine the cost for a ton of hay in your area by calling an area hay farmer. Determine how much it will cost for hay in the ration of one cow for 1 year, using the answer you calculated in Activity 10.

SELF EVALUATION

A. Multiple Choice

1. The type of mower that cuts with a scissors-type action is a _____ type mower.
   - a. rotary
   - b. flail
   - c. sickle-bar
   - d. vibrating

2. A silo is used to store
   - a. silage.
   - b. hay.
   - c. pasture.
   - d. cubes.

3. The most important legume hay in the United States is
   - a. clover.
   - b. lespedeza.
   - c. bird's-foot trefoil.
   - d. alfalfa.

4. A forage that grows in wet or poorly drained soil is
    a. reed canary grass.   c. fescue.
    b. timothy.   d. bird's-foot trefoil.

5. _____ is suitable for only pasture and lawns because it grows only 6 to 12 inches high.
    a. Dallis grass   c. Kentucky bluegrass
    b. Sweet clover   d. Orchard grass

6. Forage that is cut and allowed to dry to 40 to 55 percent moisture before storing is called
    a. silage.   c. hay.
    b. haylage.   d. cubes.

7. The harvesting of _____ requires the most labor of any of the forages.
    a. pasture   c. silage
    b. hay   d. haylage

8. The production of forages in the United States equals _____ percent of the land used for crops.
    a. 25   c. 50
    b. 35   d. 65

9. A warm-season grass is
    a. sweet clover.   c. reed canary grass.
    b. bromegrass.   d. Bermuda grass.

10. Hay is raked to
    a. allow even drying.   c. reduce its maturity.
    b. allow the hay to ferment.   d. remove leaves and weeds.

B. Matching

_____ **1.** Timothy   a. Number one grass hay crop
_____ **2.** Alfalfa   b. Excellent source of nectar
_____ **3.** Bromegrass   c. Grows very early in the spring
_____ **4.** Sweet clover   d. Number one forage crop
_____ **5.** Orchard grass   e. Popular in mixes with clover
_____ **6.** Lactic acid   f. Indicates silage is spoiled
_____ **7.** Butyric acid   g. Gives a good smell to silage

C. Completion

1. A disadvantage of _____ type mowers is that they tend to chop up the forage.

2. Sweet clover types include annual and _____.

3. Seedbeds for forages should be _____ than those for most other crops.

4. _____ is forage harvested by livestock.

5. Horizontal underground stems are called _____.

6. A _____ weed is prohibited or banned by state law.

7. Crops that are harvested for their vegetative growth are called _____.

8. Two methods of planting forages are _____ and broadcasting.

9. A _____ gathers dry forage, compresses it, and ties it with twine and wire.

10. The more mature the forage, the _____ its quality.

11. _____ converts sugars and starches to acids in the fermentation process.

12. Silage will spoil if _____ is permitted to reenter.

Ornamental Use of Plants

Ornamental Plants—An Indicator of Economic Health

When one considers the fastest growing segment of U.S. agriculture, the image of great fields of golden grain, corn "as high as an elephant's eye," or feedlots of cattle as far as the eye can see come to mind. Many would not envision greenhouses of flowers, golf courses with gorgeous turf, or suburban communities of beautifully landscaped homes. But flowers, potted plants, and landscape trees and shrubs are some of the hottest agricultural products today. They not only beautify our surroundings, but they also help purify the air.

A conservative estimate puts the worth of the floral and landscape industries in the United States at about $14 billion at the grower level in 2002, according to the U.S. Department of Agriculture's (USDA's) Economic Research Service. The value of production has grown at 5 percent per year since 1989. In 1992, flowers and house and nursery plants advanced into the sixth largest commodity group in the United States. In 21 states, this segment placed among the top five agricultural commodities.

People tend to take our decorative plants for granted and assume they are simply gifts from nature. This is far from the way it is, however. Years of painstaking research and development have gone into most ornamental plants, and research must continue on an ongoing basis to stay ahead of pests of ornamental plants and the public appetite for new decorations. For the most part, we seldom see plants from the wild—nearly all that are in common use are products of plant improvement efforts.

Consider the poinsettia, for example. Research and genetic improvements contributed greatly to this plant's phenomenal growth in commercial value during the last two decades. The poinsettia industry grew from a wholesale value of $38 million in 1976 to

$260 million by 2002. The poinsettia has become the United States's number-one potted plant, based on numbers sold, even though its annual sales period is only 6 weeks. About 68 million pots are marketed annually.

Most of the USDA's work on the poinsettia and many other ornamentals is centered at the National Arboretum in Washington, D.C., and the Agricultural Research Service's Florist and Nursery Crops Laboratory in nearby Beltsville, Maryland. Scientists there have developed many new ornamentals and are constantly improving others for commercial use. An ongoing project investigates the biochemical and physical bases for color in plants and our perception of it. Such information may help make development of new kinds of flowers easier and may provide a scientific basis for variety identification.

Applied research answers immediate needs of growers, such as solving pressing problems occurring in today's commercial ornamental enterprises. However, the long-range research goals are to stay well ahead of the current technology. From all indications, the ornamental use of plants will play an increasing role in the health and well-being of our citizens as we all pursue the great American Dream.

UNIT
23
Indoor Plants

MATERIALS LIST

- ✓ indoor plant reference books
- ✓ seed catalogs
- ✓ pencil, grid paper, and eraser
- ✓ indoor plants
- ✓ pots of various sizes
- ✓ Internet access

COMPETENCIES TO BE DEVELOPED

After studying this unit, you should be able to:

- identify plants that grow well indoors.
- select plants for various indoor uses.
- grow indoor foliage plants.
- grow indoor flowering plants.
- describe elements of design for indoor plantscapes.
- describe career opportunities in indoor plantscaping.

SUGGESTED CLASS ACTIVITIES

1. Go as a class on a field trip to a local nursery, floral shop, or the garden section of a large retail store. Observe the variety of plants that are available, and then ask the manager which plant is the most important income-producing product. Observe the tools and materials that are available for working with ornamental plants. Have each class member write a summary of the important things that were observed and discussed during the field trip.

2. Demonstrate an Internet search with a large computer monitor or in-focus machine. Search for such key words as *floriculture, research, statistics, decorative plants,* and *ornamental plants.* The class may suggest other key words that will identify sites for learning about the floral industry. Print and discuss key articles that are found during the Internet search.

3. Many indoor plants stop growing or begin to look unhealthy when they are kept in a container that is too small. Demonstrate the proper way to repot a plant. If the plant is a variety that can be divided easily into three or four individual plants, demonstrate this technique as well.

TERMS TO KNOW

floriculture	phototropism	texture	informal
succulent	relative humidity	accent	asymmetrical
foliage	rootbound	sequence	scale
variegated	plantscaping	balance	circulation
herb	basic color	formal	development limitations
foot-candle	accent color	symmetrical	

The world of indoor plants is a fascinating one. There are plants that have magnificent blooms, unusual shapes, fancy foliage, and fragrant smells. If selected and cared for properly, they can last for years and may even be passed from generation to generation. The industry of indoor plants is the floriculture part of ornamental horticulture. **Floriculture** involves the production and distribution of cut flowers, potted plants, greenery, and flowering herbaceous plants. Indoor plants present a challenge to the homeowner who wishes to invest the rooms of his or her home with living color. They also challenge those who decorate public areas (Figure 23–1).

FIGURE 23–1 Indoor plants add beauty and atmosphere to homes and public areas. *(Courtesy of Michael Dzaman)*

FIGURE 23–2 African violet *(Saintpaulia ionantha)*. *(Courtesy of Michael Dzaman)*

FIGURE 23–3 Fuchsia *(Fuchsia triphylla)*. *(Courtesy of Michael Dzaman)*

PLANTS THAT GROW INDOORS

Almost all plants can be grown indoors. There are plants, however, that favor indoor conditions. Many of the trees and shrubs do better when grown outdoors. Smaller and more succulent plants are best for indoor use. **Succulent** means having thick, fleshy leaves or stems that store moisture. There are a wide variety of shapes and sizes of indoor plants. These plants can be divided into two major groups—those that flower, and those that are grown only for their foliage. **Foliage** consists of stems and leaves.

Popular and Common Indoor Flowering Plants

African Violets   One of the most popular and common indoor flowering plants is the African violet *(Saintpaulia ionantha)*. This plant can be recognized by its small size and hairy leaves. The leaves are oval-shaped, dark green, and covered with soft, short hairs. The flowers contain four to five petals arranged in a clover pattern. They vary in color from deep purple to brilliant white (Figure 23–2).

Fuchsias   Fuchsias *(Fuchsia triphylla)* are plants with colorful flowers that cascade from the plant. Most have flowers that are two-tone pinks and reds. The foliage is dark green, and the leaves tend to be long and oval in shape, with a bronzy hint of color (Figure 23–3).

Gardenias   Particularly fragrant, flowering indoor plants, gardenias *(Gardenia jasminoides)* have deep-green shiny foliage and pure white flowers. The leaves are in clusters of three and are pointed (Figure 23–4).

Geraniums   One of the most versatile flowering indoor plants is the geranium *(Pelargonium zonale)*. These indoor plants are among the oldest. The leaves are a rounded, yellowish green with scalloped edges. The flower is borne on a stem and consists of many petals in a cluster shaped like a ball. The flower color ranges from the most popular red, to white and pink (Figure 23–5).

Impatiens   If an indoor plant with many blooms is desired, the impatiens *(Impatiens wallerana,* or *Impatiens sultanii)* is a good choice. The flowers are small and rounded, with five petals. One petal is shaped like

FIGURE 23–4 Gardenia *(Gardenia jasminoides)*. *(Courtesy of H. Edward Reiley)*

FIGURE 23–5 Geranium *(Pelargonium zonale)*. *(Courtesy of H. Edward Reiley)*

FIGURE 23–6 Impatiens *(Impatiens wallerana)*. *(Courtesy of H. Edward Reiley)*

FIGURE 23–7 Businesses and public areas are generally landscaped with foliage plants. *(Courtesy of Michael Dzaman)*

FIGURE 23–8 Boston fern *(Nephrolepis exaltata)*. *(Courtesy of H. Edward Reiley)*

a tube that protrudes from the underside of the flower. Flower colors include white, pink, salmon, coral, lavendar, purple, and red. The leaves are lance-shaped and have succulent stems (Figure 23–6).

Others   There are many other varieties of plants that can be grown indoors for their flowers. Some flowering plants are both indoor and outdoor plants, such as wax begonias, ageratums, verbenas, and petunias.

Popular and Common Indoor Foliage Plants

INTERNET KEY WORDS:

asparagus fern
emerald fern
foxtail fern

Indoor foliage plants can be divided into five groups for easy identification. These groups are ferns, indoor trees, vines, cacti and succulents, and others. A foliage plant is grown for the appearance of the leaves and stems (Figure 23–7).

Ferns   Ferns come in a variety of types. Some of the most popular indoor-type ferns are Boston fern *(Nephrolepis exaltata)* (Figure 23–8), asparagus fern *(Asparagus sprengeri;* Figure 23–9), maidenhair fern *(Adiantum capillusveneris)*, sword fern *(Nephrolepsis cordifolia)*, rabbit's-foot fern *(Davallia canariensis)*, and staghorn fern *(Platycerium biforcatum)*. Ferns are categorized by their long and often multicut leaves.

FIGURE 23–9 Asparagus fern (*Asparagus sprengeri*). (*Courtesy of H. Edward Reiley*)

INTERNET KEY WORDS:

trees, indoor

Most ferns are feathery in appearance. Some ferns are used extensively as greens in floral arrangements.

Indoor Trees An indoor tree can be an excellent accent to a room or an attractive addition to a patio or hallway. Trees can grow to be 6 to 7 feet tall. Some of the more popular indoor trees are Norfolk Island pine (*Araucaria excelsa*; Figure 23–10), fiddleleaf fig (*Ficus lyrata*), umbrella plant (*Schefflera actinophylla*; Figure 23–11), rubber plant (*Ficus elastica*; Figure 23–12), fragrant dracaena (*Dracaena fragrans*), weeping fig (*Ficus benjamina*; Figure 23–13), and croton (*Codiaeum variegatum*). Indoor trees vary in the type and size of their foliage, but all have woody-type stems.

Vines The vines are characterized by their habit of climbing or draping from the sides of the pot. Some of the more widely recognized vine-type

FIGURE 23–10 Norfolk Island pine (*Araucaria excelsa*). (*Courtesy of H. Edward Reiley*)

FIGURE 23–11 Umbrella plant (*Schefflera actinophylla*). (*Courtesy of Michael Dzaman*)

FIGURE 23–12 Rubber plant (*Ficus elastica*). (*Courtesy of Michael Dzaman*)

FIGURE 23–13 Weeping fig (*Ficus benjamina*). *(Courtesy of H. Edward Reiley)*

FIGURE 23–14 Philodendron (*Philodendron sp.*). *(Courtesy of H. Edward Reiley)*

FIGURE 23–15 Wandering Jew (*Tradescantia flumenencis*). *(Courtesy of Michael Dzaman)*

indoor plants are philodendrons (*Philodendron* sp.; Figure 23–14), Wandering Jew (*Tradescantia fluminensis* "Variegata"; Figure 23–15), grape ivy (*Rhoicissus rhomboidea*), and English ivies (*Hedera* sp.). These plants can be trained to climb up a piece of wood, around a planter box, or up a wall.

Cacti and Succulents Cacti and succulents are a unique group of indoor plants that originated in desert areas. They are among the easiest plants to grow indoors. These plants tend to hold water within their stems and leaves. Cacti and succulents survive dry heat, low humidity, and varying temperatures. Cacti usually have some type of prickly needles, whereas succulents do not. One of the most popular succulents is the jade plant (*Crassula arborescens*; Figure 23–16).

Specimen Plants There are numerous other types of indoor plants that do not fit into the categories of ferns, indoor trees, vines, and cacti and succulents. These indoor plants tend to have special characteristics or

FIGURE 23–16 Jade plant (*Crassula arborescens*). *(Courtesy of H. Edward Reiley)*

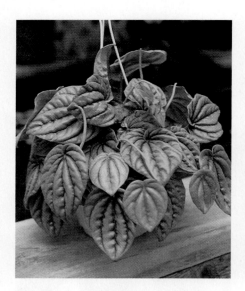

FIGURE 23–17 Peperomia *(Peperomia caperata). (Courtesy of Michael Dzaman)*

FIGURE 23–18 Passion plant *(Gynura aurantica). (Courtesy of H. Edward Reiley)*

FIGURE 23–19 Snake plant *(Sansevieria trifasciata). (Courtesy of H. Edward Reiley)*

features that people like to display, thus, they are called specimen plants. A few examples of more popular specimens are the spider plant *(Chlorophytum elatum vittatum)*, which shoots off baby plants or "spiders"; peperomia *(Peperomia caperata)*, which has pale pink to red stems with deeply grooved heart-shaped leaves (Figure 23–17); purple passion plant *(Gynura aurantica;* Figure 23–18), which has rich royal-purple leaves that are covered with velvet-type hair; bromeliads, one variety of which has a deep pink color at the center; and snake plant *(Sansevieria trifasciata)*, which has long, spikelike, thick leaves that are variegated with gold (Figure 23–19). **Variegated** means having streaks, marks, or patches of color.

SELECTING PLANTS FOR INDOOR USE

INTERNET ADDRESSES:

<http://odla.nu/garden_links/
houseplants.shtml>

<http://www.ianr.unl.edu/pubs/
horticulture/g837.htm>

There are two rules of thumb when selecting plants for use indoors. The first is to be selective when you purchase the plant. The second is to choose the right plant for the growing conditions available in the location where you want the plant to live. Careful attention to these two rules will greatly increase the chance for success with indoor plants.

Purchasing an Indoor Plant

When buying an indoor plant, look for one that appears to be healthy. Look closely at the plant for insects, being careful to check the undersides of the leaves where many insects hide or lay their eggs. Select plants with even green color, no yellowing leaves, and no spots or blotches on the leaves. Avoid plants that have spindly growth or appear wilted. If possible, purchase the plant during its growing season. Finally, look for new growth, such as leaf or flower buds.

Choosing the Right Plant

Before selecting an indoor plant, decide where the plant will be placed. Note particularly the light intensity and duration, as well as the temper-

Agri-Profile

CAREER AREAS:

**Plantscape Designer/
Plantscape Contractor/
Greenhouse and Plant
Technician/Florist**

Foliage plants, including small trees, have become popular decor for interior public areas such as shopping malls, office buildings, institutions, and housing complexes. Flowering and foliage plants have long been part of the interiors of attractive homes and have provided excellent hobbies for many people. Their extensive use in commercial settings is a relatively recent development that has stimulated many new job opportunities in ornamental horticulture.

The florist industry provides cut flowers and floral arrangements for all occasions. Interior plants are grown in greenhouses, nurseries, and fields and become part of floral and interior arrangements and plantscapes.

Youth in suburban settings may obtain jobs raising and caring for plants, installing interior plantscapes, rotating plants between growing areas and display areas, and contracting to maintain interior plantscapes. Early experiences frequently lead to life-long careers.

With more experience and education, one may become a plantscape designer, business owner–operator, grower, wholesaler, plant doctor, or extension specialist. Ornamental horticulture specialists are hired by universities, research institutes, and business firms to develop improved plants and horticultural practices. The florist industry provides cut flowers and floral arrangements for all occasions and provides jobs in design, arranging, care, and delivery.

ature of the area where the plant will be placed. Each species of indoor plants has specific conditions for optimum growth. The plant should match the location for best results.

Light   The amount of light available for a plant is a factor that is difficult to control. A shadow test can help determine this. To conduct this test, hold a piece of paper up to the light (window or lamp) and note the shadow it makes. A sharp shadow means you have bright or good light. However, if there is barely a shadow visible, the light is dim or poor. It is important to know how much light the plant needs. If a plant needs direct or full sun, then exposure of the sun is needed for at least half of the daylight hours. When a plant needs indirect or partial sun, the light should be filtered through a curtain or slats. Plants that fit into this category are Chinese evergreen (*Aglaonema modestum*; Figure 23–20) and dracaena (*Dracaena fragrans*; Figure 23–21). Even for plants preferring no direct sunlight, the room should be bright and well lit. Plants that need shade should be kept in a well-shaded part of the room. In summary, too much or too little light can greatly affect the health of the plant.

Temperature   Plants are adapted to specific temperature ranges, and they do not do well when these temperatures are exceeded. There are

FIGURE 23–20 Chinese evergreen (*Aglaonema modestum*). (*Courtesy of H. Edward Reiley*)

FIGURE 23–21 Dracaena *(Dracaena fragrans). (Courtesy of H. Edward Reiley)*

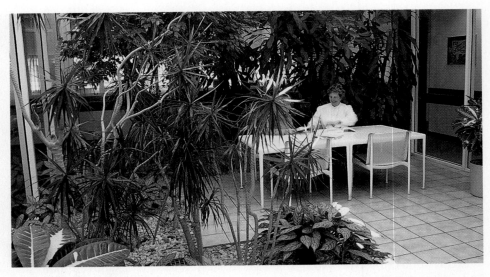

FIGURE 23–22 Decorative plants make excellent area dividers. *(Courtesy of Michael Dzaman)*

three temperature categories in which indoor plants are grouped. These are cool, moderate, and warm. The cool temperature range is 50° to 60° F, with temperatures not falling to less than 45° F. The moderate temperature range is from 60° to 70° F, with a minimum of 50° F. The warm temperature range is from 70° to 80° F, with a minimum limit of 60° F.

USES OF INDOOR PLANTS

The uses of indoor plants are varied. Plants can be used as room dividers or to brighten up a dull spot in the kitchen or bathroom. Plants can be used to divide a room into specific living areas. They may be placed in containers with special watering devices, or several plants may be grouped together to form a natural barrier. Containers are available in a number of sizes and shapes, some with castors on the bottom so they can be moved easily. Indoor trees and climbing plants are best used as living-area dividers (Figure 23–22).

Specimen or showy plants, such as the Boston fern or date palm (*Phoenix dactylifera*), are good for brightening up dull areas (Figure 23–23). Plants may be placed on ornate plant stands or in decorative pots. A series of wall shelves may also be used to display interior plants. Plants on shelves should be compact, and perhaps trailing, for a more dramatic effect.

Hanging baskets filled with indoor plants are useful when space is at a premium. Because baskets can be easily suspended from the ceiling, they can utilize the unused space above head level. Care should be used when selecting plant containers so that water does not drip on the furniture or the floor. Select baskets that are designed with drip trays attached to the bases.

Some interesting areas for hanging baskets are stairwells, offices, and foyers. One other element that is unique to hanging baskets is the hook securing the basket to the ceiling. It should be secured firmly enough to hold the basket when the soil is wet and when the plant is being cared for (Figure 23–24).

FIGURE 23–23 Date palm *(Phoenix dactylifera). (Courtesy of H. Edward Reiley)*

FIGURE 23–24 Hanging baskets are versatile and attractive. *(Courtesy of Michael Dzaman)*

FIGURE 23–25 The high humidity of bathrooms can provide an excellent environment for plants if light is adequate. *(Courtesy of Elmer Cooper)*

FIGURE 23–26 Aloe *(Aloe variegata)*. *(Courtesy of H. Edward Reiley)*

Bathrooms are excellent places for indoor plants. These rooms are humid and warm, which provides a good atmosphere for plants. They can be placed on windowsills or around the tub. Shelves can also be installed in a bathroom to display indoor plants. Plants will add a touch of color to a bathroom, especially today as the bathroom is expanding to hold a hot tub or jacuzzi. Philodendrons are excellent in bathrooms (Figure 23–25).

The kitchen is another room of the home where plants can add a beautiful touch. Refrigerators, other appliances, and kitchen windowsills all make good places for plants that prefer cooler temperatures. Most of the plants in the kitchen should be relatively small, as space is usually at a premium. Several plants suitable for the kitchen are aloes (*Aloe variegata*; Figure 23–26), maidenhair ferns, peperomias, and even a variety of herbs. **Herbs** are plants kept for aroma, medicine, or seasoning.

GROWING INDOOR PLANTS

Growing indoor plants is much like growing outdoor plants. Consideration needs to be given to the environment in both cases. Tomatoes will not grow and produce in hot, dry environments, nor will sweet corn grow in the shade of a tree. Both need full sun, but cannot survive in drought conditions or extreme temperatures. Aspects of the environment that must be considered are light, temperature, water, drainage, and feeding conditions. Although indoor plants are not harvested like tomatoes or sweet corn, they need to be similarly maintained. Some aspects of maintenance that need to be addressed are grooming, repotting, and propagation.

Light

Light is one of the most crucial factors to consider when growing indoor plants. Light is measured in foot-candles. A **foot-candle** is the amount

Science Profile

PHOTOTROPISM IN PLANTS

The tendency of plant parts to move or turn toward or away from a light source is known as phototropism. This phenomenon occurs most often when the leaves of a plant turn toward the sun to maximize the sun's intensity on the leaf surface. It is a survival mechanism that allows photosynthesis to occur in the leaf at the maximum rate. It is also observed in flowers that turn to face the sun each morning, following the track of the sun across the sky until it sets in the evening. Sometimes the petals of flowers close at night and open the next morning when the sun comes up. Each of these adaptations by plants to the sun or to other light sources is an example of phototropism.

INTERNET KEY WORDS:

plants, phototropism

of light found one foot from a burning standard candle, known as a candela. Two aspects of light that need to be determined are intensity and duration. The intensity of light varies at different times of the day and through different windows. A plant near a window with a southern exposure receives more intense sun, and for a longer period, than does a plant in a window with a northern exposure. Pulling shades across a window can reduce the intensity of light, and a tree outside a window can reduce both the intensity and duration of light that enters the window. The season of the year also affects the light's intensity and duration. Summer sun is much more intense than winter sun, because the angle of the sun to the Earth is near perpendicular during the summer.

Too little light will cause some indoor plants to grow tall and spindly and lose leaves because the long, thin, weak stems can no longer support the plants because of limited photosynthesis. Conversely, too much light causes indoor plants to wilt and lose their vibrant, green colors. The youngest leaves on plants are affected first by unfavorable conditions. Another item concerning light is the tendency of plants to grow toward the strongest source of light. If you notice that a plant is leaning toward the light, you may need to rotate the plant periodically so that it maintains a balanced shape (Figure 23–27).

There are five basic light categories for indoor plants. These are full sun, some direct sun, bright indirect light, partial shade, and shade. Full sun is a location that receives at least 5 hours of direct sun a day. This amount of sun can be found in areas that have a southern exposure. Some direct sun occurs in areas that are brightly lit but receive less than 5 hours of direct sun a day. These areas are windows facing east and west.

Bright indirect light describes areas that receive a considerable amount of light but no direct sun. An example is an area 5 feet away from a window that receives full sunlight. Partial shade refers to areas that receive indirect light of various intensities and durations. Areas 5 to 8 feet away from windows that receive direct sun are partial-shade areas. Shade refers to poorly lit sections away from windows that receive direct sun. Few plants can survive this low-intensity light.

FIGURE 23–27 Plants generally must be moved or rotated periodically to provide correct lighting to all parts of the plant. *(Courtesy of Michael Dzaman)*

FIGURE 23-28 Plants need to receive water on a regular basis, but overwatering is a frequent problem of the inexperienced owner. *(Courtesy of Michael Dzaman)*

Temperature

Plants survive best at constant temperatures. Temperatures that go up and down are not ideal for plant growth. Temperature interacts with light, humidity, and air circulation. It is best to maintain temperatures in a range from 60° to 68° F for optimum indoor plant growth. As temperatures increase beyond this range, the air gets hotter, and the available moisture in the air decreases. Although the thermostat in a house reads one temperature, each room usually varies as much as 5 degrees in one direction or the other. Because plants have various temperature preferences, it is advisable to place plants in rooms that match their specific temperature requirements.

Water

Water is an essential ingredient for growth of any living organism. The amount of water needed by indoor plants is usually not as much as one would think. The most likely problem with unhealthy indoor plants is too much water. More plants die from overwatering than from any other cause. As with other environmental factors such as light and temperature, each plant varies in its requirements for water. Unless you know the particular needs of the plant, it is best to water when the soil around the plant is a little dry. Most plants do best if allowed to dry out between waterings. When indoor plants are watered properly, the roots remain more active than if the soil becomes waterlogged or excessively wet.

There are numerous ways to water plants. The basic rule about watering is to use water that is neither hot nor cold. Water should be tepid, or a moderate temperature. One way to water is to soak the pot in a bucket of warm water for half an hour, remove the pot, and drain it. A second way is to pour the water on top of the soil slowly, filling the pot to the top with water. Allow it to absorb the water until the excess drains from the hole in the bottom of the pot. Do not rewater until the soil first becomes dry to the touch (Figure 23–28).

Besides water in the media around their roots, plants also need moisture in the air. The moisture content in the air is referred to as humidity. It is expressed as **relative humidity** (a relationship between the amount of water vapor in the air compared with the maximum moisture the air will hold at a given temperature). Almost all indoor plants prefer 50 percent relative humidity. The simplest method for humidifying the air around plants is to set pots in trays filled with gravel and add water to just cover the gravel. Misting is a good shortcut way to add humidity, but it should be done several times a day to be effective. The humidity around plants can also be increased by grouping them together. However, plants should have enough space around them to allow for adequate air circulation.

Drainage

Good drainage is achieved by using pots with porous materials in the bottom and drainage holes in the pot. Good drainage is essential for indoor plants. Adding coarse material such as sand or perlite to the soil will improve soil drainage. The addition of gravel or bits of broken clay-pot material to the bottom of a pot before adding the soil is a substantial aid to good drainage.

INTERNET ADDRESS:

<http://wwwagcomm.ads.orst.edu/agcomwebfile/garden/houseplant/potted.html>

FIGURE 23–29 Plants need to be fertilized with carefully selected materials. *(Courtesy of Michael Dzaman).*

Fertilizing

Plants need nutrients on a regular basis for good health. A balanced fertilizer, such as a 5-5-5, should be applied at regular intervals. Fertilizers with a greater proportion of nitrogen than of phosphorus and potash are often used to keep foliage plants green and healthy looking. Plants should be fertilized at 2- to 6-week intervals, depending on the type of fertilizing material. Slow-release–type fertilizers dissolve slowly and release nutrients evenly over weeks or months. In contrast, liquid fertilizers suitable for foliar application are used by the plant within a few days. They may also be applied at weekly or biweekly intervals (Figure 23–29).

Plants should be fertilized while actively growing. Fertilization should be discontinued when the plant is dormant, or in a resting stage. Flowering plants need more fertilizer. Addition of fertilizer once every 2 weeks is recommended from the time flower buds first appear until the plants stop blooming. Avoid fertilizing plants when the soil is excessively dry. Under such conditions, the fertilizer solution is likely to be too highly concentrated and may cause burned leaves or roots.

Grooming

Grooming plants is important even with the best combination of light, temperature, humidity, water, and drainage. Grooming should be done weekly. This task includes removing wilted or withering leaves, flowers, and stems using sharp scissors or shears (Figure 23–30). The plant should be observed to determine if it is getting spindly or thin. If so, pinch out new growth to force the plant to branch out. To avoid crooked stems, stake plants when they are young. Climbing or trailing plants need a stake made of bark to enable them to climb and cling to the stake.

Once a week, the plant should be dusted. Dust accumulates on the leaves and blocks the stoma so that the plant cannot breathe or transpire as well as it should. The soil around the base of the plant should be loosened with a fork or small spade to allow air to enter the soil and water to percolate through. To help control insects, mist the infested plants with a diluted solution of mild dishwashing soap and water.

FIGURE 23–30 Plants need to be examined and groomed weekly. *(Courtesy of Michael Dzaman)*

FIGURE 23–31 Repotting at appropriate times prevents the plant from becoming root-bound and promotes good plant health. *(Courtesy of Michael Dzaman)*

Repotting

In time, plants develop root systems that are restricted by the pot or container. When this happens the plant is said to be **rootbound**. There is no place for the roots to continue to grow, so repotting is necessary for plant health. In general, repotting is done in the spring or the fall. A good rule of thumb for determining pot size is to use one with a diameter at the top of the rim equal to one-third to one-half the height of the plant. This rule does not apply to plants whose growth habit is tall and slender, as opposed to a balanced top growth. When repotting, it is not desirable to move an established plant to a new pot that is more than 2 inches wider than the original pot. Excessively large pots result in wasted soil, water, and nutrients (Figure 23–31).

Flowering plants are best repotted after the flowers have faded. During repotting, check the roots for insects and root damage. Remove any roots that look or feel unhealthy.

Propagation

Propagating most indoor plants is relatively easy. The method chosen depends on the type of plant (whether it is herbaceous or woody, flowering or foliar). Both sexual and asexual methods of propagation are used.

In sexual propagation, seeds may be started in containers with a good potting soil, plenty of moisture, and adequate air circulation. Keeping pots in warmer areas of the house increases the speed of seed germination. Covering the pots with glass or plastic held up by stakes will also aid in the germination process.

Some popular methods of asexual propagation of indoor plants are leaf and stem cuttings, removal of plantlets from parent plants, and air layering. Each of these procedures results in the multiplication of the plant.

Flowering Plants

Indoor flowering plants may take some extra, special care to ensure blossoming. These plants are more sensitive to the availability of light,

Hot Topics in Agriscience

TRANSGENIC CHRYSANTHEMUM

A new floral product that shows early commercial promise is a white variety of chrysanthemum developed from a pink variety. White "mums" are preferred over pink mums by commercial florists because they sell better. DNA Plant Technology, Cinnaminson, New Jersey, produced the new flower for commercialization in Holland. The new mum has long life as a cut flower, and it flowers well under artificial light. A gene that is responsible for an enzyme that blocks pink coloring was transferred from another variety of mum. Because white is an absence of color, once the pink coloring was blocked by the enzyme, the flowers that resulted were white.

so some artificial lighting may be needed. They are also more sensitive to temperature changes. Flowering plants have particular seasons in which they flower. It is important to know when to expect the plants to bloom. Some examples of plants with specific bloom times are the Christmas cactus (*Schlumbergera x buckleyi*), which is expected to bloom between October and late January, and the Easter cactus (*Rhipsalidopsis gaertneri*), which blooms in April or May. These two plants are similar in their leaf types and flowers, but they bloom in opposite seasons of the year. Indoor plants may flower in the winter, spring, summer, or year-round.

Foliage Plants

There is such a wide variety of foliage plants grown in homes and offices that specific instructions for their care compose entire books. However, the correct management of light is probably the greatest single factor for growing foliage plants. Light is the source of energy for the process of photosynthesis, wherein the leaves produce sugars and starches to feed all parts of the plant. Different foliage plants have different light requirements. Therefore, it is recommended that a good reference book on indoor plants be used to determine the particular requirements for any given plant. Regardless of the amount and duration of light, it is desirable to rotate plants so that each side receives the same amount of light over time.

INTERIOR LANDSCAPING OR PLANTSCAPING

Plantscaping is the design and arrangement of plants and structures in indoor areas. This design and arrangement is an art. Interior plantscaping is an activity that is fun. However, it should be approached seriously, just as the arrangement of furniture or other interior decorations must be done carefully. The indoor plants should be used to complement people-oriented spaces. The more creative you are in designing, the more distinctive indoor plantscapes will be. There is no right or wrong way to design with indoor plants, but there are elements that enhance the interior plantscaping process.

Design

Before beginning a plantscape design, the designer must know the purpose or intent of the plants. Are the plants to be used as a space divider?

Science Connection

LIKE THE SEASON'S FIRST SNOWFALL

With adults not much larger than the head of a pin, whiteflies suck the life from more than 600 plants including fruits, vegetables, flowers, shrubs, and field crops. Control of and damage by the pest run an annual tab in the millions of dollars. Greenhouse growers, nurseries, and retail outlets wage a continuous battle against whiteflies in an effort to save crops from damage and infestations that would prevent the sale of these products. Similarly, homeowners and interior plantscapers must be on a constant vigil, lest unobserved eggs hatch into an infestation of damaging proportions. Like the season's first snowfall, the eggs can suddenly appear as white specs all over the plant. When the plant is moved or the insects disturbed, the masses of adults can take flight and create the appearance of a snowstorm in the air!

A serious threat to ornamentals and nonornamentals alike, whiteflies settle on plants like the season's first snow. *(Courtesy of USDA/ARS #K-4853-3)*

First observed in Florida in 1986, an especially viral strain of the whitefly spread rapidly from Florida to California. It is challenging scientists and growers alike, as it spreads nearly unchecked by either pesticides or natural enemies. Around the country, researchers are showing the textile industry how to wash off the sticky goo the insects leave on cotton fiber. They are also uncovering ways to deal with plant viruses left in the wake of the troublesome, sucking insects. They are testing legions of environmentally friendly controls such as insect parasites, predators, fungi, plant extracts, and dish detergent and vegetable oil mixtures.

Scientists have identified more than 30 predators and 25 parasites. So far, the big-eyed beetle is the most promising. It secretes a sticky substance from its mouth to stick the whitefly to the leaf, then leisurely eats the fly. Scientists have released some big-eyed beetles on a pilot basis. Another tiny but promising insect that has been identified devours an estimated 10,000 whitefly eggs or 700 nymphs in its 6- to 9-week lifetime. However, not enough is yet known about this insect to release it. Similarly, a number of parasitic wasps are under examination. One of the more promising spray materials is a biosoap, which provides some control over whiteflies, as well as other damaging insects.

INTERNET KEY WORDS:

control whitefly

Are they being used to accent existing furnishings? Will they fill empty space? Answers to these questions will result in an organized design, rather than a happenstance.

Another question to consider is the function or functions of the plants. Are they to create a specific shape, emphasize a specific area, or support a specific architectural feature of the room? If the function of the plantscape is not considered, the end result is not likely to be successful.

When selecting plants for a plantscape, consider the physical characteristics of color, form, and texture. These are determined in concert with the perceptual characteristics of accent, sequence, balance, and scale. These characteristics are the basic tools a designer uses to create the interior plantscape (Figure 23–32).

FIGURE 23–32 Interior plantscaping is an important element in shopping malls and other public areas. *(Courtesy of Michael Dzaman)*

INTERNET KEY WORDS:

interior plantscape
design principles

Color   Color is the most important physical characteristic. It can influence emotions, create specific feelings, and add beauty and harmony to the environment. There are two types of color that must be considered by the designer in creating an interior plantscape. First is the basic, or background, color. **Basic color** is the color of the walls, ceiling, and floor. These colors should influence the selection of the flowering characteristic and foliage of plant material. The second type of color is accent. **Accent color** is the color of the plants or other attention-getting objects.

Form   Form refers to shape. There are different forms that plants possess naturally. The most common shapes are round; oval; weeping, or drooping; upright; spiky; and spreading, or horizontal. These shapes can be used individually or grouped to form an artificially sculptured shape and form. Each plant shape or cluster shape adds its own particular feature to the design.

Texture   Plant **texture** refers to the visual or surface quality of the plant or plants. Texture is influenced by the arrangement and size of leaves, stems, and branches. It is described in terms of coarseness or fineness, roughness or smoothness, heaviness or lightness, and thickness or thinness. Coarse-textured plants, such as fiddleleaf fig or prickly pear cactus, should be used in large spaces, whereas a maidenhair fern should be placed in small spaces, such as on shelves.

Accent and Sequence   **Accent** means a distinctive feature or quality. An accent captures the attention of the viewer. It has a dramatic effect on the visual appearance of the room or part of the room. You can create an accent with the use of color, form, or texture. It can also be created through the use of a sequence. **Sequence** refers to a related or continuous series. Sequence is created with plant material by repeating the same color, texture, or form. The overuse of accents, however, will detract from their function to capture the attention of the viewer.

Balance and Scale   **Balance** is the state of equality and calm between items in a design. In a design, two types of balance are used: formal, or symmetrical; and informal, or asymmetrical. **Formal,** or **symmetrical,** balance occurs when the items are equal in number, size, or texture on both sides of the center of the design. **Informal,** or **asymmetrical,** balance occurs when the items are not equal in number, size, or texture on the two sides of the center. **Scale** refers to the size of items. A large patio with African violets as accents would be out of scale. Plant materials need to complement the size and, therefore, the scale of the room. A better selection for a patio would be Norfolk Island pines and dracaenas.

A Design Process

There are various types of design processes. One that is good to use has three phases. The three-phase process emphasizes the how and why of using plants for the particular space.

Phase One The first phase of the process is preplanning, which includes three steps. The first step is to develop the design objectives by determining the purpose of the plantscape. The second step is to determine the space capacities, which include the habitat and the circulation of people. It is best to know how people will move in the area, referred to as **circulation,** and the amounts of light, temperature, and humidity in the room, referred to as habitat. The third step is to determine the development limitations. **Development limitations** include the amount of money available; room and space characteristics; and habitat limitations created by light, temperature, people, pets, and other factors.

Phase Two The second phase is developing the plan. The plan will include the basic design and arrangement of plants and materials. The physical and perceptual characteristics of the plants are important in this phase. Tentative selection of plants with visual placement in the room is part of this step. The planning is done on paper, and many alternatives are examined. The final plan is drawn up with all the plants identified.

Phase Three Implementation is the third phase of the interior plantscaping process, where the design comes to life. There are three steps to this phase. The first step is the preparation of the documents. If there is any construction involved, such as platforms, decks, or planter boxes, drawings for these objects must be done. Drawings and specifications for the installation and maintenance of water lines and fixtures, electrical devices, and the plants should also be listed at this point. This step is important if you are designing for someone other than yourself. The second step in this phase is the installation of all physical modifications. Selection of plant containers and the planting of the indoor plants happens in this phase. The plants are actually put in place. The final step is evaluation of the project. A thorough look at the plants, their containers, and their position in the room or area must be done. If all looks balanced, no changes need to be made. If not, minor adjustments may be made. The interior plantscape is now complete.

Plantscape Maintenance

Maintenance of the plantscape with attention to light, moisture, humidity, and grooming is important. These factors are discussed in preceding parts of this unit. A major segment of the plantscaping picture is the nursery and greenhouse industry that provides the plants (Figure 23–33).

FIGURE 23–33 Greenhouses, nurseries, florists, and suppliers are needed to provide a constant supply of plant materials. *(Courtesy of Michael Dzaman)*

CAREERS IN INDOOR PLANTSCAPING

Interior plantscaping is a career field within the large industry of horticulture. Horticulture is the study of plants, especially garden crops, both indoor and outdoor, for human consumption, for aesthetic purposes, or for medicinal purposes. Interior plantscaping is a part of floriculture, which is a division of horticulture. Plantscaping is both a science and an art. It is a career area that allows for creativity and use of scientific knowledge and technology.

The opportunities for a career in indoor plantscaping are numerous. An individual may be involved in growing indoor plants, designing interior plantscapes, installing plantscapes, maintaining plantscapes, or selling or servicing plantscape materials. The plantscapes can be small-scale within a residence or home, or large-scale in public areas such as shopping malls or office buildings. This area of horticulture offers many individuals the chance to be entrepreneurs with relatively low investments. The education required for a career in indoor plantscaping varies. Persons with high school diplomas, two-year college degrees, or four-year college degrees can be successful in this area.

STUDENT ACTIVITIES

1. Write the Terms to Know and their meanings in your notebook.
2. Take an inventory of species, type, and number of indoor plants in your home.
3. Select indoor plants from reference books and seed or plant catalogs that are suitable for your home.
4. Develop a plan for an interior plantscape of a room in your home, the school office, or another location.
5. Repot an indoor plant.
6. Propagate indoor plants using one or more methods.
7. Perform a shadow test at several windows in your home or classroom to determine the intensity of the light available.
8. Prepare hanging baskets for use in the home or office.
9. Check with a local greenhouse or flower shop for any discarded plants you could try to save. Try to rehabilitate the plants using the knowledge you gained from this unit about fertilizer, watering, and light requirements. Donate any nice-looking plants to a nursing home, school, or community organization.

SELF EVALUATION

A. Multiple Choice

1. One of the most popular flowering indoor plants is the
   a. spider plant.
   b. petunia.
   c. African violet.
   d. Norfolk Island pine.

2. An example of an indoor tree is a
   a. philodendron.
   b. cactus.
   c. fiddleleaf fig.
   d. zebra plant.

3. When purchasing an indoor plant, check the underside of the leaves for
   a. powdery mildew.
   b. insects.
   c. price tags.
   d. brown spots.

4. The moderate temperature range for indoor plants is
   a. 60° to 70° F.
   b. 45° to 55° F.
   c. 90° to 100° F.
   d. 35° to 45° F.

5. If an indoor plant needs direct or full sun, then the plant will need sun for at least
   a. the entire day.
   b. 1 hour.
   c. a fourth of the daylight hours.
   d. half of the daylight hours.

6. The place in a home not usually considered an ideal location for an indoor plant is the
   a. bathroom.
   b. basement.
   c. kitchen.
   d. living room.

7. Which of the following is not considered an environmental factor when growing indoor plants?
   a. weather
   b. light
   c. temperature
   d. humidity

8. When watering plants, the ideal temperature for the water is
   a. ice cold.
   b. hot.
   c. tepid.
   d. cold.

9. Which of the following is not a physical characteristic of an indoor plant?
   a. texture
   b. form
   c. color
   d. balance

10. Interior plantscaping is a career area in which area of horticulture?
    a. agronomy
    b. floriculture
    c. forestry
    d. arboriculture

B. Matching

_____	**1.** *Ficus lyrata*	a. A fern-type foliage plant
_____	**2.** Gardenia	b. Has unusually colorful leaves
_____	**3.** Croton	c. A succulent-type foliage plant
_____	**4.** Wandering Jew	d. A fragrant, flowering indoor plant
_____	**5.** Jade plant	e. A tree-type indoor plant
_____	**6.** *Aspargus sprengeri*	f. A vine-type foliage plant

C. Completion

1. The two environmental factors that are most important to the survival of indoor plants are _____ and _____.

2. An indoor plant that is a tree can be identified by its usually _____ stem.

3. The bathroom is a great place for indoor plants because of the _____ usually found in a bathroom.

4. Indoor plants survive best at a _____ temperature.

5. More indoor plants die of _____ than any other cause.

6. Adding a coarse material such as sand to the soil will improve an indoor plant's _____.

7. A 5-5-5 fertilizer is an example of a _____ fertilizer.

8. When repotting an indoor plant, do not move a plant to a pot that is more than _____ wider than its original pot.

Turfgrass Use and Management

OBJECTIVE

To understand growth and development of turfgrasses and the establishment and cultural practices involved in managing these plants.

MATERIALS LIST

✓ writing materials

✓ samples of grass seed

✓ various types of turf samples

✓ brochures from lawn-care services

✓ lawn equipment catalogs

✓ Internet access

COMPETENCIES TO BE DEVELOPED

After studying this unit, you should be able to:

- identify and describe careers available in the turfgrass industry.
- identify turfgrass plant parts.
- select turfgrass species for various purposes and locations.
- state the basic cultural practices for turfgrass production and maintenance.
- list the basic steps for turfgrass establishment.

SUGGESTED CLASS ACTIVITIES

1. Establish some turfgrass demonstration plots at your school. Demonstrate the effects of different management practices, such as height of mowing and differences in fertilizer applications, on the various plots. This is a long-term project that can be extended for several years.

2. Visit a turf farm or invite a local turfgrass farmer to speak to the class about the turfgrass business. Include these topics of discussion: (1) preparing the soil for planting, (2) planting methods, (3) establishing a good stand of grass, (4) weed and pest control, (5) harvesting practices, and (6) marketing the turfgrass product.

3. Find a small area in the school yard or a nearby area that either has no grass or the grass is in poor condition. Remove any unhealthy grass that may be present. Next, break up the soil a little with a rake and evenly spread grass seed. Cover the seed with straw mulch. Protect the plot with a tape barrier or fence. Take care to water the plot daily with a light application of water until new turf is established.

TERMS TO KNOW

turfgrass	intravaginal growth	fine-texture	seed blend
playability level	seed culm	coarse-texture	seed mixture
putting green	inflorescence	bunch-type grass	seed certification programs
seminal root	induction	utility-type grass	sodding
warm-season turfgrass	sheath	fungal endophyte	turf
cool-season turfgrass	blade	transition zone	sprigging
crown	ligule	thatch	stolonizing
extravaginal growth	collar	fertilizer analysis	plugging
vegetative spreading	recuperative potential	syringing	

FIGURE 24–1 Sports turf requires maintenance practices that provide an acceptable level of playability.

THE TURFGRASS INDUSTRY

It has been estimated that the value of the turfgrass industry in the United States is $25 billion. This industry is large and diverse. Career opportunities in this field expanded rapidly in the last decade and are projected to continue growing in the years ahead. Turfgrass includes all of the grasses that are mowed frequently to maintain a short and even appearance.

Golf course superintendents and other athletic-field managers maintain turfgrass at a certain level of playability. Playability level means suitability for the intended use. It is determined by the type of sporting or recreational event. For example, putting greens are areas used for playing golf, and the turfgrass is very short. These areas are maintained to provide a surface for consistent, yet adequate, putting speeds (Figure 24–1). A football field may be managed to offer secure footing and sufficient turfgrass resiliency for player safety.

Lawn-care services offer many job positions and are one of the largest employers within the industry. Job opportunities include lawn-care specialist, branch manager, owner-operator, and others.

The maintenance of turfgrasses at large government, apartment, university, commercial, and private complexes requires personnel trained in turfgrass management.

Sod-production farms and landscaping businesses are involved in the establishment and installation of turfgrass for lawns (Figure 24–2). Turfgrass specialists are also needed for these segments of the industry (Figure 24–3).

The turfgrass industry supports a substantial sales force. Companies that produce or distribute seed, fertilizer, pesticides, and turfgrass equipment require an extensive support and sales staff (Figure 24–4).

Federal and state governments and private companies hire turf specialists and scientists with advanced degrees. Career opportunities in these areas offer challenging positions in research, teaching, and extension.

FIGURE 24–2 Turfgrass is widely used to establish new lawns that create beautiful and pleasing environments around homes, businesses, and public buildings. *(Courtesy of Elmer Cooper)*

FIGURE 24–3 Turfgrass is grown in large fields from which it is harvested for the purpose of creating "instant lawns." *(Courtesy USDA)*

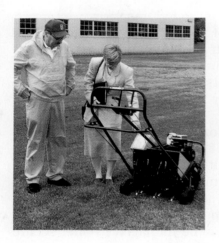

FIGURE 24–4 Equipment demonstration by a sales representative.

INTERNET KEY WORDS:

turfgrass plant anatomy

TURFGRASS GROWTH AND DEVELOPMENT

Turfgrasses are plants grouped into the Poaceae family. These grasses differ from other grass plants because they can withstand mowing at low heights. They can also tolerate vehicle and foot traffic. Turfgrasses are frequently used to hold the soil and as ornamental plants. These traits have made turfgrasses the most widely used ornamental crop in the United States. To properly maintain turfgrasses, an understanding of their growth and development is required.

THE TURFGRASS PLANT

The grass plant can be divided into two broad areas known as the root and shoot systems. The root system consists of adventitious and seminal roots. The shoot system includes the stem and leaves of the plant.

Root System

The **seminal roots** develop from the seed during seed germination. They initially anchor the seed into the soil. The seminal root system will be active for 6 to 8 weeks. The adventitious roots develop from the nodes of stem tissue. They usually compose the entire root system of a mature turfgrass stand.

Turfgrass roots are multibranching and fibrous. They are responsible for nutrient and water absorption. They also prevent soil erosion by effectively stabilizing and anchoring the plant into the soil.

Seasonal change in root growth is dependent on soil temperature and moisture. Active root growth for warm-season turfgrasses occurs in the summer. A **warm-season turfgrass** is one of a group of grasses adapted to the southern region of the United States. A **cool-season turfgrass** is a plant adapted to the northern region of the United States. These turfgrasses have active root growth in the fall and early spring. Optimum growth occurs at temperatures from 60° to 75° F.

Agri-Profile

CAREER AREAS:

Turfgrass Grower/Groundskeeper/ Landscape Maintenance Technician

In the last three decades, turfgrass production and management have become big business in the United States. In one eastern state, turfgrass recently became the number one crop based on acres covered. Starting in the 1950s, growth and development of golf courses stimulated the turfgrass industry with high-paying salaries for golf course superintendents and other turfgrass specialists.

Currently, career opportunities in turfgrass production, management, service, supervision, research, and consultation are extensive. Turfgrass technicians and specialists generally work in attractive and appealing surroundings. Many work outdoors in sunny weather and indoors when the weather is bad. In many localities, salaries have become quite attractive even for laborers.

Educational programs in turfgrass are available in high schools, technical schools, colleges, and universities. With the movement toward urbanization, interest in open spaces, concern for the environment, and increasing population, the outlook for careers in turfgrass production and management is excellent.

Turfgrass production, establishment, and maintenance provide many good jobs and create attractive surroundings for homes, businesses, and recreation. *(Courtesy of Michael Dzaman)*

Rooting depth is affected by plant species, soil factors, and cultural or maintenance practices. Average rooting depth for turfgrasses is 6 to 12 inches. The warm-season grasses have deeper root systems than the cool-season grasses. Well-drained, sandy-loam soils with neutral soil pH are best adapted for turfgrass roots.

Cultural practices that influence rooting depth and growth are mowing, fertilization, and irrigation. Frequent mowing and low mowing heights reduce rooting depth. In addition, fertilization programs that emphasize only shoot growth are likely to impair root growth. Light and frequent irrigation results in shallow-root grasses. Heavy but infrequent irrigation is more conducive to deep root growth.

Turfgrass maintenance practices should attempt to optimize the rooting potential of turfgrass plants. An extensive root system allows a plant to recover from drought and other stress conditions more rapidly.

Shoot System

The shoot system consists of stems, leaves, and seed head, or inflorescence.

INTERNET KEY WORDS:

turfgrass crown tillers rhizomes stolons

Stems Turfgrass stems include the crown, tillers, rhizomes, stolons, and seed culms. The crown is the meristem tissue of the grass plant from which new growth occurs. The crown is a stem with the nodes stacked on top of each other (Figure 24–5). All root, leaf, and other shoot growth originates from this area. The crown is located at the base of the grass plant in the soil surface area.

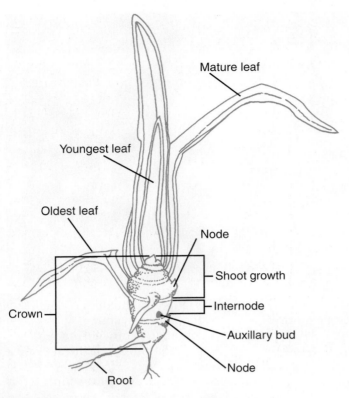

FIGURE 24–5 The crown of a turfgrass plant.

FIGURE 24–6 The major parts of a grass plant.

Rhizomes and stolons are horizontal stems. A rhizome is a creeping underground stem, whereas a stolon is an aboveground stem (Figure 24–6). They both originate from an axillary bud on the crown and will penetrate through the lower leaf sheath. This type of growth is referred to as **extravaginal growth**. Rhizome and stolon growth allows for vegetative spreading of turfgrasses. **Vegetative spreading** is reproduction from plant parts other than seed.

Tiller   Tillers are new shoots of a grass plant that develop at the axillary bud of the crown. They form within the lower leaf sheath of the plant. This type of growth is referred to as **intravaginal growth**. Increased tillering will enhance turfgrass density. All turfgrasses produce tillers. Optimum tillering for the cool-season grasses occurs in the spring and fall months. Warm-season grasses have optimum tillering during the summer. Under low-moisture and high-temperature stress conditions, tiller, rhizome, and stolon development is reduced.

Seed Culm and Inflorescence   The **seed culm**, or seed stem, supports the inflorescence of the plant. The seed culm originates at the top of the crown. **Inflorescence** is the arrangement of and the flowering parts of a grass plant. Cool-season grasses produce their inflorescence in the spring. Warm-season turfgrasses produce their inflorescence in the late summer.

Flower **induction** or initiation is caused by several environmental conditions. Temperature and photoperiod are the two induction processes for grasses. When seed heads form, they will cause a decrease

INTERNET KEY WORDS:

grass, axillary bud, crown

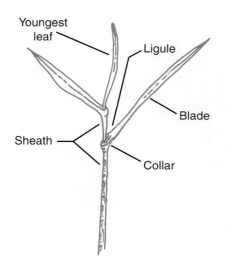

FIGURE 24–7 The leaf blade and leaf sheath of a grass plant.

INTERNET ADDRESS:

<http://www.turf.uiuc.edu/
turfspecies/tourseason/
tour.html>

in playability and appearance of the turfgrass stand. Mowing will remove the seed head, and thus improve turfgrass quality.

Leaf The turfgrass leaf consists of the sheath and blade (Figure 24–7). The **sheath** is the lower portion of the leaf and may be rolled or folded over the shoot system. The **blade** is the upper portion of the leaf. At the junction of the blade and sheath is the collar and ligule. The **ligule** is located on the inside of the leaf and is a membranous or hairy structure. The **collar** can be found on the outside of the leaf and is a light green or white-banded area (Figure 24–8). These two features are important vegetative traits for turfgrass identification.

Turfgrass growth is dependent on the production and use of carbohydrates. The turfgrass leaf is responsible for photosynthesis and, ultimately, carbohydrate production. Reserve carbohydrates will be stored in crown, rhizome, and stolon tissue.

During unfavorable growth conditions, the plant will go into dormancy. All leaf tissue will die back. However, when favorable environmental conditions occur, carbohydrate reserves from the crown and other stem tissue will be used for plant regrowth.

Turfgrass maintenance programs attempt to optimize root growth and carbohydrate accumulation. Greater recuperative potential and plant persistency occur when these two basic concepts of growth are understood and managed. **Recuperative potential** is the ability of a plant to recover after being damaged.

TURFGRASS SPECIES

There are some 7,500 plants classified as grasses. Only a few dozen are considered for turfgrass use. Turfgrasses are divided into two major groups based on climatic adaptation. The two groups are cool-season and warm-season grasses. The United States may be divided into a number of zones based on temperature and moisture conditions. It is helpful to be aware of these zones when selecting turfgrass (Figure 24–9).

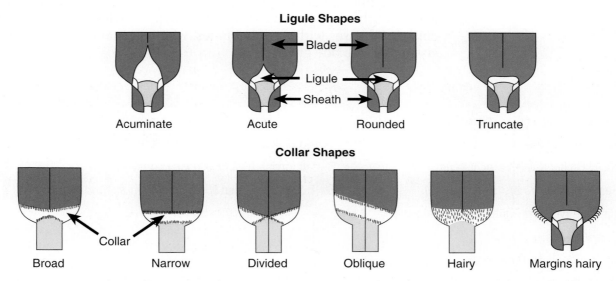

FIGURE 24–8 Grasses can be identified by using keys that eliminate incorrect identities until only the correct identity remains. The shapes of ligules and collars are used to describe plant characteristics that are included in the identification key for grasses.

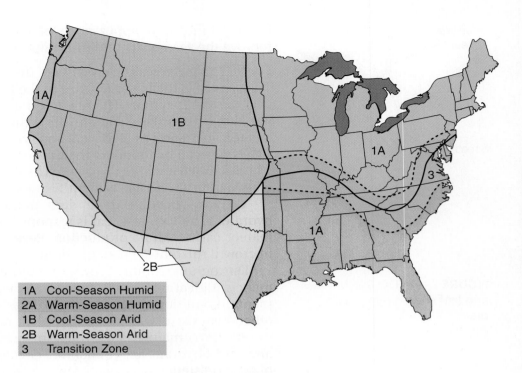

1A   Cool-Season Humid
2A   Warm-Season Humid
1B   Cool-Season Arid
2B   Warm-Season Arid
3   Transition Zone

FIGURE 24-9 The major turfgrass adaptation zones.

Cool-Season Turfgrasses

Cool-season turfgrasses originated in Europe and Asia. They have optimum growth at temperatures from 60° to 75° F. They predominate in the northern and central regions of the United States. Species adaptation within the cool-season group is determined by rainfall, soil fertility, and turf use. Major cool-season turfgrasses in the United States include Kentucky bluegrass, tall fescue, red fescue, perennial rye grass, creeping bentgrass, and crested wheatgrass (Figure 24–10).

Kentucky Bluegrass (Poa pratensis _L_)   Kentucky bluegrass is used extensively in residential and commercial lawns, in recreational facilities, and along highway rights-of-way. It performs well with moderate levels of maintenance. The plant has a medium leaf texture and an extensive rhizome system. Texture refers to leaf width. Fine-texture turf contains grasses with narrow blades, whereas coarse-texture turf consists of wide-blade grasses.

Cool-Season	Warm-Season
Colonial Bentgrass (*Agrostis tenuis*)	Bermuda grass (*Cynodon dactylon* L)
Creeping Bentgrass (*Agrostis palustris*)	Zoysia grass (*Zoysia japonica* Steud)
Kentucky Bluegrass (*Poa pratensis*)	Buffalo grass (*Buchloe dactyloides*)
Tall Fescue (*Festuca arundinacea*)	St. Augustine grass (*Stenotaphrum secundatum*)
Fine Fescue (various *Festuca* species)	
Red Fescue (*Festuca rubra* L)	
Perennial Ryegrass (*Lolium perenne*)	
Crested Wheatgrass (*Agropyron cristatum*)	

FIGURE 24–10 Some major warm- and cool-season turfgrasses in the United States.

Kentucky bluegrass grows best with full sun, moist and fertile soil, and a mowing height of 1.5 to 2.5 inches. More than 100 cultivars of Kentucky bluegrass have been developed for certain geographic areas and specific maintenance conditions. A cultivar is a plant of the same species that has been discovered and propagated because of its unique characteristics. Cultivar differences exist with respect to disease tolerance, leaf width, color, and other traits.

Tall Fescue (Festuca arundinacea Screb)

Tall fescue is a coarse-textured, bunch-type grass used in home lawns or as a utility-type turf. A **bunch-type grass** grows in clumps rather than spreading evenly over the ground. However, the bunching tendency of tall fescue can be overcome by heavy seeding, which produces thick stands of grass. A **utility-type grass** refers to a turfgrass adapted to lower maintenance levels. Tall fescues have extensive root systems and are among the most drought-tolerant, cool-season species. However, they are prone to winter injury in the northern range of the cool-season zone.

INTERNET KEY WORDS:

cool-season turf

Recent breeding work has introduced many new and improved cultivars of tall fescue. These new cultivars have medium leaf texture with more aggressive rhizome development. This turfgrass will not tolerate low mowing heights and should be mowed at 2.5 to 3 inches. Insect resistance and plant persistency is excellent for tall fescues infected with fungal endophytes. **Fungal endophytes** are microscopic plants growing within a plant. They tend to improve turfgrass.

Red Fescue (Festuca rubra L)

Red fescue is a fine-textured turfgrass well adapted to shady, dry locations. It has excellent drought tolerance and can persist on rather infertile soils and under acidic soil conditions (pH 5.5–6.0).

Red fescue is often seeded in mixtures with Kentucky bluegrass for lawn turf. It is not used on athletic fields as a permanent turf because it has poor recuperative potential. However, in the South, it may be overseeded into dormant Bermuda grass greens to provide winter play endurance and color.

Red fescue functions satisfactorily if it is mowed at 1.5 to 2.5 inches in height and is provided with minimal levels of nitrogen fertilizer and water. The major pest problem of red fescue is leaf spot disease.

Perennial Ryegrass (Lolium perenne L)

Perennial ryegrass has a medium leaf texture and is most often used in seed mixes with other turfgrasses. It has a rapid germination and excellent seedling vigor. It is often used in seed mixes to provide soil stabilization during the establishment period.

The perennial ryegrasses are used extensively for recreational turf. They have good wear tolerance and can be rapidly established if the turfgrass stand is damaged. They are also used in winter overseeding programs for warm-season turfgrasses.

Perennial ryegrasses require a moderate level of maintenance to form an attractive turf. As a group, they have poor disease resistance. Improved plant persistence and insect resistance occur if they are infected with fungal endophyte.

Creeping Bentgrass (Agrostis palustris Huds)

Used in close-cut, high-maintenance areas, this is an extremely attractive turfgrass and fits well when used on putting greens or bowling greens mowed at five

Science Connection

TOUGH TURF

Each summer, millions of people visit the museums, monuments, and Capitol Mall in Washington, DC. The daily foot traffic of Washington's employed, as well as the continuous flow of tourists, mean many thousands of feet pounding the turfgrass sections throughout the summer months. Furthermore, the seasonal use of the Mall for national rallies, concerts, inaugurations, and demonstrations pounds and punishes the turfgrass and soil like no other place. The end of summer generally leaves the turfgrass thin and vulnerable at best, and bare, muddy, and void at worst. The best of traditional turfgrasses have not been up to the task, and the most heavily traveled areas have been converted to gravel and concrete.

Turfgrass is still the preferred surface for large areas from the standpoint of both beauty and func-tion. The unique needs of the Mall have stimulated the development of a turfgrass mixture that fills the need for toughness, sustainability, and year-round green color. This is a large order in the Washington, DC, climate, which may swing from less than 0° F in the winter to more than 100° F in summer.

Enter U. S. Department of Agriculture (USDA) agronomists with an idea for a solution. Why not use one high-quality turfgrass with exceptional

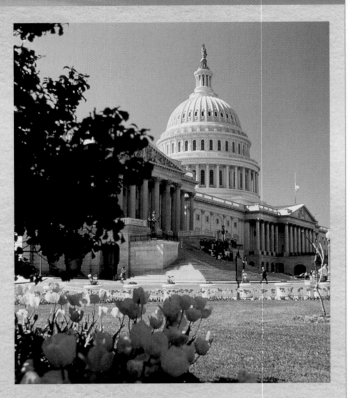

Turf created by a mixture of zoysia grass and tall fescue. *(Courtesy of the D.C. Committee to Promote Washington)*

durability in hot weather in a mixture with one having exceptional performance in cold weather resulting in a high-quality, tough, year-round turf?

For example, zoysia grass forms a dense, tough, fine-textured, attractive, and highly competitive turf from May to October. Unfortunately, it turns brown with October's first frost and stays brown until late April or early May. It is usually propagated with plugs.

In contrast, the new, improved, tall fescue turfgrasses shine in the cooler weather of fall and spring and generally stay green through Washington's winter months. They are propagated with seed. In fact, the tall fescue seed germinates with exceptional ease, and the seedlings become established and grow in competitive environments.

However, both grasses are tough competitors and tend to crowd out other plants. Could new strains, varieties, or cultivars of each species be found or bred that would survive together and combat the common enemies—excessive heat, cold, wet, dry, and the constant pounding of foot traffic in all kinds of weather?

The agronomists screened hundreds of varieties of grass in search of a fine-bladed fescue that could hold its own against zoysia grass. Eventually, they were able to bring together a mixture of fescue and zoysia grass that was worth field trials. Some 200,000 square feet, or the area of four football fields, were eventually sodded with zoysia and overseeded with tall fescue on the Washington Mall. Only time will tell if the mixture will be the answer there.

Meanwhile, a zoysia grass that will produce seed has been developed, and the turfgrass industry has picked up on the new products. Similarly, agronomists of the National Turfgrass Evaluation Program have established a testing program for zoysia grass and mixtures at more than 25 locations throughout the country. The greatest promise for the zoysia–fescue mixture is in the transition zone that stretches from New Jersey to Georgia and as far west as Kansas. Perhaps one plus one can equal one—one warm-season turfgrass plus one cool-season turfgrass equals one exceptional year-round turfgrass!

sixths of an inch. It can be maintained at higher mowing heights (0.5—0.75 inch) for use on football fields, golf course fairways and tees, lawns, and formal gardens.

The plant has extensive stolon growth and is a fine-textured turfgrass. It is best adapted to slightly acidic soil with a pH of 5.5 to 6.0 that has good internal drainage and is not prone to compaction. Creeping bentgrass has excellent cold tolerance but poor heat tolerance.

Creeping bentgrass requires a high level of maintenance to produce a quality turf. It requires proper irrigation, disease control, mowing, and cultivation practices when grown as a sports turf. Because of these high-maintenance requirements, creeping bentgrass is not recommended as a lawn turf.

Crested Wheatgrass (Agropyron cristatum) Crested wheatgrass is also known as fairway wheatgrass and is used on nonirrigated lawns and fairways in drier, colder regions. It is a coarse-textured, noncreeping, bunch grass. Crested wheatgrass is not considered a high-quality turf, but it is durable in semiarid, northern areas of the Great Plains. Its extensive, deep root system enables the plant to survive lengthy drought conditions without irrigation.

Recommended mowing height for crested wheatgrass is 1.5 to 2.5 inches, and low-to-moderate fertility is required. Heavy watering will stress the plant, therefore irrigation should be minimal.

Warm-Season Turfgrasses

INTERNET KEY WORDS:

warm-season turf

Warm-season turfgrasses originated in Africa, North and South America, and southeastern Asia, but they are adapted to the southern United States. They make optimum growth at temperatures of 80° to 95° F. These grasses go into winter dormancy as temperatures decrease to less than 50° F. There are some 14 species of warm-season turfgrasses found throughout the world.

Some major warm-season turfgrasses of the United States include Bermuda grass, zoysia grass, St. Augustine grass, and buffalo grass.

Bermuda grass (Cynodon dactylon L) Bermuda grass is considered the most important and widely used warm-season turfgrass in the United States. It is principally used as a lawn turf and sports turf. Improved breeding lines have provided fine-textured Bermuda grasses capable of being used on putting greens and fairways. The common type of Bermuda grass is a medium-textured turfgrass used for airport runways, rights-of-way, and other low-maintenance areas.

The plant spreads by both stolon and rhizome growth. It has excellent wear resistance, recuperative potential, and drought tolerance. Bermuda grass can persist on a wide range of soil types and soil pH. It has poor shade tolerance and low winter hardiness.

The improved types of Bermuda grass require a high level of maintenance and must be vegetatively established. Recommended mowing heights for these grasses range from 0.25 to 1 inch. The common type of Bermuda grass is established by seed, requires a greater mowing height than the improved cultivars, and is often considered a weed in certain situations.

Zoysia grass (Zoysia japonica Steud.) Zoysia grass has excellent low-temperature hardiness and is found in home lawns as far north as New Jersey. It can also be found on golf course fairways and tees within the transition zone. The transition zone is a geographic area of the United States where the warm-season and cool-season adaptation zones overlap.

Zoysia grass has good drought and shade tolerance. It can survive in a wide range of soil types. However, this plant will not perform well in poorly drained soils that remain waterlogged. Though zoysia grass has excellent wear resistance, it has a low recuperative potential. It is not as aggressive as Bermuda grass.

Zoysia grass requires a low-to-moderate level of maintenance. It is vegetatively established because seed germination is poor. New techniques to increase seed germination currently are being evaluated. This grass should be mowed at a height of between 1 and 2 inches. Two major pest problems of zoysia are nematodes and billbugs.

Buffalo grass (Buchloe dactyloides)   Buffalo grass is adapted to the dry, semiarid Great Plains region of the United States. It is native to Oklahoma, Texas, Arizona, Kansas, Colorado, North Dakota, South Dakota, and Montana. Regarded as the most drought-tolerant turfgrass in the United States, it survives both high and low temperature extremes and persists through droughts without irrigation. However, its shade tolerance is poor.

Buffalograss makes a gray–green turf with fine texture. Its vertical growth is slow, but it spreads by stolons, creating a continuous turf. It tolerates alkaline soil and prefers soils of the heavier, fine-textured type. It is used on nonirrigated lawns and golf fairways with cutting heights ranging from 0.5 to 2 inches. Buffalo grass is also used along roads and other low-maintenance areas for erosion control. Little or no fertilizer is necessary for survival, but turf quality may be improved by light applications of fertilizer. The grass is propagated by seed or plugs. It is virtually pest-free if not overfertilized or overirrigated.

St. Augustine grass (Stenotaphrum secundatum)   St. Augustine grass is a major lawn grass in the deep south and is used in the warmest area of the subtropical zone. It is adaptable to many soil conditions, but does best on moist, well-drained sandy soils. Drought resistance is only fair, so irrigation is required in dry weather. The grass has thick stolons and produces a coarse turf of good color, medium density, and excellent shade tolerance.

St. Augustine grass is propagated with sprigs or sod and has a good establishment rate. The grass has a vigorous growth rate with moderate maintenance requirements. Medium fertility is required, and acceptable cutting heights range from 0.5 to 3 inches. Thatch buildup, chinch bugs, and the St. Augustine decline virus are some of the problems encountered with this grass.

TURFGRASS CULTURAL PRACTICES

Mowing, irrigation, and fertilization are the most common and most important cultural practices performed to maintain turfgrass stands. These practices have tremendous effects on turfgrass quality and persistency. Improper mowing, fertilization, and irrigation are major causes of poor lawns (Figure 24–11).

INTERNET KEY WORDS:

zoysia grass
buffalo grass

Causes of Poor Quality Lawns

- Using the wrong turfgrass species or cultivars.
- Using poor quality seed.
- Mowing the lawn too closely.
- Permitting excessive growth between mowings.
- Using too little or too much lime or fertilizer.
- Improper watering.
- Too much shade.
- Droughty or poorly drained soils.
- Too much traffic.
- Damage by insects or disease.
- Improper use of chemicals.

FIGURE 24–11 Major causes of poor-quality lawns.

Recommended Mowing Heights	
Species	**Mowing height range in inches**
Bahia grass	2-4
Bermuda grass	
Common	0.5-1.5
Hybrids	0.25-1
Carpet grass	1-2
Centipede grass	1-2
St. Augustine grass	1.5-3.0
Zoysia grass	0.5-2.0
Creeping bent grass	0.2-0.5
Colonial bent grass	0.5-1.0
Fine fescue	1.5-2.5
Kentucky bluegrass	1.5-2.5
Perennial ryegrass	1.5-2.5
Tall fescue	1.5-3.0
Crested wheatgrass	1.5-2.5
Buffalo grass	0.7-2.0
Blue grama	2.0-2.5

FIGURE 24–12 Recommended mowing heights for different turfgrasses.

Mowing

Mowing will influence the functional use, persistency, and aesthetic value of a turfgrass. Grasses used for recreational purposes must be playable. The playability of an athletic-field turf is principally determined by its mowing height. A uniform turf surface, fine-leaf texture, and freedom from weed encroachment require proper mowing height and frequency.

Turfgrasses are capable of being mowed because their growing point, the crown, is located just below or at the soil surface. However, mowing does have several adverse effects on the grass plant. Reduced rooting depth and decreased carbohydrate reserves occur in mowed turfs. Improper mowing practices only accentuate these and many other adverse effects, causing a decline in turfgrass quality.

Mowing Height   The recommended ranges for mowing height are listed in Figure 24–12. If one mows below or above these ranges, various problems will occur. Mowing below the desired range will reduce photosynthesis, thus preventing carbohydrate production within the plant. This will cause a reduction in rooting and decrease the recuperative potential of the plant.

Mowing above the recommended height will increase thatch buildup and leaf texture. It will also decrease turf density and appearance. Thatch is the buildup of organic matter on the soil around the turfgrass plants. Excessive thatch will cause poor water infiltration, increased disease activity, and decreased rooting depth.

Mowing Frequency   Frequency of mowing is determined by the mowing height and the growth rate of the plant. No more than one third of the top growth should be removed per mowing (Figure 24–13). This will allow sufficient leaf area for photosynthesis after mowing. The lower the mowing height, the more frequent the mowing if the one-third rule

Science Profile

GRASS FERTILIZER

A major ecological problem that faces our high-technology society is what to do with all of the garbage that is filling up our landfills. Trash or garbage is not the only problem. A single family home can produce up to 400 lb in grass clippings each year per 1,000 square feet of lawn. This translates into 8 tons of clippings per acre each year. Many people bag their clippings into plastic trash bags and set them by the curb, never to be thought of again. However, with landfills overflowing all over the country, changes are being made to keep the grass out.

One solution that is beginning to catch on is "grasscycling," or recycling grass. Using clipped grass as an organic fertilizer in the form of mulch is a good solution. In this process, cut grass is mixed with leaves, soil, and other dead material in a plastic bin. Over time, the mixture becomes a fertile mulch that can be used anywhere commercial mulches would be used. Another option is to leave the grass as it lies on top of the lawn. When doing this, the grass must be cut into small pieces that filter through the tall grass and onto the soil surface, where it decomposes and returns nutrients to the soil. To accomplish this, you must cut grass more often or mow with mulching mowers. With a growing need to keep grass clippings out of landfills, "grasscycling" is a good alternative to curbside grass removal.

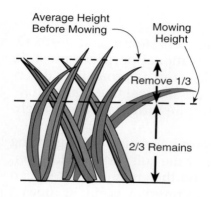

FIGURE 24–13 Mowing more than one third of the top growth of turfgrasses is likely to result in reduced rooting and less recuperative potential in the plant.

is observed. For example, creeping bentgrass mowed at a quarter inch will be mowed five to six times per week. However, tall fescue cut at 3 inches will require only one mowing per week. Under ideal growing conditions, grasses will require more frequent mowing than they will under poor growing conditions.

Fertilization

Fertility programs attempt to supply adequate levels of plant nutrients to allow for favorable plant growth. Proper fertilization should increase turfgrass density and color. Lush or succulent growth caused by excessive fertilization should be avoided.

A complete fertilizer is often recommended, with the rate of application based on the amount of nitrogen required per 1,000 square feet (Figure 24–14). A complete fertilizer consists of nitrogen, phosphorus, and potassium. The **fertilizer analysis** states the percentage of nutrients by weight in the fertilizer. A 10-6-4 fertilizer consists of 10 percent nitrogen, 6 percent phosphorus, and 4 percent potassium. A formula can be used to determine fertilizer amounts based on nitrogen-rate recommendations (Figure 24–15). Nitrogen is the most important element in turfgrass fertilization.

Proper timing and application rate are the keys to a successful fertilization program. Timing of fertilization will vary depending on turfgrass species and geographic location. For cool-season turfgrasses, the fall is the desired time period. Optimum root growth occurs during this time. In contrast, fertilization of warm-season turfgrasses should be done during late spring or the summer months (Figure 24–16).

Kentucky Bluegrass		
Time to Apply	**Nitrogen Applied Per 1,000 Square Feet**	**Acceptable Fertilizer Per 1,000 Square Feet of Area**
September	1 lb	8 to 10 lb 12-4-8 **or** 10 lb 10-6-4 **or** 5 lb 20-10-10
October	1 to 1½ lb	8 to 12 lb 12-4-8 **or** 10 to 15 lb 10-6-4 **or** 5 to 8 lb 20-10-10
November/December or February/March	1 to 1½ lb	8 to 12 lb 12-4-8 **or** 10 to 15 lb 10-6-4 **or** 5 to 8 lb 20-10-10
May-June 20	0 to ½ lb	0 to 4 lb 12-4-8 **or** 0 to 5 lb 10-6-4 **or** 0 to 3 lb 20-10-10

FIGURE 24–14 The fertility requirements for Kentucky bluegrass change with the season of the year. This is because its greatest growth occurs during the cool part of the growing season.

Percentage of nutrient(s) in the fertilizer times weight of the fertilizer equals the weight of the nutrient(s) contained.

Question #1: How much nitrogen is there in a 50 lb bag of 10-6-4 fertilizer?

Solution: % nutrient(s) x Weight of fertilizer = Weight of nutrient(s)

or

.10 x 50 lb = 5 lb of nitrogen

Question #2: How many pounds of 10-6-4 fertilizer must be used to apply 1 pound of nitrogen per 1,000 ft² on a lawn that measures 5,000 ft²

Solution: Pounds of nutrient needed divided by % nutrient equals pounds of fertilizer needed.

or

5 lb N divided by .10 = 50 lb of 10-6-4

FIGURE 24–15 Determining nutrient content of fertilizer and calculating fertilizer application rate to meet nutrient needs.

Bermuda grass and Zoysia grass		
Time to Apply	Nitrogen Applied Per 1,000 Square Feet	Acceptable Fertilizer Per 1,000 Square Feet of Area
April or May	1 lb	8 lb 12-4-8 **or** 10 lb 10-6-4 **or** 5 lb 20-10-10
June	1 lb	8 lb 12-4-8 **or** 10 lb 10-6-4 **or** 5 lb 20-10-10
July	1 lb	8 lb 12-4-8 **or** 10 lb 10-6-4 **or** 5 lb 20-10-10
August	1 lb	8 lb 12-4-8 **or** 10 lb 10-6-4 **or** 5 lb 20-10-10

FIGURE 24–16 Fertility requirements and some recommended fertilizers for Bermuda grass and zoysia grass.

Improper fertilization will reduce turfgrass quality. Turfgrasses receiving insufficient amounts of fertilizer will lack color, density, and recuperative potential. Excessive fertilization will reduce heat and drought tolerance, increase disease and insect damage, and cause excessive top growth. Appropriate nitrogen rates range from 0.25 to 1 lb nitrogen per 1,000 square feet. The rate is dependent on growth and environmental conditions. Phosphorus and potassium rates should be determined by soil testing.

Irrigation

The application of water to turf may accomplish several different objectives. First, sufficient soil moisture will allow for optimum plant growth. Irrigation is also used to establish turfgrass, reduce plant surface temperatures, and rinse in fertilizer and pesticide applications. **Syringing** is a light application of water to a turfgrass. It may be used to reduce plant temperatures or to remove dew and frost from the turfgrass leaf.

The amount of irrigation needed to maintain optimum plant growth is dependent on many factors. Some factors are turfgrass species, geographic location, soil type, weather conditions, and turfgrass use. General recommendations are to apply 1 inch of water per week during the summer months. It takes approximately 620 gallons of water per 1,000 square feet to provide 1 inch of water. This amount of water will produce a green and actively growing turfgrass stand.

Heavy but infrequent irrigation will force root growth deep into the soil. Light and frequent irrigation will keep the surface soil moist, thus encouraging shallow rooting.

The best time to irrigate is at night, when evaporation and wind are low. However, an increase in disease activity will occur at this time. Therefore, early morning watering is often selected because evaporation, wind, and disease activity are at a minimum.

TURFGRASS ESTABLISHMENT

INTERNET KEY WORDS:

turfgrass establishment

Turfgrasses may be established by seeding or vegetative propagation. Regardless of the method used, proper establishment practices,

Kentucky Bluegrass Cultivar	Quality Rating*					
	April 1	May 4	June 8	July 15	Sept. 2	Oct. 17
Adelphi	4.0*	5.6	5.4	4.5	5.9	5.7
Baron	3.8	5.1	4.8	4.4	5.4	5.5
Bensun (A-34)	3.6	5.3	6.0	5.9	6.1	5.4
Cheri	3.2	5.2	4.7	4.9	5.8	6.0
Emmundi	4.0	4.9	4.6	4.5	5.4	5.7
Glade	4.2	5.1	5.4	5.5	6.3	6.3
Majestic	3.9	6.1	6.0	5.8	6.0	5.8
Newport	3.7	3.6	3.8	4.0	4.2	4.9
Parade	4.5	6.8	5.9	4.4	5.7	5.8
Ram 1	4.3	6.9	5.9	6.0	6.2	6.4
Sydsport	3.8	6.2	6.0	5.9	6.4	6.1
Touchdown	4.1	6.3	5.9	5.7	6.2	6.5

FIGURE 24–17 Quality ratings for Kentucky bluegrass cultivars.

*It is important to note that quality ratings can vary significantly from one region or location to another. A rating of 1 = no live turf, 9 = ideal turf, >5 = acceptable quality.

including site preparation, should be performed. Correct establishment practices will ensure adequate turfgrass quality and persistency.

Turfgrass Selection

Selection of the correct turfgrass species is one of the most important decisions in the establishment process. Selecting a turfgrass seed blend or seed mixture should be based on the intended use and performance data of the turfgrasses. A seed blend is a combination of different cultivars of the same species. A seed mixture is a combination of two or more species. State universities evaluate different turfgrass species and new cultivars and provide information on their performances (Figure 24–17).

Site Preparation

Proper site preparation may include any or all of the following activities:
- debris removal;
- nonselective weed control;
- installation of a subsurface drainage or irrigation system, or both; and
- soil modification, tillage, and grading.

The removal of woody vegetation, leftover construction debris, and any large stones will reduce future maintenance problems. If woody debris is buried on the site rather than removed, the soil will settle and diseases such as fairy ring may result. Excessive amounts of stone and rock present in the seedbed will cause localized dry spots and interfere with future cultivation practices.

Nonselective weed control may be necessary if perennial grassy weeds or difficult-to-control weeds are present. A nonselective herbicide or a soil fumigant should be applied before seeding. The herbicide glyphosate (Round Up) is often used to provide nonselective weed control. A soil fumigant, such as metham, may also be used.

FIGURE 24–18 The site must be prepared properly by tilling and grading the soil before grass seed is planted.

If the site is to be used as a sports turf, the installation of a drainage and an irrigation system may be required. This should be done before final grading.

Soil modifications, followed by tillage and grading, are the last steps in site preparation (Figure 24–18). Soil amendments, such as organic matter and sand, may be incorporated to improve nutrient retention and drainage. Topsoil may also be used to provide a favorable growth medium for turfgrasses. The soil should be worked to a depth of 6 inches, and then lightly tilled to provide a uniform seedbed and adequate surface drainage (Figure 24–19). Incorporation of lime and fertilizer should be done during this time. Excessive tillage will destroy soil tilth. A seedbed with soil aggregates of 0.25 to 1 inch in diameter is ideal.

FIGURE 24–19 Tillage equipment frequently used for turfgrass establishment: (A) plow, (B) rototiller, (C) disc cultivator, and (D) harrow.

FIGURE 24–20 Some grass species that lack extensive root systems require netting to be applied to the soil surface to hold the turf together when it is harvested. *(Courtesy of Arva Burton)*

INTERNET KEY WORDS:

grass seeding practices

Additional preparation is necessary on sites where bunch grasses are selected as turfgrass crops. The lack of a strong fibrous root system makes it difficult to hold the harvested turf plugs together while they are shipped and transplanted. Netting materials are used to compensate for the lack of an extensive root system. Netting materials must be placed on the soil surface before the grass is planted. As the turfgrass matures, it becomes bound into the netting material. This allows it to be harvested and handled without causing serious damage to the turf (Figure 24–20).

Planting

Turfgrasses can be established by seeding, sodding, stolonization, sprigging, and plugging. The last four methods are vegetative means of planting. Vegetative establishment may be done if the grasses produce infertile seed or have low seed yields. Also, if quick establishment is required, sod may be installed.

Seeding Seeding is the principal method for turfgrass establishment, because it is the least expensive. Figure 24–21 shows the seeding rates for the different grasses. These rates will vary depending on seed size and percentage of seed germination.

The seed label informs the buyer or consumer of the quality and ingredients of a seed blend or mixture. Information on purity, germination, weed seed content, inert matter, and test date is present on the label (Figure 24–22). Certification programs are available for turfgrass seed. Seed certification programs are administered by state governments, and they ensure that seed is true to type. They also require that the seed meets other minimum quality standards concerning percentage of germination and weed seed content.

The best time to seed turfgrass varies, depending on the type of turfgrass. The optimum time for establishing cool-season grasses is in the fall. Ideal weather conditions and less annual weed competition are the

---

Agri-Profile

CAREER AREA:

Golf Course Maintenance Supervisor

The game of golf is one of the most popular leisure-time activities that people participate in. Many new golf courses are designed and constructed every year, and all of the existing courses must be maintained. This kind of work requires a person who knows how to establish turfgrass plantings and keep them healthy. Whether the golf course is new or well-established turf, it requires constant maintenance and care. Among the tasks that are performed as part of a golf course maintenance plan are controlling weeds, insects, and rodents; reseeding damaged turf; irrigation management; soil testing; fertilization; mowing; caring for trees; and establishing and maintaining other plantings. All of these tasks require expert knowledge.

A person who is interested in a career in golf course maintenance can obtain the required training by enrolling in a 2-year technical college program in turfgrass management or by obtaining a Bachelor's degree in plant science or a related field of study. A summer internship on a golf course would be a valuable component of either of these education options. A person who becomes a good manager of golf course resources can expect to have good employment opportunities and a reasonable income.

Seeding Rates	
Species	**Pounds of Seed per 1,000 ft^2**
Bahia grass	3-8
Bent grass	
Colonial	0.5-1.5
Creeping	0.5-1.5
Bermuda grass	
(hulled)	1-2
Bluegrass	
Kentucky	1-2
Buffalo grass	3-7*
Carpet grass	1.5-5
Centipede grass	0.25-2*
Fescue	
Fine	3-5
Tall	5-9
Grama, blue	1.5-2.5
Ryegrass	
Annual	5-9
Perennial	5-9
Wheatgrass,	
crested	3-6
Zoysia grass	1-3

*The higher rates are best, but lower rates are commonly used because the seed is expensive.

FIGURE 24–21 Seeding rates for the major turfgrass species.

principal reasons for fall establishment. The plant will also have sufficient time to develop an adequate root system before summer stress conditions. Spring seeding is less desirable because there is insufficient time to develop a mature stand capable of competing with undesirable grasses.

Warm-season turfgrasses are established during the spring and summer months. The best time is in late spring. The new grass then has the longest period for optimum growth and development after establishment.

Different types of equipment may be used to apply the seed. Fertilizer drop spreaders, overseeders, hydroseeders, and cultipacker seeders are some types of equipment used for seeding. Shallow seed placement (a quarter inch) is important for proper germination. This can be done by hand raking for small areas or the use of a drag mat for larger areas. Specialized seeding equipment, such as a hydroseeder or cultipacker, will place the seed at the right soil depth (Figure 24–23).

After seeding, the use of a mulch will provide a more favorable environment for seed germination. Mulch can conserve soil moisture and prevent soil erosion. This is extremely important, because adequate surface moisture must be present for seed germination. A straw mulch is preferred and is usually applied at the rate of 2 bales per 1,000 square feet. Other mulches are wood and paper by-products, net or fabric, and peat moss.

Sodding Sodding refers to removing a rectangular piece of grass and a slice of the soil beneath it and moving it to another location. The grass and the soil immediately beneath are referred to as turf. Sodding offers the ability to establish turf at any time of the year. It also provides an instant cover. However, the cost is extremely high, and the sod must be installed shortly after harvesting. Specialized equipment for sod

(A) Label of a Recommended Seed Mixture

Lot 410-11
FINE-TEXTURED GRASSES
29.96% South Dakota Certified
Kentucky Bluegrass
85% germination
30.06% Certified Merion Kentucky
Bluegrass
85% germination
28.92% Certified Adelphi Kentucky
Bluegrass
85% germination
8.96% Certified Pennlawn Creeping
Red Fescue
85% germination
COARSE KINDS
None Claimed

OTHER INGREDIENTS
0.05% Other Crop Seed
1.96% Inert Matter Tested: July 1995
(Must be within 90 days of purchase date)
0.09% Weed Seed

Grass Seed

(B) Label of a Poor Seed Mixture

Lot SP-24
FINE-TEXTURED GRASSES
None or None Claimed
COARSE KINDS
50.62% Kentucky 31 Tall Fescue
85% germination
19.00% Perennial Ryegrass
90% germination
25.00% Annual Ryegrass
90% germination

OTHER INGREDIENTS
3.10% Other Crop Seed
1.78% Inert Matter Tested: Jan. 1994
0.50% Weed Seed

Noxious Weeds

3 Poa annua per ounce

FIGURE 24–22 Sample labels for lawn and turf seed. (A) Label describing a recommended seed mixture of cool-season grasses. (B) Label describing a poor-quality seed mixture of cool-season grasses.

FIGURE 24–23 One method of seeding turfgrass is to use the hydroseeding technique, which mixes seed, mulch, fertilizer, and water and sprays it on the planting surface. *(Courtesy of Michael Dzaman)*

FIGURE 24–24 Cutting sod for placement in a new location as turfgrass. *(Courtesy of H. Edward Reiley)*

harvesting has been developed (Figure 24–24). Sod is cut into 12- to 24-inch widths and lengths of up to 3 feet, or in carpetlike pieces. The depth of cut will vary from 0.3 to 0.5 inch. The cut sod is then rolled or folded for transport and later placed on a carefully prepared seedbed. The new turf is then rolled to ensure good contact with the soil and watered thoroughly.

Sprigging Sprigging is the planting of a section of a rhizome or stolon, referred to as a sprig. A sprig may be up to 6 to 8 inches in length. For successful establishment, a section of the sprig with several nodes must be inserted into the soil (Figure 24–25). After sprigging, irrigation must be applied to prevent drying out. Equipment for sprigging has been developed and will plant sprigs in rows spaced 6 to 18 inches apart.

Stolonizing Stolonizing is similar to sprigging, in that 6- to 8-inch sections of rhizomes or stolons are used. However, in stolonizing, sprigs are broadcast onto the soil surface. These sprigs may be lightly top-dressed with soil and rolled before irrigation. Survival rates are lower than with sprigging. Therefore, stolonizing requires greater quantities of sprigs.

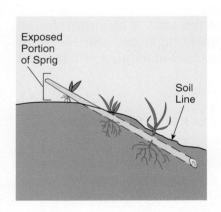

FIGURE 24–25 When sprigging, two or more nodes with shoots should be placed in the soil and watered well.

Plugging Plugging is the establishment of a turfgrass stand by using plugs, or small pieces of existing turf. Plug sizes vary, as noted in Figure 24–26. Spacing of plugs can be on 6- to 18-inch centers, depending on how quickly you need the turfgrass established. Plugging can be done by hand or with specialized equipment. It requires a longer time to cover the soil than do the other establishment techniques.

In summary, the turfgrass industry is expanding. As more space is used for nonfarm purposes, the use of turfgrass for conservation, recreation, and beautification will continue to grow. Excellent career opportunities exist, and suitable educational programs are available to prepare new workers for this increasingly popular area of agriscience.

FIGURE 24–26 Turfgrass plugs are used to plant lawns and playing fields.

STUDENT ACTIVITIES

1. Write the Terms to Know and their meanings in your notebook.

2. Determine if your community is located in a warm-season, transitional, or cool-season zone.

3. Visit or call your county cooperative extension office and obtain copies of the various publications available on turfgrass or lawn production and maintenance.

4. Determine the recommended cultural practices for establishing and maintaining a major turfgrass species or mixture for your locality. Report your findings to the class.

5. Arrange to visit a golf course and discuss with the superintendent the duties of turfgrass workers on the golf course.

6. Obtain a map of your county and attach it to the bulletin board. Place pins at the locations of all businesses, schools, government buildings, and institutions that have extensive lawn or turfgrass areas around them. Ask your classmates to help identify them. Use different colored pins to identify different types of institutions.

7. Obtain a grass identification key from your teacher or the cooperative extension service office. Collect 10 specimens of different turfgrasses and identify them by using the identification key.

8. Make your own grass clipping mulch. Inside of a compost bin, or in a plastic garbage container, mix grass clippings, leaves, soil, and any other plant material. Add water and stir the mixture each week to add moisture and oxygen to the compost. Soon you will have a useful nutrient mulch that can be used in a flower bed or a garden plot.

SELF EVALUATION

A. Multiple Choice

1. The meristem tissue of the turfgrass plant is the
   a. inflorescence.
   b. leaf blade.
   c. rhizome.
   d. crown.

2. The root system of a cool-season turfgrass is actively growing during the
   a. summer.
   b. fall and winter.
   c. fall and spring.
   d. late spring.

3. The estimated value of the turfgrass industry in the United States is

a. $25 billion.
b. $50 billion.
c. $100 billion.
d. $200 billion.

4. Which irrigation practice will develop a deeper, more extensive root system?

a. frequent and light applications
b. infrequent and heavy applications
c. frequent and heavy applications
d. infrequent and light applications

5. Turfgrasses are in the _____ family.

a. Compositae
b. Cyperacese
c. Cruciferue
d. Poacea

6. _____ is a warm-season turfgrass.

a. Kentucky bluegrass
b. Zoysia grass
c. Red fescue
d. Perennial ryegrass

7. Which cultural practice will have an adverse effect on rooting depth?

a. low mowing
b. heavy, infrequent irrigation
c. moderate fertilizer applications during the time of root growth
d. high mowing

8. A tiller is an example of

a. rhizome growth.
b. extravaginal growth.
c. stolon growth.
d. intravaginal growth.

9. Which of the following is not a turfgrass stem?

a. rhizome
b. stolon
c. leaf sheath
d. crown

10. Which cool-season turfgrass is adapted to shady, dry locations?

a. red fescue
b. Kentucky bluegrass
c. tall fescue
d. creeping bent grass

B. Matching

_____ **1.** Rhizome

_____ **2.** Stolon

_____ **3.** Creeping bent grass

_____ **4.** Sprigging

_____ **5.** Photoperiod

_____ **6.** Tall fescue

_____ **7.** Coarse texture

_____ **8.** Syringing

_____ **9.** 10-10-10

_____ **10.** Thatch

a. An induction process for inflorescence development

b. A turfgrass stem that grows horizontally aboveground

c. A cool-season turfgrass that is very drought tolerant

d. A wide-leaf blade

e. A cool-season turfgrass used on putting greens

f. A turfgrass stem that grows horizontally below ground

g. A buildup of organic matter on the soil around turfgrass plants

h. A complete fertilizer

i. Light application of water to a turfgrass

j. Vegetative establishment

C. Completion

1. _____ is a warm-season turfgrass having good winter hardiness.
2. _____ roots are present at the time of seed germination.
3. _____ is an example of intravaginal growth.
4. A seed _____ consists of different cultivars of the same species.
5. _____ is a turfgrass species with rapid seed germination.
6. As mowing heights decrease, mowing frequency will _____.
7. _____ is considered the most important element in turfgrass fertilization.
8. A planting depth of _____ is recommended for turfgrasses.
9. _____ is an establishment practice of applying the sprig to the soil surface.
10. A _____ mulch is preferred over other types of mulch for seeding turfgrasses.

UNIT
25

Trees and Shrubs

OBJECTIVE

To use trees and shrubs for beautification, wood products, improved air quality, and pollution control.

MATERIALS LIST

✓ nursery catalog
✓ pictures of plants commonly grown in an area

✓ list of insects and diseases of trees and shrubs that are normally problems in the area

✓ Internet access

COMPETENCIES TO BE DEVELOPED

After studying this unit, you should be able to:

- identify ornamental trees and shrubs.
- select trees and shrubs for appropriate landscape use.
- classify trees and shrubs according to growth habit, growth habitat needs, and other requirements.
- identify trees and shrubs by using proper nomenclature.
- purchase plant material for installation in a landscape.
- plant and maintain plant material.

SUGGESTED CLASS ACTIVITIES

1. Conduct an inventory of the trees in a park or along the streets of a small community. Identify the species and determine which species are most often found. A map of tree species may be available from the park or forestry department in some cities and towns. Teach students how to use a key to identify different species of trees.

2. Identify a potential location at the school or other community property where a landscape project could be conducted by the class. Contact the appropriate officials and obtain permission to develop a landscaping plan. Work with community groups and local government to obtain financing for the project. Install the trees, shrubs, and ornamental plants in the landscape according to the landscape design.

3. Create a hypothesis based on this question: What does a tree's age have to do with the circumference of its trunk? Take a field trip to a native tree stand in your area. Measure the circumference, at breast height, for 15 trees of the same species. Core each tree to determine its age. Calculate the average circumference and age. Write a short paper discussing how your results compare with your hypothesis.

TERMS TO KNOW

speciman plant	nomenclature	root pruning	drip line
border planting	genus	tree spade	pruning
group planting	bare-rooted plant	container grown	canopy
hardy	healing in	staking	bloom
plant hardiness zone map	balled and burlapped (B & B)	guying	

Trees and shrubs are major components of the environment. They are important in providing natural beauty and in providing oxygen to humans and animals. A well-planned landscape will improve the economic value of an area. The need for trees and shrubs has increased in recent years. This need has increased the requirement for horticulture technicians who understand the production and care of landscape plants.

TREES AND SHRUBS FOR LANDSCAPES

A well-designed landscape increases the value of a property (Figure 25–1). Residential properties are much more attractive when trees and shrubs are used to enhance their beauty and balance.

Trees and shrubs provide more than natural beauty. An acre of good, healthy trees or shrubs will produce enough oxygen to keep from 16 to 20 people alive each year. These plants are also valuable in helping to keep our air clean by using the carbon dioxide that is produced by people, automobiles, and factories.

Trees and shrubs cut noise pollution by acting as barriers to sound. They can deflect sound, as well as absorb it. When used properly in the landscape, trees and shrubs can provide shade and act as insulators, to keep the house cooler in the summer and warmer in the winter.

Many urban areas depend on trees and shrubs to soften the concrete, black top, and steel environment. In fact, many of these areas now employ urban foresters to help design, install, and maintain trees and shrubs in large cities (Figure 25–2).

FIGURE 25–1 Trees and shrubs add character and value to a property. *(Courtesy of Michael Dzaman)*

FIGURE 25-2 Trees and shrubs add beauty, provide sound insulation, purify the air, and provide privacy. *(Courtesy of Michael Dzaman)*

FOREST RESOURCES

Vast regions of North America produce forest products such as wood, cardboard, paper, solvents, medicines, fuels, and many other products. An entire industry depends on trees for the raw materials that are used to process these important products. Thirty percent of the land area in the world is forest land, and forest products are important to the economies of the developed countries. Some of the trees that are used for this purpose are produced on private land. There are also vast tracts of public land devoted to timber production.

Forests are important resources in many ways besides production of wood and forest products. A forest also functions as a biological filter system that cleans the environment by removing impurities from air and water. Forest plants restore oxygen to the atmosphere and improve watersheds that contribute to consistent supplies of fresh, pure water. Forests furnish habitat to many kinds of wild animals and birds. They also provide recreation opportunities for people who enjoy outdoor activities such as camping, going on picnics, hiking, boating, fishing, and hunting. Forests provide many kinds of resources that are important to people.

PLANT SELECTION

Trees and shrubs are distinct groups of plant materials. Trees are woody plants that produce a main trunk and grow to a height of 15 feet or more. Shrubs are woody plants with a low growth habit, produce many stems or shoots from the base, and do not reach more than 15 feet in height (Figure 25–3).

USE

Ornamental trees and shrubs in the landscape are both beautiful and functional. They are used as specimen plants, border plantings, or

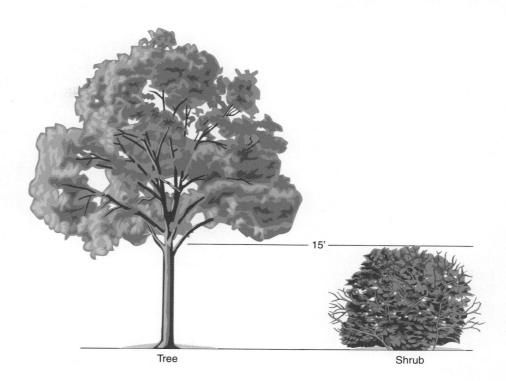

FIGURE 25–3 A tree has a single stem and a height of 15 feet or greater. A shrub has multiple stems and does not grow to a height greater than 15 feet.

Tree

Shrub

groupings. A **speciman plant** is used as a single plant to highlight or provide some other special feature to the landscape. A **border planting** is used to separate some part of the landscape from another, or to serve as a fence or a windbreak. A **group planting** consists of a number of trees or shrubs that are planted together so they point out some special feature, provide privacy, or create a small garden area.

The location of plant material in the landscape will play an important part in the selection process. Color of the leaves, texture of the plant, and color of the flowers are just some of the factors that must be considered before purchasing or planting trees or shrubs.

Geographic Location

The geographic region in which a plant will be used is an important factor in the selection of a tree or shrub. Some plants are not **hardy,** that is, they may not be able to survive or even grow properly in a given environment. The hardiness of a plant is affected by the intensity and duration of sunlight, the length of the growing season, minimum winter temperatures, the amount of rainfall, summer droughts, and humidity.

The U.S. Department of Agriculture (USDA) has issued a **plant hardiness zone map** (Figure 25–4). There are 11 zones in the United States, including Alaska and Hawaii. Each zone represents an area of winter hardiness that is based on the average annual minimum winter temperatures. It is possible that local climates may vary from the general zone map. They may be colder or warmer than is indicated on the map. Local nursery personnel are generally willing to help in the selection of the best plant material. Most nursery catalogs list the plant for the coldest zone in which it will grow normally. Two examples of such catalog listings are:

1. *Cornus florida* (Flowering Dogwood)—Zone 5: A low-branched, flat-topped tree that has a horizontal branching habit. It will grow 20 to 30 feet high and 25 to 35 feet wide. Growth rate is slow to

INTERNET KEY WORDS:

plant hardiness zone map

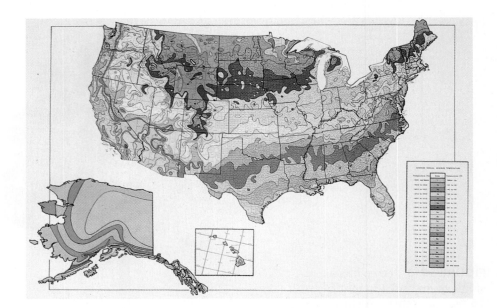

FIGURE 25–4 The U.S. Department of Agriculture plant hardiness zone map.

INTERNET KEY WORDS:

site selection, trees, shrubs

medium, and the texture is fine to medium. White flowers appear in April to May. Red berries appear in the fall.

2. *Pyrus calleryana* "Bradford" (Bradford Pear)—Zone 4: A dense, pyramidal tree that becomes brittle with age. It will grow 30 to 50 feet high and 30 to 35 feet wide. The growth rate is medium, and the texture is medium to fine. It makes an excellent all-purpose tree, useful as a shade or street tree. Profuse white flowers one-third inch in diameter appear in late April or early May in clusters up to a diameter of 3 inches.

A plant can also be expected to live in a warmer zone than indicated if rainfall, soil, and summer conditions are comparable. In some cases, it may be necessary to adjust these conditions through irrigation, correction of soil condition, wind protection, and alteration of shade or sun exposure. It is also possible to grow some plants in areas north of the indicated zone. Such plants may need special attention to protect them from wind or cold. Without such protection, they may not perform normally and are likely to suffer winter injury.

Site Location

When planning the location for planting trees and shrubs, many factors must be given consideration, including:

* flower color: Is it compatible with the house, fence, patio, and any other plants in the area?

* fruit size and type: Many trees, such as some crabapples or cherries, drop messy fruit. Therefore, they should not be placed near an area that will be walked on or used by people. Such areas include driveways, walks, patios, and swimming pools.

* other structures: Do not plant trees directly in front of doors; near wells, cesspools, or field drains; or under utility lines, where interference is likely when the plant matures.

* ornamental characteristics of the plant: Flowering time, shape, foliage texture, fall color, pest resistance, landscape suitability, and mature size are all part of any consideration when selecting a plant.

Agri-Profile

CAREER AREAS:

Landscape Architect/Landscape Technician/ Landscape Contractor/Plant Specialist

Ornamental trees and shrubs vary greatly in size, shape, temperature preference, light preference, fertility needs, and pest tolerance. This variety stimulates diversity in jobs and specialties in the area of ornamental horticulture. Career opportunities span both the arts and the sciences.

Landscape architects practice the art of design in that they plant landscapes pleasing to their clients. They must know plant species and plant materials to develop plans that are attractive yet functional in the given environment. Landscape architects generally have Bachelor's degrees in landscape design and considerable experience in ornamental horticulture.

Horticulture specialists and technicians have careers centering on nursery or greenhouse management, landscape contracting, groundskeeping, wholesaling, retailing, public information, writing, and consulting. They may specialize in just trees, just shrubs, or both. Some specialize in pruning, tree planting, pest control, or landscape maintenance. Ornamental horticulture seems to provide an endless array of career opportunities in many localities.

Plant care and management require many skills including the diagnosis of disease and insect problems. This plant specialist evaluates the extent of damage by scale insects. *(Courtesy of Michael Dzaman)*

Type and Growth Habit

Plants have many different types of growth habits. The type of growth habit is an important consideration when selecting plants (Figure 25–5). A good landscape planner or designer will be aware of the type of growth habit for a given plant used in a given area.

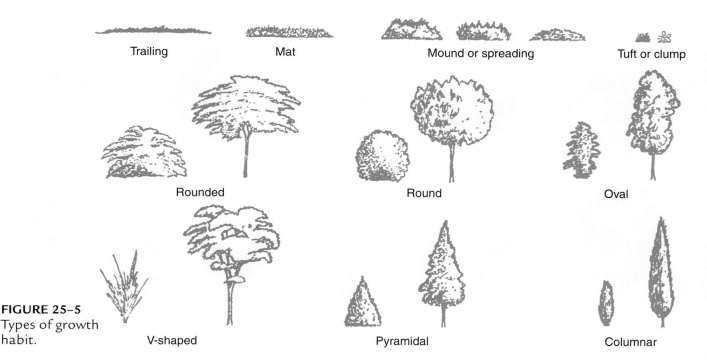

FIGURE 25–5 Types of growth habit.

Trailing Mat Mound or spreading Tuft or clump

Rounded Round Oval

V-shaped Pyramidal Columnar

Plant size in relation to the structure that is being landscaped is important. Maintenance and pruning of plants is often expensive. Pruning of shrubs is frequently done without climbing, whereas most trees must be climbed to prune them. The use of a tree that is not in proportion to nearby structures can result in an expensive, frequent pruning program. Figure 25–6 illustrates some common trees and their mature sizes. A tree that is too tall will require extra work to keep it in proper relationship to adjacent structures. Such a design error would cancel out the goal of property enhancement.

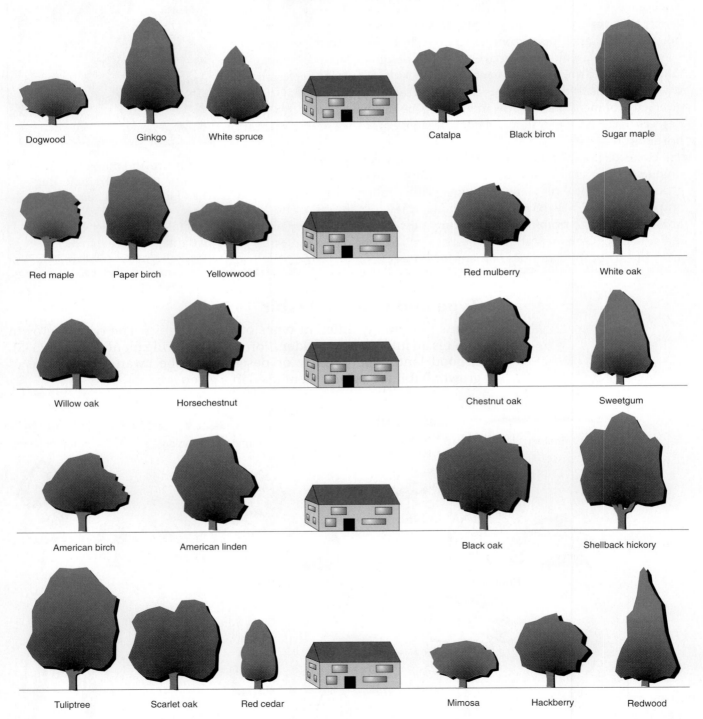

FIGURE 25–6 Tree sizes in relation to a two-story house.

FIGURE 25–7 Typical shapes of landscape trees.

Shape of the Plant

The shape of a plant is an important factor in the selection of plant material. Some important undesirable effects can be avoided by asking some basic questions. Does the plant grow straight up and give little shade? Does the plant have a trunk that divides and spreads to cast unwanted shade? Is the plant a low-growing tree that gives little shade? Consider which shapes of trees and shrubs are needed in the landscape. Then select the appropriate plant to achieve the objective (Figure 25–7).

Before plants are selected, research is necessary to better understand the types of plants used in a given area. To do this, study the types of plants sold and used in the locality and note the various characteristics of each plant. Catalogs are available from nurseries that sell plants in the locality. Also, a visit to local garden centers and nurseries will pay dividends in gathering information on plant selection.

PLANT NAMES

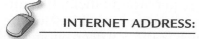

INTERNET ADDRESS:

<http://horticulture.tamu.edu:
7998/ornamentals/
207search.html>

INTERNET KEY WORDS:

plant taxonomy
nomenclature

Trees and shrubs must be ordered by their proper names. Common names such as flowering dogwood and upright juniper are not governed by any formal code of nomenclature. **Nomenclature** is a systematic method of naming plants or animals. The botanical or scientific name is recognized internationally. Scientific designations are always written in Latin and consist of two names. The first name is the genus, and the second is the species. The genus name always begins with a capital letter and is a noun. The species name is usually written in all lowercase letters and is an adjective.

The **genus** (plural is genera) is defined as a group of closely related and definable plants composing one or more species. The common, definable characteristics are fruit, flower, and leaf type and arrangement. The species (plural is also species) is the basic unit in

Bare-rooted

Twine Ties   Burlap Bag

Balled and Burlapped

Containerized

FIGURE 25–8 Methods of preparing plants for shipping and handling.

the classification system whose members have similar structures and common ancestors, and maintain their characteristics.

Variety (var.) is a subdivision of species. A variety has various heritable characteristics of form and structure that are continued through both sexual and asexual propagation. The varietal term is written in lowercase letters and <u>underlined</u> or *italicized*. Often in catalogs and on plant labels, the abbreviation var. is used. An example of how this is used might be *Cornus florida rubra* or *Cornus florida* var. *rubra.*

A cultivar (cv.) is a group of plants within a particular species that has been cultivated and is distinguished by one or more characteristics and that, through sexual or asexual propagation, will keep these characteristics. The term is written inside single quotation marks. An example is *Pyrus calleryana* 'Bradford' or *Pyrus calleryana* cv. 'Bradford.'

It is important to become familiar with this plant-naming system because common names will vary from area to area. There are no standards for creating common names. Professional horticulturists order plants by the method described above. Landscape architects place scientific names on landscape drawings and nurseries use the scientific names in their catalogs to avoid confusion by anyone who orders plant material.

OBTAINING TREES AND SHRUBS

After determining the specific plants needed in a landscape, the plants must be purchased. Plants are normally dug and shipped as bare rooted, balled and burlapped (B & B), or container grown (Figure 25–8).

Bare-Rooted Plants

Bare-rooted plants are dug, and the soil is shaken or washed from the roots. Normally only deciduous trees, shrubs, and trees with taproots are shipped this way. They are dug while the trees are dormant. Plants ordered from nursery catalogs that can survive as bare-rooted stock are shipped this way because of lower transportation costs and easier handling. It is not economical to ship soil with the roots because it is so heavy.

Bare-rooted plants are often planted while they are dormant; however, the roots must be protected to keep them from drying out. If the plant cannot be planted when it is received, the roots must be protected by putting them in a container of water, wrapping them in burlap after a good watering, or placing wet newspaper around them. If plants cannot be planted for many days, the plant may be healed in. **Healing in** is accomplished by simply digging a trench in the soil deep enough to hold the roots of the plants, placing the plants in the trench, and covering them with soil (Figure 25–9). It is important to wet the soil well after healing in. Frequently, plants that are bare rooted will need additional pruning at planting time.

Balled and Burlapped Plants

Balled and burlapped (B & B) plants are dug with a ball of soil remaining with the roots. This is wrapped with burlap and laced with twine. Normally, these plants have been root pruned. **Root pruning** is a process whereby roots are cut close to the trunk so that a good root system develops close to the trunk before the plant is dug. The result is that

HEELING-IN

1. Dig V-shaped trench in moist, shady place.

2. Break bundles and spread out evenly.

3. Fill in loose soil, and water well.

4. Complete filling in soil and firm with feet.

HANDLING SEEDLINGS IN FIELD

CORRECT AND INCORRECT DEPTHS

CORRECT
In bucket with sufficient wet moss to cover roots.

INCORRECT
Incorrect handling promotes drying of the root.

CORRECT
At same depth or 1/2" deeper than seedling grew in nursery.

INCORRECT
Too deep and roots bent.

INCORRECT
Too shallow, roots exposed.

FIGURE 25–9 Handling and planting seedling trees.

when the ball is dug, the tree or shrub will have a compact root system in the ball. This gives the plant a good chance to reestablish itself in a new environment.

It is important that transplanted trees and shrubs have a good chance to grow after transplanting. Plants typically balled and burlapped for transplanting are deciduous trees with branching root systems, conifers, azaleas, rhododendrons, and other plants that have fibrous root systems.

As is true with bare-root plants, B & B plants cannot have their root systems dry out. However, because there is soil around the roots, the drying-out process is more gradual. When purchasing B & B plants, avoid those that do not have a sound ball. Handling tends to break the ball up and strip the hair roots from the plants. Therefore, a good root ball will help substantially to avoid damage when handling. Plant material that is shipped as B & B may be planted anytime, as long as the soil can be worked.

The mechanical tree spade is becoming more popular in the industry today. A **tree spade** is an expensive piece of equipment that will dig a tree in a matter of a few minutes with a very specific-sized ball. In addition, it is used to dig holes for trees and shrubs quickly and efficiently. The tree spade saves many hours in digging and planting trees. Frequently, B & B trees and shrubs are shipped in burlap with wire cages, specially prepared baskets, or other containers. The tree spade has helped make this an efficient way to ball and burlap.

FIGURE 25–10 Nurseries package and ship many of their trees and shrubs to customers in disposable containers.

INTERNET ADDRESS:

<http://www.treefolks.org/guide/guide.html>

Bark Mulch

Set to Same Depth as it Grew Originally

Allow Roots Room to Spread Out

Soil Ring

Topsoil Mix (Added carefully to remove air pockets)

Undisturbed Soil

FIGURE 25–11 It is important to follow carefully all of the instructions for planting trees and shrubs.

These types of plants require little pruning at planting time. This makes B & B a popular way of handling plants by the mechanized professional horticulturist.

Container-Grown Plants

The use of container-grown plants is increasing in the nursery industry. **Container grown** means grown and shipped in a pot or can (Figure 25–10). Normally, the smaller types of plants are handled in this manner. They are grown for a reasonable period in the containers in which they are shipped and purchased. They are easy to handle in the field, require little or no pruning when they are planted, and can be planted any time the soil can be worked.

The disadvantage of using container-grown plants is that they need to be planted carefully. The roots of the plant have developed in a limited space and are generally rootbound. The roots may also have grown back around the trunk. To prevent the plant from strangling itself and also to encourage the root to leave the confined area, these plants must have the "container ball" broken. This is done by placing a sharp shovel through the root mass or by breaking and spreading the root mass as it is planted in the new hole.

PLANTING TREES AND SHRUBS

Planting may be done in spring, summer, or fall, as long as soil can be worked. A good practice for preparing the hole is to dig it about one-third wider than the ball or container. If the soil is hard, compacted, or of poor quality, it is advisable to dig the hole 4 to 5 inches deeper than the ball. This will allow peat moss or another soil conditioner to be added to the soil and placed under and around the root ball. Fill the hole with enough good soil to allow the ball to be the proper depth when it is put in the hole. The ball or container should be about 2 to 5 inches above the top of the hole. After cutting the twine that holds the ball together, peel back and remove the burlap. Backfill the hole around the plant, tamping the soil lightly to ensure removal of all air pockets. Continue backfilling the hole until it is level with the surrounding soil. Place a ring of soil about 4 inches high around the backfill to retain water.

In the case of a container plant, remove the plant from the container and slice the root zone with a knife or shovel. Then place the plant in the hole and backfill carefully.

After backfilling, fill the ring with several inches of water to saturate the soil to the bottom of the backfill. This will settle the soil and provide moisture for the plant (Figure 25–11).

MULCHING

Trees and shrubs will benefit from the addition of mulch around the planted area. Mulch in the form of shredded hardwood or pine straw will reduce evaporation and help hold in the moisture. A newly planted tree will take about 15 to 20 gallons of water about twice a week in a hot, dry environment. The conservation of water is important. Other advantages of mulch is that it will help provide a more even soil temperature, help control weeds, prevent erosion, and help prevent soil compaction. Some species

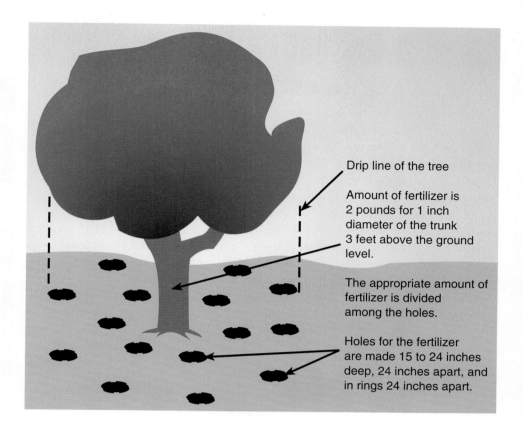

Drip line of the tree

Amount of fertilizer is 2 pounds for 1 inch diameter of the trunk 3 feet above the ground level.

The appropriate amount of fertilizer is divided among the holes.

Holes for the fertilizer are made 15 to 24 inches deep, 24 inches apart, and in rings 24 inches apart.

FIGURE 25–13 Holes for fertilizer applications should be placed within the drip line of the tree.

es apart (Figure 25–13). Each hole receives the appropriate amount of fertilizer and is sealed with soil or by closing the hole with a heel of the foot.

The amount of fertilizer needed by a tree is determined by measuring the trunk about 4.5 feet above the ground. If the trunk of the tree is less than 8 inches, multiply the number of inches of diameter by 3. If the trunk is greater than 8 inches, multiply by 6. This will give the number of pounds of fertilizer the tree needs. Divide the number of pounds of fertilizer by the number of holes to determine the amount of fertilizer to put in each hole.

For example, if the tree measures 6 inches at 4.5 feet, multiply 6×3. The answer is 18 lb fertilizer. If the tree measures 12 inches at 4.5 feet, then $12 \times 6 = 72$ lb fertilizer. If the tree has more than one trunk, combine the diameters of all the trunks and multiply by the appropriate factor.

A fertilizer with an analysis of 10-6-4 or 10-10-10 is generally acceptable. Never apply more than 100 lb fertilizer to any given tree in a year. If the rate to be applied is more than 100 lb, make the application over a 2-year period. Trees should be fertilized every 3 to 5 years.

PRUNING

INTERNET KEY WORDS:

prune trees, shrubs

Pruning is the removal of dead or undesirable limbs from a tree or shrub. Removing dead, broken, diseased, and insect-infested wood helps to protect the plant from additional damage. Trees may have waterspouts, bad-angle crotches, branches that cross over each other, or branches that form an asymmetrical habit. These need to be taken out or corrected. Sometimes it is necessary to reduce the top growth of the plant to match the root ball on a newly transplanted plant (Figure 25–14). After a tree is transplanted, the size of the **canopy,** or branches, should be reduced to equal the size of the root ball.

Science Profile

TREE GALL INFESTATION

A tree can be a host to many different kinds of creatures that do not cause any problems. Other guests are less cordial and take advantage of their host. Some organisms use trees and other plants to provide them with a place to stay and food to eat. This happens when a particular organism attacks a tree either by eating it or by laying its eggs in it. When this happens, the tree will form extra tissue called a gall. The parasitic organism will move into its newly formed home where it can live, reproduce, and eat. Galls are believed to form in response to a plant growth-regulating chemical produced by the attacking organism. The growths can be caused by many different kinds of organisms, including fungi, viruses, insects, mites, and nematode worms. They can be found on leaves, branches, trunks, and roots. Galls usually do not cause serious damage to a well-established tree and are more of a bother than a serious threat.

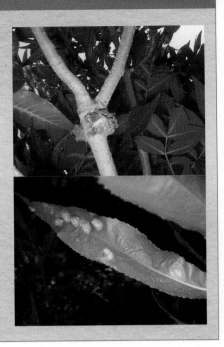

Proper pruning techniques and timing are important in growing good ornamental trees and shrubs. Incorrect pruning can leave a plant in worse condition than before. A plant in weak condition is susceptible to insects and diseases. Poor pruning can cause the loss of a season of flowers or fruit. In general, flowering plants should be pruned just after they **bloom,** or produce flowers. This is done so that the next season's flowering wood is not cut off.

It is easier to prune deciduous trees and shrubs in the late fall or winter, after the leaves drop. The framework is bare and easier to see. When removing large limbs, care should be taken to avoid splitting and tearing of the limb (Figure 25–15).

Before Pruning

After Pruning

FIGURE 25–14 Careful selection of branches for removal is important when pruning deciduous shrubs.

First cut partway through the branch at A, then cut it off at B. Make the final cut, just outside the branch collar, at C.

FIGURE 25–15 It is important to follow the recommended order of cuts when removing a large tree branch. The wrong cutting order is likely to cause damage to the tree.

Hot Topics in Agriscience

IMPROVING FOREST HEALTH

One of the great difficulties that is experienced by forest managers is court intervention by individuals who have political agendas. In recent years, many of the management decisions of forest managers have been challenged in court. Regardless of the outcomes in these cases, extra costs are added to the management of these resources, and the time delay that occurs is sometimes catastrophic. For example, an infestation of pine beetles can often be contained to a small area if it is treated with appropriate chemicals in the early stages. A challenge in court will prevent or delay treatment until a major outbreak of the beetles has occurred. In some instances, the harvest of trees that have been killed by fire has been prevented or delayed until the trees are completely dehydrated. When this occurs, the trees develop large cracks in the wood that destroy the value of the wood.

A second issue related to forest health is that some forests are simply not managed for healthy growth. Trees must be thinned to allow sufficient access to sunlight and nutrients. Forest practices must be consistent with good management practices for soil and water. New trees must be planted after timber harvests and wildfires. Pest management must be practiced, and political issues must be dealt with. All of these practices and issues have impacts on forest health. Healthy forests can sustain our needs for wood and forest products. Unhealthy forests cannot fill these needs.

FIGURE 25–16 Insect infestations can build up fast on susceptible plants. *(Courtesy of USDA/ARS #K-2849-11)*

Specific plants have specific pruning requirements. It is best to determine how each plant should be pruned before the pruning operation is started. Local plant specialists are usually willing to help or give suggestions for proper pruning. Good reference books are available with accurate information on pruning particular plants.

INSECTS AND DISEASES

Most plants are subject to damage by insects and diseases. It is easier to prevent the damage than to control it. Pest-resistant varieties should be used, and they should be kept healthy and vigorous. Weakened trees and shrubs are easily attacked and damaged by insects and diseases (Figure 25–16).

It is important that plants be selected that are suited to the environment in which they are to be planted. Plants cannot tolerate stressful conditions continuously. Plants will adapt to various conditions, but some plants adapt better than others. Some plants can withstand air pollution, drought, heat, or even wet soils. Other types of plants can tolerate infertile conditions, but not hot, dry weather. Still others may not survive in heavy shade or in full sun.

In its lifetime, any plant can be expected to experience attack by some insects and diseases. It is wise to become aware of the plants in your area and to consult the cooperative extension service for more specifics on the types of insects and diseases that can be expected to infect various plants in a given locality.

STUDENT ACTIVITIES

1. Write the Terms to Know and their meanings in your notebook.
2. Contact a local nursery and obtain a nursery catalog.
3. Visit a garden center or nursery and compile a list of trees and shrubs that are available and recommended for your community.
4. Prepare a chart of popular ornamental trees and shrubs for your community. For each plant, list the scientific name, common name, mature size of plant, time of flowering, color of flowers, spring leaf color, and fall leaf color.
5. Survey your home or other assigned area and make a list of the trees and shrubs that are present. Use books, nursery catalogs, and other resources to help identify the plants.
6. For a given area, determine the diameter of each tree 4.5 feet above the ground. Work up a recommended fertilizer program for the trees, including amount of fertilizer per tree, local fertilizer prices, cost of fertilizer for each tree, time of year to fertilize each, and the total cost of fertilizer for the area.
7. Prepare a pruning schedule for plants around your home or other area based on local recommendations. Include in the schedule the common and scientific names of each plant, time of pruning, and special requirements for pruning.
8. Prepare a sketch or a scale drawing of your home property or other assigned area. Locate the existing trees and shrubs on the sketch, using a circle with the initials of the scientific name to represent each plant. Add new plants that you believe will enhance the property you have surveyed.
9. Volunteer at a local park to help plant new trees.
10. Create a game, song, or activity that will help you remember tree names compared with their shape from Figure 25–6.

SELF EVALUATION

A. Multiple Choice

1. An acre of plant material can produce enough oxygen to keep
   a. 5 to 10 people alive each year.
   b. 10 to 16 people alive each year.
   c. 16 to 20 people alive each year.
   d. nobody alive. It does not produce enough oxygen to be of any value.

2. A border planting
   a. is used as a single plant to highlight a fence or some other special feature of the landscape.
   b. is used to separate some part of the landscape from another.
   c. is a number of trees or shrubs planted together as a point of interest.
   d. is a collection of plants that are placed in the landscape as needed.

3. When planning the location for planting trees and shrubs, which is not a major consideration?
   a. fruit size and type
   b. flower color
   c. other structures
   d. bare-rooted plants

4. The term *Cornus florida rubra* is a
    a. common name.
    b. scientific name.
    c. name that was developed in Florida.
    d. type of annual deciduous plant.

5. Planting of trees and shrubs may be done
    a. in spring, summer, or fall.
    b. in spring only.
    c. in fall only.
    d. in spring and summer.

6. Mulch should be applied
    a. 1 to 2 inches deep.
    b. 2 to 2.5 inches deep.
    c. 3 to 4 inches deep.
    d. 6 inches deep to keep out the weeds.

B. Matching

_____ **1.** Urban foresters
_____ **2.** Shrubs
_____ **3.** Specimen plant
_____ **4.** Nomenclature
_____ **5.** Pruning
_____ **6.** Canopy

a. The removal of dead, broken, unwanted, diseased, and insect-infested wood
b. Used as a single plant to highlight it or some other special feature of the landscape
c. A systematic method of naming plants
d. Woody plants that normally grow low and produce many stems or shoots from the base
e. The top of the plant; has the framework and leaves
f. Help install and maintain trees in large cities

C. Completion

1. Plant material can cut noise pollution by _____.

2. A rule of thumb is that 1 cubic yard of mulch can be spread over a _____ square foot area.

3. Newly planted trees may have to be _____ to prevent wind damage.

4. Plants should be fertilized every _____ to _____ years, starting with the growing season after the first year.

5. Proper pruning techniques and timing are important in growing good ornamental trees and _____.

6. _____ _____ is the systematic cutting of the roots by hand or machine to encourage the roots to develop close to the trunk.

D. True or False

_____ **1.** Trees and plants are valuable only for beauty.

_____ **2.** Trees can act as insulators, keeping a house cool in the summer and warm in the winter.

_____ **3.** A well-designed landscape does not increase the value of a property.

_____ **4.** All plants have the same type of growth habit.

_____ **5.** It is not a good practice to fertilize trees after middle to late July.

_____ **6.** Most plants are subject to damage from insects, diseases, or both at some time.

Animal Sciences

Animals of the Future

Making Better Sheep

Fifty years ago, desert range sheep averaged a 70 percent lamb crop—that is, each 100 female sheep averaged a total of 70 live lambs per year. Today, that figure is 85 percent, and scientists at the U.S. Sheep Experiment Station at Dubois, Idaho, believe the figure will be 150 percent in 50 years. They also project that under intensive confinement production, it is reasonable to expect a national average lamb crop of 400 percent. How will we do this? The Booroola Merino sheep, which originated in Australia, has some individual female sheep that have five or six lambs instead of the typical one or two. There is hope that the gene responsible for the multiple lambings can be isolated and bred into the genetic makeup of our most productive breeds. That would solve the numbers game, but how will the sheep and their managers keep so many newborn lambs alive and growing?

More Milk from Fewer Cows

Fifty years ago, the United States had 25 million dairy cows, and the national average milk production per cow was 4,600 lb. In 2000, the figures were 9.2 million cows averaging 18,240 lb milk. Some believe that in 50 years we will have herds averaging 40,000 lb milk per cow per year.

Will great increases in the future necessitate bigger cows? Dairy scientists at the U.S. Department of Agriculture (USDA) Dairy Forage Research Center in Madison, Wisconsin, do not believe so. Better feed could make conversion of feed to milk a more efficient process. Currently, only 30 to 40 percent of the material in plant cell walls is available to the animal. The rest simply passes through. If plant cell walls could be modified to make them more digestible without diminishing plant health and performance, then the animal could extract more nutrients from a given amount of feed.

The discovery of the hormone bovine somatotropin (BST) is affecting the way a cow's body divides calories obtained from feed between milk production and other body functions. The use of BST causes the cow to shift more of her calories into milk, thus increasing her conversion rate. The widespread use of BST causes an immediate increase in milk production by 15 percent or more. Research will bring more progress in this and other areas.

Preventing Poultry Diseases

Devastating diseases, such as avian influenza and Newcastle disease, still threaten to eliminate whole flocks in the poultry industry. Both viruses have strains with disease-producing capability that range from those not killing any birds to those that will kill 100 percent of the birds. So, on the diagnosis of a few birds with the disease, the entire flock is destroyed. The movement of people and poultry in the locality, and even within the state or region, may be restricted. The economic loss is staggering. Scientists struggle to keep ahead of such diseases by developing vaccines and practicing strict quarantines.

Veterinary researchers at the USDA Southeast Poultry Research Laboratory at Athens, Georgia, hope to find a better way to manage these poultry pathogens. The hope is to determine the fundamental genetic traits of the viruses that will enable prediction of their behavior in poultry. With genetic markers, the scientists could determine if a given virus in a bird or sample of birds is a mild one or a lethal one. The decision about destroying birds can then be made more intelligently. DNA probes for examining bird tissue promise to be a useful diagnostic tool to aid in the process.

Improvements in animal production more and more are because of genetic engineering and other laboratory procedures. *(Courtesy USDA/ARS #K-4134-6)*

Occupying the Parking Spot

Eradicating the African swine fever is a dream of the swine industry. So far, it has been an elusive dream. However, a genetic engineering technique using what is referred to as "the nonsense gene" has some promise. Certain viruses, such as that of the African swine fever, have the ability to incorporate themselves into the genetic material of an animal and be carried into the animal's offspring. However, the virus has to establish itself in a particular spot on the genetic material. If a modified version of the virus that cannot cause the disease can be engineered, then animals can be infected with the safe virus. The safe virus would become "nonsense genes," occupying the "parking spot" on the chromosome and preventing the disease-causing virus from establishing itself.

In summary, the animals of the future will continue to improve in productivity and function. Much research on farm animals focuses on increasing productivity to help feed the world's expanding population. In addition, research focuses on the feeding and management of pets and other animals for recreational uses. Some research deals with learning the needs and behavior of animals in the wild and providing improved habitat according to the unique needs of wildlife.

Animal Anatomy, Physiology, and Nutrition

MATERIALS LIST

✓ commercial feed tags from various feeds

✓ charts of various animal digestive systems

✓ Internet access

COMPETENCIES TO BE DEVELOPED

After studying this unit, you should be able to:

- compare animal digestive systems.
- understand the basics of animal physiology.
- understand how nutrients are used by animals.
- identify classes and sources of nutrients.
- identify symptoms of nutrient deficiencies.
- explain the role of feed additives in livestock nutrition.
- compare the composition of various feedstuffs.

SUGGESTED CLASS ACTIVITIES

1. Conduct a feeding trial with baby chickens using different quality rations. Weigh the chicks in each group at the beginning of the trial and at the end of the trial. Record any observations that you make as the trial is conducted. Calculate the differences in total weight gain after 2 weeks. Place all of the groups on the ration that resulted in the greatest weight gain and proceed with the trial for 2 more weeks. Record the results and offer an explanation for the results.

2. Examine the anatomy of a farm animal by dissecting a fetal pig or by butchering a lamb, hog, or beef animal. Observe each of the systems of the animal and explain the physiology or function of each system as you examine the organs that make up each system.

3. Split the class into four small groups. Each group will research the daily nutrient requirements of a different domestic animal of their choosing. The groups may use the Internet, library resources, or any other reliable source. After compiling their information, the groups will share their findings with the class.

TERMS TO KNOW

nutrition
scurvy
anorexia
obesity
ration
balanced ration
deficiency diseases
vitamin
mineral
anatomy
skeletal system
bone
bone marrow
muscular system

voluntary muscles
involuntary muscles
protein
circulatory system
lymph glands
carbohydrates
respiratory system
central nervous system
peripheral nervous system
urinary system
endocrine or hormone
    system
hormones

digestive system
ruminant
rumen
roughage
monogastric
concentrates
lactation
fructose
galactose
sucrose
maltose
lactose
starch

cellulose
fat
supplement
synthetic nutrients
oil meal
urea
feed additive
antibiotic
dry matter
TDN
by-product
green roughage
silage

Feed is animal food. It represents the largest single cost item in the production of livestock. Therefore, it is important to understand the complex nature of animal nutrition and how animals use the feed that they eat. **Nutrition** is the process by which animals eat food and use it to live, grow, and reproduce (Figure 26–1).

"You are what you eat" is an axiom that is true to a considerable extent, especially as it relates to good health of both humans and animals. This unit explores the relationship between good nutrition and good health.

FIGURE 26–1 Animal health, growth, and reproduction all are directly related to nutrition.

Agri-Profile

CAREER AREAS:

Animal Nutritionist/
Feed Formulator/Animal Manager/
Physiologist/Veterinarian

Animal nutrition is an interesting field of study that can lead to several types of career opportunities. Some careers are involved in basic research in which scientists investigate how an animal uses its food supply to grow or reproduce. They also determine the nutrient values of feeds, including how digestible they are. This career field extends to production, processing, and sale of feeds. Some people in this career field raise farm animals. Raising and managing farm animals requires an understanding of animal nutrition.

Certain elements of nutrition are basic, such as the composition of feed grains, animal by-products, and the basic nutrients needed by animals. However, the nutritive content of forages and other feedstuffs varies considerably according to stage of growth, condition, and quality. The digestive capabilities of animals vary considerably from species to species. Nutritional needs of animals vary with age, stage of development, production, and pregnancy.

Careers in nutrition may be predominantly in the basic sciences or may be in the application of science to the nutrient requirements of an animal species. They may focus on fish, small animals, pets, horses, poultry, livestock, dairy, or wild animals.

Nutrition is one of the most critical factors in animal production. Nutritional research, feed analysis, and animal evaluation are ongoing processes with production animals and pets alike. *(Courtesy of Jim Strawser)*

NUTRITION IN HUMAN AND ANIMAL HEALTH

The relationship between proper nutrition and health has long been recognized. Early sailors stocked their sailing vessels with limes when going to sea for long periods. This was to prevent the dreaded disease, scurvy. **Scurvy** is a disease of the gums and skin caused by a deficiency of vitamin C in the diet. Even today, the effects of poor nutrition are seen in the human problems of anorexia and obesity. In simple terms, **anorexia** is a result of too little nutrition, and **obesity** the result of too much or improper types of food being eaten.

In animals, proper nutrition is just as important as it is in humans. Feed efficiency, rate of gain, and days to market weight are all uppermost in the minds of those people who raise livestock for meat. Proper nutrition is just as important for animals being grown for milk, wool, or fur production. Slow growth, poor reproduction, reduced production, and poor health are generally the result of less-than-adequate animal rations. The amount and content of food eaten by an animal in 1 day is referred to as the animal's **ration**. When the amount of feed consumed by an animal in 24 hours contains all of the needed nutrients in the proper proportions and amounts, the ration is referred to as a balanced ration.

Numerous diseases may result from improper amounts or balance of vitamins and minerals. Such diseases are called deficiency diseases. Vitamins are complex chemicals and minerals are elements essential for normal body functioning of humans and animals alike. Not all types of animals require the same vitamins and minerals to maintain good health.

ANIMAL ANATOMY AND PHYSIOLOGY

The internal functions and vital processes of animals and their organs are referred to as animal physiology. The various body systems, such as the skeletal, muscular, circulatory, respiratory, nervous, urinary, endocrine, digestive, and reproductive systems, must all be properly fed and working together for the animal to be healthy and productive. To this end, proper nutrition is a must. The various parts of the body are collectively known as anatomy.

INTERNET KEY WORDS:

vitamin deficiency symptoms

Skeletal System

The skeletal system (Figure 26–2) is made up of bones joined together by cartilage and ligaments. The purpose of the skeletal system is to provide support for the body and protection for the brain and other soft organs of the body.

Bone is the main component of the skeletal system. It is composed of about 26 percent minerals. This mineral material is mostly calcium phosphate and calcium carbonate. Another 50 percent of bone is water, 20 percent is protein, and 4 percent is fat.

The material inside bones is called bone marrow, and it produces blood cells. The growth and strength of bones are greatly affected by minerals and vitamins in animal rations.

Muscular System

The muscular system (Figure 26–3) is the lean meat of the animal and the part of the body that is used for human consumption (food). The purposes of muscles are to provide for movement in cooperation with the skeletal system and to support life.

Muscles may be voluntary or involuntary, depending on whether they can be physically controlled by the animal. Voluntary muscles can be controlled by animals to do such things as walk and eat food. Involuntary muscles operate in the body without control by the will of the animal, and they function even while the animal sleeps. Examples of involuntary muscles are heart and diaphragm muscles.

Muscles are composed largely of protein. Large amounts of protein are required for the maintenance of the animal and for growth and reproduction. Proteins are nutrients made up of amino acids that are the building blocks of muscles.

Circulatory System

The heart, veins, arteries, and lymph glands compose the circulatory system (Figure 26–4). This system provides food and oxygen to the cells of the body and filters waste materials from the body. Lymph glands secrete disease-fighting materials for the body.

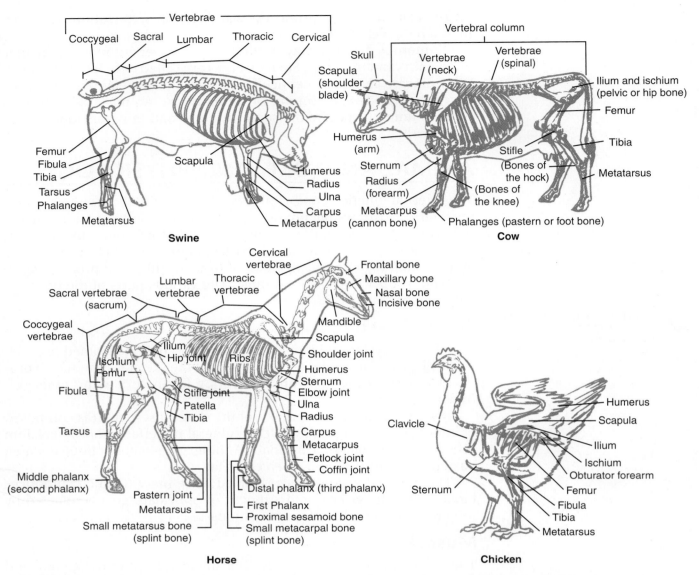

FIGURE 26–2 The skeletal system provides support for the body and protection for the soft organs. *(Courtesy of IMS, Texas A & M University)*

Vitamins, minerals, proteins, and carbohydrates are all essential for the smooth running of the circulatory system. **Carbohydrates** are sugars and starches that supply energy to the animal.

Respiratory System

The **respiratory system** provides oxygen to the blood of the animal, and removes waste gases such as carbon dioxide from the blood. The respiratory system is composed of the nostrils, nasal cavity, pharynx, larynx, trachea, and lungs (Figure 26–5). This system allows for breathing and makes use of the muscular and skeletal systems to draw air in and out of the lungs. Oxygen passes from the lungs to the blood and is distributed to the cells of the body by the circulatory system.

Muscular Structures of Domestic Animals

1. Rhomgoideus
2. Cervical cutanu
3. Serratus thoracis
4. Trapezius cervicalis
5. Trapezius thoracalis
6. Latissimus dorsi
7. External intercostal
8. Serratus dorsalis
9. Obiquus abdominis externus
10. Tensor fasciae latae
11. Gluteus superficialis
12. Biceps femoris
13. Sacro-coccygeus dorsalis
14. Scaro-coccygeus internalis
15. Coccygeus
16. Semitendinosis
17. Fascia lata
18. Aponeurosis
19. Serratus thoracis
20. Posterior deep pectoral
21. Lateral head of triceps
22. Anterior superficial pectoral
23. Long head of triceps
24. Deltoid
25. Cervical cutaneous muscle
26. Brachiocephalicus
27. Sterno-cephalicus
28. Extensor
29. Flexor

FIGURE 26–3 The well-developed muscular system distinguishes meat animals from animals that are raised for other purposes such as milk and wool production. *(Courtesy of IMS, Texas A & M University)*

Circulatory Systems of Domestic Animals

FIGURE 26–4 The circulatory system carries food and oxygen that is dissolved in the blood to all of the cells of the body. *(Courtesy of IMS, Texas A & M University)*

Respiratory Systems of Domestic Animals

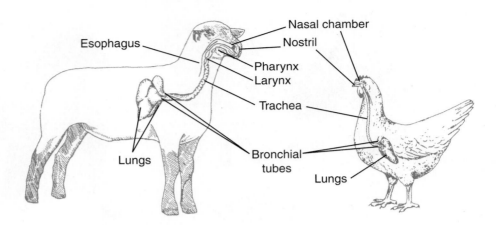

FIGURE 26–5 The respiratory system provides oxygen to the blood. *(Courtesy of IMS, Texas A & M University)*

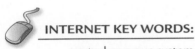

INTERNET KEY WORDS:

central nervous system
peripheral nervous system

Nervous System

The nervous system of animals is composed of the central nervous system and the peripheral nervous system.

The **central nervous system** includes the brain and the spinal cord. It is responsible for coordinating the movements of animals and also responds to all of the senses. The senses are hearing, sight, smell, touch, and taste.

The **peripheral nervous system** controls the functions of the body tissues, including the organs. The nerves transmit messages to the brain from the outer parts of the body (Figure 26–6).

Because the nervous system is composed primarily of soft tissues, proteins are particularly important in maintaining its health.

Urinary System

The function of the **urinary system** is to remove waste materials from the blood. The primary parts are the kidneys, bladder, ureters, and urethra (Figure 26–7). The kidneys also help regulate the makeup of blood and help to maintain other internal systems.

Nervous System of the Horse

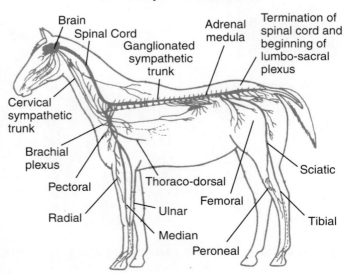

FIGURE 26–6 The nervous system consists of the brain, spinal cord, and the nerves that are distributed throughout the body. This system coordinates all of the other body systems. *(Courtesy of IMS, Texas A & M University)*

Urinary System

FIGURE 26–7 The urinary system removes waste materials from the blood. *(Courtesy of IMS, Texas A & M University)*

INTERNET KEY WORDS:

ruminant polygastric

Abnormal levels of proteins fed to animals have been known to cause stress to the urinary system, which rids the body of excess protein. Greater-than-recommended levels of minerals may also cause kidney problems.

Endocrine System

The **endocrine** or **hormone system** is a group of ductless glands that release hormones into the body. **Hormones** are chemicals that regulate many of the activities of the body. Some of these body functions are growth, reproduction, milk production, and breathing rate. Hormones are needed in only very minute amounts. For example, only 1/100,000,000 gram oxytocin hormone will stimulate the almost immediate letdown of milk in female animals. Oxytocin is a hormone from the hypothalmus gland.

Proper levels of all nutrients, especially minerals, are important for the proper functioning of the endocrine system.

Digestive System

The **digestive system** of an animal provides food for the body and for all of its systems. This system stores food temporarily, prepares food for use by the body, and removes waste products from the body.

There are three basic types of digestive systems in animals of agriscience importance: polygastric, or ruminant; monogastric; and poultry.

Polygastric, or Ruminant, System **Ruminants** are a class of animals that have stomachs with more than one compartment (Figure 26–8). Cattle and sheep are ruminants and have multicompartment stomachs. The very large compartment is called the **rumen.** The rumen can store large amounts of roughage. A **roughage** is grass, hay, silage, or other high-fiber feed. Ruminants have the ability to break down plant fibers and to use them for food far better than animals that are not ruminants.

B-complex vitamins are manufactured in the digestive systems of ruminant animals by bacteria. Such vitamins need not be added to the diets of these animals, even though they are required by the body. Notably, calves do not develop true rumens until they are several

Digestive System of the Cow

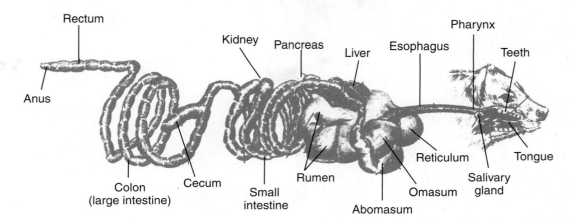

FIGURE 26–8 The ruminant digestive system can use large amounts of roughage. *(Courtesy of IMS, Texas A & M University)*

Science Profile

A FOUR-CHAMBERED STOMACH

Ruminant animals have distinct advantages over their nonruminant counterparts. They can digest roughage. The main source of energy in roughage is cellulose. Cellulose is a complex, nutrient-rich plant material that is unusable to nonruminant animals because they cannot digest it. The ruminant stomach is host to specific bacteria that produce enzymes with the capability of unlocking the complex structure of cellulose and digesting it. Not only do these microbes aid in breaking down cellulose, but they become a valuable source of protein for the animal as well.

The ruminant animal has four compartments that make up its stomach: the rumen, reticulum, omasum, and the abomasum. The compartments work together to digest food that is high in fiber. Food can flow easily between the rumen and the reticulum. Most of the roughage is broken down there. Next, food enters the omasum, where the food particles become smaller and water is reabsorbed into the body. Next, the food enters the abomasum or true stomach. Here, digestive enzymes are secreted that breakdown the food into basic nutrients. The nutrients then move into the small intestines where they are absorbed. Unused materials move into the large intestine where more water is reabsorbed, and the remaining material leaves the body as feces.

INTERNET KEY WORDS:

monogastric digestion
poultry digestion

months old; therefore, they need to be fed complete rations like non-ruminant animals until their rumens develop.

Monogastric System The digestive systems of swine, horses, and many other animals are called monogastric (Figure 26–9). Monogastric means having a stomach with one compartment. The stomachs of swine and horses are relatively small and can store only small amounts of food at any one time. Most of the digestion takes place in the small intestines. Monogastric animals are unable to breakdown large amounts of roughage. Therefore, their rations must be greater in concentrates. Concentrates are grains, low in fiber and high in total digestible nutrients. Also included in the diets of monogastric animals must be B-complex vitamins, because they cannot make such vitamins in their digestive systems.

Poultry Digestive System Although poultry have monogastric digestive systems, their digestive systems are different enough to discuss

Digestive System of Swine

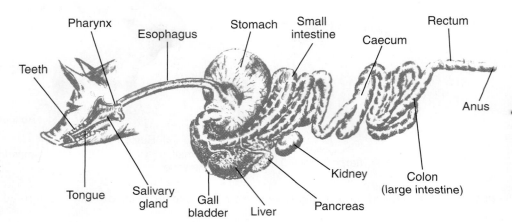

FIGURE 26–9 The monogastric digestive system has a simple stomach. *(Courtesy of IMS, Texas A & M University)*

Hot Topics in Agriscience

THE AMINO ACID "LYSINE"

Animals require nutrients of the right kinds and in the right amounts to produce the proteins that make up muscles and other body tissues. When an amino acid is unavailable, a protein that requires that amino acid can no longer be manufactured by the body.

The amino acid lysine is too low in most corn varieties to support rapid growth in hogs. Lysine must be added to corn to avoid this problem. Humans who rely on diets that are high in corn are also at risk for a lysine deficiency in the diet. However, people can supplement their diets with fish, turkey, or other high-lysine foods to correct the deficiency.

Researchers have developed new corn varieties known as "high-lysine" corn that has been shown to correct the nutrient deficiency. However, the new variety is lower in grain yield than most dent-corn varieties, and fields of "high-lysine" corn must be separated from other corn fields to prevent cross-pollination. Cross-pollinated corn yields the usual "low-lysine" corn, because the gene for "low-lysine" corn is dominant over the gene for "high-lysine" corn.

Digestive System of a Chicken

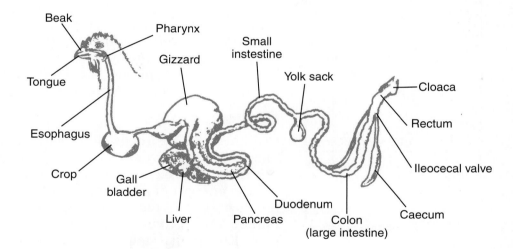

FIGURE 26–10 The poultry digestive system has no true stomach, but it does have an organ called the crop, which stores small amounts of feed. It also has a gizzard, which is the organ that grinds the seeds and other materials eaten by birds. *(Courtesy of IMS, Texas A & M University)*

them separately (Figure 26–10). Chickens have no teeth and must swallow their food whole. The food is stored in a crop and passed on to the gizzard, where it is ground up. It then passes on to the small intestine for digestion. Poultry rations must be high in food value, because poultry have no true stomachs and have little room for storage of food that has been eaten.

MAJOR CLASSES OF NUTRIENTS

Water

Water is the largest component of nearly all living things. Growing plants are usually 70 to 80 percent water. Similarly, the muscles and internal organs of animals contain 75 percent or more water.

Water is the solution in which all nutrients for animals are dissolved or suspended for transport throughout the body. Water reacts with many chemical compounds in the body to help breakdown food into products that can be used by the body.

Water provides rigidity to the body, allowing it to maintain its shape. The liquid solution in each cell is responsible for this rigidity.

Water is also important in regulating the body temperature of animals, through perspiration and evaporation. Because of the ability of water to absorb and carry heat, body temperatures of animals increase and decrease slower than would be possible otherwise.

Water is the least expensive nutrient for animals. However, most animals can live only a matter of days if they do not have access to it.

Protein

INTERNET KEY WORDS:

monogastric protein requirement
animal nutrition, amino acids

Protein is the major component of muscles and tissues. Proteins are complex materials and are made of various nitrogen compounds called *amino acids.* Some amino acids are essential for animals, and some are not. Therefore, the quality of proteins fed to animals must be considered.

Monogastric animals need specific amino acids, so it is important that they receive high-quality proteins containing the appropriate amino acids. However, in ruminant animals, quantity of protein is more important than quality. Ruminants can convert amino acids in their rumens to different amino acids to meet their needs.

Protein is used continuously by animals to maintain the body as body cells are continually dying and being replaced. In young animals, large amounts of protein are used for body growth. Protein is also important for healthful reproduction.

Carbohydrates

INTERNET KEY WORDS:

animal nutrition, carbohydrates
animal nutrition, minerals
animal nutrition, vitamins
animal nutrition, fat

Carbohydrates are a class of nutrients composed of sugars and starches. They provide energy and heat to animals. Carbohydrates are composed primarily of the elements carbon, hydrogen, and oxygen.

The energy obtained from carbohydrates is used for growth, maintenance, work, reproduction, and lactation (milk production). Carbohydrates come in several forms, with the sugars being the simplest. Examples of simple sugars used in animal feeds are glucose, fructose, and galactose. Compound sugars include sucrose, maltose, and lactose. More complex forms of carbohydrates include starch and cellulose.

Carbohydrates make up about 75 percent of most animal rations, but there is little carbohydrate in the body at any one time. Carbohydrates in the diet that are not used quickly are converted to fat and stored in the body. Fat has 2.25 times as much energy per gram as do carbohydrates.

Minerals

Minerals have many functions in the animal. The skeleton is composed mostly of minerals. They are important parts of soft tissues and fluids in the body. The endocrine system is heavily dependent on various minerals, as are the circulatory, urinary, and nervous systems.

There are 15 minerals that have been identified as being essential to the health of animals. These are calcium, phosphorus, sodium, chlorine, potassium, sulfur, iron, iodine, cobalt, copper, fluorine, manganese, molybdenum, selenium, and zinc. In the past, most of these minerals were provided naturally by feeds grown on fertile soils and by contact with the soil itself. Today, it is increasingly important to provide additional mineral matter to the diet of animals. Mineral supplements are especially important for animals that spend their lives in confinement. A mineral fed as a separate feed is called a supplement.

Vitamins

Vitamins are acquired by animals in several different ways. Some are available in roughages and concentrates, some are available in feeds made from animal by-products, and some are made by the body itself.

Vitamins are required in only minute quantities in animals. Vitamins act mostly as catalysts for other body processes. There are large variations in the necessity for vitamins in various species of animals that are important to agriscience.

Some of the specific ways that vitamins are used in animals include clotting of blood, forming bones, reproducing, keeping membranes healthy, producing milk, and preventing certain nervous system disorders.

Fat

Only small amounts of fat are required in most animal diets. The addition of fat to the diets of animals improves the palatability, flavor, texture, and energy levels of feed. The addition of small amounts of fat to the diet has also been shown to increase milk production and to aid in the fattening of meat animals. Fats are also necessary in the body as carriers of fat-soluble vitamins.

SOURCES OF NUTRIENTS

The sources of nutrients for animals are many and extremely varied. Important animal feed components include roughages, concentrates, animal by-products, minerals from mineral deposits, and nutrients made chemically, called synthetic nutrients.

Proteins

The major sources of protein for animals include oil seeds such as soybeans, peanuts, cottonseed, and linseed. These seeds are processed by cooking and other procedures to remove the bulk of the oil from them. The remainder of the seed content is then dried and ground up for feed. Feed consisting of ground oil seeds with the oil removed is called oil meal.

Cereal grains provide lesser amounts of protein than does oil meal, but they are also important protein sources. Good-quality legume hay, such as alfalfa or clover, is also a good plant source of protein for ruminant animals.

Animal protein is generally of greater quality than plant protein. More specifically, animal protein usually contains more of the essential amino acids than does protein from plants. Sources of animal protein include tankage, fish meal, blood meal, skim milk, whey, feather meal, and meat products.

Hot Topics in Agriscience

MAD COW DISEASE

Mad cow disease, also known as bovine spongiform encephalopathy or chronic wasting disease, was discovered for the first time in the United States near Yakima, Washington. Within hours, Japan, South Korea, and Taiwan had halted imports of U.S. beef, and other countries were threatening to do so. An earlier case of the disease in Canada closed the border between the United States and Canada for beef imports.

The disease is known to be spread by feeding contaminated animal products such as processed meat scraps to cattle. A ban has been placed in effect on the feeding of animal meat products to cattle. An extensive search was also implemented that traced the origin of the diseased animal and all known animals that were herd mates. Offspring of the affected cow were also located and destroyed. Government agencies have gone to great lengths to isolate the disease and eradicate it. The disease has become so widespread in Europe that thousands of animals have been killed and burned in efforts to control and eradicate it.

Nonprotein nitrogen in the form of urea can be used as a substitute for protein for ruminant animals. Urea is a synthetic source of nitrogen made from air, water, and carbon. The rumen bacteria found in the rumens of ruminant animals such as cattle and sheep can convert urea and other nitrogen sources to amino acids and proteins. These bacteria constitute a source of protein, and they are digested together with the feed. The feeding of urea should be limited to not more than 1 percent of the total dry matter in the ration. Young ruminants and all nonruminants are unable to digest urea.

Carbohydrates

Carbohydrates are found in all plant materials. The major sources of carbohydrates for animal feed are the cereal grains. Corn is the most important of these grains in the United States, followed by wheat, barley, oats, and rye.

Other sources of carbohydrates include nonlegume hays such as orchard grass, timothy, other grasses, and molasses.

Normal animal rations generally contain adequate levels of carbohydrates.

Fats

Because fats are needed in fairly small amounts in the diets of animals, it is seldom necessary to identify specific sources of dietary fat. Most sources of protein are also sources of fat. This is especially true for the oil seeds and animal by-products.

Vitamins and Minerals

Vitamins and minerals are part of all the normal feeds of animals. Ruminants manufacture B-complex vitamins in their rumens. Exposure to sunlight allows the body to manufacture vitamin D. Contact with the soil, coupled with other feeds grown on fertile land, provides most of the mineral requirements of animals that have access to pasture and high-quality feeds.

Science Connection

FEEDING FOR TWO

The nutrient needs of dairy fetuses have become the research focus of animal physiologist William House of the USDA Plant, Soil, and Nutrition Laboratory of Ithaca, New York, and of animal science professor Alan Bell of Cornell University. Their immediate goal is to produce a complete profile of fetal mineral content for dairy cattle, including a look at how the proportion of various minerals changes as the fetus progresses through gestation. Such information will help producers feed the correct amounts of minerals to pregnant dairy cows to ensure the birth of healthy calves.

Healthy cows absorb about 45 percent of the calcium they eat. Therefore, they must eat nearly twice the combined amount needed for their own bodies and that of their fetuses. Before the two researchers began their study, only calcium and phosphorus content of dairy fetuses had been measured. In their current study, they have determined the levels of potassium, sodium, magnesium, iron, zinc, copper, and manganese, as well as calcium and phosphorus taken up by the fetus, fetal membranes, fluids, and placental tissue.

Animal scientists display a 2-week-old calf born from a cow in a nutrient study. *(Courtesy of USDA/ARS #K-5177-1)*

House and Bell have confirmed that the amounts of calcium, phosphorus, and magnesium increase as the fetus grows. However, surprisingly, they discovered that the amounts of potassium and sodium decrease near the end of the term. During the latter stages of gestation, it was concluded that each day a fetus accrues an average of 0.2 g magnesium, 1 g potassium, 1.4 g sodium, 18 mg iron, 11.7 mg zinc, 1.6 mg copper, and 0.3 mg manganese. This information, combined with knowledge about the mother's body and production requirements, will enable managers to better meet the nutritional needs of dairy cows during their gestation periods.

However, it is sometimes necessary to supplement natural sources of vitamins and minerals. Commercial vitamin and mineral supplements are formulated for specific classes of animals and their special needs. Such supplements are available wherever animals exist in the developed countries of the world.

SYMPTOMS OF NUTRIENT DEFICIENCIES

Animals must be fed appropriate types and amounts of feed regularly to remain healthy and to produce milk, meat, wool, eggs, fur, work, or healthy young. Shortages, or deficiencies, of various nutrients will generally produce observable effects in animals. Some common symptoms of nutrient deficiencies are described in Figure 26–11.

Disorders Caused by:	
Vitamin Deficiency	**Mineral Deficiency**

Vitamin A
1. Night blindness
2. Loss of young
3. Poor growth
4. Nasal discharge
5. Diarrhea

Vitamin C
1. Scurvy
2. Gum inflammation
3. Hemorrhages
4. Slow healing of wounds

Vitamin D
1. Bone weakness and deformities
2. *Rickets* and *osteomalacia* are bone diseases in young and old animals, respectively
3. Thin egg shells
4. Weak, deformed young
5. Lowered milk production

Vitamin E
1. Reproductive failures
2. Degeneration of certain muscles
3. *Stiff lamb disease,* or muscle degeneration in lambs
4. *White muscle disease,* or muscle degeneration in young calves
5. Poor egg hatchability

Vitamin K
1. Poor blood clotting
2. Internal hemorrhages

Thiamine
1. Poor appetite
2. Slow growth
3. Weakness
4. Nervousness

Riboflavin
1. Slow growth
2. *Dermatitis,* or skin disorder
3. Eye abnormalities
4. Diarrhea
5. Weak legs in pigs

Niacin
1. Dermatitis
2. Retarded growth
3. Digestive troubles

Pyridoxine
1. *Anemia,* or low red-blood-cell count
2. Poor growth
3. Convulsions in pigs

Pantothenic Acid
1. "Goose-stepping" in pigs
2. Unhealthy appearance
3. Digestive problems

Biotin
1. Dermatitis
2. Loss of hair
3. Retarded growth

Choline
1. Poor coordination
2. Poor health
3. Fatty liver
4. Poor reproduction in swine

Folic acid
1. Blood disorders
2. Poor growth

Vitamin B_{12}
1. Slow growth
2. Poor coordination
3. Poor reproduction

Calcium
1. Rickets
2. Poor growth
3. Deformed bones
4. Milk fever

Phosphorus
1. Lameness
2. Stiff joints
3. Rickets
4. Poor milk production

Sodium Chloride
1. Lack of appetite
2. Unhealthy appearance
3. Slow growth

Potassium
1. Slow growth
2. Joint stiffness
3. Poor feed efficiency

Sulfur
1. General unthriftiness (Lack of strong growth)
2. Poor growth

Iron
1. Anemia
2. Labored breathing
3. Edema (swelling) of the head and shoulders
4. Flabby, wrinkled skin

Iodine
1. *Goiter,* or enlarged thyroid gland in the neck
2. Weak or dead offspring at birth
3. Hairlessness
4. Infected navels, especially in foals

Cobalt
1. Delayed sexual development
2. Poor appetite
3. Slow growth
4. Decreased milk and wool production

Copper
1. Abnormal wool growth
2. Poor muscular coordination
3. Anemia
4. Weakness at birth

Fluorine
    Poor teeth

Manganese
1. Poor fertility
2. Deformed young
3. Poor growth

Molybdenum
    Poor growth rate

Selenium
1. Muscular degeneration
2. Heart failure
3. Paralysis
4. Poor growth

Zinc
1. Poor growth
2. Unhealthy wool or hair
3. Slow healing of wounds
4. *Parakeratosis,* or a skin disease similar to mange

FIGURE 26–11 Some disorders caused by vitamin and mineral deficiencies in animals.

FEED ADDITIVES

A **feed additive** is a nonnutritive substance that is added to feed to promote more rapid growth, to increase feed efficiency, or to maintain or improve health. Feed additives fall into two major groups: growth regulators (mostly hormones) and antibiotics. **Antibiotics** are substances used to help prevent or control diseases.

Some common growth regulators include stilbestrol, progesterone, and testosterone. They increase growth rates and feed efficiency by as much as 5 percent.

There is a wide range of antibiotics that are added in low levels to the diets of animals such as swine and poultry. Antibiotics keep certain low-grade infections at bay. Antibiotics added to feed allow growing animals to gain weight at their greatest potential.

In recent years, there has been a good deal of controversy over including antibiotics and growth hormones in the feed of animals. Of major concern is the possibility of these substances remaining in the meat of animals slaughtered for human consumption. To reduce this possibility, the substances generally must be removed from the feed well before the animals are marketed.

COMPOSITION OF FEEDS

INTERNET KEY WORDS:

animal feed additives
animal feed concentrates

All feeds are composed of water and dry matter. The material left after all water has been removed from feed is **dry matter**. Water makes up 70 to 80 percent of most living things. However, dry feeds generally contain only 10 to 20 percent water.

Dry matter is made up of organic matter and ash or mineral. The organic-matter portion of animal feed consists of protein; carbohydrates, such as starch and sugar; fat; and some vitamins. The proportion of these materials varies widely among different feeds.

CLASSIFICATION OF FEED MATERIALS

In general, feed for animals is classified into two types: concentrates and roughages. Concentrates are low in fiber and high in total digestible nutrients, abbreviated as **TDN**. Total digestible nutrients include all of the digestible protein, digestible nitrogen-free extract, digestible crude fiber, and 2.25 times the digestible fat contained in the ration. Conversely, roughages are high in fiber and low in TDN.

Concentrates

Included under the classification of concentrates are the feed, or cereal, grains. These include corn, wheat, oats, barley, rye, and milo, as well as many others. These grains make up the bulk of most concentrates.

Grain by-products, such as wheat bran, wheat middlings, brewer's grain, and distiller's grain, are concentrates. They are materials left over from the production processes used in making flour and alcohol. A **by-product** is a secondary product resulting from the production of a primary commodity.

The oil meals are by-products resulting from the processing of vegetable oil from oil seeds. Both oil meals and sugar in the form of cane molasses and beet molasses are considered concentrates used in animal feeds.

Finally, animal by-products are important concentrates. They include tankage, fish meal, meat scraps, blood meal, feather meal, and dried dairy products. These products are some of the by-products of the meat- and dairy-processing industries.

Roughages

INTERNET KEY WORDS:

animal feed, roughage

Roughages can be divided into three categories: dry, green, and silage. The most important of the dry roughages is hay (Figure 26–12). Some types of legume hay are alfalfa, clover, lespedeza, soybean, and peanut. A legume is a plant in which certain bacteria can transform nitrogen obtained from the air to nitrogen that plants can use. Grass hays include timothy, orchard grass, bromegrass, Bermuda grass, and others. The hulls of cottonseed, peanuts, and rice are also included in the category of dry roughages.

Green roughages are plant materials with high moisture content, such as grasses in pastures and root plants (including sugar beets, turnips, and rutabagas). Tubers such as potatoes are also a green roughage.

Silage is the feed that results from the storage and fermentation of green crops. The fermentation takes place in the absence of air. Corn silage is the most important member of this group. Other examples are grass, legume, and small-grain silages.

The successful production of animals for fun, profit, or sport requires proper animal nutrition. A knowledge of animal physiology, feed materials, and nutrition helps keep animals healthy and productive. A knowledge of nutrition-deficiency symptoms permits the animal manager to take corrective steps when the animal experiences improper nutrition.

FIGURE 26–12 Roughages are feeds that are high in fiber, such as hay, silage, and grass.

STUDENT ACTIVITIES

1. Write the Terms to Know and their meanings in your notebook.

2. Obtain several different commercial feed tags and compare the percentages of protein, fat, TDN, and other nutrients listed on the tag. Also make note of any vitamins and mineral supplements that are part of the ingredients. Types of feed additives should also be noted. If the price is known, try to determine what causes variations in the prices of the various feeds.

3. Compare the digestive systems of ruminants, nonruminants, and poultry. Note the parts that are alike and those that are different.

4. Arrange to have the digestive tracts of ruminant and nonruminant animals dissected and compare the contents of each.

5. Make an outline of this unit. Phrases and words that are bold should be included in the outline, together with a brief description of key principles.

6. Draw and label the three different digestive systems discussed in this unit.

7. Research one of the disorders caused by a vitamin or mineral deficiency described in the unit. Write a one-page report on the disorder you have chosen. Be sure to include the symptoms, treatment, and prevention techniques associated with the illness.

8. In your own words, describe the two classifications of feed material.

SELF EVALUATION

A. Multiple Choice

1. _____ are most important for the formation of bone.

   a. Vitamins   c. Minerals

   b. Proteins   d. Carbohydrates

2. Sugar, starch, and cellulose are all examples of

   a. carbohydrates.   c. fats.

   b. minerals.   d. proteins.

3. The purpose of additional fat in the diet is to

   a. improve palatability.   c. improve feed texture.

   b. increase energy levels.   d. all of the above.

4. Night blindness may be caused by a deficiency of

   a. zinc.   c. protein.

   b. vitamin A.   d. biotin.

5. Improved feed efficiency can be accomplished by including _____ in animal feeds.

   a. fats   c. additives

   b. minerals   d. roughages

6. Grain, oil meal, molasses, and meat by-products are all forms of

   a. roughages.   c. carbohydrates.

   b. concentrates.   d. dry matter.

7. Dry matter is made up of organic matter and

   a. mineral.   c. fat.

   b. water.   d. concentrate.

B. Matching

Group I

_____ **1.** Goiter

_____ **2.** Parakeratosis

_____ **3.** Rickets

_____ **4.** White muscle disease

_____ **5.** Anemia

_____ **6.** Dermatitis

a. Iron

b. Vitamin D

c. Vitamin E

d. Zinc

e. Iodine

f. Niacin

Group II

_____ **1.** Ration

_____ **2.** Ruminant

_____ **3.** Carbohydrate

_____ **4.** Mineral

_____ **5.** Antibiotic

_____ **6.** Silage

_____ **7.** Concentrate

a. Feed high in TDN, low in fiber

b. Disease-control substance

c. Essential element

d. Amount of feed fed in one day

e. Fermented green roughage

f. Sugars and starches

g. Multicompartment stomach

Animal Health

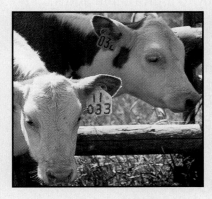

MATERIALS LIST

✓ various gauges of needles and syringes

✓ various containers or labels from containers of animal drugs

✓ discarded ears from a meat-processing plant

✓ Internet access

COMPETENCIES TO BE DEVELOPED

After studying this unit, you should be able to:

- identify signs of good and poor animal health.
- identify symptoms of animal diseases and parasites.
- understand how to prevent animal health problems.
- explain various methods of treating animal health problems.

SUGGESTED CLASS ACTIVITIES

1. Compile a list of animal diseases and identify some sources of information about each of the diseases. Assign class members to work separately or in pairs to prepare a class report on the disease they have chosen to study. (Do not assign more than one group to study the same disease.) Encourage the students to use Internet sources, as well as more traditional sources such as textbooks and extension service bulletins. Have each study group make a class presentation of their material.

2. Visit a veterinary clinic as a class, or ask a veterinarian to visit your class as a guest speaker. Ask the veterinarian to discuss the animal health problems that he or she deals with most often. Also include discussion of the career opportunities that are available in the animal health field.

3. Divide the class into six groups. Each group will have an orange, a syringe and needle (a plastic needle that cannot penetrate skin is recommended), and a cup of water. Demonstrate to each group the technique for giving intramuscular and subcutaneous injections, using the fruit. Each group member will practice this process on the orange. Administrative approval should be sought before beginning this activity.

TERMS TO KNOW

syringe	fluke	intravenous	teat
disinfectant	protozoa	intramuscular	cannula
feedlot	secondary host	subcutaneous	dipping
host animal	mange	intradermal	rectum
contagious	vaccination	intraruminal	immune
noncontagious	balling gun	intraperitoneal	veterinarian
abortion	drenching	infusion	
roundworm	injection	udder	

Maintaining animal health is the key to a profitable and satisfying animal enterprise. There are several considerations that need to be made in dealing with the health of animals. These include being able to recognize signs of good and poor health, maintaining a healthy environment, being able to identify animal diseases and parasites, and knowing how to treat health problems when they occur. These items are explored in this unit. Diseases are infective agents that result in declining health of living things. Parasites are organisms that derive their food from live organisms by ingesting body fluids such as blood or mucus.

SIGNS OF GOOD AND POOR ANIMAL HEALTH

INTERNET ADDRESS:

<http://ohioline.osu.edu/b869/b869_32.html>

Having the ability to recognize the signs of good health or the symptoms of health problems is the single most important key to being efficient in maintaining good animal health. A keen sense of observation is important, as well as the innate ability to know when something is not right with an animal (Figure 27–1).

FIGURE 27–1 The good animal manager develops a keen sense of observation and learns to recognize the symptoms of an animal under stress. *(Courtesy of USDA)*

Hot Topics in Agriscience

MODERN WAYS TO MONITOR HERD HEALTH

The earlier that a herd health problem is detected and treated, the more likely it is that a sick animal will recover. Modern electronic devices make it possible to detect health problems in animals during the early stages of infection. For example, electronic sensors are available that can take the body temperature of an animal as it walks by. These devices are placed on alley-way fences that lead to water or from milk parlors. When an animal with an abnormal body temperature approaches the sensor, a hydraulic gate sorts the animal into a holding pen to be checked by a veterinarian.

A different type of sensor is available to monitor the health of dairy cows. This sensor is placed in the milk line that carries milk from each cow to the main pipeline. The sensor measures the temperature of the milk as it comes from the cow. The sensor is useful on those dairy farms where each cow is identified by a computer chip as she enters the milking area. The sensor enters the milk temperature in the computer record for each cow, and a sorting gate is activated that sorts each cow with an abnormal milk temperature into a holding pen as she leaves the milking area. These types of devices make it possible to identify sick animals in the early stages of infection. Early detection and treatment contribute to early recovery from infections and diseases.

Class of Livestock or Poultry	Degree F Average	Degree F Range
Cattle	101.5	100.4-102.8
Sheep	102.3	100.9-103.8
Goats	103.8	101.7-105.3
Swine	102.6	102.0-103.6
Horses	100.5	99.9-100.8
Poultry	106.0	105.0-107.0

FIGURE 27-2 Normal body temperatures for animals.

INTERNET KEY WORDS:

healthy livestock environment

Signs of Good Health

One of the best signs of good health is a contented animal. Of course, a good deal of experience in dealing with animals is necessary to recognize contentment. Alertness and the chewing of the cud in ruminant animals is a good sign. A shiny hair coat, bright eyes, and pink membranes are other signs that an animal is healthy. Normal body discharges of urine and feces are further evidences that animals are not experiencing serious health problems. On the technical side, a healthy animal should have a normal body temperature, pulse rate, and respiration, or breathing, rate (Figure 27-2).

Signs of Poor Health

Often it is easier to tell when an animal is sick than to tell when it is healthy. A rough hair coat and dull, glassy eyes are often the first signs that an animal is not well. Sick animals usually stay alone with their heads down. Such animals may be drawn up and may walk slowly when forced to walk. Abnormal feces, either too hard or too soft, as well as discolored urine, may also indicate that an animal is experiencing some health problem. Loss of production, especially in dairy cattle, is often the first sign that the animal is not well. High temperatures, labored breathing, and rapid pulse rates are other indications of poor health in animals.

HEALTHFUL ENVIRONMENTS FOR ANIMALS

Maintaining a healthy environment for animals is a key factor in a complete animal-health program. It is often much less expensive to maintain a healthy environment for animals than it is to treat animals that are unhealthy because of a poor living environment (Figure 27-3).

Agri-Profile

CAREER AREAS:

Veterinarian/Animal Health Technician

Animal pathologists, animal behaviorists, physiologists, biologists, zoologists, microbiologists, geneticists, nutritionists, and others must work together to understand the complexities of animals. Animals exist as pets, production animals, work animals, pleasure animals, fish, fowl, birds, wild animals, and specimen animals in zoos. The need for health services for animals varies with the species and the primary use of the animal.

The desire to become veterinarians has been high among youths in recent years. This interest has permitted the supply of veterinarians to increase, despite the rigor of the college curriculum and tough competition to get into veterinary schools. Colleges of animal sciences offer curricula in other career areas in animal sciences, such as nutrition, breeding, education, and production.

Veterinarians and their associates work constantly to treat animals having disorders and to improve animal health.

In urban and suburban areas, veterinary clinics for pets offer many opportunities for licensed veterinarians. Animal shelters, hospitals, kennels, and pet stores provide many career opportunities for those interested in animal health at the technician level. In rural areas, large-animal veterinarian, veterinary assistant, laboratory veterinarian, and laboratory technician are typical positions in the animal-health industry.

FIGURE 27–3 A grim reminder that untreated diseases or parasites, poisonous plants, or predators can quickly take the life of an expensive animal. *(Courtesy of USDA/ARS #K-4382-2)*

Sanitation

Good sanitation is important to good health. Factors related to good sanitation include keeping facilities for animals clean. Sanitation also requires the use of clean equipment when dealing with animals. This includes feed containers, milking equipment, artificial-breeding equipment, needles and syringes, and surgical equipment. A **syringe** is an

Disposable plastic syringe   Standard metal syringe (with glass barrel)

Reusable glass or nylon syringes
Plain or metal tip   Luer lock tip

FIGURE 27–4 Syringes are instruments that are used to inject medications and vaccines into animals. Syringes are of several types, including both reusable and disposable.

instrument used to give injections of medicine or to draw body fluids from animals (Figure 27–4). Simple, on-farm surgical procedures should always be performed with the strictest sanitation possible. The liberal use of disinfectants in dealing with animals is also important. A **disinfectant** is a material that kills disease-causing organisms.

Housing

Maintenance of proper housing is an important consideration in maintaining good animal health. Housing should be clean and free from cold drafts. However, good air circulation throughout the housing is important to help decrease high temperatures in the summer and reduce humidity in the cold of winter. Extremely dry and dusty conditions are to be avoided whenever possible. Proper maintenance of animal housing also is important. Loose boards, roofing materials, and nails often pose problems in poorly maintained facilities.

Handling Manure

Piles of manure, dirty pens, and dirty feedlots are often sources of serious health problems in animals. It is important that manure not be allowed to accumulate in areas frequented by animals (Figure 27–5). Manure piles often harbor diseases and parasites. They also attract flies, which may spread diseases. Cages and pens soiled continually with animal waste products may also decrease the quality of the air breathed by

INTERNET KEY WORDS:

livestock, facilities, housing

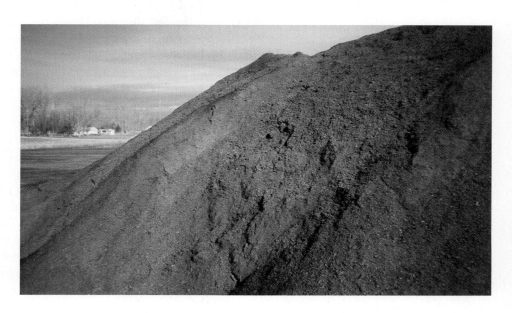

FIGURE 27–5 An efficient manure handling system helps control diseases and prevents the spread of parasites. Manure that has been composted is a valuable product for use in nurseries and gardens.

FIGURE 27–6 The drawing and analyzing of blood samples is a technique for diagnosing and treating many diseases and disorders in animals. *(Courtesy of USDA/ARS #K-4135-11)*

INTERNET ADDRESS:

<http://www.isis.vt.edu/~fanjun/text/Link_pest31.html>

INTERNET KEY WORDS:

livestock pest control
livestock parasite control

the animals. Wet, poorly drained, manure-soiled feedlots usually reduce the rate of gain in beef cattle and swine. Feedlots are areas in which large numbers of animals are grown for food. Feet and leg problems can often be traced to poorly maintained feedlots.

Controlling Pests

The control of pests and parasites is also an important consideration in the maintenance of animal health and welfare. Regular use of disinfectants to control parasites such as lice and flies is necessary in a good disease-prevention program. Regular, close observation of animals may be necessary to determine when outbreaks of parasites occur. Prevention of parasites is preferable to controlling outbreaks that occur. To that end, the development of a good prevention program is a wise decision (Figure 27–6).

The control of other pests, such as birds and wild animals, is also part of a good animal-health program. Many birds carry parasites on their bodies and in their droppings. When the parasites move from infected animals to healthy ones, they often carry diseases and parasites with them. Wild animals and pets may also cause serious health problems when allowed to roam freely around farm animals. Dogs and coyotes will often chase animals and cause them to injure themselves. Bites from these animals may also cause infection and other health problems. Just the presence of pets around farm animals may cause the farm animals to be nervous and may affect how they grow and produce.

Isolation

The isolation of animals new to a herd is an important part of any good health-prevention program. Such animals may be harboring disease or parasites that are not readily apparent. It is wise to keep them isolated from other animals for a time, usually a minimum of 30 days. This gives the new owner time to observe the isolated animals closely for health problems (Figure 27–7).

Similarly, isolation of diseased animals is important. Animals with contagious diseases that can be spread by contact should never be allowed to come into contact with healthy animals. It is difficult to treat unhealthy animals when they are living with large groups of animals.

FIGURE 27–7 It is important to isolate animals that are being added to a herd for up to 30 days. This allows time for disease symptoms to develop if they are present, and it protects the rest of the herd from exposure to the disease. *(Courtesy of USDA/ARS #K-4702-6)*

FIGURE 27–8 Pasture rotation is necessary to break the life cycles of diseases, insects, and internal parasites, and it pays additional dividends in greater pasture production. *(Courtesy of USDA/ARS #K3714-2)*

Healthy animals tend to become aggressive with unhealthy ones, making it especially difficult for such animals to regain health.

Pasture Rotation

The rotation of pastures is a consideration in maintaining a healthy environment for animals and in preventing health problems. Many disease organisms that infect animals are harbored in the soil, and they are killed only by not being able to come into contact with host animals for extended periods. A **host animal** is an animal in or on which diseases or parasites can live. Moving animals to different pastures on a regular basis also breaks the life cycle of most parasites, thus contributing to control of parasite populations (Figure 27–8).

ANIMAL DISEASES AND PARASITES

Diseases

INTERNET KEY WORDS:

animal, livestock, disease

The diseases of animals can be divided into two major classes: contagious and noncontagious. **Contagious** diseases are those that can be passed on to other animals by contact. **Noncontagious** diseases cannot be spread through contact with other animals.

The handling of these two classes of diseases varies somewhat. It is important that animals with contagious diseases be isolated from other animals in the herd as soon as the disease is identified. Because some contagious diseases of animals can be transmitted to humans, care must be taken when handling infected animals. Similarly, humans handling animals should become familiar with the proper techniques, vaccinations, and precautions to avoid human disease and parasitic infections from animals.

Noncontagious diseases pose no threat to humans or other animals, except to the animals having the diseases. Therefore, there is more leeway in dealing with animals with noncontagious diseases. It is still a good idea to isolate these animals from the herd for their own good, however.

Science Connection

THE UNSEEN HARVESTERS

The control of internal parasites in production animals and pets is probably the most persistent animal-health problem. Generally out of sight and frequently microscopic in size, parasites are small animals that live in the flesh or internal organs of larger animals where they draw nutrients from the bodies of the hosts. Animals can become infected with parasites, and they are helpless to do anything about it when symptoms start to appear. Fortunately, the body has some ways of helping to keep such intruders in check. If otherwise healthy and free of reinfection, the hosts can sometimes function fairly well even though they are infected with parasites. Sometimes they can even rid their systems of the parasites. Generally, however, the host needs outside help.

Farmers, ranchers, veterinarians, and scientists have long known that animals with parasites grow poorly and never reach their full potential. No matter how much they eat, they are undersized and underachievers. Researchers confirm that internal parasites do more than just consume nutrients intended for the host.

Researchers weigh a parasite-infected calf that has been treated. *(Courtesy of USDA/ARS #K-4736-3)*

The animal's immune system responds to invading organisms by producing chemical signals that modify the host's metabolism. These immune-response signals are small proteins called *cytokines* that manipulate the hormones that regulate feed intake, nutrient use, and, ultimately, growth of the animals. Dairy and beef calves infected with a protozoan parasite called *Sarcocystis* run fevers, lose their appetites, and become emaciated. Even after an animal has been treated and the parasites have been brought under control, some calves still do not grow normally.

Research was conducted on calves at Beltsville, MD to find some explanations. Before, during, and after acute infection with the parasite, the concentration of growth-regulating hormones in the blood was measured. It was found that after acute infection, the concentration of a hormone essential for growth decreased and the concentration of another hormone that blocks growth-hormone secretion increased. These hormone changes persisted in the infected calves even after the symptoms of infection were gone.

It is thought that the body implements a survival strategy in response to the invasion of the parasites. The strategy is one of growth restriction, so more energy can be available to fight off the intruders. It is therefore essential that every effort be made to keep animals in surroundings that are free from parasites and to use precautionary measures whenever applicable. When symptoms appear, it is important to get a professional diagnosis and promptly administer proper treatment.

Causes Contagious diseases are caused mostly by bacteria and viruses. They can be spread by direct contact with infected animals, from shared housing, or from contaminated feed or water. In some cases, the spread of infectious diseases takes place through intermediary hosts, such as birds, rodents, or insects. Diseases such as sore mouth and brucellosis are capable of infecting humans.

Noncontagious diseases may be caused by nutrient deficiencies or nutrient excesses. Poisonous plants and animals, injection of foreign material, and open wounds may cause or lead to noncontagious disease.

Symptoms General symptoms of disease are extremely varied and may include:

- poor growth, reduced production, or both;
- reduced intake of feed;
- rough, dry hair coat;
- discharge from the nose or eyes;
- coughing or gasping for breath;
- trembling, shaking, or shivering;
- unusual discharges, such as diarrhea or bloody feces or urine;
- open sores or wounds;
- unusual swelling of the body, including lumps and knots;
- **abortion**, or the loss of a fetus before it is born; and
- peculiar gait, or walking pattern, or other odd movements.

Some diseases may have little or no external symptoms and may even progress so rapidly that death of the animal precedes the occurrence of noticeable symptoms.

Parasites

INTERNET KEY WORDS:

parasite roundworm

Parasites may also be grouped into two general classifications. These are internal—inside the animal—and external—living on the outside of the animal.

The most important group of internal parasites that infest animals is the **roundworms** (slender worms that are tapered on both ends) (Figure 27–9). Other types of internal parasites include flukes and protozoa. **Flukes** are very small, flat worms, and **protozoa** are microscopic, one-celled animals. Most internal parasites spend at least some of their life cycles outside of the host animal (Figure 27–10). It is during this period that the parasite may most easily be spread to other animals. Contact with discharges from infested animals, contaminated feed, water, housing, or secondary hosts may result in the spread of internal parasites. A **secondary host** is a plant or animal that carries a disease or parasite during part of its life cycle. Some internal parasites are also spread by insects such as flies and mosquitoes.

FIGURE 27–9 Roundworm parasites come in many sizes. They infect the digestive tracts of animals and humans and are common in most parts of the world. The worms are round, as the name implies, and tapered on both ends.

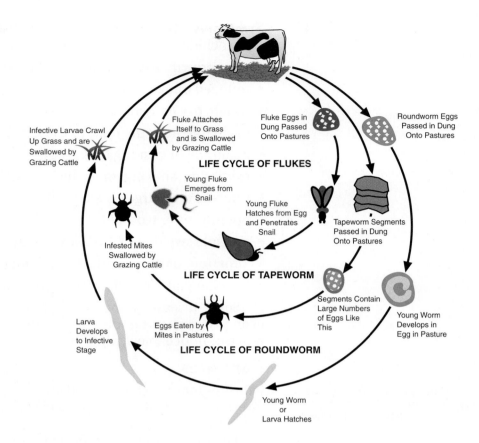

FIGURE 27–10 Life cycles of some common internal parasites.

External parasites include flies, ticks, mites, and fleas. They are spread from animal to animal through physical contact.

Symptoms of parasite infestation may include:

- poor growth;
- weight loss;
- constant coughing and gagging;
- anemia;
- reduced production and reproduction;
- diarrhea or bloody feces;
- worms in the feces;
- swelling under the neck;
- poor stamina;
- loss of hair and the occurrence of **mange**, resulting in a crusty skin condition caused by mites; and
- visibility of the parasite itself.

PREVENTING AND TREATING ANIMAL HEALTH PROBLEMS

There are a number of activities and procedures that are used to prevent and treat health problems found in animals. Some of these include administering drugs, dipping, and restraining animals. The roles of feed additives and vaccination will also be explored.

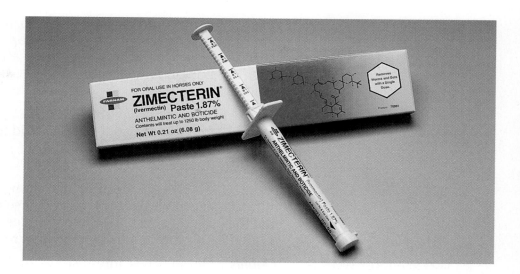

FIGURE 27–11 Pills, liquids, powders, and pastes are methods of getting medicine to animals. It is important that the drug manufacturer's recommendations be followed closely. *(Courtesy of Michael Dzaman)*

Vaccination is the injection of a modified disease organism into an animal to prevent disease.

Practical Veterinary Skills

All owners of animals should have a few basic veterinary skills. Farmers and ranchers would spend far more on animal medical costs if they had a veterinarian administer all of the routine vaccinations required by their animals. Some procedures that are commonly performed by animal owners are birthing assistance, castration, vaccination, treating for lice or mites, minor cut repair, mastitis treatment, and many more. Although there are many tasks that can be done by animal caregivers, there are still many more serious procedures that must be completed by a licensed veterinarian.

Many common animal illnesses have distinct symptoms and can be treated initially in the field. Medicines for these illnesses can be purchased at a farm supply store or from a veterinarian. Some knowledge on how to provide medical care to animals can be learned through observation and experience. Other skills are learned through organized education classes. Local college classes, agricultural extension workshops, clinics, and veterinary handbooks can also teach basic veterinary skills that are needed on the farm or ranch.

Administering Drugs

There are several factors to be considered before administering drugs to an animal. These include determination of the amount to be administered, type of drug to use, purpose of the drug, site of administration of the drug, and type of animal to be treated. Most of this information can be found on the drug container. It is important that the drug manufacturer's recommendations be followed closely. Another factor that needs to be considered is the amount of time that the drug remains in the animal. This is important when determining how long milk from the animal will be contaminated by the drug. Contaminated milk must be discarded to make sure it is not in the human food supply. Also, it must be determined how long to wait before a treated animal can be slaughtered for meat.

Drugs may be manufactured and sold as pills, powder, paste, or liquid (Figure 27–11).

Pills   The procedure for giving a pill to an animal is to restrain the animal and lift its head so that the mouth opens. Gently place the pill as far back on the tongue as possible, using a balling gun (a device used to place a pill in an animal's throat). Massage the animal's throat until it swallows the pill.

Powders   Powder drugs are normally mixed in the feed or water of the animal. Often it is necessary to withhold feed or water for a time before administering the drug. Otherwise, the animal may refuse to eat or drink the drugged food.

Paste   Paste is normally used for treating horses for worms. The preparation is placed on the back of the horse's tongue with a caulking gun, and the horse is forced to swallow. Pastes are used for horses because it is often nearly impossible to treat them for worms by any other method.

Liquids   Liquid drugs administered orally (by mouth) are often placed directly in the animal's stomach by drenching. Drenching is the process of administering fairly large amounts of liquid to an animal. A syringe or drenching gun is used. In the process, the animal is restrained, with the head held level. The upper lip of the animal is lifted, and the tube is inserted along the side of the tongue. The drug is released, and the animal is allowed to swallow. Care must be taken not to get the drug into the animal's lungs.

The injection of drugs into animals takes many forms, based on the location of the injection. Injection is the process of administering drugs by needle and syringe. Some of the injection sites include intravenous (in a vein), intramuscular (in a muscle), subcutaneous (under the skin), intradermal (between layers of skin), intraruminal (in the rumen), and intraperitoneal (in the abdominal cavity). One determining factor as to where injections are made is how fast the drug needs to work. A drug injected into the blood is available faster than one injected under the skin. Often, it is desirable for drugs to be released slowly over a long period. Growth hormones are generally administered in this way.

The procedure for giving an injection is to:

1. fill the syringe, making sure that all air is removed;
2. restrain the animal;
3. select the location for the injection;
4. disinfect the area to be injected;
5. if the injection is to be made intradermally, clip the hair from the area to be injected;
6. Pop the needle into the desired area without the syringe attached (this prevents the loss of the drug if the animal jumps); and
7. attach the syringe to the needle and inject the liquid.

Infusion   Infusion is another method of getting drugs to the site of the infection. It is used most often to treat dairy animals with udder and teat problems. The udder is the milk-secreting gland of the animal. Teats are the appendages of an udder. A sterile cannula (blunt needle) is inserted into the opening of the teat, and the drug is forced into the teat canal (Figure 27–12).

Dipping   Dipping is a process for treating animals, mostly cattle and sheep, for external parasites. It involves filling a vat with medicated

INTERNET KEY WORDS:

animal injection, intramuscular animal injection, subcutaneous

INTERNET ADDRESS:

<http://www.ianr.unl.edu/pubs/ Beef/g1314.htm>

FIGURE 27–12 Infusion is a method of treating cows and other large animals for infections in the udder.

water and forcing the animal to walk or swim through it. This process is also used to treat dogs for ticks and fleas. Dipping is popular in cases where large numbers of animals must be completely covered with the medication.

Taking Temperatures

Taking an animal's temperature is basically the same as taking a human's temperature. It is usually taken in the rectum. The rectum is the last organ in the digestive tract. Animal thermometers are normally longer and heavier than those used in human medicine. An animal thermometer also has an eye at one end and should have a string attached to it to prevent loss of the thermometer in the body cavity of the animal.

To use an animal thermometer, first shake down the column of mercury. Coat the thermometer with sterile jelly to make insertion easier. Do not force the thermometer into the rectum. If there is resistance, injury may result. Correct the conditions that are causing the resistance, and then reinsert the thermometer. After several minutes, remove the thermometer and read the temperature on the scale. Thermometers with digital readouts are also available.

Determining Pulse and Respiration Rates

The pulse rate for a large animal can be taken by holding your ear against the animal's chest and listening to the heartbeat. The number of heartbeats in 1 minute is the pulse rate.

The respiration rate of an animal can be determined by watching its rib cage move. Counting the number of breaths that the animal takes in 1 minute will indicate the rate of respiration.

Restraining Animals

There are a number of ways to restrain animals for observation and treatment of diseases and parasites. These include head gates, squeeze chutes, halters, twitches, nose leads, and casting harnesses. Head gates trap the heads of large animals, whereas squeeze chutes hold the whole animal. When halters are used, they are usually tied to a post or something else substantial to hold the animal. Twitches hold the tender lip of a horse (Figure 27–13). Nose leads hold cattle by the nose (Figure 27–14). Sometimes, large animals must be lying down to be examined. An easy way to accomplish this is by using a casting harness (Figure 27–15). Properly applied, a casting harness will cause an animal to fall down when steady pressure is applied by pulling the rope.

Vaccination

The prevention of a disease is nearly always less expensive than treating animals once they have become infected by a disease. A good disease-prevention program should include vaccination of animals. Vaccination is the injection of an agent such as a modified disease organism into an animal to immunize it against a specific disease. The agent causes the animal's body to become immune to the disease. An animal that is immune to a disease is not affected by the disease organism. Vaccination programs are usually part of the services of a veterinarian (an animal doctor). Vaccination programs vary widely with the type of animal, the disease organism involved, and the area of the country where the program is being implemented.

FIGURE 27–13 A twitch is used to obtain the attention of the horse so it will hold still for diagnosis or treatment.

Feed Additives

Feed additives are used primarily to control the incidence of low-level infections in growing animals. The materials used are primarily antibiotics that help increase feed efficiency and rate of gain, as well as control of diseases. Feed additives are sometimes used to control internal parasites. Caution must be taken to always follow the manufacturer's recommendations concerning the use of these materials. Failure to do so may lead to illegal contamination of animal products that are used for human consumption.

INTERNET KEY WORDS:

livestock feed additives

FIGURE 27–14 Cattle can be restrained during examination or treatment by using a nose lead to hold them in position. The nose lead distracts the animal's attention from the procedure by applying moderate pressure on the tissue between the nostrils when the animal struggles.

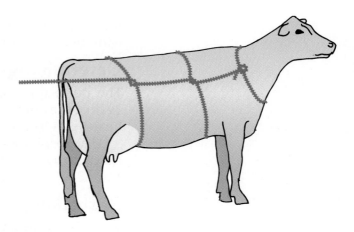

FIGURE 27–15 A properly used casting harness makes putting a large animal down a simple task.

Science Profile

FIGHTING "FIRE" WITH "FIRE"

The number-one cause of death in newborn pigs is an infection caused by the *Escherichia coli* bacteria. As a result of many deaths, the swine industry loses millions of dollars every year. Up to this point, antibiotics have been used to treat the sick animals. Unfortunately, current techniques are becoming less and less effective because of the ability of bacteria to develop resistances to the antibiotics. When a bacterium has developed a resistance, it means that an antibiotic no longer kills it. The scientist Roger Harver heads a research project committed to finding an alternative to antibiotics.

A promising alternative currently is being tested. "Good" bacteria from the digestive systems of healthy pigs are cultured, and then introduced to newborns. When this is done, the good bacteria attach themselves to the intestines, and therefore prevent *E. coli* from causing an intestinal infection. Using "good" bacteria may prevent the bacteria that are responsible for so many infant pig deaths.

Medication Withdrawal

Drugs that are used to medicate food animals such as dairy cows, poultry, hogs, beef, and sheep must be withdrawn before the animal products can be used for human consumption. It often requires several days for drugs to leave an animal's system. If the withdrawal period has not been observed before slaughtering a meat animal or consuming eggs or milk from treated livestock, the medicine may end up in human food. People who are allergic to antibiotics may experience serious allergic reactions to these drug residues. Frequent use of foods that are contaminated with drugs can also build up a tolerance for antibiotics. Such tolerances for antibiotics make them less effective as medicines when they are needed to treat human infections.

VETERINARY SERVICES

The veterinarian is an essential part of any good health program for animals. It is important to know when to call the veterinarian for help and when to deal with a problem yourself. There are no hard-and-fast rules in this regard, and it will vary greatly depending on the experience of the individual. In general, it pays to call the veterinarian any time that you are not absolutely sure of the problem and how to handle it.

A veterinarian should normally be consulted when you are planning and executing a disease-prevention program. Any time that an animal is having reproductive problems, a veterinarian should be consulted; such problems include failure to conceive, abortion, or great difficulty in giving birth. When an animal dies suddenly and there is no apparent reason, a veterinarian should also be consulted to determine the cause of the death. A veterinarian should also be contacted when animals have symptoms of a contagious disease. This will help minimize spread of the disease.

STUDENT ACTIVITIES

1. Write the Terms to Know and their meanings in your notebook.

2. Compare the methods of administration, dosage rates, and times of withdrawal of several drugs for animals.

3. Practice giving injections of water using discarded animal ears from a meat processing plant.

4. Develop a complete disease-prevention program for a livestock operation.

5. Check the housing of production animals for animal safety and proper sanitation at home, on someone else's farm, or at someone's business.

6. Visit a farm supply store or a pet center and record the names of animal medicines, animal pest controls, and disinfectants that are for sale there. Also list the use(s) of each item.

7. Interview the manager of a small animal or livestock operation concerning the disease-prevention and health-maintenance practices used. Report your findings to the class.

8. At your home, a friend's home, or a willing farmer's home, practice some restraint techniques on an animal. Determine what works best for you and for the animal.

9. Contact a local veterinarian and ask to spend a few hours observing him or her as he or she works.

SELF EVALUATION

A. Multiple Choice

1. Subcutaneous injections are made
   a. in a vein.
   b. under the skin.
   c. between layers of skin.
   d. in the body cavity.

2. An animal that lives and feeds on other animals is a
   a. parasite.
   b. disease.
   c. vaccination.
   d. host.

3. When taking the temperature of an animal, use a/an
   a. catheter.
   b. oral thermometer.
   c. rectal thermometer.
   d. syringe.

4. Which of the following is *not* a means of administering drugs orally?
   a. drench
   b. balling gun
   c. pills
   d. infusion

5. The purpose of vaccination is to
   a. prevent parasites.
   b. prevent diseases.
   c. control parasites.
   d. treat diseases.

6. Which of the following is *not* a sign of good health in animals?
   a. smooth hair coat
   b. increased pulse rate
   c. alertness
   d. contentment

7. The purpose of isolating sick animals is to
   a. prevent spread of contagious diseases.
   b. keep other animals from hurting the sick animal.
   c. allow for easier treatment of the problem.
   d. all of the above.

8. Which of the following is *not* an internal parasite?

   a. fluke

   b. mite

   c. stomach worm

   d. protozoan

9. Call a veterinarian when the animal

   a. has bright eyes.

   b. chews its cud.

   c. aborts.

   d. starts to lose hair.

10. A twitch is used to

    a. restrain horses.

    b. treat internal parasites.

    c. give injections.

    d. isolate infected animals.

B. Completion

1. _____ diseases cannot be passed from one animal to another.

2. An _____ injection is one that is made in the vein of an animal.

3. Parasites that are very small, flat worms are called _____.

4. _____ refers to the number of heartbeats in 1 minute.

5. An agent that prevents the growth of a disease organism is called an _____.

6. Protozoa are a form of _____ parasites.

7. Oral administration of medicine refers to administering the drug through the animal's _____.

8. Vaccination is a form of disease _____.

9. An animal doctor is called a _____.

10. A nose lead is used to _____ animals.

Genetics, Breeding, and Reproduction

MATERIALS LIST

✓ copies of illustrations of male and female reproductive systems

✓ animal magazines
✓ bulletin board materials

✓ Internet access

COMPETENCIES TO BE DEVELOPED

After studying this unit, you should be able to:

- define terms associated with genetics and reproduction.
- understand the principles of genetics.
- identify systems of breeding.
- identify parts of reproductive systems.
- understand new technologies in animal reproduction.
- evaluate animals for type and production.
- understand types of testing programs.

OBJECTIVE

To determine the role of genetics, breeding, and reproduction in animal agriscience today.

SUGGESTED CLASS ACTIVITIES

1. Obtain a video that describes the processes by which artificial insemination (AI) and/or embryo transfer are accomplished in livestock. View the video as a class. Take time to discuss various aspects of the process by stopping the video at strategic points. Discuss key questions such as: What are the advantages of AI and embryo transfer procedures as opposed to maintaining a herd sire on the farm? Why are embryos collected only from the most productive female animals?

2. Obtain female reproductive tracts from a biological supply house or other reliable source and use them to teach students how AI is performed. Obtain some disposable gloves and plastic pipettes. Teach the students to thread the pipette through the cervix of a cow to simulate the AI process.

3. In groups of two or three, search the Internet for information involving animal genetics. Choose information based on current research from a reputable source. Prepare a 2- to 4-minute class presentation of your findings.

TERMS TO KNOW

artificial insemination (AI)	meiosis	estrogen	grading up
egg	DNA (deoxyribonucleic acid)	estrus	inbreeding
ova		progesterone	sire
sperm	homozygous	mammary system	dam
embryo transfer	heterozygous	ovulation	closebreeding
zygote	incomplete dominance	embryo	linebreeding
geneticist	genotype	parturition	outcrossing
breed	phenotype	fetus	natural service
gene	sex-linked	gestation	pasture mating
chromosome	mutation	sterility	hand mating
gamete	polled	freemartin	semen
dominant	lethal	purebred	ejaculate
recessive	heritability	registration papers	in vitro
cell	testes	crossbreeding	pedigree
mitosis	testosterone	grade	

Probably the fastest growing area of technology in agriscience is in genetics and reproduction. Artificial insemination has allowed for more improvement in milk production in the last 30 years than had occurred in the previous 200 years. **Artificial insemination (AI)** is the placing of sperm in proximity with female reproductive cells, called **eggs** or **ova,** by a method other than natural mating. **Sperm** consists of male reproductive cells. Artificial insemination allows for use of a

Agri-Profile

CAREER AREAS:

Geneticist/Genetic Engineer/ Animal Physiologist/Veterinarian/ Insemination Technician

The greatest improvements in animal production and performance have come about as a result of selection and breeding. Recent advances in the field of genetics and the ability to stimulate multiple ovulation, to transplant embryos to surrogate mothers, and to freeze embryos for future incubation have greatly increased the genetic output of superior individuals.

However, genetic engineering, in which the genetic coding of cells can be manipulated, holds forth the greatest opportunities for changing animal characteristics and capabilities in the future. These changes are products of biotechnology.

A technician removes frozen semen from liquid nitrogen for artificial insemination of a farm animal.

In the foreseeable future, the greatest number of new positions in agriscience is expected to be for scientists, engineers, and related specialists. The animal industry will provide many of these positions.

FIGURE 28-1 Successful embryo transfer requires the services of a highly trained veterinarian backed up by sophisticated laboratory equipment.

INTERNET KEY WORDS:

livestock embryo transfer

INTERNET ADDRESS:

<http://www.equigen.net/key>

superior male animal to father many times more offspring than would be possible naturally. Embryo transfers have made it possible for superior female animals to produce far more offspring than would be possible otherwise.

Embryo transfer is a process that removes fertilized eggs, or zygotes, from a female animal and places them in another female animal who carries them until birth. Embryo transfers are even being used to reproduce endangered species of animals faster than would normally be possible (Figure 28–1).

Genetic engineering has made it possible to increase resistance to diseases, improve production, and improve efficiency of animals. Gene splicing, recombinant DNA, and biotechnology are common terms used by geneticists today. A geneticist studies genetics, or heredity. Heredity is the passing on of traits or characteristics from parents to offspring.

This unit explores some new technologies, as well as the basics of animal breeding. Also notable are current directions of research in animal breeding.

ROLE OF BREEDING AND SELECTION IN ANIMAL IMPROVEMENT

The English geneticist Robert Bakewell is generally credited with being the father of animal husbandry. His work in selection of Merino sheep for fine wool production and quality encouraged other farmers of his era to try to improve their livestock. Bakewell and others took great pains to always mate the most desirable female animals with the best male animals, with the expectation that the offspring would be as good as or superior to their parents. These practices have continued through the years and have resulted in major advances in animal agriscience (Figure 28–2).

FIGURE 28–2 Mating the most desirable female animals with the best male animals usually results in offspring as good as or superior to their parents. *(Courtesy of USDA/ARS #K-2681-13)*

INTERNET KEY WORDS:

livestock breed associations
horse breed associations
sheep breed associations
cattle breed associations
swine breed associations

By continually selecting animals for a specific type or characteristic, the resulting generations of animals tend to conform to the characteristics for which they were selected. For example, 200 years ago, cattle were not separated into dairy and beef types. Through careful selection of those animals with superior milk production and those types with excellent meat production, two distinct types of animals emerged from the same ancestors. The many breeds of animals have been developed in the same way. A **breed** is a population of animals having similar physical characteristics that are passed on to their offspring. It also should be noted that selection is an extremely important part of animal agriscience today. This is especially true as consumer demand for animal products changes, and the margin of profit continues to decline.

Science Profile

HUMAN GENOME PROJECT

In 1990, a team of international scientists began the Human Genome Project. Their goal was to identify the sequence of genes on each chromosome, including the entire DNA in a human. Ten years later, in June of 2000, these scientists announced that their work was essentially complete. Before mapping a human's genome, the scientists sequenced the genomes of other organisms to perfect their techniques. They started with bacteria, which has much less DNA than a human. Researchers also worked on the cattle genome. Once the human genome was finished, comparisons could be made between organisms of different species. The results of one comparison involved the DNA of cattle and a human. This comparison showed that the two species share many of the same genes. They also found that four whole chromosomes shared identical genes. Scientists can identify a gene in one organism and apply that knowledge to another. The application of this kind of information is immeasurable. Could the gene responsible for some cancers be identified and eliminated? Could the lactation gene be manipulated in such a way that 1 cow will do the work of 20 on less feed? The future of animal genetics is compelling, and only time will tell where these types of discoveries will lead agriscientists.

PRINCIPLES OF GENETICS

Gregor Mendel, an Austrian monk, is generally given credit for having discovered the basic principles of genetics. He did this through keen observation as he raised peas in his garden. These principles have become the foundation of modern genetics. They are summarized as follows:

- In every living organism there is a pair of genes in every cell that determines every trait in that individual. A **gene** is a unit of hereditary material located on a chromosome. A **chromosome** is the rodlike structure that functions as a carrier for genes.

- Individuals receive one gene for each trait from each parent.

- Genes are transmitted from parent to offspring as unchanging units.

- In the production of reproductive cells, known as gametes, gene pairs separate; only one gene for each trait is contained in each gamete. A **gamete** is a reproductive cell.

- When different genes are present for a single genetic trait, in most instances, only one trait will be expressed. The trait that is expressed is known as the **dominant** trait, whereas the trait that is masked is called the **recessive** trait.

Cells and Cell Division

Cells are the basis of all genetic activity. A **cell** is a unit of protoplasmic material with distinct structures (Figure 28–3). It is the basic structure of all living things. Cells are microscopic in size. All plant and animal life begins as a single cell. The nucleus of the cell contains pairs of chromosomes on which genes are located at specific locations. Each gene controls a genetic trait. The gene for a specific trait is always located in the same place on the same pair of chromosomes in a given plant or animal species.

Animal growth and reproduction takes place through cell division. Simple cell division for growth is called **mitosis**. The process begins as

INTERNET ADDRESSES:

http://www.accessexcellence.
org/AB/GG/mitosis.html

http://www.accessexcellence.
org/AB/GG/meiosis.html

FIGURE 28–3 Each animal cell consists of distinct structures that carry out the functions of the cell.

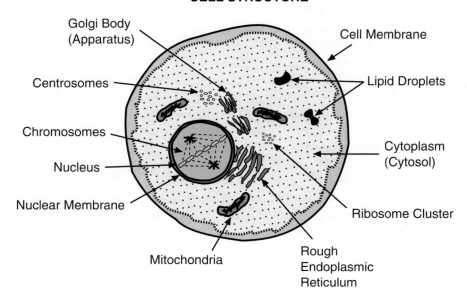

CELL STRUCTURE

Golgi Body (Apparatus)

Centrosomes

Chromosomes

Nucleus

Nuclear Membrane

Mitochondria

Cell Membrane

Lipid Droplets

Cytoplasm (Cytosol)

Ribosome Cluster

Rough Endoplasmic Reticulum

MITOSIS

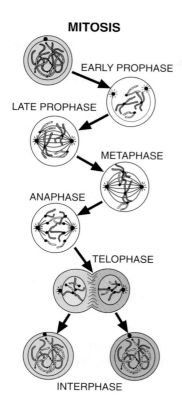

EARLY PROPHASE

LATE PROPHASE

METAPHASE

ANAPHASE

TELOPHASE

INTERPHASE

FIGURE 28–4 Mitosis is the process of division or duplication of a typical cell.

INTERNET KEY WORDS:

deoxyribonucleic acid
DNA heredity

each chromosome first divides in two. The wall of the nucleus disappears, and the chromosomes move to opposite sides of the cell. A new nucleus wall forms around each of the groups of chromosomes. The walls on the two sides of the cell then move toward each other until they divide the cell into two new cells complete with nuclei and pairs of chromosomes (Figure 28–4).

The cell division that occurs during reproduction resulting in the formation of gametes is called meiosis (Figure 28–5). It differs from mitosis primarily in that instead of the chromosomes dividing and moving in pairs to the opposite sides of the cell, they separate and move individually to the cell walls. When the new cells are formed, each cell contains only one of each chromosome pair rather than both chromosomes. Meiosis occurs only in the reproductive organs of animals. When an egg is fertilized, the sperm contributes one of each chromosome pair; the egg contributes the other.

Genes

Genes are the units of genetic material that are responsible for all of the traits or characteristics that an animal inherits from its parents. Genes occur at specific locations on structures called *chromosomes.* Chromosomes control certain enzyme and protein production that influences physical traits in animals. The chromosomes themselves are composed of a protein covering surrounding two chains of DNA, or deoxyribonucleic acid. This substance serves as the coding mechanism for heredity.

In pairs of genes on matching chromosomes, the genes may be either alike or different. Pairs of genes that are alike are said to be homozygous, whereas those pairs that are different are called heterozygous. When the two genes in a pair are different, one gene usually expresses itself, and the expression of the other remains hidden. The gene that expresses

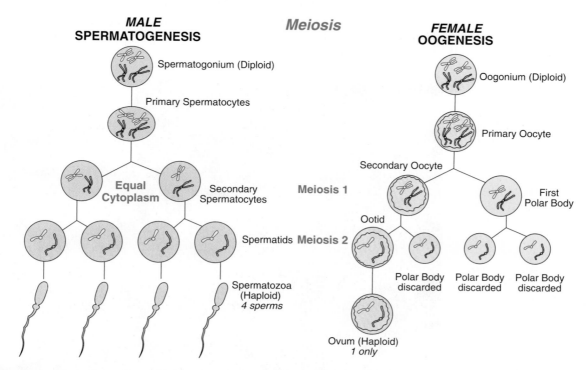

FIGURE 28–5 Meiosis is the division or duplication of egg and sperm cells.

itself is referred to as dominant. The gene that remains hidden and expresses itself only in the absence of a dominant gene is called recessive. Sometimes neither gene of a pair expresses itself to the exclusion of the other. When this happens, the gene pair is referred to as expressing partial dominance, or incomplete dominance. The actual configuration of genes in an animal is called the genotype, whereas the physical appearance of the animal is referred to as phenotype. All of this is important when exploring the basics of genetics and the use of genetics in animal breeding.

Some traits are controlled by genes that are located on the chromosomes that determine the sex of the animal. These are called sex-linked traits. The chromosomes that control sex in most animals are not perfectly matched. The result is that not all of the genes on these chromosomes occur in pairs. When this happens, some traits show only in male animals.

Genes normally duplicate themselves accurately. However, sometimes accidents or changes occur. These genetic accidents or changes in genes are called mutations. Sometimes these mutations result in desirable changes in animals. One such example is the polled characteristic in breeds of cattle that are normally horned. Polled means naturally or genetically hornless. In other cases, the mutation results in a lethal characteristic, which causes an animal to be born dead or to die shortly after birth.

Genetics in the Improvement of Animals

The improvement of animals through genetics can be either natural or planned. In natural selection, the "survival of the fittest" occurs. In other words, as changes in genes occur naturally, only those animals experiencing gene changes that make them better adapted to their environment will survive. Popular examples include protective colorations, ability to digest certain feeds, and ability to survive in extreme heat or cold.

When a fertilized egg divides into two identical eggs, identical twins result (Figure 28–6).

FIGURE 28–6 These two polled calves are identical twins that resulted when a fertilized egg was divided into two identical cells or masses of cells in the early stages of embryonic development. *(Courtesy USDA/ARS #K-4323-18)*

In planned or artificial selection, people decide which traits they want in animals. They then use the animals with the desirable traits in the breeding program. Over a period, the animals that result from such selection show more and more of the desired traits. Heritability is the capacity of a trait to be passed down from a parent to offspring.

Unfortunately, most of the traits for which people are selecting animals are the result of a combination of many pairs of genes. Because of this, few traits are 100 percent heritable from parents. For example, the extent of heritability for loin-eye size in pigs is 50 percent. A boar with a 6-inch loin-eye is crossed with a sow that has a 5-inch loin-eye. The expected average loin-eye size for the resulting offspring would be 5.5-inch if loin-eye size were 100 percent heritable. However, because loin-eye size is only 50 percent heritable, the offspring can only be expected to have 5.25-inch loin-eyes.

The heritability rates of other traits can be found in Figure 28–7. These rates should be used as a guide only when attempting to improve animals through genetics.

Environmental factors often play a part in the expression of genetic traits, masking to some extent the true potential of the animal. For example, an animal that is improperly fed or cared for may never reach the size or weight that its genetic potential would permit.

INTERNET ADDRESS:

<http://www.bovineengineering.com/heritabi.htm>

Estimated Percent Heritability						
Trait	Cattle	Sheep	Swine	Poultry	Rabbits	Horses
Fertility	0-10	0-15	0-15	0-15	–	Low
Number of young weaned	10-15	10-15	10-15	–	3	–
Weight of young at weaning	15-30	15-20	15-20	–	35	–
Postweaning rate of gain	50-55	50-60	25-30	–	60	–
Postweaning gain efficiency	40-57	20-30	30-35	–	–	–
Fat thickness over loin	40-50	–	40-50	–	–	–
Loin-eye area	50-70	–	45-50	–	60	–
Percent lean cuts	40-50	–	30-40	–	60	–
Milk production (lb)	25-30	–	–	–	–	–
Milk fat (lb)	25-30	–	–	–	–	–
Milk solids, nonfat (lb)	30-35	–	–	–	–	–
Total milk solids (lb)	30-35	–	–	–	–	–
Body weight	–	–	–	35-45	40	–
Feed efficiency	–	–	–	20-25	–	–
Total egg production	–	–	–	20-30	–	–
Age at sexual maturity	–	–	–	30-40	–	–
Viability	–	–	–	5-10	–	–
Speed	–	–	–	–	–	25-50
Wither height	–	–	–	–	–	25-60
Body length	–	–	–	–	–	25
Heart girth circumference	–	–	–	–	–	34
Cannon bone circumference	–	–	–	–	–	19
Points for movement	–	–	–	–	–	40
Temperament	–	–	–	–	–	25

FIGURE 28–7 The rates of heritability for certain traits of domestic animals.

Genetic Engineering

Genetic engineering has much potential for improving animals for the use of humans. Genetic engineering is the process of transferring genes from one individual to another individual or organism by inserting them in the chromosomes of the new individual or organism. Geneticists have been able to link specific genes to specific traits. They also have developed procedures for removing the genes from the cells of one animal and inserting them into the cells of another animal.

The potential for change in animals using genetic engineering is tremendous. For example, if an animal species is genetically resistant to a certain disease, genes that make that animal resistant could be inserted into cells of an animal species that is not resistant. Because genes are passed on to offspring from parents, resulting generations of animals would be resistant to that disease.

Some of the areas being explored by geneticists working with genetic engineering include disease resistance, cancer research, vaccines, increased growth and production, and immunology.

REPRODUCTIVE SYSTEMS OF ANIMALS

Male Reproductive System

The male reproductive system functions to produce, store, and deposit sperm cells. Secondary functions include production of male sex hormones and elimination of urine from the body (Figure 28–8).

The actual structural makeup of the male reproductive system varies widely with different species of animals. The **testes** are the organs that produce sperm cells. They also produce **testosterone**, the male sex hormone. The testes are attached to the body by the spermatic cord and are protected by the scrotum. A coiled tube called the epididymis stores and transports the sperm cells. The vas deferens carries the sperm cells to the urethra, which extends through the penis. Other glands of the male reproductive system include the seminal vesicles and the prostate and Cowper's glands. The seminal vesicles secrete seminal fluid.

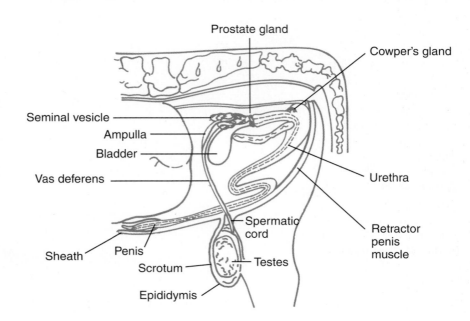

FIGURE 28–8 Reproductive system of a bull. *(Courtesy of Maryland State Instructional Guide, University of Maryland)*

The prostate gland provides nutrition to the sperm cells, and the Cowper's gland prepares the urethra for the passage of the sperm cells. The penis serves to deposit the sperm cells into the female reproductive tract.

Female Reproductive System

The female reproductive system produces the ova, or egg, together with the female sex hormones estrogen and progesterone (Figure 28–9). Estrogen regulates the heat period, estrus, whereas progesterone prevents estrus during pregnancy and causes development of the mammary system. The mammary system produces milk.

The parts of the female reproductive system include two ovaries, which produce ova, also referred to as eggs. The infundibulum, a funnel-shaped device, catches the eggs during ovulation (the process of releasing a mature egg from the ovaries). The egg then passes to the fallopian tube, which is also called the oviduct. The fallopian tubes are the site where fertilization of the egg takes place. The fertilized egg, or embryo, then travels to the uterine horn where it attaches to the wall of the uterus, and remains until birth, or parturition. When the embryo attaches itself to the uterine wall, it becomes known as a fetus. The period between fertilization of the egg by a sperm cell and birth is called the gestation period. The vagina, which is separated from the uterine horn by the cervix, serves as the passageway for the sperm cells. The external opening, or vulva, protects the rest of the female reproductive system from infection from the outside.

Reproductive Problems

Sterility is the inability of an animal to reproduce. There are a number of conditions that may result in reproductive problems or failures. Some of these problems are physical, but others are genetic. Examples of

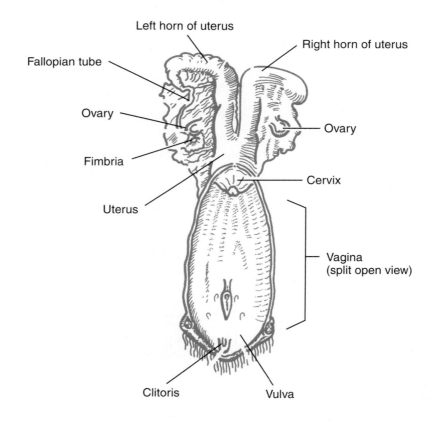

FIGURE 28–9 Reproductive system of a cow. *(Courtesy of Maryland State Instructional Guide, University of Maryland)*

genetic reproductive problems include sterile female cattle (also some female lambs) that are born as a twin or triplet with male offspring. These female animals are exposed to male hormones before birth that interfere with the development of female organs. A female animal that exhibits this problem is called a freemartin. Another problem of a genetic nature includes the scrotal hernia. This is a muscle tear that is often caused by a hereditary weakness in the muscle. Undeveloped or missing ovaries are often thought to be the result of genetic or developmental accidents.

Infections and diseases are important causes of sterility of a physical nature. Physical damage to the reproductive system and nutritional deficiencies are known to contribute to reproductive failures.

SYSTEMS OF BREEDING

There are a number of breeding systems that are important to animal agriscience. Which system or systems to use depends on many factors. Some of the considerations include type of operation, markets available, resources available, climatic conditions, size of operation, goals of the breeder, and personal preference.

Commonly recognized systems of breeding include purebreeding, crossbreeding, grading up, inbreeding, and outcrossing (Figure 28–10).

Purebreeding

Purebreeding occurs when a purebred animal is bred to another purebred animal. A purebred animal is one of a recognized breed and one having registration papers. Registration papers are records of ancestry.

Although there are no guarantees, purebred animals are usually considered to be superior to nonpurebreds. They are used as show animals and are important parts of the crossbreeding and grading-up systems of breeding.

People who elect to use the purebreeding system of breeding should have ample resources and a good knowledge of genetics. They should also be good salespersons capable of marketing the purebred animals at premium prices that reflect the superior qualities of the animals.

Crossbreeding

Crossbreeding is the breeding of one recognized breed of animals to another recognized breed. The resulting offspring are called hybrids.

There are a number of advantages to hybrid animals. They tend to be faster growing, stronger, and greater producing as a result of the combination of desirable traits from the two breeds. This is called hybrid vigor. They also tend to be more fertile and more disease resistant.

Crossbreeding is generally used by commercial producers who are more interested in offspring that are efficient producers than in maintaining a specific breed of animals.

Grading Up

Livestock producers who are raising animals that are not purebred often use grading up to improve their herds. When a nonpurebred female animal, called a grade, is mated with a purebred male animal, the process is called grading up. The idea is that the purebred male animal should

INTERNET KEY WORDS:

livestock closebreeding
livestock inbreeding
livestock linebreeding
livestock outcrossing

	Relationship of Mates	**Advantages**	**Disadvantages**
Purebreeding	Unrelated	1. Concentration of selected traits 2. Breed ass'n to create demand	1. May result in less desirable traits 2. Loss of hybrid vigor
Crossbreeding	Unrelated	1. Increased growth 2. Increased prod. 3. Increased hybrid vigor 4. Higher fertility 5. Disease resistance	1. Less uniformity of offspring 2. Not eligible for registry
Grading up	Unrelated	1. Herd improvement w/o purchasing purebreds 2. Develop uniformity	Slow process of improvement
Closebreeding	Sire-daughter Son-dam Brother-sister	Concentrates desirable traits	1. Concentrates undesirable traits 2. Expression of abnormal traits
Linebreeding	Not closer than ½ brother to ½ sister	Concentrates desirable traits of one individual	1. Concentrates desirable traits 2. May result in expression of abnormal traits
Outcrossing	Unrelated	Produces hybrid vigor within a breed	

FIGURE 28–10 Systems of breeding for animal improvement.

be superior to the grade female animal and that the resulting offspring should be superior to their mother. Succeeding generations of female animals are also mated to purebred male animals of superior quality.

The purposes of grading up include the improvement of quality and production in the offspring. Offspring are also called progeny. The development of uniformity in the herd is also a reason for grading up animals.

Inbreeding

Inbreeding is the mating of animals that are related. The purpose of inbreeding is to intensify the desirable characteristics of a particular animal or family of animals. Unfortunately, inbreeding also intensifies the undesirable and abnormal characteristics.

There are various degrees of inbreeding based on how closely related are the individuals being mated. When a father, or sire, is mated with his daughter, a son is mated with his mother, or dam, or a brother is mated to his sister, the term closebreeding is used.

Linebreeding is the mating of less closely related individuals that can be traced back to one common ancestor. Normally, the most closely related cross made in linebreeding is half brother to half sister.

Inbreeding must be used carefully, because inbred animals tend to exhibit more undesirable characteristics than animals produced by other systems. Unless a breeder is willing to carefully select the outstanding individuals resulting from inbreeding, this system should be avoided.

Outcrossing

Outcrossing is the mating of unrelated animal families in the same breed. This is probably the most popular system of breeding used in purebred herds of animals. It also has many of the advantages of cross-breeding, including increased production and improved type.

METHODS OF BREEDING

There are three general methods of breeding animals: natural service, artificial insemination, and in vitro. Natural service occurs when the male animal is allowed to mate directly with the female animal. There are several ways in which this may be accomplished, depending on the type of operation, amount of labor available, number of animals in the herd, requirements of breed associations, and personal preference.

Bio-Tech Connection

EMBRYO CO-CULTURE

Sheep and cattle embryos can be kept alive and growing outside of the mother's womb for up to 6 days when tissue-cultured oviduct cells are used for nourishment. This technique was first developed for sheep and is known as *co-culture*. The technique permits embryos to be held for observation and checked for success of genetic engineering before being implanted into surrogate mothers.

Single-celled embryos that have had a gene inserted are placed in cultures of cells from the oviduct. Scientists believe that certain nutrients from the cultured cells keep the embryos alive and growing. Microscopic examination of the cultured embryos is used to determine if they are developing properly. An embryo cultured for 3 days should have more than eight cells. If there are eight or less cells, it signals that development is not progressing normally and implanting should not be performed.

Another technique that can be used is to place gene-implanted embryos into rabbits for 5 days, and then remove them for implanting. Greater labor requirements and other disadvantages related to the use of rabbits makes the use of oviduct cells the more promising of the techniques, however.

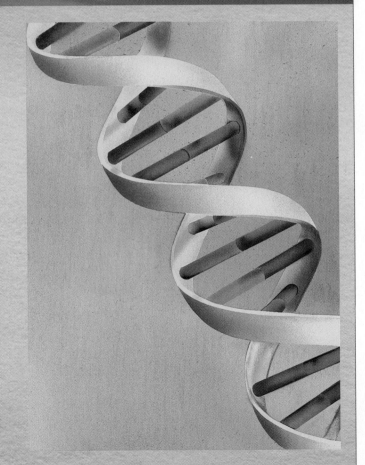

Various genetic-engineering techniques are being used for embryonic development.

Pasture mating is a system of natural service in which the male animal is allowed to roam freely with the female animals in the herd. The male animal is responsible for detecting the heat period of each female animal in the herd and for mating with her at the appropriate time. There are some disadvantages to this system. One male can mate with only a limited number of females. Also, if more than one female is in heat at the same time, all may not get bred. Similarly, breeding records are more difficult to keep. Finally, the male animal may become sterile and thus the female animals may not get bred at all.

Hand mating is the bringing of the female animal to the male animal for mating. More labor and better management is required, because someone must determine when the female animal is in heat and get the female to the male animal so mating can take place. Advantages of hand mating include being able to keep more accurate breeding records and being able to mate more females to a single male.

There are numerous advantages of mating animals through AI. There are a few disadvantages as well. Artificial insemination involves collecting semen from a male animal and placing it in the reproductive tract of the female animal. Semen is composed of the sperm cells and accompanying fluids. The technique of AI has been responsible for tremendous increases in animal productivity in recent years.

Advantages of AI include that semen that is collected from the male animal can be used fresh or stored frozen in liquid nitrogen for later use. One ejaculate, or the amount of semen produced at one time by a male animal, can be diluted and used to breed many female animals. Because semen can be frozen and stored for long periods,

Hot Topics in Agriscience

EMBRYO CLONING— MORE OFFSPRING FROM OUTSTANDING FEMALE ANIMALS

Outstanding female cattle are frequently given hormone treatments to increase the number of ova they produce. This allows outstanding cows to produce more than one calf per year when "extra" embryos are placed in other cows of lesser quality. In addition, the number of embryos can be expanded by splitting the cells that make up each embryo mass. In this instance, each of the cell masses that is obtained is capable of producing identical individuals, or clones. An embryo that is in the 16- to 32-cell stage of development can be separated into as many as 4 to 8 different clones.

When these cloned embryos mature in the uterus of a cow that is not the genetic mother, they retain the genetic makeup that they inherited from the cow and the sire that produced the embryos. Each of these calves may be nourished by its surrogate mother, but it is the genetic offspring of the cow from which the embryos were collected. In this manner, numerous offspring can be obtained from a single outstanding cow.

Embryo Splitting

16 – 32 cells

8 – 16 cells

Genetically identical embryos

Embryos in the 16- to 32-cell stage can be successfully split into genetically identical "clones." These embryos are placed in surrogate mothers, but they retain the genetic characteristics of the "mother" that produced the original ovum.

the use of an outstanding male can be greatly extended. Artificial insemination also greatly reduces the need to keep male animals and reduces the danger of having male animals around. Other advantages include less chance of injury to breeders and their animals, reduced spread of reproductive diseases, and improved record keeping.

There are some disadvantages to AI as a method of mating animals. A person trained to inseminate female animals must be available when the animal is in heat. Semen collection from the male animal can be a dangerous activity requiring the services of trained technicians, special equipment and facilities, and excellent management. Finally, genetic defects of the male can be spread faster and more broadly.

In vitro mating occurs outside of the animal's body. Mature eggs are flushed from the female animal and fertilized by sperm cells collected from the male animal. The fertilized eggs are then placed in host female animals for development into offspring. Although this process is very exacting in its requirements and facilities, there are times when this is the only way that a viable fetus can be obtained. This means of mating is often used as part of the new technology of genetic engineering.

INTERNET ADDRESS:

<http://www.hyperionfarm.com/
invitro.htm>

SELECTION OF ANIMALS

There are a number of methods by which animals may be selected. These methods fall into two major types: selection based on physical appearance, and selection based on performance or production of either the individual or its progeny.

Selection based on physical appearance is generally used when choosing purebred animals. Often, the sole criterion for selection of an animal is how well he or she performs in the show ring. This is an acceptable means of selection when animal breeders are raising animals for show and to sell to others for the same purpose. However, this method often leaves much to be desired when the animal so chosen is expected to produce a product or perform a desired activity. Animals fitted to perform or look their best in the show ring often fall sadly short of expectations in the milking parlor or in other performance settings.

The use of comparative judging helps individuals to develop skills in evaluating animal appearance. In livestock judging events, one animal is compared with another and frequently judged in groups of four (Figure 28–11).

Selection based on production or performance is usually a more reliable means of choosing animals. If you are selecting dairy cows to produce milk, it makes more sense to select cows with high-production records and high-production relatives. In meat animals, progeny testing, or the testing of performance of the offspring, is often the only way to predict the breeding value of the parents. Other measures of production on which various animals may be selected include rate of gain, feed efficiency, butterfat production, back-fat thickness, loin-eye area, yearly egg production, and pounds of wool produced.

Sometimes, selection of animals is based on pedigree. A **pedigree** is a record of an animal's ancestry and is included on registration papers for purebred animals (Figure 28–12). Although the consideration of pedigree can be important in the selection process, it should always be used in combination with other methods of selecting animals.

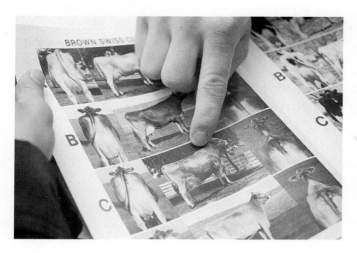

FIGURE 28–11 Comparative judging is a good technique for studying the details of the animal, such as feet, legs, udder, head, body, and overall appearance. *(Courtesy of FFA)*

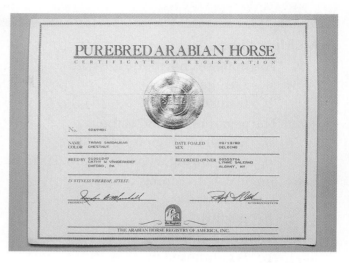

FIGURE 28–12 Pedigree and registration papers are valuable documents when buying purebred animals. *(Courtesy of Michael Dzaman)*

In summary, tremendous gains have occurred in the productivity of animals in the last 100 years. Fewer and fewer animals are providing the needs of an ever-growing human population. Genetics and gains in animal breeding have been responsible for much of the increase in productivity. This will continue as technology in the field of animal agriscience continues to advance.

STUDENT ACTIVITIES

1. Write the Terms to Know and their meanings in your notebook.
2. Label the parts of the male and female reproductive systems.
3. Sketch and label the various stages in mitosis and meiosis.
4. Suppose you mate a Hereford bull that has horns to six female animals that were born without horns (polled). If the female animals were homozygous for the polled characteristic, how many would probably bear calves having no horns? Why?
5. Select a species of animal and determine the origin of the popular breeds in that species.
6. Clip pictures of animals from magazines and newspapers and make a collage of popular breeds for the bulletin board.
7. Develop a bulletin board showing the popular breeds of sheep, cattle, horses, swine, dogs, rabbits, and cats in your community.
8. Suppose that you mated a black male rabbit with a white female rabbit and that the female gave birth to eight bunnies. If white is dominant to black, and the female is heterozygous for hair color, how many of the bunnies are likely to be black?
9. If a red pig with floppy ears was crossed with a white pig with erect ears and all possible characteristics were homozygous, what is the probability that the offspring will be red with erect ears? In pigs, white hair color and erect ears are dominant.
10. Dissect the reproductive organs of a female animal and identify the major parts.
11. Visit a farm or other animal facility and observe artificial insemination or other reproductive techniques being conducted.

SELF EVALUATION

A. Multiple Choice

1. The male sex hormone is called
- a. estrogen.
- b. testosterone.
- c. progesterone.
- d. oxytocin.

2. The gestation period is the
- a. length of pregnancy.
- b. time during which an animal is in heat.
- c. period when an animal is fertile.
- d. time it takes the egg to mature.

3. Single cell division is called
- a. meiosis.
- b. parturition.
- c. mitosis.
- d. estrus.

4. A chromosome
- a. is composed of DNA.
- b. is a carrier of genes.
- c. occurs in pairs.
- d. is all of the above.

5. A pair of genes with characteristics that are alike is said to be
- a. heterozygous.
- b. homozygous.
- c. genotype.
- d. recessive.

6. Mating of an animal of one breed to an animal of another breed is
- a. purebreeding.
- b. grading up.
- c. crossbreeding.
- d. outcrossing.

7. The phenotype of an individual is
- a. what the genes look like.
- b. the physical appearance.
- c. the type of animal.
- d. the expected production.

8. When genes are transferred from one individual to another other than through mating, it is referred to as
- a. artificial insemination.
- b. in vitro fertilization.
- c. genetic engineering.
- d. hybrid vigor.

9. The inability to reproduce is
- a. ovulation.
- b. meiosis.
- c. biotechnology.
- d. sterility.

10. Mating a sire to his daughter is
- a. inbreeding.
- b. crossbreeding.
- c. purebreeding.
- d. grading up.

B. Matching

_____ **1.** Gamete
_____ **2.** Genotype
_____ **3.** Fetus
_____ **4.** Progeny
_____ **5.** Dam

- a. Developing embryo
- b. Mother
- c. Sex cell
- d. Offspring
- e. Configuration of genes

C. Completion

1. The _____ manufacture male sex cells.

2. A _____ studies genetics.

3. The passing of traits from parents to offspring is _____.

4. The crossing of half brother to half sister is _____.

5. _____ is the coding mechanism for heredity.

Small Animal Care and Management

✓ an observation beehive

✓ bulletin board materials

✓ magazines and other materials with pictures of poultry and rabbits

✓ labels from poultry and rabbit feed bags

✓ Internet access

OBJECTIVE

To determine the types, uses, care, and management of small animals.

COMPETENCIES TO BE DEVELOPED

After studying this unit, you should be able to:

- describe the domestication and history of small animals.
- determine the economic importance of the various classes of small animals.
- list the types and uses of the various classes of small animals.
- describe the approved practices in feeding and caring for small animals.

SUGGESTED CLASS ACTIVITIES

1. Obtain a merit badge booklet for small animal care from the local center for the Boy Scouts of America. Review the requirements for this merit badge carefully, and use the booklet to organize the class discussion on small animal care and management.

2. Invite a person who raises rabbits to bring some animals to class to discuss the care and management of rabbits. Prepare class members to ask key questions of their guest during the presentation. Discuss rabbits as a source of meat that can be produced with minimal space in the backyard. Identify appropriate breeds of rabbits that might be used for this purpose. Ask the guest lecturer to address these topics in his or her presentation.

3. Invite class members who own small animals to bring one of them to class. Have these students describe the care needed to keep their animal healthy. Allow class members to ask questions at the end of each presentation. Be sure to get approval for this activity with the school administration first, so that there is a clear understanding about where the animals will be kept while students are in other classes.

TERMS TO KNOW

domesticate	pullet	hatchery	comb
jungle fowl	hen	flock	apiculture
waterfowl	bantam	pelt	swarm
poultry	tom	angora	queen
broiler	poult	buck	super
layer	drake	doe	worker
chick	duckling	hutch	drone
roaster	gander	litter	apiary
cockerel	goose	tattoo	
cock	gosling	honey	
rooster	squab	hive	

INTERNET KEY WORDS:

domesticate chicken,
duck, turkey

As our world population continues to expand and there is less and less room for humans and large animals to coexist, the importance of small animals continues to increase. They are more efficient than larger animals in converting the feed they eat into usable food and other products for humans. They are less intrusive on the lives of people, and more people are keeping animals as pets (Figure 29–1). There are many important species of small animals. This unit explores poultry, rabbits, pets, and honeybees, and their contributions to our daily lives.

POULTRY

History and Domestication

The domestication of chickens occurred about 4,000 BC in Southeast Asia. To domesticate means to tame for the use of people. The jungle fowl is an ancestor of our modern chickens. Its association with humans benefitted both the jungle fowl and its human neighbors. Humans made small clearings in the jungle that attracted insects and other food for the jungle fowl. The jungle fowl provided some eggs and meat for humans. This association over centuries gradually led to the domesticated chicken of today. Chickens came to the New World with the earliest settlers, and the people of Jamestown settlement had their pens of chickens.

Turkeys are the only domesticated animals of agricultural importance to have originated and been domesticated in the New World. When early explorers arrived in the New World, they found that the natives of Central America had highly domesticated turkeys that were being grown for food for animals and humans. Although modern turkeys are direct descendants of the wild turkey of the Americas, their physical makeup has changed to such a degree that they are totally dependent on humans and cannot survive in the wild.

The various types and breeds of ducks and geese have originated from places all over the world. Ducks and geese are also known as waterfowl.

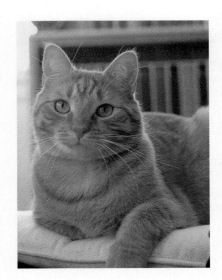

FIGURE 29–1 Pet ownership and the animal care industry have grown extensively in the last two decades. *(Courtesy of Sandy Clark)*

ECONOMIC IMPORTANCE

The consumption of red meat has increased by 5 percent between 1997 and 2002. Negative publicity regarding fat and cholesterol was probably responsible for declining consumption before 1997. Consumption of poultry and poultry products has been increasing. Poultry is a group name given to all domesticated birds. United States consumption of chicken meat increased from 26.4 lb to 81.4 lb per person from 1975 to 2003, and turkey meat consumption increased from 6.5 lb in 1975 to 17.6 lb in 2003. People in the United States also eat about 250 eggs per person each year. As a result, the poultry industry is and will continue to be an important part of the U.S. agriscience industry. Currently, poultry production ranks first ahead of beef and swine production in pounds of meat consumed. Some of the largest farms in the United States are poultry operations.

Major centers of production in the United States for broilers (young chickens grown for meat) produced each year are Arkansas, Georgia, Alabama, North Carolina, Mississippi, Texas, and Maryland. Including export production, 32.6 billion pounds of broilers were produced in 2003. California, Indiana, and Pennsylvania are the leading producers of eggs in the United States, followed by Ohio and Georgia.

The production of turkeys is spread over a wide area, with North Carolina, Minnesota, California, Arkansas, Missouri, and Virginia being leading states. In 2002, approximately 272 million turkeys were produced. Nearly 60 percent of the more than 10 million ducks produced in the United States each year comes from Long Island, New York. New York, Missouri, Iowa, South Dakota, and Minnesota are major producers of geese.

Most poultry farms are large and have thousands of birds in production. This is because there is greater potential to earn a profit when fixed costs such as buildings are divided among more animals. Despite this, it is still attractive to many people to keep a few chickens or other poultry for their own use. Chickens, ducks, geese, and turkeys can be raised in small numbers in nearly any environment, including backyards and small acreage properties. They tend to control some insects, such as grasshoppers. In outdoor settings, they are capable of finding some of their own food. Care must be taken, however, to protect them from predators such as dogs and skunks.

Types and Uses of Poultry

The types of poultry can be divided into the following general groups: chickens, turkeys, ducks, geese, and captive game birds.

Chickens are usually classified as either layers or broilers, depending on their intended use. A layer is a chicken that was developed to produce large numbers of eggs (Figure 29–2). They may produce either white or brown eggs, depending on the breed. Laying chickens are also maintained to produce eggs to be hatched for the production of broiler chicks. A chick is a baby chicken.

Chickens produced for meat are usually classified according to age. Broilers are young-meat chickens usually not more than 8 weeks old (Figure 29–3). A roaster is a mature chicken used for meat. Game or Cornish chickens are popular breeds of chickens raised for meat.

A young male chicken is called a cockerel, whereas an adult male is called a cock or rooster. These terms also apply to male pheasants.

FIGURE 29–2 Laying hens are bred to convert feed into eggs rather than body flesh. *(Courtesy of Bill Muir, Purdue University)*

FIGURE 29–3 Broilers are chikens that are bred for meat production. They are produced in large numbers, especially in the southern states. *(Courtesy USDA)*

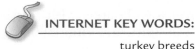

INTERNET KEY WORDS:

turkey breeds
breeds of ducks

A young female chicken is called a **pullet**, and an adult female chicken is called a **hen**. Adult female turkeys, ducks, and pheasants are also called hens.

Other classes of chicken include the **bantam**, or miniature chicken, and ornamental chickens, which are of value as hobby chickens.

There are more than 200 recognized breeds of chickens in the United States. However, nearly all of the layer and broiler types used are the result of crossbreeding to maximize production. The foundation breed of most laying-type chickens is the White Leghorn. Most broilers can trace their ancestors back to Cornish or game chickens.

Turkeys are of a single breed but eight varieties, including Beltsville Small White, Black, Bourbon Red, Bronze, Narragansett, Royal Palm, Slate, and White Holland. Of these, only the White Holland, which includes the Broad-Breasted White (Figure 29-4), and the Bronze, which includes the Broad-Breasted Bronze, are used extensively for commercial production. The broad-breasted birds have been intensively bred for superior meat qualities. They are produced by artificial breeding because they are no longer capable of breeding naturally. However, these two varieties account for nearly all of the turkey meat that is produced in the United States each year. A male turkey is called a **tom**, a female turkey is called a hen, and a young turkey is called a **poult**.

Ducks can be classified as meat producers or egg producers. The primary meat breed is the Pekin (Figure 29–5). They reach a market weight of about 7 lb in 8 weeks. This makes them faster growing than broilers, which reach 4 lb in the same period. Other breeds of duck used for meat production are Aylesbury, Muscovy, Rouen, and Call.

Egg-laying ducks are generally either Khaki Campbell or Indian Runner. The Khaki Campbell is the champion egg layer of the bird world, often averaging more than 350 eggs per year (Figure 29–6). This compares to an average of about 250 eggs laid per year for laying chickens. A male duck is called a **drake**, and a young duck is a **duckling**.

Geese are raised primarily for meat. There is also a limited market for geese used for weeding certain crops. The Chinese breed is popular for this use. Other breeds of goose are Toulouse, Emden, Pilgrim, and African. A male goose is called a **gander**, a female is a **goose**, and a young goose is a **gosling**.

FIGURE 29–4 The most popular commercial variety of turkey in the United States is the Broad-Breasted White. *(Courtesy USDA)*

FIGURE 29–5 Pekin ducks are the most popular breed of duck used for meat. *(Courtesy of Jurgielewicz Duck Farm)*

FIGURE 29–6 Khaki Campbell ducks are the champion egg layers of the bird world, producing as many as 350 eggs per year compared with an average of approximately 250 eggs per year for chickens.

Captive game birds include pheasants, quail, chukar partridge, and pigeons (Figure 29–7). The uses of game birds include meat and eggs. Some game birds are also raised to release to the wild or on game preserves for hunting. Pigeons may be raised for sport or for meat. A young pigeon is called a squab. It is used for meat before it learns to fly.

Some Approved Practices for Poultry Production

Up-to-date information on poultry and other small animals is available from suppliers of animals, equipment, medicine, and building materials. Recommended or approved practices are available from your state university. Approved practices for the production of poultry include the following:

- Purchase young poultry with a specific use in mind.

- Purchase young poultry or eggs for hatching from reputable hatcheries or breeders only. A hatchery is a business that hatches young poultry from eggs.

A

B

FIGURE 29–7 Two types of captive game birds are pictured here: (A) Chinese Ringnecked Pheasant and (B) Ruffed Grouse. *(Courtesy of U.S. Fish and Wildlife Services)*

Science Profile

GOT ENERGY?

Hunger leaves an individual feeling tired and weak. This is because the brain has ways of letting the body know that it needs food. Food provides the basic building blocks for the energy supply of the body. Inside of cells, the mitochondria (the cell's power plant) works tirelessly producing energy for the cell. The process whereby food is transformed into chemical energy is known as the Kreb cycle. Before the Kreb cycle can begin, another process breaks a sugar molecule called glucose in half, creating two energy molecules called adenosine triphosphate (ATP). ATP is the molecule that cells use for energy. Next, the two sugars are sent into the Kreb cycle where they go through a number of chemical reactions that results in 34 more ATPs.

Animals require a lot of energy. From the beating of a heart to running, every function of the body takes energy. Which animal do you think needs more energy, a horse or a rabbit? The answer may surprise you . . . it's the rabbit. Small animals need more energy per ounce than larger animals. In general terms, the bigger an animal is, the less energy it needs. Eighty to 90 percent of the energy of the body is spent regulating the body's temperature. Small animals lose heat much faster than large animals, so small animals use more energy or ATP to maintain their bodies at the right temperature.

- Purchase chicks, pullets, or poults that are immune to and free from disease.
- Purchase young poultry at the proper time for hitting a target market date. Broilers should be 7 to 8 weeks old before marketing them; ducks, 7 to 8 weeks; turkeys, 12 to 14 weeks; and geese, 12 to 14 weeks old. Layer chicks should be purchased 20 to 22 weeks before you expect them to produce eggs.
- Ensure that proper housing is available for the type and number of poultry you are planning to raise. Housing considerations include size, ventilation, ease of cleaning, lighting, heating and cooling, feed storage, and maintenance requirements.
- Secure and maintain the proper equipment for the type of poultry operation planned. Consider feeder and waterer space, and brooder size.
- Feed a balanced ration designed especially for the type of poultry being grown.
- Plan and follow a flock health program. A flock is a group of birds.
- Plan for marketing at the optimum time.
- Properly clean and disinfect facilities between flocks of poultry.

RABBITS

History and Domestication

Much of the early history of the rabbit is obscure. It is believed that the Phoenicians brought rabbits to Spain about 1,100 BC. They are also given credit for having introduced rabbits to most of the world that was known at that time.

Romans kept rabbits in special enclosures. Roman women were known to have eaten large quantities of rabbit meat. They believed that it enhanced their beauty.

FIGURE 29–8 Rabbit production can be done on a small basis with very little capital and space. *(Courtesy of FFA)*

Early monasteries produced large amounts of rabbit meat and fur. These religious institutions are given credit for having domesticated the rabbit. It is known that great pride was taken in producing good-quality rabbits and that much rabbit trading existed between monasteries.

Rabbit meat has long been an important component of the diets of people in densely populated countries of Europe. Rabbits are efficient converters of feed to meat. They take up relatively little space and reproduce rapidly.

Rabbits have been raised in the United States since the time of the early settlers, but serious rabbit production did not begin until the turn of the 20th century. An intense advertising campaign was conducted for "Belgian Hare" at that time to promote commercial production of rabbits.

Rabbit production also got a boost in the United States during the two world wars. At a time when shortages and rationing of food products occurred, rabbits became an inexpensive source of lean, red meat.

Economic Importance

Rabbit production is an important agriscience enterprise in the United States. Each year, 6 to 8 million rabbits are raised. The U.S. population consumes 25 to 30 million lb rabbit meat each year. Biomedical teaching and research use another 600,000 rabbits yearly.

Rabbit production is an ideal enterprise for a young person because it can be started with limited capital (Figure 29–8). With only a small investment in housing and equipment, a person with one pair of rabbits can produce 50 to 60 rabbits each year to eat or to sell. Because they are small and generally accepted by people, rabbits are better adapted for production in more urbanized areas than are most other types of animals. Rabbit meat is low in fat (4 percent), sodium, and cholesterol and is high in protein (25 percent).

The outlook for rabbit production in the future is variable with a need for good market analysis, good market development, and careful production management. As the competition between humans and animals for grain products increases, the rabbit will play an important role in meeting the protein needs of humans in the future.

INTERNET KEY WORDS:

rabbit care husbandry

A   B   C

FIGURE 29–9 Important meat-producing breeds of rabbit include the (A) New Zealand, (B) Californian, and (C) Champage D'Argent. *(Courtesy of Isabelle Francais)*

Types and Uses

Animals commonly known as rabbits come from two families and three genera, with distinct differences. These include rabbits, cottontails, and hares. Rabbits bear their young in underground burrows in the wild. The young are born blind, hairless, and completely helpless. In contrast, cottontails and hares usually give birth in nests aboveground. The young are born with their eyes open and with hair. They are able to fend for themselves shortly after birth. Hares also have larger hind legs and longer ears.

Hares and cottontails include the jackrabbit, arctic hare, and snowshoe rabbit. Because they belong to different genera, cottontails, hares, and rabbits cannot interbreed.

Domestic rabbits can be divided into a number of groups based on use. These groups include meat, fur, pets, show, and laboratory use. Many of the breeds will fall into several of these use groups.

The primary use of rabbits in the United States is for the production of meat, with pelts being a by-product. A pelt is an animal skin with the hair attached. Almost 100 million rabbit pelts are used in the United States each year. Most of these are imported from other countries because, in the United States, rabbits are slaughtered at too young an age to have desirable pelts.

Although all of the breeds of rabbit will produce meat, some breeds are far more efficient in producing desirable-quality meat. The New Zealand White is the most popular breed of rabbit in the United States for meat production. This rabbit occurs in three colors: white, red, and black. It is of medium size. It may be grown into a 4-lb rabbit at 8 weeks of age, using about 4 lb feed for each pound of rabbit produced. The Californian and Champage D'Argent are also popular breeds used for meat production (Figure 29–9).

Some breeds of rabbit are grown for their lustrous fur for use in the manufacture of fur coats and many other rabbit-fur products. The Satin,

FIGURE 29–10 Important fur-producing breeds of rabbit include the (A) Satin, (B) Rex, and (C) Havana. *(Courtesy of Isabelle Francais)*

FIGURE 29–11 The (A) Dutch and (B) Florida White join the (C) New Zealand White as important breeds in biomedical research and education. *(Courtesy of Isabelle Francais)*

Rex, and Havana are examples of rabbit breeds grown for their fur (Figure 29–10).

Rabbits have long been important for use in laboratory work. They are used for research in the development of drugs for treating a wide range of diseases. They are also important in nutritional studies and in genetic research. Commonly used breeds of rabbit for laboratory work include New Zealand White, Dutch, and Florida White. Producers who breed rabbits for laboratory work should be aware that many laboratories will use only white rabbits of medium size (Figure 29–11).

FIGURE 29–12 Angora rabbits, such as the French Angora (pictured here), are raised for the production of angora wool. *(Courtesy of Isabelle Francais)*

FIGURE 29–13 Netherland Dwarf rabbits, generally weighing in at less than 2 lb, are popular pets. *(Courtesy of Isabelle Francais)*

The Angora rabbit is used strictly for the production of wool called angora. The wool from Angora rabbits is sheared or pulled from the rabbit about every 10 to 12 weeks. Mature Angora bucks may produce 1 to 1.5 lb wool each year. A **buck** is a male rabbit. A female rabbit is called a **doe**. There are two breeds of Angora rabbit: French and English (Figure 29–12).

All of the 40 breeds of rabbit recognized by the American Rabbit Breeders Association can be used for pets, for show, or for both. They range in size from the Flemish Giant, which can weigh nearly 20 lb, to the Netherland Dwarf, which seldom weighs more than 2 lb and makes a popular pet (Figure 29–13).

Personal preference and the availability of breeding stock usually determine what breed or breeds of rabbit to raise for pets or show.

Approved Practices for Rabbit Production

Approved practices for the production of rabbits include the following:

- Select the correct breed for the intended use.
- Use purebred stock if you plan to sell breeding stock and to maintain uniformity in your herd.
- Purchase breeding stock only from reputable breeders with accurate records.
- Build or choose a hutch of the proper size for the breed of rabbit that you are growing. A **hutch** is a cage or house for a rabbit. For small- and medium-sized breeds, provide hutches 30 inches wide × 36 inches long × 18 inches high. For large-sized breeds, provide hutches 30 inches wide × 48 inches long × 18 inches high.
- Place the hutch where the rabbit will have adequate ventilation and be protected from heat, wind, rain, sleet, and snow.

Hot Topics in Agriscience

THE ROLE OF SMALL ANIMALS IN MEDICAL RESEARCH

Medical researchers have identified the causes and many of the cures for a variety of diseases and infections that afflict people and animals. This has been accomplished by studying the organisms that cause these infections. Once such an organism has been identified, it is cultured, modified, weakened, and tested in animals in an effort to create a vaccine that will prevent the infection from becoming established. Fertile eggs are often used to culture such organisms in live tissue. These eggs are supplied to medical research facilities from farms that specialize in producing the quality of eggs that is needed for this purpose.

Rabbits and other small animals are used in medical research to test the effects of potential vaccines and medications on animal life. Much of the progress that medical science has made in controlling diseases and infections of various kinds is dependant on the use of animals for research purposes. Laws prevent researchers from using human subjects until medicines and vaccines have been proved to be effective in laboratory animals. Without research animals, we would not have the ability to control many of the diseases and infections that are known to afflict humans. The average life span of humans would be much less in the absence of research animals.

- Provide adequate feeder and waterer space. Rabbits should have access to fresh, clean water at all times.
- Provide a separate hutch or cage for each mature rabbit.
- Breed rabbits when does are 5 to 8 months old, depending on the breed, and when bucks are 6 to 7 months old. It may be wise to delay breeding of large rabbits, because they are slower in reaching sexual maturity.
- Take the doe to the buck's cage for breeding and return her to her own cage immediately after breeding.
- Maintain 1 mature buck for every 10 to 25 does, depending on the breed and mating management.
- Place a nesting box in the doe's cage 25 days after mating occurs.
- Keep the handling of rabbits to a minimum to avoid injury. When handling rabbits, hold them by the skin on the back of the neck, with the other hand supporting the weight of the rabbit.
- Feed a commercial pelleted-feed free choice (feed available at all times) to does and litters (a group of young born at one time to the same parents). Feed single bucks and does 3 to 6 ounces feed each day. Rabbits need to be fed only once a day, and they should be fed in the evening if possible.
- Maintain accurate breeding, production, and health records for all rabbits.
- Tattoo all breeding rabbits for identification. A tattoo is a means of marking rabbits and other animals for identification. Rabbits are tattooed in the ear.
- Plan for and maintain a strict herd health program.
- Dispose of sick and dead rabbits promptly.
- Market rabbits as soon as they reach market size or weight.

HONEYBEES AND APICULTURE

INTERNET KEY WORDS:

honeybee
honeybee pollination

History and Domestication

Honeybees are an important part of human history. Cavemen drew pictures of bees and honey collection on cave walls. Honey is a thick, sweet substance made by bees from the nectar of flowers. In Egypt, mummies were embalmed and stored in a liquid based on honey. Jars of honey have been found in many of the Egyptian tombs.

The Bible makes many references to honey and the use of honey for food. During biblical times, honey was not produced in nice, neat combs as it is today. Rather, in most cases, the hive was destroyed in the process of removing the honey and comb. A hive is a home for honeybees. Comb is the wax foundation in which bees store honey.

Greeks and Romans were very familiar with honeybees and honey. Pompey used poisoned honey to defeat his enemies in at least one engagement. Aristotle wrote in great detail about bees and their production of honey.

Most early civilizations considered honey to be the food of gods. Athletes competing in Olympic games often ate honey before their events to gain extra strength and endurance.

Early beekeepers kept their bees in hollow logs, straw hives, or even in crude clay cylinders. All of these containers had to be destroyed to remove the honey.

With the invention in the 1850s of movable combs with wax foundations to encourage bees to make neat, straight honeycombs, the whole beekeeping industry changed. Honey was finally a commodity to be enjoyed by nearly everyone (Figure 29–14). Soon after the development of movable comb, the discovery that honey could be whirled out of the comb led to the invention of the honey extractor. It was no longer necessary to destroy the comb to get to the honey. The comb, after having been emptied of honey, can now be placed back in the hive to be refilled by the bees.

Today, the production of honey in the United States is a large and profitable business. Far more important than the production of honey is the work that honeybees do in the pollination of crops that are important to agriculture. Modern beekeeping is known as apiculture.

Economic Importance

It is difficult to accurately gauge the true economic importance of honeybees. They are responsible for about 80 percent of insect pollination of plants. Without honeybees, many of the crops important to agriscience would simply disappear from the earth.

Pollination of orchard crops such as citrus, peaches, and apples by honeybees is so important that many beekeepers rent their bees to orchardists when these trees are in bloom. With $10 to $30 in rent per hive, and flatbed trailers to move hives, commercial beekeepers make more money from bee rental than from honey. Such beekeepers operate from Florida to Maine and from Texas to Washington State (Figure 29–15).

There are about 300,000 beekeepers in the United States, of which about 99 percent are hobby or part-time beekeepers. These 300,000

FIGURE 29–14 Biological technician Gary Delatte and graduate student Lilia de Guzman examine a honeycomb and check the health of the honeybees. *(Courtesy of USDA/ARS #K-5064-2)*

Science Connection

BEES IN BIOCONTROL BUSINESS

An old adage says, "If you want to get a job done, ask a busy person to do it." Noting another saying, "busy as a bee," entomologists have found a way to have busy honeybees spread biocontrol agents to plants as they pollinate billions of dollars worth of crops each year. Scientists are using unsuspecting honeybees to deliver biocontrol agents right where the agents are needed on plants. Two serious pests are targeted for control by the process. One is the fire blight bacterium, which infects pear and apple trees; and the other is the corn earworm, one of the worst insect pests of corn and cotton.

Fire blight is caused by *Erwinia amylovora*, a bacterium that first colonizes a flower's stigma, the part that receives pollen grains during pollination. The bacteria then multiply and spread quickly. The disease causes cankers on twigs and branches and causes the leaves to have a burnt appearance; hence, the name "fire blight." The disorder weakens or kills the trees. On the positive side, scientists have found that spraying blossoms with beneficial bacteria helps to prevent the fire blight disease. The beneficial bacteria apparently compete with the harmful bacteria by consuming the nutrient-rich stigma, and the pathogen cannot get a foothold and invade the tree.

Fire blight is caused by a bacterium that enters apple and pear trees by colonizing the stigma of the fruit blossom. Honeybees, carrying a fire blight–fighting bacteria, are used to colonize the stigma with beneficial bacteria, as well as pollinate the fruit blossoms.

Rather than spraying with the beneficial bacteria, entomologists and plant pathologists at Utah State University and Oregon State University are doing reseach in which honeybees deliver the beneficial bacteria to the point-of-entry for the fire blight pathogen—the stigma. Such a strategy maximizes the use of the biocontrol material. The procedure is being field-tested, wherein numerous beehives are fitted with devices called pollen inserts. The pollen insert is a slotted passageway for bees that automatically dusts the bees with the beneficial bacterium, which is the biocontrol in this case, each time they leave the hive. A hive contains from 10,000 to 100,000 bees, each visiting perhaps 100 blossoms per hour. The beneficial bacteria do no apparent harm to the bees.

Meanwhile, entomologists in Tifton, Georgia, are using honeybees to carry a natural virus, *Heliothis*, that destroys the larval stage of the corn earworm. The device used to "load" the bees is one in which the bees use a different entrance and exit to the hive. The exit device includes a metal tray that contains the virus material. The exiting bees carry the material on their feet, legs, and undersides and deposit it on their rounds. The technique reduced earworm infestations in clover, and it should be effective on other flowering plants.

people care for about 6 million hives of bees. In a normal year, a hive of bees will produce 100 to 150 lb honey in excess of the approximately 150 lb needed by the bees to live. At $1.50 to $2.00 per pound of honey retail, it is easy to see that the production of honey is big business.

FIGURE 29–15 Bees are responsible for up to 80 percent of all insect pollination of plants, and they are absolutely necessary for pollination of fruit crops such as citrus, peaches, apples, and other orchard crops. *(Courtesy of USDA/ARS #K-4715-1)*

Approved Practices for Beekeeping

The following is a list of approved practices to be used in the keeping of bees:

- Check local regulations before starting a beekeeping operation.
- Locate bees out of direct contact with people and neighbors' yards and gardens.
- Place hives facing away from prevailing winds. They should also be protected from hot summer sun.
- Thoroughly clean and disinfect hives before allowing new groups of bees to use them.
- Purchase bees from reputable sources. It is usually far more profitable to purchase a 3-lb package of bees with a purebred queen than to rely on a swarm to populate a new beehive. A swarm is a group of bees complete with a queen that leaves an overcrowded hive to find a new home.
- Replace queens every 2 years. A queen bee is the only fertile, egg-laying female bee in each hive (Figure 29–16).
- Have your bees inspected for contagious diseases by a federal bee inspector each year.
- Make sure that each hive of bees has a store of at least 75 lb honey for the winter. Hives that do not have enough surplus honey stored for winter should be fed a sugar–water mixture to supplement their own honey stores.
- Always be sure that bees have ample room to store the honey being produced.
- Remove surplus honey as soon as the bees have capped it over with wax.

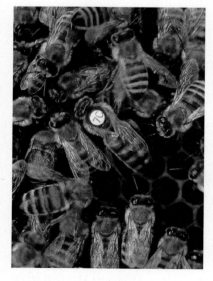

FIGURE 29–16 Worker bees tend the queen (marked), the hive, and the brood, as well as making and storing honey for the future. *(Courtesy of USDA/ARS #K-5069-22)*

- Remove honey in the evening or at night when nearly all of the bees are in the hive. Supers containing surplus honey can be freed from bees by blowing cool smoke over the bees and brushing them off the comb with a bee brush. A **super** is a box filled with a movable foundation that is used by the bees to store honey. You can also use a bee excluder between the honey to be removed and the hive body.

INTERNET KEY WORDS:

beekeeping

- After moving a hive, put a deflector in the entrance of the hive so the bees will notice that they have been moved. Hives must be moved at least 5 miles to prevent bees from returning to the former site of their hive.

- Inspect beehives at least monthly to determine the strength of the hive and the queen. Be sure to observe the number of eggs being layed by the queen. Also note whether the worker bees are building drone or queen cells in the hive. Drone and queen cells look like peanuts. Such cells should be destroyed. **Worker** bees are undeveloped female bees and constitute all of the working force of the hive. A **drone** is a male bee whose only purpose in life is to fertilize the queen once in her life span.

- Reduce or prevent swarming of bees by providing ample hive space for the bees and eliminating queen cells as they are found. Overcrowding often causes bees to develop a second queen. A new queen will attract a group of worker bees and leave the hive to start a new colony. This process is called swarming. Bees will not swarm without a queen because she is their only hope of survival.

- Be aware of pesticides being used in the area that could kill bees or be stored in the honey.

- Secure the proper equipment before starting an apiary. An **apiary** is an area for the keeping of beehives.

- Keep honey that has been removed from bees in an area that bees cannot get to. Otherwise, they will steal all of the honey in a short time.

- Extract honey from the comb as soon as possible after harvesting it. Honey stored for long periods in the comb may granulate, which makes it impossible to extract.

- Develop a market for your honey.

PET CARE AND MANAGEMENT

Animals make excellent companions and can provide specialized functions for children and adults alike. For many animals, the line between pet, companion animal, and work animal is a fine one. For example, guide dogs are essential "eyes" for the blind and actually must make intelligent choices for their masters when crossing streets (Figure 29–17). Similarly, dogs and other pets provide companionship for the elderly and may warn of intruders. Good dogs may ward off attackers and provide other protection when appropriately trained. Guard dogs are essential for herding and controlling sheep and protecting sheep from predators.

Many kinds of caged birds and reptiles are used for pets. Their needs are rather specialized and advice for their care is usually obtained from pet shops and other suppliers. Similarly, a large assortment of fish and

FIGURE 29–17 Guide dogs serve dual purposes. When at work, they act as "eyes" for the blind. When at rest or play, they provide companionship. *(Courtesy of Guide Dog Foundation for the Blind, Inc.)*

FIGURE 29–18 Aquatic animal and plant projects create interest and provide excellent opportunities for learning and profit. *(Courtesy of FFA, Photo by Michael Wilson)*

FIGURE 29–19 The care of exotic animals creates new challenges for experienced pet handlers. *(Courtesy of FFA)*

other aquatic creatures make attractive and challenging aquarium projects for all ages (Figure 29–18). Even exotic animals such as llamas and kangaroos find their ways onto farms and into the hearts of those in the United States (Figure 29–19).

Dogs and cats are probably the most prevalent companion animals in the United States. Although they have different moods and behave in different ways, they have similar needs (Figure 29–20). The following is

FIGURE 29–20 Cats make great pets that provide companionship to people. They also have needs, such as food and veterinary care, that must be provided by their owners. *(Courtesy of FFA)*

Agri-Profile

CAREER AREAS:

Animal Technician/Grower/Beekeeper/Manager

The career opportunities for small-animal care and management are good in many areas. The extensive use of laboratory animals for research, small animals for pets, fish for home aquariums and garden pools, small animals for fur, and animals for zoological parks ensures attractive jobs in the future.

The operation of animal hospitals, kennels, grooming services, pet stores, training programs, boarding facilities, public aquariums, animal shelters, and humane societies provides opportunities in a number of fields. These include animal nutrition, facilities construction and maintenance, feeds, health services, care and management, production, breeding, and marketing.

Curricula in small-animal care and management have been added to many high school agriscience programs. Similarly, programs are available in many technical schools, community colleges, and universities.

Animal trainers teach animals to obey commands and to perform tricks for audiences.
(Courtesy of Tracy and Pedro Navarro)

INTERNET KEY WORDS:

care for animals

a list of approved practices for the care of dogs and cats (courtesy of Pennsylvania State University):

- Select animals that are alert and healthy.
- Vaccinate for rabies at age-determined intervals.
- Select a breed adapted to your situation.
- Prepare a clean, draft-free living area.
- Provide an adequate number of feed dishes.
- Provide a diet of high-quality food for the breed and age.
- Provide toys, treats, or both.
- Clean feed and water dishes daily.
- Keep clean, fresh bedding in sleeping area.
- Provide plenty of clean, fresh water.
- Clean pen or box and exercise area daily.
- Develop and implement a sound health plan.
- Consult with your veterinarian to prevent and control internal and external parasites.
- Vaccinate animals routinely at proper times.
- Train animals for show.
- Properly bathe and groom the animal.
- Use a proper carrier for transport.
- Use proper restraint procedures.
- Maintain accurate breeding and production records.
- Select a male animal to mate with your female animal.

FIGURE 29–21 Businesses are needed to supply specialized care, food, health supplies, and equipment for pets and small-animal projects. *Courtesy of Michael Dzaman)*

- Use a superior proven male animal.
- Take extra precautions for female animals in heat.
- Prepare a clean area for whelping (giving birth).
- Train properly.
- Market and sell young animals.
- Develop a private market for the animals.
- Complete a registration application if animals are purebred.
- Neuter animals not involved in breeding.
- Summarize and analyze records.

What is one person's pet always becomes someone else's business in providing replacement animals, feed, supplies, and equipment for that pet (Figure 29–21).

Raising small animals provides the opportunity for individuals with limited capital and facilities to get a start in animal agriscience. Most small animals are better adapted to production in urban and suburban areas than are larger animals. The same experiences in planning for, caring for, managing, and marketing can be learned with small-animal enterprises without the large outlay of cash needed for the production of large animals.

STUDENT ACTIVITIES

1. Write the Terms to Know and their meanings in your notebook.
2. Make a bulletin board display of breeds and types of poultry and rabbits.
3. Attend a fair or show and record the names of the breeds of poultry and rabbits shown there.
4. Interview a local beekeeper about beekeeping practices in your area.
5. Compare the label from a bag of poultry feed with one from a bag of rabbit feed. Determine the differences in ingredients, percentage of protein, additives, and fiber.
6. Set up an observation beehive in the school.
7. Make a list of local crops that are of importance to agriscience and that bees pollinate.
8. Develop a crossword puzzle or word search using the Terms to Know.
9. Take your small animal to the veterinarian for a checkup. While there, have all of the scheduled vaccinations given.
10. Ask your veterinarian to discuss with you the proper care that your animal needs. Be sure to talk about how much food and water it should be getting, as well as any special care it may require.

SELF EVALUATION

A. Multiple Choice

1. A hive of honeybees needs about _____ lb honey stored to live during the winter.
   - a. 25
   - b. 50
   - c. 75
   - d. 100

2. A gosling is a baby
   - a. chicken.
   - b. duck.
   - c. pigeon.
   - d. goose.

3. A pair of rabbits can produce _____ lb meat per year.
   - a. 10 to 30
   - b. 30 to 50
   - c. 50 to 80
   - d. 200 or more

4. A male duck is a
   - a. drake.
   - b. capon.
   - c. cockerel.
   - d. rooster.

5. The only bee in a hive capable of laying eggs is the
   - a. king.
   - b. queen.
   - c. worker.
   - d. drone.

6. Honeybees account for about _____ percent of all insect pollination.
   - a. 20
   - b. 40
   - c. 60
   - d. 80

7. The _____ breed of chicken is the foundation of nearly all types of laying hens.
   - a. Cornish
   - b. Pekin
   - c. Leghorn
   - d. game

8. A type of rabbit that produces wool is
   - a. pelt.
   - b. fur.
   - c. angora.
   - d. mohair.

9. The champion egg-laying breed of bird is the
   - a. Leghorn.
   - b. quail.
   - c. Khaki Campbell.
   - d. Toulouse.

10. More than _____ lb rabbit meat are produced in the United States each year.
    - a. 1 million
    - b. 10 million
    - c. 25 million
    - d. 50 million

B. Matching

_____	**1.** Drone	a.	Male goose
_____	**2.** Cockerel	b.	Young chicken
_____	**3.** Gander	c.	Male bee
_____	**4.** Doe	d.	Female bee
_____	**5.** Drake	e.	Female pheasant
_____	**6.** Chick	f.	Young turkey
_____	**7.** Buck	g.	Male rabbit
_____	**8.** Worker	h.	Female rabbit
_____	**9.** Hen	i.	Male chicken
_____	**10.** Poult	j.	Male duck

C. Completion

1. More than 90 percent of the turkeys produced in the United States are _____.

2. Ducks raised for meat reach a weight of 7 lb in about _____ weeks.

3. The most popular breed of rabbit for meat production is the _____.

4. A _____ is a home for honeybees.

5. The average laying hen produces about _____ eggs each year.

6. Another word for beekeeping is _____.

7. _____ and _____ are probably the most common companion animals in the United States.

8. Whelping means _____.

UNIT 30 | Dairy and Livestock Management

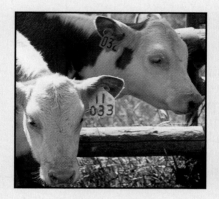

OBJECTIVE

To determine the history, types, uses, care, and management of dairy and livestock.

MATERIALS LIST

✓ copies of various livestock breed magazines

✓ bulletin board materials

✓ paper glue

✓ scissors

✓ paper for notebook covers

✓ examples of various animal products or things made from animal products

✓ Internet access

COMPETENCIES TO BE DEVELOPED

After studying this unit, you should be able to:

- describe the history and economic importance of dairy and livestock.
- recognize major types and classes of livestock.
- list major uses of livestock.
- understand basic approved practices in the care and management of dairy and livestock.

SUGGESTED CLASS ACTIVITIES

1. Take a class field trip to a livestock farm located near your community. Ask the farm owner or manager to talk to the class about the critical management decisions that he or she makes that affect the profitability of the farm. Be prepared to ask leading questions about the nutrition, health, breeding, and marketing strategies that are used on the farm. Prepare individual written student reports that address these and other important issues that were discussed during the visit.

2. Study the breeds of livestock by assigning each class member to write to a breed association to obtain information about the breed. Have students prepare an oral presentation and report their information to the class. Provide class members with a worksheet on which they record key points about each livestock breed as the reports are given.

3. Invite a dairy, sheep, swine, or goat judge to come to the class and give a presentation outlining the physical characteristics that make a good show animal, as well as a good producer. You may also want him or her to provide a presentation on showmanship. Students should be ready to ask questions and participate in the discussion.

TERMS TO KNOW

mammal	calve	sow	lamb
milk	cow	piglet	mutton
veal	dry period	boar	ewe
beef	cull	gilt	ram
butterfat	draft	farrow	lambing
calf	bull	farrowing crate	shearing
colostrum	implant	dock	mohair
milk replacer	steer	ear notching	chevon
free choice	pork	creep feed	lactation period
wean	lard	shepherd	kid
heifer	render	wool	

Large animals, including dairy, beef, sheep, goats, and swine, are the backbone of animal agriculture. Keeping wild animals captured made it necessary for early humans to provide feed for them. It also changed their lives, in that it forced humans to tend the animals. The capturing and domestication of animals changed humans from hunters to farmers. Today, much of agriscience is centered around the production of animals and animal products and the production of feed for those animals.

DAIRY CATTLE

Origin and History

In the wild, mammals normally produce only enough milk to feed their offspring. A **mammal** is an animal that produces milk. **Milk** is white or yellowish liquid secreted by the mammary glands of animals for the purpose of feeding young. When early humans realized that milk was good to drink and that some types of animals produced more milk than others, the domestication of milk animals began. Although the cow, buffalo, goat, ewe, and mare, have been and are currently being used for the production of milk in various parts of the world, this unit concentrates primarily on cattle in discussing dairy production.

There are early historical records of the use of cattle to produce milk, and there are also a number of references in the Bible to milk and the production of milk. Even Hippocrates, the father of modern medicine, recommended milk as a medicine in his writings around 400 BC.

INTERNET KEY WORDS:

milk production, U.S.

There were no dairy cows in the New World until Columbus brought them with him on his second voyage in 1493. Cattle also came to the New World in 1611 with the Jamestown colonists. The production of milk was limited to a few cows per family during the colonial period. It was not until the late 1800s that dairying became an important agricultural industry in the United States. Since then, efforts have been intensive in breeding and developing the dairy type (Figure 30–1).

PARTS OF A DAIRY COW

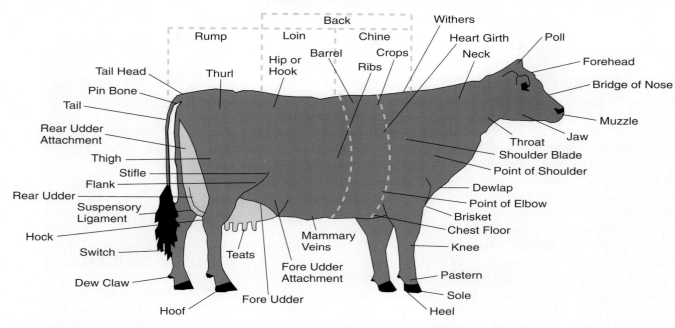

FIGURE 30–1 Dairy cattle were developed to convert nutrients into milk rather than meat.

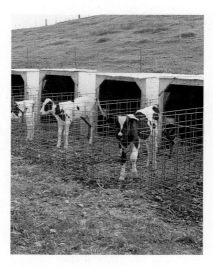

FIGURE 30–2 Today's female calves become tomorrow's dairy cows, so good calf care and management are top priorities on dairy farms. *(Courtesy of FFA)*

INTERNET KEY WORDS:

Holstein cattle
Jersey cattle
Guernsey cattle
Brown Swiss cattle

Economic Importance

The production of milk is the second most important animal enterprise in the United States, if sales dollars are the criterion for importance. The consumption of milk and other dairy products has remained nearly steady during the past several years, after several decades of steady decline. The average person in the United States uses slightly more than 580 lb of milk and dairy products each year. There are about 9.12 million cows in the United States producing about 170.3 billion lb milk annually. Average production per cow is more than 18,700 lb milk per cow per year. Dairy is truly big business.

The production of milk is not the only income-generating part of dairy production. Calves not needed as replacement for the dairy herd may be sold as **veal** (the meat of young calves). Similarly, cows that are no longer profitable producers of milk and the male offspring of dairy cows are sold for **beef** (meat from cattle). The female calves become tomorrow's cows (Figure 30–2).

Dairy-cattle operations are generally divided into two types: Class A and Class B. The class refers to the intended use for the milk that is produced. Class A milk is produced under strict standards and is intended for consumption as fluid milk. Fluid milk includes whole milk, reduced-fat milk, and cream. Class B milk, which can be produced under less strict standards, is intended to be used to make butter, cheese, ice cream, nonfat dry milk, and other manufactured dairy products.

Types and Breeds

More than 90 percent of all dairy cattle in the United States are of the Holstein breed. The familiar black and white cattle are the greatest average producers of milk of any breed in the country. Because of the large

Agri-Profile

CAREER AREAS:

Farm Manager/Herd Manager/Farmer/Rancher/ Breed Association Representative

Science and production can no longer be separated. Individuals and teams of scientists visit farms, ranches, and feedlots frequently to gather data, analyze problems, and seek solutions. Farmers, ranchers, and feedlot managers are frequently graduates of agricultural colleges and may have advanced degrees in business, management, or animal sciences.

In addition to scientists and managers, technicians find meaningful careers in the animal production businesses of the nation. Livestock enterprises include beef, dairy, sheep, and swine. There are many nonfarm career opportunities as field representatives for breed associations, feed companies, marketing cooperatives, supply companies, animal-health products, herd-improvement associations, and financial-management firms.

Many individuals use their farming and ranching backgrounds in agriscience writing, publishing, and telecasting careers. Farm and ranch experience provides excellent background for careers in agriscience teaching and extension work.

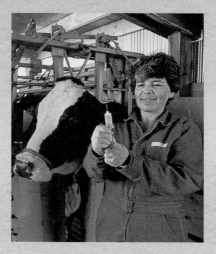

Farming, ranching, and feedlot management have become big business, and they are based on scientific research. *(Courtesy of USDA)*

numbers of cattle involved, the breed also has been able to make the most genetic improvement in recent years.

The second most popular breed of dairy cattle is the Jersey. Even though they are the smallest of the dairy breeds, they rank number one in butterfat production. Butterfat is the fat in milk. Their rich milk is also high in milk solids such as protein. Another popular breed of dairy cattle is the Guernsey. This breed is known for the yellow color of its milk. Ayrshire and Brown Swiss round out the top five breeds of dairy cattle in the United States (Figure 30–3).

Approved Practices

Raising Calves General approved practices for raising dairy calves include the following:

- Make sure the newborn calf receives colostrum as its first food and for at least the first 36 hours of its life. Colostrum is the milk that a cow produces for a short time after calving. It contains antibodies that protect the newborn animal from diseases until the animal can build up its own natural defenses.

- Feed milk or milk replacer at 8 to 10 percent of the calf's weight daily until the calf is 4 weeks old. Milk replacers are dry dairy or vegetable products that are mixed with warm water and fed to young calves in place of milk.

- Start feeding calf starter, a grain mixture, free choice at about 10 days. Free choice means making feed available at all times.

FIGURE 30–3 Major breeds of dairy cattle in the United States include (A) Holstein, (B) Jersey, (C) Guernsey, (D) Ayrshire, and (E) Brown Swiss.

- Wean calves from milk when they are eating 1.5 lb calf starter per day. To **wean** means to stop feeding milk.
- Feed calves green, leafy hay and water free choice at 2 to 9 months of age. Up to 4 lb grain can be fed daily.
- Make up the bulk of the ration with forages fed free choice after 9 months of age.
- Remove horns at an early age, preferably as soon as the horns begin to develop.
- Remove extra teats at an early age.
- Identify calves with ear tags or tattoos as soon as possible after birth.
- Prevent calves from sucking each other.
- Keep hooves properly trimmed.
- Vaccinate for calfhood diseases at the recommended times.
- Maintain the calf in clean and sanitary conditions.
- Plan for and maintain a disease- and parasite-prevention and control program.
- Breed heifers to calve at 20 to 24 months of age. A **heifer** is a female that has not given birth to a calf. With cattle, to **calve** means to give birth.
- Maintain heifers and calves in uniform groups according to size and weight.

INTERNET KEY WORDS:

dairy cow management

Dairy Cows Some approved practices for dairy cows include the following:

- Rebreed cows 45 to 60 days after calving. A **cow** is a female animal of the cattle family that has given birth. Cows should be bred to calve once every 12 months.
- Observe cows for evidence of heat period twice daily (Figure 30–4).
- Check cows to determine whether they are pregnant. This should be done 45 to 60 days after breeding them.
- Provide a dry period of about 60 days before calving to allow the cow to rebuild her body. The **dry period** refers to the time when a cow is not producing milk.
- Feed dairy cows according to their levels of production and stages of pregnancy.
- Maintain complete health, breeding, and production records for every cow in the herd (Figure 30–5).
- Establish and maintain a disease- and parasite-prevention and control program.
- Contact a veterinarian anytime that you are unsure of how to treat a dairy-herd problem.
- Milk dairy cows at regular intervals each day. Two milkings each day, approximately 12 hours apart, is a normal routine. However, milking three times per day is common in many high-producing herds.

COMING INTO HEAT

1. Smells other cows
2. Attempts to nose and ride other cows.
3. Vulva moist, red, and slightly swollen.

TRUE HEAT

1. Nervous and easily excited.
2. Bawls and bellows frequently.
3. Attempts to ride other cows.
4. Stands to be ridden.
5. Wet, shiny pin bones, with strings of clear mucus discharge from vulva (late in period).
6. Some animals "hold up" their milk at milking time while in heat.

AFTER HEAT

1. Smells other cows.
2. Will not stand to be ridden.
3. Clear mucus discharge on tail and pin bones.

Poor | Fair | Good | Excellent Time to Breed | Good | Egg Release From Ovary

6–10 HOURS   0   3   6   9   18   24 27   30

NOTE: Suggestions are based on average conditions. Many exceptions are true of normal cows. Cows noticed in "True Heat" at milking time in the morning probably have been in heat for several hours. The numbers 3, 6, 9, etc. indicate hours of heat.

FIGURE 30–4 Breeding at the proper stage of the heat period is essential to maximize pregnancy rates.

Factors Determining Efficiency			
Factor	Results	Goals	State Average
_____ 1. Number of cows in herd			
_____ 2. Total lb of milk produced			
_____ 3. Total lb of butter fat produced			
_____ 4. Average annual milk production/cow (lb)			
_____ 5. Average annual butter fat production/cow (lb)			
_____ 6. Average annual % butter fat for student's herd			
_____ 7. Feed cost/lb of butterfat produced			
_____ 8. Average feed cost/cow			
_____ 9. % of calves sold or kept until 3 months of age			
_____ 10. Profit or loss			
_____ 11. Lbs. of milk sold/hour of total labor			
_____ 12. Dollar returns/hour of self labor			
_____ 13. Production cost/lb of milk			

FIGURE 30–5 Efficiency factors are good indicators of how the dairy business is operating.

- Maintain a regular routine in handling dairy cattle to maintain maximum production.
- Cull unprofitable dairy cows. To **cull** means to remove from the herd.
- Properly maintain dairy housing and milking equipment.

BEEF CATTLE

Origin and History

INTERNET ADDRESS:

<http://www.beef.org>

It is likely that cattle were domesticated in Europe and Asia at some time during the New Stone Age. Domesticated cattle of today are probably all descended from one of two wild species, *Bos taurus* or *Bos indicus* (Figure 30–6). Some of these wild cattle stood as tall as 7 feet at the shoulders. They were domesticated for meat, milk, and draft. **Draft** means for work.

Owning cattle was a symbol of wealth in early times. They were worshipped, used, and often abused. Several early civilizations had "gods" fashioned to look like cattle. The use of cattle in sports such as bullfighting also developed during early times.

Cattle came to the New World with the earliest settlers, who were more interested in animals that could do heavy work than in those that could produce meat. As the European explorers came to the New World, they also brought cattle with them. Spanish settlers brought longhorn-type cattle as a source of food for Christian missions in the Southwest. As demand increased for beef, the cattle industry developed on the frontier, where grass and the large open spaces required for cattle were abundant. Great cattle drives originated in these areas, as cattlemen drove their cattle to human population centers to market them.

Today, the beef cattle industry is concentrated in the Midwest and South, where plenty of feed is available and production costs are less than in some areas of the country.

FIGURE 30–6 *Bos taurus* was one of the wild ancestors of today's cattle.

Science Connection

OF COWS AND CARS

Cows and cars are sometimes the recipients of the same technology. Transponders are small devices that can be attached to the ears of cows and other livestock to identify them. As animals come into the electronic fields of certain other devices, their identities can be noted and recorded by computer. The computer can be programmed to simply note and remember what the animal does, or it can direct other devices to provide selected kinds and amounts of feed, water, medicine, pesticide, or other treatment. This technology is becoming common in modern milking parlors and feeding systems.

At the U.S. Department of Agriculture (USDA) Fort Keogh Livestock and Range Laboratory in southeastern Montana, scientists found transponders to be the answer to a weighty problem! To conduct feeding, grazing, pasture-improvement, and pasture-management experiments, they had to weigh their research subjects, range cattle, at frequent intervals. Animal response to treatment generally shows up in the form of weight gain or loss. At the same time, any disturbance or change in an animal's routine, such as

Transponders and other sensing units permit better research and management. The unit being applied here will record motion and other data so scientists can determine how much time the animal spends grazing. *(Courtesy of USDA/ARS #K-4235-8)*

cattle roundup, forced weighing, or catching in a head gate, will influence an animal's weight. Good research required a system that could weigh the animals at frequent intervals without disturbing them—that is, they needed to weigh themselves!

The answer was found in a system using a scale, a computer, and a transponder. Cattle need water. A continuous supply of fresh water must be readily available if body functions are to be optimal. Therefore, the researchers set up a scale and guide railing so the animal had to cross over the scale to drink. An ear-tag transponder on the animal identifies the animal as it enters the range of the pickup device and steps on the scale. The electronic scale sends continuous weight readings to the computer, indicating the animal's weight before, during, and after drinking. The computer records the weight data, together with time, temperature, weather, and other information monitored at the site.

As drivers zip across California's Golden Gate Bridge and blissfully bypass the toll booth, they are probably legal motorists simply using the same transponder device that was pioneered by the USDA's Agricultural Research Service. If so, the car-mounted transponder in their car gave the computer the auto's account number, and the computer will have noted the time, date, and auto number for future billing to the motorist. The system is being used by many toll facilities across the country.

Economic Importance

The beef cattle industry is the number-one red-meat production industry in the United States. The U.S. population eats about 76 lb beef per person per year. This is part of a total consumption of about 190 lb meat, poultry, and seafood.

The beef cattle business is one that will likely remain important for many years to come. There are 101.6 million head of cattle and calves on U.S. farms today. Leading states in the sale of beef are Texas, Nebraska, Kansas, Colorado, Iowa, and California.

By-products from the Production of Meat	
• Bone for bone china.	• Drumheads and violin strings.
• Horn and bone handles for carving sets.	• Animal fats for soap and feed.
• Hides and skins for leather goods.	• Wool for clothing.
• Rennet for cheese making.	• Camel's hair (actually from cattle ears) for artists' brushes.
• Gelatin for marshmallows, photographic film, printers' rollers.	• Cutting oils and other special industrial lubricants.
• Stearin for making chewing gum and candies.	• Bone charcoal for high-grade steel, such as ball bearings.
• Glycerin for explosives used in mining and blasting.	• Special glues for marine plywoods, paper, matches, window shades.
• Lanolin for cosmetics.	• Curled hair for upholstery. Leather for covering fine furniture.
• Chemicals for tires that run cooler.	• High-protein livestock feeds.
• Binders for asphalt paving.	
• Medicines such as various hormones and glandular extracts, insulin, pepsin, epinephrine, ACTH, cortisone, and surgical sutures.	

FIGURE 30-7 Many important substances are obtained from the meat-processing industry. They range from raw materials for cosmetics to medicines and many other products that are used for a variety of purposes. ACTH, adenocorticotropic hormone.

The production of meat is not the only use for beef cattle. Cattle convert inedible grasses into food for people. Cattle manure provides fertilizer for crops, and meat by-products are made into many nonfood products that we use everyday (Figure 30–7).

Types and Breeds

The general types of beef-cattle operations include purebred breeders, cow-calf operations, and slaughter-cattle, or feedlot, operations. In a purebred operation, only cattle of a single, pure breed are raised. Operations are geared to produce purebred bulls for cow–calf operations and to produce animals to be sold to other purebred operations. A **bull** is a male animal of the cattle family. Breeders of purebred cattle have been responsible for much of the genetic improvement in beef cattle in recent years.

Cow–calf operations serve to produce feeder calves for slaughter-cattle producers. They are located mostly in the upper Great Plains states and in the western range states, where grass is in abundance and much of the land is unsuited to produce many other crops. Usually calves are born in the spring, stay with their mothers during the summer, and are weaned in the fall. They are then sold to feedlot operations. Most cow–calf producers use purebred bulls on the grade cows in the herd to produce superior calves.

The feedlot operator buys calves from cow–calf operators and feeds them until they reach slaughter weight. These operations are concentrated in the Midwest, where there is an abundance of corn and other grain available for feed. There are large feedlot operations, located wherever an abundance of cheap feed exists.

Until about 40 years ago, there were basically three breeds of beef cattle in the United States: Hereford, Angus, and Shorthorn. Although these three breeds are still important today, they have been joined by more than 50 other breeds from all over the world (Figure 30–8). These breeds can be divided into many classifications. However, for our purposes, they are designated as English, exotic, and American.

English breeds originated in the British Isles and include Hereford, Angus, Shorthorn, Galloway, Devon, and Red Poll. In general, English

INTERNET ADDRESS:

<http://www.miobuffer.com.tw/foreigener/breed/02.htm>

INTERNET KEY WORDS:

beef cattle breeds

FIGURE 30–8 Today's predominant beef breeds in the United States are (A) Hereford *(Courtesy American Hereford Association)* and (B) Angus *(Courtesy American Angus Association)*. (C) Simmental *(Courtesy American Simmental Association)*. (D) Limousin *(Courtesy North American Limousine Foundation)*. (E) Beefmaster *(Courtesy Beefmaster Association Universal)*. (F) Charolais *(Courtesy American-International Charolais Association)*

G   H

FIGURE 30-8 *(Continued)* More of today's predominant beef breeds in the United States are (G) Brahman *(Courtesy American Brahman Breeders' Association)* and (H) Shorthorn *(Courtesy American Milking Shorthorn Society).*

FIGURE 30-9 Many commercial producers have crossed and mixed breeds of cattle, such as the Brangus bull pictured here. *(Courtesy American Red Brangus Association)*

breeds of cattle are of medium size and are noted for the excellent quality of meat that they produce.

Exotic breeds of cattle were first imported to the United States from all over the world when consumers began demanding leaner beef. Beef producers also became aware that calves from these breeds of cattle grew faster and much more efficiently than most of the English breeds. Exotic-breed bulls are often used in grade cow–calf operations to increase the weight of calves being produced. Examples of exotic breeds of cattle include Charolais, Limousin, Simmental, Blond D'Aquitaine, and Maine Anjou.

U.S. breeds of beef cattle were developed from necessity. To grow beef cattle in the South and Southwest, heat tolerance and disease and parasite resistance were needed. To develop cattle that met these requirements and still had desirable quality meat, breeders crossed Brahman cattle from India with the English breeds. Examples of U.S. breeds of beef cattle are Brangus, Beefmaster, Santa Gertrudis, and Barzona. Many commercial herds are of crossbred, or mixed, breeds (Figure 30–9).

INTERNET KEY WORDS:

growth implants, beef
beef management practices

Approved Practices for Beef Production

The following are some of the considerations in the production of beef cattle:

- Select breeds and individual beef cattle according to intended use, area of the country, and personal preference.

- Buy cattle from only reputable breeders.

- Select purebred cattle according to physical appearance, pedigree, and available records (Figure 30–10).

- Isolate new animals from the herd for at least 30 days to observe for possible diseases and parasites.

- Provide enough human contact so that beef cattle can be handled when necessary.

- Break calves to lead as soon as possible, if they are going to be exhibited at fairs and shows.

- Plan a complete herd-health program and follow through with it.

- Use the services of a veterinarian for serious or unknown health problems.

- Vaccinate to prevent diseases of local concern.

- Castrate, dehorn, and permanently identify calves at an early age.

- Breed heifers to calve at 2 years of age.

- Implant steers and heifers grown for slaughter with approved growth hormones. To **implant** is to place a substance under the skin. The substance is released slowly over a long period. A **steer** is a castrated male animal of the cattle family.

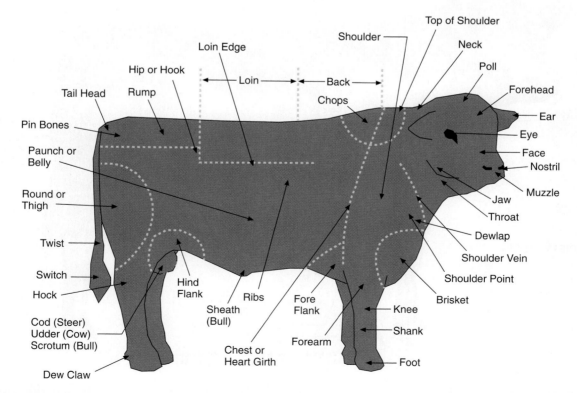

FIGURE 30–10 Learning the proper names of cattle parts is necessary for discussing and evaluating animals.

- Wean calves at 205 days of age and at 450 to 500 lb.
- Provide supplemental nutrition to cattle when natural forages are in short supply.
- In a pasture-breeding system, allow one mature bull for every 25 to 30 cows. With pen mating, one bull can be used for every 30 to 50 cows.
- Restrict the length of the breeding season so that all calves are born in a 30- to 60-day period.
- Assure that cows calve in clean stalls or pastures.
- Group the cattle being fed for slaughter according to size and sex.
- Provide shelter from inclement weather.
- Provide access to clean, fresh water at all times.
- Feed slaughter cattle to reach market weight at 15 to 24 months of age.
- Have proper facilities and equipment available for the type of operation planned.
- Market animals at the optimum time to maximize profits (Figure 30–11).
- Maintain complete and accurate records.

FIGURE 30–11 Animals should be bred, fed, managed, and marketed in a manner that delivers the best product for the consumer. *(Courtesy of American Angus Association)*

SWINE

INTERNET KEY WORDS:

National Pork Producers
annual pork consumption

Origin and History

Swine were apparently domesticated in China during the Neolithic Age. The first written record of keeping swine appears in 4900 BC. Mention of swine occurs in the Bible as early as 1500 BC.

Domestic swine originate from two wild stocks: the European wild boar, *Sus scrofa,* and the East Indian pig, *Sus vittatus.* There are several other wild types of swine that exist today. Even domesticated swine can revert quickly to the wild when the opportunity or need arises.

Swine came to the New World in 1493 on the second voyage of Columbus. As European explorers came to what is now the United States, they brought swine with them as a food supply. The original 13 head of swine brought to the New World by the explorer de Soto multiplied to more than 700 head in just 3 years and provided food for the explorers.

European settlers also brought swine with them as they settled on the East Coast of the United States. Because swine could find food on their own and reproduce rapidly, they were soon known as "mortgage lifters." Swine in excess of local needs were exported as pork and lard. **Pork** is meat from swine. **Lard** is rendered pork fat. **Render** means to cook and press the oil from fatty tissue.

As westward expansion occurred, swine production followed. In time, the center of swine production settled in the Corn Belt. This area provided large amounts of corn and other grains necessary for large-scale pork production.

Economic Importance

Pork production is the number-two red-meat industry in the United States. It generates more than $38 billion in retail sales each year. With a population of about 97 million hogs, the United States ranks second in the world in swine production, behind China. There are about 780 million head of swine in the world today, with more than 45 percent of them in China alone. U.S. production of pork makes up approximately 10 percent of the world supply.

Science Connection

OH, THE SWEET SMELL OF SUCCESS

The suburbs are moving farther and farther away from the cities. Farms and subdivisions find themselves becoming neighbors. As a result, conflicts are becoming more frequent because of the smells produced by animal wastes. Fortunately, researchers at the University of Florida have designed a piece of equipment called a fixed-film digester that may be a solution that will keep everyone breathing deeply. A system of plastic pipes line a 100,000-gallon tank. Bacteria inside of the pipes transform smelly waste into compounds that do not create an odor problem. The system can reduce offensive smells by 90 percent. Besides clearing the air, the digester also produces a free fuel source for the farmer, biogas (methane). Another benefit is that water can be cleaned and recycled. The system currently has a high price tag, but as it becomes commercially available, the cost is expected to come down and make the system a reasonable farm expense that will keep both farmers and their neighbors breathing the sweet smell of success.

Since 1997, people in the United States have consumed about 56 lb pork per person annually. The consumption of red meat has been slowly decreasing, but the consumption of pork has increased.

The modern hog of today is vastly different from its ancestors. It is lean and trim compared with the lard-type hog in demand 50 years ago, and even the meat-type hog of 20 years ago. Hogs are the most efficient converters of feed into meat among the large red-meat animals. It takes about 3.5 lb feed to 1 lb gain.

Leading states in the United States in the production of swine include Iowa, Illinois, Minnesota, Nebraska, and Indiana. More than 25 percent of the swine production in the country takes place in Iowa alone.

Swine Operations

The basic types of swine operations in the United States are feeder-pig producers and market-hog producers. Operations that produce feeder pigs usually maintain large herds of sows that produce 2 to 2.5 litters of piglets each year. A **sow** is a female animal of the swine family that has given birth. A **piglet** is a young member of the swine family. Piglets are often sold to other producers, who feed them until they reach market weight. In most operations, crossbred sows are bred to purebred boars to produce offspring with more hybrid vigor than would otherwise result (Figure 30–12). A **boar** is a male member of the swine family.

Market-hog operations normally purchase pigs at 5 to 8 weeks of age from feeder-pig producers. They feed the pigs until they reach a market weight of about 240 to 260 lb. They are then marketed and sent to slaughter plants for processing.

Another type of swine operation is the purebred producer. This kind of swine operation is responsible for producing high-quality boars for feeder-pig operations and purebred stock for other purebred farms. They contribute much to the genetic improvement of swine in general.

Types and Breeds

Purebred swine in the United States consist of meat-type hogs. With the decreased demand for lard and the demand for lean pork nearly devoid of fat, the lard-type hog has been bred out of existence.

Popular breeds of swine that were developed in the United States include Duroc, Hampshire, Chester White, Poland China, and Spotted

INTERNET ADDRESS:

<http://www.nppc.org>

INTERNET ADDRESS:

<http://www.ansi.okstate.edu/
breeds/swine>

FIGURE 30–12 Feeder-pig producers keep brood sows and boars, which provide the young stock for the swine industry. *(Courtesy Coles Swine Farms, Inc.)*

Hog. The Berkshire, Yorkshire, and Tamworth breeds were developed in England, and the Landrace originated in Denmark. Duroc, Hampshire, and Yorkshire are the most popular breeds of swine in the United States today (Figure 30–13).

Approved Practices in Swine Production

INTERNET KEY WORDS:

swine management

Some of the approved practices in the production of swine include:

- Buy pigs from only reputable producers or at certified feeder-pig sales.

- Observe newly purchased animals for signs of disease and parasites.

- Group pigs according to size in groups of not more than 20 to 25 animals.

- Feed a complete, balanced ration based on the age and weight of the animal being fed.

- Ensure that access to an unlimited supply of fresh water is available at all times.

- Keep facilities and equipment clean and sanitary.

- Clean and disinfect all facilities and equipment after each group of animals leaves and before the next group arrives.

- Select replacement gilts for the breeding herd at an early age and raise them separately from market hogs. A gilt is a female pig that has not given birth.

- Breed gilts at 8 months of age or 250 to 300 lb, so they farrow at approximately 1 year of age. To farrow means to give birth to baby pigs.

- Use a hand-mating system to breed gilts and sows. In a hand-mating system, the boar and sow are kept separate except during mating. Use a boar to check for animals in heat.

- Put bred gilts or sows in farrowing facilities or farrowing crates 3 days before they are due to farrow. Farrowing crates are specially made cages or pens in which swine give birth. This 3-day period gives the female pig time to adjust to new surroundings before the piglets are born.

- Perform the following to the piglets at birth:
    1. Clip needle or wolf teeth.
    2. Clip or tie navel cord and dip the end in iodine.
    3. Provide supplemental iron.
    4. Dock tails of pigs to be marketed for meat. To dock means to remove all but about 1 inch of the tail.
    5. Weigh all pigs in the litter.
    6. Ear notch all pigs for identification. Ear notching is a system of permanently marking animals for identification by cutting notches in their ears at specific locations.

- Provide creep feed for the baby pigs by the time they are 1 week old. Creep feed is feed provided especially for young animals to supplement milk from their mothers.

- Castrate male pigs at an early age.

- Wean pigs at 4 to 6 weeks of age. Weaning at about 6 weeks is normal in most herds of swine.

FIGURE 30–13 Major breeds of swine in the United States include (A) Duroc, (B) Yorkshire, (C) Hampshire, (D) Spotted, and (E) Chester White.

- Rebreed sows on the first heat period after weaning the pigs. This usually occurs about 3 days after the pigs are weaned.

- Limit the feed for sows to prevent them from getting too fat.

- Provide protection from heat and cold, especially heat. Swine have no sweat glands, and care must be taken to keep them cool in hot weather.

- Maintain complete health and production records for each animal in the breeding herd.

- Set realistic production goals and cull animals not meeting the goals.

SHEEP

Origin and History

The domestication of sheep apparently occurred before the time of recorded history. Fibers of wool have been found in ruins of villages in Switzerland more than 20,000 years old. Egyptian sculptures showing the importance of sheep date back 7,000 years. Even ancient historical texts are filled with mentions of sheep and shepherds (those who take care of sheep).

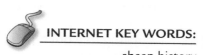

INTERNET KEY WORDS:

sheep history
sheep rangeland

Sheep were probably domesticated from wild types in Europe and Asia by early humans to use for meat, wool, pelts, and milk. Wool is a modified hair with superior insulating qualities. It is usually obtained from sheep, but it is also obtained from some other animals such as llamas. Wool has been an important product of international trade for many centuries, and it is still one of the most important fibers known to humans.

As civilization advanced, the production of wool became a priority in sheep production. As a result, specific wool-producing breeds were developed in Europe. Most of today's breeds can be traced back to these breeds developed 500 to 1,000 years ago.

Columbus had sheep with him on his second voyage to the West Indies in 1493. Cortez brought sheep when he explored Mexico in 1519. Spanish missionaries also kept sheep and taught American Indians of the Southwest how to weave wool into cloth.

English settlers on the East Coast also raised sheep for the production of wool. Lamb and mutton were of secondary concern. The term lamb refers to a young sheep, as well as the meat of young sheep. Mutton is meat from mature sheep.

Centers of sheep population gradually moved from the Northeast to the West as populations expanded. Areas of open spaces and abundant grasses such as western rangelands have proved to be ideal for the production of sheep.

Economic Importance

Production of sheep in the United States is relatively unimportant compared with dairy, beef, and swine. The U.S. population eats only about 1.7 lb lamb and mutton per person each year. This amount is increasing slowly, but is not expected to seriously challenge beef and pork producers. Each person in the United States also uses about 0.8 lb wool each year.

There are approximately 7.8 million sheep on U.S. farms. Sales of U.S. sheep and lamb total nearly $490 million each year. The sale of wool adds about another $70 million to the income of sheep producers.

FIGURE 30–14 Sheep can use low-quality forage, graze land unsuited for other purposes, survive severe weather conditions, and produce meat and wool. *(Courtesy of USDA/ARS #K-4166-5)*

INTERNET ADDRESS:

<http://www.ansi.okstate.edu/
breeds/sheep>

INTERNET KEY WORDS:

sheep management

Because sheep have the ability to survive in areas of limited feed and harsh climates, they are of more economic importance than would be expected. In many such areas, only sheep can survive, and they constitute a major animal enterprise (Figure 30–14).

Types and Breeds

Sheep operations can be divided into two basic types: farm flocks and range operations. A flock is a group of sheep. Farm-flock operations are generally small and are often part of diversified agricultural operations. They may raise either purebred or grade sheep. These flocks usually average less than 150 animals. They are responsible for about one third of the sheep and wool produced in the United States.

The other two thirds of the approximately 11 million sheep in the United States are produced on range operations. Many flocks contain 1,000 to 1,500 head. They are nearly 100 percent grade sheep. Range production is concentrated in the 12 western states. Texas, California, Wyoming, South Dakota, and Colorado are leading states in sheep and wool production.

There are five basic classifications of sheep according to wool type in the United States. They are fine wool, medium wool, long wool, crossbred wool, and fur sheep.

Fine-wool breeds of sheep produce wool that is fine in texture with a long staple length. It has a wavy texture, is very dense, and is used to make fine-quality garments. Fine-wool sheep often produce as much as 20 lb wool per sheep per year. Breeds of fine-wool sheep all originated from the Spanish Merino breed. Fine-wool breeds in the United States include the American Merino, Delaine Merino, Debouillet, and Rambouillet.

Medium-wool breeds of sheep were developed for meat, and little emphasis was placed on the production of wool. Popular medium-wool breeds include Suffolk, Shropshire, Dorset, Hampshire, and Southdown.

The long-wool breeds of sheep were developed in England. They tend to be larger than most of the other breeds. The wool produced tends to be long and coarse in texture. Long-wool breeds in the United States include Leicester, Lincoln, Romney, and Cotswold.

Crossbred-wool breeds are the result of crossing fine-wool breeds of sheep with long-wool breeds. They were developed to combine good-quality wool with good-quality meat. Because they tend to stay together as a group better than other breeds of sheep, they are popular in the western range states. Crossbred-wool breeds include Corriedale, Columbia, Panama, and Targhee.

There is only one breed of fur sheep in the United States. The Karakul is grown for the pelts of its lambs. The pelts are taken from young lambs and made into expensive Persian lamb coats. The production of wool and meat is of little importance in this breed.

Approved Practices in Sheep Production

Some of the approved practices in the production of sheep and wool include the following:

- Select lambs that are large and growthy for their age.
- Select purebred stock based on physical appearance and pedigree.
- Select breeding stock with a history of multiple births.
- Provide shelter from severe weather conditions.

- Provide good-quality forages and unlimited fresh, clean water.
- Vaccinate for disease problems of local concern.
- Treat regularly for internal and external parasites.
- Breed ewes to lamb at no more than 2 years of age. A **ewe** is a female animal of the sheep family.
- Use marking harnesses on rams to tell when ewes have been bred. A **ram** is a male animal of the sheep family.
- Use a system of identification to distinguish ewes from one another.
- Do not disturb ewes during lambing. With sheep, **lambing** means to give birth.
- Shear ewes at least 1 month before lambing. **Shearing** is the process of removing wool from sheep.
- Provide clean, warm, dry stalls for lambing.
- Make sure that ewes accept their new lambs.
- Dock (remove) lambs' tails at 7 to 10 days of age.
- Castrate ram lambs to be marketed.
- Keep hooves properly trimmed.
- Maintain a complete flock disease- and parasite-prevention and control program.
- Cull ewes that do not lamb or those that have health problems.
- Maintain a complete and accurate record-keeping system.

GOATS

Origin and History

The domestication of goats probably took place in western Asia during the Neolithic Age, between 7000 and 3000 BC. Remains of goats have been found in Swiss lake villages of that period. Mention of the use of mohair from goats is made in the Bible. **Mohair** is hair from Angora goats that is used to make a shiny, heavy, wooly fabric.

Goats were imported to the United States from Switzerland for milk production early in the colonial period. Angora goats, to be used for mohair production, were also imported from Turkey. Nearly 95 percent of the mohair-producing goats in the United States are located in Texas.

Milk or dairy-type goats can be found all over the United States. However, there are concentrations on the East and West coasts.

Economic Importance

Goats are of relatively low economic importance in the United States. Most dairy goats are raised in small numbers by suburbanites and small farmers to produce milk and meat for their own families. There are few large herds of milk goats. Finding a processor to bottle goat's milk is often difficult.

The United States produces nearly 60 percent of the mohair in the world. Texas provides more than 95 percent of the 17 million pounds produced in the United States annually.

Goats provide little competition for food for cattle and sheep. They prefer to eat twigs and leaves from woody plants rather than grass.

FIGURE 30–15 Dairy goats may use high or remote mountain pastures or may be raised in relatively small confinement operations. *(Courtesy American Goat Association)*

INTERNET ADDRESS:

<http://www.ansi.okstate.edu/breeds/goats>

INTERNET KEY WORDS:

goat production management

Types and Breeds

There are two types of goats in the United States: hair producing and milk producing.

The only hair-producing goat is the Angora. It produces 6 to 7 lb mohair per year. Mohair ranges in length from 6 to 12 inches. Angora goats are best adapted to a dry climate with moderate temperatures. In addition to producing mohair, Angora goats are used for meat and to help control weeds and brush. The current interest in Boer, Kiko, and Spanish meat goat breeds has resulted in a market for meat-producing goats. Goat meat is called chevon.

Milk-producing, or dairy, goats are found in every state in the United States. Normal production (per goat) averages 3 to 4 quarts milk per day during a 10-month lactation period. The lactation period is the time during which an animal produces milk. The common breeds of dairy goat are Nubian, Alpine, Saanen, and Toggenburg (Figure 30–15).

Approved Practices in Goat Production

Some approved practices in the production of goats and goat products include the following:

- Select goats according to intended use.
- Use physical appearance, pedigree, and records as a basis of selection.
- Purchase replacement animals from reputable breeders.
- Provide adequate feed for hair goats on the range in winter.
- Feed dairy goats supplemental grains based on the amount of milk production.
- Breed goats in the fall to have kids (young goats) in the spring.
- Breed does (female goats) for the first time at 10 to 18 months of age.
- Use one buck (male goat) for every 20 to 50 does.
- Shear or clip hair goats twice each year.
- Maintain clean and sanitary conditions for the production of milk.
- Castrate bucks that are not to be used for breeding at an early age.

Hot Topics in Agriscience

THE "GENE FOR LEANNESS"

Lean, tender beef is the product that is in demand from cattle producers. These qualities have been pursued in the past by attempts to breed cattle that will develop marbling in the meat without putting on too much fat on the outside of the carcass. Now science has identified a new approach to producing lean, tender beef. A gene has been isolated that codes for a protein called *myostatin*. In its active form, this gene controls the number of muscle fibers that develop in the fetus of a calf. Another form of the gene produces inactive myostatin. When inactive myostatin genes are paired in an animal, a condition known as *double-muscling* occurs.

Double-muscling is considered to be undesirable because the extra muscle mass tends to produce stress in animals, and it has been shown to cause difficult births. When this gene is paired with another form of the gene, more extensive muscling occurs without the related stress. The animal produces a carcass with approximately 7 percent more beef and 14 percent less overall carcass fat than cattle that have active myostatin.

- Maintain a herd-health program.
- Milk dairy goats twice daily.
- Dehorn dairy-type goats at an early age.
- Maintain complete and accurate records of reproduction, production, and health.

The production of dairy and livestock in the United States is big business. Receipts from sales of dairy, beef, swine, sheep, and goat products in the United States total more than $70 billion annually. There are types of livestock production that are adapted to almost every locality and situation. Nearly every facet of life is affected in one way or another by animals and animal products. Although food from animals is more expensive than food from crops, animal products add variety and quality to the human diet. Similarly, animals are sources of high-quality fabrics, leather, and many other products. Therefore, animals and the production of animals will be important enterprises in the future.

STUDENT ACTIVITIES

1. Write the Terms to Know and their meanings in your notebook.
2. Develop a word search or other puzzle using the Terms to Know. Trade your puzzle with someone else in class and solve his or her puzzle.
3. Make a chart showing when and where the various types of animals were domesticated.
4. Participate in a class discussion on how and why certain animals were domesticated and others were not.
5. Take a class survey of all of the breeds of animals owned by students and their families.
6. Invite a breeder of purebred livestock and a breeder of commercial, or grade, livestock to class to discuss advantages and disadvantages of each type of operation.
7. Conduct a survey of your local school district to determine the types of livestock being raised, the breeds, and the numbers of each.
8. Make a bulletin board showing the various types of livestock in your area and the importance of each.

9. Visit a livestock operation to determine the types of jobs that need to be performed there and what training would be necessary to perform the tasks.

10. Survey various purebred livestock magazines to determine prices of animals being sold at various purebred sales.

11. Make a notebook-cover collage showing as many breeds as possible of the particular type of livestock in which you are interested.

12. Make a bulletin board showing some items made from animal products.

SELF EVALUATION

A. Multiple Choice

1. The number-one livestock industry in the United States is _____ production.
   a. dairy
   b. beef
   c. swine
   d. sheep

2. Milk for the purpose of making cheese is produced in _____ operations.
   a. Grade A
   b. Grade B
   c. Class A
   d. Class B

3. Merino sheep belong to the _____ -wool type.
   a. fine
   b. medium
   c. long
   d. crossbred

4. The average Angora goat produces about _____ lb mohair each year.
   a. 2 to 3
   b. 3 to 5
   c. 6 to 7
   d. 10 to 15

5. The most popular breed of dairy cattle in the United States is
   a. Jersey.
   b. Ayrshire.
   c. Brown Swiss.
   d. Holstein.

6. The leading state in swine production is
   a. California.
   b. Texas.
   c. Iowa.
   d. Pennsylvania.

7. Docking refers to
   a. fees charged for selling animals.
   b. removal of tails.
   c. backing a cattle trailer to a loading ramp.
   d. none of the above.

8. Sows normally have _____ litter(s) of piglets per year.
   a. one
   b. two
   c. three
   d. four

9. The average person in the United States eats about _____ lb beef each year.
   a. 12.1
   b. 55.5
   c. 62.5
   d. 76

10. Dairy heifers should be bred to calve when they are _____ year(s) old.
    a. 1
    b. 2
    c. 3
    d. 4

B. Matching

_____ **1.** Buck   a. Give birth to cattle

_____ **2.** Mutton   b. Remove wool from sheep

_____ **3.** Milk   c. Calf meat

_____ **4.** Calve   d. Male goat

_____ **5.** Veal   e. One who takes care of sheep

_____ **6.** Chevon   f. Meat from mature sheep

_____ **7.** Farrow   g. Fat in milk

_____ **8.** Shepherd   h. Give birth to piglets

_____ **9.** Shear   i. Liquid from mammary glands

_____ **10.** Butterfat   j. Goat meat

C. Completion

1. _____ were domesticated in China.

2. Goats are of two types: hair and _____.

3. _____ is the most common system of swine identification.

4. _____ is rendered pork fat.

5. Milk-producing animals are called _____.

6. The _____ period is the time when cows are not producing milk.

7. Three classifications of beef-cattle breeds are English, exotic, and _____.

8. Consumption of red meat in the United States has been _____ in recent years.

9. _____ feed is feed provided for young animals to supplement milk from their mothers.

10. Breeds of sheep grown especially for meat belong to the _____ wool class.

MATERIALS LIST

- ✓ bulletin board materials
- ✓ horse breed magazines
- ✓ examples of horse tack
- ✓ Internet access

COMPETENCIES TO BE DEVELOPED

After studying this unit, you should be able to:

- understand the origin and history of the horse.
- determine the economic importance of the horse in the United States.
- recognize the various types and breeds of horse.
- understand approved practices for the care and management of horses.
- understand the basics of English and western riding.
- list rules of safety for handling horses.
- understand the vocabulary generally associated with horses.

SUGGESTED CLASS ACTIVITIES

1. Attend a horse show or other event that involves horses. Observe the horses and record evidence of training that the horses have received to perform as they are expected. Discuss some of the methods that are used to teach horses to perform on farms and ranches, as well as in horse shows.

2. Assign class members to research the history of the horse and to prepare a written paper reporting what they have learned. This assignment could be modified to include the history of a particular breed of horse. If breeds of horses are researched, give the assignment in time for class members to request and receive printed information from the individual horse breed associations.

3. Invite a horse judge to come and talk to the class about the physical characteristics that make a good show animal. You may also want him or her to provide a demonstration on showmanship. Students should be ready to ask questions and participate in the discussion.

TERMS TO KNOW

equine	jack	walk	lope
light horse	mare	trot	filly
draft horse	hinny	canter	foal
donkey	stallion	gallop	colt
coach horse	jennet	rack	gelding
horse	tack	pace	shod
pony	gait	reins	farrier
hand	equitation	bit	blemish
color breed	saddle	horsemanship	unsoundness
mule	stirrup	jog	vice

Horses have been closely associated with humans for much of recorded history. One important way they have served humankind is by providing transportation for humans and materials. Horses and mules have filled important roles in military operations, making it possible to move soldiers and supplies quickly. They provided power for farming, and teamsters used them to haul goods and materials, much as our modern trucks move the products of commerce today.

Horses continue to be used extensively on working cattle and sheep ranches, particularly in areas where other forms of transport are difficult or impossible. In addition, they are the foundation of a significant racehorse industry that generates millions of dollars every year. Horses are used in competitive events such as rodeo, jumping, polo, and competitive riding events. But, most importantly, most horses are owned by people who take pleasure in training, riding, and caring for them (Figure 31–1). The term **equine** means horse or horselike. From this term come the names of activities associated with horses, such as equestrian events that feature horses and riders performing skills for which they are highly trained.

ORIGIN AND HISTORY

There are fossil remains of the horse family that have been found in the United States dating back nearly 58 million years. However, by the time of the discovery of America, no horses remained in the New World. They apparently had all died out only a few thousand years before. It should be noted that these early horses were only about the size of a collie dog.

Even though horses became extinct in the New World a few thousand years ago, it is a common belief that they had migrated to Europe and Asia when Alaska and Siberia were connected. It is from these horses that humans domesticated the horses of today.

The horse was probably one of the last animals to be domesticated. This occurred first in Central Asia or Persia about 5,000 years ago. Horses came to Egypt in about 1680 BC from Asia. From there, the Egyptians introduced horses over the known world.

The Arab horse is the ancestor of most modern light breeds of horse. **Light horses** are breeds used for riding. However, horses were not used to any great extent until AD 500 to 600.

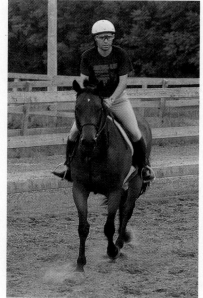

FIGURE 31–1 Horses are friends, working companions, and pets to youths and adults. *(Courtesy of Michael Dzaman)*

Science Profile

THE ORIGIN OF THE HORSE

The fossil record of the horse is probably the most complete record science has documenting the physical changes that horses have undergone. Since the time horses appeared on earth, they have been successful. Because of their survival success, many fossils have been found of animals that appear to be ancestors to the modern horse. The story of the horse begins about 60 million years ago with an animal that has been called both Hyracotherium and Eohippus.

The first ancestor is a small, dog-sized animal that is thought to be from the horse family because it had teeth like the unique ridges seen in molars of the modern horse. Over time, many different horse ancestors appeared. Probably the most significant change that took place was the adaptation of the foot. Early horses had between three and four toes. As mutations occurred and the most successful species were left to reproduce, gradual changes took place. The modern horse has only one toe with a thick toenail or hoof on the end. Studying the fossil record left behind by the ancient relatives of the horse can provide an interesting glimpse of what our modern horse once might have been.

INTERNET ADDRESSES:

<http://www.equinenet.org/heroes>

<http://www.pbs.org/wildhorses/wh_origin/wh_orgin2.html>

INTERNET KEY WORDS:

horse origin development

Draft horses probably originated from the heavy Flanders horse of Europe. **Draft horses** are used for work. At the time of domestication they were often used in religious rites for sacrifice or for food.

Donkeys were domesticated in Egypt some time before 3400 BC, when they appeared on slates of the First Dynasty. A **donkey** is a member of the horse family with long ears and a short, erect mane. Mention of the use of donkeys appears in many places in the Bible. They were generally used as saddle animals or beasts of burden.

Horses were imported to the New World by Columbus in 1493. When Cortez came to Mexico in 1519, he brought nearly 1,000 horses with him. The death of the explorer de Soto in the upper Mississippi region led to the abandonment of many of his horses. These horses, coupled with some of those lost or stolen from Spanish missionaries, probably formed the nucleus of the horses used by the American Indians of the Plains.

The introduction of the horse to the American Indians of the Plains completely changed their culture. Horses permitted easier hunting of buffalo in wider areas. This led to competition between various tribes of Plains Indians and a nearly constant state of war in an area where peace had previously prevailed.

The European settlers on the East Coast were far less dependent on the horse than were the explorers. It was of little advantage to have a riding horse with no place to ride. As draft animals, oxen or cattle were more popular because they were stronger and produced meat and milk, as well as work.

It was only as Americans prospered that horses became an important part of their lifestyles. Horses have had an illustrious past, and a bright future looms ahead.

ECONOMIC IMPORTANCE

There are approximately 6.9 million equines in the United States today. These include horses, ponies, mules, burros, and donkeys. Most horses are owned by nonfarmers and nonranchers. They are used primarily for pleasure, racing, breeding, and companionship.

Agri-Profile

WILD HORSES IN THE UNITED STATES

Scattered across 43 million acres of wild desert rangelands in the western United States are approximately 48,500 wild horses and burros. Most of these animals live on federal lands where the ranges are shared with livestock and other wildlife. The current rate of increase in the wild horse population is approximately 15 percent per year, and agencies managing the herds have been hard-pressed to maintain the population at a consistent level that can be supported by the available supply of forage. Legislation mandates that the herds may not be reduced by lethal means; therefore, horses are rounded up from time to time, and excess animals are offered to the public for adoption.

It is generally believed that the wild herds are the remnants of horses that escaped from the Spanish conquistadors in the years after America was discovered. The wild horses are smaller than most domestic light horse breeds. Wild stallions weigh about 1,000 lb, whereas stallions of domestic breeds generally weigh close to 1,200 lb. Controversy has surrounded these wild animals in recent times because of the tendency of the horses to overrun the ranges as their numbers expand. Adoptions are expensive, with costs approaching $100 million since the adoption program began.

FIGURE 31–2 Horses provide opportunities for families to participate in outdoor activities together, and they provide companionship. *(Courtesy of Michael Dzaman)*

Horse owners spend more than $16 billion on horses and their use each year. Horse racing is one of the most popular spectator sports in the United States. Individual racing events may offer as much as $1 million in purses.

Horse shows are also popular events, especially for young people. One breed association alone recently sponsored more than 4,000 shows in 1 year.

There are numerous benefits of horses besides economics. They help develop a sense of responsibility in young people. They provide physical activity for the young and old alike. They provide opportunities for families to participate in outdoor activities together, and they provide companions when needed (Figure 31–2). Horseback riding is also therapeutic for the healing of certain injuries and disabilities. And, horses are still essential for many cattle roundups.

TYPES AND BREEDS

For the purpose of this discussion, members of the equine family are divided into *horses, ponies, donkeys,* and *mules.* Horses are further divided into *light horses, coach horses,* and *draft horses.* Coach horses were developed to pull heavy coaches and freight wagons.

Members of the horse family that are 14.2 hands or more tall are called horses. Animals of the same family that are less than 14.2 hands tall are ponies. A hand measures 4 inches and is used as a unit of measure for members of the horse family to determine the size of horses, mules, donkeys, and ponies.

The light breeds of horse comprise by far the most popular horses in the United States. Light breeds are further divided into true breeds, or

Agri-Profile

CAREER AREA:

Horse Breeder/Trainer/Rancher/Manager/Jockey/Farrier

Careers in horse management are available in a variety of settings. Urban police forces and park services have rediscovered the advantages of patrolling on horseback, and horse-drawn carriages have captured the fancy of sightseers in many areas. In addition, many families have adopted the horse for teenage animal-production projects in semirural, as well as rural, settings. These frequently involve the entire family in fairs, shows, or rodeos. Horses are used extensively for herding and handling cattle on ranches and in cattle feedlot operations. Pack mules and riding horses are used extensively for hunting and other recreational pursuits. Finally, the horse-racing industry is huge in the United States, involving many people and extensive career possibilities.

Although many people raise horses as an avocation, there are rich and varied career opportunities in raising, managing, or caring for horses. Veterinarians, trainers, managers, attendants, jockeys, and farriers are common around racetracks and breeding farms. Tack, feed, and equipment supply centers provide business opportunities for those wishing to interface with the horse industry but preferring not to handle horses directly.

Horse health, soundness, breeding, behavior, and performance command the attention of those in research, veterinarian services, and education. Horse-related jobs may be indoor or outdoor and run the spectrum from laborer to scientist.

The number of horses in the United States is increasing. The services of the farrier are still in demand, as demonstrated by Mike Woods in Fort Keogh's old cavalry remount barn in Montana. Maintaining working horses in peak condition is vital to the daily operations of many enterprises. *(Courtesy of USDA/ARS K-3906-7)*

INTERNET ADDRESS:

<http://www.ansi.okstate.edu/breeds/horses>

purebreds, and color breeds. True breeds have registration papers and parents of the same breed. Color breeds need only to be specific colors or color patterns, although other characteristics may be considered for registration. They need not have purebred parents.

Examples of the nearly 50 true breeds of horse in the United States include Arabian, Morgan, Thoroughbred, Quarter Horse, Standardbred, Tennessee Walking Horse, and American Saddle Horse (Figure 31–3). The most popular of the light breeds of horse in the United States is the Quarter Horse. The Quarter Horse is used for riding, hunting, and sports, and on cattle ranches to work with cattle (Figure 31–4).

Some of the color breeds of horse include the Palamino, American White, American Creme, and Pinto. They may or may not breed true to color. The uses of the color breeds are the same as those of the true breeds of light horse.

When coaches disappeared from the U.S. scene, coach horses became rare in the United States because there was little need for them. While they were used, though, they combined the qualities of both the light horse and draft breeds. They were fast (like the light horse) and strong enough (like a draft horse) to pull the heavy stage coaches of their

FIGURE 31–3 The most popular breeds of light horses in the United States, based on number registered, are (A) Quarter Horse, (B) Thoroughbred, (C) Arabian, (D) Standardbred, (E) Paint Horse, and (F) Appaloosa. *(Photo provided by the Appaloosa Horse Club)*

Hot Topics in Agriscience

"A HORSE OF A DIFFERENT COLOR"

The color of a horse is determined by the combination of several genes and the presence or absence of two important pigments in the hair and skin. Two primary pigments are found in the coats of mammals. These pigments are eumelanin, a black or brown pigment, and phaeomelanin, which is red. All of the colors are combinations of these pigments. In many instances, these pigments are diluted by certain gene combinations, resulting in different degrees or shades of color. Spots of white indicate the absence of the pigments. Genes control the distribution of the pigments in the skin and hair. This explains the white and colored spots of the Appaloosa and Paint breeds of horses.

Overriding all other genes for color is the dominant gene *A*, which results in the coloring of bay or one of its shades. Every time that the *A* gene is present, the horse will be bay in color. Another interesting color combination is "roan." In this instance, hair of different colors is mixed together. A "sooty" or "smutty" color occurs when black is mixed with hair of other colors.

Many different colors are possible in horses. However, once the offspring of a stallion has been observed for color, it is possible to work out the genetics that are involved. When the color genetics of both the stallion and the mare are known, the possible color combinations in the foal can be predicted with reasonable accuracy. However, it is beyond the scope of this book to expand on the specific genetic combinations and the colors that they produce.

INTERNET KEY WORDS:

equine, horse, color

INTERNET ADDRESS:

<http://www.vgl.ucdavis.edu/
~lvmillon/coatclr.html>

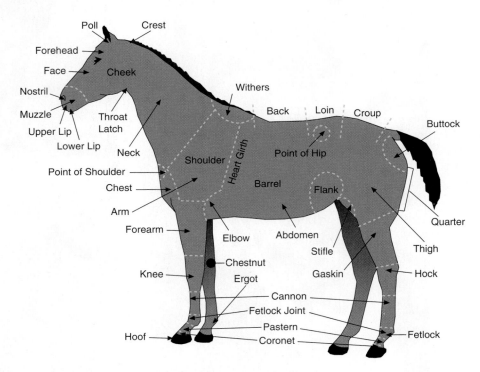

FIGURE 31–4 Learning the proper names of the parts of a horse aids in discussing, evaluating, and caring for horses.

time. The only example of the old coach-style horse in the United States today is the Cleveland Bay.

Draft, or work, horses were once the backbone of U.S. agriculture. Today, they are used primarily for show and recreation. Their doom was the invention of the internal combustion engine, which only had to be

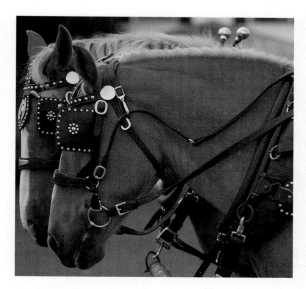

FIGURE 31-5 Draft horses are kept primarily for historic interest, show, and recreation. *(Courtesy of Michael Dzaman)*

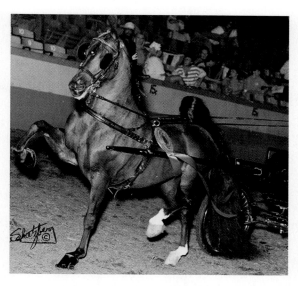

FIGURE 31-6 The Shetland is the most popular breed of pony in the United States. *(Courtesy Multi-World Champion Shetland Harness Pony)*

FIGURE 31-7 The mule is a cross between a jack and a mare. *(Mules owned by Jack and Laura Schreiner of West Sand Lake, NY. Courtesy of S & R Photo Acquisitions.)*

INTERNET KEY WORDS:

mule
hinny

fed when it was actually working and could do the work of many horses. Breeds of draft horse include Belgian, Clydesdale, Percheron, Shire, and Suffolk (Figure 31-5).

Ponies are smaller versions of the horse. They are used for riding and driving and as pets. Some breeds were originally developed to work in coal mines and in other pursuits where small size is important. The common breeds of pony include Shetland, Welsh, Gotland, Appaloosa, Connemara, and Pony of the Americas. The Shetland is the most popular breed of pony in the United States (Figure 31-6).

Donkeys are used in the United States for work and as pack animals. Miniature donkeys also make excellent pets. The male donkey, called a jack, is used as the male parent of mules. A **mule** is a cross between a jack and a mare. A **jack** is a male donkey or mule. **Mares** are mature female horses or ponies. The opposite cross, a stallion with a jennet, results in a **hinny**. A **stallion** is an adult male horse or pony. A **jennet** is a female donkey or mule (Figure 31-7).

Science Connection

AN INTERNATIONAL SOLUTION

U.S. veterinarians and other animal health officials are constantly on the alert for a lurking menace that could threaten the horse industry if allowed to get out of hand. A malaria-like disease of horses, equine babesiosis, infects the horses of the world in varying degrees. Horses of Brazil, Argentina, Russia, and Poland have lived with the parasite so consistently that they have developed a level of immunity that protects them from serious consequences of the disease.

Most U.S. horses have not been exposed to the parasite; therefore, they have developed little or no resistance to it. Such horses are vulnerable, and a bout with equine babesiosis can be fatal. Babesiosis resembles malaria in humans because, in both diseases, insects transmit the infectious agent, which then attacks and destroys red blood cells. Sick animals become feverish and lethargic and refuse to eat. Microscopic, single-celled parasites, called *Babesia equii,* enter the horse through infected blood-sucking ticks. The parasites then multiply in the bloodstream and form tiny pear- or ring-shaped bodies, called *merozoites.* The merozoites then invade the red blood cells. The horse's immune system reacts by forming antibodies to attempt to fend off the invaders.

Most U.S. horses can become targets of equine babesiosis. A new test promises to help keep infected horses from entering the United States. *(Courtesy of Barbara Lee Jensen)*

The United States tries to protect its horse population by keeping the parasite out of the country. However, horses coming into the United States from any other part of the world may be carriers of the parasite, even though they appear healthy themselves. There has been a test for the parasite in general use, but the test has yielded uncertain results.

Donald Knowles of the Agricultural Research Service Animal Disease Research Unit in Pullman, Washington, has found what may be a good test. By studying the blood of infected horses, Dr. Knowles discovered specific proteins on the surface of the merozoites that were present regardless of the origin of the horse. One of the proteins produced an antibody response in 200 horses from 20 different countries. His colleague, Lance Perryman of Washington State University, made a monoclonal antibody that can accurately detect the presence of this protein. The antibodies are produced by a laboratory cell culture of fused mouse spleen and myeloma cells, called a *hybridoma.* The antibodies are then studied as a component of a diagnostic test.

The National Veterinary Services Laboratory in Ames, Iowa, screens all blood samples from animals attempting entry. However, it is expensive to ship a horse to an international border only to be required to ship it back home because of rejection for health reasons. If tests are economical and reliable, courtesy tests can be run on blood samples taken before the animal is shipped. If no problems are indicated, the animal can be shipped, and then retested at the point of entry. The new procedure developed by Knowles and Perryman should be a step forward in solving the old international dilemma of how to move horses across borders economically without the danger of spreading the dreaded equine babesiosis.

Mules are thought to be more intelligent than horses and can do more work than can comparable-sized horses. There are several types and sizes of mule, depending on the breed or type of horse that serves as the female parent. Mules may be used for work, sports, or pleasure.

RIDING HORSES

The styles of riding horses can be divided into two general classifications: English and western. The style of riding and the **tack** (horse equipment) required vary greatly between the two classifications. Even the **gaits** (the ways an animal moves) of the horses involved have different names.

English Equitation

In English equitation, the saddle is smaller and lighter and has shorter stirrups than the western saddle (Figure 31–8). **Equitation** is the art of riding on horseback. A **saddle** is the padded leather seat for the rider of a horse. **Stirrups** are footrests hung from the saddle. The English riding clothing includes closefitting breeches, jodhpurs, jacket, and hard hat or derby (Figure 31–9).

The gaits of the English-equitation horse are walk, trot, canter, gallop, rack, and pace. The **walk** is a slow, four-beat gait. The **trot** is a fast, two-beat diagonal gait. The **canter** is a slow, three-beat gait, whereas the **gallop** is a fast, three-beat gait. The **rack** is a fast, four-beat gait, and the **pace** is a side-to-side, two-beat gait (Figure 31–10).

A

B

FIGURE 31–8 The English saddle is lighter and is structured differently than the western saddle. *(Courtesy of Michael Dzaman)*

A. Western Riding Attire

Chaps

Frontier shirt

Wide-brimmed hat

Bell-bottom jeans or riding pants

Cowboy boots

B. Pleasure Riding Attire

Breeches   Jodhpurs

Schooling chaps

Light jacket

Casual jodhpur boots

Soft hat

Field boots

NOTE: Clothing selection is very flexible for park hacking & trail riding

C. 5-Gaited Show Horse Riding Attire

Turn-over collar

5-GAITED

Check waist-coat

Gaited riding whip

Derby (5-GAITED)

Top hat (3-GAITED)

Patent jodhpur boots

Walk-trot stick

Riding habit-change accessories as shown for the respective gaits

D. Hunt Riding Attire

Stock   Hunt club colors

Hard hat

Hunt coat

Staghorn hunting crop

Top hat or derby

Breeches

Hunt boots

English boots

The hunting habit may be varied according to individual taste and club approval.

FIGURE 31–9 Proper riding attire for riders includes western, pleasure, 5-gaited show horse, and hunt.

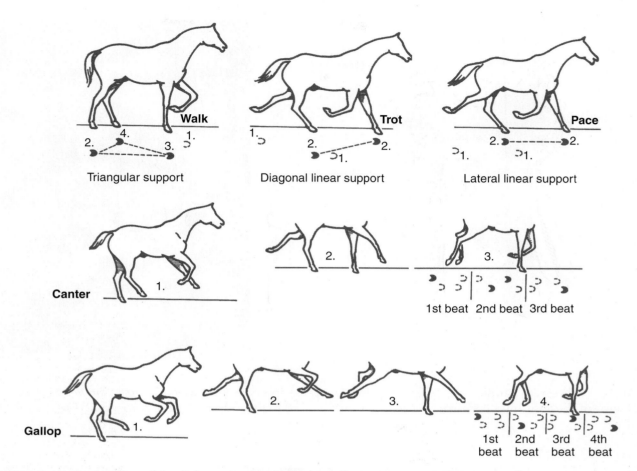

FIGURE 31–10 Basic gaits of English equitation include walk, trot, pace, canter, and gallop.

The English-equitation horse is controlled by reins that may be held either in one or both hands. **Reins** are leather or rope lines attached to the bit. The **bit** is a metal mouthpiece used to control the horse.

Western Horsemanship

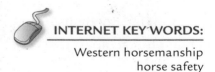

Western horsemanship differs from English equitation in many ways. **Horsemanship** is the art of riding and knowing the needs of the horse. The western saddle is heavy and has a horn that is used when roping livestock. This saddle has longer stirrups than the English saddle and may be plain or have fancy ornamentation. The western saddle was designed for the comfort of cowboys, who spend much of their time in a saddle (Figure 31–11).

Western riding attire was also designed for comfort. It consists of jeans and a western shirt and hat. Cowboy boots and chaps complete the typical western outfit.

The gaits of the western horse are walk, jog, lope, and gallop. The **jog** is a slow, smooth, two-beat diagonal gait. The **lope** is a slow canter.

The western rider controls the horse with the reins held in one hand. The hand holding the reins cannot be changed during competition riding.

A

B

FIGURE 31–11 Western saddles were originally developed for cowboys who worked cattle. However, present-day designs of these saddles vary from basic and functional to highly ornamental. *(Courtesy of Michael Dzaman)*

HORSE SAFETY RULES

Some of the safety rules to observe when riding and caring for members of the horse family include:

- Always approach a horse from the front left side.
- Never do anything to startle or scare a horse. It may kick or rear up.
- Be sure that the horse always knows exactly what you intend to do to or with it.
- Pet the horse by putting your hands on its shoulder, not on its nose.
- Tie horses that are strangers to each other far enough apart that they cannot fight.
- Walk beside the horse when leading it, not in front of or behind it.
- Never wrap the lead rope around your hand when leading a horse.
- Use care in adjusting the saddle on the horse. Be sure that it is tight enough so that it will not slip or slide.
- Always mount the horse from the left side (Figure 31–12).
- Always keep the horse under control.
- Do not allow the horse to misbehave without disciplining it.
- Walk the horse up and down steep slopes, on rough ground, and across paved roads.
- Reduce speed when riding on rough terrain or in wooded areas.
- In groups, ride in single file and on the right side of the road.

FIGURE 31–12 Always mount a horse from the left side. *(Courtesy of Michael Dzaman)*

- Be calm and gentle when dealing with your horse.
- Always wear appropriate riding attire, including hard hats when jumping obstacles.
- Never tease the horse.
- Walk the horse to and from the barn or holding area to prevent it from developing a habit of riding to the area when it comes into sight.
- Know the temperament and vices of the horse.
- Be especially aware of barking dogs and other things that may frighten the horse.
- Do not allow other people to ride your horse unsupervised.

APPROVED PRACTICES FOR HORSES

Breeding Horses

Some of the approved practices for breeding horses are:

- Breed fillies so they will foal for the first time at 3 to 4 years of age. A **filly** is a young female horse or pony. A **foal** is a newborn horse or pony. To foal is to give birth in the horse family.
- Breed mares in April, May, or June, when their fertility is normally greatest. Breeding at this time also allows the mare to foal in the spring, when pastures are actively growing.
- Rebreed mares 25 to 30 days after foaling, if they are in good condition with no reproductive-tract problems.
- Make sure that the pregnant mares and fillies get plenty of exercise.
- Allow the mare to foal in a clean pasture free from internal parasites, or in a large, clean pen or stall (Figure 31–13).
- Put the mare in the facility in which she will foal several days before she is expected to foal. This gives her some time to get accustomed to the facility.

- Remain out of the mare's sight when she is foaling. Mares prefer to be alone with no interruptions during the foaling process.

- Contract the services of a veterinarian if the mare shows signs of difficulty in foaling or if the position of the foal is abnormal for delivery.

- At birth, the following actions should be performed:
  1. Make sure that the foal is breathing. Tickling the foal's nose with a piece of straw often will stimulate it to start breathing after it is born. More drastic measures, such as artificial respiration, may be needed in more serious situations.
  2. Be sure to remove any mucus from the nose of the foal, and dry the foal off.
  3. Dip the end of the navel cord in iodine after tying it off. This prevents infection from entering the foal through the open navel cord.
  4. Make sure that the foal nurses as soon as it can stand. It should be allowed to stand on its own. Normally, nursing first takes place within one-half hour after birth.
  5. Check to see that the foal has a bowel movement within the first 12 hours of birth. The foal may need an enema if the bowel movement does not occur naturally.
  6. If the foal shows signs of diarrhea, reduce the amount of milk that it is allowed to drink.

- Provide creep feed for the nursing foal when it reaches 10 days to 2 weeks of age.

- Begin training the foal as soon as possible. A foal that is trained from an early age seldom needs to be broken.

- Do not mistreat the mare or foal at any time.

FIGURE 31–13 Mares and foals should be kept in clean, parasite-free surroundings. *(Courtesy of Barbara Lee Jensen)*

- Wean the foal at 4 to 6 months of age. The foal and the mare should not be allowed to see each other for several weeks after weaning to break the bond between mother and offspring.

- Castrate colts that are not to be used for breeding purposes. A **colt** is a young male horse or pony. A castrated male horse is called a **gelding**. Geldings are usually much more docile and easier to handle than stallions.

Care and Management of All Horses

Approved practices for the care and management of all horses and ponies include the following:

- Groom the horse daily or weekly at a minimum (Figure 31–14).

- Always use soft brushes when grooming a horse. Its skin is sensitive and easily damaged by rough treatment.

- Shorten the mane and tail when needed by pulling the hairs from the underside.

- Be especially careful not to get water in the horse's ears when washing. Be sure to dry off the horse quickly so that it does not catch a cold in cool weather.

- Inspect and clean the horse's hooves daily and always before and after riding.

- Trim the horse's hooves every 4 to 6 weeks if the horse is not shod. **Shod** means wearing shoes.

- Replace the shoes of a shod horse every 4 to 6 weeks. A **farrier** (a person who shoes horses and cares for their feet) should be engaged to perform this task.

A

B

FIGURE 31–14 (A) Grooming a horse helps it to develop trust for people. It is also important to keep the horse clean and to provide an opportunity to make a complete inspection of the horse. (B) Frequent inspection and care of a horse's feet are important because a lame horse has no value. It is much easier to prevent lameness than it is to cure it. *(Courtesy of Michael Dzaman)*

- Shoe a horse for the first time when it is 2 years of age or when it starts being worked.
- Cool out the horse thoroughly after every exercise period and after riding. This should be done before the horse is allowed to drink large quantities of water.
- Do not overfeed a horse. The total daily consumption of concentrates and roughages should not total more than 2 to 2.5 percent of the horse's weight.
- Feed a horse on a regular routine. The number of times that the horse is fed per day makes little difference, as long as it is the same everyday.
- Do not abruptly change the ration being fed. The stomach of the horse is temperamental and slowly adjusts to changes in feeding practices.
- Never feed moldy feed to a horse.
- Provide unlimited access to fresh, clean water at all times.
- Maintain a strict health-care program for all horses.
- Maintain clean and sanitary facilities at all times.
- When buying a horse, be sure to:
  1. Note blemishes and unsoundnesses of the horse. A blemish is an abnormality that does not affect the use of the horse. An unsoundness is an abnormality that does affect the use of the horse.
  2. Check the age of the horse by looking at its teeth (Figure 31–15). Examining the teeth can also indicate how long a horse might be usable.
  3. Determine that the horse is not blind or having other problems that may affect its value and use.
  4. Check for evidence of good health.
  5. Try to determine if the horse has undesirable vices. A vice is a bad habit.
  6. Try to determine the personality and spirit of the horse.
  7. Be sure to consider the price of the horse and whether you can afford the cost of owning a horse.
  8. Check the pedigree of the horse, if it is a purebred.

Owning and caring for horses and ponies is an excellent way to get outdoor exercise. Horses have given many young people companionship during the trials of youth. Competitors find horses and racing excellent outlets for their excess energy. Cattle producers find horses to be an essential part of a team when working with cattle. The horse was, is, and will continue to be an important part of the world of many people in the United States.

A. Pincers (nippers)
B. Intermediates (middles)
C. Corners

Change in Shape of Tooth Surface

Young

Middle age

Old

Changes in Angle and Shape of Teeth

Note difference in length and width of teeth

Young   Old

Birth to 10 days
- Temporary nippers erupt

6 weeks
- Temporary intermediate erupt

6 to 10 months
- Temporary corners erupt
- Full set of temporary incisors, both upper and lower

2 to 2-1/2 years
- Permanent nippers erupt
- All foal teeth show surface wear

3 to 3-1/2 years
- Permanent intermediates erupt
- Outer edges of nippers show wear
- Corner foal teeth blunted

4-1/4 years
- Permanent corners erupt
- Male hooks erupt
- Nippers worn on outer and inner edges

5 years
- All permanent teeth up and same height
- No wear shows on corners

6 years
- Cups disappear in lower nippers

7 years
- Cups disappear in lower intermediates
- 7-year hook evident (see below)
- Note angle of teeth as shown below

8 years
- Cups disappear in lower corners
- Dark spots will show where all cups have been
- Do not mistake spots for cups

9 years
- Cups disappear from nippers in upper teeth

10 years
- Cups disappear from nippers in upper intermediates

11 years
- Cups disappear from upper corners
- Horse is now smooth mouth

FIGURE 31–15 The age of a horse can be determined by the number and shape of its teeth.

STUDENT ACTIVITIES

1. Write the Terms To Know and their meanings in your notebook.
2. Survey the students in your school regarding how many have horses and what type and breed of horse they own.
3. Develop a bulletin board showing as many breeds and types of horse as you can.
4. Write a report about a breed or type of horse that interests you.
5. Have a horse breeder or owner talk to the class about the care and management of horses.
6. Have a horse veterinarian talk to the class about health-care and first-aid procedures for horses.
7. Develop a word search using the Terms to Know.
8. Visit a tack shop and make a list of the various types of horse equipment sold there.
9. Look at horse-breed-and-care magazines to determine the uses of the various breeds of horse. Also look for the types of job opportunities that may be available in the horse industry.
10. Demonstrate to the class the techniques of grooming a horse.
11. Research the history and development of the horse. Write a paper, create a chart, or draw pictures outlining the physical changes that occurred in the fossil record of the modern horse.

SELF EVALUATION

A. Multiple Choice

1. Horses were probably domesticated in
   a. Russia.
   b. the United States.
   c. Persia.
   d. China.

2. Light horses are used for
   a. riding.
   b. rodeos.
   c. racing.
   d. all of the above.

3. A mule is a cross between a
   a. jack and a mare.
   b. stallion and a jennet.
   c. jack and a jennet.
   d. stallion and a mare.

4. The number-one spectator sport in the United States is
   a. rodeos.
   b. horse racing.
   c. horse shows.
   d. riding.

5. Fillies should be bred to foal at _____ years of age.
   a. 1 to 2
   b. 2 to 3
   c. 3 to 4
   d. 4 to 5

6. Always approach a horse from the
   a. rear.
   b. right side.
   c. left side.
   d. none of the above.

7. Horses should be reshod every _____ weeks.
   a. 2 to 4
   b. 4 to 6
   c. 6 to 8
   d. 8 to 10

8. The most popular breed of light horse in the United States is the
   a. Thoroughbred.
   b. Appaloosa.
   c. Morgan.
   d. Quarter Horse.

9. A fast, three-beat gait is the
   a. gallop.
   b. canter.
   c. pace.
   d. trot.

10. There are approximately _____ horses in the United States.
    a. 100,000
    b. 1 million
    c. 7 million
    d. 100 million

B. Matching

_____	**1.** Stallion	a. Way of moving
_____	**2.** Jennet	b. Young female horse
_____	**3.** Gelding	c. Female donkey
_____	**4.** Mare	d. Newborn horse
_____	**5.** Gait	e. Adult male horse
_____	**6.** Filly	f. Adult female pony
_____	**7.** Tack	g. Male donkey
_____	**8.** Jack	h. Horse equipment
_____	**9.** Foal	i. Young male horse
_____	**10.** Colt	j. Castrated male horse

C. Completion

1. A _____ is a person who shoes horses.

2. Horses originated in what is now the _____.

3. _____ horses were developed for work.

4. _____ breeds of horse need only to be certain colors or color patterns to be registered.

5. The art of riding on horseback is called _____.

6. An abnormality that affects the use of a horse is called a(an) _____.

7. Bad habits in horses are called _____.

8. The result of the cross between a stallion and a jennet is a _____.

9. The most popular breed of pony is the _____.

10. The unit of measurement for determining the size of a horse is the _____.

Food Science and Technology

Wave of the Future

Protein-rich, edible-food coatings and biodegradable food packaging are a limited reality now and a wave of the future. Chemist Franklin Shih of the Agricultural Research Service's Southern Research Center in New Orleans is conducting research on soybean film types that promises a variety of new products. These films may have uses ranging from films on citrus and other fruits, to keep them fresh, to fast-food wrappers that can be eaten. Plastics from soybeans have been around since the 1940s, but plastic-like digestible and biodegradable films are new and still developing.

The process developed at the New Orleans facility involves the separation and isolation of proteins of the soybean by freeze-drying protein to remove the water and grinding the protein into fine powder. One product is soy concentrate containing 70 percent protein; another is soy isolate containing 90 percent protein. The proteins then can be mixed with various ingredients and additives before being cast into films for coatings for food products. By using enzymes and other treatments, the protein can be modified for films and coatings with specific uses. Other researchers have developed films and coatings from corn and wheat starches, but soy-protein films promise some additional adaptability.

Protein mixes with oils, and this quality is important in processing products with moisture-resistant characteristics. Soy protein can be mixed with starch and gums to form films that keep moisture in, but keep oxygen out. This creates an ideal barrier for maintaining food quality. By adjusting other factors, films can be tailored for effective food packaging and preservation. For instance, fat-containing foods, such as meats, can develop off-flavors after being cooked. This creates problems in institutional cooking, such as in restaurants and school lunch programs. Coating such foods with the appropriate protein film could solve the problem of off-flavors. The use of protein and other edible films must be approved by the Food and Drug Administration before being put into commercial practice. The development and use of such "agriplastics" will surely continue because they present the advantages over petroleum-based films of being edible, nutritionally valuable, and environmentally safe.

A

B

C

(A) Chemist Fred Shih analyzes the permeability of various soybean film types, (B) treating fruits and vegetables with protective film, and (C) entomologist Jennifer Sharp evaluates the effectiveness of insect deaths by heat treatment of shrink-wrapped grapefruit. *(Courtesy of (A) USDA/ARS #K-4472-8; (B) USDA/ARS #K-3517-5; and (C) USDA/ARS #K-3183-10)*

UNIT
32

The Food Industry

OBJECTIVE

To explore elements, trends, and career opportunities in the food industry.

MATERIALS LIST

✓ bulletin board materials

✓ State Department of Agriculture reports on commodities grown and foods processed in particular states

✓ Internet access

COMPETENCIES TO BE DEVELOPED

After studying this unit, you should be able to:

- explain what is meant by the term *food industry*.
- determine the importance of the food industry to the consumer.
- describe the economic scope of the food industry.
- identify government requirements and other assurances of food quality and sanitation.
- compare the major crop and animal commodity production areas in this nation and in the world.
- discuss the major food commodity groups and their predominant origins.
- explain the major operations that occur in the food industry.
- describe career opportunities in food science.
- discuss future developments predicted for the food industry.

SUGGESTED CLASS ACTIVITIES

1. Invite a representative of a food-processing plant or business in your area to speak to the class about the food-processing industry. Key topics for discussion should include food safety laws and regulations, food science careers, quality-control methods, sources of raw products, transportation, and marketing.

2. Create a map that shows the major food production areas in your state or region. Obtain statistics from your state department of agriculture or the Economic Research Service of the United States Department of Agriculture that indicate the ranking of your state or region in the production of the foods that you identified. Create a chart with the state rankings for each of these food items.

SUGGESTED CLASS ACTIVITIES—*continued*

3. Follow one locally grown product on its trip from the field to the shelf. For this task, you will need to contact producers and harvesters, processing plants, distributors, wholesalers, and grocery store retailers. When interviewing individuals, plan the questions that you will ask before starting the interview. End the project with a class discussion and project summary.

TERMS TO KNOW

gourmet	packer	harvesting	migratory labor
food industry	trucker	maturity	processing
retailer	harvester	underripe	bran
wholesaler	producer	overripe	endosperm
distributor	grade	spoiled	germ
processor	climatic condition	microorganism	edible
grader			

Food is all around us—in the school cafeteria, on the dinner table at home, at the fast-food chains that dot our nation's highways, and in our nation's supermarkets. Less visible are the gourmet and specialty food stores and restaurants that satisfy special dietary needs and tastes. Gourmet means sensitive and discriminating taste in food preferences. Learning about other cultures is frequently accomplished by tasting their foods and learning how those foods are prepared.

The food industry is involved in the production, processing, storage, preparation, and distribution of food for consumption by living things. Pet and animal food, as well as human food, requires a chain of people, places, equipment, regulations, and resources to change farm products into edible foods (Figure 32–1).

This unit explores the many operations in the food-processing industry. Careers are plentiful in this industry. As the different opera-

FIGURE 32–1 Most modern food products are prepared for the consumer, as seen in ready-to-heat-and-eat grocery items. *(Courtesy of USDA/ARS #K-3550-1)*

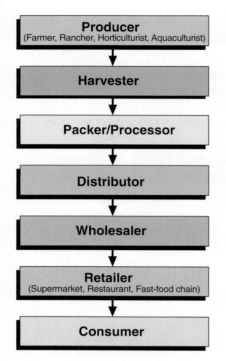

FIGURE 32–2 Producers, such as farmers, ranchers, and aquaculturists who actually produce food, are joined by many others before most food items reach the consumer.

tions are explored, remember that new people are needed to maintain and expand the vital functions that keep the abundant food before us.

ECONOMIC SCOPE OF THE FOOD INDUSTRY

When you purchase groceries in the food store or a hamburger at the local fast-food restaurant, does most of your food dollar go directly to the farmer who grew the beef? How about the lettuce, tomato, pickle, bun, and sesame seeds that adorn your hamburger? There are many businesses and individuals that join the farmer in dividing your food dollar. The economic chain reaction that begins with your food purchase sends signals to the retailer, wholesaler, distributor, processor, grader, packer, trucker, harvester, producer, and others to replace that food for your next purchase (Figure 32–2).

A **retailer** is the person or store that sells directly to the consumer. The retailer is the end of the marketing chain, whereas a **wholesaler** is a person who sells to the retailer, having purchased fresh or processed food in large quantities. A **distributor** stores the food until a request is received to transport the food to a regional market. A **processor** is anyone involved in cleaning, separating, handling, and preparing a food product before it is ready to be sold to the distributor. A **grader** is the person who inspects the food for freshness, size, and quality, and determines under what criteria it will be sold and consumed. A **packer** is the person or firm responsible for putting the food into containers, such as boxes, crates, bags, or bins, for shipment to the processing plant; and a **trucker** is the person responsible for transporting the product anywhere along the way from farm to consumer. A **harvester** is the person who removes the edible portions from plants in the field. Finally, the **producer** grows the crop and determines its readiness for harvest. It seems like everyone gets a piece of your food dollar (Figure 32–3)! Moreover, the dollars spent on food and fiber in the United States provide jobs for approximately 20 percent of the country's working population (Figure 32–4).

FIGURE 32–3 The farmer's share of the food dollar is less than 25 percent. Numerous other services and value-added inputs to a product account for 76 percent of each dollar spent for food.

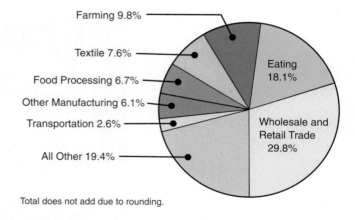

FIGURE 32–4 Career sectors in the food and fiber system.

Proportion of Income Used for Food

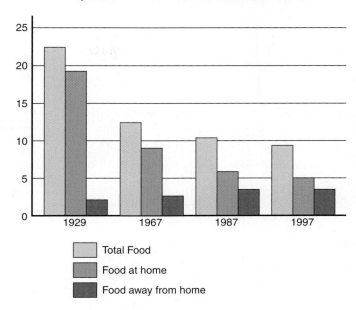

FIGURE 32–5 Proportion of disposable income used for food in the United States.

Where you spend your food dollar also has an influence on who gets how much of your dollar. A meal purchased in a restaurant costs considerably more than a meal prepared from raw food products at home. How has our change in lifestyles and the shift to families with two or more people employed outside the home influenced how and what we eat? More meals are eaten outside the home than was the case a generation ago. Convenience foods for use at home are more in demand today. In 1997, nearly 38 percent of the food dollar in the United States was spent on meals and snacks away from home. This was up from 26 percent in 1970. By 2003, the proportion of income needed for food was 12.1 percent of disposable income in the United States. (Figure 32–5).

QUALITY ASSURANCE

Grading and Inspecting

In the United States, we have become accustomed to high-quality food in every state and every store. The grading system established by the U.S. Department of Agriculture (USDA) has provided a uniform set of trading terms known as grades. Grades are based on quality standards. These improve acceptability of products by the consumer.

Grade standards are established for the following commodities: meat, cattle, wool, poultry, eggs, and dairy products; fresh, frozen, canned, and dried fruits and vegetables; cotton, tobacco, and spirits of turpentine; and rosin. Grades indicate freshness, potential flavor, texture, and uniformity in size and weight, depending on the commodity (Figure 32–6).

Sanitation

Additional quality-assurance programs administered by the USDA include inspection of slaughtering houses and processing plants and oversight of processing operations (Figure 32–7). The USDA oversees

FOOD	CURRENT GRADE NAMES	WHAT THE GRADES MEAN
MEAT		
Beef	*USDA PRIME	Very tender, juicy, flavorful; has abundant marbling (flecks of fat within the lean).
	*USDA CHOICE	Quite tender and juicy, good flavor; slightly less marbling than Prime.
	USDA SELECT	Fairly tender; not as juicy and flavorful as Prime and Choice; has least marbling of the three.
Lamb	*USDA PRIME	Very tender, juicy, flavorful; has generous marbling.
	*USDA CHOICE	Tender, juicy; flavorful, has less marbling than Prime.
Veal	*USDA PRIME	Juicy and flavorful; little marbling.
	*USDA CHOICE	Quite juicy and flavorful; less marbling than Prime.
POULTRY		
Chickens Turkeys Ducks Geese	*U.S. Grade A	Fully fleshed and meaty; uniform fat covering; well formed; good, clean appearance.
	U.S. Grade B	Not quite as meaty as A; may have occasional cut or tear in skin; not as attractive as A.
	U.S. Grade C	May have cuts, tears, or bruises; wings may be removed and moderate amounts of trimming of the breast and

* Indicates grades most often seen at retail.

FOOD	CURRENT GRADE NAMES	WHAT THE GRADES MEAN
EGGS		
	*U.S. Grade AA	Clean, sound shells; clear and firm whites; yolks practically free of defects; egg covers small area when broken out — yolk is firm and high and white is thick and stands high.
	*U.S. Grade A	The same as AA except egg may cover slightly larger area when broken out and white is not quite as thick.
	*U.S. Grade B	Sound shells, may have some stains or shape may be abnormal; white may be weak and yolk enlarged and flattened; egg spreads when broken out.
DAIRY PRODUCTS		
Instant nonfat dry milk	*U.S. Extra Grade	Sweet, pleasing flavor; natural color; dissolves readily in water.
Butter	*U.S. Grade AA	Delicate sweet flavor and smooth texture; made from high-quality fresh sweet cream.
	*U.S. Grade A	Pleasing flavor; fairly smooth texture; made from fresh cream.
	U.S. Grade B	May have slightly acid flavor or other flavor or body defects.
Cheddar cheese	*U.S. Grade AA	Fine, pleasing Cheddar flavor; smooth, compact texture; uniform color.
	U.S. Grade A	Pleasing flavor; more variation in flavor and texture than AA.

* Indicates grades most often seen at retail.

FOOD	CURRENT GRADE NAMES	WHAT THE GRADES MEAN
FISH		
	*U.S. Grade A	Uniform in size, practically free of blemishes and defects, in excellent condition, and having good flavor for the species.
	U.S. Grade B	May not be as uniform in size or as free from blemishes or defects as grade A products; general commercial grade.
	U.S. Grade C	Just as wholesome and nutritious as higher grades; a definate value as thrifty buy for use where appearance is not an important factor.
FRESH FRUITS AND VEGETABLES		
The grade is more likely to be found without the shield.	*U.S. Fancy	Premium quality; only a few fruits and vegetables are packed in this grade.
	*U.S. No. 1	Good quality; chief grade for most fruits and vegetables.
	U.S. No. 2	Intermediate quality between No. 1 and No. 3.
	U.S. No. 3	Lowest marketable quality.
PROCESSED FRUITS AND VEGETABLES AND RELATED PRODUCTS		
Canned and frozen fruits and vegetables	*U.S. Grade A	Tender vegetables and well-ripened fruits with excellent flavor, uniform color and size, and few defects.
	U.S. Grade B	Slightly mature vegetables; both fruits and vegetables have good flavor but are slightly less uniform in color and size and may have more defects than A.
	U.S. Grade C	Mature vegetables; both fruits and vegetables vary more in flavor, color, and size and have more defects than B.
Dried or dehydrated fruits. Fruit and vegetable juices, canned and frozen. Jams, jellies, preserves. Peanut butter, honey, catsup, tomato paste.	*U.S. Grade A	Very good flavor and color and few defects.
	U.S. Grade B	Good flavor and color but not as uniform as A.
	U.S. Grade C	Less flavor than B, color not as bright, and more defects.

* Indicates grades most often seen at retail.

FIGURE 32–6 Grades of food commodities, as established by the U.S. Department of Agriculture (USDA). *(Courtesy of USDA)*

food labeling and enforces regulations regarding representation on such labels (Figure 32–8). The National Shellfish Sanitation Program, the U.S. Public Health Service, and the U.S. Food and Drug Administration work with USDA to ensure the safety of food and food products. States, counties, and municipalities also have inspectors. They regulate local conditions to ensure sanitation and safe food handling, especially in restaurants and food-preparation areas.

Nutrition Facts		
Serving Size		1 Cup (55g/2.0 oz)
Servings per Container		13

	Cereal	Cereal with 1/2 Cup Vitamins A & D Skim Milk
Amount Per Serving		
Calories	170	210
Fat Calories	10	10

	% Daily Values**	
Total Fat 1.0g*	2%	2%
Sat. Fat 0g	0%	0%
Cholesterol 0mg	0%	0%
Sodium 300 mg	13%	15%
Potassium 340mg	10%	16%
Total Carbohydrate 43g	14%	16%
Dietary Fiber 7g	28%	28%
Sugars 17g		
Other Carbohydrate 19g		
Protein 4g		

Vitamin A	15%	20%
Vitamin C	0%	2%
Calcium	2%	15%
Iron	45%	45%
Vitamin D	10%	25%
Thiamin	25%	30%
Riboflavin	25%	35%
Niacin	25%	25%
Vitamin B$_6$	25%	25%
Folate	25%	25%
Vitamin B$_{12}$	25%	35%
Phosphorus	20%	30%
Magnesium	20%	25%
Zinc	25%	25%
Copper	15%	15%

*Amount in cereal. One-half cup skim milk contributes an additional 40 calories, 65mg sodium, 6g total carbohydrate (6g sugars), and 4g protein.

**Percent Daily Values are based on a 2,000-calorie diet. Your daily values may be higher or lower depending on your calorie needs:

	Calories	2,000	2,500
Total Fat	Less than	65g	80g
Sat. Fat	Less than	20g	25g
Cholesterol	Less than	300mg	300mg
Sodium	Less than	2,400mg	2,400mg
Potassium		3,500mg	3,500mg
Total Carbohydrate		300g	375g
Dietary Fiber		25g	30g

Calories per gram:
Fat 9 • Carbohydrate 4 • Protein 4

FIGURE 32–8 The commodity label is the consumer's best assurance of food quality and value.

FIGURE 32–7 Quality-control personnel, as well as government inspectors, use U.S. Department of Agriculture standards to monitor foods for cleanliness, wholesomeness, and quality. *(Courtesy of USDA/ARS #K-3512-3)*

COMMODITY GROUPS AND THEIR ORIGINS

What Foods Are Grown Where

Food is grown all over the world. Climatic conditions and available technology dictate that some foods grow better and in greater abundance in certain areas of the world. **Climatic conditions** refer to average temperature, number of days with a certain temperature range, length of growing season, and amount of precipitation for a given geographic area. Technology refers to the equipment and scientific expertise available to cultivate, store, process, and transport the crop for consumption in a variety of forms after harvest. Food production in the United States has always been influenced by geography and climate (Figure 32–9).

Since early times, when humans traveled and traded, foods have been introduced outside the areas where they are grown naturally. The origins of the soybean, for example, can be traced back 3,000 years to China, where it is still produced and consumed today. However, major growing areas for the soybean today include the United States, Brazil, and western Europe.

Modern technology has allowed producers to raise crops somewhat artificially with irrigation and in greenhouses where the conditions of temperature and moisture are controlled (Figure 32–10). Different varieties of food have been developed to grow under different climatic conditions, such as extreme heat or cold. Similarly, in aquaculture, seafood and fish are produced under controlled conditions (Figure 32–11).

In the United States, we are accustomed to having almost every food available fresh at any time of the year; but all foods are not grown in all parts of the country. Citrus fruit, including oranges and grapefruit, require warm climates, such as those found in Texas, California, and

AGRICULTURAL REGIONS

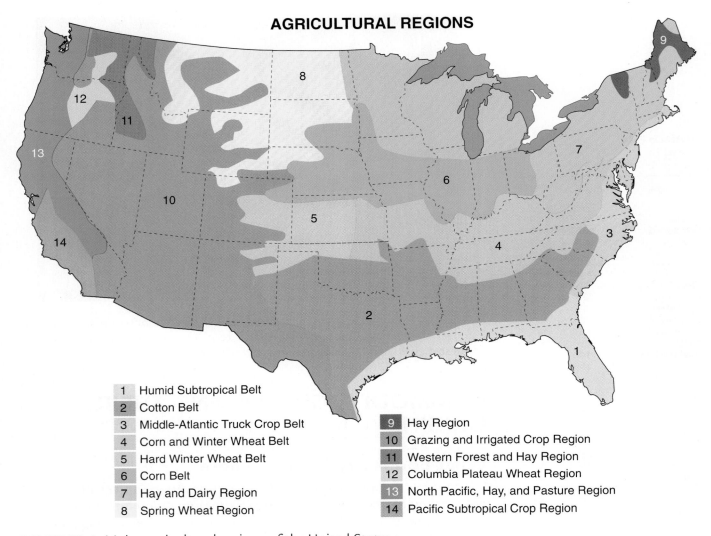

1 Humid Subtropical Belt
2 Cotton Belt
3 Middle-Atlantic Truck Crop Belt
4 Corn and Winter Wheat Belt
5 Hard Winter Wheat Belt
6 Corn Belt
7 Hay and Dairy Region
8 Spring Wheat Region

9 Hay Region
10 Grazing and Irrigated Crop Region
11 Western Forest and Hay Region
12 Columbia Plateau Wheat Region
13 North Pacific, Hay, and Pasture Region
14 Pacific Subtropical Crop Region

FIGURE 32–9 Major agricultural regions of the United States.

FIGURE 32–10 Technologies such as modern irrigation systems have made it possible to produce crops in environments such as deserts.

FIGURE 32–11 Controlled living environments such as fish runs make it possible to produce fish under controlled conditions. The aquaculture industry is the product of this technology.

Florida. More than 70 percent of the fresh vegetables that are consumed in the United States are grown in California, Florida, and Arizona. However, people in North Dakota and Maine enjoy the nutrients and good taste of fresh or processed citrus and vegetable products, such as orange juice, daily. Our country is not only "America the Beautiful," it is America the bountiful.

The many operations of the food industry that are discussed later in this unit will explain how we can enjoy foods that are not grown naturally in our particular region of the country and world. Realizing where some of our food products originate also makes you appreciate the size and scope of the food industry.

Crop Commodities

Grains Various grains have different growing requirements. Therefore, they are produced in different parts of the United States and the world. Wheat, which originated in Asia, is grown in the cooler climates of the United States. Different varieties have been developed to accommodate different growing seasons and climatic conditions around the world. Corn is a warm-weather crop, but the many varieties and types permit its growth in every state in the United States. Rice, however, has special moisture requirements; therefore, its production is limited to specific areas of the country.

Oil Crops Oil crops are sometimes thought of as the invisible food product. Soybeans, corn, cotton, flax, sunflowers, coconut, peppermint, and spearmint are all significant oil crops in the United States. The United States is a world leader in the production of soybeans, corn, cotton, and peanut oils. Soybean products are referred to in ancient Chinese literature, and the origin of the peanut may be traced to Brazil and Paraguay. Sunflowers are native to the United States and are growing in importance. Sunflowers are also grown in Spain, China, Russia, and Mediterranean areas. Safflower oil is a relatively minor oil in terms of proportion to the total oil crop worldwide. It originated in northern

INTERNET KEY WORDS:

grain crops
oil crops
fruit crops
vegetable crops

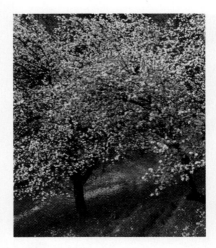

FIGURE 32–12 Apples are produced in areas with a relatively cool season. Apples are one of our favorite fresh fruits, and large volumes of apples are consumed every year.

INTERNET KEY WORDS:

meat, beef, pork, lamb products

India, North Africa, and the Middle East. Its origin is indicative of its drought tolerance.

Sugar Crops   Sugar beets and sugar cane are the principal sugar crops in the United States. Corn is a secondary source of sugar. Sugar beets are grown in temperate areas such as Minnesota, Colorado, and California, whereas sugar cane is grown in tropical and subtropical locations around the world. Florida, Hawaii, and Louisiana are the three largest producers of cane sugar in the United States.

Citrus   Oranges, limes, lemons, and grapefruit all require warm temperatures to survive. Consequently, the southern states with warmer climates, such as Florida, Texas, Arizona, and California, are the major producers of citrus.

Tree Fruits   The many varieties of fruit that grow on trees require different weather conditions. Therefore, various fruits are adapted to different parts of the country. Apples and pears require cooler temperatures and do well in mountainous areas (Figure 32–12). Bananas require very warm conditions and grow best in tropical areas.

Vegetables and Berries   Vegetables and berries are consumed shortly after harvest or they are processed by canning, drying, or freezing for future consumption. Some vegetables require warm climates and some require cool climates. Vegetables that require cooler climates include cabbage, broccoli, potatoes, and cauliflower. Vegetables requiring warmer environments include beans, tomatoes, and sweet corn.

Meat Commodities

Animals, like crops, are typically raised in locations with some regard to climatic conditions. Artificial cooling or heating of livestock is expensive. Large amounts of water must be available for livestock. Where fewer artificial conditions are introduced, the cost of production is minimized. The type, cost, and availability of livestock feed is an important factor influencing where animals are raised in the United States and around the world.

Beef   Most beef is raised near corn, the main feed source of cattle. More than half of the corn in the United States is grown in Nebraska, Iowa, Illinois, and Minnesota. Therefore, beef is raised extensively in the Midwest. The open ranges in the western part of the United States provide other important areas where beef is raised.

Pork   Corn is also the primary food of hogs. Similar to beef, the primary area where hogs are raised is the midwestern part of the United States. The mid-Atlantic and southern states are also important hog-producing areas.

Lamb   Sheep are animals that require large amounts of grazing pasture area. Therefore, they are raised extensively in the range states of the far West. However, as is true with beef and pork, lamb products are produced in other states.

Dairy Products   Wisconsin has long been called the "Dairy State." Indeed, many of our dairy products are produced in Wisconsin and other northern-tier states. However, California has a large dairy industry with many cooperatives and processing plants. These provide the dairy products to consumers nationwide. Dairy animals prefer cooler

Agri-Profile

CAREER AREAS:

Scientist/Inspector/Quality Controller/Buyer/
Seller/Processor/Trucker/Wholesaler/Retailer

The food industry is massive and includes both plant and animal products. It includes the producers, processors, distributors, wholesalers, retailers, fast-food establishments, restaurants, and the home kitchens where food is prepared.

Career opportunities include many that have been discussed previously and some new ones, too. The meat processing-and-packaging industry is massive and employs a large number of individuals in the United States. Field supervisors and coordinators direct the work of crews to harvest crops at the peak of their quality and transport them to processing or packing plants. Similarly, crews take fish, oysters, clams, lobsters, and other seafood from production habitat to processing centers. In many cases, huge packing and/or processing machines are used in the field, orchard, or boat. Quality-control personnel collect food specimens, label them, test them, and maintain records to ensure quality control on each batch of food coming out of the plant.

Careers in food science, store management, produce management, meat cutting, laboratory testing, field supervision, research, diet and nutrition, health and fitness, and promotion are all possibilities in the food industry.

A poultry research scientist investigates ways of improving the hatchability of eggs.
(Courtesy of James Strawser; Cooperative Extension Service, The University of Georgia)

environments, so the industry is extensive in the northern part of the United States. During recent years, however, the industry has expanded significantly in Idaho, New Mexico, and other western states.

Game Each state has native game. Whether it is harvested as an agricultural product depends on the demand for the product. Venison is the tasty and popular meat of deer. In some states, deer are being raised in captivity to help meet the market demand for venison.

Seafood States that border the Atlantic and Pacific oceans and the Gulf of Mexico are considered the primary suppliers of seafood in the United States. However, the science of aquaculture permits the production of fish in interior states.

Poultry Poultry can be raised in a variety of settings. Typically, chickens and turkeys are raised indoors, where ventilation and temperature are carefully controlled. Most poultry is raised in the mid-Atlantic and southern states; however, important poultry-producing areas are also found in California and other states.

OPERATIONS WITHIN THE FOOD INDUSTRY

The food industry begins with the process of photosynthesis in plants. It progresses through plant and animal growth and on to commodity processing and distribution. The discussion in this unit focuses primarily on the food industry after the crops and animals have been grown.

Harvesting

Harvesting means taking a product from the plant where it was grown or produced. This may involve taking potatoes out of the ground, picking oranges off a tree, removing bean pods from bean plants, or removing and threshing grain from stalks (Figure 32–13). It is most important that the crop be harvested in a timely and careful fashion. The plant should be harvested at the correct stage of maturity. **Maturity** means the state or quality of being fully mature. Harvesting a crop when it is of proper maturity means that it is not underripe, overripe, or spoiled. **Underripe** means that it has not reached maturity. **Overripe** means that the plant is past the optimum maturity; stalks or limbs can easily break or shatter, or fruit can drop. **Spoiled** means that chemical changes have taken place in the food or food product that either reduce its nutritional value or render it unfit to eat.

Spoilage is usually caused by microorganisms. **Microorganisms** that contribute to food spoilage are bacteria, fungi, and nematodes. Bacteria are a group of one-celled plants. Fungi are plants that lack chlorophyll and obtain their nourishment from other plants, thus causing rot, mold, and plant diseases. Nematodes are small worms that feed on or in plants or animals. They live in moist soil, water, or decaying matter and pierce cells of plants and suck the juices. The moisture content of a product determines many changes that the product can undergo. It also dictates the types of handling procedures and storage facilities that are required for certain foods. Certain health considerations are influenced by the various levels of crop maturity. Some fruits, such as bananas and tomatoes, continue to ripen after they have been picked from the tree and vine. Other foods, such as beans and oranges, do not continue to ripen once picked, and they must be handled accordingly. Knowledge of the complete growth process of a commodity is essential for the producer and harvester.

Harvesting involves the use of equipment and labor. Because timing is critical, migratory labor is often used at harvest time. **Migratory labor**

FIGURE 32–13 Harvesting is the timely removal of crops from the field. Most crops are harvested mechanically using specialized machines, such as this potato-harvesting equipment.

FIGURE 32–14 Mechanical harvesters replace the back-breaking and difficult work of hand harvesting—for example, a cranberry harvester in action. *(Courtesy of USDA/ARS #K-4416-14)*

refers to workers who move from place to place where harvesting is occurring throughout the year. As crops ripen, laborers migrate to new geographic locations to harvest the crops. Workers are not always available. They may be poorly trained and may not harvest crops as efficiently and economically as do modern machines.

Therefore, engineers and business people have developed and now sell machines to perform many harvesting operations. Such machines make harvesting easier than it was in the past (Figure 32–14). However, machines may not be as gentle as human hands. Consequently, plant breeders have developed new crop varieties that better lend themselves to mechanical harvesting. Such varieties may not be as good to the human touch or taste. For instance, tomato varieties that have skins tough enough for mechanical harvesting are harder to slice and may not be as juicy as varieties suitable for the home garden.

Processing and Handling

INTERNET KEY WORDS:

methods of food processing

The steps involved in turning the raw agricultural product into an attractive and consumable food are collectively known as **processing**. Processing factories or plants clean, dry, weigh, refrigerate, preserve, store, and turn the commodity into a variety of other products (Figure 32–15).

Wheat is cleaned, dried, weighed, and graded for quality. It is then ground into flour. However, before this occurs, it may be separated into bran and germ. The skin or covering of a wheat kernel is known as **bran**. Inside the bran is the **endosperm**, which will become flour, and the **germ**, which is a new wheat plant inside the kernel. Wheat flour is used for breads, cereals, cakes, and pasta. Other grains have similar parts and are used to make similar products.

The processing of tomatoes results in a variety of products (Figure 32–16). Some people claim that the fresh tomato defines summer. The peak of the North American growing season occurs when the tomato is ripe. To the gardener, this means picking tomatoes directly from the vine in a backyard garden and consuming them immediately. However, tomatoes are harvested year-round somewhere in the world, and the food industry can get them to us in edible form. **Edible** means fit to eat. Processing permits tomatoes to be kept for future use. This begins after the tomato is harvested in the field or greenhouse and arrives at the processing plant.

Science Connection

DO NOT CHUCK ROTTEN TOMATOES

The vitamin and mineral industry may spell big money for tomato growers. A valuable antioxidant, lycopene, is found in tomatoes. It is the substance responsible for the red color of the tomato. Lycopene helps protect cells from harm that can be caused by the oxidation process. It has been found to reduce the risk for prostate cancer and other diseases. Until now, the substance has been expensive, costing $2,500 a kilo in its pure form.

Agriscientists at the University of Florida have created an inexpensive way to extract lycopene from tomatoes. This new discovery may earn the tomato industry millions of dollars. Every year tomato growers lose money because some of their tomatoes are unfit for market because of blemishes that are unappealing to consumers. With the new extraction technique, these blemished tomatoes can be marketed to the health industry. The technique will provide consumers with a less expensive product, and it will provide growers with a profit, instead of a costly loss.

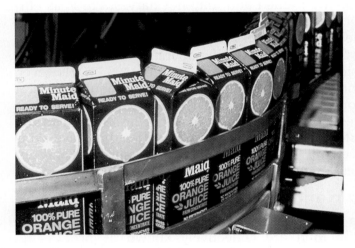

FIGURE 32–15 Processing is one of many intermediate steps between the producer and the consumer. *(Courtesy of FFA)*

FIGURE 32–16 Tomatoes can be processed into a variety of products. *(Courtesy of W. Altee Burpee Company)*

After tomatoes are cleaned and separated for size and quality, they may be canned whole; chopped, cooked, and strained for juice; or made into other products. Such products include spaghetti and hamburger sauces, relishes, and many other foods.

Transporting

INTERNET KEY WORDS:

importance of food transport

Trucks, planes, boats, cars, trains, carts, and bicycles are vehicles used by the food industry in various parts of the world (Figure 32–17). The transporting of fresh and processed food products composes 5.5 percent of the marketing cost within the food industry in the United States. Timing and the distance foods must travel contribute to the ultimate cost of the foods. The efficiency of transportation can influence food quality in terms of freshness and spoilage. Insulated and refrigerated trucks enable food products to move in fresh form to most parts of the

FIGURE 32–17 Food commodities move around the world by every mode of transportation.

FIGURE 32–18 Refrigerated trucks permit year-round supplies of fresh fruits, vegetables, milk, eggs, and meats to every part of the United States. *(Courtesy of Michael Dzaman)*

country year-round (Figure 32–18). This luxury is not available to most people of the world.

Approximately 90 percent of our perishable food is shipped by truck. Much of the less-perishable foods, such as wheat, potatoes, and beets, are shipped by rail. Air transportation allows us to enjoy perishable foods from distant regions and different countries. For example, pineapples and papayas from Hawaii are enjoyed all over the world because of air transportation. Because of modern transportation, fresh seafood is enjoyed in many areas where fish are not bred.

The consumer provides the final link from farm to table. How far the food is shipped, how the food was wrapped for transportation, how long the food was in transit, and how warm the food became during transport all influence ultimate food quality. From the milk truck driver taking milk from farm to processing center, to the trucker delivering products to the local store, transportation is a key component in the food industry. Knowledgeable and competent employees are great assets when the cash value of loads on their trucks or rail cars is considered. A delay could be costly for both business and the consumer.

Even if food is in perfect condition when you buy it, the quality can decrease substantially before you get it home if precautions are not taken. Perishable food should be packaged correctly in the store, kept cool in the car or truck, and refrigerated or frozen on arrival home.

Marketing

INTERNET KEY WORDS:

marketing wholesale foods

Wholesalers purchase food products from packing houses, processors, fish markets, and produce terminals. They, in turn, sell to retailers and institutions such as hospitals, schools, restaurants, and retail stores (Figure 32–19). Grocery stores and fast-food chains are important links in the food chain before the food is purchased by the consumer.

There are many types of retail stores from which consumers can purchase their food items. Such marketing sources meet the needs of consumers in different locations and situations. Superstores carrying 15,000 or more items, conventional supermarkets, limited-assortment

Science Connection

KEEPING THE LID ON DRUGS

It is generally agreed by authorities that the food supply in the United States is among the safest in the world. Federal and state food inspectors work to ensure that our nation's food supply is wholesome and is not contaminated beyond acceptable limits. Avoiding contamination by common bacteria and other living organisms is impossible in any place except a sterile environment. Therefore, we produce our food in sanitary, but not sterile, environments and remove any residual external biological contamination by washing or peeling. We then heat the food to sufficient temperatures to kill internal biological organisms.

Inspectors must monitor the quality of milk and dairy products to ensure that the products are safe and wholesome. *(Courtesy of James Strawser, Cooperative Extension Service, The University of Georgia)*

Contamination by chemicals is quite a different matter. Plants grown in soils and fish grown in water containing poisonous chemicals, heavy metals, and other pollutants can result in chemical contamination of fruits, vegetables, nuts, and fish products. Similarly, animals grazing on contaminated pastures or eating contaminated feeds can have chemical contaminants in their meat and milk. Drugs used for medicines to keep animals free from diseases, parasites, and insects can contaminate meat and milk if not used according to label instructions. Unfortunately, heating does not remove or neutralize most chemical contaminants. Therefore, food inspectors and other health inspectors keep a vigilant watch with all the tools of modern science to monitor foods from source-to-table and divert unacceptable products from the food stream. In actual practice, reputable food producers, processors, and handlers protect the food supply through quality-control measures to protect their businesses and to avoid penalties by the government monitoring authorities.

Residues of pesticides and medicines in meat and milk receive continuous scrutiny. Laboratory and other scientific procedures are being improved constantly to help. A case in point is the development of new, high-tech probes to speed-up tests for benzimidazoles on meats. These are medicines used to protect cattle, sheep, pigs, chickens, and goats from internal parasites. If the drugs are administered in incorrect doses or with inappropriate timing, some can remain in the meat.

Meat specimens are routinely scrutinized using a high-performance liquid chromatograph. However, future chemists may complement this approach with monoclonal antibody assays. Such assays hold the promise of dramatically speeding up testing without compromising accuracy. Already, monoclonal antibodies have been developed that will inexpensively detect benzimidazoles in meat at concentrations as low as one part per billion—that is the equivalent of 1 second in 32 years!

and box stores, convenience stores, nonconventional food stores, small stores, corner stores, food cooperatives, farmers markets, roadside stands, pick-your-own businesses, and other farm outlets are the most common places that consumers purchase their food items (Figures 32–20 and 32–21). The major differences between the various types of stores are the number of items stocked and the physical size of the facilities.

FIGURE 32–19 Wholesale terminals provide the facilities for trucks and trains to bring food commodities together, allowing buyers to obtain commodities for their retail outlets. *(Courtesy of FFA, Photo by Bill Stagg)*

FIGURE 32–20 Superstores may stock as many as 15,000 items or more. *(Courtesy of FFA)*

CAREER OPPORTUNITIES IN THE FOOD INDUSTRY

Each of the areas discussed in this unit and in Unit 33 require people to manage, operate, and carry out the many and varied elements of the food industry. As with all careers in agriscience, those in the food industry present many challenges and rewards. When you consider the chain-reaction nature of the food industry, career opportunities await individuals at the local, county, state, national, and international levels. Careers in the food industry can be divided into seven, often overlapping, categories. Career opportunities in each area are numerous (Figure 32–22).

THE FOOD INDUSTRY OF THE FUTURE

The food industry is ever changing, with new developments occuring each day. Some areas that may attract the food researcher include new food products, new processing and preserving techniques, and new equipment for harvesting labor-intensive crops.

Aquaculture is meeting the increasing demand for fish and will continue to supplement the catches of commercial fishermen. The use of extreme heat and cold in processing has contributed to the development of many food items that meet the demand for convenience foods. The convenience food store will continue to play a larger role in the food chain. Economic efficiency in convenience foods and convenience stores is under constant review. In addition, the USDA and other agencies continue their vigilance regarding safety and nutritional standards at all steps of the food chain.

Improved harvesting equipment for products such as grapes is being tested to reduce the labor costs of such crops. Fuel alternatives for cost-effective transportation, and refrigeration with carbon dioxide

FIGURE 32–21 Farmers' markets provide fruits, vegetables, honey, and other farm products that are fresh from the fields to consumers. They have become popular marketing outlets in many areas. *(Courtesy USDA)*

Some Careers in Food Science and the Food Industry

Business

Accountant
Buyer
Distributor
Financial analyst
Loan officer
Marketing specialist
Salesperson
Statistician

Communications

Advertising specialist
Broadcaster
Media specialist
TV Producer/Demonstrator
Writer

Education

College professor
Extension specialist
Industry educator
Dietician
Teacher

Processing

Butcher
Efficiency expert
Engineer
Plant line worker
Plant supervisor
Refrigeration specialist
Safety expert

Quality Assurance

Food analyst
Grader
Inspector
Lab technician
Quality-control supervisor
Quarantine officer

Research and Development

Distribution analyst
Biochemist
Microbiologist
Packaging specialist
Process engineer

Retailing/Food Service

Baker
Cook/Pizza maker
Counter salesperson
Deli operator
Meat cutter
Nutritionist
Produce specialist
Restaurant owner/operator
Waiter/waitress

Transportation

Dispatcher
Trucker
Rail operator
Merchant marine

FIGURE 32-22 Career opportunities in food science and the food industry are many and varied.

snow instead of conventional diesel-powered mechanical refrigeration are some of the many developments under constant review in the food industry. This industry must continue to meet the demands for high-quality food in the United States and the world through effective research and qualified employees.

Hot Topics in Agriscience

IRRADIATION OF FOODS

Harmful insects, such as the Mediterranean fruit fly, enter the United States on imported fruits and vegetables. One method of reducing the threat of these and other invading insects is to subject incoming fruits and vegetables to radiation. In high doses, this treatment is capable of killing many insects. At lower exposure levels, many insects become sterile or unable to reproduce. Irradiated insect eggs do not hatch, thus irradiated produce is essentially free of dangerous insects even though some of them may survive the radiation treatment. The process works by breaking the chromosomes in the cells of the insect. Irradiation is expected to play a major role is controlling insects in the future.

Countries such as Japan require fumigation treatments of cherries and other produce from the United States before they are allowed into the country. Irradiation is less damaging to the produce than is fumigation, and it is hoped that wider acceptance of the irradiated fruit and vegetables will lead to acceptance of the irradiation process on U.S. exports of fruits and vegetables. United States military installations have used irradiated foods, and it is hoped that consumers can be educated to accept widespread use of this technology for protecting fruits and vegetables.

STUDENT ACTIVITIES

1. Write the Terms to Know and their meanings in your notebook.

2. Keep a food-dollar diary to document where your food dollars are spent. Record the cost of meals and snacks eaten in and outside of the home.

3. Do a cost comparison of meals prepared at home and similar meals consumed at fast-food places and restaurants.

4. Trace the activities that occur in transforming wheat in the field to a hamburger roll consumed in your home.

5. Draw a diagram tracing the individual food components of a deluxe hamburger back to the places where the components were produced. Label each component, process, and commodity along the way.

6. Ask your instructor to arrange a field trip to a butcher shop or supermarket to observe demonstrations on meat cutting and packaging.

7. Make a collage illustrating the various activities of the food industry.

8. Prepare a report on some of the safety regulations that are in place in the United States to prevent illness caused by food contamination.

9. Refer to Figure 32–22, which lists a variety of career opportunities in the food industry. Choose a career that interests you. Find out what steps you will have to take to get the job you are most interested in. Be sure to include all training and education that will be required.

SELF EVALUATION

A. Multiple Choice

1. Approximately what percentage of the U.S. food dollar is spent on meals away from home?
   a. 15 percent
   b. 25 percent
   c. 35 percent
   d. 45 percent

2. Which of the following products is native to North America?
   a. soybeans
   b. wheat
   c. sunflowers
   d. peanuts

3. In the United States, more than one-half of the fresh fruits and vegetables are grown in which states?
   a. Montana, Oregon, and Washington
   b. California, Florida, and Texas
   c. New Jersey, North Carolina, and Georgia
   d. Arizona, Nebraska, and Ohio

4. Migratory workers would harvest wheat last in which state?
   a. Arizona
   b. Nebraska
   c. Montana
   d. Ohio

5. When you spend one dollar for food, approximately how much goes into the labor required to harvest, and then process that food after it leaves the farm?
   a. $0.94
   b. $0.64
   c. $0.34
   d. $0.04

6. Approximately what percentage of all the jobs in the food and fiber system is related to wholesale and retail sales?
   a. 60 percent
   b. 50 percent
   c. 40 percent
   d. 30 percent

7. Which product is consumed away from home the most?
   a. fruits
   b. beverages other than milk
   c. vegetables
   d. meat

8. Superstores are likely to carry how many items?
   a. 15,000
   b. 1,500
   c. 150
   d. 15

B. Matching

_____   **1.** Harvester   a. Purchases food in large quantities

_____   **2.** Grader   b. Stores food until requested

_____   **3.** Retailer   c. Follows crop harvesting geographically

_____   **4.** Wholesaler   d. Is involved in the transportation of food

_____   **5.** Migrant worker   e. Takes the crop from the field

_____   **6.** Trucker   f. Is involved in cleaning, sorting, and preparing a product

_____   **7.** Processor   g. Inspects food and determines how it will be sold

_____   **8.** Distributor   h. Sells directly to the public

C. Completion

1. Three careers that you could pursue in the quality-assurance area of the food industry include _____, _____, and _____.

2. Differences in how grocery stores are categorized are primarily determined by _____ and _____.

3. When buying food at a pick-your-own farm, the producer is also the _____.

4. A beef grade of _____ would indicate very tender, juicy, and flavorful, with abundant marbling.

5. Grocers purchase their supplies through _____.

UNIT 33

Food Science

MATERIALS LIST

✓ bulletin board materials

✓ Internet access

COMPETENCIES TO BE DEVELOPED

After studying this unit, you should be able to:

- discuss nutritional needs of humans and the food groups that meet these needs.
- categorize foods in the U.S. Department of Agriculture nutrition pyramid.
- discuss food customs of major world populations.
- relate methods used in processing and preserving foods.
- list the major steps used in slaughtering meat animals.
- list the major cuts of red-meat animals.
- identify methods of processing fish.
- describe techniques used to enhance retail sales of food commodities.

SUGGESTED CLASS ACTIVITIES

1. Conduct a class competition to identify the greatest number of processed food products. Divide the class into teams of four to five students. Assign each team to make a collage of photos or graphics that depict food items, using old magazines and other printed materials. Each food item must be a completely different product, not just different brand names of the same thing. Ten points are scored when the product is represented in the collage by a graphic from its package. Five points are earned when the product is represented in the collage by a picture from an advertisement. One point is scored when the product name is printed. Only the highest category is scored for each item. An additional 50 points is awarded for neatness and imagination. Use the collages for decorating bulletin boards.

2. Have a class discussion about possible new food products that you have recently heard about or even some new ideas for foods that do not now appear on the market. Encourage students to use their creativity to imagine some new way to prepare a food or to combine foods to make new products. For example, consider the fruit flavors and colors for single-serving cartons of milk. Create a list of the ideas that come from the class.

SUGGESTED CLASS ACTIVITIES—*continued*

3. Have the students keep a food journal for up to 1 week before beginning this unit. Introduce them to the Food Guide Pyramid. Be sure the students understand how much one serving is for each of the food groups. Next, have the students reflect on their food choices for the last week. Students should then write a short essay highlighting some positive food choices they made and areas in which they could improve.

TERMS TO KNOW

food	shelf life	vacuum pan	disassembly process
nutrients	dry-heat cooking	slaughter	fabrication and boxing
fermentation	moist-heat cooking	rendering insensible	processed meat
controlled atmosphere	conventional oven	shackle	evisceration
refrigeration	microwave oven	hoist	singe
blanching	convection oven	bleeding out	leaf fat
canning	dehydrator	hide	giblet
dehydration	smoker	viscera	kosher
freeze-drying	casein	carcass	dressing percentage
oxidative deterioration	cream	offal	sweetbreads
dehydrofrozen product	cheese	split carcass	tripe
humidity	cottage cheese	shroud	tankage
retortable pouch	condensed milk	age (ripen)	collagen
irradiation	evaporated milk	block beef	

You are what you eat. Have you ever stopped to consider what that statement means? Food is defined by Webster as a material containing or consisting of carbohydrates, fats, proteins, and supplementary substances, such as minerals, used in the body of an organism to sustain growth, repair, and vital processes and to furnish energy, especially parts of the bodies of animals and plants consumed by humans and animals. This unit explores the foods that humans need to maintain health and sustain growth (Figure 33–1). In addition, it explores how those foods reach our tables from their beginning as raw products.

FIGURE 33–1 The future well-being of infants is largely dependent on the nutrition they receive. Good nutrition, coupled with good health habits, will keep a body productive and vibrant for a lifetime. *(Courtesy of USDA)*

| Carbohydrates |
| Minerals |
| Proteins |
| Water |
| Fats |
| Vitamins |

FIGURE 33–2 Nutrients are classified into six categories. Each category of nutrients supports different functions in the body.

NUTRITIONAL NEEDS

The body is a complex system that has many nutritional demands. Nutrients are substances that are necessary for the functioning of an organism. There are more than 50 specific nutrients required for bodily functions. Nutrition involves the combination of processes by which all body parts receive and use materials necessary for function, growth, and renewal. Nutrition includes the release of energy, the building up of body tissues (both hard and soft), and the regulating of body processes.

After food is digested and in the blood system, nutrients are able to do their work. Nutrients are classified into six major groups. The different groups support different functions in the body (Figure 33–2).

Carbohydrates

Carbohydrates serve as the main source of energy for the body. Carbohydrates have four calories per gram. There are three different types of carbohydrates: sugars, starches, and fiber (Figure 33–3). Sugars are simple carbohydrates and are found naturally in many foods, such as fruit, milk, and peas. Refined sugar, or the substance used in many households, comes from sugar beets and sugar cane. Starch is a complex carbohydrate that is found in foods such as bread, potatoes, rice, and vegetables. Starches and sugars are converted to glucose in the body and serve as the major body fuel. Some of the fuel that is generated is stored by the body for later use. However, when the glucose is not used by the body, it is changed to fat. Fiber is also a complex carbohydrate and is found in the walls of plant cells. Humans are unable to digest fiber, yet it plays an important role in moving food through the body and expelling waste after digestion.

Fats

Fats are another source of energy for the body. Fats have nine calories per gram. They are considered to be a more compact source of energy because they have 2.25 times the number of calories as the other two energy sources—carbohydrates and proteins. Some vitamins are

FIGURE 33–3 Fruits, vegetables, and grains are healthful sources of carbohydrates. *(Courtesy of Price Chopper Supermarkets)*

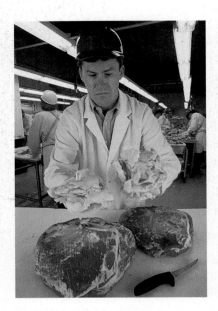

FIGURE 33–4 Animal products are good for their protein; but their fatty parts, such as skin and fatty layers, should be removed to avoid excess fat in the diet. *(Courtesy of USDA/ARS #K-4285-2)*

retained in the body by becoming dissolved in fats. They require fats to carry them to the parts of the body where they are needed. These vitamins are classed as fat-soluble vitamins. Although fat is necessary in the body, too much fat results in obesity and serious diseases, such as heart problems and high blood pressure. Fats are present in differing amounts in most foods. Foods that are known to be high in fat content include cheese, meat, poultry skin, avocados, and numerous others. Some foods that we are accustomed to require fats as part of their preparation. Baked goods, such as cakes and cookies; salad dressing; and fried foods acquire fats through preparation (Figure 33–4).

Proteins

The body needs food with proteins to build and rebuild its cells. Proteins have four calories per gram. Hair, skin, teeth, and bones are all parts of your body that require protein. Proteins are in a continuous cycle of building up and breaking down. Approximately 3 to 5 percent of your body's protein is rebuilt each day. Beans, peanut butter, meats, eggs, and cheese are high-protein foods.

Vitamins

Vitamins are also essential to the functions of the body. Some vitamins are dissolved in body fat and are stored in the body. Fat-soluble vitamins are not required in the diet each day, because they can be stored in the body. These include vitamins A, D, E, and K. Nine other vitamins—vitamin C and eight B vitamins—are water soluble and must be replenished daily. Specific vitamins have specific jobs in the body, and some foods are known to be richer in specific vitamins (Figure 33–5).

VITAMINS

Functions and Sources					
A	**B**	**C**	**D**	**E**	**K**
Functions					
vision bones skin healing wound	using protein, carbohydrates, and fats to keep eyes, skin, and mouth healthy brain nervous system	wound healing blood vessels bones teeth other tissues works with minerals	needed for using calcium and phosphorus bones teeth	preserve cell tissue	blood clotting
Sources					
yellow, orange, and green vegetables	whole-grain and enriched cereals, breads, meats, beans	citrus fruits, melons, berries, leafy green vegetables, broccoli, cabbage, spinach	fatty fish, liver, eggs, butter, added to most milk	vegetable oils, whole-grain cereals	leafy green vegetables, peas, cauliflower, whole grains

FIGURE 33–5 Vitamins have specific functions, and it is important to choose foods that supply adequate vitamin content.

Functions of Minerals			
Bone Development	Fluid Regulation	Materials for Cells	Trigger Other Reactions
Sources of Minerals			
calcium milk products *magnesium* nuts, seeds, dark-green vegetables, whole-grain products *phosphorus* no specific food group *fluorine* some seafood, some plants, may be added to drinking water	*sodium* salt *potassium* bananas *chlorine* salt	*iron* meats, liver, beans, leafy green vegetables, grains *works with vitamin C* *iodine* iodized salt added to salt	*zinc* whole-grain breads and cereals, beans, meats, shellfish, eggs *copper* fish, meats, nuts, raisins, oils, grains

FIGURE 33–6 The careful selection of foods can ensure a correct balance of minerals in our diets.

Minerals

There are more than 20 minerals that are needed by the body. The amounts needed may be small, but they are required nonetheless. The 20 minerals are divided into 4 major groups. Some minerals are parts of bones, others regulate bodily functions, some are needed to make special materials for cells, and others trigger chemical reactions in the body (Figure 33–6).

Water

The human body is more than 50 percent water. Water carries nutrients to cells, removes waste, and maintains the body's proper temperature. Fluid foods, such as milk and juice, obviously help to supply the body with water. However, foods such as meat and bread also provide water.

Food Groups that Meet Needs

INTERNET KEY WORDS:

food guide pyramid

Each food is different in the types of nutrients it contains and ultimately provides to the body. Foods are divided into five major food groups, which represent the nutritional needs of the body. The five food groups are the bread, cereal, rice and pasta group; vegetable group; fruit group; milk, yogurt, and cheese group; and meat, poultry, fish, dry beans, eggs, and nut group. The U.S. Department of Agriculture (USDA) represents these groups in a pyramid format to help us remember the relative proportions that each group should make up in a daily diet (Figure 33–7).

You generally should not seek foods in the fats, oils, and sweets group at the top of the pyramid, because plenty of these are used to prepare foods in the other groups, and they show up in adequate amounts in most people's diets in the United States. The names of the groups suggest some of the typical foods that they include. However, the quality of diet can be increased by selecting the most nutritious items from each group (Figure 33–8).

Science Connection

NEW KID ON THE BLOCK

After a lengthy period of inquiry and research, the U.S. Department of Agriculture (USDA) approved the irradiation of raw packaged poultry as a safe and effective procedure for protection against food-borne illness. Irradiation of food is being done in more than 35 countries.

Proper cooking is still the last stage in food handling to eliminate microorganisms that can cause food poisoning or other illness to those who consume the food. The use of water for washing and cooling fruits, vegetables, and meats during and after processing may spread harmful microorganisms throughout the batch and leave every item with low concentrations of the microorganisms. That is, all of the food has low-level contamination, and the organisms will multiply and increase their hazard if conditions are right to do so. Therefore, the handling of raw meat having pathogens on the surface carries the threat of contamination of hands, countertops, cooking areas, and, ultimately, cooked food.

As early as January 1992, irradiated Florida strawberries were sold at a northern Miami supermarket. *(Courtesy of USDA)*

Science has long sought to find a way to eliminate food-borne pathogens without altering or damaging the product or leaving poisonous chemicals on the food. Irradiation seems to be the answer. Irradiation provides the same benefits as when food is processed by heat, refrigeration, or freezing or is treated with chemicals—to destroy insects, fungi, or bacteria that cause food to spoil or cause human disease. Irradiated foods are wholesome and nutritious. The treatment process involves passing food through an irradiation field, but the food itself never contacts a radioactive substance. The process has been subjected to extensive study and declared safe by special scientific committees in Denmark, Sweden, the United Kingdom, and Canada, and it has received official endorsement from the World Health Organization and the International Atomic Energy Agency. No known risks exist from consuming food that is irradiated by legal procedure.

Notably, the approved level of irradiation does not kill all microorganisms—in other words, it does not sterilize food. Therefore, other food-preserving measures must be followed when storing irradiated food and preparing it for consumption. Irradiation does not alter the food in any perceptible way for the consumer. The Cordex Alimentarius, an international committee on food safety, has developed a green irradiation logo. Irradiated foods from the United States will bear this green logo along with the words "Treated with Radiation" or "Treated by Irradiation."

Eating from each of the major food groups daily will ensure a well-balanced diet and will provide the essential nutrients needed for growth and development. The number of portions consumed per day differs with each group. At different stages in life, requirements within each group may vary to some degree, but no food group should dominate or be eliminated from the diet. More information on nutrition is provided in Unit 26.

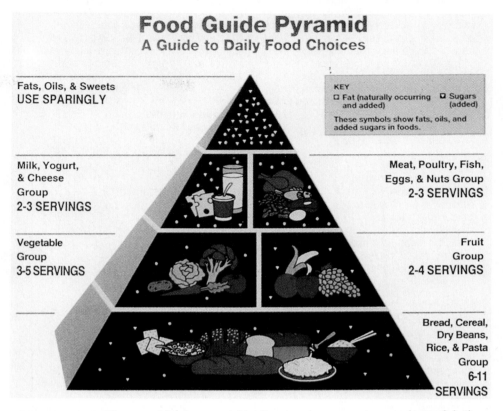

FIGURE 33–7 The nutrition pyramid indicates the range in number of daily servings for each major food group as required for a healthful diet. *(Courtesy of USDA)*

FOOD CUSTOMS OF MAJOR WORLD POPULATIONS

What is eaten from each of the major food groups varies around the world. Food habits reflect what is most readily available.

In hot and wet climates such as in Southeast Asia, a great deal of rice is consumed. In areas of the world where corn grows well, many food items contain corn in some form. For example, cornmeal may be used to make tortillas or pancakes and may be mixed as a cereal in countries such as Mexico. Availability of food and technology to prepare that food has dictated eating habits over the years. For instance, introducing dairy products in countries where dairy cows are not raised presents educational, as well as transportation and processing, challenges.

METHODS OF PROCESSING, PRESERVING, AND STORING FOODS

INTERNET KEY WORDS:

food preservation
refrigeration

One of the oldest ways to preserve food for delayed use is fermenting and pickling. Fermentation is a chemical change that involves foaming as gas is released. Long ago, it was determined that some foods did not spoil when allowed to ferment naturally or when fermented liquids were added to the foods. Controlled fermentation is now

POULTRY, FISH, MEAT, AND EGGS

Lower-scoring foods tend to be high in calories, cholesterol (eggs), fat (red meat), or sodium (processed meats). The foods near the top are relatively low in fat. Most of the foods are rich in protein and iron. (All servings are 4 ounces broiled, baked, or roasted, unless noted otherwise.)

	NUTRITION SCORE
clams, steamed	19
turkey breast, skinless	10
tuna canned in water (3 oz.)	6
cod	1
egg white (1 large)	1
salmon, canned (3 oz.)	1
scallops, steamed	1
flounder	-1
lobster meat, boiled	-4
salmon fillet	-5
blue crab meat, steamed	-6
chicken breast, skinless	-8
turkey breast luncheon meat, 3 slices (2 oz.)	-13
tuna canned in oil (3 oz.)	-18
shrimp, steamed	-21
chicken breast with skin	-23
Canadian bacon, fried, 2 slices (2 oz.)	-24
veal cutlet, breaded, pan-fried	-25
round steak, trimmed (5 oz.)	-29
ham, luncheon meat, 2 slices (2 oz.)	-32
pork chops	-48
bacon, fried, 4½ slices (1 oz.)	-54
egg (1 large)	-59
shrimp, fried	-63
bologna, 2 slices (2 oz.)	-70
leg of lamb	-76
salami, luncheon meat (2 oz.)	-80
hamburger, lean	-81
sirloin steak (5 oz.)	-86
hamburger, regular	-92
round steak, untrimmed (5 oz.)	-97
chicken thighs, fried, home recipe (2)	-103
sausage links, 2 (2 oz.)	-112
pot roast	-114

GRAIN FOODS

Contrary to myth, starchy grain foods are not fattening. Most people would do well to eat more grain foods in place of meat. Grains, especially whole grains, are a nicely balanced, low-fat source of carbohydrate, vitamins, minerals, and protein. (All serving sizes are 1 cup cooked, unless noted otherwise.)

	NUTRITION SCORE
bulgur (cracked wheat)	69
wheat germ (¼ cup)	61
pearled barley	60
brown rice	45
spaghetti or macaroni	45
oatmeal	38
hominy grits	35
whole-wheat bread (2 slices)	31
hamburger or hotdog roll (1)	18
corn muffin (1)	1

FRUITS

Fruits can give you, naturally, all the sweetness you want, plus fiber, vitamins A and C, and other nutrients. Go easy on the dried fruits! Their sugars are sticky and promote tooth decay. (All servings are one medium piece, unless noted otherwise.)

	NUTRITION SCORE
papaya (½ medium)	74
cantaloupe (¼ medium)	67
strawberries (1 cup)	65
orange	62
prunes, uncooked (5)	51
dried apricots (5)	49
tangerine	41
watermelon (2 cups cubed)	36
apple	36
pear	36
blueberries (1 cup)	36
pink grapefruit (½)	35
pineapple, fresh (1 cup)	34
banana	32
cherries (1 cup)	32
honeydew melon (1/10 melon)	31
raisins (1 oz.)	28
plums (2)	27
applesauce, unsweetened (½ cup)	18
peach	17
grapes (30)	16
peaches in heavy syrup (½ cup)	2

DAIRY

While most dairy foods are rich in protein and calcium, the lower-scoring foods are high in saturated fat, cholesterol, and sodium.

	NUTRITION SCORE
yogurt, nonfat (1 cup)	58
milk, skim (8 oz.)	40
yogurt, plain lowfat (1 cup)	36
milk, 1% lowfat (8 oz.)	28
milk, 2% lowfat (8 oz.)	16
yogurt, fruit-flavored lowfat (1 cup)	13
chocolate milk, 2% lowfat (8 oz.)	6
cottage cheese, 1% fat (½ cup)	3
sour cream, lowfat (2 Tbsp.)	-2
ricotta cheese, part skim (1 oz.)	-3
yogurt, plain (1 cup)	-5
mozzarella cheese, part skim (1 oz.)	-5
milk, whole (8 oz.)	-7
cheddar cheese, reduced fat (1 oz.)	-7
nondairy powder coffee creamer (2 tsp.)	-12
half and half cream (2 Tbsp.)	-15
Swiss cheese (1 oz.)	-15
cottage cheese, 4% fat (½ cup)	-16
mozzarella cheese (1 oz.)	-19
sour cream (2 Tbsp.)	-26
cheddar cheese (1 oz.)	-32
American cheese (1 oz.)	-34
whipped cream (2 Tbsp.)	-59

VEGETABLES

Most vegetables are great sources of vitamins—especially A and C—and minerals. Try a new vegetable today! (All serving sizes are ½ cup cooked, unless noted otherwise.)

	NUTRITION SCORE
sweet potato, baked (1 medium)	184
potato, baked (1 medium)	83
spinach	76
kale	55
mixed vegetables, frozen	52
broccoli	52
winter squash (acorn, butternut), baked	44
Brussels sprouts	37
cabbage, chopped, raw (1 cup)	34
green peas	33
carrot (1)	30
okra	30
corn on the cob (1 ear)	27
tomato (1 medium)	27
green pepper (½)	26
cauliflower, raw	25
artichoke (½)	24
romaine lettuce, raw (1 cup)	24
collard greens	23
asparagus	22
celery (four 5" pieces)	19
green beans	18
turnips	16
sauerkraut	15
summer squash (zucchini)	12
green beans, canned	10
iceberg lettuce, raw (1 cup)	8
bean sprouts (¼ cup)	7
onion, chopped, raw (¼ cup)	7
eggplant	6
cucumber slices, raw	4
mushrooms, raw (¼ cup)	2
dill pickle (½ large)	-3
avocado (½ medium)	-25

LEGUMES

Beans are excellent sources of dietary fiber, protein, vitamins, and minerals. They are also very low in fat. (All serving sizes are ¾ cup cooked, unless noted otherwise.)

	NUTRITION SCORE
kidney beans	91
navy beans	82
black beans, black-eyed peas, or lima beans	78
lentils	74
chickpeas	68
split peas	56
tofu/bean curd (4 oz.)	33

DESSERTS

Most desserts are high in fat, sugar, and calories. Next time, try fresh fruit or nonfat frozen yogurt for a change. (Serving sizes are 1 cup, unless noted otherwise.)

	NUTRITION SCORE
angelfood cake (2 oz.)	1
chocolate pudding (½ cup)	-2
Jell-O (½ cup)	-7
brownie with nuts (1¾" square)	-23
sherbet	-37
vanilla ice cream	-73
cheesecake (4½ oz.)	-161

FIGURE 33–8 Estimated relative nutritional values of selected food items.

FIGURE 33-9 Fermentation is a food preservation process whereby bacteria convert sugars to acids or alcohol that protects the food from spoilage. *(Courtesy of USDA)*

FIGURE 33-10 Food storage and preservation through ice, refrigeration, and freezing is the cornerstone of milk, meat, fish, fruit, and vegetable handling today. *(Courtesy of USDA)*

used to produce cheeses, wines, beers, vinegars, pickles, and sauerkraut (Figure 33–9).

Today, the primary objective of processing and preserving is to change raw commodities into stable forms. With refrigeration and various processing techniques, we now expect almost all foods at any time during the year and in any part of the world.

Slowing deterioration is the primary goal in food preservation. Tomatoes and cucumbers that will be sold raw are waxed to retard shriveling while they are in the grocery store. Apples may be treated with a decay inhibitor. Table grapes are fumigated with sulfur dioxide to control mold. Similarly, silos where grains are stored are purged with 60 percent carbon dioxide to control insects.

Carbon dioxide inhibits the growth of bacteria. Controlled atmosphere (CA) is the process whereby oxygen and carbon dioxide are adjusted to preserve or enhance particular foods. An example of food preservation by controlled atmosphere is the transporting of cut lettuce in controlled atmosphere to prevent the edges from turning brown.

There are many other processing and preservation techniques. These techniques slow deterioration and allow one to enjoy foods in a variety of forms throughout the year and around the globe.

Refrigeration is an important key to many processing and preservation techniques. Refrigeration is the process of chilling or keeping cool. Low temperatures reduce or stop processes that contribute to the deterioration of products. Refrigeration retards respiration, aging, ripening, textural and color change, moisture loss and shriveling, insect activity, and spoilage from bacteria, fungi, and yeasts. Refrigeration is expensive and is used widely only in well-developed countries.

When crops are harvested in the field, their temperatures are between 70° and 80° F. The goal of refrigeration is to quickly reduce that temperature to near 32° F. How quickly the food is cooled varies depending on the type of cooling technique used. Some vegetables are precooled in the field by cold air blasts, hydrocooling, or vacuum. Hydrocooling means cooling with water. Milk is cooled in refrigerated lines and tanks on the farm (Figure 33–10).

The logical step after a product is cooled is to continue cooling it until it is frozen (Figure 33–11). When foods are kept at 0° F or less, little

Science Connection

MILK STAYS COOL UNDER PRESSURE

Alexander Lin, a graduate student at Texas A&M University, has improved an old technique for processing milk. The result is a longer shelf life for milk products. During processing milk is homogenized. Homogenization is performed by pumping milk into a chamber that applies between 2,000 and 3,000 pounds of pressure per inch (psi). The pressure causes the fat in the milk to break into tiny random-shaped particles that stay suspended in the milk instead of rising to the surface like it does in nonhomogenized milk. Lin decided to find out what would happen if the pressure was increased to 14,000 psi. The result of the greater pressure was a smaller, more uniform fat globule disbursed throughout the milk. This made the final product more stable with an increased shelf life. This research may change the way dairy products are processed and may help keep them fresher longer. The improved high-pressure homogenization will benefit milk processors, grocers, and consumers.

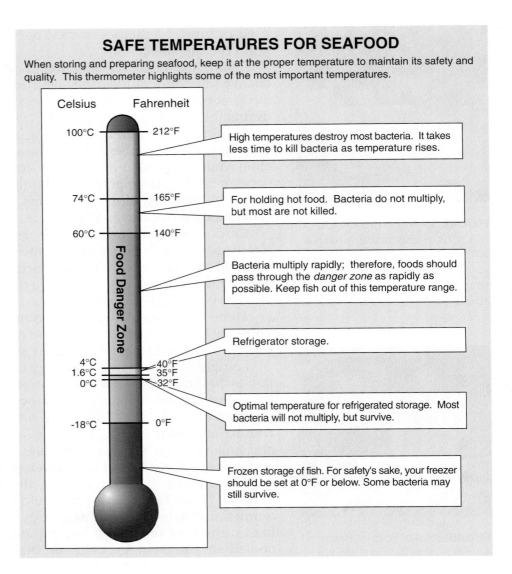

SAFE TEMPERATURES FOR SEAFOOD

When storing and preparing seafood, keep it at the proper temperature to maintain its safety and quality. This thermometer highlights some of the most important temperatures.

Celsius	Fahrenheit	
100°C	212°F	High temperatures destroy most bacteria. It takes less time to kill bacteria as temperature rises.
74°C	165°F	For holding hot food. Bacteria do not multiply, but most are not killed.
60°C	140°F	Bacteria multiply rapidly; therefore, foods should pass through the *danger zone* as rapidly as possible. Keep fish out of this temperature range.
		Refrigerator storage.
4°C	40°F	
1.6°C	35°F	
0°C	32°F	Optimal temperature for refrigerated storage. Most bacteria will not multiply, but survive.
-18°C	0°F	Frozen storage of fish. For safety's sake, your freezer should be set at 0°F or below. Some bacteria may still survive.

Food Danger Zone

FIGURE 33–11 It is important to choose the appropriate heat intensity and duration for the intended purpose.

deterioration occurs. However, even frozen foods have storage limits. Fruits and vegetables should be consumed within 1 year after freezing. Meats should be consumed within 3 to 6 months. Vegetables that are to be frozen often require blanching before freezing. **Blanching** is the scalding of food for a brief time before freezing it. This process inactivates enzymes that cause undesirable changes when plant cells are frozen.

Another popular preservation technique is canning. **Canning** involves putting food in airtight containers and sterilizing the food to kill all living microorganisms that could cause spoilage. Temperatures of 212° to 250° F are required to successfully can food products. Metal cans are coated to reduce chemical reactions between the can and its contents. A 2-year shelf life for canned food is considered normal.

Another popular way to process and preserve food is dehydration. **Dehydration** means reducing the moisture content to inhibit growth of microorganisms. Moisture can be removed by the sun, by indoor tunnel or cabinet dehydrators, or by freeze-drying. **Freeze-drying** is the newest

Agri-Profile

CAREER AREAS:

Food Scientist/Food Technician/Nutritionist/Dietician

Food science programs are available in high schools, colleges, technical institutes, and universities. Careers in food science may be in specialty areas, such as meats, fruits, vegetables, baked goods, dairy products, wine, or other beverages.

Food companies and consumers rely on food scientists and technicians to develop new products to meet the ever-changing needs of a busy world. Convenience products, such as instant and freeze-dried coffee, dried and vacuum-packed fruits and meats, processed chicken tenders, boneless rolled meat, yogurt, ice cream, pasteurized and homogenized milk, shelf-safe milk, low-calorie and low-fat food, space-age food, unrefrigerated foods, and many other products are the handiworks of food science and technology.

Food scientists work for food companies, universities, research centers, food chain stores, dairies, radio and television stations, and government agencies. Some become authors and publish journal and magazine articles, cookbooks, and recipe books.

Are these young visionaries contemplating careers in food science? Food scientists develop better ways to process, handle, package, prepare, and market food products. *(Courtesy of USDA/ARS #K-48191)*

method of dehydration. It involves the removal of moisture by rapid freezing at very low temperatures. When foods are dehydrated, they have a 2- to 10-percent moisture content. The shelf life of dehydrated foods is as long as 2 years. Oxidative deterioration is loss of quality because of a reaction with oxygen. This can occur when air reaches dried foods. Glass or metal containers are more airtight than plastic ones. Dried soup mixes, packaged salad dressing, spices, and dried fruits are foods that have been dehydrated. Dehydrated foods are lighter in weight and lower in volume than the same whole foods.

Another processing technique is dehydrofreezing. A dehydrofrozen product is processed by precooking, evaporating water, and freezing. Potatoes are being used in the development of this technique. They are precooked as cubes or slices, and water is evaporated to reduce the weight by 50 percent. The potatoes are then frozen.

Another factor to consider in the processing and handling of foods is humidity during storage and transportation. Humidity is the amount of moisture in the air. A humidity level of 90 to 95 percent is required for high-moisture products such as meats and vegetables. Dry onions require only 75 percent humidity for optimum storage.

Various wrappings are used in the processing, preservation, and transportation of foods. Cardboard boxes, wood boxes, molded pulp trays to reduce bruising, and plastic wraps in a variety of thicknesses or plies all meet different needs. Newer retortable pouches provide protection from light, heat, moisture, and oxygen transfer, all in one wrapping. Retortable pouches are flexible packages consisting of two layers of film

FIGURE 33–12 Irradiation is now an approved method for controlling Salmonella bacteria in chicken meat. *(Courtesy of USDA/ARS #K-3783-20)*

(or plastic) with a layer of foil between them. Although the cost is now high, the benefits of a 1- to 2-year shelf life are appealing.

Irradiation is a relatively new procedure using gamma rays to kill insects, bacteria, fungi, and other organisms in food products. Gamma rays pass through food without heating or cooking. Therefore, no heat-sensitive nutrients are lost in the process (Figure 33–12).

The cost of each of the preceding processes and techniques is different and must be taken into account by processing plants. Where energy costs are low, more sophisticated techniques may be used without incurring excessive costs. Research is always looking at ways of preserving food more economically in relation to energy costs.

Food Additives to Enhance Sales of Food Commodities

INTERNET KEY WORDS:

food additives

The processing of some foods may reduce their natural nutritional value. To compensate for that loss, vitamins and minerals are added back into foods to restore their nutritional value. Bread, noodles, and rice have vitamins and minerals added to them before they are packaged. Vitamins A and D may be added to fluid milk before its sale to the public.

A food additive is anything that is added to a food during processing and before it goes into a package. In addition to food additives that restore nutritional value, preservatives are also added to food to extend its shelf life. Shelf life refers to the amount of time before spoilage begins. Food additives also enhance the color or appearance of foods. Other food additives reduce the cooking time of foods such as oatmeal.

Sugar is probably one of the most widely used food additives. It is found on the labels of many cereals and beverages. Food labels identify the contents of food products. The order of ingredients on a food label indicates the proportions of each, in descending order (Figure 33–13).

The government is continually testing and evaluating the positive and negative effects of food additives. It has been determined that some food additives may cause harm. Such additives are usually banned by the government.

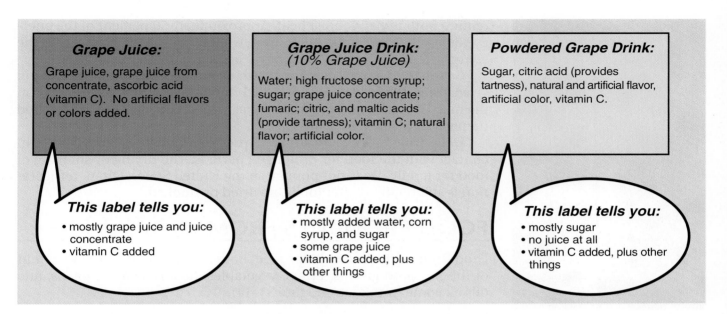

FIGURE 33–13 Products may look the same, but read the labels for the real differences. Ingredients are listed in order from the most to the least amount found in the product.

FOOD PREPARATION TECHNIQUES

Some foods can be eaten as they come from the package. Crackers and raisins are examples of foods that do not require additional processing at home to ensure safe eating. However, cooking is often required for other foods, whereas heating may only improve flavor in others. Raw meat should be cooked before it is eaten to ensure safety. How food is cooked is determined by the type of food and the appliances available.

There are two basic methods of cooking foods. The appliances used to accomplish these methods vary from household to household. **Dry-heat cooking** involves surrounding the food with dry air in the oven or under the broiler. This method is usually used for tender cuts of meat having little connective tissue and for vegetables having a high-moisture content, such as potatoes. **Moist-heat cooking** involves surrounding the food with hot liquid or by steaming, braising, boiling, or stewing the food. The warm moisture breaks down the connective tissues. Moist-heat cooking is a popular method used for less tender cuts of meat and for vegetables with a low-moisture content.

Appliances have been designed to accommodate different types of food, as well as different time and energy demands. The goal of preparing good food in a short period, without expending excessive energy, has resulted in the development of several appliances that supplement the traditional gas or electric range and oven. Pressure cookers concentrate moisture by sealing it in, thus reducing cooking time. Crockpots and oven bags allow slow, moist cooking without the operator continually being near the cooking process.

Tabletop **conventional ovens** are used to cook small amounts of foods by dry heat. This requires less energy than do large, traditional ovens. The **microwave oven** uses electromagnetic waves to heat and cook food. It offers a more energy-efficient way to cook foods that

FIGURE 33–14 Baked goods from grains form the basic diet in many cultures. *(Courtesy of Price Chopper Supermarkets)*

require both dry and moist heat. Approximately 50 percent of the energy goes to the food in the microwave, whereas only 6 to 14 percent of the energy used by a conventional range actually goes to the food.

Convection ovens heat food with the forced movement of hot air. Ovens that combine convection and microwave functions are now on the market. Again, energy efficiency, as well as convenience for the food preparer, is a constant goal.

Dehydrators dry food; **smokers** preserve food by keeping smoke in contact with the food for prolonged periods. The drying or smoking of food for family use is not popular in the United States. Often, refrigeration is still required after foods are dried or smoked.

FOOD PRODUCTS FROM CROPS

Food from plant sources helps meet body requirements for food in four of the five food groups. Fruits, vegetables, breads, cereals, beans, and oils all come from crops (Figure 33–14).

Fruits, Vegetables, and Nuts

Fruits, vegetables, and nuts are nearly ready to eat when they are harvested. This can be as simple as pick, wash, and eat for items such as leafy vegetables, berries, and fruits. The process could include picking, cutting, peeling, shelling, washing, trimming, cooking the food, or eating it raw. These products may be processed for storage for as short as a few hours to as long as many years. Food processing may be done at home for the family or commercially for the billions of people in the world market.

For some food products the journey from field to table may involve many processes. The first stage may be to simply cool the product and hold it in a temperature- and humidity-controlled environment until the next stage, or it may be to process the food without first precooling. Where processing occurs immediately, the product typically passes through washing equipment designed to avoid injury to the product. Washing may involve flotation, rotary, water-jet, or other cleaning procedures. Skins and hulls may be removed by hot water bath, steam, flame, or cutters. The product may then be trimmed, halved, quartered, sectioned, sliced, diced, or crushed for drying, canning, freezing, or processing into ready-to-cook or precooked products.

Processed foods must be packaged and correctly labeled. All along the way, nutritionists, chemists, inspectors, public health officials, and others monitor the process and product. The law requires that food be safe for consumption and of the weight, quality, and grade that is specified on the label.

Cereal Grains

Cereal grains compose the major diet of most of the world's people. Rice, wheat, corn, barley, oats, and other grains are consumed as whole grains, cracked or rolled grains, flour, bran, and many other products. Grains are economical to process because they can be left on the plant until nearly dry enough to prevent spoilage. If stored in a cool, dry place, properly dried grains can remain edible for many years. However, much of the world's food grain is lost every year because of rotting in the field and in storage, or being consumed or ruined by insects, birds, and rodents.

Processing grain for human consumption generally means separating or milling the grain into its basic components—hulls, bran, flour, and germ. These components are then used to make breakfast cereals, breads, pastries, pasta, and thousands of products that adorn the grocery shelves.

Oil Crops

Soybeans, cottonseed, peanuts, rape, palm nuts, coconuts, olives, and corn are examples of crops that are rich in vegetable oils. These are used for cooking, frying, baking, and other food processes, as well as for many food products such as dressings, coffee creamers, and shortening. These oils are also used in the manufacture of paints, lacquers, plastics, and many products of industry. The seeds or nuts and other oil-rich plant parts are crushed or ground and heated. The oil is then extracted by solvents and purified for food and industrial uses. The meal is dried and ground mostly for livestock feed (Figure 33–15).

Those who do not eat meat products can still meet the requirements for food from all five food groups. People with allergies to milk products can meet their nutritional needs for food from soybeans and other plant sources.

FIGURE 33–15 Processing of grains and oil-crop seeds releases the basic building blocks for many products.

FIGURE 33–16 Meats, fish, poultry, and dairy products are nutritious foods from animal sources. *(Courtesy of Price Chopper Supermarkets)*

FIGURE 33–17 Milk is pasteurized to kill bacteria, homogenized to keep the milk fat evenly suspended, and bottled for the convenience of the customer. *(Courtesy Marilyn Parker)*

FOOD PRODUCTS FROM ANIMALS

Meats, fish, poultry, dairy products, and the proteins group are nutritious foods from animal sources (Figure 33–16).

Dairy Products

Milk is used and consumed in a variety of ways. Approximately 37 percent of all milk consumed in the United States is in fluid form. The remainder is used to make many products, including cheese, butter, frozen foods, dried whole milk, cottage cheese, evaporated milk, and condensed milk. The processing of milk in different ways results in these many different products (Figure 33–17).

Milk and Milk Products

Although recommendations for milk consumption are greatly influenced by calcium needs, milk contains some of all the essential food nutrients needed by the body. The major components of milk are water, fat, protein, sugar, and minerals. However, there are numerous other highly important components in milk.

Water   Although cow's milk is a fluid product containing about 88 percent water, it also contains 12 to 13 percent total solids. This is comparable to the solids content of many other foods. Because milk is a food specifically prepared by nature for the nourishment of the very young, it provides the water necessary for life. The water also acts as a carrier for dissolved, suspended, and emulsified components.

Protein   Milk provides a substantial proportion of the total protein in our food supply. Casein is the predominate protein in milk. It is found only in milk and makes up about 82 percent of the total milk protein. Casein exists in suspended form and is easily coagulated by the action of acids and enzymes. Milk proteins are high-quality proteins that contain all of the essential amino acids in proper balance for good nutrition.

INTERNET KEY WORDS:

dairy products nutrition

Milk Fat   Fats are concentrated sources of energy. There are many different fatty acids in milk that give it the distinctive, pleasing flavor that complements many prepared foods. It also contains those vitamins that are fat soluble—vitamins A, D, E, and K. Milk fat exists in a highly emulsified state, which facilitates its digestion.

Lactose   Lactose is milk's major carbohydrate and accounts for about half of the nonfat solids in milk. The relative sweetening power of lactose is about one-sixth that of sucrose (common table sugar).

Minerals   Milk contains seven minerals as major constituents and many more in minor or trace amounts. Calcium and phosphorus are essential in human nutrition for building bony structures and for certain metabolic processes. Milk is the chief source of food calcium in the diets of people in the United States. It has the added advantage of containing phosphorus in the same biological relationship to calcium as occurs in the growing skeleton. It is difficult to provide the recommended daily amounts of dietary calcium without using milk or milk products because calcium is poorly distributed among other foods.

Vitamins   All of the vitamins known to be required by humans are found in milk. Some are fat soluble and are associated with butterfat, whereas those that are water soluble are found in the nonfat portion of milk. Vitamin A and carotene are present in high concentration in milk fat. Carotene, from which vitamin A is formed in the body, gives milk fat its characteristic color. The vitamin D content of fresh milk is low. However, most commercially pasteurized milk is fortified with vitamin D to balance the product for best nutrition.

Milk is an abundant source of riboflavin (vitamin B_2) and an important source of niacin. Although the niacin content of milk is low, it is in a fully available form. Milk contains significant amounts of thiamine (vitamin B_1). Other vitamins of the vitamin B complex occurring in milk include pantothenic acid, pyridoxine (vitamin B_6), biotin, vitamin B_{12}, folic acid, and choline.

INTERNET KEY WORDS:

making cheese, butter, yogurt
making ice cream, sherbet

Processed Milk Products   The nutritional completeness of milk has led to its reputation as nature's most nearly perfect food. Where it falls short in minor ways, science and the food-processing industry have intervened to improve the options for the consumer.

Fluid milk sold in the United States is pasteurized to ensure safety from disease-causing organisms, and it is homogenized to keep the milk fat in suspension so it does not require stirring before each use. Whole milk is marketed with 3.5 percent milk fat. However, for those desiring less fat intake, milk can also be purchased with 2 percent, 1 percent, or no milk fat. The latter is called skim milk. Milk may also be purchased as fortified milk with vitamins A and D added.

The fat in milk is called butterfat. Cream is a component of milk that contains up to 40 percent butterfat. Butter, which is about 80 percent fat, is made from cream. Cream is made by concentrating the fat portion of the milk. This is accomplished by passing milk through a cream separator. Whipping cream contains about 40 percent fat, table cream 18 to 20 percent, and "half-and-half" approximately 12 percent.

Ice cream, ice milk, and sherbet are also dairy products. They account for most of the frozen desserts in the United States.

Nonfat dried milk is used both as human food and animal feed. It is frequently used as an ingredient in dairy and other food products.

FIGURE 33–18 Cheese is manufactured from milk by coagulating the milk solids to form curd. The curd is ground and pressed into larger blocks of cheese. Bacteria act on the cheese to give it the characteristic flavor that makes each variety unique.

Cheese is made by exposing milk to certain bacterial fermentations or by treating it with enzymes. Both methods are designed to coagulate some of the proteins found in milk. There are many types and varieties of cheese (Figure 33–18). **Cottage cheese** is made from skimmed milk.

Condensed and evaporated milks are canned milk products. **Condensed milk** and **evaporated milk** are both produced by removing large portions of water from the whole milk through a machine called a **vacuum pan**. Condensed milk is further treated by adding large amounts of sugar. The sugar content makes condensed milk an important ingredient in the baking and ice cream industries. The many components of milk have resulted in the evolution of an extensive dairy industry. Consumers now can choose from a great array of dairy products (Figure 33–19).

Meat Products

Meat products, like milk, are enjoyed in a variety of forms. Animal species vary somewhat, but the procedures for slaughtering and processing each are similar. There are common by-products from the processing of animals, as well as the familiar meats that are important to U.S. eating habits.

Beef

Meats of all kinds come from slaughtered animals. **Slaughter** means to kill and process or dress animals for market purposes. The procedure is as follows for beef.

The first step is to render insensible. **Rendering insensible** means making the animal unable to sense pain. There are several methods of accomplishing this that comply with the Humane Slaughter Act of 1958. Packers that do not comply with provisions of this act are unable to sell meat to the federal government. The approved methods must be rapid and effective. Methods used in rendering insensible include a single blow or gunshot, electrical current, or use of carbon dioxide gas. In addition, rendering insensible may be accomplished by following the ritual requirements of religious faiths.

When the animal is insensible to pain, it is shackled, hoisted, and stuck to permit bleeding. **Shackles** are mechanical devices that confine

INTERNET KEY WORDS:

home butchering
slaughter meat
processing meat

FIGURE 33–19 Milk is the basic ingredient of many food items. *(Courtesy of USDA)*

FIGURE 33–20 After slaughter, the carcass is inspected and cooled before being cut into block or retail cuts. *(Courtesy of USDA/ARS #K-4284-12)*

the legs and prevent movement; hoist means to raise into position. To *stick* means to cut a major artery to permit the blood to drain from the body. A large artery is severed for efficient bleeding out. The head of the animal is removed during or after the bleeding-out process.

The removal of the hide or skin is the next step in the process. The hide is cut open at the median, or midline, of the belly of the animal, and hide pullers are used to remove the hide in one piece. The breast and rump bones are split at this time by sawing.

The term viscera refers to organs located in the cavity of the animal. Organs that are removed include the heart, liver, and intestines. The kidneys are not removed at this time. Plants that are regulated by the U.S. Department of Agriculture must have the carcass and viscera inspected to confirm good animal health. Carcass refers to the body meat of the animal—the part that is left after the offal has been removed. Offal consists of the nonmeat material that is converted to by-products. It includes the blood, head, shanks, tail, viscera, hide, and loose fat.

The next step is to split the carcass by cutting through the center of the backbone and removing the tail. The split carcass (sides of the animal) is then washed with warm water under pressure.

A high-quality carcass should have a smooth appearance after it is cooled. To accomplish this, the hot carcass is wrapped with a large cloth called a shroud. Sides are cooled for a minimum of 24 hours before ribbing and further processing (Figure 33–20). The meat is kept at 34° F until it is sold and consumed.

Fresh beef must be aged, or ripened. To age, or ripen, means to leave undisturbed for a period so that minor biological changes can take place while the beef cools. Fresh beef is not in its most tender state immediately after slaughter. While the beef is aging, evaporation (the loss of moisture) and discoloration are kept to a minimum. A fairly thick covering of fat on the carcass helps the aging process. There are three methods used to age beef—traditional aging, fast aging, and vacuum packaging. Time, temperature, and technology help define these three methods.

Beef carcasses are generally disposed of in three ways: as block beef; fabricated, boxed beef; and processed meats. Traditionally, meat is shipped in exposed halves, quarters, or wholesale cuts to be cut into retail cuts in supermarkets. In this condition, it is referred to as block beef. It is ready for sale "over the block" or counter. This traditional method creates concern over sanitation, shrinkage, spoilage, and discoloration. Therefore, more packers are using the disassembly process. The disassembly process means that the carcass is divided into smaller cuts, vacuum sealed, boxed, moved into storage, and shipped to retailers. This process is also known as fabrication and boxing. Processed meats are made from scraps of meat that are not in suitable form for sale over the block. Such meats have the bones removed and are sold as boneless cuts. They can also be canned, made into sausage, dried, or smoked.

Sheep

Sheep are slaughtered in much the same way as beef. They are first rendered insensible and bled. Next, the front feet are removed, the pelt or furlike covering is removed, and the hind feet and head are removed. The opening of the carcass and removal of the viscera, called evisceration, are similar to the procedures outlined for beef animals. In view of the small size of a lamb, the forelegs are folded at the knees and are held in place by a skewer after evisceration. Washing and cooling procedures are similar to those used for beef.

Hogs

The procedure for slaughtering hogs is a little different than for cattle and sheep. Hogs are rendered insensible, shackled, hoisted, and bled. The carcasses are then plunged into water at 150° F for about 4 minutes. This process is required to loosen the hair and dry skin.

The hair of hogs is removed by mechanical scraping. A dehairing machine can remove the hair from about 500 hogs per hour. After the hair is removed, the hog is returned to overhead racks and processing continues.

The hog is washed and singed before the removal of the head. **Singe** means to burn lightly to remove hair. Next, the carcass is opened and eviscerated, before being split or halved with a cleaver or electric saw. The leaf fat is removed. **Leaf fat** is layers of fat inside the body cavity. Before the carcass is washed, the kidneys and facing hams are inspected. After washing, the carcass is sent to coolers at 34° F.

Unlike the fat from beef cattle, lard is considered a product along with the meat. Lard is the fat from hogs. It is used for a variety of cooking and baking products. Lard is often combined with other animal fats and with vegetable oils, such as cottonseed, soybean, peanut, and coconut. Such mixes are used extensively for baking, cooking, frying, and other food preparations, and for commercial products.

Poultry

The steps in poultry processing are similar in many ways to those required for other animals. The feathers that cover a bird, like the hair that covers the hog, must be removed before evisceration.

The process begins by securing the bird on a conveyor belt and bleeding out. Next, the bird is scalded before feather removal or picking. Singeing or lightly burning the skin is required to remove the fine hairs that cover a bird under its feathers. After the feathers and hair have been removed, the bird is washed and eviscerated, and the giblets cleaned. **Giblets** are the heart, liver, and gizzard of a bird. The bird is then cooled to 40° F, usually with ice (Figure 33–21).

FIGURE 33–21 Modern processing plants make it possible to process and package the millions of broilers that are consumed fresh or frozen for later use. *(Courtesy of USDA)*

Fish

After fish are caught or harvested, they too must be prepared for processing and consumption. The procedure depends on the type of fish. Evisceration is usually done after the scales and head are removed. Washing and cooling follow.

Fish, like other foods, are consumed in a variety of forms. Whether a tuna is to be consumed whole or processed to be eaten later will determine whether the fish is left whole or is cut up. Some shellfish, such as crab and lobster, are kept alive until they are cooked. The heat of cooking kills them. The meat can then be removed and consumed or processed.

Kosher Slaughter

Kosher means right and proper. This type of animal processing is based on the religious ritual of the Jewish faith. It requires that animals be killed by a rabbi or a specially trained representative. The methods, as well as the time by which the meat of the animal must be sold, are based on ritual that relates to concerns for sanitation. Meat must be consumed quickly. Neither packers nor retailers are permitted to hold kosher meat for more than 216 hours (9 days). Washing is required every 72 hours.

Major Cuts of Meat

INTERNET KEY WORDS:

cuts of meat
dressing percentage beef

After the meat is slaughtered, it is further prepared for use. Different areas of the animal are useful for different purposes. Cuts of beef, lamb, and pork, with wholesale and retail terms, are shown in Figures 33–22, 33–23, and 33–24.

By-products—How Waste Products Are Used

Although most of the animal is consumed by humans, all of the animal is not edible. The dressing percentage is a term used to indicate the percentage or yield of hot carcass weight to the weight of the animal on foot. The offal is removed from the live animal to arrive at the dressing percentage. The formula is: hot carcass weight divided by the live weight times 100. The offal may be 40 percent of the live weight of the animal.

What happens to the offal accounts for many products that are used daily (Figure 33–25). By-products can generally be divided into 12 categories:

1. Hides—Leather from animal hides is used to make a variety of consumer products, such as shoes, harnesses, saddles, belting, clothing, sports equipment, hats, and gloves (Figure 33–26).
2. Fats—These are used to make products such as oleomargarine, soaps, animal feeds, lubricants, leather dressing, candles, and fertilizers.
3. Variety meats—The heart, liver, brains, kidneys, tongue, cheek meat, tail, feet, sweetbreads (thymus and pancreatic glands), and tripe (pickled rumen, or stomach, of cattle and sheep) are sold over the counter as variety or fancy meats.
4. Hair—Brushes for artists are made from the fine hairs on the inside of the ears of cattle. Other hair from cattle and hogs is used for toothbrushes, paintbrushes, mattresses, upholstery for residential and commercial furniture, air filters, and baseball mitts.
5. Horns and hoofs—These items are used as a carving medium and are fashioned into decorative knife and umbrella handles, goblets, combs, and buttons.

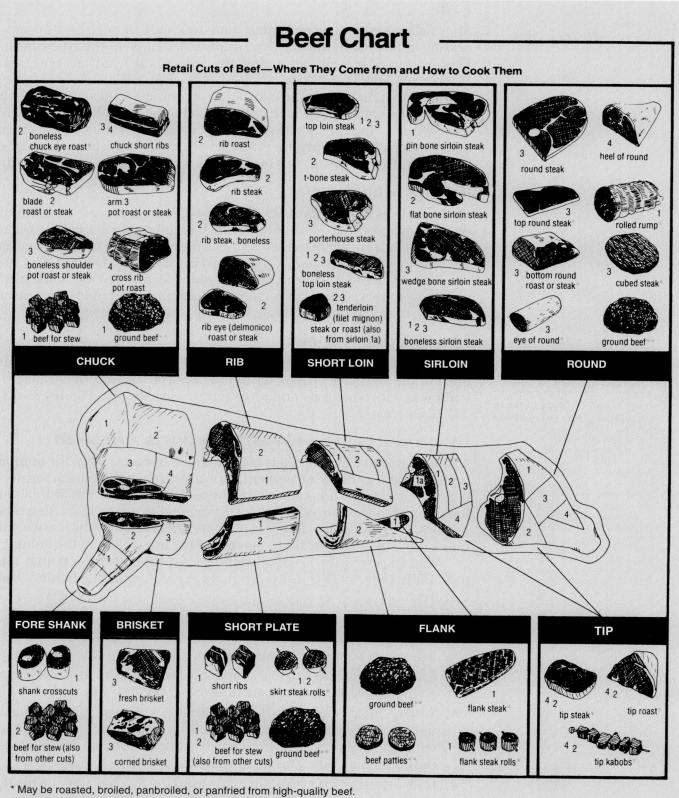

Beef Chart

Retail Cuts of Beef—Where They Come from and How to Cook Them

CHUCK
- 2 boneless chuck eye roast
- 3 4 chuck short ribs
- blade 2 roast or steak
- arm 3 pot roast or steak
- 3 boneless shoulder pot roast or steak
- 4 cross rib pot roast
- 1 beef for stew
- 1 ground beef **

RIB
- 2 rib roast
- 2 rib steak
- 2 rib steak, boneless
- 2 rib eye (delmonico) roast or steak

SHORT LOIN
- top loin steak 1 2 3
- 2 t-bone steak
- 3 porterhouse steak
- 1 2 3 boneless top loin steak
- 2 3 tenderloin (filet mignon) steak or roast (also from sirloin 1a)

SIRLOIN
- 1 pin bone sirloin steak
- 2 flat bone sirloin steak
- 3 wedge bone sirloin steak
- 1 2 3 boneless sirloin steak

ROUND
- 3 round steak
- 4 heel of round
- 3 top round steak*
- 1 rolled rump
- 3 bottom round roast or steak*
- 3 cubed steak
- 3 eye of round
- 3 ground beef **

FORE SHANK
- 1 shank crosscuts
- 2 beef for stew (also from other cuts)

BRISKET
- 3 fresh brisket
- 3 corned brisket

SHORT PLATE
- 1 short ribs
- 1 2 skirt steak rolls*
- 1 2 beef for stew (also from other cuts)
- ground beef **

FLANK
- ground beef **
- 1 flank steak*
- beef patties **
- 1 flank steak rolls*

TIP
- 4 2 tip steak*
- 4 2 tip roast*
- 4 2 tip kabobs*

* May be roasted, broiled, panbroiled, or panfried from high-quality beef.
** May be roasted, baked, broiled, panbroiled, or panfried.

FIGURE 33–22 Retail cuts of beef—where they come from and how to cook them.

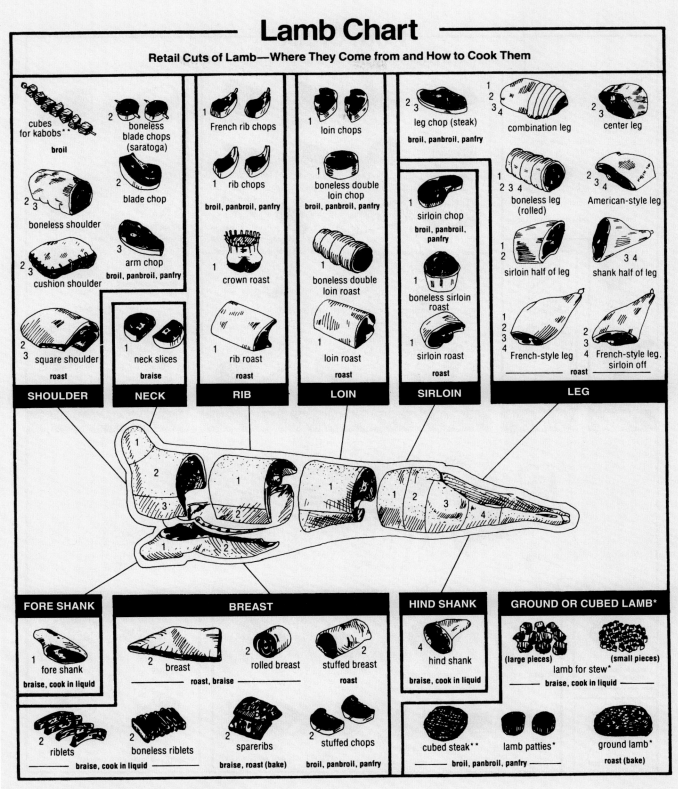

FIGURE 33–23 Retail cuts of lamb—where they come from and how to cook them.

Pork Chart

Retail Cuts of Pork—Where They Come from and How to Cook Them

BOSTON SHOULDER

cubed steak*

pork cubes

braise, cook in liquid, broil

2 blade steak — braise, panfry

2 smoked shoulder roll — roast (bake), cook in liquid

2 boneless blade boston roast

2 blade boston roast — braise, roast

1 CLEAR PLATE 4 FATBACK

4 fatback — panfry, cook in liquid

1 4 lard — pastry, cookies, quick breads, cakes, frying

LOIN

1 blade chop   2 rib chop   2 loin chop   3 sirloin chop

cubed steak*   2 3 butterfly chop   2 top loin chop   3 sirloin cutlet

braise, broil, panbroil, panfry

1 country-style ribs   1 2 back ribs   2 smoked loin chop   1 2 3 Canadian-style bacon

roast (bake), braise, cook in liquid   roast (bake), broil, panbroil, panfry

1 2 3 boneless top loin roast   1 2 3 boneless top loin roast (double)   2 3a tenderloin roast (bake), braise, panfry

roast

1 blade loin   2 center loin   3 sirloin

roast

LEG (Fresh or Smoked Ham)

1 2 3 boneless leg (fresh ham) — roast

1 2 3 sliced cooked "boiled" ham — heat or serve cold

1 2 3 boneless smoked ham   1 2 3 canned ham

roast (bake)

2 boneless smoked ham slices   center smoked ham slice

broil, panbroil, panfry

1 2 smoked ham, rump (butt) portion   3 smoked ham, shank portion

roast (bake), cook in liquid

JOWL

1 smoked jowl — cook in liquid, broil, panbroil, panfry

1 pig's feet — cook in liquid, braise

PICNIC SHOULDER

3 4 fresh arm picnic — roast

smoked arm picnic 3 4 — roast (bake), cook in liquid

3 arm roast — roast

ground pork* — roast (bake) panbroil, panfry

1 fresh hock   smoked hock   2 3 neck bones — cook in liquid   3 arm steak — braise, panfry   sausage* link roll — panfry, braise, bake

braise, cook in liquid

1 SPARE RIBS 2 BACON (Side Pork)

1 spareribs   2 slab bacon

1 salt pork — bake, broil, panbroil, panfry, cook in liquid   sliced bacon — bake, broil, panbroil, panfry

*May be made from Boston shoulder, picnic shoulder, loin, or leg.

FIGURE 33–24 Retail cuts of pork—where they come from and how to cook them.

GOOD THINGS FROM CATTLE
(Prepared by the Beef Industry Council of the Meat Board)

432 lb Retail Beef
• Steaks
• Roasts
• Ground Beef

Pharmaceuticals
• Rennet   • TSH
• Epinephrine   • ACTH
• Thrombin   • Cholesterol
• Insulin   • Estrogen
• Heparin   • Thyroid Extract

Variety Meats
• Liver   • Kidneys
• Brains   • Tripe
• Tongue   • Sweetbreads
• Ox Joints

1,000-lb STEER

Inedible By-Products
• Leather   • China
• Sports Equipment • Photographic Film
• Surgical Sutures • Sandpaper
• Soap   • Violin Strings
• Cosmetics   • "Camel Hair" Brushes
• Buttons   • Explosives

Edible By-Products
• Oleo Stock   • Canned Meat
• Oleo Oil   • Candies
• Gelatin   • Marshmallows
• Natural Sausage Casings

FIGURE 33–25 There are many useful by-products that come from animals.

FIGURE 33–26 Leather is a by-product of the meat-processing industry. It is used for many purposes, including the manufacture of shoes, furniture, upholstery, belts, coats, saddles, and numerous other items. *(Courtesy of Michael Dzaman)*

6. Blood—The blood from animals is used in making animal feeds, fertilizers, and shoe polish.

7. Meat scraps and muscle tissue—After separation from the fat, meat scraps and muscle tissue are most often made into meat-meal or tankage. **Tankage** is the dried animal residue used as fertilizer and feed.

8. Bones—Some bones are put to the same uses mentioned for horns and hoofs. In addition, bones are converted into stock feed, fertilizers, and glue.

9. Intestines and bladders—Sausage, lard, cheese, and snuff all use the intestines and bladders from cattle. Strings for musical instruments and tennis rackets are also made from these by-products.

10. Glands—The pharmaceutical industry relies heavily on animal glands for many medical drugs.

11. Collagen—**Collagen** is the chief constituent of the connective tissues. Glue and gelatin are made from collagen. These products, in various forms, are used in the furniture, photography, medical, and baking industries.

12. Contents of the stomach—Stomach contents of slaughtered animals are used primarily in the production of feed and fertilizer.

NEW FOOD PRODUCTS ON THE HORIZON

The foods that we eat and how they reach us is an exciting area of agriscience. New foods, and new versions of familiar foods, are arriving on the market daily. Research is being conducted on improving the nutritional values of foods and keeping our food costs at a minimum.

The fruit industry uses an electronic, fruit-shaped beeswax sensor to log the bumps and bruises sustained by fruit during shipment.

Hot Topics in Agriscience

LIFE EXPECTANCY—AN INDICATOR OF FOOD SAFETY

A newborn baby in the United States is entering the most dangerous period in life—infancy and early childhood. U.S. boys, on average, can expect to live to the age of 73.6 years, whereas U.S. girls can expect to live to the age of 79.4 years. These calculations are based on the records and charts of life insurance companies, which insure against death on the basis of their charts. Each year that you survive, these numbers will go up unless you take up habits (such as smoking) that are known to reduce survival rates.

The average life expectancy of an individual depends on adequate medical care, clean drinking water, sanitary facilities, and food of sufficient amount and quality. Countries in which these resources are unavailable usually experience high death rates, especially among infants and small children. Other factors such as wars also influence the average number of years a person may reasonably expect to live. Those countries with the greatest life expectancy are those that are wealthy enough to feed and care for their people.

The average life expectancy at birth has been increasing in many countries of the world for quite a few years. The following examples of life expectancy at birth illustrate the importance of providing access to plenty of good-quality food, medical care, clean water, and sanitary facilities: Ethiopia—35 to 39 years, Mozambique—46.5 years, Indonesia—64 years, United States and Western Europe—70 to 74 years, and Japan, Norway, Iceland, and Sweden—75+ years.

Improvements in handling equipment are resulting in better apples and are reducing loss for producers.

Eggs are used in the preparation of other foods, as well as being eaten by themselves. Eggs have sometimes been taken out of diets because of their high cholesterol content, which leads to heart problems. Low-carbohydrate diets have brought about a resurgence in consumption of eggs. Researchers are investigating ways of reducing the cholesterol level in the egg without negating its nutritional value. In addition, the poultry industry is looking at factors that make an animal fat and at ways to reduce the high-cholesterol component of poultry products. Perhaps we will soon be able to benefit from low-carbohydrate, low-cholesterol, high-protein eggs.

Evaluation of different processing techniques with regard to cost, digestibility, nutrient retention, and flavor is constant. For example, nutrient loss in poultry has been found to be less with freeze-drying than with dehydration.

The conveniences that everyone enjoys will continue to be improved, as there continue to be new developments in the food science field of agriscience.

STUDENT ACTIVITIES

1. Write the Terms to Know and their meanings in your notebook.
2. Keep a diary of what you eat for a week. Note which foods represent each of the five food groups.
3. With your teacher, arrange a visit to a meat department of a local grocery store to observe meat processing.
4. Compare the end result when a food has been processed in a variety of ways—for example, fresh, canned, frozen, dehydrated, and freeze-dried potatoes.
5. Make a collage of items that exist because of animal by-products.
6. Using an outline of an animal, identify the major cuts of meat and where they come from on the animal.
7. Research the steps that must be taken to prepare kosher foods from food products other than meats. Write a brief summary outlining your findings.
8. Create a game that will help you remember the number of servings a person should eat daily from each food group. Use the Food Guide Pyramid as your model.

SELF EVALUATION

A. Multiple Choice

1. Which of the following is *not* a carbohydrate?
    a. sugar
    b. starch
    c. fiber
    d. meat

2. The number of minerals needed by the body is approximately
    a. 2.
    b. 12.
    c. 20.
    d. 200.

3. The suggested number of daily portions from the meat, fish, poultry, eggs, and bean group is
    a. two 2- to 3-oz servings.
    b. four 2- to 3-oz servings.
    c. six 2- to 3-oz servings.
    d. eight 2- to 3-oz servings.

4. The butterfat content of table cream is approximately
    a. 10 to 12 percent.
    b. 18 to 20 percent.
    c. 30 to 32 percent.
    d. 38 to 40 percent.

5. To which milk product is sugar added during processing?
    a. evaporated milk
    b. condensed milk
    c. skimmed milk
    d. dried milk

6. Which mineral is added to most drinking water in the United States to assist in tooth development?
    a. zinc
    b. iron
    c. fluorine
    d. chlorine

7. Iron is important in the diet for the development of
    a. bones.
    b. vision.
    c. blood.
    d. hair.

8. Traditional aging of meat takes approximately
    a. 1 to 6 hours.
    b. 1 to 6 days.
    c. 1 to 6 weeks.
    d. 1 to 6 months.

B. Matching

_____ **1.** Canning
_____ **2.** Shrouding
_____ **3.** Dehydration
_____ **4.** Viscera
_____ **5.** Sweetbreads
_____ **6.** Food additive

a. Thymus and pancreatic glands
b. Internal organs
c. Added before packaging
d. Stored in airtight containers
e. Reduced moisture content
f. Wrap in cloth

C. Completion

1. Vitamins _____, _____, _____, and _____ are fat soluble.
2. Minerals that assist in fluid regulation are _____, _____, and _____.
3. Meat from animals that have been slaughtered as prescribed by Jewish ritual is called _____.
4. The method of precooking food, removing moisture, and then freezing is known as _____.

Communications and Management in Agriscience

Putting It All Together

Looking at the broad picture, agriscience is the industry that feeds, clothes, shelters, and keeps the nation healthy. These benefits spill over to many other nations of the world. The United States shares the results and products of research in agriscience and donates or sells things such as food, fiber, crops, livestock, plant oils, lumber, medicines, breeding stock, germ plasm, seed, agricultural chemicals, machines, and an endless array of agriscience materials for producing food and fiber, as

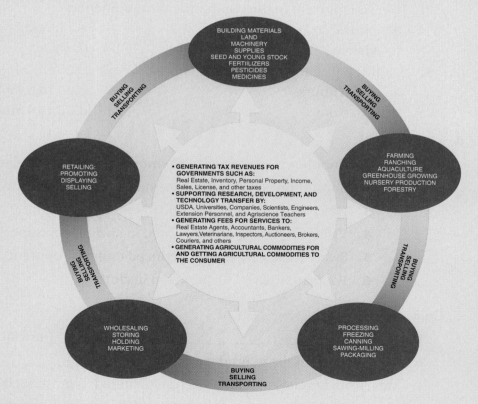

BUILDING MATERIALS
LAND
MACHINERY
SUPPLIES
SEED AND YOUNG STOCK
FERTIILIZERS
PESTICIDES
MEDICINES

BUYING
SELLING
TRANSPORTING

BUYING
SELLING
TRANSPORTING

RETAILING:
PROMOTING
DISPLAYING
SELLING

FARMING
RANCHING
AQUACULTURE
GREENHOUSE GROWING
NURSERY PRODUCTION
FORESTRY

- **GENERATING TAX REVENUES FOR GOVERNMENTS SUCH AS:**
  Real Estate, Inventory, Personal Property, Income, Sales, License, and other taxes
- **SUPPORTING RESEARCH, DEVELOPMENT, AND TECHNOLOGY TRANSFER BY:**
  USDA, Universities, Companies, Scientists, Engineers, Extension Personnel, and Agriscience Teachers
- **GENERATING FEES FOR SERVICES TO:**
  Real Estate Agents, Accountants, Bankers, Lawyers,Veterinarians, Inspectors, Auctioneers, Brokers, Couriers, and others
- **GENERATING AGRICULTURAL COMMODITIES FOR AND GETTING AGRICULTURAL COMMODITIES TO THE CONSUMER**

BUYING
SELLING
TRANSPORTING

BUYING
SELLING
TRANSPORTING

WHOLESALING
STORING
HOLDING
MARKETING

PROCESSING
FREEZING
CANNING
SAWING-MILLING
PACKAGING

BUYING
SELLING
TRANSPORTING

Agriscience has many facets, and communications and management transcends the entire industry.

well as the end products themselves. Our land-grant system of colleges and universities has provided a system of research in agriscience that is the envy of the world. In many countries of the world, the greatest prize in education is to be admitted to and graduate from a land-grant college or university in the United States.

The agriscience industry in the United States provides the nation with its playgrounds, outdoor fitness centers, and a source of renewal for a nation boxed in by modern skyscrapers, large industrial facilities, and clogged highways. These take the form of open spaces for softball, baseball, soccer, hockey, horseback riding, golf, hiking, hunting, biking, walking, fishing, bird watching, recreational farming, and gardening. A significant number of people make their livings as farmers, ranchers, greenhouse operators, aquaculturists, landscapers, nursery operators, zoo workers, and others in the outdoor segment of the food, fiber, and ornamental plant and exotic animal segments of the industry.

Agriculture employs about 20 percent of this nation's workers and provides the income for these workers to enjoy the fruits of our free-enterprise system. In contrast to many nations, where most of the income is consumed by the purchase of food, we spend only 11 percent of our income on food and have the remaining 89 percent for clothing, housing, transportation, recreation, health, automobiles, appliances, communications, education, leisure pursuits, and gadgets. No other nation has the abundance of opportunities that the United States offers under an agricultural industry that produces such affordable food and frees so much income for other needs and preferences.

The driving force for this abundance is planning and management. In no other sector of the economy is the profit incentive of the free-enterprise system in greater evidence. In a system in which one can receive and use or accumulate the proceeds from work or innovation, the human spirit and initiative accelerates; the results are demonstrated in greater productivity. The old adage that "where there's a will, there's a way" is proven correct every day as people apply their education, inventiveness, innovation, and energy in a workplace where a substantial amount of the proceeds come back to the individual. Whether one is the owner or a stockholder in line for the profits of the organization; a salesperson or manager in line for bonuses or commissions; a pieceworker or assembly-line worker whose income is tied directly to output; or a per-hour worker with the option to seek a better paying position, the free-enterprise system rewards productivity.

This section addresses the application of communication and management skills in agriscience. These skills transcend agriscience and culminate in the marketplace where the fruits of our labor are sold. The marketplace yields the cash to buy labor and other inputs for the business and yields profits for the inventors, entrepreneurs, and other risk-takers in the system.

34 | Marketing in Agriscience

MATERIALS LIST

✓ various newspaper and marketing reports

✓ Internet access

COMPETENCIES TO BE DEVELOPED

After studying this unit, you should be able to:

- describe the marketing strategies that maximize profits.
- describe various pricing strategies.
- distinguish between wholesale and retail marketing.
- describe some methods of marketing at farms, roadside stands, and farmers' markets.
- discuss advantages and disadvantages of terminal markets, auctions, and direct marketing.
- recognize fees, commissions, and other costs of marketing.
- list the procedures for handling livestock to minimize losses during marketing.
- understand the grades of some popular agriscience commodities.
- recognize marketing trends and cycles.
- describe the use of futures in agriscience marketing.

SUGGESTED CLASS ACTIVITIES

1. Visit a livestock auction or a fruit and vegetable wholesale market. Observe how the price is established for the product that is offered for sale. Make arrangements ahead of time for the manager of the market to speak to the class about the services that are offered at the place of business. Ask about such business issues as insurance, bonding, financing, storage or yarding practices, and transportation. Have each class member prepare a written report that covers the basic points that were learned. Limit the report to one to two pages in length.

2. Obtain a videotape that is used in a video auction. Obtain permission to listen in to the phone call when a sale is actually made. If the time of the sale is not convenient for the class, request permission to record the audio portion of the sale, and then play it as you observe the videotape. Discuss the advantages and disadvantages of using a video auction to market livestock.

SUGGESTED CLASS ACTIVITIES

3. Divide the class into small groups. Have each group design a new marketing campaign for a locally produced agricultural product. Make a commercial. Prepare a videotape or present it live to the class.

TERMS TO KNOW

supply	consumer	vertical integration	wether
demand	middlemen	pencil shrink	commodity exchange
consumer demographics	wholesale marketing	feeder calf	futures market
product advertising	terminal market	yearling	futures
institutional advertising	yardage fee	slaughter cattle	opening a position
psychological pricing	commission	quality grade	offsetting a position
penetration pricing	auction markets	finish	out-of-the-market
skimming	auctioneer	yield grade	profit
loss-leader pricing	video merchandising	pig	
prestige pricing	direct sales	hog	
retail marketing	cooperative	barrow	

In our free-enterprise system, goods and services may be produced, modified, or improved. The expectation is to sell or market the commodity or service as value is added along the way to the final user or consumer. Many people receive financial benefits in the form of salaries, profits, or "in kind" benefits. Governments receive tax revenues from the various transactions. This unit explores the basics of marketing agriscience commodities to maximize profits.

There are a number of factors that need to be considered when deciding on how to market agriscience products. Some of these factors include the following:

- demand for the product to be produced;
- supply of the product already available;
- types and availability of markets in the area;
- competition from similar products;
- buying power of intended consumers; and
- seasonal variations in demand.

Supply and demand have long been the factors that determine whether the production of a product can be profitable. **Supply** is the amount of a product available at a specific time and price. Some of the things that may determine supply are how many people there are in the business of producing that product in the market area, how much of the product is coming into the market area from other areas, and the history of profitability for that product in your area.

Demand for a product is also determined by numerous factors. **Demand** is the amount of a product wanted at a specific time and price. It is often determined largely by price. The less expensive a product is, the more of it that is wanted. However, there are other factors that influence demand for a product. The amount of money available to

FIGURE 34–1 In a free-market system, the relationship between supply and demand will determine price.

FIGURE 34–2 Knowing who the customers are, as well as knowing their needs and preferences, is essential to developing an effective marketing plan.

consumers to buy that product is a factor that is often overlooked. Competition from similar products may reduce demand. Seasonal variations in demand also need to be considered. It is easier to sell ice cream in the summer than in the winter. Advertising, personal contacts, and product samples to create demand are also factors to be taken into consideration (Figure 34–1).

FOCUS ON CONSUMERS

Consumers ultimately determine marketing success. Commodities or services may be on the doorsteps of consumers, but if the consumer chooses not to buy and does not submit to persuasion or coercion, then there is no sale. In our democratic, capitalistic system, we have the right to produce and market goods and services for personal profit within the state and federal legal frameworks. Similarly, the right of refusal to buy goods and services is a basic right of consumers. The marketing system involves the interplay between producers and consumers, and all of the people, processes, and jobs that come into play between them.

Consumer Demographics

One of the first marketing functions is to decide who will be the target population. The target population consists of the people who will be given deliberate exposure to the product. Some factors to consider when determining target populations are: Who are the consumers? How many consumers are there? What are their preferences? Where are they located? Categories of information about consumers or potential consumers are known as **consumer demographics**. A knowledge of consumer demographics is helpful in determining target populations and marketing strategies (Figure 34–2).

Some useful categories of consumer demographics for the marketing of agriscience products are population density, ethnic makeup of the population, family income, discretionary money held by individuals within the family, family size, eating preferences, who makes buying decisions in the household, occupation and work locations, tendencies to eat at home or away, clothing preferences, styles, and recreational preferences. People generally satisfy their needs for basic food, clothing, and shelter first, then move on to purchases that reflect group and individual tastes. Because the purchase of food in the United States requires less than 12 percent of the average family income, there is considerable room for choice in the purchase of additional goods and services.

ADVERTISING AND PROMOTION

Advertising is any form of nonpersonal presentation of a product or service. For maximum effectiveness, advertising must be coordinated with other marketing techniques, such as public relations, promotional programs, pricing, personal selling, product description and characteristics, sales promotions, and warranties. Effective advertising performs a variety of functions. These include creating an awareness of the product, motivating a customer to seek out the product, educating the potential customer, and reinforcing the value of purchases already made.

Product advertising focuses on the product itself. This might include the product's usefulness, durability, price, value, and the customer's need

INTERNET KEY WORDS:

beef checkoff
pork checkoff
sheep checkoff
milk checkoff

INTERNET ADDRESS

<http://www.gotmilk.com>

for the item (Figure 34–3). In contrast, institutional advertising is designed to create a favorable image of the firm or institution offering the products or services. Such advertising is likely to claim financial stability, longevity, reliability, and integrity of the company or institution and its employees.

Advertising efforts that seek to find new users for a given product, such as motor boats, are trying to create primary demand. Similarly, advertising intended to attract buyers to a specific brand of the product is attempting to create selective demand. The promotion of any one agricultural commodity from farms, ranches, ponds, nurseries, or greenhouses is expensive and difficult because there are so many small-volume producers. This is in contrast to other industries, such as automobiles, where there are relatively few manufacturers, and nationwide advertising is practical and affordable.

To address the problem of many small-volume producers of individual agricultural commodities, the United States government has instituted a system that generates a pool of money from producers of a given commodity such as wheat, rice, cotton, tobacco, peanuts, beef, pork, or milk. Here, the seller contributes a small amount for each unit sold to a money pool to be used to purchase advertising for the specific commodity. The program is referred to as the *check-off program* for the given commodity. It is estimated that about 90 percent of all U.S. farmers contribute to some 300 federal and state generic promotion programs covering approximately 80 commodities. The spending for research and promotion by these programs is reportedly near the half-billion-dollar mark annually. Media advertising for agricultural goods and services typically includes the use of newspapers, radio, television, outdoor billboards, signs, pamphlets, transportation advertising, direct mail, telephone book classified section, breed journals, product catalogs, and magazines.

The promotion of commodities may be done in numerous ways. Product displays are common in stores and other retail outlets. Similarly, displays and sales of machinery, animals, food, flowers, ornaments, landscape designs, and other commodities and services are common at fairs, shows, open houses, trade shows, and professional meetings. Free

FIGURE 34–3 Product advertising focuses on some aspect of the product itself.

Science Profile

HOW MUCH WILL YOU PAY FOR GOOD HEALTH?

People are willing to pay more for health-promoting genetically modified (GM) food, according to a University of Purdue study. Jayson Lusk, an associate professor of agricultural economics at the University of Purdue, sent out a mail survey that asked consumers how much they would pay for golden rice. Golden rice is a GM food that has a gene from a daffodil in it. This gene triggers rice to produce a chemical that the human body can change into vitamin A. In his questionnaire, Lusk outlined the benefit of golden rice from a consumer's standpoint. Earlier studies indicated that consumers would pay more for foods that were not GM foods. In these studies, the benefits to farmers were outlined, but the potential benefits for consumers were not mentioned. The results of Lusk's study show that consumers will pay a greater price for golden rice if they understand the benefit. A valuable lesson can be learned from this study. As products like these become available to consumers, a shift in marketing must follow. Consumers need to be informed about the personal benefits they will get from the GM food. If they are not informed of the benefit, they will have no incentive to buy the food. Without proper marketing, GM foods may not sell well. Conversely, if advertising is done correctly, foods such as golden rice can be a big success.

INTERNET ADDRESSES:

<http://www.ams.usda.gov/marketnews.htm>

<http://ianrwww.unl.edu/markets>

FIGURE 34–4 Modern marketing cooperatives have given small producers much more power to influence the prices they receive for their commodities. (*Courtesy of Michael Dzaman*)

samples may even be offered in stores, at sidewalk promotionals, fairs, shows, public auctions, and wherever prospective customers may gather.

COMMODITY PRICING

Final determination of prices for agricultural commodities is not always under the control of the producer or representative of the producer. This is due to a large extent to the perishable nature of nonprocessed foods; the relationship between peak quality and price of fruits and vegetables; and the lack of control over production of crops, eggs, milk, and meat after the production cycle is started. Therefore, the producer may not have much control over the supply that is being collectively generated and may have to accept lower prices or loss of the product because of spoilage.

However, the growth of producer-owned marketing cooperatives has enabled producers to negotiate prices on a large-volume basis. Another benefit of cooperatives is that of owning processing plants so fresh commodities can be processed and withheld, if necessary, until better prices are offered. Cooperatives, privately owned businesses, and individuals all use a variety of strategies when setting prices. These strategies are applied according to the benefits the price setter sees in effecting more sales and greater profits (Figure 34–4).

Psychological pricing is a strategy designed to make the price seem lower or less significant than it is. For instance, the pricing of an item at $1.99 seems to attract many buyers that would reject the same item priced at $2.00. Needless to say, the profit from selling more items will easily offset the loss of one cent per item. Another technique is the offering of multiple items for a given price, such as 3 for 23 cents. Here the customer is lured into buying three to save a small amount, when one item may be all the customer needs. Discount pricing and bulk selling are also used as techniques of psychological pricing. Sometimes customers are lured into buying more than they would normally buy because of the discount attraction, and then consume more than they would otherwise consume. Some consumers believe that larger packages are

FIGURE 34–5 Modern computers and scanning devices enable retailers to change prices quickly in accordance with promotional activities or changing market conditions. *(Courtesy of USDA/ARS #K-2656-2)*

automatically priced at less cost per unit. Occasionally, sellers exploit this perception by pricing commodities in larger containers at greater costs per unit than in the standard package (Figure 34–5).

Penetration pricing is a strategy wherein price is set below that of competitors to entice customers to try an item. Although profits will be low temporarily, they can be regained if new customers continue to buy and the price is gradually increased to new levels. Skimming is the opposite of the penetration pricing strategy. With this method, the price of a new product is set for unusually high profits at first, when affluent and willing customers are available. This enables the producer to recoup development expenses more quickly and gradually reduce the price as sales volume increases or other competing products enter the market. Skimming is also known as "sliding down the demand curve."

Loss-leader pricing is a procedure wherein a popular commodity is offered for sale in a special sales promotion at prices less than the cost level—in other words, sold at a loss. The purpose is to attract customers into the business establishment with the hopes that they will buy other items priced at profitable levels. Prestige pricing is a procedure used to target buyers with special desires for quality, fashion, or image. The customer's notion or perception that a certain brand is better may be further reinforced by a higher price tag. Prestige pricing seems to work well in the garment and cosmetic industries, and it may be quite profitable.

In general, in a nonregulated, free-enterprise system, there is a tendency to price commodities according to what the market will bear, rather than what it costs to produce the items. This pricing strategy is generally regarded as fair, because our system rewards risk-taking with the potential for profit. If one must be prepared to shoulder losses when offering a commodity that does not reach sales expectations, then it only seems reasonable to permit that individual or company to benefit from the occasional high-profit opportunities.

MARKETING STRATEGIES FOR MAXIMIZING PROFITS

To make the most money from a production enterprise, successful marketing is a must. Strategies that can be used to market agricultural products most profitably include:

• Determine what types of markets are available to you.

• Determine the costs of various types of marketing.

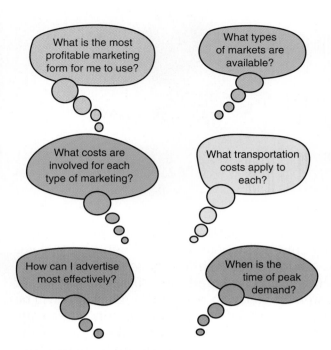

FIGURE 34–6 Good marketing strategies are developed by asking key questions about potential markets and the cost of accessing them.

FIGURE 34–7 Retail marketing by the producer adds extra dimensions to the business. It allows the producer to capture additional shares of the profits from the commodity.

- Determine transportation costs to market at each of the markets available to you, and sell where transportation costs are favorable.
- Determine the most profitable form in which to market your product (age, size, weight, degree of preparation).
- Advertise to create markets where none existed before.
- In seasonal markets, market your product at the peak of demand (Figure 34–6).

RETAIL MARKETING

Retail marketing is selling a product directly to **consumers** (people who use a product). This may take place on the farm, at roadside markets, or at the farmers' markets that have developed around many centers of population (Figure 34–7).

Retail marketing is generally used for ready-to-eat products. For example, ham, sausage, and fresh fruits and vegetables, rather than live pigs, are usually sold at a retail outlet. This creates many other problems that need to be dealt with and that have an effect on the profitability of a production enterprise. When an animal product is processed for retail sales, various standards and state and federal regulations must be met. Also, changes in facilities are often necessary. Sanitary storage of processed products and the increased labor costs associated with retail customers must also be considered.

Retailing at the Farm

On-the-farm retail sales present special problems. If retail marketing is to take place on the farm where the product is produced, the farm must

Agri-Profile

CAREER AREAS:

Auctioneer/Coop Manager/Dealer/Grower/ Packer/Grader/Meat Cutter/Produce Manager/ Communications Specialist

The marketing of agricultural products occurs in all segments of our society on a continuing basis. The year-round displays of produce, dairy, meat, seafood, bread, frozen foods, canned goods, deli foods, and ready-to-eat foods in supermarkets is a testimony to the strength of the agriscience industry. The producers and breeders of vegetable and fruit crops, grains, flowers, ornamental shrubs, turfgrass, trees, pets, laboratory animals, livestock, horses, dairy animals, poultry, fish, wildlife, milk, eggs, wool, and fur, all market the products of their businesses. This is only the beginning in the process of agricultural marketing. From the farm, ranch, pond, fish tank, feedlot, orchard, nursery, greenhouse, turf farm, or other production unit, there exists a whole system of professional and scientific workers.

Marketing cooperatives, dealers, auction markets, auctioneers, livestock handlers, jobbers, brokers, clerks, truckers, railroad personnel, commodity market managers, futures traders, writers, and broadcasters all get in the marketing act. Government, as well as company inspectors, monitor the products as they make the numerous transitions from production site to finishing site, pet shop, by-product, or dinner table. Teachers and extension workers provide instruction and professional development programs for personnel.

Marketing careers may be launched through programs in technical schools, colleges, or universities in the plant and animal sciences, food science, agricultural economics, marketing, or communications. Studies in the biological sciences are appropriate for laboratory work associated with agricultural products and marketing.

New products that are developed by agriscience researchers are rapidly solving environmental problems. Chemist Sevim Erhan evaluates inks from soybeans that are biodegradable. They are quickly replacing inks of the past. *(Courtesy of USDA/ARS #K-3998-5)*

---

be well maintained and attractive to the customers buying the product. Facilities for parking are necessary. Animal and sales facilities must be clean, and the animals present must be well cared for and contented. An educational program to make consumers aware of good management techniques is often necessary so that they understand what is happening on the farm. Salespersons or the farm family themselves must be available to assist customers. Privacy is often difficult to achieve, because retail customers often feel that when products are sold at the farm, it is acceptable to drop by at any time.

On the positive side, profits from farm sales are often greater because retail prices are higher. Also, middlemen are eliminated in the marketing procedure. **Middlemen** are people who handle an agricultural product between the farm and the consumer. Examples of middlemen include buyers, processors, and salespeople. Greater prices can often be charged when a superior product is produced. For farm families who like to meet new people, farm retail sales create excellent opportunities to do so. These types of sales provide opportunities for urban and suburban people to see the agricultural way of life firsthand.

Roadside Markets

Roadside marketing retains nearly all of the advantages of on-the-farm retail sales, while eliminating some of the disadvantages. The sales unit is usually somewhat removed from the actual farm operation when roadside marketing is used. By removing the customers from the production area, more efficient use of labor can take place. Less perfect care and maintenance of facilities can be tolerated.

On the less positive side, separate facilities for the retail sales unit must be maintained. More labor may be required because it may be inconvenient for the farmer to staff the roadside stand. Furthermore, the positive effects of the consumer seeing the actual farm in operation are diminished.

Farmers' Markets

Farmers' markets have appeared in many large metropolitan areas to cater to the demands of urban consumers. They are normally operated 1 or 2 days per week and give urban and suburban consumers access to fresh agricultural products directly from the producers. Farmers' markets give the producer access to markets that would seldom be available otherwise. The producer also has the opportunity to educate the consumer concerning the value of good products (Figure 34–8).

There are several costs and inconveniences associated with marketing agricultural products at farmers' markets. There are often fees to be paid for the privilege of participating in farmers' markets. Competition may be greater, especially if several producers are selling the same type of product. Vehicles with heating, cooling, or both may be necessary to get products to the market as fresh as possible. Products may be subject to certain food regulations and packaging requirements to meet state and federal standards. Facilities to hold and display products must also be available at the market.

In summary, the decision whether to market agricultural products by retail means requires much consideration. Although the returns are generally greater when products are marketed retail, the costs are also higher. Consideration must be given to the product being produced, availability of markets, labor availability, and personal desires before a decision concerning marketing method can be made.

WHOLESALE MARKETING

Wholesale marketing is the marketing of a product through a middleman, who eventually forwards the product to the consumer. Most agricultural products in the United States are marketed in this way. Wholesale markets allow for marketing large volumes of products with comparatively little labor. Types of wholesale markets include terminal markets, auction markets, and direct sales.

Terminal Markets

A **terminal market** is usually a stockyard that acts as a place to hold animals until they are sold to another party. The terminal market never actually owns the animals. The animals that are delivered to the terminal market are consigned to a selling agent, who does the actual selling

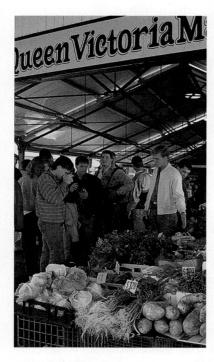

FIGURE 34–8 Farmers' markets provide opportunities for producers to sell at retail without opening their production facilities to the general public. *(Courtesy of FFA)*

Hot Topics in Agriscience

VIDEO MERCHANDISING

Video merchandising is a marketing tool that allows buyers at different locations to bid in a competitive auction for a product that they have already viewed or that they are simultaneously viewing on video. The buyers base their bids on their own judgment from viewing the video and on the description of the product by an expert judge of quality. Video merchandising has become especially popular as a way of selling livestock. Many feedlot operators in the Midwest Corn Belt now make use of video merchandising to purchase feeder calves and cattle from the western range states and other regions of the country.

An element of trust must be preserved among the buyer, the seller, and the on-site evaluator that the quality of the product is represented accurately. Many of the buyers will not actually see the product until it has been delivered to them by a trucker several days later. Some buyers may never see the product at all because they are acting as brokers by purchasing the product for one of their customers. They must be able to accurately represent the description and quality of the product to their clients. Video merchandising has found a market "niche" in the agricultural industry.

of the animals. The terminal market charges the seller a fee for caring for the animals until they are sold. This is called a **yardage fee**. The selling agent also receives a fee, called a **commission**, for the service of selling the animals.

The use of terminal markets to market animals has always been mostly confined to the midwestern and western states. In recent years, the use of terminal markets to market animals has greatly decreased, and most livestock are currently being marketed by other methods.

Auction Markets

At **auction markets**, animals are sold by public bidding on individual lots of animals. An **auctioneer** generally conducts the sales at an auction market. These markets are widespread and are convenient to most local communities. They have grown in popularity and now represent the most common means of marketing animals of importance to agriculture. Auction markets are usually most practical for the small livestock producer.

Video merchandising is a recent marketing innovation. **Video merchandising** is usually done by recording the livestock or other commodity on a videotape. The tapes are distributed to potential buyers together with a time when interested buyers can join a conference call to bid on the consignment of merchandise. This type of marketing strategy has become popular in recent years as a good way to sell livestock and other agricultural commodities. It has the advantage of engaging potential buyers from a much larger geographic area than traditional livestock auctions are capable of doing.

As is true when selling in terminal markets, commissions are charged for selling animals at auction markets. The amount of commission charged varies with the type and size of the animal. Because many auction markets are small, there may not be the competition that is present at terminal markets to buy smaller numbers of animals (Figure 34–9).

INTERNET ADDRESS:

<http://www.cattlepages.com/services/videoauction.asp>

FIGURE 34–9 Auction markets are popular for selling livestock in many communities. *(Courtesy of Bill Angell)*

Direct Sales

Direct sales refers to the producer selling crops and animals directly to processors. This method of marketing has several distinct advantages. There are no commission fees to be paid to selling agents, and there are no yardage fees. Transportation costs are kept to a minimum because the buyer generally comes to the farm to make purchases. The animals look their best for the buyer because they have not been exposed to the stresses of hauling and contact with strange facilities and animals.

There are a few negative aspects to direct marketing of commodities. A producer must have fairly large numbers of animals to be marketed at one time before the purchaser comes to the farm or ranch to buy. The producer is generally at the mercy of the buyer concerning the price received for the animals sold, and the producer often is not paid for the animals until after they have been processed.

Direct sales of animals, especially beef cattle, have increased significantly in recent years.

Other Wholesale Marketing Techniques

Approximately 75 percent of the milk marketed in the United States is sold through farmer-owned, milk-marketing cooperatives. Cooperatives are groups of producers who join together to market a commodity. The cooperatives then either process the milk and sell it directly to consumers or sell it to other large processing plants. Marketing cooperatives have the ability to maintain product quality, arrange for transportation of products from farm to market, balance supply and demand of agricultural products, and plan advertising to increase sales.

In the poultry market, almost 99 percent of the chickens produced for meat are grown under a system called vertical integration (several steps in the production, marketing, and processing of animals are joined together). The use of vertical integration in the production and marketing of agricultural products allows for extremely large systems of production that can be efficient. Only the anticipated number of animals that will be demanded by consumers are produced. There is less competition from other producers, and all phases of production can be controlled (Figure 34–10).

FIGURE 34–10 Vertical integration in the broiler industry takes the dressed bird directly to the retailer and eliminates the need for live-bird markets. *(Courtesy of USDA)*

Science Connection

RAINBOW COTTONS, HIGH-TECH LEATHER, AND BIODEGRADABLE DETERGENTS

Agriscience research is turning out new products. For example, new colors of no-wrinkle fabrics are made with cotton, leather is tanned by less-polluting methods, and a new detergent made from animal fats is biodegradable. Cotton has always been a preferred fabric, but its tendency to wrinkle has limited its use. In the past, cotton fabric was dyed before a no-wrinkle finish was applied, because the chemical bond created before the refinishing process would repel dye if applied after the process. The cotton fabric must swell to accommodate the molecules of dye, but this cannot normally occur after the fabric has been heated and treated with a no-wrinkle finish. The fabric industry needs to be able to dye fabric just before clothing is manufactured to keep up with the rapid changes in clothing fashions. After all, who wants to get stuck with a warehouse full of fabric dyed in the hues and colors that are out of fashion?

Frank Scholnick and James Chen of the U. S. Department of Agriculture examine the results of experimental treatments for leather. *(Courtesy of USDA/ARS #K-4388-1)*

To solve the problem, scientists at the Agricultural Research Service (ARS) Textile Finishing Chemistry Research Unit in New Orleans have developed techniques that allow industry to apply a no-wrinkle finish to cotton fabric before dying. By adding a variety of quaternary ammonium salts to the no-wrinkle treatment solution and a positive charge to the fabric, they have developed a procedure that produces a no-wrinkle fabric that will accept dye. The procedure also broadens the choice of dyes that can be used and gives industry more options for supplies and a wider range of shades with deeper, more vibrant colors.

On another front, scientists at the ARS Eastern Regional Research Center in Philadelphia are looking for more environmentally friendly ways to decrease bacterial decay in hides, and thus preserve leather. One of the most promising techniques is the use of electron beam irradiation. The technique promises to replace the salt-and-brine method of curing and extend the qualities of strength and elasticity for some new leather products.

On yet another front, scientists have long sought laundry detergents that will not pollute groundwater when they are discharged with waste water. Tallow is beef fat and a by-product of the beef industry. Agriscientists have developed a tallow-laced soap that is environmentally friendly. The soap contains no phosphates; will not harm humans, domestic animals, or wildlife; and will usually biodegrade in 24 hours. The product is as effective as phosphate detergents and is economical, too.

The oldest of all marketing strategies is bartering, or trading, agricultural commodities for other products or services. This system is used widely in developing countries of the world, and it is still used to a limited extent in rural communities across the United States. Another obvious method for marketing agricultural produce and livestock is the cash sale. It is still a common practice for people to buy beef or a hog for the freezer or to buy vegetables or fruit directly from the farm. Although this method would not be feasible for large farming operations, it does account for a significant number of sales on some farms.

There are a number of methods of marketing agricultural commodities. The methods chosen by an individual producer are often a matter of what is available and what the producer prefers. Care should be taken to carefully choose the means of marketing. Intelligent marketing practices may well make the difference between profit and loss in a competitive business.

MARKETING FEES AND COMMISSIONS

Because there are numerous methods of marketing animals, the fees and commissions vary fairly widely. Livestock are usually sold by the head, and a fee is charged for each animal sold. This fee varies from $1.50 to $3.00 per animal. At many purebred livestock sales, up to 10 percent of the selling price goes for fees and commissions. If animals are kept at a terminal market, yardage fees are also charged for the feed and care of the animals. Charges for insurance may also be deducted from the seller's check.

For animals that are purchased on the farm for slaughter, a fee called pencil shrink is sometimes deducted from the selling price of the animals. Pencil shrink is the estimated amount of weight that an animal will lose when it is transported to market. This usually varies from 1 to 5 percent.

Fees may also be charged for the sale of commodities based on a percentage of the gross amount received for the product.

MARKETING PROCEDURES

How produce is packaged and how animals are prepared for and transported to markets can often mean the difference between profit and loss. Some procedures to follow when shipping animals are:

- Always handle live animals quietly and carefully. Animals that are calm and quiet lose less weight during transportation.

- Move animals when temperatures are moderate, if possible. Usually this means at night during the summer.

- Do not overcrowd animals on trailers or in lots while they are waiting to go to market.

- Do not overfeed animals just before hauling them to market. Animals do not travel well on full stomachs. Do provide ample fresh, clean water.

- Avoid injuring or bruising animals when loading and unloading them. Dead or injured animals are worth little at the market.

- Sort animals according to sex and size before shipping them to market.

- Precondition animals for several days before marketing them.

GRADES AND MARKET CLASSES OF ANIMALS

Cattle

Beef animals are usually classified as either calves or cattle. Calves are younger than 1 year, whereas cattle are older than 1 year. Calves are further divided into veal calves, feeder calves, and slaughter calves. Veal calves are younger than 3 months and are slaughtered for meat that is also called *veal*. They usually weigh less than 200 lb.

Calves that are between 3 months and 1 year old and are marketed for meat are called *slaughter calves.* They have usually been fed at least some grain. Feeder calves are 6 months to 1 year old and are sold to people who feed them to market weight as slaughter cattle. The sex classes for feeder and slaughter calves are steers, heifers, and bulls. Steers are castrated male calves, heifers are young female calves, and bulls are unaltered male calves.

Cattle are divided into feeder cattle and slaughter cattle. Feeder cattle are further categorized into age groups of yearlings and 2-year-olds and older. Yearlings are between 1 and 2 years old, and 2-year-olds are 2 or more years old. These two classes of cattle can also be divided into five sex classes—steer, heifer, bull, cow, and stag. A cow is a female animal that has had a calf, and a stag is a male animal that was castrated after reaching sexual maturity.

Slaughter cattle are marketed for the purpose of being slaughtered for meat. They are divided into the same age and sex classes as are feeder cattle. They are also divided into quality and yield grades. Quality grades refer to the amount and distribution of finish (fat) on the animal. The quality grades for cattle are prime, choice, select, standard, commercial, utility, canner, and cutter (Figure 34–11). Yield grades are based on the amount of lean meat an animal will yield in relation to fat and bone (Figure 34–12). The yield grades are 1 through 5, with yield grade 1 producing the largest amount of lean meat.

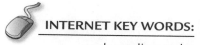

INTERNET KEY WORDS:

cattle quality grades
cattle yield grades

Swine

The use classes of swine are feeder pigs and slaughter hogs. In general, pigs are swine younger than 4 months old, whereas hogs are swine older than 4 months. Feeder pigs are sold to be fed to higher weights before being slaughtered. Slaughter hogs are sold for immediate slaughter.

The sex classes of swine are gilt, barrow, boar, sow, and stag. Gilts are young female swine. Barrows are castrated male swine. Boars are unaltered male swine, stags are mature male swine that have been castrated, and sows are mature female swine.

Swine are also graded according to quality, with the official USDA grades being U.S. No. 1 through U.S. No. 4. Animals with lean meat of an unacceptable quality are graded U.S. Utility. The highest grade of swine is U.S. No. 1 (Figures 34–13 and 34–14).

Sheep

Sheep that are marketed are classified according to age, use, sex, and weight. The age classes are lambs, yearlings, and sheep. Lambs are young sheep and can be divided into hothouse lambs, spring lambs, and lambs. Hothouse lambs are marketed under 3 months old, usually for the Christmas or Easter holidays. Spring lambs are 3 to 7 months old, and lambs are 7 to 12 months old. Lambs can also be classified as feeder lambs or slaughter lambs, depending on whether they are to be fed to heavier weights or slaughtered immediately.

FIGURE 34–11 Quality grades of slaughter cattle are based on the amount and distribution of finish.

FIGURE 34–12 Yield grades of slaughter cattle are based on the yield of lean meat in proportion with the amount of fat and bone, with grade 1 being the highest and grade 5 being the lowest.

Yearlings are between 1 and 2 years of age, and sheep are older than 2 years. Sheep are divided into three sex classes: ram, ewe, and wether. Rams are unaltered male sheep, ewes are female sheep, and **wethers** are castrated male sheep.

Sheep are also graded according to yield and quality. The yield grades are 1 through 5, with 1 being the most desirable. Quality grades are prime, choice, good, utility, and cull (Figure 34–15).

A review of useful terms regarding cattle, swine, and sheep is provided in Figure 34–16.

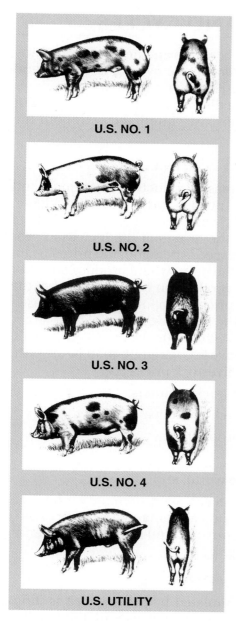

FIGURE 34–13 Quality grades of feeder pigs.

FIGURE 34–14 Quality grades of slaughter hogs.

FIGURE 34–15 Quality grades of slaughter lambs.

FIGURE 34–16 Terms given to meat animals of various sexes and maturities.

	Cattle	Swine/Hogs	Sheep
Young, immature	Calf	Pig	Lamb
Young, but maturing	Feeder or yearling	Feeder	Yearling
Castrated male	Steer or stag	Barrow	Wether
Mature breeding male	Bull	Boar	Ram
Mature breeding female	Cow	Gilt or sow	Ewe
Meat of the young	Veal	Pig or pork	Lamb
Meat of the mature	Beef	Pork	Mutton

MARKETING TRENDS AND CYCLES

Agricultural commodity prices tend to increase and decrease on a fairly well-defined cycle. The cause of general price cycling is supply and demand or government intervention. However, cycles are sometimes interrupted, accelerated, or delayed by disasters or political events. For example, the discovery of "mad cow disease" interfered with both the beef and milk price cycles.

Trends and Cycles in Animal Markets

INTERNET KEY WORDS:

beef cycle

When prices are high, producers increase production of animals and animal products. At first, this action pushes the prices even higher, because female animals that normally would have been marketed are retained in the breeding herd. When the results of the increased production reach the market and supply exceeds demand, the prices begin to decrease. When prices decrease to the point where production is not profitable, livestock producers reduce production by selling some of the breeding herd. Of course, this pushes prices even lower. With decreased supplies, the demand increases, and the cycle begins again.

Today, there are many other factors that influence prices. As a result, the traditional livestock price cycles have less variation than in the past. Some factors that currently influence livestock prices are:

- importation and exportation of livestock products;
- development of new uses for livestock products;
- increased advertising of livestock products;
- world weather conditions;
- general economic conditions in the world; and
- changes in consumer demands. In recent years, demand for beef and pork has eased up and down, whereas demand for poultry has been increasing dramatically. Egg consumption is fairly steady, and milk consumption is increasing slightly.

Trends and Cycles in Crop Marketing

The availability of a given crop is stimulated by price, which is influenced by demand. Many grain crops have so many uses that the market for the items is complex. For instance, corn is the number-one choice for most farm-animal and poultry feeds. It is frequently used in pet foods, breakfast cereals for humans, and as the source of corn oil, plastics, alcohol, and ingredients for baked goods.

However, in many of these commodities, corn can be replaced by substitutes. For instance, barley and other grains will be substituted if the price of corn increases to more than a certain level. Similarly, when the price of crude oil decreases to less than a certain level, the price of corn may no longer be competitive. Gasoline refiners will then use more petroleum and less grain alcohol in the fuel. Weather and other production factors will influence the size of the world grain supply and influence prices in the United States and abroad. Trade policies greatly influence prices of a commodity, because the flow of commodities into the world market or the policy of permitting world supplies to flow into the United States greatly complicates the formulas that determine prices.

FIGURE 34–17 Large trucks and modern highways move agricultural commodities quickly from farm to the processor or market. *(Courtesy of FFA)*

A classic example of how U.S. government policy can affect prices, supplies, and farm conditions occurred in 2003. The U.S. government closed the border to shipments of cattle from Canada due to the first case of mad cow disease in North America, which occurred in Canada. Beef prices in the United States hit record highs in the following months. When a case of the disease was later found in Washington state, a temporary decline in consumer demand impacted the market again.

Many factors make agricultural marketing unpredictable. The use of rapid transportation on a modern system of interstate highways (Figure 34–17); air shipping of highly perishable foods; refrigerated shipping; food processing procedures that lock in freshness; extensive, specialized, high-technology storage facilities; and an excellent marketing system all contribute to the availability of food and fiber commodities to most any part of the nation on a year-round basis. These factors help to maintain markets and prices that are fairly stable and predictable.

GLOBAL MARKETING IN AGRISCIENCE

Historically, agricultural producers have focused on producing commodities and managing their businesses in ways that held the best promise for acceptable profits. Although producers always dream of favorable prices and fair profits, these do not always materialize. Telephones, radios, televisions, fax machines, computers, and other modern communication devices have permitted many changes in marketing during the last two decades. There have always been many middlemen in processing and marketing. The number and types are increasing (Figure 34–18).

The Agricultural Commodity Futures Market

Relatively new on the scene is the use of futures contracts in the agriculture commodity marketing complex. The Chicago Board of Trade, Chicago Mercantile Exchange, Winnipeg Commodity Exchange, and London International Financial Futures Exchange are all commodity exchanges. A **commodity exchange** is an organization licensed to manage the process of buying and selling commodities under specific laws using a system of licensed brokers. Commodity exchanges also manage futures markets. A **futures market** is a procedure conducted by commodity exchanges to

FIGURE 34-18 Despite all of the high tech marketing strategies that are available, direct face-to-face marketing is still used to introduce selected products to potential buyers.

INTERNET ADDRESS:

<http://www.cbot.com/index.html>

provide networks and legal frameworks for sellers and buyers to work through brokers in making contracts called *futures contracts,* or simply *futures.*

Buying and Selling Futures Futures are defined as legally binding agreements, made on the trading floor of a futures exchange, to buy or sell something in the future. Futures prices are determined by competitive bidding of brokers for prospective buyers and sellers from all over the world. Anyone with appropriate assets can work through a broker and buy futures or sell futures on the commodity market. The futures market floats up and down by the moment while the futures exchange is "open," or in session. Telephone and computer networks around the world permit agricultural managers, buyers, sellers, and brokers to communicate almost instantly to place their bids to buy or sell, and the commodity exchange handles the legal work to complete the transactions. Many agricultural commodities, as well as other commodities, are bought and sold in this manner (Figure 34–19).

FIGURE 34–19 Commodity exchanges make possible the use of futures trading as valuable marketing and management tools. *(Courtesy of Chicago Board of Trade)*

Farmers, ranchers, cattle feeders, crop growers, and other agricultural owners and managers use the futures market as a tool to stabilize their businesses. The wise buying and selling of futures can protect them from losses caused by excessive price fluctuations. For instance, before deciding how much corn to grow, how much to spend on fertilizer and other inputs, or how much corn to keep for feeding out cattle or hogs on the farm, the farmer should know the price per bushel for which the corn could be sold after it is grown. The other decisions would then be easier to make.

By looking in the commodity-price sections of major business newspapers, such as the Wall Street Journal, the seller can determine the futures price for corn delivered in a given month. If the farmer regards that price as high enough to commit a certain number of bushels for sale, then he will inform his broker to "sell" that many bushels at the rate of the futures price for the month selected. The broker will negotiate a contract that obligates the farmer to deliver that many bushels of corn on the specified date for the amount of the futures price. Now the farmer is guaranteed the price for the corn, and other decisions about the farming operation can be based on this selling price. If the price of corn is lower in December than the $1.80 futures contract that was made, the payment on delivery will still be $1.80 per bushel; and the farmer will have successfully protected the business against a price drop. However, if the price of corn is greater than $1.80 per bushel in December, the farmer might wonder if the purchase of a futures contract was a wise decision.

Here is how a futures contract might be made for 5,000 bushels of corn. Suppose the settlement price on the futures market for December is $1.80 per bushel. The value of the contract for 5,000 bushels is 1.80 × 5,000, or $9,000. The farmer is obligated to deliver the corn for $9,000 regardless of the market price of corn at the expiration of the contract.

Opening and Offsetting a Position The purchase of a futures contract does not have to be a final transaction that plays through to the actual buying or selling of the commodity. In fact, in most cases, the goods are not delivered because the participant makes another contract that offsets the first one. The initial step in the futures market is called opening a position. At a later date, before the futures contract expires, the participant can take a second position to offset the first. Such action is known as offsetting a position. When the position is offset, the participant is said to be out-of-the-market and has no obligation to deliver or take delivery. To offset the position described previously, the farmer would need to buy 5,000 bushels of corn. This would offset the contract to sell 5,000 bushels. If the farmer were fortunate enough to find a day when the futures price was less than $1.80, and buy futures for 5,000 bushels for the specified date of delivery at that lower price, then the farmer could pocket the difference as income, or profit, from the business transaction.

INTERNET KEY WORDS:

agriculture, export, markets

EXPORT MARKETING

As transportation and communications improve, the world seems to get smaller. More people are familiar with more places and people around the world, so more business is being conducted between and among people and nations of different continents. Technology and computer languages are spreading rapidly, so different cultures are able to communicate in business, commerce, engineering, and other fields. Fiber optics; lasers; satellites; computers; nationwide and worldwide telephone, television, and computer networks; international airlines and air routes; trade agreements; and improved education all contribute to the increase in international investment, business, trade, and commerce.

The United States now exports an estimated 40 percent of its agricultural products. The livestock, meats, hides, dairy products, grain, lumber, fruits, vegetables, ornamentals, and processed plant and animal products that we export help to offset our trade deficit. This trade activity has great implications for marketing products. In recent years there has been a substantial increase in travel abroad. State delegations, including governors and staff officials, legislators, educators, farmers, ranchers, horticulturists, and consultants of all kinds are traveling abroad. The purposes of such delegations typically are to develop good will; promote professional linkages, partnerships, and exchanges; and develop markets for commodities. When preparing for careers in agriscience or developing strategies for marketing agricultural products, it is important to consider the international and global opportunities.

The marketing of products is a complex operation if maximum profits are to be realized. There are many different markets for plants and animals and many ways of marketing them and their products. Planning for marketing should take place before entering into production, and it should be kept in mind during the entire production process.

STUDENT ACTIVITIES

1. Write the Terms to Know and their meanings in your notebook.
2. Obtain several copies of marketing reports and newspapers that have marketing reports. Compare the prices received for the various classes and grades of animals listed. Also compare the prices received from various markets.
3. Interview a livestock buyer. Ask how prices are determined and where animals are bought and sold. Also ask how to go about getting a job as a livestock buyer. Report your findings to the class.
4. Talk to a livestock grader about how to determine the various grades and classes of livestock. Try to grade some animals yourself.
5. Invite the operator of a retail market to class to talk about the issues involved in marketing animals and animal products retail.
6. List the types of markets for agricultural commodities in your area.
7. Write a brief paper on the "best philosophy for pricing."
8. Develop an ad or other promotional plan for an agricultural commodity.
9. Build a display for retailing an agricultural commodity.
10. Research the procedures used to buy and sell in the futures market.
11. Do research on one or more major international trade agreements that influence the marketing of agriscience products.

SELF EVALUATION

A. Multiple Choice

1. Consumer demographics include:
    a. populations, preferences, and product cost.
    b. family size, eating preferences, and discretionary money.
    c. customer location and tax rates.
    d. transportation costs.

2. The grading system based on the amount and distribution of finish on an animal is called
    a. prime.
    b. quality.
    c. yield.
    d. commercial.

3. The amount of a product that is available at a specific time and price is the
    a. supply.
    b. demand.
    c. commission.
    d. production.

4. The estimated amount of weight that an animal loses during transportation to market is
    a. yardage.
    b. stress.
    c. loss.
    d. pencil shrink.

5. Yardage is a fee paid to terminal markets for
    a. feed.
    b. insurance.
    c. selling.
    d. none of the above.

6. Lambs that are slaughtered when they are younger than 3 months old are called
    a. mutton.
    b. chevon.
    c. feeder lambs.
    d. hothouse lambs.

7. The least desirable yield grade for cattle is
    a. U.S. No. 1.
    b. Utility.
    c. U.S. No. 4.
    d. 5.

8. The sex classes for sheep are
    a. stag, boar, and sow.
    b. lamb, sheep, and mutton.
    c. ram, ewe, and wether.
    d. none of the above.

9. Roadside stands are
    a. wholesale markets.
    b. terminal markets.
    c. retail markets.
    d. direct markets.

10. A male animal castrated after reaching sexual maturity is a
    a. wether.
    b. steer.
    c. stag.
    d. barrow.

11. An example of a wholesale market is a/an
    a. terminal market.
    b. roadside stand.
    c. on-the-farm market.
    d. farmers' market.

12. A definite trend in agriscience marketing is
    a. cost per unit keeps going up.
    b. pricing is always based on cost.
    c. number of commodities is decreasing.
    d. marketing is more global.

B. Matching

_____ **1.** Prime

_____ **2.** Ewe

_____ **3.** U.S. No. 1

_____ **4.** Barrow

_____ **5.** Veal

_____ **6.** Futures purchase

_____ **7.** Offset position

a. Castrated male pig

b. Female sheep

c. Quality grade of cattle

d. Quality grade of swine

e. Calf meat

f. Out-of-the-market

g. Opening a position

C. Completion

1. The lowest quality grade of slaughter hogs is _____.

2. _____ is the amount of a product desired at a specific place and time.

3. A professional livestock seller receives a _____ for selling animals for producers.

4. _____ markets are those where animals are sold by competitive bidding.

5. Futures for agricultural commodities are purchased through a _____.

6. Rather than delivering goods sold on the futures market, most people take a(n) _____ position.

Agribusiness Planning

OBJECTIVE

To define management, determine management performance, determine how decisions are made, and describe economic principles that affect management.

MATERIALS LIST

✓ writing materials

✓ calculator

✓ income statement

✓ newspapers

✓ Internet access

COMPETENCIES TO BE DEVELOPED

After studying this unit, you should be able to:

- define management.
- describe the importance of management.
- describe kinds of agriscience management decisions.
- list eight steps in decision making.
- describe the economic principles of supply and demand, diminishing returns, comparative advantage, and resource substitutions.
- use capital and credit wisely in business management.

SUGGESTED CLASS ACTIVITIES

1. Invite the owner or manager of an agribusiness to make a presentation to the class on the subject of managing a successful business. Ask him or her to identify the key elements of successful agribusiness management. Also discuss the barriers to managing a successful business, and invite the guest to share examples of ways that he or she has overcome these barriers.

2. Develop a business plan for a school business such as producing flowers in a school greenhouse. Implement the plan by electing a board of directors, hiring a manager, and producing or obtaining a marketable product that can be sold to the public. Conduct the business enterprise for a specific period, and determine profits or losses for that time period. What steps were the most difficult to implement? Which elements of the business plan were most critical to the success of the business?

3. Invite a banker or other agricultural lender to talk to the class about obtaining credit. Ask him or her to discuss small business loans, prequalifications for credit, interest rates, and repayment plans. The students should be prepared to ask appropriate questions and participate in the discussion.

AGRIBUSINESS MANAGEMENT

Agribusiness management is the human element that carries out a plan to meet goals and objectives in an agricultural business or enterprise, generally referred to as an agribusiness. Management decides the types of business or production activities included in the agribusiness, such as horticulture, aquaculture, and farm supplies. Agribusinesses include farming, ranching, nursery operations, landscaping, retail stores, service enterprises, marketing businesses, lending institutions, veterinary services, and consulting activities.

According to Dun and Bradstreet, 88 percent of all businesses that fail do so because of poor management. Poor management of an agribusiness frequently results from no set objectives or goals. Management is usually considered to be good when maximum profits are achieved from the available resources. Generally, good management has established goals and objectives as guidelines, which translate into good action plans (Figure 35–1).

Influences on Agribusiness Management

Agribusiness management is influenced by the members who make up the boards of directors in corporations and cooperatives. Land, labor, and capital, being in limited supply, also influence management decisions. For instance, limited **capital**, which is money or property, may prevent the agribusiness manager from buying a new piece of equipment that would make the operation more efficient or protect the operation from excessive losses (Figure 35–2). Similarly, limited land will influence the agribusiness manager's selection of enterprises. For example, a cow–calf beef enterprise will need more land than will a feedlot operation. Limited labor will influence the manager's selection of enterprises. For instance, if labor is limited, it will prevent the use of crop enterprises with high labor requirements, such as watermelons or tobacco. In contrast, grain production requires minimal labor.

Estimating a Manager's Performance

The performance or ability of the agribusiness manager can be estimated in several ways. Dollar income is the measurement most often used. According to a study conducted in Ohio, the managerial ability of an individual can be determined by the manager's economic orientation, decisiveness, ability to look at alternatives, and extent of social activities or distractions from business goals (Figure 35–3).

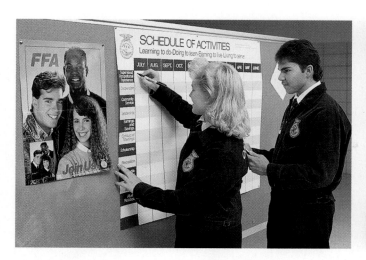

FIGURE 35–1 Planning skills that are learned in an agriscience class can also be used to develop the goals and objectives for a business plan. Many businesses that fail did not have clearly defined goals and objectives. *(Courtesy of FFA)*

FIGURE 35–2 Modern irrigation equipment permits profitable farming in areas that would otherwise be inefficient or too risky because of crop loss by drought. *(Courtesy of FFA; Photo by Ken Whitmore Association)*

Agribusiness is constantly changing, and it is expected to change even faster in the future. As agribusiness becomes more complex, errors in management will be more costly. Therefore, the need for people who are well educated in agribusiness management is increasing.

CHARACTERISTICS OF DECISIONS IN AGRIBUSINESS

Although there are various types of decisions in agribusiness, decisions of managers are usually organizational or operational (Figure 35–4). Both types of decisions affect the success of the business. The degree of influence on the business is determined in part by the characteristic of the decision. The following discussion covers four characteristics of a decision, as well as steps in decision making.

- Income goal (Economic orientation)
- Willingness or tendency to make decisions (Decisiveness)
- Manager's ability to recognize alternatives and opportunities
- The extent of social activities or distractions from business goals

FIGURE 35–3 Factors for predicting the managerial performance of agribusiness managers.

Agribusiness Decisions

Organizational
- Should I rent more land or should I borrow money and purchase land?
- Should the business be operated as a partnership or corporation?
- What lender offers the best source of borrowed capital?

Operational
- Should cattle be sold this week or next?
- Should I use high-magnesium lime or regular lime?
- When should I start planting corn? Soybeans? Small grain?
- When should I change the photoperiod for my poinsettias?

FIGURE 35–4 Kinds of agribusiness decisions.

Agri-Profile

CAREER AREAS:

Owner/Manager/Consultant

Agribusiness activities cut across the food and non-food spectrums of agriscience. This includes the supply side, as well as the output side, of production. Career opportunities exist in every sector where goods are bought or sold.

Agribusinesses sell products such as feed, seed, pesticides, fertilizer, tools, equipment, plants, or animals. They also sell services such as animal care, crop spraying, or recreational fishing privileges. Careful planning is important before starting an agribusiness, because many new businesses end in failure. Careful planning increases the chances of success.

Agribusinesses are the backbones of many communities. Here, a consultant uses computer and communication resources to predict future business outcomes. *(Courtesy National FFA)*

Career opportunities in agribusiness planning include financial services such as banking, accounting services, management services, teaching, university extension service, marketing, market analysis, product specialization, product engineering, sales, ownership, and management. Agricultural economics, business management, accounting, finance, and personnel management are college programs that can lead to careers in agribusiness.

Importance

The importance of a decision may be determined by measuring the potential loss or gain that is likely to occur as a consequence of the decision. For example, the selection of an agribusiness enterprise or business venture is likely to be more important than the selection of a brand of a given commodity. The manager must spend a sufficient amount of time selecting a type of business, because this decision cannot be changed easily after it is made. In contrast, if sales of a certain brand of wire or species of plant are not successful, a different brand or species can be added or substituted at little additional cost.

Frequency

The frequency, or how often a decision is made, varies greatly. Determining the cost per day to rent a truck may be used as an example. The cost of renting the truck for only 1 day, for a one-time job, might not be an important decision. However, the cost of renting the truck every few weeks as an ongoing expense makes the decision more important.

Urgency

It is important that certain decisions be made immediately. Some decisions can be delayed. The urgency of a decision depends on the cost of waiting. Two examples of decisions with different degrees of urgency are

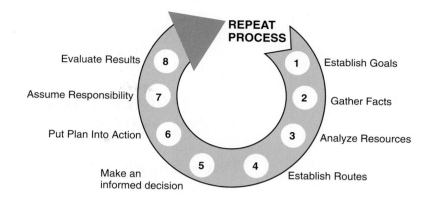

FIGURE 35-5 Eight steps in the decision-making process.

hiring a new employee and buying a new office radio. If it is the busy time of year, an additional person may be needed immediately and should be hired. If you delay, you may lose sales. However, the delay in buying an office radio is not likely to affect the overall management and profit of the agribusiness.

Available Choices

INTERNET KEY WORDS:

decision-making skills

Some situations offer several choices. If several choices are available, the manager should delete the least likely courses of action and focus on the more promising ones. The manager should take time to gather important information regarding the options that are helpful in making the correct decision. If no choices are available, there is no decision to be made.

Steps in Decision Making

Making decisions based on facts will increase your management ability. The agribusiness manager must determine the process to be used for making better decisions. The following eight steps should be helpful in developing a management process (Figure 35–5).

1. Start the management process with a situation or established goal.
2. Gather all available facts and information for accurate analysis of the problem.
3. Analyze the available resources. Human resources are labor, time, skills, and interest. Material resources are land, equipment, and capital. Reevaluate your goal and adjust it if appropriate.
4. Determine the possible ways or established routes of accomplishing the goal or solving the problem.
5. Make an informed decision concerning the route(s) that will be taken.
6. Follow through with a plan of action.
7. Assume responsibility for implementing the plan.
8. Evaluate the results to determine whether the goals were accomplished. If not, could a better route have been selected? Should the goals be modified?

After going through these eight steps, the manager should determine whether the process worked well. If not, where were the weak links? What changes should be made to improve the process? These and other questions will enable the manager to fine-tune the process until it fits his or her style, the personalities of the staff, and the structure of the organization.

FUNDAMENTAL PRINCIPLES OF ECONOMICS

INTERNET KEY WORDS:

price, supply, demand
law of diminishing returns

INTERNET ADDRESS:
<http://www.stfx.ca/people/
mgerriet/SDDETER.html>

Price, Supply, and Demand

Price is the amount received for an item that is sold. The price is determined by three price-making factors: (1) the supply of the item, (2) the demand for the item, and (3) the general price level. No one factor can be used to explain all price changes.

The general price level of a commodity is influenced by the supply or availability of the item, the demand for it, wars, depressions, and many other factors. When supply and demand are in balance, a general price is established. An increase or decrease in either supply or demand is likely to influence the prevailing price.

The quantity of a product that is available to buyers at a given time is called supply. A producer of vegetables has control over some factors that influence the vegetable supply. However, one factor that the manager cannot control is the weather. If the rains do not come, the supply of vegetables will be reduced. The ability to buy items needed to produce the vegetables can also increase or reduce the supply. The ability to buy these items is also influenced by their prices. Cost and availability of credit may also influence supply. Credit is money loaned. It permits the agribusiness manager to buy items needed for production or for business operation.

Demand may be defined as the quantity of a product that buyers will purchase at varying prices at a given time. The quantity that a buyer is willing to purchase depends on the quantity available and the price. Both the desire and the ability to purchase influence the extent of demand. The population growth may also influence demand. As the population of the country changes, the demand for various foods changes.

Diminishing Returns

The term diminishing returns is often used in economics. It refers to the amount of profit generated by additional inputs. Most of the time, the term is not completely understood. An understanding of the law of diminishing returns can be extremely helpful to the agribusiness manager in decision making.

The principle of diminishing returns has two parts: physical returns and economic returns. An explanation of both physical and economic returns follows.

Satisfaction from eating may be used to illustrate diminishing physical returns (Figure 35–6). When eating, each bite of food represents an additional unit of input. Each bite helps satisfy a part of the hunger. However, the added amount of satisfaction of hunger diminishes as you eat more. Each mouthful results in less satisfaction. This tendency is known as *diminishing physical returns*. At a certain point, the amount of hunger satisfied becomes negative with each bite taken. This means

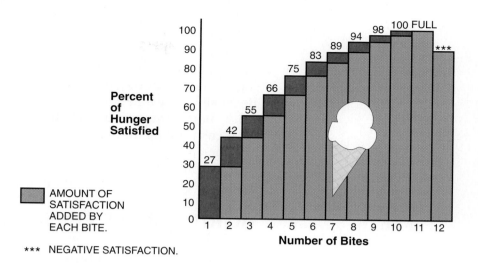

Satisfaction from Eating Ice Cream

FIGURE 35–6 The law of diminishing returns.

*** NEGATIVE SATISFACTION.

Effect of Nitrogen on Corn Yields*		
Nitrogen Applied (lb)	Yield (bu.)	Yield Increase (bu.)
0	67	—
60	103	+36
120	133	+30
180	154	+21
240	173	+19
300	169	−4

*Hypothetical and not based on research

FIGURE 35–7 Effect of increased nitrogen inputs on corn yields.

that each additional bite takes away from the satisfaction gained by previous bites.

An example of diminishing economic returns is demonstrated with the addition of units of nitrogen fertilizer as inputs and the effect on corn yields as outputs (Figure 35–7). Each additional input requires an additional cost. It may be observed that adding the first unit of nitrogen produced the greatest increase in yield. However, the rate of increase in yield diminished as more units of input were added. The decision that the manager needs to make is how many units of nitrogen should be applied to obtain the greatest profit from growing the corn. If the number of inputs is too low or too high, the greatest potential income will not be achieved.

Comparative Advantage

The United States has nine major farming areas that have developed over a period of years. These areas have developed because of changes in demand or other factors. Within each area, different commodities are produced. Most operations in an area produce similar commodities and have similar systems of production. The reason for the similarity in operations is the comparative advantage found in following the programs that have evolved in that area.

Comparative advantage is the emphasis in a given area where the most returns can be achieved. Comparative advantage may be illustrated in several ways. A recent example has been the change in livestock production in New York State. At one time, most of the farms in New York raised sheep. Today, these farms are producing milk because of the high demand and favorable prices for milk in New York. It has become more advantageous to produce milk than to produce lambs and wool.

A similar situation exists for broiler production on farms. Fifty years ago, most farms raised enough chickens to supply their families with eggs and chicken for Sunday dinner. Few farms raise chickens anymore. The production of eggs and broilers has become so competitive that it is usually less expensive for farm families to buy eggs and broilers at the grocery store than it is to produce them in small numbers on their farm.

Resource Substitution

Resource substitution refers to the use of one resource or item to replace another, when the results are the same. It is often possible to substitute a less expensive item for a more expensive one.

For example, it may be possible to substitute barley for corn in making a cheaper dairy feed. This substitution can be done without affecting total milk production. Barley is frequently less expensive than corn. Because barley and corn have slightly different feeding values for dairy cattle, barley must sell for $4.50 or less per 100 lb (cwt) to be a better buy than corn selling for $5.00 per cwt.

The application of the principle of resource substitution is not limited to feeding livestock. This principle can be used in most management situations. Can you think of other examples of resource substitutions?

AGRIBUSINESS FINANCE

Importance and Uses of Credit

The management of finances is the most important single function of the agribusiness manager. Possessing a great deal of technical knowledge and know-how is important. However, this know-how will not make one a successful agribusiness manager unless he or she is a competent money manager. Careful planning of finances is probably more important than planning other aspects of the agribusiness. Some good agribusinesses have failed because the manager was not a good money manager.

Credit is borrowed money. At one time, the manager who sought credit was considered to be a poor manager. Today, this is no longer true. The manager approaching a lending institution to negotiate for credit should not do so in an apologetic manner. The lending institution must sell its commodity, money, to stay in business. The use of credit is a privilege that must not be abused, however. Once labeled as a poor credit risk, the reputation is difficult to overcome. Credit used wisely, however, is a valuable tool in business.

Classifications of Credit

Credit is classified according to its period of use. Loans may be classified as long-term, intermediate-term, or short-term (Figure 35–8).

Long-term loans are used to purchase land and buildings. The loan period ranges from 8 to 40 years. Interest rates for this type of loan are usually less than they are for other types. These loans are made by

Classifications of Credit		
Long-Term Loans	**Intermediate-Term Loans**	**Short-Term Loans**
Used to purchase land and buildings, and are for a period of eight to forty years.	Extend for a period of one to seven years for the purchase of breeding livestock, farm equipment, tractors, and similar items.	Usually written for a period of one year or less to cover the cost of feeder livestock, feed, fertilizer, seed, fuel, etc.

FIGURE 35–8 Classifications of credit.

federal land banks, the Farmers Home Administration (FHA), insurance companies, local banks, and individuals.

Intermediate-term loans are for periods of 1 to 7 years. Capital investments are usually made with the money from this type of loan. **Capital investment** is money spent on commodities that are kept 6 months or longer. Examples are breeding stock, tractors, store equipment, and warehouse equipment. Production credit associations, the FHA, finance companies, and local banks are sources of this type of credit.

Short-term loans are for a period of 1 year or less. They are often referred to as *production loans*. Some managers are able to borrow short-term money without security. This is particularly true of managers and businesses with good reputations. Money borrowed without security is referred to as a *signature loan*. The borrower signs a promissory note that the loan will be paid on or before a specified date. Production credit associations and local banks are sources of short-term loans. It is not uncommon for local agribusinesses to permit their reliable customers to make purchases on a short-term credit basis by simply signing a sales slip.

Types of Credit

INTERNET KEY WORDS:

credit, short-term, intermediate-term, long-term

Two types of credit are productive credit and consumptive credit (Figure 35–9). Productive credit is used to increase production or income. This is justifiable when the estimated increase in production will increase profits. This type of credit is used to purchase supplies,

EXAMPLES

PRODUCTIVE CREDIT

Livestock

Plants/Flowers

Equipment

CONSUMPTIVE CREDIT

Vehicles

Clothing/Shoes

Entertainment

FIGURE 35–9 Types of credit based on how the loans are used. Productive credit is used to pay for the cost of raising crops and livestock or to purchase materials for resale. Consumptive credit is used to purchase goods or services that are used up with no expectation of earning a profit.

plants, flowers, livestock, land, equipment, storage facilities, seed, fertilizer, labor, and other materials.

Consumptive credit is used to purchase consumable items used by the individual; it does not contribute to the business income. It is relatively easy for a family to abuse the use of consumptive credit. This type of credit can also limit the amount of productive credit available to a family business.

A form of consumer credit growing in popularity is credit card purchases. Most credit card purchases paid in 30 days or less are interest free. However, the cost of consumer credit is high if the balances are not paid off within the first billing cycle. Federal statutes require firms charging interest on credit card purchases to publish their interest rates.

Sources of Credit

There are many sources of credit available and differences in lending policies. Because of this, a knowledge and understanding of credit and credit practices should be helpful in securing credit. Lenders differ in the interest rates they charge, the length of the loan period, and the purposes for which money is loaned.

When seeking a loan, remember that the interest rate is not the only important factor. Of substantial importance is the lender's willingness to extend a line of credit should an unexpected event occur. Another important factor is the lender's knowledge of the agribusiness.

Some agencies and institutions make only certain types of loans (Figure 35–10). Commercial banks are the most important source of credit. About 25 percent of the total agricultural debt is owed to banks. This percentage represents about 13.5 percent of the real-estate loans and 45 percent of other loans. Commercial banks lead in loan volume because they make all types of loans.

Individuals are the next most important source of agribusiness credit. About 36 percent of agricultural real estate mortgages are held by individuals. Interest rates, tax deferments and reductions, and payment options are some of the reasons that individuals are willing to finance real estate.

Agri-Profile

GRANTS . . . FREE MONEY?

A grant is money that is given to a person or organization to finance a particular activity. Although this money rarely has to be repaid, a grant really is not free money. The organizations or individuals providing the funding must see a success story. They need to know that their money has been used for a good purpose. Often, the grantor asks for documentation that shows that grant money has been used wisely.

There are many grants available to individuals in agriculture. Some common ones are educational, livestock promotion, value-added, and rural rehabilitation. These grants can be of great benefit to people whose goals match those of the grantor and to those who are willing to work hard to obtain results. Finding a grant for which you qualify will take time and effort. Grant information is widely available on the Internet, in government publications, and in trade magazines. Once a person finds a grant for which they qualify, an application will need to be completed. This process is the most important step in getting a grant. Many applications are usually accepted for each available grant. Educating yourself on grant writing will increase your chances of getting a grant. There are many resources that provide valuable information on how to write a successful grant proposal. Local colleges offer classes and workshops for grant writing.

Sources of Agricultural Credit and Typical Rates				
Source	**Length of Loan**	**Annual Interest Rate***	**Percent of Appraisal Loan Value**	**Purpose of Loan**
Commercial Banks & Trust Companies	6-12 Months	8-13%	To 100%	Farm Production Items
	2-3 Years	9-12 1/2%	70-80%	Machinery, Equipment, and Livestock
	10-20 Years	9-11%	60-75%	Real Estate
Federal Land Banks	20-35 Years	8 1/2 % Variable	Up to 85%	Real Estate
Production Credit Association	1 Year	9% Variable	To 100%	Production Items
	3-7 Years	9%	Varies	Machinery, Equipment, and Livestock
Farmers Home Administration	40 Years	5%	To 100%	Real Estate
	7 Years	7-9%	To 100%	Machinery, Equipment, and Livestock
Insurance Companies	20-35 Years	Varies	To 75%	Real Estate
Individuals	15-30 Years	6-7%	To 75%	Real Estate
	2-5 Years	6-8%	To 80%	Machinery, Equipment, and Livestock
Equipment Manufacturers	1 Month-5 Years	17.5%	100%	Machinery and Equipment

*Annual interest rates fluctuate based on the current prime rate. These percentages are for example only and should not be considered current.

FIGURE 35-10 Sources of agribusiness credit.

INTERNET KEY WORDS:

federal land bank
farm credit system
production credit associations
Commodity Credit Corporations
Farmers Home Administration
Small Business Administration

Retail merchants also supply non–real-estate credit. Their credit is usually in the form of an open account. Sometimes cash loans are made. The interest rates are usually higher from these sources than they are from commercial banks. The popularity of this type of credit is based on its convenience and availability.

Federal land banks were created by Congress in 1916 to provide long-term credit for agriculture. These banks are excellent sources of credit for the purchase of real estate because of the favorable interest rates.

Production Credit Associations (PCA) were established by an Act of Congress in 1933. The purpose was to provide favorable, short-term credit for agriculture. These funds are secured from the federal intermediate credit banks, who in turn secure their money from private lenders.

Life insurance companies have been excellent sources of credit for financing long-term real estate loans. Loans are made through brokers, correspondents, and company representatives. Insurance companies hold about 15 percent of agricultural real estate mortgages.

The **Commodity Credit Corporation (CCC)** was administered as an agency of the U. S. Department of Agriculture (USDA). The CCC is a government-owned corporation. Loans are made on eligible

A Comparison of the Different Methods of Calculating Interest		
Example: A Loan of $600.00 for 12 Months at 10%. Repaid in 12 Equal Payments.		
	FORMULAS FOR CALCULATING INTEREST	INTEREST CALCULATED
SIMPLE INTEREST	Interest (I) = Principal X Rate X Time (I = P X R X T)	$600.00 X .10 X 1 = $60.00 = 10% True Interest
DISCOUNTED LOAN	Rate (R) = $\frac{\text{Interest (dollar cost)}}{\text{Principal X Time}}$ $\left(R = \frac{I}{P \times T}\right)$	$\frac{60.00}{540.00}$ X 1 = .11, or 11% True Interest
ADD-ON LOAN	Rate (R) = $\frac{2 \text{ X No. of Payments X Interest Charged in \$}}{\text{Beginning Principal X Years X (No. of Payments + 1)}}$	$600.00 X .10 = $60.00 660.00 = Total Payments $\frac{2 \text{ X } 12 \text{ X } 60}{600 \text{ X } 1 \text{ X } 13}$ = .184, or 18.4% True Interest

FIGURE 35-11 Formulas for calculating interest.

commodities, such as grain, cotton, peanuts, and tobacco. Farmers use the commodities as security for loans. Payment of loans is made by purchasing the loans or delivering the commodities to the CCC.

Farmers Home Administration (FHA) was created by the government during the depression. The FHA's original purpose was to assist tenant farmers in becoming landowners. The FHA provides financing to farmers who are unable to secure credit from any other sources. The advantages of borrowing money from the FHA are as follows:

- A large percentage of the total cost of the property can be borrowed.
- The repayment plan is based on the borrower's ability to repay.
- Supervision of the loan and assistance with planning are provided.

The **Small Business Administration (SBA)** also provides loans to agribusinesses. In 1976, Congress passed legislation that permits the SBA to make agricultural production loans. These loans have a favorable interest rate.

Cost of Credit

Interest is the greatest expense in borrowing money. However, borrowers are also required to pay other fees. These fees include commissions, recording fees, title certification charges, insurance, and service charges. There are many formulas used for calculating interest (Figure 35-11).

A **simple-interest loan** refers to one in which the full amount of a loan is received by the borrower and is paid back with interest after a short period. The borrower generally signs a **promissory note** agreeing to the terms of the loan. If payments are made several times throughout the duration of the loan, the interest is paid to date and interest is charged on the remaining balance of the principal.

In the case of a **discount loan**, interest is subtracted from the principal at the time the loan is made. For example, on a $1,000 loan, the borrower would receive only $900 of the $1,000 if the interest were 10 percent. Because the full amount of the principal is not received, the resulting true interest rate is 11 percent.

The **add-on loan** method is used for calculating interest on consumer loans. The interest is charged for the entire amount of the principal for the entire length of time. The total principal plus total interest is divided into combined equal installments. This results in a very high rate of true interest. For example, a $1,000 loan at 10 percent interest would require repayment of $1,100 to be repaid in monthly installments of $91.67 ($1,000 principal + $100 interest = $1,100/12 = $91.67).

The amortized loan is generally used for the purchase of land, buildings, and other expensive items. The payments are made monthly and are computed so the interest owed plus the payment on the principal is equal throughout the repayment time. By this method, almost all of the payment at the beginning of the repayment period is for interest, whereas most of the payment near the end is for principal. In other words, the amount being paid on the principal increases proportionately as the amount due for interest decreases. A careful analysis of amortization schedules for a loan over various lengths of time will help the manager determine the cost of borrowing and will aid in making wise decisions on when, how much, at what rate, and for how long to borrow capital (Figure 35–12).

Seeking a Loan

Most agribusinesses must borrow money to expand their operations or increase income. The decision to incur debt is not easily reached. A careful examination of available alternatives must be made. Then the best alternative must be chosen.

When seeking a loan, the agribusiness manager must be prepared to fully explain the benefits and risks to potential lenders. Many loan applications are not approved because the borrower does not present the details of the agribusiness. Other loan applications are not approved because the borrower does not effectively present the advantages of expanding. To be effective, the presentation must be organized, and the details of the operation should be put in writing. Lenders are concerned about the benefits and risks associated with providing a loan. Lenders are also concerned about a logical presentation of facts and the desires concerning specific agreements in the loan contract.

Information that should be included in the presentation includes: (1) an agribusiness plan; (2) business records (income statements, expense records, net worth statement, and financial history); (3) terms of the loan; and (4) method of repayment.

A borrower should investigate several sources of credit before selecting a lender. The final selection should be one that best meets the needs of the borrower.

Selecting a Lender

Factors that should be considered in selecting a lender include the following:

- lending institution representative's knowledge of agribusiness problems and practices;
- lending institution representative's experience in handling agricultural credit of a similar nature;
- reputation of the lending institution;
- loan policies (interest rate, repayment schedule, closing costs, penalty clause, optional prepayment clause, and policy regarding failure to meet payment because of circumstances beyond borrower's control);

A. 48 PAYMENTS OF $368.67 EACH

Pmt	Principal	Interest	Balance	Total Interest	Pmt	Principal	Interest	Balance	Total Interest
		12.000%	14,000.00				12.000%	14,000.00	
1	228.67	140.00	13,771.33	140.00	25	290.35	78.32	7,541.63	2,758.38
2	230.96	137.71	13,540.37	277.71	26	293.25	75.42	7,248.38	2,833.80
3	233.27	135.40	13,307.10	413.11	27	296.19	72.48	6,952.19	2,906.28
4	235.60	133.07	13,071.50	546.18	28	299.15	69.52	6,653.04	2,975.80
5	237.95	130.72	12,833.55	676.90	29	302.14	66.53	6,350.90	3,042.33
6	240.33	128.34	12,593.22	805.24	30	305.16	63.51	6,045.74	3,105.84
7	242.74	125.93	12,350.48	931.17	31	308.21	60.46	5,737.53	3,166.30
8	245.17	123.50	12,105.31	1,054.67	32	311.29	57.38	5,426.24	3,223.68
9	247.62	121.05	11,857.69	1,175.72	33	314.41	54.26	5,111.83	3,227.94
10	250.09	118.58	11,607.60	1,294.30	34	317.55	51.12	4,794.28	3,329.06
11	252.59	116.08	11,355.01	1,410.38	35	320.73	47.94	4,473.55	3,377.00
12	255.12	113.55	11,099.89	1,523.93	36	323.93	44.74	4,149.62	3,421.74
13	257.67	111.00	10,842.22	1,634.93	37	327.17	41.50	3,822.45	3,463.24
14	260.25	108.42	10,581.97	1,743.35	38	330.45	38.22	3,492.00	3,501.46
15	262.85	105.82	10,319.12	1,849.17	39	333.75	34.92	3,158.25	3,536.38
16	265.48	103.19	10,053.64	1,952.36	40	337.09	31.58	2,821.16	3,567.96
17	268.13	100.54	9,785.51	2,052.90	41	340.46	28.21	2,480.70	3,596.17
18	270.81	97.86	9,514.70	2,150.76	42	343.86	24.81	2,136.84	3,620.98
19	273.52	95.15	9,241.18	2,245.91	43	347.30	21.37	1,789.54	3,642.35
20	276.26	92.41	8,964.92	2,338.32	44	350.77	17.90	1,438.77	3,660.25
21	279.02	89.65	8,685.90	2,427.97	45	354.28	14.39	1,084.49	3,674.64
22	281.81	86.86	8,404.09	2,514.83	46	357.83	10.84	726.66	3,685.48
23	284.63	84.04	8,119.46	2,598.87	47	361.40	7.27	365.26	3,692.75
24	287.48	81.19	7,831.98	2,680.06	48	365.26	3.65	0.00	3,696.40

B. 24 PAYMENTS OF $659.03 EACH

Pmt	Principal	Interest	Balance	Total Interest	Pmt	Principal	Interest	Balance	Total Interest
		12.000%	14,000.00				12.000%	14,000.00	
1	519.03	140.00	13,480.97	140.00	13	584.86	74.17	6,832.53	1,399.92
2	524.22	134.81	12,956.75	274.81	14	590.70	68.33	6,241.83	1,468.25
3	529.46	129.57	12,427.29	404.38	15	596.61	62.42	5,645.22	1,530.67
4	534.76	124.27	11,892.53	528.65	16	602.58	56.45	5,042.64	1,587.12
5	540.10	118.93	11,352.43	647.58	17	608.60	50.43	4,434.04	1,637.55
6	545.51	113.52	10,806.92	761.10	18	614.69	44.34	3,819.35	1,681.89
7	550.96	108.07	10,225.96	869.17	19	620.84	38.19	3,198.51	1,720.08
8	556.47	102.56	9,699.49	971.73	20	627.04	31.99	2,571.47	1,752.07
9	562.04	96.99	9,137.45	1,068.72	21	633.32	25.71	1,938.15	1,777.78
10	567.66	91.37	8,569.79	1,160.09	22	639.65	19.38	1,298.50	1,797.16
11	573.33	85.70	7,996.46	1,245.79	23	646.04	12.99	652.46	1,810.15
12	579.07	79.96	7,417.39	1,325.75	24	652.46	6.52	0.00	1,816.67

FIGURE 35–12 Comparison of a loan for $14,000 at 12 percent interest paid over 4 years versus the same loan repaid in 2 years.

Science Connection

WORKHORSES OF MODERN BUSINESS PLANNING

The microcomputer has become the workhorse of modern agribusiness planning, as well as the workhorse of business operations. Capable of split-second computations and tremendous storage capacity, the microcomputer is found in nearly every business and generally serves as the nerve center for the business. After a small investment in software and a modest investment in accessories, the computer can be used to evaluate marketing alternatives, predict financial outcomes for proposed business ventures, and accumulate data on prospective customers. By using a good computer system, the manager and staff can handle complex mailing lists, individualize communications to large numbers of patrons, track sales and inventories, keep appropriate financial accounts, and develop presentations for the owners, board of directors, or patrons.

Computer programs and computer models help the business planner to test alternative practices without risking time and capital. *(Courtesy of Michael Dzaman)*

The very small business may have just one computer with a high-quality printer and monitor. However, such businesses usually graduate to multiple computers. Color monitors, printers, and scanners are replacing black-and-white systems, as the price for computer hardware continues to decline even as memory capacity increases. The software for a business may be an integrated software package, typically with word processing, database, spreadsheet, and communications software. Communications software permits the computer to use telephone lines to access outside networks and download data and programs from outside sources. This permits the business access to market quotations, continuous status reports on inventories, up-to-date marketing information, and databases on world business and commerce.

Data for tax computation, annual inventory, business reports, and other government obligations are readily available. Specialized computer programs can be purchased to calculate taxes and prepare tax returns, thereby cutting costs for services and consultants for the business.

Salespeople, buyers, company executives, and other on-the-road personnel use portable computers to gather and synthesize information from sales contacts and conferences and to phone the information to the home computer for staff to act on before the traveler returns to the office. With the arrival of voice-input capacity and computers that talk back, typical agribusinesses will continue to find new ways to use the computer, and its role as a business "workhorse" will undoubtedly expand.

- date the loan would be advanced;
- possibility of increasing the loan; and
- availability of credit for other purposes.

Using Credit Wisely

There is a chance that the borrower will fail to meet the financial obligations of a loan. In this case, the borrower may lose part or all that is owned. Because of this, the borrower takes a much greater risk than

Hot Topics in Agriscience

INTERNET SALES

Many businesses are taking advantage of the Internet to market and sell their products. This is a new way to connect a customer with a product, and nearly anything a person may want is available for sale at an Internet site. The Internet is particularly adapted to the sale of unique products that may be hard to find, but it is also a source of the products that are easily obtained at a store. For example, you can buy tires, groceries, insurance, and financing for your home on the Internet. Shopping is easy, and you can compare prices among competitors quickly.

A major effort has been made to protect the security of credit card information that is provided to companies with products for sale. The credit card is the key medium of exchange for Internet sales because it makes it possible to collect the payment instantly before an item is shipped or delivered. It also provides some protection to the buyer against fraudulent sales. Most major credit card companies will work with their cardholders to reverse a payment when there is evidence that a product was misrepresented. Internet sales are here to stay.

does the lender. The borrower, however, can minimize risk by applying certain rules (Figure 35–13) as follows:

- Production loans should be used to increase income.
- The borrower should limit the amount borrowed on new or unfamiliar business ventures.
- The borrower should keep debt as low as possible while still maintaining efficiency.
- The borrower should keep abreast of markets and trends.
- A debt-to-net-worth ratio of 1:1 or less should be maintained.
- A proper debt-to-income relationship should be maintained. The income must be greater than the principal and interest payments.
- Dependability and terms of the loan should be considered when selecting a lender.
- The borrower should have a definite repayment plan and schedule supported by evidence of an adequate cash flow.
- The borrower should be businesslike, fair, and frank.

Ways Whereby Borrower Can Minimize Risk

- Use production loans to increase income
- Limit amount borrowed on new or unfamiliar enterprises
- Keep debt as low as possible while still maintaining efficiency
- Keep abreast of markets and trends
- Maintain proper debt-net worth ratio
- Maintain proper debt-income relationship
- Select lender on dependability and terms
- Have a definite repayment schedule
- Be business-like, fair, and frank
- Reduce risk by carrying adequate insurance

FIGURE 35–13 Ways in which a borrower can minimize risk.

• The borrower should have adequate insurance to reduce the lender's risk. Property, liability, and crop and life insurance provide partial protection against risk.

The use of management in an agribusiness is extensive. Management is a vital part of the business and commerce associated with supplies, services, production, processing, distribution, and selling of plant, animal, and natural-resource commodities. Careful planning and the application of sound agribusiness principles and procedures are necessary for success in many agriscience careers.

STUDENT ACTIVITIES

1. Write the Terms to Know and their meanings in your notebook.
2. Explain how the eight steps in decision making can be used in planning an FFA or other school money-raising activity.
3. Explain how the principal of resource substitution may be used in making an agricultural mechanics project.
4. Discuss with a banker the procedures used in applying for an agribusiness loan.
5. Arrange to have a banker discuss agribusiness capital and credit in a class presentation.
6. Arrange for an accountant to talk to the class about taxes to consider when planning an agribusiness.
7. Make an outline of this unit. Phrases and words that are bold should be included, together with a brief description of the key points of the text.
8. Search the Internet for grants for which you may qualify. Discuss with your teacher what steps you will need to take to apply for a grant.

SELF EVALUATION

A. Multiple Choice

1. Most businesses fail because of
   a. death of the manager.
   b. lack of capital.
   c. labor problems.
   d. poor management.

2. Management is considered good if
   a. labor is adequate at all times.
   b. maximum profits are achieved.
   c. the business survives for 5 years.
   d. the business survives for 10 years.

3. The importance of a decision may be measured by the
   a. availability of the manager.
   b. inventory.
   c. potential for gain or loss.
   d. time of year.

4. The amount of profit generated by additional inputs is known as
   a. capital.
   b. diminishing returns.
   c. margin.
   d. profit.

5. Using a resource in the place of another is known as
   a. bailing out.
   b. integration.
   c. resource management.
   d. resource substitution.

6. Loans for 8 to 10 years are called
   a. amortized loans.
   b. intermediate-term loans.
   c. long-term loans.
   d. short-term loans.

7. The most important source of credit is
   a. commercial banks.
   b. individuals.
   c. insurance companies.
   d. land banks.

8. The formula for calculating simple interest is
   a. $I = P \times R \times T$.
   b. $I = P/R \times T$.
   c. $I = R/P \times T$.
   d. $I = P \times T$.

9. A debt-to-net-worth ratio should not exceed
   a. 5:1.
   b. 3:1.
   c. 2:1.
   d. 1:1.

10. The type of insurance that a borrower should have is
    a. liability.
    b. life.
    c. property.
    d. all of the above.

B. Matching

_____ **1.** Capital   a. Money loaned

_____ **2.** Price   b. Quantity desired

_____ **3.** Supply   c. 1 year or less

_____ **4.** Credit   d. Quantity available

_____ **5.** Demand   e. 1 to 7 years

_____ **6.** Short-term loan   f. Money property

_____ **7.** Intermediate-term loan   g. Pay interest in the beginning

_____ **8.** Generated income   h. Consumptive credit

_____ **9.** Used for the individual   i. Amount received

_____ **10.** Add-on loan   j. Productive credit

C. Completion

1. List the eight steps in decision making, in their correct order.

2. Study Figure 35–12, and then answer the following questions.
   a. What is the total interest paid on the loan if repaid in 4 years (48 payments)? $_____
   b. What is the total interest paid on the loan if repaid in 2 years (24 payments)? $_____
   c. How much more does the loan cost if repaid in 4 years rather than 2 years? $_____
   d. What are your observations about stretching a loan out over more time?

UNIT
36

Entrepreneurship in Agriscience

MATERIALS LIST

✓ variety of publications with pictures

✓ materials to make a collage or bulletin board

✓ Internet access

COMPETENCIES TO BE DEVELOPED

After studying this unit, you should be able to:

- define and describe entrepreneurship.
- describe steps in planning a business venture.
- state five basic functions performed in the operation of a small business.
- select a product or service for a personal or group enterprise.
- determine the basic functions performed by small-business managers.
- analyze the outcome of a business venture.
- use small-business financial records.
- analyze the benefits of self-employment versus other types of employment.

SUGGESTED CLASS ACTIVITIES

1. Schedule a class period in a computer laboratory that has Internet access. Explore the Web for evidence of entrepreneur activities and have class members share the locations with each other. Print the best examples and compile the pages into a reference booklet.

2. Divide the class into groups of four to five students. Assign each group to brainstorm about a possible product or service that might be provided to a specific target population by an entrepreneur. Design a plan for establishing a business to market the new venture following the steps that are provided in this unit.

3. Invite a local successful entrepreneur to discuss with the class what steps he or she took to start his or her own business. The students should be prepared to ask appropriate questions and participate in the discussion.

TERMS TO KNOW

entrepreneur	promoting function	actuating	profit and loss statement
entrepreneurship	distribution function	balance sheet	cash flow statement
inventor	financing function	asset	inventory report
buying function	short-term plan	liability	
selling function	long-term plan	net worth	

Entrepreneurship in agriscience provides extensive career possibilities at all levels of effort. It provides opportunities for the inventor, risk taker, profit seeker, owner, manager, and employee. The products and services of entrepreneurs include fresh food, processing plants, equipment sales, fishing, guide services, veterinarian work, forestry work, farming, ranching, teaching, research, mechanics, environmental efforts, law, insurance, real estate, and finance. Opportunities in agriscience exist worldwide (Figure 36–1).

THE ENTREPRENEUR

The **entrepreneur** is the person who organizes a business or trade, or improves an idea. The word is taken from the French word, *entreprendre,* which means to "undertake." **Entrepreneurship** is the process of planning and organizing a small-business venture. It also involves managing people and resources to create, develop, and implement solutions to problems to meet the needs of people. The **inventor** is the person responsible for devising something new or making an improvement to an existing idea or product.

The entrepreneur can be the inventor, as well as the small-business manager. However, in many cases, they are different people, each with distinct talents. The entrepreneur functions as the liaison between the inventor and the manager or management team. The entrepreneur brings these two groups together for the purpose of getting the invention to the individuals it will serve.

It is the entrepreneur who visualizes the venture strategy and is willing to take the risk to get the venture off the ground. Inventors or business managers are not entrepreneurs unless they organize the venture. An understanding of the differences, as well as the similarities, among these roles is important to comprehending entrepreneurship as a career option.

ENTREPRENEURSHIP

Various types of business enterprises are part of the entrepreneurship system. The element of individual, partner, or corporate ownership means private control as opposed to government ownership control. The profits (or losses) from entrepreneurship go to the owners. However, owners hire managers and other employees; therefore, entrepreneurships provide jobs for everyone (Figures 36–2 and 36–3).

A sales project conducted by a school organization such as the FFA is a type of entrepreneurship. However, it differs from a typical small-business venture in the following ways:

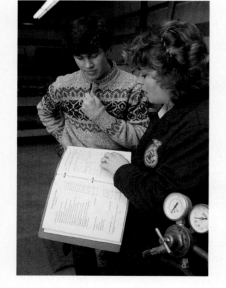

FIGURE 36–1 An entrepreneur must learn to recognize and take advantage of business opportunities when they occur. Agribusiness opportunities are found throughout the world, and today's young entrepreneurs will find opportunities to do business in the international arena. *(Courtesy National FFA)*

FIGURE 36–2 Many businesses start with high school occupational-experience programs in agriscience. *(Courtesy of FFA; Photo by Bill Stagg)*

FIGURE 36–3 Successful businesses generate job opportunities for managers and other employees. *(Courtesy of FFA)*

INTERNET KEY WORDS:

help entrepreneur
marketing tips

1. The sales project should be a valuable learning experience as well as a money-making activity.
2. The sales project is usually of a limited duration. It is completed within a short period. A small business typically is designed to operate indefinitely.
3. The sales project involves the voluntary participation of the class. The small business must hire and pay its employees.
4. The sales project usually involves little risk for financial loss to the individual. The small-business owner (entrepreneur) could lose his or her investment.
5. The sales project may need to be approved by the school administration, but it usually does not need to be licensed by the government, pay taxes, or file employee reports.

With these distinctions in mind, you can plan, organize, and implement a sales project and gain insight into the world of entrepreneurship at the same time (Figure 36–4).

Operating Businesses

Five basic functions are performed in the operation of both a small business and a sales project. These functions represent the basic steps in moving a product or service from the supplier to the consumer. They are the buying function, selling function, promoting function, distribution function, and financing function.

Buying Function The buying function involves selecting a product or service to be marketed or sold for profit. The selection of a product or service is based on thorough marketing research, which determines consumer (customer) needs and wants. It also determines who, if anyone, is already promoting the product or service.

Selling Function The selling function includes studying the product or service to determine the reasons that customers will want and need the product or service. This function also involves developing suitable customer approaches. Planning sales presentations, determining methods of overcoming objections, and planning for the close of a sale are all part of this function (Figure 36–5).

FIGURE 36–4 Group projects in school can help students develop valuable entrepreneurial skills. *(Courtesy of FFA; Photo by Michael Wilson)*

Hot Topics in Agriscience

INTERNET MARKETING

Internet marketing is a new marketing strategy that has allowed entrepreneurs to market products directly to anyone who has access to a computer with an Internet connection. It allows a person to run a business from his or her own home, and it allows business organizations of all sizes to serve Internet shoppers. Internet marketing has opened up new advertising and promotion opportunities, and virtually anyone who has a product or service to sell can do so using the Internet.

Internet marketing has also introduced some new problems to the business world. Payment is made by credit card, but how do you keep credit card numbers secure from discovery by dishonest people? How do you protect Internet shoppers from being exploited by cybercrooks and others who develop lists of names by tracing the Internet sites that are visited? What privacy rights should be protected by Internet providers? What new laws are needed to protect consumers who use the Internet? These are only a few of many issues that are developing together with this new marketing opportunity.

Promoting Function The promoting function involves developing a plan to identify ways to make potential customers aware of the product or service to be offered. Examples of promotional activities include newspaper, TV, radio, and outdoor advertising.

Distribution Function The distribution function involves physically organizing and delivering the selected product or service. For products, this includes receiving, storing, and distributing the merchandise. Many of the same activities are associated with services, particularly distributing to the customer (Figure 36–6).

Financing Function The financing function includes obtaining capital for the initial inventory, recording sales, maintaining inventory, computing profit or loss, and reporting the results of the venture. A venture is an activity with some risk involved.

FIGURE 36–5 Effective selling and promotion is absolutely essential to the success of most businesses. How a product is displayed is important. The flowers at this greenhouse are eye-catching and visually appealing. This type of presentation is sure to attract customers. *(Courtesy of FFA)*

FIGURE 36–6 Product distribution may be made to retail outlets or to other businesses, homes, farms, and many other destinations. *(Courtesy of Michael Dzaman)*

Agri-Profile

CAREER AREAS:

Owner/Manager/Assistant Manager/ Management Trainee

Agribusiness management is seen by many as an ideal career. This career has the advantage of continuing in the work of the family where one gains experience as he or she grows up. A person may also start a business and work in a locality where there may not be jobs in specialized areas.

Many high school and college students are well established in agribusinesses before they finish their educations. Some popular agribusinesses include lawn services, logging, lumber business, greenhouse or nursery operations, machinery repair, agricultural supplies, home and garden centers, florist shops, retail flower sales, livestock sales, farming, and ranching.

Preparation for careers in agribusiness management includes both formal education and on-the-job training. High school agribusiness and agriscience programs provide excellent training for business. Such programs should include classroom, laboratory, supervised agriscience experience, and leadership development. Advanced agribusiness programs may be taken at technical schools, colleges, or universities to obtain appropriate training in economics, finance, and management.

The entrepreneur is constantly seeking ways to provide better goods and services to the clientele. *(Courtesy of Michael Dzaman)*

SELECTING A PRODUCT OR SERVICE

The first function in the process of establishing a new business or developing a fund-raising project is the selection of a product or service (buying function). This involves analyzing potential products/services in terms of the potential demand (number of customers). The analysis of potential products or services should answer the following questions:

- Who are the potential purchasers of the proposed product or service? Customers can often be viewed in groups that have similar interests.

- On what basis does each group make decisions on the product or services? Typical responses to this question include price, quality, and service.

- What are the competing products or services? How do they compare with your product or service relative to price, quality, and service?

- What is the total estimated demand for your product or service?

- How much profit would you make on the sale of this product or service? (Multiply the number of units sold by your markup per unit and subtract any expenses you incur in the purchase, sale, and delivery of the product or service.)

ORGANIZATION AND MANAGEMENT

The role of a venture's organization and management is to make things happen so the venture can achieve its goals. To accomplish this, managers work mainly with data and people. Managing involves getting all the parts of the business—including personnel, marketing strategies, finances, and records—to function together to achieve the venture's goals.

No two managers have jobs that are the same. The jobs are shaped by the type of venture and the personality of the individual manager. However, managers perform the same functions. These functions are:

- planning work;
- organizing people and resources for work;
- actuating work; and
- controlling and evaluating work.

Planning

INTERNET KEY WORDS:

business plan

INTERNET ADDRESS:
<http://www.onepagebizplan.com>

When managers make plans, they set objectives or goals. They recommend policies to achieve the goals, and they develop procedures, methods, or programs to implement those policies. Plans must be constantly reviewed and updated. No matter how thorough, the plans themselves do not guarantee success. In planning, managers must make short-term and long-term plans. **Short-term plans** are accomplished in several days or weeks. **Long-term plans** are accomplished over several months or years.

Organizing

After a plan is developed, the work must be organized. The manager must identify the people needed to carry out the plan and arrange for the necessary equipment and supplies. Staff development may also be required.

Actuating

Actuating simply means putting the plan into action. The manager needs to inform employees of the plan. All persons involved must understand their roles. Actuation also includes motivating people to work efficiently and effectively and to want to get the job done.

Controlling and Evaluating

Managers must carefully control implementation of plans after the work begins. The quality and quantity of all results must be evaluated. If the results are satisfactory, work can continue. If problems arise, changes must be made and alternate plans may need to be developed. Managers must make adjustments in personnel, equipment, policies, or procedures whenever necessary (Figure 36–7).

In a very small venture, all management functions may be carried out by the same person. As a venture grows and its goals and objectives become more involved, additional managers may be required. When they are hired, organizational lines of responsibility must be developed and job descriptions written to identify the functions that are performed by each manager.

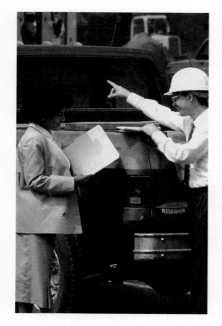

FIGURE 36–7 The manager must be familiar with all of the functions and able to work with all of the employees of the business.

Hot Topics in Agriscience

COOPERATION THROUGH AGRISCIENCE

A cooperative, or co-op, is a type of business organization that is created, owned, and controlled by its members. The function of a co-op is to ensure that the objectives of its members are met. There are three main types of co-ops: market, purchasing, and service based. A co-op that is organized for the purpose of marketing receives products from its members and then resells those products for the best possible price. Marketing co-ops are common in agriculture. One producer alone may not be able to produce enough crops or other agricultural commodities to obtain a large contract. When many producers band together, they can ensure that they will have the ability to compete for large-order contracts.

Purchasing co-ops can make large-quantity purchases for its members. By purchasing in bulk, the members are able to get the best possible prices for products they need. Items such as farm equipment, seed, fertilizers, dairy supplies, and much more can be purchased at reduced prices for members. Service co-ops provide low-cost service items to their members. For example, health insurance can be purchased at a much lower cost when a large number of people are becoming insured as opposed to a single family. Co-ops are valuable tools to the farm and ranch families that elect to participate in them. Members have discovered that much more can be accomplished when working together than could be done by themselves.

INTERNET KEY WORDS:

financial records needed, small business

SMALL-BUSINESS FINANCIAL RECORDS

Financial records are invaluable tools to the entrepreneur and manager. Good financial management will allow the manager to maintain control of the business venture and to increase profits or reduce losses.

Financial records can reveal which items are selling and which are not. These records can also show the extent of success of each person on the sales force. Financial records should indicate the amount of inventory on hand and how much has been sold. How much profit is made and the venture's total value at any time can also be computed using appropriate financial records.

The small-business owner or entrepreneur needs and uses more detailed financial records than are needed to maintain the typical sales project for a club or organization. Financial records for items such as buildings, fixtures and equipment, credit, debt, and return on investments are examples of records maintained by small businesses. However, balance sheets, profit and loss statements, and inventory and sales reports may be useful for a group project, as well as for a small business. Some typical financial records are discussed in the following sections (Figure 36–8).

Balance Sheet

A **balance sheet** is like a photograph of the business at a given time. It shows the assets, liabilities, and owner's investment on a particular date. The equation for the balance sheet is: assets = liabilities + net worth. **Assets** are anything the venture owns, including cash on hand, equipment, and inventory. **Liabilities** are both current and long-term debts. **Net worth** is the owner's investment in the business, including profits as they occur. A sample balance sheet is shown in Figure 36–9.

FIGURE 36–8 Financial records are fundamental management tools. *(Courtesy of FFA)*

Balance Sheet	
	Date: _____
ASSETS	
Current: Cash	_____
Merchandise	_____
Accounts Receivable	_____
Fixed: Land	_____
Building	_____
Machinery	_____
Equipment	_____
Other:	
1.	_____
2.	_____
3.	_____
TOTAL ASSETS:	_____
LIABILITIES	
Current: Notes Payable	_____
Accounts Payable	_____
Noncurrent:	
Debts more than one-	
year maturity	_____
TOTAL:	
NET WORTH:	_____
TOTAL LIABILITIES AND NET WORTH:	_____

FIGURE 36–9 A balance sheet is a statement of a business owner's assets, liabilities, and net worth.

Science Connection

GOLDEN OPPORTUNITIES WAITING

It is widely known that research in agriscience plays a continuous role in keeping the food, fiber, and natural resources sectors in the United States productive and efficient. What is not so well known is the impact of agriscience research on the businesses and industries that are not seen by the general public as being agribusinesses. In a recent tally, the USDA Agricultural Research Service documented nearly 75 companies that were manufacturing products or using processes from 53 new technologies developed by that agency. Some of these companies were formed to take advantage of new opportunities to meet apparent needs with emerging technologies. Others were established companies that were able to expand their product lines as a result of the research (Figure 36–10).

The U.S. government is looking for new ways to transfer newly developed technologies to private firms. Any decrease in time required for technology transfer from laboratory to product development and distribution reduces the pay-back time for the cost of research. New technology releases provide new opportunities for prospective entrepreneurs. Inquiries should be sent to: USDA Alternative Agricultural and Commercialization Center, 12th and C Street SW, Washington, DC 20250, or telephone (202) 401-4860.

PRODUCT	COMPANY	LOCATION	PRODUCT	COMPANY	LOCATION
No-calorie, high-fiber flour	Mt. Pulaski Products Canadian Harvest	Mt. Pulaski, IL Cambridge, MN	Controlled bulk vegetable fermentation	Trumark, Inc.	Roseville, NJ
Plant virus test kit	Agdia, Inc.	Elkhart, IN	Reducing water content of emulsions, suspensions, and dispersions with highly absorbent starch-containing polymeric compositions	Super Absorbent Co. Grain Processing Co. Worne Biotech Promar, Inc.	Lumberton, NC Muscatine, IA Medford, NJ Milwaukie, OR
Super-Slurper (starch-derived absorbent material)	Super Absorbent Co. Grain Processing Co. Henkel Corporation	Lumberton, NC Muscatine, IA Kankakee, IL			
Traps for stable flies	Tiger Farm Products	Hopedale, MA	Removal of heavy metal ions from waste water	Tetrahedron, Inc.	Stanhope, NJ
Turkey hemorrhagic enteritis virus vaccine	Arko Laboratories Oxford Laboratories Willmar Poultry Co.	Jewell, IA Worthington, MN Willmar, MN	In-ovo vaccination	Embrex, Inc.	Morrisville, NC
Dietary supplement	Monarch Nutri. Labs	Ogden, UT	Super Slurper dewatering cartridge	Central Illinois Manufacturing Co.	Bement, IL
Southwestern corn borer pheromone trap	Great Lakes IPM	Vestaburg, MI	Turkey semen extender	Continental Plastic Corp.	Delavan, WI
Microbial insecticides	Reuter Labs	Haymarket, VA	Cotton and soybean seed	Delta and Pineland Seed Co.	Scott, MI
Improved fire ant insecticide	Griffin Corp.	Valdosta, GA	Recirculating wiper for agricultural chemicals	Apple Machine Co.	Fort Pierce, FL
Japanese beetle trap	Consep Membranes	Bend, OR	Biopesticide for stone fruit	Fermenta	Painesville, OH
Direct marketing of fruits and vegetables	Three Rivers Produce	Southeastern, OK	Marek's disease vaccine	Select Laboratories, Inc. Tri Bio Laboratories	Gainesville, GA State College, PA
Biochemical and physiological factors influencing turkey egg hatchability	Agrimatic Corp.	Paramount, CA	Biodegradable plastic	Agri-Tech	Peoria, IL
Improve flavor, texture, and juiciness of processed poultry meat	Continental Grain Co.	Pendergrass, GA	Vaccination of chicken embryos for protection against avian coccidiosis	Embrex, Inc.	Morrisville, NC
Discover and develop mycoparasites for biocontrol of selected soilborne plant pathogens	Agracetus, Inc.	Middleton, WI	Methods that affect molecular genetic transfer and transformation in crop plants	Pioneer Hi-Bred, Inc.	Johnston, IA
Bee breeding and genetics for bee stock improvement	Weaver Apiaries, Inc.	Navasota, TX	Physiology of agriculturally important spiroplasmas and mycoplasmas	Agdia, Inc.	Elkhart, IN
Modification of starch to provide controlled delivery of agricultural chemicals	Illinois Cereal Mills, Inc.	Paris, IL	High-solids tomatoes and low-sugar potatoes using tissue culture	Northrup King Co.	Gilroy, CA
Cotton gin system design and evaluation to maximize product quality and minimize processing costs	Lummus Industries, Inc.	Columbus, GA	Construction of expression vectors from Marek's disease virus	Select Laboratories, Inc.	Gainesville, GA
Systems to apply irrigation water efficiently, control intake, and reduce nitrate leaching	Elvert Edgar Oest and Co.	Fruita, CO	Computerized control and management of center pivot irrigation systems	Valmont Industries	Valley, NV
Biological control of weeds using exotic and endemic plant pathogens	Confederated Tribes, Coleville Reservation	Nespelem, WA	Colonization factors expressed by *Campylobacter jejuni* in chicken	Southeastern Poultry and Egg Association	Decatur, GA
New engineering concepts for deciduous fruit production, harvesting, sorting	Agri-Tech	Woodstock, VA	New technologies in cotton ginning	Moisture Systems Corp.	Hopkinton, MA
Molecular biology of orbiviruses to diagnose and characterize bluetongue virus	Veterinary Diagnostic Technology, Inc.	Wheat Ridge, CO	Integrated strategies for managing filth-breeding flies on dairy farms	Olson Products, Inc.	Medina, OH
Apparatus and method for rapid analyses of multiple samples	Lachat Chemicals Alpkem Corp.	Mequon, WI Clackamas, OR	Monoclonal antibodies for diagnosis of Marek's disease	Vineland Laboratories Intervet, Inc.	Vineland, NH Millsboro, DE
Novel durable press finishing of textiles	Duro Finishing Corp.	Fall River, MI	Development of avian expression vectors using serotypes 1, 2, and 3 of Marek's disease virus	Solvay Animal Health, Inc.	Charles City, IA
Milk-like products from peanuts	Seabrook Blanching	Edenton, NC	Techniques for automated nondestructive quality evaluation of horticultural crops	Process Equipment Co.	Tipp City, OH
Highly absorbent polymeric compositions derived from flour	Illinois Cereal Mills Industrial Services Venture Chemicals Polysorb	Paris, IL Bradenton, FL Lafayette, LA Smelterville, ID	Development of methods to protect packaged agricultural products from insect infestation	Trécé	Salinas, CA
Rope wick applicator	Rear's Mfg. Co. Brothers Equip. Co. BCP Manufacturing Co. Agman, Inc. Rodgers Sales Stanley Resser	Eugene, OR Friend, NV Winters, TX Riverside, MO Clarksdale, MI Weldon, IL	Effects of residue decomposition in low-input sustainable agricultural systems	Crops Genetics Intl.	Hanover, MD
Starch-based semi-permeable films	Uni-Star Industries	Cuba, IL	Retention of quality of lightly processed fruits and vegetables	Tree Top, Inc.	Selah, WA

FIGURE 36–10 Some new technologies developed by the Agricultural Research Service of USDA, and the companies that have marketed products using those technologies. *[Adapted from USDA/ARS]*

Profit and Loss Statement

The **profit and loss statement** projects costs and other expenses against sales and revenue over time. The five basic sections in a profit and loss statement are (Figure 36–11):

1. total sales
2. cost of goods sold
3. gross profit
4. expenses
5. net profit

Cash Flow Statement

A **cash flow statement** describes the availability of funds for the purposes of running a business (Figure 36–12). When a positive cash flow exists, the income that is received by the business exceeds the expenses, and there is cash on hand. A negative cash flow indicates that the expenses that are generated by the business exceed the income at a given time. Nearly every business deals with periods of negative cash flow, and it is at these times that an operating loan may be needed to tide the business over until income is received. The cash flow statement allows a business owner to estimate the need for credit and to establish when the credit will be needed. For example, a beef ranch operates with a negative cash

Profit and Loss Statement	
For period ending _____	
TOTAL SALES	$75,000
COST OF GOODS SOLD	
Beginning Inventory	55,000
(Plus) Purchases	10,000
(Less) Ending Inventory	15,000
TOTAL COST OF GOODS SOLD	50,000
(Beginning Inventory plus Purchases	
minus Ending Inventory)	25,000
GROSS PROFIT...............................	25,000
(Total Sales minus Total Cost of Goods Sold)	
EXPENSES	
Salaries	18,000
Payroll Taxes.............................	2,500
Rent...	4,500
Advertising	1,000
TOTAL EXPENSES...........................	26,000
NET PROFIT (LOSS) (before taxes)	(1,000)
(Gross Profit minus Total Expenses)	

FIGURE 36–11 Sample profit and loss statement.

		Beginning Inventory	Items Sold	Ending Inventory	Total Cost Beg./Inv.	Total Cost End/Inv.	Total Sales	Total Profit
Item No.	Description							
000	Tick Tack Tote	45	45	0	33.75	.00	45.00	11.25
001	Brown Tray	32	32	0	25.60	.00	48.00	22.40
002	Fish Tray	20	20	0	15.00	.00	30.00	15.00
003	Lion Tray	6	6	0	4.80	.00	9.00	4.20
004	Guitar Tray	8	8	0	6.40	.00	12.00	5.60
005	Asst. Animal Tray	34	34	0	25.50	.00	51.00	25.50
006	Elephant Tray	10	10	0	5.00	.00	15.00	10.00
007	Screwdriver Set	22	22	0	11.00	.00	33.00	22.00
008	Brush & Shoehorn	36	36	0	32.40	.00	81.00	48.60
009	Fully Auto. Umbrella	9	9	0	20.70	.00	31.50	10.80
010	Auto. Umbrella	8	8	0	16.00	.00	24.00	8.00
011	Dad's No. 1 Keychain	200	190	10	90.00	4.50	190.00	104.50
012	Grandpa Keychain	100	79	21	45.00	13.95	79.00	47.95
013	Dad's Pad	48	30	18	28.80	10.80	45.00	27.00
014	Mini Screwdriver Set	75	75	0	60.00	.00	150.00	90.00
015	Dad's Plaque	72	72	0	28.80	.00	108.00	79.20
016	Tic Tac Toe	36	36	0	21.60	.00	36.00	14.40
017	Dad's Pen	1	1	0	.80	.00	1.50	.70

Inventory Report

FIGURE 36–12 Sample cash flow statement.

flow for most of the year. Once the calves are ready for market, they are sold together with cows that are not being retained in the breeding herd. Most of the income is received within a few days or weeks. In contrast, a dairy farm has a fairly constant cash flow throughout the year from the sale of milk and excess animals. The cash flow statement is intended to help the business manager and the loan officer determine how much operating cash is needed, when it is needed, and when it will be repaid.

Inventory Report

The **inventory report** includes how many units of each product are on hand, how many were sold, each item's cost, total sales, and profit. The inventory report can be used to make decisions such as reducing the price of slow-selling items, reordering other items, and learning which items are yielding the best profits.

SELF-EMPLOYMENT VERSUS OTHER FORMS OF EMPLOYMENT

The decision to continue working for someone else or to open a business is a difficult one to make. A way to help make that decision is to look at the advantages and disadvantages of both working for someone else and working for yourself.

Being an Employee

The advantages of being an employee working for someone else center on security. The salaried employee has no personal financial risk or responsibility to the company for which he or she works. Employees generally put in regular hours. If they work additional hours, they may be paid overtime for those hours. In addition, employees are often guaranteed vacation time and fringe benefits such as life and health insurance. In many cases, they are offered retirement plans. An employee can count on a somewhat stable lifestyle, with a fairly accurate idea of what the income will be from year to year. In addition, the employee may move up the career ladder within the company.

However, there are some disadvantages to working for someone else. The company does not have any financial responsibility to the employee should there be a recession resulting in a cutback in personnel. Some companies have a predetermined salary scale, so the employee could remain at a particular salary level until the right combination of years or experience is met. This is also true of promotions that could be based on years accumulated rather than on merit. The work pattern could become fairly routine. If no positions open within the company, the employee must wait until a position opens or look for a position with another company. The company management controls these decisions, not the employee.

Being Self-employed

INTERNET KEY WORDS:

become self-employed

One of the major reasons generally cited for opening a business is that it gives the owner control over his or her own destiny. The owner has the opportunity to set personal goals and recruit a team to help carry out those goals. Successfully meeting those goals results in a sense of achievement and, hopefully, financial reward.

Owning the business and being responsible for the decisions related to the business gives the entrepreneur a sense of independence. The owner's ability to make money is not restricted to a particular level.

The disadvantages of owning a business are not as obvious as the advantages. For instance, the necessary capital outlay may jeopardize family savings or even the family home. The number of hours required to run the business will mean a definite commitment from the owner, as well as the family members involved. Should the company have difficulty, the responsibility for both the management and the financial problems rests with the owner.

CONTRIBUTIONS OF SMALL BUSINESSES

INTERNET KEY WORDS:

economic development entrepreneur

Entrepreneurs have been credited with being the cornerstone of the U.S. enterprise system. Furthermore, many see entrepreneurs as the self-renewing agents of the economic environment in the United States. In recent years, numerous socialistic countries of the world, such as China, Russia, and eastern European countries, have encouraged entrepreneurship after forbidding capitalism for decades.

The role of entrepreneurs in the United States is highlighted by the following:

- Most businesses in the United States (95 percent) are classified as small by the Small Business Administration.

- New businesses are formed at a rapid rate (about 600,000 per year).

- Small businesses generate almost half (48 percent) of the U.S. gross national product (GNP).

- Small businesses employ almost half (48 percent) of all U.S. workers.

- Small businesses created 60 percent of the new jobs in our economy in the last two decades.

FIGURE 36–13 The small businesses that dot the landscape of the United States are highly productive and responsive to changing needs of communities. *(Courtesy of Michael Dzaman)*

- Small businesses produce 2.5 times as many innovations (products, services, techniques) as large businesses (Figure 36–13).

As the U.S. economy continues toward an emphasis on services, the role and importance of small business and entrepreneurship should increase. This is expected to occur because small businesses are especially dominant in the service sector of the economy.

New businesses can be the center of innovation because they are not generally tied to existing ways of doing things. They have a sense of energy and vitality that comes from the entrepreneurial spirit.

The contribution of small businesses to the U.S. private-enterprise system is important. It is likely to remain so in the future.

STUDENT ACTIVITIES

1. Write the Terms to Know and their meanings in your notebook.
2. Develop a collage on the bulletin board illustrating entrepreneurship opportunities in agriscience.
3. Make a table showing the advantages and disadvantages of being (1) an employee and (2) an entrepreneur. The following format is suggested:

Being an Employee	**Being Self-employed**
Advantages	*Advantages*
1.	1.
2.	2.
etc.	etc.
Disadvantages	*Disadvantages*
1.	1.
2.	2.
etc.	etc.

4. Participate in a group sales project in the FFA, some other school group, 4-H, or some other community group. Encourage the group to operate the project to help all participants obtain useful business skills.

5. Organize a cooperative business within your class or FFA for buying or marketing a product of interest to the group.

6. Become familiar with the record book and record-keeping system used by members of your local FFA chapter or 4-H club for keeping records on individual projects.

7. Become an entrepreneur—own and operate your own business venture.

8. Get a summer or after-school job with a company similar to one you may be considering for your own future career. While working for that company, learn as much about running a business as you can.

9. Create an individual business plan using the information you have learned from this unit. Discuss with your parents and trusted advisors what steps you will need to take to make your business plan a reality.

SELF EVALUATION

A. Multiple Choice

1. The selection of a product or service to be sold for profit is called the
   a. buying function.
   b. distribution function.
   c. promoting function.
   d. selling function.

2. Determining reasons that customers may wish to buy a product or service is called the
   a. buying function.
   b. distribution function.
   c. promoting function.
   d. selling function.

3. Development of a plan to identify ways to make potential customers aware of a product or service is called the
   a. buying function.
   b. distribution function.
   c. promoting function.
   d. selling function.

4. When analyzing competing products or services, which is *not* a factor?
   a. price
   b. quality
   c. service
   d. supply

5. Which is *not* a function of managers?
   a. actuating
   b. controlling and evaluating
   c. organizing people and resources
   d. setting policy

6. A disadvantage of being an employee is
   a. overtime pay.
   b. regular hours.
   c. little or no control over future job.
   d. fringe benefits.

7. A major advantage of owning a business is
   a. control over your destiny.
   b. financial responsibility.
   c. the relationship of business income and family finances.
   d. a shorter work week.

B. Matching

_____ 1. Entrepreneur   a. Business venture

_____ 2. Entrepreneurship   b. Organizer; risk taker

_____ 3. Balance sheet   c. Expenses versus revenues

_____ 4. Profit and loss statement   d. One who devises

_____ 5. Inventor   e. Photograph of a business

Glossary

A horizon – layer near the soil surface consisting of mineral and organic matter.

    horizonte A – capa cerca de la superficie que consta de materia mineral y orgánica.

abiotic – nonliving.

    abiótico – no viviente.

abortion – loss of a fetus before it is viable.

    aborto – la muerte de un feto antes de que esté viable.

accent – distinctive feature or quality.

    acento – rasgo o calidad distintiva.

accent color – attention-getting color.

    color de enfasis – color llamativo.

acid – pH of less than 7.0.

    ácido – que tiene valor pH de menos que 7.0.

acidity – sourness or the tendency toward a low pH less than 7.0.

    ácidez – agrio o la tendencia a pH bajo 7.0.

active ingredient – a component that achieves one or more purposes of the mixture.

    principo activo – componente que alcanza uno o más de los objetivos de la mixtura.

actuating – putting a plan into action.

    accionar – ponerse en acción de un programa.

acute toxicity – a measurement of the immediate effects of a single exposure to a chemical.

    toxidad grave – una medida de los efectos inmediatos de una sola exposición a una sustancia química.

adaptation – a process that occurs as heritable traits favoring the survival of an organism are passed from one generation to the next.

    adaptación – un proceso que ocurre mientras que los rasgos hereditarios que favorecen la supervivencia de un organismo se pasan a partir de una generación al siguiente.

add-on-loan – the method used for calculating interest on consumer loans. The loan is repaid in installments of equal payments. Interest is added on at the beginning of the payments.

    préstamo adicional – el método que se usa para calcular el interés en los préstamos de consumidor. El préstamo se paga a plazos de pagos iguales. El interés se agrega al comienzo de los pagos.

Adenine (A) – a base in genes designated by the letter *A*.

    adenino – una base en los genes nominado por la letra «A».

adjourn – a motion used to close a meeting.

    suspender – una moción que se usa para levantar una junta.

adventitious root – root other than the primary root or a branch of a primary root.

    raíz adventicia – una raíz diferente de la raíz fundamental o de una rama de la raíz fundamental.

aeration – the mixing of air into water or soil to improve the oxygen supply of plants and other organisms.

    airear – el mezclar del aire con agua o tierra para hacer mejoras en la provisión de oxígeno de las platas y de otros organismos.

aeroponics – the growing of plants in which their roots hang in the air and are misted regularly with a nutrient solution.

    cultivo aeropónico – el cultivo de plantas donde las raíces de las plantas flotan en el aire y son rociadas regularmente.

aerosol – a can with contents under pressure.

    aerosol – lata con el contenido bajo presión.

agar – a plant nutrient medium in which plant tissues are placed during the tissue-culture process.

    agar – medio de nutrición de plantas donde los cultivos son colocados durante el proceso.

age (ripen) – to leave undisturbed for a period.

    madurarse – dejar sin tocar por un período.

aggregate culture – a material such as sand, gravel, or marbles that supports the plant roots.

    cultivo árido – una materia como la arena, el cascajo, o las canicas que sostiene las raíces vegetales.

aggregate – soil unit containing mostly clay, silt, and sand particles held together by a gel-type substance formed by organic matter.

    árido – elemento de tierra que incluyen principalmente partículas de tierra arcillosa, légamo y arena juntados por una sustancia de tipo de gel formada por materia orgánica.

agribusiness – commercial firms that have developed with or stem from agriculture.

    agroindustria – las firmas comerciales que han desarrollado con o se derivan de la agricultura.

agribusiness management – the human element that carries out a plan to meet goals and objectives in an agriscience business.

administración de agroindustria – la parte humana que lleva a cabo un programa para cumplir con las metas y los objetivos de un negocio de agriciencia.

agricultural – business, employment, or trade in agriculture, agribusiness, or renewable natural resources.

agrícola – negocio, empleo, o industria de la agricultura, de la agroindustria, o de los recursos naturales renovables.

agricultural economics – management of agricultural resources, including farms and agribusinesses.

economía agropecuaria – administración de recursos agrícolos, inclusive las fincas y las agroindustrias.

agricultural education – teaching and program management in agriculture.

educación agrícola – enseñanza y dirección de programa en la agricultura.

agricultural engineering – application of engineering principles in agricultural settings.

ingeniería agrícola – aplicación de los principios de ingeniería dentro del ambiente agrícola.

agricultural mechanics – design, operation, maintenance, service, selling, and use of power units, machinery, equipment, structures, and utilities in agriscience.

mecanía de agrícola – la industria que acarrea, clasifica, transforma, envasa, y comercializa los productos desde los orígenes de producción.

agricultural processing, products, and distribution – industry that hauls, grades, processes, packages, and markets commodities from production sources.

transformación, productos, y distribución de agrociencia – diseño, manejo, mantenimiento, reviso, venta, y uso de los motores, maquinaria, herramientas, estructuras y utilidades de la agrociencia.

agricultural supplies and services – businesses that sell supplies and agencies that provide services for people in agriscience.

provisiones y servicios de agrociencia – los negocios que venden provisiones, y las agencias que prestan servicios para las personas de agrociencia.

agriculture – activities concerned with the production of plants and animals, and the related supplies, services, mechanics, products, processing, and marketing.

agricultura – actividades que tratan de la producción de plantas y de animales, y los artículos, servicios, mecanismo, productos, transformación, y comercio relativos.

agriculture/agribusiness and renewable natural resources – broad range of activities in agriculture.

agricultura/agroindustria y los recursos naturales renovables – una gran variedad de actividades de la agricultura.

agriscience – the application of scientific principles and new technologies to agriculture.

agriciencia – la aplicación de los principios científicos y la tecnología reciente a la agricultura.

agriscience literacy – education in or understanding about agriscience.

educación agrocientífica – el estudio o la comprensión de la agrociencia.

agriscience professions – professional jobs dealing with agriscience situations.

profesiones de agrociencia – puestos profesionales que se ocupan de los asuntos de agrociencia.

agriscience research project – an original agricultural research project consisting of identifying a problem, a review of scientific literature, conducting a research project, and reporting the results.

proyecto de investigación de agrociencia – un proyecto de investigación original que consiste en identificar un problema, una crítica de literatura científica, conducir un proyecto de investigación y reportar los resultados.

agriscience skills profile – a record of skills that have been developed, and the level of competence in each.

perfil de pericia agrocientífica – un record de las habilidades que han sido desarrolladas y el nivel de competencia de cada una.

agronomy – science and economics of managing land and field crops.

agronomía – ciencia y economía del manejo de tierras y cultivos del campo.

air – colorless, odorless, and tasteless mixture of gases.
aire – mezcla de gases incolora, inodora e insípida.

air layering – plant propagation by girdling a plant stem, wrapping with sphagnum peat, and protecting with plastic.

acodo al aire – propagación de las plantas por anular el tallo de una planta, envolverlo con turba de esfagno, y protegerlo con plástico.

alkaline – pH of more than 7.0.
alcalino – que tiene valor pH de más que 7.0.

alkalinity – sweetness, or the tendency toward a high pH greater than 7.0.
alcalinidad – dulce o la tendencia a un alto nivel de pH sobre 7.0.

alluvial deposit – soil transported by streams.
depósito aluviale – depósito de tierras transportadas que quedan de arroyos de agua.

amend – a type of motion used to add to, subtract from, or strike out words in a main motion.

enmendar – un tipo de moción que se usa para aumentar, substraer, o quitar las palabras de una moción principal.

amendment – in addition to; change in.
enmienda – además de; rectificación en o cambio de.

ammonia/nitrite/nitrate – the chemical components generated during biological breakdown of animal wastes.

amoníaco/nitrito/nitrato – los componentes químicos que se producen durante la descomposición biológica de los desechos animales.

amortized loan – loan repaid in equal installments of principal and interest.

préstamo amoritzado – préstamo pagado en cuotas iguales del capital principal e interés.

amphibian(s) – organisms that complete part of their life cycle in water and part on land.

anfibios – organismos que cumplen parte del ciclo de la vida en el agua y otra parte en la tierra.

anatomy – the structure and arrangement of the various parts of the body of an animal or plant.

anatomía – la estructura y composición de varias partes del cuerpo, de un animal o una planta.

angiosperm – a plant with its seeds enclosed in a pod or seed case.

angiosperma – planta cuya semilla está envuelta por una vaina o una cápsula.

angora – wool from wool-producing rabbits or goats.

angora – lana de los conejos que dan la lana.

animal science – animal growth, care, and management.

ciencia de animales – cría, cuido, y manejo de animales.

animal science technology – use of modern principles and practices in animal growth and management.

tecnología de las ciencias de animales – uso de principios y prácticas modernos en la cría y el manejo de animales.

anion – ion that is negatively charged.

anione – ione que son cargados negativamente.

annual – a plant with a life cycle that is completed in one growing season.

anual – una planta cuyo ciclo de vida es completo en una estación de crecimiento.

annual ring – ring in a cross section of a tree root, trunk, or limb representing 1 year's growth.

capa cortical – capa de una sección transversal de raíz, tronco, o rama de un árbol que representa el crecimiento de un año.

annual weed – a weed that completes its life cycle within 1 year.

mala hierba anual – una mala hierba que cumple el ciclo de vida dentro de un año.

anorexia – the result of too little nutrition.

anorexia – el resultado de la falta de nutrición.

anther – portion of the male part that contains the pollen.

antera – parte de lo macho que contiene el polen.

antibiotic – substance used to help prevent or control infections and diseases of animals.

antibiótico – sustancia que se usa para evitar o controlar ciertas infecciones y enfermedades de los animales.

apiary – area where beehives are kept.

apiario – sitio en donde se mantienen las colmenas.

apiculture – beekeeping.

apicultura – arte de criar abejas.

aquaculturalist – trained professional involved in the production of aquatic plants and animals.

piscicultor – profesional educado, dedicado a la producción de plantas y animales acuáticos.

aquaculture – raising of finfish, shellfish, and other aquatic animals under controlled conditions. Also, the management of the aquatic environment for production of plants and animals.

piscicultura – la crianza de peces, mariscos, crustáceos, y otros animales acuáticos bajo condiciones controladas. Adicionalmente, la dirección del ambiente acuático para la producción de plantas y animales.

aquifer – water-bearing rock formation.

acuífero – formación terciaria que embebe agua.

arachnid – a living creature such as a spider or mite that is distinguished from insects by having light legs.

arácnido – creatura viva que chupa coma araña ácaro que se distingue de insectos por tener ocho patas.

arboriculture – care and management of trees for ornamental purposes.

arboricultura – cuidado y manejo de árboles para fines de adorno.

area of cell division – area of the root or stem tip in plants where new cells are formed.

área de división celular – área de la raíz o punta del tallo en plantas donde nuevas células son formadas.

area of cell elongation – portion of the new root where the cells start to become specialized and begin their function.

área de estiramiento celular – parte de la nueva raíz en donde las células se ponen a diferenciarse y desempeñar su función.

area of cell maturation – area where cells mature.

área de maduración celular – área en donde se maduran las células.

arid – an area deficient in rainfall; dry.

árido – un área que carece de precipitación.

arthropod – eight-legged animals such as spiders and mites.

artrópodo – animales de cuatro patas como arañas y acáridos.

artificial insemination (AI) – the placing of sperm cells in contact with female reproductive cells by a method other than natural mating.

inseminación artificial – el meter de las células de esperma en contacto con células reproductivas de la hembra mediante método de otra manera que el apareamiento natural.

asbestos – heat- and friction-resistant material.

asbesto – materia que es resistente a calor y fricción.

asexual reproduction – propagation using a part or parts of one parent plant.

reproducción asexual – reproducción que utiliza una o mas partes de una sola planta madre.

assets – anything the business owns.

activos – total de lo que posee una empresa.

asymmetrical – not equal on both sides of center.

asimétrico – que ambos lados no son iguales relativos al centro.

auctioneer – person who conducts the sales at auction markets.

subastador – persona que dirige las ventas en los mercados de subasta.

auction market – market where products are sold by public bidding.

mercado de subasta – mercado donde se venden los bienes por hacer pujas.

axil – the upper angle between the leaf or flower stem and the stalk of the plant.

axila – el ángulo superior entre la hoja o tallo de flor y el tallo de la planta.

axillary bud – bud that is located in the axil of the leaf.

botón axilar – botón que se encuentra en la axila de la hoja.

B horizon – soil below the A horizon or topsoil and generally referred to as subsoil.

horizonte B – la tierra debajo del horizonte A o de la capa superficial del suelo, y generalmente es conocido a nombre de subsuelo.

bacteria – one-celled, microscopic organisms.

bacteria – microorganismos vegetales unicelulares.

balance – state of quality and calm between items.

equilibrio – estado de bienestar y tranquilidad entre los cuerpos.

balance sheet – a statement of the assets and liabilities of a business on a specific date.

hoja de balance – un estado de bienes y deudas de un negocio en una fecha específica.

balanced ration – the daily supply of feed for an animal containing all of the needed nutrients in the proper proportions and amounts.

ración balanceada – el suplemento diário de alimento para un animal conteniendo todos los nutrientes necesarios en cantidades y proporciones apropiadas.

balled and burlapped (B&B) – plants that have been dug in the field, wrapped in burlap, and laced with a heavy twine.

empacado y harpillerado – plantas que han sido arrancados del campo, envueltas en arpillera, y atados con bramante grueso.

balling gun – a device used to place a pill in an animal's throat.

pistola de «balling» – aparato que se usa para meter una pastilla en la garganta de un animal.

band application – placing fertilizer about 2 inches to one side and slightly below the seed.

abonado en fajas – meter el abono más o menos a os pulgadas de un lado y un poco debajo de la semilla.

bantam – a small breed of chicken.

gallina bántum – raza de gallina miniatura.

barbed wire – wire with sharp points used to discourage livestock from touching fences.

alambre de espino – alambre con puntas agudas que se usa para que el ganado se haga desistir de tocar las cercas.

bare-rooted plant – plant that is dug for transplanting with little soil remaining on the roots.

planta de raíces expuestas – planta que son desenterradas para el transplante con sólo un poco de tierra que queda en las raíces.

barrow – castrated male animal of the swine family.

cerdo castrado jóven – macho castrado de la familia cochina.

bases – genetic material that connects strands of DNA.

bases – nombre original por materia genética que conecta fibras de ADN.

basic color – background color.

color básico – color de fondo.

bay – a natural indentation along a coastline with a small inlet and a protected area behind it where the water is relatively calm.

bahías – indentación natural en la costa con una pequeña entrada y una área protegida atrás donde el agua es relativamente calmada.

bedrock – the area below horizon C consisting of large soil particles.

roca de fondo – el área debajo del horizonte C que consiste en partículas de tierra más grandes.

beef – meat from cattle.

carne del ganado – carne del ganado.
Note: In Spanish, there is no equivalent to "beef"—it is just called "cattle meat."

BelRus – superior baking potato bred to grow well in the Northeast.

«BelRus» – patata al horno que es producida para cultivarse bien en el noroeste de los EE.UU.

Beltsville Small White – breed of turkey that weighs only 8 to 12 lb at maturity.

«Beltsville Small White» – raza de pavo que pesa solamente 8 a 12 libras al madurarse.

biennial – a plant that takes two growing seasons or 2 years from seed to complete its life cycle.

bianual – una planta que toma dos estaciones de crecimiento o dos años desde semilla para completar su ciclo de vida.

biennial weed – a weed that will live for 2 years.

mala hierba bienal – una mala hierba que vive por dos años.

binomial – having two names.

binomio – que tiene dos nombres.

bio – life or living.

bio – vida o viviente.

biochemistry – chemistry as it applies to living matter.

bioquímica – química que se aplica a la materia viviente.

biological control – pest control that uses natural control agents.

control biológica – control de los insectos o animales nocivos que utiliza los agentes de control natural.

biology – basic science of the plant and animal kingdoms.

biología – la ciencia básica de los reinos vegetal y animal.

biotechnology – use of cells or components of cells to produce products or processes.

biotecnología – uso de células o componentes de células para producir productos o procesos.

biotic disease – disease caused by living organisms.

enfermedade biótica – enfermedade causada por organismos vivientes.

bit – metal mouthpiece used to control a horse.

bocado – aparato que se mete sobre la boca del caballo para frenarlo.

blade – the upper portion of the grass leaf.

brizna – la parte superior de la hoja de pasto.

blanching – the brief scalding of food before freezing.

escaldar – el sumergir de los alimentos por un tiempo breve en agua hirviendo antes de congelación.

bleeding out – draining blood from an animal.

hacer sangrías – drenar la sangre de un animal.

blemish – in horses, any abnormality that does not affect the use of the horse.
    defecto – de los caballos, cualquiera abnormalidad que no afecta el uso del caballo.

block beef – meat sold over the counter to consumers.
    carne descuartizada – carne de res que se vende al por menor a los consumidores.

bloom – production of a flower by a plant.
    floración – producciòn de una flor por una planta.

boar – male animal of the swine family.
    verraco – el macho de la familia cochina.

board foot – a unit of measurement for lumber that equals $1 \times 12 \times 12$ inches.
    «board foot» (b.ft.) – unidad de medida de los maderos que es igual a 1″ por 1″ por 12″.

bog – water-logged area.
    ciénaga – área de aguas estancadas.

bone – the main component of the skeletal system.
    hueso – el componente principal del esqueleto.

bone marrow – material inside of bones that makes blood cells.
    médula de hueso – materia dentro de los huesos que fabrica los glóbulos.

border planting – a planting that is used to separate some part of the landscape from another. It might also be used as a fence or a windbreak.
    arriate – un cuadro de plantas que se usa para separar alguna parte del paisaje de la otra. Se puede usar como cerca o también protección contra el viento.

bovine somatotropin (BST) – hormone that stimulates increased milk production in cows.
    somatotropino bovino – una hormona que estimula un aumento de secreción láctea en las vacas.

brackish water – waters influenced by tide and river flow with intermediate salinity of 3 to 22 percent.
    agua salobre – aguas influidas por la marea y el flujo de río, que tienen salinidad de 3 a 22 por ciento.

bract – modified leaf that is often brightly colored and showy.
    brácteras – hoja modificada que muchas veces es colorada brillantemente y es llamativa.

bran – skin or covering of a wheat kernel.
    salvado – cascarilla del grano de los cereales.

breed – a group of animals having similar physical characteristics that are passed on to their offspring.
    raza – grupo de animales que tienen características físicas semejantes que se pasan para adelante a su descendencia.

broadcasting – scattering seeds rather than sowing them in rows.
    sembrar al voleo – esparciendo al aire las semillas en vez de sembrarlas en hileras.

broadcast planter(s) – scatters seed in a random pattern.
    sembrador al voleo – éste esparce las semillas en una pauta hecha al azar.

broiler – young chicken grown for meat.
    pollo tomatero – gallina jóven que se cría para comer.

buck – male animal of the goat, deer, or rabbit family.
    cabrón – macho de la cabra.
    conejo – macho de la coneja.

bud grafting – the union of a small piece of plant tissue containing a bud and a plant rootstock.
    injerto de brote – la unión de un pequeño pedazo de corteza que contiene un brote y una rama de una planta enraízada.

budding rubber – a piece of rubber band that is used to wrap the location where a bud has been inserting into the stock in a process called bud grafting.
    una liga de brote – una tira de hule usada para envolver el lugar donde un brote ha sido insertado en la planta en el proceso llamado injerto de brote.

buffer – a substance in a solution that tends to stabilize the pH.
    estabilizador – substancia en una solución que tiende a estabilizar el pH.

bulb – short underground stem surrounded by many overlapping, fleshy leaves.
    bulbo – tallo corto, subterráneo envuelto por muchas hojas carnosas y traslapadas.

bull – male animal of the cattle family.
    toro – macho del ganado vacuno.

bunch-type grass – a turfgrass that grows in clumps.
    césped de manojo – un tipo de césped que crece en manojos.

business meeting – a gathering of people working together to make decisions.
    sesión (de negocio) – asamblea de personas trabajando juntas para hacer decisiones.

butterfat – fat found in milk.
    grasa de la leche – grasa que se encuentra en la leche.

buying function – selection of a product or service to be marketed or sold for a profit.
    función adquisitivo – selección de un producto o servicio para que se ponga en venta o se venda algo con ganancia.

by-product – a secondary product left from the production of a primary commodity.
    subproducto – un producto solbrante que queda después de la producción de un mercancía primaria.

C horizon – soil below the B horizon; it is important for storing and releasing water to the upper layers of the soil.
    horizonte C – la tierra debajo del horizonte B; es importante para la guarda y suministración del agua a las capas superiores de la tierra.

calf – young member of the cattle family.
    becerro – un jóven del ganado vacuno.

calve – to give birth in cattle.
    parir – en el ganado, dar a luz a la cría.

calyx – group of sepals of a flower.
    cáliz – grupo de los sépalos de la flor.

cambium – growth layer in a tree root, trunk, or limb.
　cambium – capa de crecimiento de raíz, tronco, o ramo del árbol.

cane cutting(s) – stems are canelike and cuttings are cut into sections that have one or two eyes, or nodes.
　esqueje a la caña – los tallos son parecidos a la caña y se cortan las talas en secciones que tienen uno o dos yemas o nudos.

cane – mature wood in grape vines and some other fruits that has produced fruit and lost the leaves. The next year's new shoots and fruit grow from the cane.
　caña – madera madura en viñedos u otras frutas que han producido y perdido sus hojas. Los retoños del próximo año crecen de las cañas.

canning – storing food in airtight containers.
　enlatado – guardar los comestibles en recipientes herméticos.

cannula – blunt needle.
　cánula – aguja desafilada.

canopy – the top of the plant that has the framework and leaves.
　partes aéreas – la parte superior de la planta que tiene la estructura y las hojas.

canter – in horses, a fast three-beat gait.
　medio galope – de caballos, un paso rápido de tres tiempos.

capability class – soil classification indicating the most intensive but safe land use.
　clase de capacidad – clasificación de suelos indicando el uso más intensivo pero seguro de la tierra.

capability subclass – soil group within a class designated by a small letter.
　subclase de capacidad – grupo de suelos dentro de una clase a que se denomina por letra minúscula.

capability unit – soil group within a subclass.
　unidad de capacidad – grupo de suelos dentro de una subclase.

capillary water – water held by soil particles and available for plant use.
　agua capilar – agua sujetada por partículas de tierra y disponible para el uso de las plantas.

capital – money or property.
　capital – dinero o propiedad.

capital investment – money spent on commodities that are kept 6 months or longer.
　inversión de capital – dinero gastado en mercancía que se guarda por seis meses o más.

carbohydrate – starches and sugars that provide energy in the diet.
　carbohidrato – féculas y azúcares que proporcionan energía en la dieta.

carbon monoxide – colorless, odorless, and highly poisonous gas; carbon dioxide and water combine to make plant food and release oxygen.
　monóxido de carbono – gas incoloro, inodoro, y sumamente tóxico; bióxido de carbono y agua se combinan para hacer alimento vegetal y soltar oxígeno.

carcass – body of meat after the animal has been eviscerated.
　res abierta en canal – el cuerpo de carne después de haber sido destripado el animal.

carcinogen – a chemical capable of producing a tumor.
　agente cancerigeno – una sustancia química capaz de producir un tumor.

career – a person's occupation or profession.
　carrera – la profesión u ocupación de una persona.

career exploration – learning about occupations and jobs as part of the process of choosing one's life work.
　exploración de carrera – aprender sobre ocupaciones y trabajos como parte del proceso de elejir el trabajo de la vida de un individuo.

carrying capacity – the number of animals that a pasture will provide feed for.
　capacidad de carga – la cantidad de animales a las que el pasto proporcionará pienso.

casein – predominant protein in milk.
　caseína – proteina prevalente de la leche.

cash crop – a crop grown for cash sale.
　cultivo comercial – una cosecha cultivada para vender por pago al contado.

cash flow statement – a financial document that describes the availability of operating funds for a business at different points in time.
　estado de dinero efectivo en caja – documento financiero que describe los fondos disponibles para negocios en diferentes puntos de tiempo.

cation – ion that is positively charged.
　catione – ione que son cargados positivamente.

causal agent – an organism that produces a disease.
　agente causal – un organismo que produce una enfermedad.

cell – a unit of protoplasmic material with a nucleus and cell walls.
　célula – unidad de materia protoplásmica con núcleo y paredes.

cellulose – woody fiber parts that make up plant cell walls.
　celulosa – partes leñosas de fibra que constituyen las paredes de célula vegetal.

central nervous system – the brain and the spinal cord.
　sistema nervioso central – el cerebro y la médula espinal.

cereal crop – grasses grown for their edible seeds.
　cultivo de cereal – hierbas que son cultivadas por sus semillas comestibles.

cheese – milk that is exposed to bacterial fermentation.
　queso – leche que está expuesta a la fermentación bacterial.

chemical control – the use of pesticides for pest control.
　control químico – el uso de pesticidas para control de los insectos y animales nocivos.

chemistry – science dealing with the characteristics of elements or simple substances.
　química – la ciencia que trata de las características de los elementos o sustancias sencillas.

chevon – meat from goats.
　carne de chivo – carne de ganado caprino.

chick – newborn chicken or pheasant.
　polluelo – gallina or faisán recién nacido.

chlorofluorocarbons – group of compounds consisting of chlorine, fluorine, carbon, and hydrogen used as aerosol propellants and refrigeration gas.

cloroflurocarbonos – grupo de compuestos que se compone de cloro, flúor, carbono, e hidrógeno usados como propulsores de aerosol y de gas de refrigeración.

chlorophyll – green pigment in leaves.
clorofila – pigmento verde de las hojas.

chloroplast – membrane-bound body inside a cell containing chlorophyll pigment; necessary for photosynthesis.
cloroplasto – dentro de la célula, cuerpo tapado con membrana que contiene la clorofila; necesario para la fotosíntesis.

chlorosis – yellowing of the leaf.
clorosis – el volverse amarillo de la hoja.

chromosome – the rodlike carrier for genes.
cromosoma – cuerpo en forma de bastoncillo que lleva los genes.

chronic toxicity – a measurement of the effect of a chemical over a long period and under lower exposure doses.
toxidad crónica – una medida del efecto de una sustancia química durante un período largo de tiempo y bajo dosises más pequeños de exposición.

circulation – activity in a plantscaped area.
circulación – actividad en un área ajardinado.

circulatory system – the system that provides food and oxygen to the cells of the body and filters waste materials from the body.
aparato circulatorio – el conjunto de órganos que proporciona los alimentos y oxígeno a las células del cuerpo, y filtra los residuos desde el cuerpo.

citizenship – functioning and interacting in a society in a positive way.
ciudadanía – funcionamiento e interacción en una sociedad en una forma positiva.

clay – smallest of soil particles; less than 0.002 mm.
arcilla – las partículas más pequeñas del suelo; menos de 0.002 mm.

clean culture – any practice that removes breeding or overwintering sites of a pest.
cultivo limpio – cualquiera práctica que elimina los criaderos o sitios de hibernación de los animales o insectos nocivos.

clear cut – removal of all marketable trees from an area.
corte raso – el cortar y traslado de todos los árboles comerciales de un área.

climate – the weather conditions of a specific region.
clima – las condiciones atmosféricas de una región específica.

climatic conditions – temperature, temperature range, and precipitation.
condiciones climáticas – temperatura, variedad de temperatura, y precipitación.

clod – a lump or mass of earth.
terrón – masa de tierra.

clone – exact duplicate.
clon – una duplicación exacta.

closebreeding – mating of father to daughter, mother to son, or brother to sister.
procreación en consanguinidad – acoplamiento de padre a hija, madre a hijo, o hermano a hermana.

coach horse – type of horse developed to pull stagecoaches.
caballo de coche – tipo de caballo desarrollado para arrastrar diligencias.

coarse texture – a turfgrass with a wide leaf blade.
tierra de textura gruesa – césped que tiene una brizna ancha.

coarse textured (sandy) soil – loose and single grained soil.
suelo textured grueso (de sandy) – suelo granuloso flojo y solo.

coccidiosis – disease of poultry that costs growers nearly $300 million a year.
coccidiosis – enfermedad de los aves de corral que les cuesta a los agricultores casi 300 milliones de dólares al año.

cock – adult male chicken or pheasant.
gallo – macho adulto de los aves de corral.

cockerel – young male chicken or pheasant.
gallo jóven – gallo jóven.
faisán jóven – faisán jóven.
Note: Spanish does not have an equivalent to "cockerel"—a cockerel is referred to as either a "young male chicken" or "young male pheasant."

cold frame – a bottomless wood box with a sloping glass in which early season plants are protected from freezing temperatures.
bastidor frío – caja de madera sin fondo con un vidrio inclinado de base en el cual plantas tempranas de la estación son protejidas de temperaturas congelantes.

collagen – chief component of connective tissue.
colagen – componente principal de tejido conjuntivo.

collar – light-green or white banded area on the outside of a leaf blade.
cuello – parte alargada y estrecha de color blanco o verde ligero en el exterior del limbo de una hoja.

colluvial deposit – soil deposited by gravity.
coluvial – suelo depositado por la gravedad.

color breed – breed of horses based on color.
cría selectiva basado en su color – cría selectiva basado en su color.
Note: The translation for «color breed» is the same as that of the English definition provided.

colostrum – first milk produced by mammals; high in antibodies.
calostro – primera leche producida por mamíferas; tiene niveles elevados de anticuerpos.

colt – young male horse or pony.
potro – caballo macho jóven.

comb – wax material on and in which bees store honey.
panal – material de cera en y sobre la cual las abejas depositan la miel.

combine – machine that is used to cut and thresh seed crops such as grain.
segadora trilladora – máquina que siega y trilla el grano en el campo.

commensalism – one type of wildlife living in, on, or with another without either harming or helping it.

comensalismo – una clase de fauna que vive en, encima de, o por, otra clase sin dañarla ni tampoco ayudarla.

commission – fee for selling a product.

comisión – honorario por la venta de un producto.

Commodity Credit Corporation (CCC) – institution that lends money for production of farm commodities.

corporación de crédito para mercancía – institución que presta dinero para la producción de mercancías agrícolas.

commodity exchange – organization licensed to manage the buying and selling of commodities.

la bolsa – organización autorizada para administrar la compra y la venta de productos.

common name – name given to a pesticide by a recognized committee on pesticide nomenclature.

nombre genérico – nombre dado a un pesticida por un cómite reconocido de nomenclatura de pesticida.

comparative advantage – the ability to produce a commodity with greater returns than those of competitor because of a favorable condition such as climate.

ventaja comparativa – la abilidad de producir mercancías con más grandes lucros que los del competidor debido a condiciones favorables tales como el clima.

competition – two types of wildlife eating the same source of food.

competición – dos clases de fauna comiendo de la misma fuente de alimentación.

complete fertilizer – a fertilizer having nitrogen, phosphorus, and potassium.

abono completo – abono que tiene nitrógeno, fósforo, y potasio.

compost – mixture of partially decayed organic matter.

estiércol vegetal – mixtura de materia orgánica descompuesta en parte.

compound – a chemical substance that is composed of more than one element.

compuesto – una sustancia química que se compone de más de un elemento.

compound leaf – two or more leaves arising from the same part of the stem.

hoja digitada – doso más hojas brotando del mismo rabo.

concentrate – feed high in total digestible nutrients and low in fiber.

concentrado – pienso con alto nivel de alimentos nutritivos digestibles totales y bajo en fibra vegetal.

condensed milk – milk that has had water removed and sugar added.

leche condensada – leche de la que se ha quitado el agua y añadido el azúcar.

condominium – apartment building or unit in which the apartments are individually owned.

condominio – casa de pisos o unidad en la cual tiene su propio dueño cada apartamento.

conifer – evergreen tree with needlelike leaves and that produce cones.

conífera – árboles de hoja perene como aguja que produce conos.

conservation plan – plan developed to use land for its maximum production and water conservation without unacceptable damage to the land.

plan de conservación – plan desarrollado para el uso de producción máxima de la tierra y conservación de agua sin inaceptable daño a la tierra.

conservation tillage – techniques of soil preparation, planting, and cultivation that disturb the soil the least and leave the maximum amount of plant residue on the surface.

labranza de conservación – técnicas de preparación del terreno, siembra y cultivación que el menos perturba la tierra y que deja la cantidad máxima de residuos vegetales en la superficie.

consumer demographics – categories of information about preferences of consumers or potential consumers.

demográfica del consumidor – categorías de información sobre preferencias de consumidores o clientes potenciales.

consumer – person who uses a product.

consumidore – persona que usan un producto.

contact herbicide – a herbicide that will not move or translocate within the plant.

herbicida de contacto – un herbicida que no se mueve ni se desplaza dentro de la planta.

contagious – diseases that can be spread by contact.

contagiosas – enfermedades que se pueden transmitir por contacto.

container grown – plants that are grown in pots or other type of container and are shipped in the container.

crecido de contenedor – plantas que se crecen en macetas u otros tipos de contenedor y son transportadas en el contenedor.

contaminate – to add material that will change the purity or usefulness of a substance.

contaminar – añadir material que cambiará la pureza o utilidad de una sustancia.

continuous-flow system – the nutrient solution flows constantly over the plant roots.

sistemas de flujo continuo – la solución nutritiva fluye constantemente sobre las raíces de las plantas.

contour – level line around a hill in which all points along the line are the same elevation.

contorno – linea de nivel alrededor de una colina en la cual todos los puntos a lo largo de la colina tienen la misma elevación.

contour practice – operations such as plowing, discing, planting, cultivating, and harvesting across the slope and on the level.

trabajos de cosecha a lo largo del declive – actividades así como arar, abrir los surcos con discos, sembrar, cultivar, y cosechar a lo largo del declive y sobre la llanura.

controlled atmosphere – adjusting oxygen and carbon dioxide where food is stored and while it is being transported.

atmósfera controlada – que ajusta los oxígeno y dióxido de carbono donde se almacena la comida, y mientras está transportada.

convection oven – an oven that heats food by circulating hot air.

horno de convección – un horno que calenta la comida por circular aire caliente.

conventional tillage – land is plowed, turning over all crop residues.

labranza clásica – se ara la tierra, volteando todos los residuos de la cosecha.

cool-season turfgrass – turfgrass adapted to the northern regions of the U.S. that grow best at 60° to 75° F.

tepe de temporada fresca – tepe adaptado a las regiones norteñas de los Estados Unidos, las cuales crecen mejor a los 60° hasta 65° F.

Cooperative Extension System – an educational agency of the U.S. Department of Agriculture and an arm of land-grant state universities.

System Cooperativo de Extensión – una agencia educativa del Departamento de Agricultura de los Estados Unidos, y un ramo de las universidades estatales de concesión de terrenos.

cooperative – group of producers who join together to market a commodity and/or to purchase supplies.

cooperativa – grupo de productores que se reunen para mercar mercancía y/o para comprar abastecimientos.

corm – short, flattened underground stem surrounded by scaly leaves.

bulbos – tallo corto aplanado, subterréaneo envuelto por hojas escamosas.

corn picker – machine that removes ears of corn from stalks.

recolectora de maíz – máquina que recoge las espigas de maíz de los tallos.

corolla – collectively, all of the petals of the flower.

corola – colectivamente, todos los pétalos de una flor.

cottage cheese – a product made of skimmed milk.

requesón – un producto hecho de leche desnatada.

cotton gin – machine that removes cotton seed from cotton fiber.

desmotadora – máquina que saca la semilla a la fibra del algodón.

courage – willingness to proceed under difficult conditions.

valor – la voluntad de seguir en adelante a pesar de condiciones duras.

cover crop – close-growing crop planted to protect the soil and prevent erosion.

planta protectora – cultivo que crece muy junto y se planta para proteger las tierras de la erosión.

cow – female of the cattle family that has given birth.

vaca – hembra de la familia vacuna que ha parido.

cream – milk containing 40 percent butterfat.

nata – leche que tiene 40 por ciento de grasa de leche.

credit – money borrowed.

crédito – dinero prestado.

creep feed – feed provided to young animals to supplement their mother's milk.

alimentación suplementaria – pienso suministrado a la cría de animales para suplementar la leche de amamantamiento.

crop rotation – planting of different crops in a given field every year or every several years.

rotación de cultivos – el sembrar de cultivos diferentes en un campo determinado cada año o cada dos o tres años.

crop science – use of modern principles in growing and managing crops.

ciencia de cultivos – el uso de principios modernos en el cultivar y manejar de cultivos.

crossbreeding – mating of parents of two different breeds.

híbrido – el cruce de animales con padres de dos razas diferentes.

crown – an unelongated stem of major meristematic tissue of turfgrass.

corona – un tallo no alargado de tejido meristemático mayor de césped.

crumbs – aggregates.

gránulos – áridos.

crustacean – aquatic organism with an exoskeleton that molts during growth.

crustáceo – organismo aquático con dermatoesqueleto que mudan durante el crecimiento.

cull – remove from the herd.

entresacar – sacar de entre el rebaño.

cultivar – a group of plants with a particular species that has been cultivated and is distinguished by one or more characteristics; through sexual or asexual propagation, it will keep these characteristics.

variedad obtenida por selección – un grupo de plantas con una especie particular que ha sido cultivada y que se distingüe por una o más características; por la propagación sexual o asexual mantendrá estas características.

cultivation – the act of preparing and working soil.

cultivo – la acción de preparar y labrar la tierra.

cultural control – pest control that adapts farming practices to better control pests.

compaña contra los animales dañinos – control de los animales o insectos nocivos que adapta las prácticas agrícolas para mejorar el control de ellos.

cuticle – top-most layer of the leaf; waxy protective covering of the leaf.

cutícula – la superficie de la hoja; capa protectora cerosa de la hoja.

cutting – vegetative part removed from the parent plant and managed so it will regenerate itself.

tala – la parte vegetativa que se quita de la planta progenitora para que se regenere.

Cytosine (c) – a base in genes designated by the letter *C*.

citosino – una base en los genes nominado por la letra «C».

dam – mother.

progenitora – madre.

deciduous – plants that lose their leaves every year.

caduco – plantas cuyas hojas se caen cada año.

decompose(d) – decay into soil-building materials and nutrients.

descomponerse – marchitarse y corromperse, convirtiéndose en materiales y alimentos nutritivos que fortalecen la tierra.

decomposer – an organism that is capable of breaking down dead plant and animal matter into soil components.

decomposer – un organismo que es capaz de analizar la materia muerta de la planta y del animal en componentes del suelo.

deficiency – less available than needed for optimum growth.

carencia – cuando no se obtiene lo necesario para el crecimiento óptimo.

deficiency disease – condition resulting from improper levels or balances of nutrients.

enfermedade por carencia – estado de salud resultando de niveles bajos o equilibrios malos de alimentos nutritivos.

defoliate – to strip a plant of its leaves.

deshojar – quitar las hojas a una planta.

dehydration – removing moisture with heat.

deshidratación – quitar el agua a algo, usando calefacción.

dehydrator – a device for drying food.

deshidratador – un aparato para secar alimentos.

dehydrofrozen product – removing moisture after partial cooking, and then freezing.

producto desidrocongelado – parcialmente y después congelarlo.

demand – the amount of a product wanted at a specific time and price.

demanda – conjunto de productos pedidos en un tiempo específico y por un precio específico.

deoxyribonucleic acid (DNA) – coded genetic material in a cell.

ácido desoxirribonucleico (ADN) – la materia genética codificada de una célula.

desert – large areas of the world having an adequate balance of soil particles and minerals for plant growth, but having continuous, severe water shortages.

desierto – grandes áreas del mundo con un balance adecuado de partículas de tierra y minerales para el crecimiento de las plantas, pero con continua escasez de agua.

development limitations – limitations imposed by resources or use in a plantscape.

límites de explotación – límites impuestos a un paisaje por recursos o utilidad.

dicotyledon (dicot) – plant with two seed leaves.

dicotiledónea – plantas con dos hojas de semilla.

digestive system – system that provides food for the body of the animal and for all of its systems.

aparato digestivo – el conjunto de órganos que proporciona los alimentos al cuerpo del animal y para todos sus sistemas.

diminishing returns – point at which each additional unit of input decreases the output or returns.

rendimientos decrecientes – el punto en donde cada unidad adicional de insumo disminuye el rendimiento o el reembolso.

dipping – the process of treating animals for external parasites by walking or swimming them through a medicated bath.

desinfección por inmersión – el proceso de curar animales de los parásitos exteriores por recorrerles por un balneario medicinal.

direct sale – the selling of crops or animals directly to a processor by the producer.

venta sin intermediario – a la venta de animales directamente a los transformadores por el productor.

disassembly process – dividing the carcass into smaller cuts.

carnear – tomar la res abierta en canal y descuartizarla.

discount loan – a loan in which interest is subtracted from the principal at the time the loan is made.

empréstito de descuento – un empréstito en que se substraye el interés del principal en el momento en que se presta.

disease triangle – the term applied to the relationship of the host, pathogen, and the environment in disease development.

triángulo de enfermedad – aplícase al desarrollo de una enfermedad: la relación entre el huésped, el agente patógeno, y el medio ambiente.

disease – infective agent that results in decreased health in living things.

enfermedades – agente infecciosos que resultan en el mal estado de salud en los cuerpos vivientes.

disinfectant – material that destroys infective agents such as bacteria and viruses.

desinfectante – materiale que matan a los agentes infecciosos como bacterias y viruses.

dissolved oxygen – oxygen that is dissolved in water.

oxígeno disuelto – oxígeno que es disuelto en agua.

distribution function – physical organization and delivery of product or service.

función de distribución – organización física y expedición de un producto o servicio.

distributor – person or business storing food for transport to regional markets.

distribuidor – persona o empresa almacenando alimentos para transportarlos a mercados regionales.

DNA – deoxyribonucleic acid.

ADN – ácido desoxirribonucleico.

dock – remove or shorten tails of certain animals.

cortar la cola – quitar o acortar las colas de ciertos animales.

doe – female of the goat, deer, or rabbit family.

cabra – hembra del cabrón.
coneja – hembra del conejo.

domestic – of or pertaining to the household.

doméstico – del hogar o perteneciendo al hogar.

domesticate – the act of taming an animal for use of humans.

domesticar – el acto de amansar un animal para el uso de los humanos.

dominant – gene that expresses itself to the exclusion of other genes.

dominante – el gene que se manifiesta con exclusión de otros genes.

donkey – member of the horse family with long ears and a short erect mane.

burro – miembro de la familia de caballos con orejas grandes y crines rapados.

dormant – resting stage, no active growth.

 dormición – estado de descanso; no hay crecimiento activo.

double-eye cutting – used when the plant has leaves that are opposite.

 esqueje de yema doble – empleado cuando la planta tiene hojas simétricas.

draft – animals used for work.

 tiro – animales que se usan para cargar.

draft horse – type of horse bred for work.

 caballo de tiro – tipo de caballo criado para cargar.

drake – male duck.

 pato – el macho de los patos.

drenching – a process of administering drugs orally to animals.

 administrar una pócima – el proceso de administrar las drogas oralmente a los animales.

dressing percentage – the proportion of the live weight of an animal to the weight of the carcass before cooling.

 rendimiento a la canal – la proporción del peso vivo del animal con el peso de la canal antes de enfriarse.

drift – movement of a pesticide through the air to nontarget sites.

 arrastre – movimiento de un pesticida por el aire hasta sitios intencionales.

drill planter – machine that plants seed in narrow rows at a high population.

 sembradora a chorillo – máquina que siembra semillas en surcos angostos con una alta concentraciòn.

drip irrigation – the use of small tubes to deliver irrigation water to the roots of crop plants in a uniform manner.

 irrigación de goteo – el uso de pequeños tubos que llevan agua de irrigación a las raíces de pantas de cultivo.

drip line – the edge of the tree where the branches stop.

 lindero para viento – el límite del árbol donde terminan las ramas.

drone – male bee.

 zángano – macho de la abeja.

drupe – a stone fruit.

 drupa – fruta que contiene un hueso.

dry-heat cooking – surrounding food with dry air while cooking.

 cocinar en estufa de secado – rodeando la comida con aire seco mientras cocinarla.

dry matter – material left after all water is removed from a feed material.

 materia seca – materia que queda después de quitar toda el agua de material de pienso.

dry period – period when a cow is not producing milk.

 período seco – período de tiempo cuando la vaca no está secretando leche.

duckling – young duck.

 patito – cría del pato.

dwarf – tree that has rootstock that limits growth to 10 feet or less.

 enano – árbol que tiene el rizoma, lo cual que limita el crecimiento a 10 pies o menos.

ear notching – system of permanently marking animals by cutting notches from their ears.

 poner una señal – práctica de señalar animales permanentemente por hacer una muesca en las orejas.

ease of working – refers to the difficulty in cutting, shaping, nailing, and finishing wood.

 facilidad de tallar – se refiere a la dificultad en cortar, labrar, clavar, y acabar madera.

economic threshold level – the level of pest damage to justify the cost of a control measure.

 «precio de umbral» – el nivel de daño a causa de animales o insectos nocivos que justifica el costo de una medida de control.

edible – fit to eat or consume by mouth.

 comestible – apropiado para comerse o tomar por la boca.

egg – female reproductive cell.

 óvulo – células reproductores de la hembra.

ejaculate – the amount of semen produced at one time by a male.

 ejaculación – la cantidad de semen expulsada a una vez por el macho.

element – a uniform substance that cannot be further decomposed by ordinary means.

 elemento – una sustancia uniforme que no puede descomponerse por medios ordinarios.

embryo – fertilized egg.

 embrión – óvulo fecundado.

embryo transfer – a process that removes fertilized eggs from a female and places them in another female who carries them until birth.

 traslado de embrión – un proceso que quita óvulos fecundados de la hembra y los coloca en otra hembra, que los lleva hasta el parto.

endocrine or hormone system – a group of ductless glands that release hormones into the body.

 sistema endocrino u hormonal – conjunto de glándulas de secreción interna las cuales secretan hormonas por el cuerpo.

endosperm – interior of a wheat kernel that will become wheat flour.

 endosperma – parte interior de un grano de trigo que llegará a ser harina de trigo.

enterprise – commercial business to generate profits.

 empresa – establecimiento comercial que produce ganancias.

enthusiasm – energy to do a job and inspiration to encourage others.

 entusiasmo – energía para hacer una obra y inspiración para animar a otras personas.

entomology – science of insect life.

    entomología – ciencia de la vida de insectos.

entomophagous – insects that feed on other insects.

    entomófagos – insectos que se alimentan con otros insectos.

entrepreneur – person organizing a business, trade, or entertainment.

    empresario – persona que organiza un negocio, comercio, o espectáculo.

entrepreneurship – process of planning and organizing a business.

    empresiales – el proceso de planear y organizar un negocio.

entrepreneurship supervised agricultural experience – enterprises that are supervised by teachers and conducted by students as owners or managers of businesses based on agriscience.

    experiencia agrocientífica supervisada por empresarios – empresas que están supervisadas por maestros y conducidas por estudiantes como dueños o gerentes de negocios basados en agrociencia.

environment – All the conditions, circumstances, and influences surrounding and affecting an organism.

    medio ambiente – espacio y masa que nos rodea.

epidermis – surface layer on the lower and upper sides of the leaf.

    epidermis – capa de las partes superior e inferior de la hoja.

equine – having characteristics of a horse.

    equino – tener características de un caballo.

equitation – the art of riding on horseback.

    equitación – el arte de montar caballo.

eradicant fungicide – a fungicide applied after disease infection has occurred.

    fungicida desarraigante – un fungicida aplicado después de que se ha ocurrido la infección.

eradication – complete control or removal of a pest from a given area.

    desarraigo – control completo o extirpación de un animal o insecto dañino de un área determinada.

erosion – wearing away.

    erosión – desgasto o deterioro lento.

estrogen – hormone that regulates the heat period.

    estrógeno – hormona que regula el período en que el animal está en celo.

estrus – heat period or time when female animal is receptive to the male animal.

    celo – período o tiempo fecundo cuando la hembra está receptiva al macho.

estuary – ecological system including bays, streams, and tidal areas influenced by brackish water.

    estuario – sistema ecológico inclusive bahías, arroyos, y tierras bajas del litoral que son afectados por agua salobre.

evaporated milk – milk that has had water removed.

    leche deshidratada – leche a que el agua se ha quitado.

evaporation – changing from a liquid to a vapor or gas.

    evaporación – transformación de un líquido o vapor a gas.

evergreen – plants that do not lose their leaves on a yearly basis.

    planta de hoja perenne – plantas a que no se caen las hojas cada año.

evisceration – removal of the viscera.

    destripación – extirpación de las vísceras.

ewe – female animal of the sheep family.

    oveja – hembra del carnero.

executive meeting – meeting of the officers to conduct business of the organization between regular meetings.

    reunión del consejo de dirección – reunión de los directores para dirigir el negocio de la organización en medio de reuniones corrientes.

exoskeleton – the external body wall of an insect.

    dermatoesqueleto – caparazón o pared exterior del cuerpo de un insecto.

experience – anything and everything observed, done, or lived through.

    experiencia – cada y toda cosa observada, hecha, o vivida.

exploratory supervised agricultural experience – a program in which the teacher conducts activities that allow students to become involved in learning a variety of subjects about agriscience and careers related to agriculture.

    experiencia agrocientífica exploratoria supervisada – un programa en el cual el maestro conduce actividades que permiten a los estudiantes envolverse en el aprendizaje de una variedad de materials sobre agrociencia y carreras relacionadas a la agricultura.

extemporaneous – speech delivered with little or no preparation.

    extemporáneo – dícese de un discurso pronunciado con poca preparación.

extemporaneous speaking – a form of public speaking wherein some preparation is made, but the speech is not written out or memorized.

    discurso improvisado – una forma de discurso público en donde se hace una cierta preparación, solamente el discurso no se pone en escrito ni se memoriza.

extravaginal growth – turfgrass growth in which growth originates from an axillary bud on the crown.

    crecimiento extravaginal – crecimiento de césped en que brota el crecimiento del botón axilar en la corona.

fabrication and boxing – vacuum sealing of meat in boxes before shipment.

    fabricación y encajamiento – envasar al vacío la carne en cajas antes de transportarla.

famine – widespread starvation.

    escasez – inanición o hambre general.

Farmers Home Administration – government agency that assists farmers to become landowners.

    Administración de Hogares de Agricultores – agencia de gobierno que ayuda a los cultivadores a volverse propietarios de tierra.

farrier – a person who shoes horses.

    herrador de caballos – persona que clava y ajusta las herraduras a un caballo.

farrow – giving birth in the swine family.

    parir a cerdos – parir a cerdos.

    Note: In Spanish, the word *parir* means "to give birth in the non-human mammal." Thus, there is no other word that differentiates the birth process of a particular species of animal.

farrowing crate – special cage or pen used for swine to give birth in.

    box de parición – jaula o pocilga especial dentro de la cual paren las cerdas.

fat – nutrients that have 2.25 times as much energy as carbohydrates.

    grasa – alimentos nutritivos que tienen 2.25 veces más la energía que los carbohidratos.

feces – solid body waste.

    heces fecales – excrementos sólidos del cuerpo.

federal land bank – lending institution that provides long-term credit for agriculture.

    Banco Federale de Tierra – ésto dan crédito a plazo largo para la agricultura.

feed additive – a nonnutritive substance added to feed to improve growth, increase feed efficiency, or maintain health.

    aditivos a pienso – una sustancia no nutritiva añadida al pienso para mejorar el crecimiento, para aumentar la eficacia del pienso, o para mantener la salud.

feeder calf – calf intended to be fed to a heavier weight before slaughter.

    becerros de engorde – becerro a que les da tanto para comer para que se engorden mucho antes de ser matados.

feedlot – area in which large numbers of animals are grown for food.

    «feedlote» – área en que muchos animales son cultivados para que sirvan como carne.

feedstuff – any edible material used for animal feed.

    forraje – cualquiera material comestible que sirve para pienso.

fermentation – a chemical change that results in gas release.

    fermentación – un cambio químico que resulta en desprendimiento de gases.

fertility – amount and type of nutrients in the soil.

    feracidad – cantidad y clase de alimentos nutritivos en la tierra.

fertilizer – material that supplies nutrients for plants.

    abono – material que proporciona los alimentos nutritivos a las plantas.

fertilizer analysis – the percentage of nutrients by weight in the fertilizer.

    análisis de abono – por peso, es el porcentaje de alimentos nutritivos que hay en el abono.

fertilizer grade – percentages of primary nutrients in fertilizer.

    grado del abono – proporción de alimentos nutritivos en el abono.

fetus – embryo from the time of attachment to the uterine wall until birth.

    feto – embrión desde la hora de la unión al útero hasta el nacimiento.

FFA – a national intracurricular organization for students enrolled in agriscience programs in their schools.

    AFF – una organización dentro del programa para estudiantes matriculados en los programas de agriciencia.

fibrous root – one of the two major root systems, consisting of many fine, hairlike roots.

    raíz fasfciculada – uno de dos sistemas principales de raíz, que consiste de muchas raíces finas como el pelo.

field crop – a class of crop that includes grains, oil crops, and specialty crops.

    cultivo de campo – una clase de cultivo que incluye granos, cultivos de aceite y cultivos de especialización.

filament – structure that supports the anther.

    filamento – base que sostiene la antera.

filly – young female horse or pony.

    potra – hembra jóven del caballo o poney.

financing function – obtaining capital for a business.

    función de financiar – obtenier capital para un negocio.

fine texture – a turfgrass with narrow leaves.

    textura fina – césped con hoja delgada.

fine-textured (clay) soil – soil that usually forms very hard lumps or clods when dry; plastic when wet.

    tierra fina – usualmente forma terrones duros cuando seca; como plástico cuando mojada.

finish – fat covering on a market animal.

    acabado de engorde – envultura de gordo en el animal de mercado.

flock – group of birds or sheep.

    bandada – grupo de aves juntas.

    rebaño – grupo de ovejas juntas.

floriculture – production and distribution of cut flowers, potted plants, greenery, and flowering herbaceous plants.

    floricultura – producción y distribución de flores cortadas, de plantas en maceta, ramas y hojas verdes, y plantas herbáceas florecientes.

flower – reproductive part of the plant.

    flor – parte reproductiva de la planta.

flowering bud – a terminal bud on a plant that produces flowers.

    brote de flor – un botón terminal en una planta que produce flores.

fluke – very small, flat worm that is a parasite.

    trematodos – gusanos de cuerpo plano muy pequeños que son parásitos.

foal – newborn horse or pony. Also, to give birth in horses.

    potro – caballo o poney recién nacido.

    parir – dar a luz a los animales.

    See note under farrow.

foliage – stems and leaves.

    follaje – ramas y hojas.

foliar spray – liquid fertilizer sprayed directly to the leaves of plants.

    abonado foliar – abono líquido que se pulveriza directamente a las hojas de las plantas.

food – material needed by the body to sustain life.

    alimento – material que necesita el cuerpo para sostener la vida.

food additive – anything added to food before packaging.

    suplemento aditivo – cualquiera cosa añadida a la alimentación antes de envasarla.

food chain – interdependence of plants or animals on each other for food.

    cadena trófica – dependencia reciproca de plantas o animales unos a otros para alimentarse.

food industry – production, processing, storage, preparation, and distribution of food.

    industrias alimenticias – producción, tratamiento, almacenaje, preparación, y distribución de alimentos.

food pyramid – a visual illustration of the kinds and number of servings of foods that are required for proper human nutrition.

    el pirámide de alimento – una ilustración visual de las clases y del número de porciones de los alimentos que se requieren para la nutrición humana apropiada.

foot-candle – amount of light found 1 foot from a standard candle or candela.

    candela por pie cuadrado – cantidad de luz que se encuentra de un pie de una candela o vela estándar.

forage – crop plants grown for their vegetative growth and fed to animals.

    forraje – plantas cultivadas por su vegetación y dadas de comer a los animales.

forest – large group of trees and shrubs.

    bosque – gran conjunto de árboles y arbustos.

forester – a person who studies and manages forests.

    silvicultor – una persona que estudia y maneja los bosques.

forest land – land at least 10 percent stocked by forest trees.

    tierra forestal – tierra poblada de por lo menos 10 por ciento de árboles forestales.

forestry – industry that grows, manages, and harvests trees for lumber, poles, posts, panels, paper, and many other commodities.

    silvicultura – industria que cultiva, maneja, y cosecha los árboles para maderos, varas, palos, tableros, y muchos otros productos.

formal – equal in size and number.

    formal – igual en cantidad, tamaño y textura.

formulation – the physical properties of the pesticide and its inert ingredients.

    formulación – las propiedades físicas del pesticida y sus ingredientes inertes.

4-H – network of youth clubs directed by Cooperative Extension System personnel to enhance personal development and provide skill development in many areas.

    4-H – una red de asociaciones dirigida por el personal del Servicio Cooperativo de Extensión para fomentar el desarrollo personal y enseñar técnicas en muchas áreas.

free choice – making feed available at all times.

    alimentación a elección – dejando que está disponible todo el pienso que quiera el animal todo el tiempo.

freemartin – sexually imperfect female calf (usually sterile) born twin to a male in cattle.

    ternera estéril – ternera hembra que es sexualmente defectuosa, nacida gemela a un macho del ganado.

    Note: The Spanish term for "freemartin" implies sterility (*estéril*), so the "(usually sterile)" part of the definition was deleted in Spanish.

free water – water that drains out of soil after it has been wetted.

    agua libre – agua que avena del suelo después de haber sido mojado el suelo.

freeze-drying – removing moisture with cold.

    deshidratar por congelación – quitar a algo el agua por el frío.

fresh water – water that flows from the land to oceans and contains little or no salt.

    agua dulce – agua que corre desde la tierra hasta los mares y que tiene poco o no sal.

fructose – simple fruit sugar.

    fructosa – azúcar simple de frutas.

fruit – mature ovary; seed.

    fruto – óvulo maduro, semilla.

fry – small, newly hatched fish.

    Alevines – peces recién salidos del huevo.

fungal endophyte – microscopic plant growing within a plant.

    endofito fungoso – planta microscópica creciendo dentro de una planta.

fungi – members of a major group of lower plant life that lack chlorophyll and obtain nourishment from either live or decaying organic matter.

    hongos – miembros de un grande grupo de plantas de vida inferior que no producen clorofila y obtienen nutrición de organismos vivos o podridos de materias orgánicas.

fungicide – a material used to destroy fungi or protect plants against their attack.

    fungicida – una sustancia que se usa para desarraigar hongos o proteger las plantas de su ataque.

furrow – groove made in the soil.

    surco – hendidura hechas en la tierra.

futures – legally binding agreements made on the trading floor of a futures exchange.

    futuros – convenios obligatorios legalmente, hechos en el parquet de un cambio a futuro.

futures market – legal frameworks for sellers and buyers to buy and sell futures contracts.

    mercado afuturo – estructuras legales para vendedores y compradores para comprar y vender contratos a futuro.

gait – way of moving.

    paso – manera de pasar.

galactose – simple milk sugar.

    galactosa – azúcar simple de leche.

gallop – fast, three-beat gait.
    galope – paso rápido de tres tiempos.

gamete – a reproductive cell.
    gameto – una célula reproductiva.

gander – male goose.
    ganso – macho de la gansa.

gavel – a small wooden hammer-like tool that is used by the presiding officer to direct a meeting.
    martillo – un pequeño martillo de madera usado por un oficial que dirige una reunión.

gelding – castrated male animal of the horse family.
    caballo castrado – caballo cuyos órganos necesarios a la generación han sido extirpados.
    Note: Because there is no term for "gelding" in Spanish other than "castrated horse," the Spanish definition reads, "castrated horse — horse whose organs necessary for procreation have been removed."

gene – a unit of hereditary material located on a chromosome.
    gene – unidad de materia hereditaria ubicada en un cromosoma.

gene mapping – finding and recording the locations of genes in a cell.
    trazar un mapa de genes – localizando y recordando las posiciones de genes en una célula.

general-use pesticide – a pesticide that poses a minimal amount of risk when applied according to label directions.
    pesticida para uso general – un pesticida que plantea un riesgo mínimo cuando se aplica según las instrucciones de etiqueta.

generation – the collective offspring of common parents.
    generación – la descendencia colectiva de padres comunes.

gene splicing – the process of removing and inserting genes in cells.
    empalme de genes – el proceso de extirpar y meter las genes en las células.

genetic control – pest control using resistant varieties of crops.
    control genético – control de animales o insectos dañinos usando variedades resistentes de cultivos.

genetic engineering – movement of genes from one cell to another.
    ingeniería genética – traslado de genes de una célula a otra.

geneticist – a person who studies genetics.
    genetista – persona que especializa en genética.

genetics – the biology of heredity.
    genética – la biología de la herencia.

genotype – the genetic makeup of an individual or group of organisms.
    genotipo – cómo se manifiestan los genes.

genus – (plural, *genera*) a closely related and definable group of animals or plants comprising one or more species.
    genus – un grupo definible y estrechamente relacionado de plantas que consta de úna o más especies.

germ – new wheat plant inside the kernel.
    germen – progenie de trigo dentro del grano.

germinate – a seed sprouting or starting to grow.
    germinar – una semilla brotando o empezando a crecer.

germination rate – percentage of seed that sprouts and begins to grow.
    proporción de germinación – porcentaje de semilla que brota y comienza a crecer.

gestation – length of pregnancy.
    gestación – tiempo que dura en el estado de embarazo.

giblets – heart, liver, and gizzard of poultry.
    menudillos – corazón, higadillo, y molleja de las aves.

gills – organ of aquatic animal that absorbs oxygen from the water.
    agalla – órgano de animal acuático que absorbe el oxígeno del agua.

gilt – female of the swine family that has not given birth.
    cerda jóven – hembra de la familia cochina que no ha parido.

ginning – the process of removing the seeds from cotton.
    desmotar – el proceso de sacar la semilla al algodón.

Girl Scouts and Boy Scouts – youth organizations that provide opportunities for leadership development and skill development.
    Niñas Exploradoras y Niños Exploradores – asociaciónes que da oportunidades del desarrollo de liderazgo y de técnica.

glacial deposit – soil deposited by ice.
    depósitos glaciales – tierras depositadas por hielo.

glucose – simple sugar and the building blocks for other nutrients.
    glucosa – azúcar simple y los fundamentos para otros alimentos nutritivos.

goose – female of the goose family.
    gansa – hembra del ganso.

gosling – young goose.
    ansarino – ganso jóven.

gourmet – sensitive and discriminating taste in food preferences.
    gastronómico – gusto sensible y muy bueno en cuanto a las preferencias de comida.

grade – animal that is not purebred or is of unknown ancestry.
    cruce – animal que no es de pura sangre o raza; o de linaje desconocido.

grader – person who inspects the food for freshness, size, and quality.
    clasificadora – persona que inspecciona los alimentos en cuanto a su frescura, tamaño, y calidad.

grade – quality standard.
    grado – estándare de calidad.

gradient – a measurable change in an amount over time or distance.
    índice – un cambio mensurable de una cantidad por un tiempo o una distancia.

grading up – crossing a purebred male animal with a grade female animal.

procreación por sustitución – Acoplar un macho de pura raza y una hembra de cruce.

grafting – joining two plant parts together so that they will grow as one.

injertar – juntando dos partes de planta para que crezcan así como úna.

graft union – the location where a scion and rootstock meet when parts of two plants are grafted together.

unión de injerto – sitio donde una vara y un tallo de dos plantas diferentes se juntan.

grain crop – crop grown for its edible seeds.

cultivo de cereale – planta cultivada por sus semillas comestibles.

grass waterway – strip of grass growing in the low area of a field where water can gather and cause erosion.

estepa de gramíneas – franja de hierba creciendo en la parte baja del campo donde el agua puede acumularse y resultar en erosión.

gravitational water – water that drains out of soil after it has been wetted.

agua de gravitación – agua que avena del suelo después de haber sido mojado el suelo.

greenhouse effect – a buildup of heat at the earth's surface caused by energy from sunlight becoming trapped in the atmosphere.

efecto de invernadero – concentración de calor en la superficie de la tierra debido a la energía de la luz del sol que se encierra en la atmósfera.

green manure (crop) – crop grown to be plowed under to provide organic matter to the soil.

plantas para abono verde – plantas cultivadas para ser enterradas con el objetivo de proporcionar materia orgánica al suelo.

Green Revolution – process in which many countries became self-sufficient in food production.

revolución verde – proceso en donde muchos países logran a ser autosuficientes en la producción de alimentos.

green roughage – plant material with high moisture content such as pastures and root plants.

forraje grosero verde – materias vegetales con alto contenido de humedad así como pastos y tubérculos.

group planting – trees or shrubs planted together so as to point out some special feature or to provide privacy or a small garden area.

siembra por agrupación – árboles o arbustos están plantados juntos para señalar alguna característica especial o para proveer intimidad o un jardín pequeño.

Guanine (G) – a base in genes designated by the letter *G*.

guanino (G) – una base en los genes nominado por la letra «G».

guard cells – cells that surround the stoma.

células guardias – células que rodean el estoma.

gully erosion – removal of soil to form relatively narrow and deep trenches known as gullies.

erosión en cárcavas – extirpación de tierra para formar acequias relativamente hondas y estrechas que son conocidas por el nombre de cárcavas.

guying – the process of tying a tree or other plant to stakes using wire or rope to support the plant.

sujetar – el proceso de atar un árbol u otra planta a varillas usando alambre o cuerda para sostener la planta.

gypsum – a soil amendment that is often used to reduce the alkalinity of agricultural soils.

yeso – un aditivo de tierra frecuentemente usado para reducir la alcalinidad de tierras agriculturales.

habitat – area or type of environment in which an organism or biological population normally lives.

hábitat – área o tipo de medio ambiente en que normalmente vive un organismo o una población biológica.

hand – unit of measurement for horses equal to 4 inches.

palmo – unidad de medir a los caballos que es igual a cuatro pulgadas.

hand mating – system of breeding in which the male and female animals are kept apart except during mating.

cubrición a mano – el sistema de acoplamiento en donde el macho y la hembra son separados excepto durante la época de celo.

hardness – wood's resistance to compression.

dureza de la madera – la resistencia de madera a la compresión.

hardwood – wood from deciduous trees.

madera dura – madera de los árboles caducos.

hardy – the ability of a plant to survive and grow in a given environment.

resistente – la capacidad de una planta de sobrevivir y crecer en un ambiente determinado.

harvester – person responsible for taking products from plants in the field.

cosechadora – persona responsable por recoger el fruto del campo.

harvesting – taking a product from the plant where it was grown or produced.

cosechar – recoger un producto de la planta donde fue cultivado o producido.

hatchery – a place where eggs are hatched.

criadero – sitio donde sale de los huevos la cría.

hay – forages that have been cut and dried to a low level of moisture, used for animal feed.

heno – forrajes que han sido cortados y secados a un nivel bajo de humedad, usado para pienso.

haylage – silage made from forages dried to 40 to 55 percent moisture.

hierba presecada y ensilada – ensilaje hecho de forajes secados a 40 o 55 por ciento de humedad.

healing in – placing plant roots in a trench that is deep enough to cover the roots with moist soil.

recuperación – el colocar las raíces en una sanja que es suficientemente profunda para cubrirlas con tierra húmeda.

heartwood – inactive core of a tree trunk or limb.

duramen – núcleo inactivo de un tronco o rama de árbol.

heel cutting – a shield-shaped cut made about halfway through the wood around the leaf and the axial bud.

escudete – un corte en forma de escudo hecho a medio camino a través de la madera alrededor de la hoja y el botón axilar.

heifer – female animal of the cattle family that has not given birth.

vaquilla – hembra del vacuno que no ha parido.

heritability – the capacity to be passed down from parent to offspring.

capacidad para heredar – la capacidad para ser transmitido del progenitor a la progenitura.

hen – adult female chicken, duck, turkey, and pheasant.

gallina – adulta hembra del gallo.

herb – plant kept for aroma, medicine, or seasoning.

hierba – planta guardada para aroma, medicina, o condimiento.

herbaceous – a plant that has a stem that does not turn woody; it is more or less soft and succulent, and often lacks winter hardiness.

herbáceo – una planta que tiene el tallo que no vuelve a ser leñoso; es más o menos blando y carnoso, tampoco es resistente al invierno.

herbicide – a substance for killing weeds.

herbicida – una sustancia empleada para matar las malas hierbas.

heredity – transmission of characteristics from parent to offspring.

herencia – la transmisión de rasgos de los progenitores a la progenie.

heterozygous – pairs of genes that are different.

hetercigótico – pares de genes que son diferentes.

hide – skin of an animal.

cuero – piel de los animales.

high technology – use of electronics and ultra-modern equipment to perform tasks and control machinery and processes.

alta tecnología – el uso de electrónica y equipo ultra-moderno para desempeñar funciones y controlar maquinería y procesos.

hinny – cross between a stallion and a jennet.

burdégano – un cruce entre un semental y una burra.

hive – place or box in which bees are kept.

colmena – sitio o caja en donde se guardan las abejas.

hog – swine older than 4 months.

cerdo, marrano – cerdo que tiene más de cuatro meses de edad.

hoist – lift into position.

subir – llevar a un lugar o posición más alta.

home gardening – production of vegetables for use by a single family.

hortaliza – producción de vegetales para el uso de una sola familia.

homozygous – pairs of genes that are alike.

homocigótico – pares de genes que son semejantes.

honey – a thick, sweet substance made by bees from the nectar of flowers.

miel – una sustancia viscosa y dulce hecha por las abejas con el néctar de las flores.

horizon – layer.

horizonte – capa.

hormone – chemical that regulates activities of the body.

hormona – sustancia química que regulan muchas de las actividades del cuerpo.

horse – member of the horse family 14.2 or more hands tall.

caballo – miembro de la familia de los caballos que mide 14.2 o más de palmos de alto.

horsemanship – the art of riding and knowing the needs of the horse.

equitación – el arte de montar a caballo y conocerlo en cuanto a sus necesidades.

horticulture – the science of producing, processing, and marketing fruits, vegetables, and ornamental plants.

horticultura – la ciencia de producir, procesar, y poner en venta las frutas, verduras, y plantas de adorno.

host animal – a species of animal in or on which diseases or parasites can live.

huésped animal – una especie de animal en o sobre el cual viven parásitos o enfermedades.

hotbed – a cold frame that has a source of artificial heat.

superficie caliente – un bastidor frío que tiene un surtidor de calor artificial.

humidity – moisture in the air.

humedad – impregnación de agua en el aire.

hutch – a rabbit cage or house.

conejera – jaula o morada para conejos.

hybrid – plant or animal offspring from crossing two different species or varieties.

híbrido – progenitura vegetal o animal que procede del cruzar dos distintas especies o razas.

hybrid vigor – offspring of greater strength and potential resulting when two different breeds or varieties are crossed.

vigor híbrido – progenitura de más fuerza y potencial resultando cuando se cruzan dos distintas razas o cruces.

hydrocarbon – organic compound containing hydrogen and carbon.

hidrocarburo – compuesto orgánico que contienen hidrógeno y carbono.

hydrocooling – removal of heat by immersing in cold water.

enfriamiento del agua – quitarse del calor por inmersión en agua fría.

hydroponics – the practice of growing plants without soil.

cultivo hidropónico – la práctica de cultivar plantas sin tierra.

hygroscopic water – water that is held too tightly by soil particles for plant roots to absorb.

agua higroscópico – agua que está sujetada con demasiado rigor por las partículas de tierra para que las raíces de planta la absorban.

hyphae – the threadlike vegetative structure of fungi.

hifas – la estructura vegetal filiforme de los hongos y mohos.

I

ice-minus – bacteria that retard frost formation on plant leaves.

«ice-minus» – bacteria que retrasan la formación de escarcha en las hojas vegetales.

imbibition – the absorption of water into the cell causing it to swell.

embebición – la absorción de agua en la célula, resultando en que se hincha.

immune – not affected by.

inmune – que no se afecta.

impatiens – popular, easy-to-grow, summer flowering plant.

balsamina nicaragua – planta de floración del verano que es popular y fácil para cultivar.

imperfect flower – flower that is missing one or more of the following parts: stamen, pistil, petals, or sepals.

flor imperfecta – flor que falta de úna o más de las partes siguientes: estambre, pistilo, pétalos, o sépalos.

implant – to place a substance under the skin to improve growth of animals.

implantar – meter una sustancia bajo la piel para mejorar el crecimiento de animales.

improvement activity – project that improves beauty, convenience, safety, value, or efficiency learned outside of the regularly scheduled classroom or laboratory classes.

proyecto de hacer mejoras – proyecto que hacen mejoras en la belleza, conveniencia, seguridad, valor, o eficacia y que los aprende uno fuera de las clases de escuela o laboratorios.

improvement by selection – picking the best parents for the next generation.

cultivo mejoramiento – seleccionando los mejores padres para la generación siguiente.

inbreeding – mating of animals that are related.

procrear en consanguinidad – apareamiento de animales del mismo parentesco.

incomplete dominance – neither gene expresses itself to the exclusion of the other.

dominancia incompleta – ninguno de los genes se manifiesta con exclusión al otro.

incorporated – mixed into the soil by spading, plowing, or discing.

incorporación del abono al suelo – metido en la tierra y mezclado por arar, usar la pala, o usar los discos abresurcos.

induction – a specific set of conditions must occur to cause flower development.

inducción – tiene que suceder unas condiciones específicas para causar el desarrollo de la flor.

industrial technology – the application of science to industrial uses.

tecnología industrial – la aplicación de la ciencia para los usos industriales.

inert – inactive.

inerte – inactivo.

inflorescence – the flowers and ultimately the seed area of the plant.

florescencia – son las flores y ultimamente el área de las semillas de la planta.

informal – in landscaping, planting areas that contain elements that are not equal in number, size, or texture.

informal – cuando se refiere al ajardinar, se refiere a las áreas de plantar que contienen elementos que no son iguales en cantidad, tamaño, ni textura.

infusion – the process for treating udder problems through the teat canal.

infusión – el proceso usado para tratar problemas de la ubre por el canal de la teta.

injection – the process of administering drugs by needle and syringe.

inyección – el proceso de administrar medicamento por medio de jeringuilla hipodérmica.

inner bark – contains the cambium from which the tree grows and the bark is renewed; transports food within a tree.

corteza interna – contiene la sustancia de la que crece el árbol y la corteza se renova; transporta alimentos dentro del árbol.

inorganic compound – a compound that does not contain carbon.

compuesto inorgánico – un compuesto que no contiene el carbono.

insect – a six-legged animal with three body segments.

insecto – un animal de seis patas con tres sementos de cuerpo.

insecticide – a material used to kill insects or protect against their attack.

insecticida – un material que se usa para matar insectos o proteger contra su ataque.

insemination – the placement of semen in the female reproductive tract.

inseminación – la introducción de semen en las vías genitales de la hembra.

instar – the state of the insect during the period between molts.

etapas de crecimiento – la fase del insecto durante el período en medio de las mudas.

institutional advertising – advertising designed to create a favorable image of the firm or institution.

publicidad institucional – publicidad creada para establecer un concepto favorable de la empresa o institución.

insulin – chemical used to control blood sugar levels.

insulina – sustancia química que se emplea en la regulación de niveles de la cantidad de glucosa.

integrated pest management (IPM) – pest-control program based on multiple-control practices.

lucha integrada – programa de control de los animales o insectos nocivos basado en ejercicios de varios controles.

integrity – capable of personally upholding a high moral standard.

integridad – capaz de personalmente sostener un estandar de alta moral.

intermediate-term loan – loan with payment periods that range from 1 to 7 years.

préstamo de término intermedio – préstamo que tienen plazos de pagos que oscilan entre uno y siete años.

internode – area between two nodes.
    entrenudos – área entre dos nudos.

internship – a career experience that places a student in a business for a specific period to learn technical skills through actual work experiences.
    internado – una experiencia que coloca al estudiante en un negocio por un término específico de tiempo para aprender habilidades técnicas a través de experiencias reales en el trabajo.

intradermal – between layers of skin.
    intradérmico – entre las capas de piel.

intramuscular – in a muscle.
    intramuscular – en el interior de los músculos.

intraperitoneal – in the abdominal cavity.
    intraperitoneal – en el interior de la cavidad abdominal.

intraruminal – in the rumen.
    intraherbarial – en el interior del herbario.

intravaginal growth – a type of growth in which shoots develop within the lower leaf sheath at a crown's axillary bud.
    crecimiento intravaginal – un tipo de creimiento en que desarrollan los brotes en el interior de la vaina de la hoja exterior, en la corona del botón axilar.

intravenous – in a vein.
    intravenoso – en el interior de la vena.

inventor – person devising something new or improving an existing idea or product.
    inventor – persona que crea algo nuevo o mejora una idea o producto existente.

inventory report – units received and sold, unit cost, total sales, and profit.
    inventario – unidades recibidas y vendidas, coste por unidad, ventas totales, y ganancia.

in vitro – fertilization of an ovum or egg outside the body of the mother.
    en vitro – fertilización de un óvulo o huevo fuera del cuerpo de la madre.

involuntary muscle – muscle that operates in the body without control by the will of the animal.
    músculo involuntario – músculo que funcionan en el cuerpo sin control por la voluntad del animal.

ion – atom or a group of atoms that has an electrical charge.
    iones – átomo o grupo de átomos que tienen una carga eléctrica.

irradiation – treating food with gamma rays.
    irradiación – tratamiento de comestibles con rayos gamma.

irrigation – addition of water to plants to supplement that provided by rain or snow.
    irrigación – adición de agua a los cultivos para suplementar la proporcionada por la lluvia o la nieve.

jack – male donkey or mule.
    burro – macho de la burra o mula.

jennet – female donkey or mule.
    burra – hembra del burro o mula.

jog – in horses, a slow, smooth, two-beat, diagonal gait.
    trote corto – de caballos, paso diagonal, lento y suave, de dos tiempos.

jungle fowl – wild ancestor of the chicken.
    ave de corral de la selva – antepasado salvaje de la gallina.

Katahdin – popular potato variety of the 1930s.
    «Katahdin» – variedad popular de patata durante los 1930.

key pest – a pest that occurs on a regular basis for a given crop.
    animal o insecto dañino regular – animal o insecto dañino que ocurre en una basis regular para un cultivo determinado.

kid – young goat.
    chivo – cabrío jóven.

knife application – injection of fertilizer into soil in gaseous form.
    fertilización por inyección – inyección del abono al suelo en forma gaseosa.

knowledge – familiarity, awareness, and understanding.
    conocimiento – familiaridad, conciencia, y entendimiento.

kosher – prepared in accordance with Jewish dietary laws.
    «Kosher» – preparado según las leyes dietéticas judaicas.

lactation – milk production.
    lactancia – secreción de leche.

lactation period – period when mammals are producing milk.
    período de amamantamiento – época en que los mamíferos están secretando leche.

lactose – compound milk sugar.
    lactosa – compuesto de azúcar de leche.

lacustrine deposit – soil deposited by lakes.
    depósito lacustre – los suelo depositados por lagos.

lamb – member of the sheep family younger than 1 year; also, meat from young sheep.
    cordero – miembro de la familia de ovejas que tiene menos de un año de edad; también la carne de la oveja jóven.

lambing – act of giving birth in sheep.
    parir a corderos – parir a corderos.
    See note under farrow.

lard – rendered pork fat.
    manteca de cerdo – grasa rendida de cerdo.

larvae – mobile organisms that become fixed and grow into non-mobile adults.
    larvas – organismos movibles que logran a ser fijados y crecen a ser adultos no movibles.

laser – an intense, narrow beam of light that is used to measure nutrient deficiencies in plants from an orbiting satellite.

laser – un haz de luz intenso, estrecho que se utiliza para medir deficiencias nutrientes en plantas de un satélite que se mueve en órbita

lay on the table – a motion used to stop discussion on a motion until the next meeting. The way to table a motion is to say, "I move to table the motion."

presentar una moción – una moción usada para terminar discusión sobre otra moción hasta la próxima reunión. La manera de presentar una moción es decir, "Yo propongo que se presente la moción."

layer – chicken developed for the purpose of laying eggs.

gallina ponedora – una gallina criada para el objetivo de poner huevos.

layering – a practice inducing formation of roots for new plants while maintaining the health and vigor of the parent plant.

acodar – una práctica que induce la formación de raíces para plantas nuevas mientras manteniendo la salud y vigor de la planta madre.

LC$_{50}$ – lethal concentration of a pesticide in the air required to kill 50 percent of the test population.

CM$_{50}$ – concentración mortal de un pesticida en el aire necesaria para matar 50 por ciento de una población prueba.

LD$_{50}$ – lethal dose of a pesticide required to kill 50 percent of a test population.

DM$_{50}$ – dosis mortal de un pesticida necesaria para matar 50 por ciento de una población prueba.

leached – certain elements have been washed out of the soil.

lixiviado – ciertos elementos han sido derrubiados del suelo.

lead – to show the way by going in advance or to guide the action or opinion of.

guiar – mostrar por medio de ir delante o dirigir la acción o la opinión de una.

leadership – the capacity or ability to lead.

dirección – la capacidad de guiar.

leaf – plant part consisting of a stipule, petiole, and blade.

hoja – parte vegetal consistiendo de estípulo, pecíolo, y brizna.

leaf blade – the wide portion of a leaf in which photosynthesis occurs.

brizna – la porción de una hoja donde ocurre la fotosíntesis

leaf cutting – a cutting made from a leaf without a petiole.

corte de hoja – un corte hecho de una hoja sin tallo.

leaf fat – loose fat on hogs.

gordo suelto – grasa flácida en el cuerpo de los cerdos.

leaf mold – partially decomposed plant leaves.

mantillo – hojas vegetales parcialmente descompuest.

leaf petiole cutting – a plant consisting of a leaf and petiole on which a rooting compound is applied and the cutting is placed in a soil medium to develop into a new plant.

corte de tallos de hojas – una planta que consiste de una hoja y un tallo en la cual un compuesto para enrraizar se aplica donde el corte es colocado en un medio de tierra parar desarrollar una nueva planta.

leaf section cutting – a section of a plant leaf containing a vein that is placed in a plant growth medium to generate a new plant.

sección de hoja de corte – una sección de una hoja de la planta que contiene una vena, que es colocada en un medio de crecimiento para plantas para generar una nueva planta.

legume – plant in which certain nitrogen-fixing bacteria use nitrogen gas from the air and convert it to nitrates that the plant can use as food.

legumbre – planta en la cual ciertas bacterias fijadores del nitrógeno utilizan el gas nitrogenado del aire y lo convierten a nitratos para que los puede usar la planta como alimentación.

lenticels – pores in the stem that allow the passage of gases in and out of the plant.

lenticelas – poros en el tallo o el pedúnculo que permiten que los gases entren a y salgan de la planta.

lethal – genetic characteristic that causes an animal to be born dead or to die shortly after birth.

mortal – característica genética que causa que un animal nazca muerto o muera poco depués del nacimiento.

liabilities – current and long-term debts.

pasivos – deudas actuales y a largo plazo.

light horse – type of horse developed for riding.

caballo de silla – tipo de caballo criado para montar.

light intensity – brightness of light.

intensidad luminosa – el brillo de la luz.

ligule – a membranous or hairy structure located on the inside of the leaf.

lígula – está ubicada en la vaina foliar interior y es una estructura membranosa y pelosa.

lime – material that reduces the acid content of soil and supplies nutrients such as calcium and magnesium to improve plant growth.

cal – material que reduce el contenido de ácido del suelo y proporciona a minerales así como calcio y magnesio para mejorar el crecimiento de las plantas.

linebreeding – mating of individuals with a common ancestor.

cría por líneas puras – apareamiento de animales que tienen antepasado común.

linen – fabric produced from fibers in flax plants.

lino – tejido producido de las fibras de linácea.

linseed oil – oil produced from flax seed.

aceite de linaza – aceite producido de semilla de lino.

litter – group of young born at the same time with the same parents.

cama de paja – la cría nacida a la misma vez de los mismos padres.

loamy – a granular soil containing a good balance of sand, silt, and clay.

marga – un suelo granular que contiene un buen equilibrio de arena, légamo, y arcilla.

loess deposit – soil deposited by wind.

loess – tierra depositada por el viento.

long-term loans – loans with payment periods that range from 8 to 40 years.

préstamos a largo plazo – préstamos cuyos plazos de pago oscilan entre 8 a 40 años.

long-term plans – plans accomplished over months or years.

planes a largo plazo – planes realizados por meses o años.

lope – in horses, a very slow canter.

paso largo – de caballos, medio galope muy lento.

loss-leader pricing – commodity offered for sale at prices less than the cost level.

valoración del artículo barato para atraer clientes – se ofrece en venta la mercancía a precios por debajos del nivel de coste.

loyalty – reliable support for an individual, group, or cause.

lealtad – apoyo seguro para un individuo, un grupo, o una causa.

lumber – boards cut from trees.

maderos – trozos cortados de la sustancia dura de los árboles.

lymph gland – gland that secretes disease-fighting materials for the body.

ganglio linfático – glándula que secretan materiales que combaten las enfermedades del cuerpo.

macronutrient – element used in relatively large quantities.

elemento nutritivo esenciale – elemento usado en cantidades relativamente grandes.

main motion – a basic motion used to present a proposal for the first time. The way to state it is to get recognized and then say, "I move . . ."

moción mayor – la moción básica para presentar una propuesta por la primera vez. La manera de declararla es para ser reconocido y luego decir, «Yo propongo . . .»

malting – process of preparing grain for the production of beer and alcohol.

malteado – proceso de preparar el grano para la producción de cerveza y alcohol.

maltose – compound milk sugar.

maltosa – azúcar de malta compuesto.

mammal – milk-producing animal.

mamífero – animal que secrete la leche.

mammary system – milk-production system in the female.

aparato mamario – aparato de producción de leche de la hembra.

manage – to use people, resources, and processes to reach a goal.

administrar – utilizar a personas, recursos, y procesos para lograr una meta.

mange – crusty skin condition caused by mites.

sarna – enfermedad de la piel que es causada por los acáridos.

mapping – the process of identifying the location of a specific gene on a chromosome.

trazar el mapa – el proceso de identificar el lugar de un gene específico en un cromosoma.

mare – adult female horse or pony.

vegua – hembra adulta del caballo o poney.

marginal burn – dead spots or edges on a leaf that usually indicate deficiencies in one or more nutrients.

quemada marginal – manchas o bordes muertos en una hoja que usualmente indican deficiencias en uno u más nutrientes.

margin – edge of the leaf.

linde – el límite de la hoja.

market gardening – production of a wide variety of vegetables that are generally sold in local roadside markets.

hortaliza para mercado – producción de una amplia variedad de vegetales que son generalmente rendidos en puestos locales por caminos.

mastitis – infection of the milk-secreting glands of cattle, goats, and other milk-production animals.

mastitis – infección de las glándulas que secretan la leche del ganado, cabras y otros mamíferos.

maturity – the state or quality of being fully grown.

madurez – estado o cualidad del desarrollo o crecimiento completo.

mechanical pest control – pest control that affects the pest's environment or the pest itself.

control mecánico de pestes – control de los animales o insectos nocivos que afecta el animal o insecto nocivo mismo o el ambiente de éso.

media – material that provides plants with nourishment and support through their root systems.

medio – material que proporciona alimentación y sostenimiento a las plantas por sus raíces.

medium – surrounding environment in which something functions and thrives.

medio ambiente – los alrededores en los cuales funciona y crece alguna cosa.

medium-textured (loamy) soil – a relatively even mixture of sand, silt, and clay.

suelo limoso – una mixtura relativamente igual de arena, légamo, y arcilla.

meiosis – cell division that results in the formation of gametes.

meiosis – división celular que resulta en la formación de los gametos.

mentor – a person who teaches their career skills to students using hands-on experiences in work settings.

mentor – una persona que enseña habilidades profesionales a estudiantes usando experiencias prácticas en el lugar de trabajo.

meristematic tissue – plant tissue responsible for plant growth.

tejido meristemático – tejido vegetal que causa el crecimiento de la planta.

mesophyll – tissue of the leaf where photosynthesis occurs.

mesofilo – tejido de la hoja en donde ocurre el fotosíntesis.

metamorphosis – the change in growth stages of an insect.

metamorfosis – la mudanza de forma en las etapas de crecimiento del insecto.

microbe – living organism that requires the aid of a microscope to be seen.

microbio – organismo viviente que requiere la ayuda del microscopio para que se vea.

micronutrient – element used in very small quantities.

micronutriente – elemento usado en cantidades pequeñas.

microorganism – tiny plant or animal that may contribute to food spoilage.

microorganismo – microbio animale o vegetale que pueden contribuir a la putrefacción de los comestibles.

microwave oven – an oven that heats food by using an electromagnetic wave.

horno de microonda – un horno de cocina que calenta la comida por medio de usar ondas electromagnéticas.

middlemen – people who handle an agricultural product between the farm and the consumer.

intermediarios – personas que manejan un producto agrícolo entre la finca y el consumidor.

migratory labor – workers who move to places where harvesting is occurring.

obreros migratorios – trabajadores del campo que se mudan a lugares donde se está ocurriendo el cosechar.

milk – white or yellowing liquid produced by the mammary glands for the purpose of feeding young.

leche – líquido blanco o amarillento que es segregado por las glándulas mamarias para el objetivo de alimentar la cría.

milk replacer – dried dairy or vegetable product that is mixed in warm water and fed to calves in place of milk.

sustitivo de leche – producto secados lácteos o vegetale que son mezclados con agua tibia y dado de comer a los becerros en lugar de leche.

milking machine – machine that milks cows and goats.

máquina de ordeñar – aparato o máquina que extrae la leche de la ubre de las vacas y cabras.

mineral – element essential for normal body functions.

mineral – elemento esencial para las funciones normales de cuerpo.

mineral matter – nonliving items such as rocks.

materia mineral – cosas no vivientes como rocas.

minimum tillage – soil is worked only enough so that seed will germinate.

labranza mínima – la tierra es labrada justo lo suficiente para que germinen las semillas.

minutes – the official written record of a business meeting.

actas – el nombre del record escrito oficial de una junta directiva.

misuse statement – pesticide label information that reminds the user to follow the label directions when using the products.

aviso de mal manejo – información en la etiqueta que avisa al usuario que haya de seguir las direcciones de la etiqueta mientras usando los productos.

mitosis – simple cell division for growth.

mitosis – división celular simple que resulta en el crecimiento.

mode of action – how a pesticide affects its target.

manera de afectar – cómo el pesticida afecta su objetivo.

mohair – hair from Angora goats used to make a shiny, heavy, woolly fabric.

mohair – pelo de la cabra de Angora usado para hacer un tejido lanoso, grueso, y con brillo.

moist-heat cooking – surrounding the food with liquid while cooking.

cocinar por humectación – rodeando a la comida con líquido mientras cocinando.

moldboard plow – plow with a curved bottom that will turn prairie soils.

arado de vertedera – arado con orejera la cual puede voltear el suelo de llanura.

molt – softening and cracking of the exoskeleton so that crustaceans may escape and grow a larger exoskeleton.

mudar – ablandarse y agrietarse del dermatoesquéleto para que los crustáceos puedan escaparse y dejarse crecer un dermatoesquéleto más grande.

monitoring – to check or observe pest numbers and damage.

observar – ver u observar la cantidad de animales o insectos nocivos y el daño resultando de ellos.

monoclonal antibody – natural substance in blood that fights diseases and infections.

anticuerpo monoclonale – substancia naturale en la sangre que luchan contra enfermedades e infecciones.

monocotyoledon (monocot) – plant with one seed leaf.

monocotiledóneo – planta con solamente una hoja de semilla.

monogastric – animal with a single-compartment stomach.

monogástrico – animal con solamente una cavidad en el estómago.

mosaic – a virus disease of plants in which a leaf shows a symptom of light- and dark-green mottling of the foliage.

mosaico – una enfermedad viral de las plantas en donde la hoja manifiesta un síntoma de vetas manchas de verde claro y oscuro en el follaje.

motion – a proposal for group action that is presented in a meeting to be acted on by an organization.

movimiento – una propuesta para acción de un grupo que es presentada en una junta para tomar acción por la organización.

mulch – material placed on soil to break the fall of raindrops (preventing erosion), prevent weeds from growing, or improve the appearance of the area.

pajote – material puesto encima del suelo para cortar la caída de las gotas de lluvia (para prevenir la erosión), para prevenir el crecimiento de malas hierbas, o para hacer mejoras del aspecto del área.

mule – cross between a jack and a mare.

mulo – una cruce entre el burro macho y la yegua.

muscular system – the lean meat of the animal.

sistema muscular – la carne sin grasa del animal.

mutation – change in genes.

mutación – cambios en los genes.

mutton – meat from mature sheep.

cordero – carne de cordero mayor.

mutualism – two types of wildlife living together for the mutual benefit of both.

mutualismo – dos clases de fauna que viven juntas para el beneficio mutuo de las dos.

mycelium – a collection of fungal hyphae.

micelio – un grupo de hifa fungosa.

natural fishery – existing breeding ground of wild fishes, that is, oceans, continental shelves, lakes, and bays.

criadero naturale de peces – criadero existente de la cría de peces en estado salvaje; por ejemplo, mares, el plataformo continental, lagos, y bahías.

natural service – male mates directly with the female.

servicio natural – cuando el macho acopla directamente con la hembra.

nematode – a type of tiny roundworm that causes damage to specific kinds of plants.

nematodo – un pequeño tip a de ascáride que causa daños a un tipo específico de plantas.

net contents – the amount of pesticide in the container.

contenidos netos – la cantidad de pesticida en el contenido.

net worth – the value of an owner's assets minus liabilities (debts).

valor total – el valor de los bienes de un dueño menos las deudas.

neutral – neither acidic nor alkaline.

neutro – ni ácido ni siquiera alcalino.

nitrate – a form of nitrogen used by plants.

nitrato – una forma de nitrógeno usado por plantas.

nitrogen fixation – conversion of nitrogen gas to nitrate by bacteria.

fijación del nitrógeno – la conversión de gas de nitrógeno en nitrato por bacteria.

nitrous oxides – compounds containing nitrogen and oxygen; they make up about 5 percent of the pollutants in automobile exhaust.

óxido nitroso – compuestos que contienen nitrógeno y oxígeno; constituyen aproximadamente 5 por ciento de los agentes contaminadores en el gas de escape de los automóviles.

node – portion of the stem that is swollen or slightly enlarged that gives rise to buds.

nudo – la parte del tallo o pedúnculo que es hinchada o una poca expandida que echa renuevos.

nomenclature – a systematic method of naming plants or animals.

nomenclatura – un método sistemático de dar nombres a las plantas o los animales.

noncontagious – diseases that cannot be spread to other animals.

no contagiosas – enfermedades que se transmiten a otros animales por contacto.

nonselective herbicide – a chemical that kills all plants with which it comes in contact.

herbicida no-selectivo – una sustancia quimica que mata todas las plantas a las que toca.

no-till – seed is planted directly into the residue of the previous crop, without exposing the soil surface.

sin labranza – dícese de cuando se siembra la semilla directamente en el residuo de la cosecha previa, sin exponer la superficie del suelo.

noxious weed – plant that is highly damaging to an environment that is controlled under authority of state law.

hierba tóxica – planta que es sumamente dañina al medio ambiente que es controlada bajo la autoridad de la ley estatal.

nucleic acid – original name for deoxyribonucleic acid.

ácido nucleico – nombre original por ácido desoxirribonucleico.

nurse crop – a crop used to protect another crop until it can get established.

planta protectora – un cultivo que se usa para proteger a otro cultivo hasta que éste pueda establecerse.

nursery – a place where young trees, shrubs, and other plants are grown.

vivero – un lugar donde se recrían árboles, arbustos, y otras plantas jóvenes.

nut – a fruit or seed contained within a removable outer cover.

nuez – fruto o semilla envuelto en una cáscara o cubierta exterior a que se quita.

nutriculture – water culture.

nutricultura – cultivo de agua.

nutrient – substance necessary for the functioning of an organism.

alimento nutritivo – sustancia necesaria para el funcionar del organismo.

nutrition – the process whereby all body parts receive materials needed for function, growth, and renewal.

nutrición – el proceso en que todas las partes del cuerpo reciben materiales necesarios para función, crecimiento, y renovación.

O horizon – the soil layer that is on the surface that is composed of organic matter and a small amount of mineral matter.

horizonte O – la capa de tierra que está en la superficie de la tierra que está compuesta de substancias orgánicas y una pequeña cantidad de substancias minerales.

obesity – the result of too much or improper types of food being eaten.

obesidad – el resultado de comer demasiada comida o de comer demasiado de alimentos de tipo inadecuado.

offal – materials thrown away during slaughter, including the head, blood, viscera, loose fat, and others.

asaduras – materias desechadas durante el sacrificio, incluso la cabeza, la sangre, las vísceras, el gordo suelto, etc.

offsetting a position – taking a second position on the futures market to offset a first.

compensar una posición – tomar una posición secundaria en el mercado a futuro para compensar una primera.

oil meal – by-product of the production of vegetable oil.

harina de aceite – subproducto de aceite vegetal.

oilseed crop – crop produced for the oil content of their seeds.

cultivo de semilla oleaginosa – cultivo cultivado para el contenido de aceite de sus semillas.

olericulturalist – one who studies the cultivation of vegetables.

oleicultor – uno que estudia la cultivación de verduras.

olericulture – the cultivation of vegetables.

oleicultura – la cultivación de verduras.

on-the-job training – experience obtained while working in an actual job setting.

aprendizaje – experiencia obtenida mientras trabajando en un empleo verdadero.

opening a position – initial step in the futures market.

una posición inicial – paso inicial en el mercado a futuro.

order of business – the items and sequence of activities conducted at a meeting.

orden de negocio – a los puntos y la sucesión de actividades conducidos en una reunión.

organic compound – a compound that contains carbon.

compuesto orgánico – un compuesto que contiene carbono.

organic food – food that has been grown without the use of certain chemical pesticides.

alimentos orgánicos – alimentos que han sido cultivados sin el uso de ciertos pesticidas químicos.

organic matter – dead plant and animal tissue that originates from living sources such as plants, animals, insects, and microbes.

materia orgánica – tejido muerto animal o vegetal que origina de orígenes vivientes como plantas, animales, insectos, y microbios.

ornamental plant – used to improve the appearance of a structure or an area.

de adorno – usada para hacer mejoras al aspecto o estructura de un área.

ornamental – fruits, vegetables, and ornamental plants.

cultivo de planta de adorno – ciencia de frutas, verduras, y plantas de adorno.

osmosis – the flow of a fluid through a semipermeable membrane separating two solutions, which permits the passage of the solvent but not the dissolved substance. The liquid will flow from a weaker to a stronger solution, thus tending to equalize concentrations.

osmosis – el flujo de un líquido a través de una membrana semipermeable que separa dos soluciones, que permite el paso del solvente pero no de la sustancia disuelta. El líquido fluirá de un más débil a una solución más fuerte, así tendiendo para igualar concentraciones.

outcrossing – mating of unrelated animal families within one breed.

«outbreeding» – acoplamiento de familias animales no relacionadas dentro de la misma raza.

out-of-the-market – when a futures position is offset by a second position.

fuera del mercado – cuando una posición a futuro es compensado por una posición segunda.

ova – female reproductive cells or eggs from the ovary of an animal.

óvulo – células reproductivas femeninas o huevos del ovario de un animal.

ovary – female organ that produces eggs or female sex cells; also, that portion of the flower that contains the ovules or seeds.

ovario – órgano de la hembra que produce los óvulos o células sexuales femeninas; además aquella parte de la flor que consta de los óvulos o semillas.

overgrazing – damage to plants or soil due to animals eating too many of the plants at one time or reducing the plants' ability to hold soil or recover after grazing.

sobrepastorear – daño a las plantas o al suelo causado por el comer demasiado de parte de los animales a las plantas de una vez o el reducir de la capacidad de la planta de guardar tierra o recuperar después del apacentamiento.

overripe – beyond maturity.

demasiado maduro – más allá de madurez.

overseeding – seeding a second crop into one that is already growing.

sobresembrar – sembrar un cultivo segundo encima de úno que ya está creciendo.

ovulation – process of releasing mature eggs from the ovary.

ovulación – el proceso de desprender óvulos maduros del ovario.

ovule – unfertilized seed.

óvulo (vegetal) – semilla ya no fecundada.

oxidative deterioration – decay resulting from exposure to air.

deterioro oxidativo – deterioro resultando de la exposición al aire.

ozone – compound that exists in limited quantities about 15 miles above the Earth's surface.

ozono – compuesto que existe en cantidades limitadas hacia quince millas encima de la superficie de la tierra.

pace – two-beat, side-to-side gait.

amblar – paso de dos tiempos, un lado tras otro lado.

packer – person or firm responsible for preparing commodities for shipment.

empaquetador – persona o firme responsable de preparar productos para envío.

palisade cells – elongated, vertical cells that give the leaf strength and manufacture food.

células empalizadas – células verticales, alaragadas que dan fuerza a la hoja y fabrican alimentación.

parasite – organism that lives in or on another organism with no benefit to the host.

parásito – organismo que vive en o encima de otro organismo sin beneficio al huésped.

parasitism – one type of wildlife living and feeding on another without killing it.

parasitismo – una clase de fauna viviendo asido y alimentándose con otra sin matarla.

parent material – the horizon of unconsolidated material from which soil develops.

materiales madres del suelo – el horizonte de material descomprimido desde lo cual desarrolla el suelo.

parliamentary procedure – a system of guidelines or rules for making group decisions in business meetings.

procedimiento parlamentario – un sistema de principios o normas para hacer decisiones de grupo en reuniones de negocio.

particulates – a form of pollution consisting of small particles that are suspended in the air.

particulatas – una forma de contaminación que consiste de pequeñas partículas que están suspendidas en el aire.

parturition – birth.

parto – nacimiento.

pasture – forages that are harvested by the livestock itself.

pasto – forrajes que se cosechan por el ganado mismo.

pasture mating – male allowed to roam freely with the herd and mate at random.

servicio a campo – cuando se permite que el macho ronde libremente por el rebaño y acople al azar.

pathogen – organism that produces disease.

agente patógeno – organismos que producen las enfermedades.

peat moss – a type of organic matter made from spagnum moss.

turba – un tipo de materia orgánica hecha de esfagno.

pedigree – record of ancestry.

carta de origen – registro de linaje.

pelt – skin of an animal with the hair attached.

pellejo – la piel de un animal, inclusive el pelo.

pencil shrink – the estimated amount of loss that an animal will sustain during the trip to market.

contracción – la cantidad estimada de pérdida que sostendrá el animal durante el transporte al mercado.

penetration pricing – a strategy in which price is set less than that of competitors.

valoración psicológica – una estrategia donde se fija el precio más bajo que lo de los competidores.

percolation – movement of water through the soil.

filtración – paso del agua a través de la tierra.

perennial – a plant that lives from year to year.

perene – una planta que vive de año a año.

perennial weed – a weed that lives for more than 2 years.

mala hierba perenne – una mala hierba que vive más de dos años.

perfect flower – flower containing all of the parts: stamen, pistil, petals, and sepals.

flor perfecta – flor que consta de todas las partes: estambre, pistilo, pétalos, y sépalos.

peripheral nervous system – system that controls the functions of the body tissues including the organs.

sistema nervioso periférico – sistema que controla las funciones de los tejidos de cuerpo inclusive los órganos.

perlite – natural volcanic glass material having water-holding capabilities.

perlita – material vítreo volcánico natural que tiene capacidades de aguantar el agua.

permeable – permitting movement.

permeable – que permite movimiento a través de si mismo.

pest – any organism that adversely affects human activities.

insecto, animal dañino – cualquier organismo que afecta adversamente las actividades del hombre.

pesticide – chemical used to control pests.

pesticida – sustancia química empleada para combatir los animales e insectos dañinos.

pesticide resistance – the ability of a pest to tolerate a lethal level of a pesticide.

resistencia de pesticida – la capacidad de un animal o insecto dañino para tolerar un nivel mortal de pesticida.

pest population equilibrium – a condition that occurs when the number of pests stabilizes or remains steady.

equilibrio de población de pestes – una condición que ocurre cuando el número de pestes se estabiliza o queda estable.

pest resurgence – ability of a pest population to recover.

ensurgimiento renacimiento – la capacidad de una población de insectos o animales dañinos para recuperar.

petal – brightly colored, sometimes fragrant portion of the flower.

pétalo – parte de la flor de color vivo y a veces fragrante.

petiole – the slender leaf stock that supports the blade, attaching it to the stem of a plant.

pecíolo – el esbelto rabo de la hoja que sostiene la brizna, uniéndola al tallo de la planta.

pH – measurement of acidity or alkalinity from 1 to 14.

valor pH – medida de la acidez o alcalinidad desde 1 a 14.

phenotype – physical appearance of an individual.

fenotipo – características externas, o aspecto de un individuo.

pheromone – a chemical secreted by an organism to cause a specific reaction by another organism of the same species.

feromona – una sustancia química secretada por un organismo para causar una reacción específica por otro organismo de la misma especie.

phloem – cells that carry manufactured food to areas of the plant where it is stored or used.

líber – células que llevan alimentación fabricada a las áreas de la planta donde se conserva o se usa.

phloem tissue – conductive tissue responsible for transport of food downward in a tree.

líber – tejido conductivo que es responsable del transporte de los alimentos nutritivos hacia abajo en un árbol.

photodecomposition – chemical breakdown caused by exposure to light.

fotodescomposición – la descomposición química causada por la exposición a la luz.

photosynthesis – process in which chlorophyll in green plants enables those plants to use light to manufacture sugar from carbon dioxide and water.

fotosíntesis – proceso en donde el clorafilo de las plantas verdes las capacita utilizar la luz, fabricar azúcar de dióxido de carbono y agua.

phototropism – a process by which a plant leaf is capable of adjusting its angle of exposure to the sun.

fototropismo – un proceso por el cual la hoja de una planta es capaz de ajustar su ángulo al exposición al sol.

physiology – study of the functions and vital processes of living creatures and their organs.

fisiología – el estudio de las funciones y procesos vitales de criaturas vivientes y sus órganos.

pig – swine younger than 4 months.

gorrino – cerdo que tiene menos de cuatro meses de edad.

piglet – newborn of the swine family.

lechón – el recién nacido del cerdo.

pistil – female part of the flower consisting of stigma, style, ovary, and ovule.

pistilo – parte femenina de las flores que consta de estigma, estilo, ovario, y óvulo.

placement supervised agricultural experience – career experiences that place students with employers who are conducting agricultural business such as farming ranching, greenhouses, and others.

colocación supervisada de experiencia agrocientífica – experiencias de carrera que coloca a estudiantes con empleadores que conducen negocios agriculturales tales como agricultura,ranchería, viveros y otros.

plan – to think through.

planear – pensar muy bien en cómo se va a hacer algo.

plant disease – any abnormal plant growth.

enfermedad de planta – cualquier crecimiento vegetal anormal.

plant fertilization – addition of elements to a plant-growing environment.

fertilización de planta – adición de elementos a un ambiente en que se crecen las plantas.

plant hardiness zone map – a map developed by the U.S. Department of Agriculture, dividing the country into 10 zones based on average winter temperatures.

Mapa de Zonas de Plantas Resistentes al Frío – mapa desarrollado por el Departamento de Agricultura de los EE.UU., que divide el país en diez zonas basado en las temperaturas medias de invierno.

plant nutrition – provision of elements to plants.

alimentación nutritiva vegetal – el suministro de elementos a las plantas.

plant residue – plant material that remains when a plant dies or is harvested.

residuo de planta – material de planta que queda cuando una planta muere o es cosechada.

plantscaping – the design and arrangement of plants and structures in an indoor area.

el plantscaping – el diseño y el arreglo de plantas y de estructuras en un área de interior.

playability level – the suitability for the intended recreational or sporting use of the turf.

nivel de resistencia – la conveniencia del uso intencional recreativo o deportivo del césped.

plugging – establishment of turf by using small pieces of existing turfgrass.

cubrir de trasplantes de tepe – iniciando el césped por plantear tepes ya existentes.

plywood – construction material made of thin layers of wood glued together.

contrachapado – material de construcción hecho de hojas delgadas de madera que son pegadas juntas.

point of order – a procedure used to object to some item in or about a meeting that is not being done properly. The procedure to use is to say, "point of order." The presiding officer should then recognize the member by saying, "State your point."

cuestión de orden – un procedimiento que sirve para oponerse a algún punto de o sobre la reunión que no está haciéndose adecuadamente. El procedimiento de usar es decir, «¡Cuestión de orden!» Entonces el director ha de reconocer al miembro por decir, «Exprese su cuestión.»

polled – hornless, either naturally or otherwise.

mocho – sin cuernos, naturalmente o de otro modo.

pollen – small male sperm or grains that are necessary for fertilization in the flower.

polen – esperma pequeño o polvillo macho que es necesario para la fecundación de la flor.

pollination – transfer of pollen from anther to stigma.

polinización – transporte del polen del antero hasta el estigma.

polluted – containing harmful chemicals or organisms.

contaminado – que contiene sustancias químicas u organismos dañosos.

pome – fleshy fruits with embedded core and seeds.

pomo – frutos carnosos con corazón fijado y semillas.

pomologist – a fruit grower or scientist.

pomólogo – cientista o cultivador de frutos.

pony – member of the horse family less than 14.2 hands tall.

jaca – miembro de la familia de los caballos menos de 14.2 palmos de alto.

porcine somatotropin (PST) – hormone that increases meat production in swine.

somatotropino porcino – una hormona que aumento la producción de carne en los cerdos.

pore – spaces between soil particles through which plant roots penetrate, and in which air, water, and nutrients are stored.

poro – espacio entre partículas de tierra en los que las raíces penetran a través y en cuales el aire, agua, y nutrientes son almacenados.

pork – meat from swine.

puerco – carne del cerdo.

port – town having a harbor for ships to take on cargo.

puerto – ciudad o pueblo que tiene un lugar en la costa o el río que es dispuesto para seguridad y las carga y descarga de las naves.

postemergence herbicide – a herbicide applied after the weed or crop is present.

herbicida postemergente – un herbicida aplicado después de que sea presente la mala hierba o el cultivo.

potable – drinkable, that is, free of harmful chemicals and organisms.

potable – que puede beberse, o sea que es libre de organismos y sustancias químicas dañosos.

poult – young turkey.

pavipollo – pavo jóven.

poultry – group name for domesticated birds.

aves de corral – nombre de grupo para los aves domesticados.

ppm – parts per million.

partes por millón – partes por millón.

ppt – parts per thousand.

partes por mil – partes por mil.
Note: In Spanish, abbreviations are not used for "PPM" and "PPT," as they would both be "PPM," translated verbatim.

precipitate – a chemical action that results in the dropout of solids in a solution.

precipitado – una reacción química que resulta en el depositarse de sedimento en una solución.

precipitation – formation of rain and snow.

precipitación – formación de lluvia y nieve.

precooling – rapid removal of heat before storage or shipment.

prerefrigeración – el quitarse del calor rápidamente antes de almacenamiento o transporte.

predation – a way of life where one type of wildlife eats another type.

rapacidad – una manera de vivir en que un tipo de fauna se alimenta con otro tipo.

predator – an animal that feeds on a smaller or weaker animal.

animal de rapiña – un animal que se alimenta con otro animal más pequeño o más débil.

preemergence herbicide – a herbicide applied before weed or crop germination.

herbicida pre-emergente – un herbicida aplicado antes de la germinación de malas hierbas o cultivos.

presiding officer – a president, vice president, or chairperson who is designated to lead a business meeting.

presidente – el presidente, vicepresidente, o presidente interino designado para dirigir una sesión de negocio.

prestige pricing – pricing to buyers with special desires for quality, fashion, or image.

valoración de prestigio – valoración a los compradores que tienen deseos especiales para calidad, moda, or reputación.

prey – animal eaten by another animal.

presa – animal comido por otro animal.

price – amount received for an item that is sold.

precio – cantidad recibido por un artículo que es vendido.

primary nutrients – in agriculture, nitrogen, phosphorus, and potash.

alimentos nutritivos mayores – en la agricultura, nitrógeno, fósforo, y potasa.

processed meat – meat that is not suitable for sale over the counter.

carne elaborada – carne que no es adecuado para venderse al por menor.

processing – turning raw agricultural products into consumable food.

elaboración – el convertirse de los productos agrícolos en bruto en alimentos comestibles.

processor – person or business cleaning, separating, handling, and preparing a product before it is sold to the distributor.

elaborador – persona o empresa que limpia, separa, maneja, y prepara un producto antes de que se venda al distribuidor.

producer – person who grows a crop.

productor – persona que cultiva el cultivo.

product advertising – advertising that focuses on the product itself.

publicidad de producto – publicidad que fija en el producto si mismo.

production agriculture – farming and ranching.

agricultura de producción – labrar y llevar un rancho.

Production Credit Association(s) (PCA) – lending institution that provides short-term credit for agriculture.

Asociación de Crédito de Producción – institución de préstamo que provee crédito para agricultura a corto plazo.

production or productive enterprise – project conducted for wages or profit.

producción o proyecto productivo – proyecto conducido por salarios o lucro.

profession – occupation requiring an education, especially law, medicine, teaching, or the ministry.

profesión – vocación que exige una enseñanza superior, especialmente el derecho, la medicina, la enseñanza, o el ministerio.

profile – a cross-sectional view of soil.

pérfil – una vista del suelo por corte transversal.

profit – proceeds from a business transaction.

beneficio – ganancias de una transacción comercial.

profit and loss statement – projection of costs and expenses against sales and revenue over time.

estado de ganancias y pérdidas – la previsión de los costos y gastos en contra a las ventas e ingresos por un tiempo.

progeny – offspring.

progenie – descendecia.

progesterone – hormone that prevents estrus during pregnancy and causes development of the mammary system.

progesterona – hormona que impide el estro durante el embarazo y causa el desarrollamiento del aparato mamario.

program – total plans, activities, experiences, and records.

programa – planes, actividades, experiencias, y registros totales.

project – a series of activities related to a single objective or enterprise.

proyecto – una serie de actividades relativas a un sólo objetivo o empresa.

promissory note – note signed by borrower agreeing to the terms of the loan.

pagaré – obligación escrita firmada por el prestatario, consintiendo a los plazos del préstamo.

promoting function – plan to make potential customers aware of the product or service.

función de promocionar – plan para hacer que los clientes se den cuenta del producto o servicio.

propagation – process of increasing the numbers of a species or perpetuating a species.

propagación – el proceso de aumentar la población de una especie o de perpetuar una especie.

protectant fungicide – a fungicide applied before disease infection.

fungicida preventivo – un fungicida aplicado antes de infeccionarse.

protein – nutrient made up of amino acids and essential for maintenance, growth, and reproduction.

proteina – alimento nutritivo que consta de aminoácidos y que es esencial para el mantenimiento, el crecimiento, y la reproducción.

protozoa – microscopic, one-celled animals that are parasites of animals.

protozario – microorganismos unicelulares que son parásitos de otros animales.

pruning – the removal of dead, broken, unwanted, diseased, or insect-infested wood.

podar – el quitar de madera muerta, trozada, no deseada, enferma o plagada por los insectos.

psychological pricing – a strategy designed to make a price seem lower or less significant.

valoración psicológica – una estrategia creada para hacer que un precio parezca más bajo o menos significativo.

pullet – young female chicken.

polla – hembra jóven del pollo.

pulpwood – wood used for making fiber for paper and other products.

madera para pasta de papel – madera usada para hacer fibras para papel y otros productos.

purebred – animal with both parents of the same breed and with registration papers.

animal de pura raza – animal cuyos dos padres son de la misma raza.

purify – removal of all foreign material.

purificar – extirpación de toda materia extranjera.

putting green – golf-playing area with turfgrass mowed very short.

campo más pequeño que es utilizado para jugar al golf – área para jugar al golf que tiene el césped muy segado.

Q

quality grade – grade based on amount and distribution of finish on an animal.

grado de calidad – grado basado en la cantidad y distribución del acabado de engorde de un animal.

quarantine – isolation of pest-infested material.

cuarentena – aislamiento de material plagado.

queen – only fertile, egg-laying female bee in the hive.

reina – la sóla abeja hembra que es fecunda y pone huevos.

R

rack – in horses, a fast, four-beat gait.

pasitrote – de los caballos, un paso rápido de cuatro tiempos.

radioactive material – material that is emitting radiation.

materia radioactivo – materia que emite la radiación.

radon – colorless, radioactive gas formed by disintegration of radium.

radón – gas radioactivo incoloro, formado por la desintegración de radium.

ram – male member of the sheep family.

carnero – macho de la oveja.

ration – the amount of feed fed in one day.

ración – la porción de pienso que se da de comer en un día.

real-world experience – an activity conducted in the daily routine of our society.

experiencia de la vida diaria – una actividad conducida dentro de la rutina diaria de la sociedad.

reaper – machine that cuts grain.

segador – máquina que corta mieses.

recessive – a gene that remains hidden and expresses itself only in the absence of a dominant gene.

recesivo – un gene que queda oculto y se expresa sólo en la ausencia de un gene dominante.

recombinant DNA technology – gene splicing.

tecnología de ADN preparado en el laboratorio – empalme de genes.

rectum – the last organ in the digestive tract.

recto – el último órgano del aparato digestivo.

recuperative potential – the ability of a plant to recover after being damaged.

potencialidad de recuperar – la capacidad de la planta para reestablecerse después de ser dañada.

refer – a motion used to refer some other motion to a committee or person for finding more information and/or taking action on the motion on behalf of the members. The way to state a referral is to say, "I move to refer this motion to . . ."

referir – una moción usada para referir a otra moción a una comisión para averigüar más información y/o para tomar acción sobre la moción en nombre de los socios. La manera de declarar una referencia es para decir, «Yo propongo referir esta moción a . . .»

refrigeration – keeping cool.

refrigeración – mantenerse congelado.

registration papers – records proving that an animal is purebred.

cartas de registro – registros probando que un animal determinado es de pura raza.

reins – leather or rope lines attached to the bit for control of a horse.

riendas – cuerdas de cuero o soga fijadas al bocado y empleada para controlar el caballo.

relative humidity – the relationship between the actual amount of water vapor in the air to the greatest amount possible, at a given temperature.

humedad relativa – relación entre la cantidad verdadera de vapor de agua en el aire a la cantidad más grande que sea posible, a una temperatura determinada.

render – to cook and press the oil from.

derretir – cocinar y extraer el óleo de algo.

rendering insensible – making an animal unable to feel pain.

rendir insensible – hacer que el animal no pueda sentir el dolor.

renewable natural resources – resources provided by nature that can replace themselves.

recursos naturales renovables – recursos suministrados por la naturaleza que pueden volver a llenarse de nuevo.

reproduction – the making of a new plant or animal.

reproducción – el hacerse de planta o animal nuevo.

residual soil – parent material formed in place.

suelo residuo – materiales madres del suelo formados en lugar.

resource substitution – the use of a resource or item for another when the results are the same.

sustitución de recursos – el uso de un recurso o cuerpo en vez de otro cuando los resultados salen semejantes.

Resource Inventory – summary of the resources that may be available for conducting a SAEP.

inventario de recursos – resumen de los recursos que pueden ser disponibles para conducir un PEAS.

respiration – a process in which energy and carbon dioxide are released due to digestion or the breakdown of plant tissues during periods of darkness.

respiración – un proceso en el cual la energía el bióxido de carbono son liberados debido a la digestión o la descomposición de tejidos de plantas durante períodos de oscuridad.

respiratory system – system that provides oxygen to the blood of the animal.

aparato respiratorio – conjunto de órganos que proporciona el oxígeno a la sangre del animal.

restricted-use pesticide – a pesticide that poses a greater risk to humans and the environment than one that is not so labeled.

pesticida de uso restringido – un pesticida que posa un mayor riesgo a humanos y al medio ambiente que uno que no está rotulado.

resume – a summary of a person's education, technical skills, and career experiences.

resumen – un sumario de la educación de una persona, habilidades técnicas y experiencias de carrera.

retailer – person or store that sells directly to the consumer.

minorista – persona o tienda que vende directamente al consumidor.

retail marketing – the selling of a product directly to consumers.

mercadeo al por menor – la venta de un producto directamente al consumidor.

retortable pouch – package with foil between two layers of plastic.

embalaje a devolver – paquete que tienen hoja de aluminio entre dos hojas de plástico.

rhizome – horizontal, underground stems.

rizoma – tallo subterráneo, horizontal.

roaster – old chicken used for meat.

pollo para asar – pollo viejo usado para carne.

rolls over – rapid change in water quality because of a change in the relative position of cold and warm water.

derribo – cambio rápido de la calidad del agua.

rooster – adult male chicken or pheasant.

gallo – adulto macho del pollo.

faisán – adulto macho de la faisana.

rootbound – roots restricted by a container.

raíces restringidas por contenidor – raíces restringidas por un contenidor.
Note: The translation of "rootbound" is the same as that of the English definition provided.

root cap – the outermost part of a root that protects the tender tip of a growing root as it penetrates the soil.

capucha de raíz – la parte más externa de una raíz que protege la tierna punta de una raíz en crecimiento mientras penetra la tierra.

root crop – crop grown for the thick, fleshy storage root that it produces.

raíz planta – cultivo cultivado por el tubérculo grueso carnoso que produce.

root cutting – section of a root cut used for propagation purposes.

estaca de raíz – cuando se corte parte de la raíz y se usa para fines de propagación.

root hairs – small microscopic roots that arise from the cells located on the surface of a plant root.

peluza de raíz – pequeñas raíces microscópicas que surgen de las células localizadas en la superficie de la ríz de una planta.

rooting hormone – chemical used to stimulate root formation on a cutting.

hormona radicular – sustancia química usada para estimular la formación de raíz en una tala.

root pruning – systematic cutting of the roots by hand or machine to encourage the roots to develop within the root ball range.

poda de restauración – el podar sistematicamente de las raíces a mano o a máquina para apoyar que las raíces desarrollen dentro del área del cepillón.

rootstock – root system and stem of a plant on which another plant is grafted.

portainjerto – sistema radicular y tallo de una planta sobre la cual se injerta otra planta.

roughage – grass, hay, or silage and other feeds high in fiber and low in TDN.

forraje – hierba, heno, o ensilaje y otros piensos altos en forraje vegetal y bajos en NDT.

roundworm – slender worm that is tapered on both ends.

ascáride – lombrice delgado que tienen aspecto afilado en dos extremos.

row crop planter – machine that plants seed in precise rows and with even spacing within the row.

sembradora a hilera – máquina que planta semillas en hileras precisas y a distancias uniformes dentro del surco.

rumen – one of the compartments of the stomach in cattle, deer, and sheep that is responsible for the breakdown of cellulose in the feed that is consumed.

herbario – uno de los componentes del estómago de ganado, venados, y ovejas que es responsable por la descomposición de celulosa en el alimento que es consumido.

ruminant – animal that has a stomach with four digestive compartments.

rumiante – animal que tiene estómago con más de una cavidad.

russet – a baking potato variety best adapted for growth in sandy vocanic soils.

russet – una variedad de la patata de la hornada se adaptó lo más mejor posible para el crecimiento en suelos vocanic arenosos.

S

saddle – padded leather seat for the rider of a horse.

silla – aparejo de asiento rellenado y de cuero que usa el jinete.

safe water – water that is free of harmful chemicals and disease-causing organisms.

agua potable – agua que es libre de las químicas dañosas y de los organismos que causan las enfermedades.

salinity – a measure of total dissolved mineral solids in water.

salinidad – una medida de minerales disueltos totales que hay en el agua.

salmonid – any of the family of soft-finned fish such as salmon or trout that have the last vertebrae upturned.

salmonides – cualquieres de la familia de peces de alas suaves como salmón o trucha que tienen el vértebra última respingona.

saltwater marsh – the ecological system of plants and animals influenced by tidal waters with salinity of 15 to 34 parts per million.

salina – sistemas ecológicos de plantas y animales influidos por mareas con salinidades de 15 a 34 partes por mil.

sand – largest soil particles; 1 to 0.05 mm.

arena – las partículas de tierra más grandes; de 1 a 0.05 mm.

sapwood – transports water and minerals upward in tree roots, trunks, and stems.

albura – traslada los agua y minerales hacia arriba en raices, troncos, y tallos de árbol.

saturated – water added until all the spaces or pores are filled.

saturado – el añadido del agua hasta que se llenen todos espacios o poros.

scale – size of items.

escala – tamaño de un cuerpo.

scarify – to soak, keep moist, or mechanically scrape a seed coat to aid germination.

escarificar – empapar, mantener húmedo, o raspar a máquina de la capa de una semilla para apoyar la germinación.

scientific method of inquiry – a procedure for investigating problems of a scientific nature.

método científico de investigación – un procedimiento para investigar problemas de naturaleza científica.

scion – the top of the stem and leaves of a plant that is to be propagated.

codo o varilla de retoño – la punta del tallo y hojas de una planta que se va a propagar.

scorch – dead spots or edges on a leaf that usually indicate deficiencies in one or more nutrients.

quemadura – manchas o bordes muertos en la hoja que usualmente indican deficiencias en uno u más nutrientes.

scurvy – vitamin C deficiency disease.

escrobuto – una enfermedad por la carencia de vitamina C.

seasonal – pertaining to a particular season of the year.

estacional – que pertenece a una estación determinada del año.

secondary host – a plant or animal that carries a disease or parasite during part of the life cycle.

huésped secundario – una planta o un animal que lleva una enfermedad o parásito durante parte del ciclo de la vida.

secretary – a person elected or appointed to take notes and prepare minutes of the meeting.

secretaria – una persona elegida o apuntada para recordar y preparar las actas de la sesión.

seed blend – combination of different cultivars of the same species.

mezcla de semilla – combinación de cultivos diferentes de la misma especie.

seed certification program – program administered by state governments to ensure seed quality.

programa de certificación de semilla – programa administrado por gobiernos estatales para asegurar la calidad de la semilla.

seed culm – stem that supports the inflorescence of the plant.

caña de semilla – tallo que sostiene la florescencia de la planta.

seed legume crop – crop that is nitrogen-fixing and produces edible seeds.

legumbre – cultivo que es fijador del nitrógeno y produce semillas comestibles.

seedling – young plant grown from seed.

planta de semillero – plantón cultivado de semilla.

seed mixture – combination of two or more species.

mixtura de semilla – combinación de dos o más especies.

seed piece – vegetative means of propagating sugar cane.

multiplicación por estacas – medida vegetativa de propagar la caña de azúcar.

seining – physical netting of fish in open water.

pescar con jábega – el coger de los peces con red en las aguas abiertas.

selective breeding – mating adults who have characteristics desired in the offspring.

crianza selectiva – selección de padres para apareamiento que tienen los rasgos que se quiere que tenga la descendencia.

selective herbicide – a chemical that kills or affects certain types of plants.

herbicida selectivo – una sustancia química que mata o afecta ciertos tipos de plantas.

selflessness – placing the desires and welfare of others above oneself.

desinterés – poniendo los deseos y el bienestar de los demás mas allá de los de si mismo.

selling function – market research, sales plans, and sales closures.

función de vender – las investigaciones del mercado, las ventas, los planes, y cierres de venta.

semen – the sperm cells and accompanying fluids.

semen – las espermatazoides y los líquidos que los acompañan.

semiarid – an area partially deficient in rainfall; dry.

semiárido – un área que parcialmente carece de la precipitación.

semidwarf – tree that has rootstock that limits growth to 15 feet or less.

semienano – árbol que tiene el rizoma, lo cual que límite el crecimiento a lo de 15 pies o menos.

seminal root – root formed at the time of seed germination to anchor the seed in soil.

raice seminale – raice formada en momento de la germinación de semilla para fijar la semilla en el suelo.

semipermeable membrane – membrane that permits a solution to move through it.

membrana semipermeable – membrana que permite que la penetre una solución.

sepal – small, green, leaflike structure found at the base of the flower.

sépalo – cuerpo verde pequeño y de aspecto de hoja que se encuentran al base de la flor.

sequence – related or continuous series.

sucesión – series continuas o relacionadas.

sewage system – receives and treats human waste.

alcantarillado – recibe y trata los desechos humanos.

sex-linked – genes carried on chromosomes that determine sex.

(herencia) ligada al sexo – genes que se transmiten por los cromosomas que determinan el sexo.

sexual reproduction – union of an egg and sperm to produce a seed or fertilized egg.

reproducción sexual – unión de un óvulo y un esperma para producir una semilla u óvulo fecundado.

shackles – mechanical devices that restrict movement.

trabas – dispositivos mecánicos que restringen el movimiento.

shearing – to remove wool from sheep.

esquilar – quitar la lana a la oveja.

sheath – lower portion of the turfgrass leaf.

vaina foliar – parte a la base de la hoja de césped.

sheet erosion – removal of soil from broad areas of the land.

erosión laminar – eliminación del suelo de áreas amplias del terreno.

shelf life – time between packaging and spoilage.

tiempo de conservación de un producto en venta – el tiempo entremedias de envasar y de putrefacción.

shellfish – any aquatic animal having a shell or shell-like exoskeleton.

mariscos – cualquier animal acuático que tiene cáscara o dermatoesqueleto parecido a la cáscara.

shepherd – one who cares for sheep.

pastor – persona que guarda las ovejas.

shod – wearing shoes.

herrado – dícese del caballo llevando herraduras.

short-term loan – loan with a payment period of 1 year or less.

préstamo a corto plazo – préstamo con plazos de pago de un año o menos.

short-term plan – plan accomplished in days or weeks.

plane a corto plazo – plane realizado dentro de días o semanas.

shrinkage – changes in dimensions of wood as it reacts to changes in humidity and temperature.

contracción – disminución de volumen de la madera mientras reacciona a los cambios de humedad y temperatura.

shrouded – wrapped tightly with a cloth.

amortajar – envolver ajustadamente con un paño.

shrub – a woody perennial plant that normally produces many stems or shoots from the base and does not reach more than 15 feet in height.

arbustos – plantas perennes leñosos que normalmente crecen bajo, producen muchas ramas o brotes desde la base, y no alcanzan a más que 15 pies de alto.

side-dressing – placing fertilizer in bands about 8 inches from the sides of growing plants.

abonado en fajasa lado – meter el abono más o menos a ocho pulgadas de los lados de los cultivos crecientes.

signal word – required words on the label that denotes the relative toxicity of the product.

palabra señalada – la palabra requerido en la etiqueta que indica la toxicidad relativa del producto.

silage – feed resulting from the storage and fermentation of green crops in the absence of air.

ensilaje – pienso formado a causa del ensilar y fermentación de las plantas verdes debido a la ausencia del aire.

silo – air-tight storage facility for silage.

silo – almacenamiento estanca al aire para el ensilaje.

silt – intermediate soil particles; 0.05 to 0.002 mm.

légamo – partículas de tierra intermedias; 0.05 a 0.002 mm.

silviculture – the scientific management of forests.

silvicultura – el manejo científico de los bosques.

simple-interest loan – the full amount of a loan is received by the borrower and paid back with interest after a short period.

préstamo de interés simple – la cantidad total del préstamo es recibido por el prestatario y pagado con interés después de un período corto.

simple layering – stem is bent to the ground, held in place, and covered with soil.

acodo simple – el tallo es metido debajo de la tierra, sujetado y cubierto. con tierra.

simple leaf – a single leaf arising from a plant stem.

hoja simple – una hoja singular que se levanta de un tallo de planta.

simulate – to look or act like.
    simular – dar la apariencia de algo o imitar algo.

singe – burn lightly.
    chamuscar – quemar superficialmente.

single-eye cutting – cutting made with the node of a plant with alternate leaves.
    codos o cortes con un solo retoño – tala hecha con la yema de una planta que tiene hojas alternativas.

sire – father.
    semental – padre.

skeletal system – bones joined together by cartilage and ligaments.
    esquelético – huesos juntados por el cartílago y los ligamentos.

skimming – setting the price of a new product for unusually high profits at first when affluent and willing customers are available.
    importe de los derechos – fijar el precio de un producto nuevo para que se saquen ganancias extraordinariamente altas al principio, cuando hay clientes opulentos y dispuestos.

slaughter – the killing of animals for market.
    sacrificar – la matanza de animales para el mercadeo.

slaughter cattle – cattle intended for slaughter.
    ganado de sacrificio – ganado para el matadero.

Small Business Administration (SBA) – institution that provides loans to agribusinesses.
    Administración de Comercios Pequeños – hace empréstitos a las agroindustrias.

smoker – device used to add smoke flavor and taste to food.
    secador al humo – dispositivo que se usa para añadir el sabor ahumado a los comestibles.

sodding – removal of a portion of the soil and turf-grass plant for vegetative establishment purposes.
    cubrir con césped – excavación de una parte del suelo y césped para fines de establecimiento vegetal.

softwood – wood from conifers.
    madera blanda – la madera de las coníferas.

soil – top layer of the earth's surface suitable for the growth of plant life.
    tierra, suelo – capa de la superficie de la tierra que es conveniente para el crecimiento de seres vegetales.

soil science – study of the properties and management of soil to grow plants.
    ciencia de estudio de los suelos – el estudio de las propiedades y administración de los suelos para cultivar plantas.

solution culture – same as water culture.
    cultivo de solución – el mismo que cultivo hidropónico.

sow – female member of the swine family that has given birth.
    cerda – hembra del cerdo que ha parido.

spat – the young shellfish just after it attaches to underwater structures.
    freza – los mariscos jóvenes, inmediatamente después de ligarse a estructuras subterráneas.

spawn – egg-laying process of fish.
    frezar – de los peces, el proceso de poner huevos.

species – the basic unit in the classification system whose members have similar structure, common ancestors, and maintain their characteristics; subgroup of genus.
    especies – la unidad básica del sistema de clasificación cuyos miembros tienen estructura semejante, antepasados comunes, y mantienen sus características; subgrupo de genus.

specimen plant – plant that is used as a single plant to highlight it or some other special feature of the landscape.
    planta maestra – planta que se usa como planta singular para demostrar a ella o a otro rasgo especial del paisaje.

sperm – male reproductive units.
    células de esperma – unidades reproductivas del macho.

sphagnum – pale and ashy mosses used to condition soil.
    esfagno – musgo pálido y ceniciento usado para acondicionar el suelo.

split carcass – halves of the animal after it is killed.
    res abierta en canal – las mitades del animal después del sacrificio.

split-vein cutting – cutting made by slitting a large leaf on the veins before placing in a rooting medium.
    esqueje del nervio partido – una tala hecha por hender una gran hoja en los nervios antes de meterla en un medio para echar raíces.

spoiled – chemical changes that reduce nutritional value or render food unfit to eat.
    estropeado – cambios químicos que reducen el valor nutritivo o render el comestible inadecuado para comer.

spongy layer – lower, irregular layer in the leaf that allows the veins, or vascular bundle, to extend into the leaf.
    capa esponjosa – capa del vaina exterior desigual, que permite que los nervios, o el haz vascular, se extienda a la hoja.

sprigging – planting a section of a rhizome or stolon (sprig) in the soil.
    propagación por estacas de raíz – plantar una parte de un rizoma o estolón (ramito) en el suelo.

sprinkler – device that sprays water on crops; its effect is much like rain.
    regadera – dispositivos que espuman agua al aire para regar los cultivos; tiene casi el mismo efecto como la lluvia.

sprinkler irrigation – the use of pipes to deliver water under pressure to a sprinkler head for the purpose of watering plants that are remote from a water source.
    irrigación de regadera – el uso de pipas de entregar el agua bajo presión a una cabeza de regadera con el fin de las plantas de riego que están alejadas de una fuente de agua.

squab – young pigeon that is butchered for meat before it is old enough to fly.
    pichón – aves jóvenes que son matadas por su carne antes de ser suficientemente maduras para volar.

square foot – a unit of area measuring 1 foot on each side of a square.
    pie cuadrado – una unidad de medir un pie en cada lado del cuadro.

staking – driving a wooden or metal rod into the ground near the plant and tying the plant to the rod.

estacar – clavar una estaca o varilla de metal en el suelo cerca de la planta y amarrar la planta a la varilla o estaca.

stallion – adult male horse or pony.

semental – adulto macho del caballo o poney.

stamen – male part of the flower that contains the pollen, anther, and filament.

estambre – órgano sexual masculino de la flor que consiste de los polen, antero, y filamento.

starch – major energy source in livestock feeds.

fécula – gran fuente de energía en los piensos.

starter solution – diluted mixture of single or complete fertilizers used when plants are transplanted.

abono de material inicial – mixtura diluida de abonos individuos o completos que se usan cuando se trasplantan las plantas.

steer – castrated male member of the cattle family.

bueyezuelo – macho castrado de la familia vacuna.

stem – supports the leaves and conducts the flow of water and nutrients.

tallo – sostiene las hojas y conduce el flujo de agua y alimentos nutritivos.

stem section cutting – a section of plant stem containing a node that is placed in a plant medium to generate a new plant.

corte de sección del tallo – una sección del tallo de la planta que contiene un nudo que es colocado en un medio de planta para generar una nueva planta.

stem tip cutting – cutting taken from the end of a stem or branch, normally including the terminal bud.

esqueje de punta de tallo – una tala sacada del extremo de un tallo o ramo, normalmente inclusive el botón terminal.

sterility – inability to reproduce.

esterilidad – incapacidad para reproducir.

stigma – part of the pistil that receives the pollen.

estigma – parte del pistilo que recibe el polen.

stimulant crop – crop that stimulates the senses of users.

cosecha estimulante – cosecha que estimula los sentidos de los usuarios.

stirrup – attached to the saddle by the stirrup leather; this device is a footrest used when riding horses.

estribo – atado a la silla por la barriguera, este aparato es un descanso para el pie que se usa al montar caballos.

stock – sometimes called the rootstock; a section of a plant consisting of the root and the base of the stem to which a scion is grafted.

cepa o tronco – a veces llamado raíz de tronco, una sección de una planta que consiste en una raíz y la base del tallo en el cual el codo es injertado.

stolon – a stem that grows aboveground.

estolón – un tallo que crece encima de la tierra.

stolonizing – type of vegetative establishment in which sprigs are broadcast onto the soil surface.

sembrar estolones al voleo – tipo de establecimiento vegetal en que los ramitos son sembrados al voleo a la superficie del suelo.

stoma – small openings, usually on the lower side of the leaf, that control movement of gases.

estoma – aberturas pequeñas, usualmente en la vaina exterior de la hoja, que controlan el movimiento de gases.

strip cropping – alternating strips of row crops with strips of close-growing crops.

cultivo en fajas – fajas alternas de cultivos a hilera con fajas de cultivos que crecen muy juntos.

structure – the tendency of soil particles to cluster together and function as soil units.

estructura – refiere a la tendencia de las partículas de tierra de agrupar y funcionar como unidades de tierra.

style – enlarged terminal part of the pistil.

estilo – parte terminal engrandizado del pistilo.

subcutaneous – under the skin.

subcutáneo – por debajo de la piel.

subsoil – a soil layer that corresponds to the B horizon that is composed almost entirely of mineral with generally large chunky soil structure.

subsuelo – una capa del suelo que corresponde al horizonte B que está compuesto de casi todo mineral y generalmente con grandes trozos de tierra en su estructura.

successive – following one right after the other.

sucesivo – en seguida.

succulent – in horticulture, thick, fleshy leaves or stems that store moisture.

planta carnosa – de la horticultura: hojas gruesas y carnosas que almacena agua.

sucrose – compound cane sugar.

sucrosa – azúcar de caña compuesto.

sugar crop – crop grown for the sugar content of its stems or roots.

cultivo de azúcar – un cultivo cultivado por el contenido de azúcar en los tallos o raíces.

sulfur – pale-yellow element occurring widely in nature.

azufre – elemento amarillo blanco que aparece mucho en la naturaleza.

super – box in which bees store excess honey.

super – parte superior del conjunto de la colmena.

supervised – looked after and directed.

supervisado – guiado y dirigido.

supervised agricultural experience program – SAEP; all supervised agriscience experiences.

programa de experiencia agrícola supervisado – PEAS; todas experiencias agrícolas supervisadas.

supplement – extra nutrient added to animal feed.

suplemento – alimento nutritivo añadido al pienso.

supply – the quantity of a product that is available to buyers at a given time.

oferta – la cantidad de un producto que está disponible a los consumidores.

surface irrigation – water flows over the soil surface to the crop.

riego de superficie – el agua fluye sobre la superficie del suelo al cultivo.

swarm – group of bees with a queen that leaves an overcrowded hive to find a new home.

enjambre – grupo de abejas con una reina que sale de una colmena atestada para hallar una morada nueva.

sweetbreads – thymus and pancreatic glands of animals.

mollejas – glándulas de timo y páncreas de los animales.

symbol – a warning illustration such as a skull and crossbones located on a chemical container that warns of chemical toxicity to humans and animals.

símbolo – una ilustración de advertencia como una calavera y huesos localizados en un recipiente de químicos que advierte de toxicidad química para animales y humanos.

symmetrical – in landscaping, a planting that is equal in number, size, and texture on both sides.

simétrico – en labrar el paisaje, cuando hay plantas iguales en número tamaño, y textura en ambos lados.

symptom – the visible change to the host caused by a pest.

síntoma – el cambio visible al huésped causado por un animal o insecto nocivo.

synthetic nutrient – nutrient manufactured chemically.

alimentos nutritivos sintéticos – alimentos nutritivos que son fabricados quimicamente.

syringe – an instrument used to give injections of medicine or to draw body fluids from animals.

jeringuilla – un instrumento usado para administrar inyecciones de medicina o para sacar fluidos corporales de los animales.

syringing – a light application of water to a turfgrass.

aplicación con mango de riego – una aplicación ligera de agua a un césped.

systemic herbicide – a herbicide that is absorbed by the roots of a plant and translocated throughout the plant.

herbicida sistemático – un herbicida que es absorbido por las raíces de la planta y desplazado por la planta.

tack – equipment or gear for horses.

arreo – guarniciones o aparejo para caballos.

tact – skill of encouraging others in positive ways.

tacto – característica o facilidad de animar a los demás en maneras positivas.

TAN – measurement of total ammonial nitrogen.

NAT – medida de nitrógeno amoniacal total.

tankage – dried animal residue after slaughter.

fertilizante orgánico – los desechos secados del animal después del sacrificio.

taproot – large main root of the system; usually has little or no branch roots.

raíz eje – gran raíz primaria del sistema radicular; usualmente no tiene pocos o ningunos otros raices secundarias.

targeted pest – identified pest that, if introduced, poses a major economic threat.

animal o insecto nocivo objetivo – animal o insecto nocivo que, si se introduce, representa una amenaza económica mayor.

tattoo – means of marking animals for identification.

tatuaje – medios para marcar los animales para fines de identificación.

taxonomy – systematic classification of plants and animals.

taxonomía – clasificación sistemática de plantas y animales.

T-budding – a grafting method in which a bud is placed on a root-stock using a vertical cut in the stem with a horizontal cut across the top forming a slit into which the bud is placed.

retoños-T – un método de injerto en el cual un codo es unido a un tallo usando un corte vertical en el tallo con un corte horizontal a lo largo de la parte superior formando una ranura dentro de la cual el codo es colocado.

TDN – total digestible nutrients; the measure of digestibility of feed.

NDT – nutrientes digestibles totales.

teat – the appendage of an udder.

tetas – los apéndices del ubre.

technology – application of science to an industrial or commercial objective; also, the equipment and expertise to cultivate, harvest, store, process, and transport crops for consumption.

tecnología – la aplicación de ciencia a un objetivo industrial o comercial; también, el equipo y pericia para cultivar, cosechar, almacenar, procesar, y transportar las cosechas para consumpción.

terminal – growing at the end of a branch or stem.

terminal – dícese de algo que está creciendo en el extremo de una rama o un rabo.

terminal bud – bud at the end of a twig or branch.

botón terminal – yema que está en el extremo de una rama o ramita.

terminal market – a stockyard that acts as a place to hold animals until they are sold to another party.

mercado terminal – un corral que sirve para guardar los animales hasta que se vendan a otro individuo.

terrace – soil or wall structure built across the slope to capture water and move it safely to areas where it will not cause erosion.

bancal – estructura de tierra o pared construida a lo largo del declive para coger el agua y transportarla seguramente a los áreas en donde no causarán la erosión.

terrestrial – land organism.

terrícola – organismo que vive en la tierra.

testes – organs that produce sperm cells.

testes – órganos que producen los espermatozoides.

testosterone – male sex hormone.

testosterona – hormona sexual masculino.

tetraethyl lead – colorless, poisonous, and oily liquid.

plomo tetraetílico – líquido incoloro, tóxico, y aceitoso.

texture – visual or surface quality.

textura – cualidad visual o del superficie de algo.

thatch – building of organic matter on the soil and around turf-grass plants.

fertilización del suelo – la fomentación de materia orgánica en el suelo y alrededor de las plantas de césped.

thermo – classification of plants according to growing season required.

termo – clasificación de las plantas según la temporada de cultivación que se exigen.

threshing – removal of grain from a plant.

trilla – acción de separar del grano de la paja de la planta.

Thymine (T) – a base in genes designated by the letter *T*.

timino – una base en los genes nominado por la letra «T».

tidewater – water that flows up the mouth of a river as the ocean tide rises and comes in.

agua de marea – agua que fluye a la boca de un río mientras la marea sube y la entra.

tillable – soil that is workable with tools and equipment.

arable – tierra que se puede labrar con herramientas y equipo.

tiller – a young plant shoot that grows from the crown of a plant.

retoños – brotes de joven planta que emergen del botón axilar de la corona de una planta.

timberland – forest land capable of producing more than 20 cubic feet of industrial wood per year.

bosque maderable – tierra forestal capaz de producir más de 20 ft³ de maderos industriales por año.

tip layering – tip of a shoot place in media and covered.

acodo de punta – la punta del vástago es metido en la media y tapada.

tissue culture – plant reproduction using small, actively growing plant parts under sterile conditions and medium.

cultivo de tejidos – reproducción vegetal usando partes pequeñas y activamente crecientes de la planta bajo condiciones y medios estériles.

tofu – food made by boiling and crushing soybeans and letting it coagulate into curds.

tofu – comestible hecho por hervir y triturar la soja y dejándo que se coagule para formar requesón.

tom – male turkey.

pavo – macho de la pava.

top-dressing – a procedure in which fertilizer is broadcast lightly over close-growing plants.

abonado de cobertera – un procedimiento en lo cual se siembra al voleo el abono ligeramente sobre los cultivos que crecen muy juntos.

topsoil – desirable proportions of plant nutrients, chemicals, and living organisms located near the surface that support good plant growth.

superficie – proporciones deseosas de alimentos nutritivos, sustancias químicas vegetales y organismos ubicados cerca de la superficie, que sostienen buen crecimiento vegetal.

townhouse – one of a row of houses connected by common side walls.

casa de ciudad de tres o más pisos – una de una filera de casas conectadas por paredes laterales comunes.

toxic – a pesticide or other material that is poisonous or harmful to an organism other than the target organism.

tóxico – el pesticida o el otro material que es venenoso o dañoso a un organismo otro ese el organismo de la blanco.

toxicity – a measurement of how poisonous a chemical is.

toxicidad – una medida de la noxidad de una sustancia química.

toxin(s) – a poisonous substance, causing injury to animals or plants.

toxina – una sustancia capaz de producir efectos tóxicos, resultando en daño a animales o plantas.

tractor – source of power for belt-driven machines, as well as for pulling.

locomotora de tracción – fuente de potencia para máquinas propulsadas por correa, tanta como para arrastrar.

trade name – a manufacturer's name for its pesticide product.

marca registrada – el nombre del fabricante por su producto de pesticida.

transition zone – a geographic area where warm-season and cool-season adaptation zones overlap.

zona de transición – una área geográfica donde las zonas de adaptación de la estación fría y de la estación cálida se enlazan.

transpiration – process by which a plant loses water vapor.

transpiración – proceso por lo cual la planta se pierde de su vapor.

transplant – plant grown from seed in a special container.

trasplante – planta cultivada de semilla en estructuras especiales.

trap crop – a susceptible crop planted to attract a pest into a localized area.

planta-tampa – un cultivo susceptible sembrado para atraer a un animal o insecto dañino a un área localizada.

tree – a woody plant that produces a main trunk and has a more or less distinct and elevated head (a height of 15 feet or more).

árbol – planta leñosa que produce un tronco principal y que tiene una copa más o menos distinta y elevada (de 15 pies o más de alto).

tree spade – a piece of equipment, operated by hydraulics, that will dig a tree in a matter of a few minutes to a specific size ball.

pala mecánica – un aparato, operado por la hidráulica, que arrancará a un árbol en meros minutos a un cepillón muy específico.

tripe – the pickled rumen of cattle and sheep.

callos – el herbario adobado de ganado y corderos.

trot – in horses, fast, two-beat diagonal gait.

trote – de los caballos, un paso rápido diagonal de dos tiempos.

truck cropping – large-scale production of a few selected vegetable crops that are shipped to wholesale markets.

transporte camionero agrícola – producción de grande escala de pocos vegetales seleccionados de cultivos que son enviados a mercados de abasto.

trucker – person transporting commodities from farm to consumer.

camionero – persona que transporta la mercancía de la finca al consumidor.

tuber – specialized food-storage stem that grows underground.

tubérculos radiculares – raíz especializada de almacenamiento de alimentación que crece subterráneamente.

tuber crop – crop grown for its fleshy, underground storage stem.

tubérculo – cultivo cultivado por su raíz carnosa de almacenamiento.

turbidity – measure of suspended solids.

turbiedad – medida de sólidos suspendidos.

turf – the turfgrass plant and soil immediately below it; also, grass used for decorative or soil-holding purposes.

tepe – la planta de césped y la tierra inmediatamente por debajo de éste; también la hierba usada para fines de adorno o de sujetar el suelo.

turfgrass – grass that is mowed frequently for a short and even appearance.

césped – hierbas que se siegan frecuentemente para obtener un aspecto corto y plano.

turgor – swollen or still condition as a result of being filled with liquid.

turgencia – condición hinchada o rígida como resulto de llenarse con líquido.

udder – the milk-secreting glands of the animal.

ubre – las glándulas del animal que secretan la leche.

underripe – as applied to vegetation, any that is not mature.

insuficientemente madura – como se aplica a la vegetación, cualquiera que no es madura.

universal solvent – material that dissolves or otherwise changes most other materials.

solvente universal – material que disuelve o de otro modo cambia la mayoría de los otros materiales.

unsoundness – in horses, any abnormality that affects the use of the horse.

defectuoso – de los caballos, cualquiera abnormalidad que afecta el uso del caballo.

urea – synthetic source of nitrogen made from air, water, and carbon.

urea – fuente sintética de nitrógeno hecho de aire, agua, y carbono.

urinary system – system that removes waste materials from the blood.

aparato urinario – conjunto de órganos que elimina los desechos de la sangre.

urine – liquid body waste.

orina – desechos líquidos corporales.

utility-type grass – a turfgrass adapted to lower maintenance levels.

césped de utilidad – un césped adaptado a niveles bajos de mantenimiento.

vaccination – the injection of an agent into an animal to prevent disease.

vacunación – la inyección de un agente a un animal para evitar la enfermedad.

vacuum pan – a device used to remove water from milk.

tacho de vacío – un dispositivo usado para quitar el agua de la leche.

vapor drift – movement of pesticide vapors caused by chemical volatilization of the product.

arrastre de vapor – el traslado de vapores de pesticidas debido a la volatilización química del producto.

variegated – a leaf having streaks, marks, or patches of color.

jaspeado – una hoja que tiene rayas, marcas, o manchas de color.

variety – a subdivision of a species with various inheritable characteristics of form and structure that are continued through both sexual and asexual propagation.

variedad – una subdivisión de una especie, tiene varias características capaces de transmitir por herencia, de forma y estructura que se transmiten por la propagación sexual y asexual.

vascular bundle – food, water, and nutrient-carrying tissue extending from roots to tips of plants.

haz vascular – tejido que conduce la alimentación, el agua y los nutrientes, que extiende desde las raíces hasta las puntas de las plantas.

veal – young calves or the meat from young calves.

ternura – terneros jóvenes o el carne de ellos.

vector – a living organism that transmits a disease.

portador – un organismo viviente que transmite una enfermedad.

vegetable – any herbaceous plant whose fruit, seeds, roots, tubers, bulbs, stems, leaves, or flower parts are used as food.

verdura – cualquiera planta herbácea cuyos frutos, semillas, raíces, bulbos, tallos, hojas, o partes de flor, son usados como comestibles.

vegetative – plant parts such as stems, leaves, roots, and buds, but not flowers, fruits, and seeds.

vegetativo – partes de la planta como los tallos, las hojas, las raíces, y los botones, pero no las flores, los frutos, y las semillas.

vegetative bud – a terminal plant bud that produces stem and leaf growth.

botón vegetativo – el botón de una planta terminal que produce crecimiento de tallo y hoja.

vegetative spreading – reproduction in plants other than by seed.

propagación vegetativa – reproducción de las plantas de otro modo de por semilla.

veneer – thin sheet of wood used in paneling and furniture.

chapa – hoja delgada de madera que se usa en revestimiento de madera y en mueblería.

vermiculite – mineral matter used for starting plant seeds and cuttings.

vermiculita – materia mineral usada para empezar semillas y talas de plantas.

vertebrate – an animal with a backbone.

    vertebrados – animales que tienen columnas vertebrales.

vertical integration – occurs when several steps in the production, marketing, and processing of animals are joined together.

    integración vertical – ocurre cuando varios pasos en la producción, mercadeo, y procesamiento de animales son unidos.

veterinarian – animal doctor.

    veterinario – doctor de animales.

viable – alive and capable of functioning.

    viable – vivo y capaz de funcionar.

vice – bad habit in horses.

    resabio – mala costumbre de un caballo.

video merchandising – the use of video recordings of livestock that are offered for sale at a distant location, thus enabling a potential buyer to bid for the livestock in a telephone auction.

    ventas por video – el uso de grabaciones de videos de ganado en venta ofrecidos en lugares distantes, posibilita a un comprador potencial a ofrecer un precio por el ganado en una subasta telefónica.

virgin – a forest that has never been harvested.

    selva virgen – un bosque que nunca jamás ha sido cosechado.

virus – pathogenic entity consisting of nucleic acid and a protein sheath.

    virus – cuerpos patogénicos que consisten de ácido nucléico y una vaina de proteina.

viscera – internal organs of an animal including heart, liver, and intestines.

    víscera – órganos internos del animal inclusive los corazón, higado e intestinos.

vitamin – complex chemical essential for normal body functions.

    vitamina – sustancias químicas esenciales para funciones normales del cuerpo.

volatilization – process in which a liquid or solid changes to the gaseous phase.

    volatilización – proceso en que un líquido o sólido transforma en el estado gaseoso.

voluntary muscle – muscle that can be controlled by animals to do things such as walk and eat food.

    músculo voluntario – músculo que pueden ser controlados por los animales para hacer cosas como pasar y comer alimentos.

walk – in horses, a slow, four-beat gait.

    ponerse al paso – de los caballos, un paso lento de cuatro tiempos.

warm-blooded animal – animal with the ability to regulate its body temperature.

    animal de sangre caliente – animal con la capacidad de regular su temperatura de cuerpo.

warm-season turfgrass – turfgrass adapted to the southern region of the United States; grow best at 80° to 95° F.

    césped de temporada de verano – césped adaptado a la región sureña de los Estados Unidos que crece mejor a temperaturas de 80° a 95° F.

warp – the tendency of wood to bend permanently because of moisture change.

    alabearse – la tendencia de la madera de torcer permanentamente debido al cambio de la humedad.

water – clear, colorless, tasteless, and nearly odorless liquid.

    agua – líquido claro, incoloro, insípido, y casi inodoro.

water culture – growing of plants with roots in a nutrient solution.

    cultivo acuático – la cultivación de plantas con raices en una solución nutritiva.

water cycle – movement of water to surface, to atmosphere, to surface.

    ciclo de agua – movimiento de agua desde la superficie, hasta la atmósfera, y regreso a la superficie.

waterfowl – ducks and geese.

    aves acuáticos – patos y gansos.

water hardness – measured as the amount of calcium (ppm) in the solution.

    dureza de agua – medida como la cantidad de calcio (partes por mil) en la solución.

water-logged – soaked or saturated with water.

    anegado – mojado o empapado.

water quality – measurement of factors that affect the utilization of water.

    cualidad de agua – medida de los factores que afectan la utilización de agua.

water resources – all aspects of water conservation and management.

    recursos de agua – todos aspectos de conservación y administración de agua.

water table – level below which soil is saturated or filled with water.

    nivel hidrostático – nivel debajo de lo cual el suelo se llena o se empapa de tierra.

watershed – a large land area from which water flows or from which it is absorbed from rain or melting snow, and from which water drains as it emerges to the surface through springs.

    cuenca – una área grande de tierra de donde fluye el agua o donde es absorvida de la lluvia o nieve derritida, y de donde el agua desemboca mientras emerge a la superficie por manantiales.

wean – to accustom an animal to take food other than by nursing.

    destetar – hacer que el animal se acostumbre alimentarse de otro modo de mamar.

weed – a plant that is not wanted.

    mala hierba – una planta no deseable.

wether – castrated male member of the sheep family.

    carnero castrado – carnero castrado.

wetland – a lowland area often associated with ponds or creeks that is saturated with freshwater.

    tierras mojadas – una región de tierra baja que muchas veces es asociada con charcos o arroyos, que es saturada con agua dulce.

wholesale marketing – the marketing of a product through a middleman.

    mercado al por mayor – el mercadeo de un producto por un intermediario.

wholesaler – person who sells to the retailer.

mayorista – persona que vende al minorista.

wildlife – animals that are adapted to live in a natural environment without the help of humans.

fauna – animales que se han adaptado a vivir en un ambiente natural sin la ayuda de los seres humanos.

woodlot – small, privately owned forest.

rodal pequeño – bosque pequeño que tiene propietario individual.

woody – hard, stiff, dark-colored growth of plants; they are winter hardy.

leñoso – crecimiento duro, resistente, y de color oscuro de las plantas; son resistentes al invierno.

wool – modified hair obtained from sheep and some other animals. It is a fiber with good insulating qualities that is used to make cloth.

lana – pelo modificado obtenido de borregos y otros animales. Es una fibra con muy buenas cualidades de aislamiento que se usa para hacer telas.

worker – female bee that does the work in the hive.

obrera – abeja hembra que hace el trabajo de la colmena.

X-Gal – compound that causes marked bacteria to turn blue.

«X-Gal» – una compuesta que causa que se vuelva azúl la bacteria marcada.

xylem – vessels of the vascular bundle that carry the water and nutrients from roots to leaves.

xilema – vasos del haz vascular que llevan el agua y los alimentos nutritivos desde las raices a las hojas.

xylem layer – cell structures that transport water, minerals, and nutrients within plants from the roots to the leaves.

capa del xylem – estructuras de la célula que transportan el agua, los minerales, y los alimentos dentro de las plantas de las raíces a las hojas.

xylem tissue – conductive tissue responsible for transport of water and nutrients within the plant.

tejido de xilema – tejido conductivo responsable de la transportación de agua y alimentos nutritivos dentro de la planta.

yardage fee – a fee for caring for animals until they are sold.

gastos del encierro – un honrario por cuidar los animales hasta que sean vendidos.

yearling – cattle between 1 and 2 years of age.

añal – ganado entre uno y dos años de edad.

yield grade – grade based on the amount of lean meat in relation to the amount of fat and bone in cattle and sheep.

proporción carne/grasa – grados basados en la cantidad de carne magra en proporción con la cantidad de grasa y hueso en el ganado vacuno y las ovejas.

zygote – fertilized egg.

cigoto – óvulo fecundado.

Appendix A

DEVELOPING A PERSONAL BUDGET

One of the most important parts of becoming an adult is that of creating a budget to handle your personal finances. A budget is a plan of action that includes projections of income and expenses for all or part of business or personal expenses. The process in this appendix is designed to help you understand what a personal budget is. Many of the same principles can be used when developing your supervised agricultural experience program (SAEP).

There are several reasons for developing and using a budget. Following is a list of some of these reasons:

- helps you plan for life of assets
- excellent device for organizing
- is useful to obtain credit
- allows experimenting with different outcomes
- identifies costs and income
- helps refine and organize
- is a good management tool for anyone with limited experience

LIMITATIONS OF BUDGETS

There are several reasons why people do not budget. Being aware of these limitations will help you avoid the mistake of not carefully budgeting. Following is a list of some of these reasons:

- budgeting takes time
- must search for accurate information
- sometimes difficult to predict future actions
- managers tend to underestimate costs and overestimate income

THE BUDGET PROCESS

Budgets do not need to be complicated. The process can be as simple as determining, listing, and summarizing all of your anticipated expenses and all of your anticipated income. Obviously, the goal should be for your income to exceed your expenses. If this is not the case, you will need to either change the level of income or the expenses.

Listed on the next page is a simple worksheet that may help you in organizing a budget.

MONTHLY EXPENSES

Home: **Rent/Own**_____
              Maintenance_____

Utilities: **Phone** _____
              Electricity_____
              Cable_____
              Gas_____

Insurance: **Home** _____
              Vehicle_____
              Life_____
              Health_____
              Disability_____

Vehicle: **Payment**_____
              Gas_____
              Oil/Tires/Maintenance_____

Food:_____
Clothing:_____
Savings:_____
Entertainment/Recreation:_____
Emergencies:_____
Cleaning/Toiletry/Beauty:_____
Gifts/Contributions:_____
Medical Care:_____

Installment Payments: **Credit Cards**_____
                            Bank Notes_____
                            Department Stores_____

MONTHLY INCOME

Salary:_____
Odd jobs:_____
Gifts:_____
Sales:_____
Other Income:_____
Total Income:_____

BALANCE: (*income minus expenses*):_____

Portions of Appendix A were adapted from The Georgia Agricultural Education Curriculum Guide, Atlanta, GA.

Appendix B

PLAN SUPERVISED AGRICULTURAL EXPERIENCE PROGRAMS

This appendix includes plans for building birdhouses and nesting boxes. Both of these units can be constructed using various types of wood. Because the houses of the structures are constantly exposed to the elements, it is essential that wood used in the construction be carefully selected. Use your problem-solving skills to determine which materials are best suited for the various parts. Comparisons of different types of woods are included to aid in your decisions. Safety is a constant concern when constructing projects or conducting experimentation. Rules governing safety are included in this appendix.

GUIDELINES FOR SAFETY IN THE LABORATORY

The following rules must be observed to ensure your safety and the safety of others in the agricultural education classroom and laboratory.

1. Your concern for safety should begin even before the first laboratory or shop activity. Always read and think about safety before starting an activity.

2. Perform laboratory work only when your teacher is present. Unauthorized or unsupervised laboratory experimenting is not allowed.

3. Know the location and use of all safety equipment in your laboratory. These should include the eye wash station, first-aid kit, fire extinguisher, and safety shower (if available).

4. Wear safety glasses or goggles at all times in the wood and metal laboratory. Wear protective clothing and eye protection at all times as directed by the instructor. Tie back loose hair and secure loose clothing.

5. Check chemical labels twice to make sure you have the correct substance. Pay attention to the hazard classification shown on the label.

6. Avoid unnecessary movement and talk in the laboratory.

7. Gum, food, or drinks should not be brought into the laboratory.

8. Never sniff chemicals and do not place your nose near the opening of a chemical container.

9. Any laboratory accident, however small, should be reported immediately to your teacher.

10. In case of a chemical spill on your skin or clothing, rinse the affected area with plenty of water. If the eyes are affected, water-washing must begin immediately and continue for 10 to 15 minutes or until professional assistance is obtained.

11. When discarding used chemicals, or hazardous materials, carefully follow the instructions provided.

12. Know where the Material Safety Data Sheet forms are located and how to use them.

13. No horseplay, running, and so on is allowed at any time in the laboratory/shop or classroom.

14. Do not use equipment unless you have been instructed on its use and safety precautions and have been given permission to do so.

15. Keep all tools stored in their correct locations while not in use.

16. Make sure all safety devices, guards, and other items of protective equipment are in place and working at all times.

17. Make sure compressed gas cylinders are kept chained to a wall or on a secure rack with caps on tightly.

18. Before making adjustments on any piece of equipment, make sure it is turned off and unplugged.

19. Keep your work area clean and clear of hazards.

20. If water has been spilled on the floor, clean it up immediately. If a hazardous chemical has been spilled, notify your teacher but do not attempt to clean it up until you are told to do so.

21. Do not put hands and fingers closer than 2 inches to a blade while using saws, drills, and other similar equipment.

22. Know where all tools, chemicals, and equipment are located in the shop.

23. Always use the proper tool for the job.

24. Do not stand closer than 6 feet to power equipment being operated by someone else.

25. Recognize the color coding for shops and observe all regulations regarding color coding:

Safety RED	=	*DANGER*
Safety ORANGE	=	*WARNING*
Safety YELLOW	=	*CAUTION*
Safety BLUE	=	*INFORMATION*
Safety GREEN	=	*SAFETY*
Safety WHITE	=	*TRAFFIC MARKINGS*

26. Lift heavy objects safely using proper lifting techniques.

27. Never throw anything in the classroom or laboratory.

28. Take responsibility for the safety of yourself and others. Follow tool and equipment safety manuals. Obey rules and guidelines, as well as the instructions of your teacher. Use good judgment for safety sake. If in doubt, ask!

Portions of Appendix B were adapted from the Georgia Agricultural Education Curriculum Guide, Atlanta, GA.

CONSTRUCTION PROJECT: FLICKER/WOODPECKER HOUSES

Plan for Flicker/Woodpecker Houses

DIMENSIONS

FLOOR OF HOUSE:	7" x 7½"
DEPTH OF HOUSE:	7½"
DIAMETER OF HOLE:	2½"
CENTER OF HOLE ABOVE FLOOR:	14½" (15¼" OUTSIDE)
HEIGHT ABOVE GROUND:	18' TO 20'
MATERIAL	¾"

MODERN HOUSE

TREE NEST BOX

FLAT FRONT NEST BOX

FRONT   SIDE   ROOF   BOTTOM   BACK

Courtesy of Cooperative Extension Service, Virginia Polytechnic Institute and State University, Blacksburg, VA.

Bill of Materials (for modern house)

Materials Needed	Quantity	Dimensions	Description or Use
Lumber (rough and rustic)	1	3⁄4" × 8-1⁄2" × 17-1⁄2"	Front (A)
''	2	3⁄4" × 7-1⁄2" × 19"	Sides (B)
''	1	3⁄4" × 7" × 7-1⁄2"	Bottom (C)
''	1	3⁄4" × 7-1⁄2" × 8-1⁄2"	Roof (D)
''	1	3⁄4" × 9-1⁄2" × 22-1⁄2"	Back (E)
Box nails or finishing nails		4d or 6d	Attach

The flicker will nest readily in a well-placed birdhouse of the proper dimensions. Roughen the interior of the nest box to assist the young woodpeckers to reach the entrance hole. Cover the bottom with sawdust so that the mother bird can shape the nest for eggs and the young birds. Material may be one-half inch instead of three-fourths inch thick, in which case the width of the front and roof will be 8 inches.

Construction Procedure for Flicker/Woodpecker Houses

1. Select pine, redwood, or other lumber that will withstand weather without paint. Rough lumber is generally preferred and will provide a rustic appearance. Planed lumber should be used if the house is to be painted for decorative purposes, but this is not recommended for most species except Martin houses.

2. Cut all parts from a 3⁄4″ × 9-1⁄2″ × 8′ board.

3. Cut the sides to the proper angle.

4. Bevel the top edge of the front to conform to the angle of the sides.

5. Bevel the upper edge of the roof to conform to the back and sides.

6. Nail the front, sides, and bottom together.

7. Nail the back to the sides and bottom. (Center the assembly on the back for a good appearance. Leave 1 inch at the bottom and 2.5 inches at the top.)

8. Attach the roof with two hinges or nail it lightly so the roof is easily removed for periodic cleaning of the nest area.

CONSTRUCTION PROJECT: NESTING AND DEN BOXES

Plan for Nesting and Den Boxes

Courtesy of Maryland Department of Natural Resources, Public Information Services.

Bill of Materials

Bills of materials are not provided with these nesting and den boxes because of the variety of types pictured. Nesting and den boxes should be planned from the information provided. An appropriate bill of materials may then be developed for the plan.

Construction Procedure for Nesting and Den Boxes

Construction procedures will vary with the nesting or den box being constructed. However, the procedure outlined for the flicker/woodpecker houses may be helpful.

CHARACTERISTICS OF COMMON WOODS

Species	Hardness	Known for	Some Major Uses
Birch	Hard	Surface veneer for panels	Cabinets and doors
Cedar, red	Medium	Pleasant odor	Furniture, chests, and birdhouses
Cherry	Hard	Red grain	Fine furniture
Cypress	Medium	Rot resistance	Structural material in wet places, birdhouses
Fir and Hemlock	Soft	Light, straight, strong	Construction framing, siding, sheathing
Locust, black	Hard	Rot resistance	Fence posts, birdhouses
Maple	Hard	Light grain	Floors, bowling alleys, durable furniture
Mahogany	Medium	Reddish color	Fine furniture
Oak	Hard	Toughness, strength	Floors, barrels, wagon bodies, feeders, farm buildings
Pine, yellow	Medium	Wear resistance, tough	Floors, stairs, trim
Pine, white	Soft	Easy to work, straight	Shelving, siding, trim
Redwood	Soft	Excellent rot resistance	Yard posts, fences, birdhouses
Walnut, black	Hard	Brown grain	Fine furniture
Willow, black	Soft	Brown grain, easy to work, walnut look	Furniture

TYPES OF NAILS

STAPLE   HINGE   PLASTER BOARD   ROOFING

LEAD HEAD   CUT   FINISHING   BOX   COMMON   DUPLEX

LENGTH OF NAILS

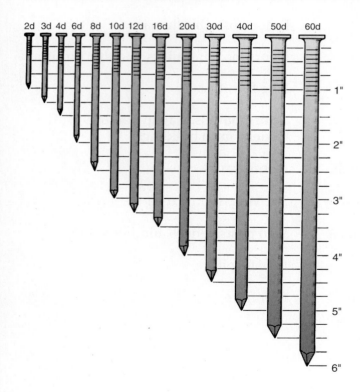

2d 3d 4d 6d 8d 10d 12d 16d 20d 30d 40d 50d 60d

1"
2"
3"
4"
5"
6"

GAUGES AND TYPES OF WOOD SCREWS

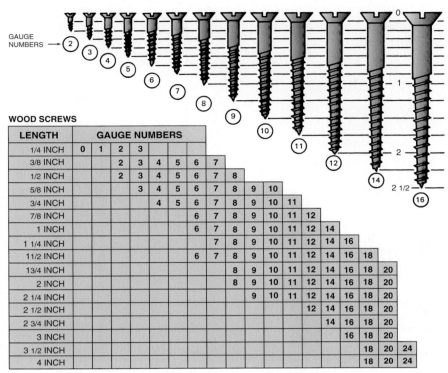

WOOD SCREWS

LENGTH	GAUGE NUMBERS															
1/4 INCH	0	1	2	3												
3/8 INCH			2	3	4	5	6	7								
1/2 INCH			2	3	4	5	6	7	8							
5/8 INCH				3	4	5	6	7	8	9	10					
3/4 INCH					4	5	6	7	8	9	10	11				
7/8 INCH							6	7	8	9	10	11	12			
1 INCH							6	7	8	9	10	11	12	14		
1 1/4 INCH								7	8	9	10	11	12	14	16	
1 1/2 INCH							6	7	8	9	10	11	12	14	16	18
1 3/4 INCH									8	9	10	11	12	14	16	18	20
2 INCH									8	9	10	11	12	14	16	18	20
2 1/4 INCH										9	10	11	12	14	16	18	20
2 1/2 INCH													12	14	16	18	20
2 3/4 INCH														14	16	18	20
3 INCH															16	18	20
3 1/2 INCH																18	20	24
4 INCH																18	20	24

When You Buy Screws Specify (1) Length, (2) Gauge Number, (3) Type of Head–Flat, Round, or Oval, (4) Material–Steel, Brass, Bronze, etc., (5) Finish–Bright, Steel Blued, Cadmium, Nickel, or Chromium Plated.

Index

Page numbers in *italics* refer to figures

A

Abiotic diseases, 261
Acetanilides, 276
Acidity ranges, 325
Adaptation, 242
Additives, 668
Add-on loans, 720–721
Adenine, 49
Adenosine triphosphate (ATP), 576
Advertising, 688–690
Aeroponics, 185
Aerosols, 37
African swine fever, 517
African violets, 350, 458
Age, horses, *633*
Age profiles, 32
Aggregate cultures, 185
Aggregates, 156, 175–176
Agribusiness. *See* Business
Agriculture
  -agribusiness, 6–7
  careers, 63–64, 66–67
  definition, 6
  distribution, 63–64
  economics, 14
  engineering, 8
  environmental impact, 24
  global aspects, 11
  green revolution, 38
  jobs in, *7*
  mechanization, 33–35
  processing, 63–64
  production, 6–7, 63
  publications, 12
  services, 66–67
  supervised experience, 14
  supplies, 66–67
  worker shortage, 15
Agriscience
  aspects, 8–9, 13–14
  benefits, 684–685
  contributions of, 33
  definition, 6–10, 61
  education, 59
  everyday influence, 10–12
  facets, *684*
  food supply, 7
  future advances, 41–42
  impact of, 6
  literacy, 82

  opportunities, 15, 734, *735*
  project, 14–15
  skills profile, 85
Agronomists, 426, 444
Agronomy, 13
Agropyron cristatum, 485
Agrostis palustris, 483, 485
A horizon, 173
AI. *See* Artificial insemination
Air
  consumption, 135
  currents, 126
  definition, 125
  layering, 343–344
  plants and, 321
  pollution, 126–130, 135
  seed germination, 337
Air quality
  control, 128
  improving, 136
  maintaining, 136
  threats to, 125–130
Alfalfa
  bacteria, *439*
  characterization, 438–439
  high-protein, 447
Algae, 144, 177
Alkalinity
  definition, 177
  excessive, correcting, 180–181
  ranges, 325
Alluvial deposits, 166
Aluminum sulfate, 180
American beech, *206*
American Star Awards, 81
American sycamore, *206*
Amino acids, 528
Ammonia, 245
Amortized loans, 721
Amphibians, 240
Angora, 580
Animals. *See also* Wildlife; *specific*
     species
  anatomy, 521–527
  aquatic, 240, 242–243
  breeding, 35–38
  byproducts, 677
  circulatory system, 521–522, *523*
  companion, 585–587
  competition, 223–224
  digestive system, 525–527
  domestication, 47
  endocrine system, 525

  environment and, 26
  foods from, 672–677, 682
  gene changes, 560–561
  genetic manipulation, 51
  manager, 520
  manure, 184
  market trends, 702
  medical treatment, basic, 547–550
  microscopic, 52–53
  muscular system, 521
  nervous system, 524
  nutritional function, 10–11
  physiology, 521–527
  predators, 222
  prey, 222
  respiration, 135
  respiratory system, 522
  restraining, 549–450
  science, 8, 14
  selective breeding, 47, 568–569
  skeletal system, 521–523
  tallow, 39–40
  technician, 587
  transgenic, 52
  urinary system, 524–525
  vertebrate, 221
  waste, 605
  woodlot grazing, 215
Anions, 323
Annual wormwood, 434
Anther, 311
Antibiotics, 533
Aphids, 267, 354
Apiarys, 582–583, 585
Apiculture. *See* Honeybees
Apples
  cross-section, *311*
  fruiting spur, *405*
  harvesting, *412*
  storage, 412
  tree pruning, *409*
Applied science, 13
Aquaculture
  biology, 242–243
  caged, 246–247
  competition in, 242–243
  crop selection, 245–246
  definition, 8, 239
  food chain, 242
  hatcheries, 248
  production, 243–244, 246
  research, 249
  resource management, 249

salinity gradient, 240–241
trends, 250
Aquaculturists, 239, 247
Aquifer, 155
Arboriculture, 212
Artemisinin, 434
Artificial insemination
advantages, 555–556, 568
definition, 555
disadvantages, 568
process, 556, 567
Artificial nesting, 231
Asbestos, 129–130
Aseptic technique
description, 350
hood for, 351
materials preparation, 352
practice activities, 351
procedure, 353–354
Asexual propagation
division, 344
grafting, 345–346
layering, 343–344
leaf cuttings, 340–342
root cuttings, 342
stem cuttings, 338–339
tissue cultures, 346–350
Ash, *206*, 207
Asian lady beetle, 509
Aspen, *205*, 207
Assets, 733
Auctioneers, 693
Auction markets, 695
Avian influenza, 517
Axil, 304

B

BA. *See* Small Business Association
Babesia equii, 624
Babesiosis, equine, 624
Bacillus popilliae, 267
Bacteria. *See also specific species*
alfalfa roots, *439*
characterization, 262
soil, 177
in waste management, 54
Bakewell, Robert, 556
Balance sheets, 733
Baldwin, Anna, 34
Balled plants, 506–507
Balsam fir, *203*, 204
Banks, 718–719
Bare-rooted plants, 506
Barley, 418–419
Bausher, Michael, 329
Beech, *206*, 207
Beef
aging, 675
block, 675
boxing, 675
fabrication, 675
retail cuts, *678*
ripening, 675
Beef cattle
associated careers, 599
breeds, 600–602
byproducts, *600*, 677
carcasses, 675

classification, 698–699
commodities, 646
disassembly process, 675
economics, 599–600
embryo co-culture, 566
hides, 675
history, 598–599
lean gene, 613
mad cow disease, 530
origins, 598–599
processing, 675
production practices, 603–604
products, 674–675, *681*
reproductive system, *562*
restraining, 550
slaughter, 674–675
species, 598
types, 600–602
Beekeepers, 582–583, 587
Bees. *See* Honeybees
Beetles, 276
Beets, sugar, 41
Bell, Alan, 531
Bell, Marion, 275
Beneficial Insects Introduction
Research Laboratory, 373
Bermuda grass, 442, 485
Berry, Edwin, 386
B horizon, 173
Bighorn sheep, 226
Binomial systems, 313
Biochemistry, 13
Biodegradable detergents, 697
Bio-diesel, 39–40, 423
Biodiversity, 213
Biological attractants, 38
Biological control, 267
Biology, 13
Biosphere 2, *3*
Biotechnology. *See also* Genetic
engineering
benefits, 36
definition, 9–10
designer foods, 429
environmental cleanup, 229
ethics in, 55
history, 46
improvements from, 46–47
mums from, 470
pest control application, 268
pesticides, 284
reproduction (*See* Artificial
insemination)
safety, 55
waste management, 53–54
Biotic diseases, 262
Birch, *205*, 207
Birds, artificial nesting, 231
Bird's-foot trefoil, 439–440
Bitternut hickory, *206*
Black cherry, *206*
Black walnut, *206*, 208
Black willow, *206*
Bladder products, 681
Blanching, 666
Block beef, 675
Blood products, 681
Blueberries, 38
Bone marrow, 521

Bones, 681
Border plantings, 501
Botany, 301
Bottomland Hardwoods Forest, 200
Bovine somatotropin
consumer resistance to, 55
discovery, 516–517
production, 51
Boxing, beef, 675
Boy Scouts of American, 105–106
Brackish water, 241
Bracts, 310
Bradford pear, 502
Breeding
AI, 555–556, 567
dairy cattle, *597*
genetic engineering and, 562
horses, 629–631
methods, 566–568
natural selection, 560–561
papers, 564
planned selection, 560–561
selective, 556–557
systems of, 564–566
Britz, Steven, 359
Broadcast planters, 430
Brokers, 426
Bromegrass, 440
Bromus tectorum, 257
BST. *See* Bovine somatotropin
Buchloe dactyloides, 486
Buds, grafting, 345
Buffalo grass, 486
Bulbs, 304, 328
Burlapped plants, 506–507
Business. *See also* Entrepreneurship
careers, 712, 731
decisions, 711–714
economics, 714–716
employment options, 737–738
enterprise, 83
finance, 716–724, 730
management, 710–711
operations, 729–730
owning, 736
records, 733–734
small, contributions, 738–739
software, 723
Byproducts, 677

C

Cacti, 461
Calcium, 327
Calcium nitrate, 325
Calgene, 38
Calves, 698–699
Canada goose, 227
Cane cuttings, 339
Canning, 666
Canola, *421*
Canopy, 511
Capillary water, 148
Carbamates, 279
Carbohydrates
characterization, 528
function, 522
nutritional role, 659
sources, 530

Carbon dioxide
  in aquaculture, 242
  characterization, 134
  global warming and, *132*
  measuring, *318*
  in photosynthesis, 135, *136*
  shortage, 318
Carbon monoxide, 127
Carcasses, 675
Carcinogens, 290
Careers
  advisors, 72–74
  agriculture, 63–64, 66–67
  agriscience, 9, 15
  agronomist, 426
  animal technician, 587
  aquaculturists, 247
  beekeepers, 587
  biology, 301
  botany, 301
  broker, 426
  business, 712, 731
  cattle industry, 599
  education, 69
  elevator manager, 426
  entomology, 261
  environmental management, 31
  FFA events, 108–109
  foliage plant-associated, 463
  food industry, 639–640, 647, 653, *654*
  forage crops, 444
  forestry, 64–66, 208
  fruit/nut production, 400
  gardening, 368
  genetics, 48, 555
  golf course-associated, 492
  grain handler, 426
  home garden, 379
  horse-associated, 620
  horticulture, 64
  hydroponics, 189
  internships, 84
  landscaping, 503
  marketing, 693
  mechanics, 67
  natural resources, 66
  nutrition, 667
  nutrition-associated, 520
  opportunities, 61–62
  personal assessment, 72–73
  pest control, 277
  planning for, 70
  plant breeding, 339
  plant pathology, 261
  plant physiologists, 322
  plantscaping, 473–474
  portfolio, 69–70
  professions, 68
  requirements, 61
  SAEP, 78–81
  skills, 71
  taxonomy, 301
  teaching, 59
  truck cropping, 383
  turfgrass-associated, 479
  vacancies, 61
  veterinarians, 520, 540
  wildlife biologist, 229
Carson, Rachel, 28, 263
Cash flow statement, 736–737

Catfish, 245
Cations, 323
Cats, 586–587
Cattle. *See* Beef cattle; Dairy cattle
Causal agents, 261
CCC. *See* Commodity Credit
    Corporation
Cedar, *203*, 204
Cells
  definition, 558
  division, 302, 558–559
  maturation, 302
  structure, 558
Cellulose, 526
CELSS. *See* Controlled ecological life
    support system
Central Broad-leaved, 200
Cereal crops, 426
CFCs. *See* Chlorofluorocarbons
Chain saws, *134*
Cheatgrass, 257
Check-off program, 689
Cheese, 674
Chemicals, impact, 28
Chemicals, spills, 152
Chemistry, description, 13
Cherry, *206*, 207
Chickens. *See* Poultry
Chilocorus kuwande, 509
China
  forests, 29
  populations, 32–33
  soil erosion, 154
Chinch bus, *260*
Chlorinated hydrocarbons, 279
Chlorofluorocarbons
  characterization, 129
  ozone hole and, 132
  recapturing, 134
Chlorophyll, *135*, 318
Chloroplasts, 307, 318
Chlorosis, 326
C horizon, 173
Chromatids, 335
Chromosomes, *51*, 558, 559
Chrysanthemum, 470
Circulatory system, 521–522, *523*
Citrus blight, 329
Clay soils, 174–175
Clear cutting, 214
Climate
  change, 131–134
  crops and, 365, 643
  definition, 24
  fruit/nut crops, 403
  soil and, 166
Clones
  African violets, 350
  aseptic technique, 350–354
  definition, 50
  embryo, 567
  potato, 348
Closebreeding, 565
Clovers, 439
Coccidiosis, 40–41
Cockroaches, 26
Cold frames, 371–372
Collagen, 681
Colluvial deposits, 166
Colorado potato beetles, 52, 276

Colorado River Basin, *149*
Colors, plantscaping, 472
Columbus, Christopher, 592, 609
Columbus Center for Marine Sciences,
    249
Commensalism, 223
Commodities
  exchanges, 703–704
  fruits, 646
  futures market, 703–705
  grains, 645
  meat, 646–647
  oil, 645–646
  pricing, 690–691
  promotion of, 688–690
  safety, 652
  vegetables, 646
Commodity Credit Corporation,
    719–720
Communication, software, 723
Communication, specialists, 693
Community, influence, 23–24
Community gardens, 362
Comparative advantage, 715
Composts
  building, *164*
  definition, *10*
  description, 163–164
  materials, *165*
Computers, 68, 723
Condominium, 22–23
Conservation
  soil hydrology, 150
  tillage, 156
  water, 148, 150–153
Consumers
  demographics, 688
  direct sales, 692
  power of, 688
Container growth, 508
Contamination, 26
Continuous-flow systems, 186
Contour practice, 156
Controlled ecological life support
    system, 358–359
Convection ovens, 670
Cooking methods, 669–670
Cooperative extension system, 367
Cooperatives
  description, 733
  managers, 693
  pricing, 690
Corms, 304, 328
Corn
  characterization, 416
  classifications, 416
  crop, 11
  cultivation, 417
  as forage, 442
  high-lysine, 527
  nutrient deficiencies, *182*
  origins, 416
  picker, 34
  taxonomy, *313*
Cornus florida, 501–502
Corolla, 311
Cortez, Hernán, 618
Cotton
  boll, *255*
  characterization, 422

importance, 422
rainbow, 697
seeds, 422
Cotton gin, 34
Crawfish, 243, 245
Cream, 673
Credit. *See also* Loans
classification, 716–717
cost of, 720–721
obtaining, 721, 723
rates, *719*
risk, minimizing, 724
sources, 718–720
types, 717–718
use of, 716, 723–725
Creed speaking, 111
Creeping bentgrass, 483, 485
Crested wheatgrass, 485
Creutzfeldt-Jakob disease, 27, 530
Crockpots, 669
Crops. *See also* Foods
aquaculture, 245–246
climate and, 643
commodities, 645–646
cool-season, 427
edible, 649
field (*See* Field crops)
foods from, 670–671
frost susceptibility, 403–404
handling, 650
harvest temperatures, 665
high-value, *187*
major regions, *644*
marketing, 651–652
market trends, 702–703
maturity, 648
nurse, 443
processing, 650
rotation, 156, 184–185
science, 8
sod, 185
soil pH range, 177–178
spoilage, 648
transporting, 650–651
truck, 378, 382–383, 387–389
wildlife and, 228
Crossbreeding, 564
Crowns, 479
Crustaceans, 243
Cucurbitaceae, 381
Culm, 480
Cultures, 185
Cuscuta campestris, 301
Cuticles, plant, 307
Cuttings
cane, 339
leaf, 340–342
roots, 342
stem, 338–339
Cynodon dactylon, 442, 485
Cytosine, 49

D

Dairy cattle
breeding, *597*
breeds, 593–595
BST and, 516–517
calf raising, 596

cars risk to, 594
culling, 597
dry period, 597
economics, 593
history, 592
industry, 90
milk production, 516
nutrient needs, 531
origin, 592
production practices, 596–597
transponders for, 594
types, 593–595
Dairy products. *See* Milk
Dallis grass, 442
Dams, 565
DDT, 28, 279
Dealers, 693
Decompsers, 177
Decorative plants. *See* Ornamentals
Deere, John, 34
Deer management, *230*
Dehydration, 666–667
Dehydrofrozen, 667
Demand, determining, 687–688
Demand, principle, 714
Demographics, 688
Deserts, 147
Diameter limit cutting, 214
Dicotyledons, 379
Diet. *See* Nutrients
Dietitians, 667
Digestive system, 525–527
Diminishing returns, 714–715
Dinitroanilines, 276
Dipping, 548–549
Direct sales, 696
Disassembly process, 675
Discount pricing, 690
Diseases. *See also* Health
abiotic, 261
animal, spread of, 26
biotic, 262
classification, 543–544
common symptoms, 545
contagious, 543–544
control, 255
deficiency, 521
insect, spread of, 26–27
noncontagious, 543–544
parasites-induced, 545–546
plant, 261–263
poultry, 517
resistance to, 52
stalks, *324*
trees, 513
triangle, 261
whirling, 248
Dissolved oxygen, 244
Distribution, 63–64, 730
Division, 344
DNA
bases, 49
components, 49–50
definition, 49
fingerprinting, 56
recombinant, 50
shape, 49
strands, 49–50, 51
traits, 49
Dodder, 301

Dogs, *585*, 586–587
Dogwood, flowering, 501–502
Domestication
honeybees, 582
poultry, 572
rabbits, 576–577
Donkeys, 618, 623
Dosage, lethal, 284
Douglas fir, *203*, 204
Drainage, 467–468
Drea, John J., 509
Drenching, 548
Drill planters, 430
Drip irrigation, 391, 431
Drugs
administering, 547–548
forms of, 548–549
withdrawal, 550
Dry-heat cooking, 669
Dual, 276
Ducks, 231
Dust, radioactive, 128–129
Dust bowl, 147
Dutch Elm disease, 51

E

Earth
atmosphere gases, 131
circumference, 126
crust, 146
sun proximity to, 130
Earthworms, 386
Ecology, definition, *224*
Economics
beef production, 599–600
dairy production, 593
definition, 14
food industry, 640–461
goat production, 611
honey production, 582
horse breeding, 618–619
poultry production, 573
principles, 714–716
rabbit production, 577
sheep production, 609–610
small business, 738–739
swine production, 605–606
threshold level, 264–265
Eggs
characterization, 555
preparation, 682
production, 563
Electrical conductivity meter, 190–191
Electricity, rural, 34
Electron microscopes, 305
Elevator manager, 426
Embryos
cloning, 567
co-culture, 566
definition, 563
transfer, 556
Employment options, 737–738
Endocrine system, 525
English ivy, 300
Enterprise production, 83
Entomologists, 261
Entomology, 13
Entomophagous insects, 259

Entrepreneurship. *See also* Business
  buying/selling, 729
  definition, 728
  management, 732
  operating, 729–730
  opportunities, 734
  organization, 732
  product selection, 731–732
  profits/losses, 728
  SAE, 83
  sales projects, 729
  service selection, 731–732
  types, 728–730
Environment
  agricultural impact, 24
  animals and, 26
  chemicals and, 28
  climate and, 24–25
  community, 23–24
  definition, *10*
  family, 23
  food supply, 23
  health and, 539–543
  homes, 20–23
  humans and, 26
  influencing factors, 26–28
  insects and, 26–27
  manipulation, 264–265
  neighborhood, 23–24
  sharing, 29
  topography and, 24–25
  variety in, 19–20
Environmental management, 31
Environmental Protection Agency, 284
Epidermis, plant, 307–308, 310
Epidermis, roots, 320
Equipment
  agricultural, 66–67
  computers, 68, 723
  garden, 134, *364*
  harvesting, 34
  planters, 445–446
  riding, 625, 628
  spades, 507
  sterilizing, 348–439
  tillage, *429, 491*
  turfgrass, 480
Eradication, 266
Erosion
  controlling, 152
  gully, 155
  prevention, 155–157
  reducing, 156–157
  sheet, 155
  types, 155
Erwinia amylovora, 583
Escherichia coli, 53, 551
Esters, 423
Euonymus scale, 509
Evisceration, 675–676
Export markets, 705–706

F

Fabrication, beef, 675
Fabric industry, 697
Family, impact, 23
Farmers, 693

Farmers Home Association, 720
Farmers' markets, 394–395, 694
Farms. *See also specific crops*
  contour practice, 156
  crop rotation, 156
  migratory labor, 648–649
  retailing at, 692–693
  slash and burn, 154
  sod-production, 477
  strip cropping, 156
  turf, *65*
  wildlife, 226–227, 228
Farriers, 620, 631
Fats
  animal, 677
  bio-diesel from, 39–40
  characterization, 529
  nutritional role, 659–660
  reducing, 40
  sources, 530
Federal land banks, 719
Feed. *See also* Nutrition
  additives, 533, 550
  classification, 533–534
  composition, 533
  concentrates, 533
  definition, 519
  formulator, 520
  roughage, 525, 534–535
Fermentation
  definition, 663
  function, *665*
  process, 450–451
Ferns, 459–460
Fertilization
  definition, 322
  fruit/nut crops, 407–408
  ornamentals, 468
  trees, 510–511
  turfgrass, 488–489
  vegetable gardens, 388
Fertilizers
  applications, 183
  complete, 181
  by crop rotation, 184–185
  definition, 156
  function, 182
  grades, 181
  holes, 510, *511*
  ingredients, 181
  inorganic, 181–183
  manure, 184
  nutrient content, *183, 488*
  organic, 181
  unmixed, 183
Fescues, 442, 483
Festuca arundinacea, 483, 484
Fetus, definition, 563
FFA
  aim, 106–107
  career development, 108–109
  colors, 107
  degree requirements, 108–109
  description, 105
  emblem, 107
  instruction, 80–81
  member profile, 108
  motto, 107–108

proficiency awards, 81
  programs, 80–81
  purpose, 106–107
  SAEP and, 80
  salute, 108
  speech competitions, 111
  student handbook, 119
FHA. *See* Farmers Home Association
Field crops
  classification, 426–427
  grains, 416–420
  harvesting, 433–434
  irrigation, 430–431
  medical applications, 434
  oilseed, 420–422
  planting, 429–430
  seedbed preparation, 428
  seeds, 425–426
  specialty, 422–425
  spoilage, 432
  storage, 434
Finance. *See also* Credit
  balance sheet, 733
  business, 716–724, 730
  cash flow statement, 736–737
  inventory report, 737
  profit/loss statement, 736
  records, 733–734, 736–737
Fire ant control, 40
Fire blight, 583
Fires, 215, 216
Fish
  adaptation, 242
  artificial stocks, 233
  in caged culture, 246–247
  freshwater, 241–242
  fry, 241
  genes, 249
  growth, control, 249
  migrating, 243
  oxygen needs, 244
  -plants, relationship, 242
  processing, 677
  saltwater, 241–242
  shellfish, 243, 245, 247
  sport, *234*
  supplies, 647
  temperatures, 666
  trout, 244, *245,* 248
Fisheries
  catfish, 247
  crop selection, 245–246
  natural, 239–240
  trout, *245*
Flail-type harvesters, *449*
Flax, 422
Floriculture, 457
Flowering dogwood, 501–502
Flowers
  beauty, 310
  buds, 304
  characterization, 298–299
  exotic, 41
  growing, 457
  indoor, 458–459, 469–470
  induction, 480–481
  ovary, *311*
  parts, *311*

perfect, 311
removal, 338
structure, 311
varieties, selecting, 367
Flukes, 545
Foliage
definition, 459
light and, 470
types, 459–462
Food chains
aquatic, 242
description, 143
models, *144*
origins, 144
plants role, 299
Food guide pyramid, 663
Food industry
careers, 639–640, 647, 653, *654*
dairy products, 672–274
economics, 640–641
function, 639
future developments, 653–654
grading, 641
handling, 649–650
harvesting, 648–649
inspecting, 641
marketing, 651–652
meat products, 674–675
new products, 681–682
operations within, 647–652
processing, 649–650
quality assurance, 641–642
sanitation, 641–642
Foods. *See also* Crops
additives, 668
animal-based, 672–677, 682
borne pathogens, 662
contamination, 652
convenience, 653
cooking methods, 669–670
crop-based, 670–671
customs, 663
designer, 39, 429
function, 576
genetically modified, 39, 690
global aspects, 643, 645
groups, 661–662
irradiation, 654, 662
labels, 669
life expectancy and, 682
microorganisms in, 46
money spent on, 6
nutritional values, 38, 664
organic, 10, 289
plant storage, 327–329
poisoning, 56
preparation, 669–670
preservation, 371, 663, 665–668
processing, 663, 665–668
protein-rich, 636–637
safety, 652
science, 636–637
scientist, 667
storing, 663, 665–668
supply, 7, 23
surpluses, 10
technician, 667
varieties, 643, 645

Foot-candles, 465–466
Forage crops
associated careers, 444
corn as, 442
fermentation, 450
function, 427
harvesting, 446–447
importance, 438
manager, 444
maturity, 447
oak as, 442
pest control, 446
planting, 445–446
rangeland, 442
seedbeds, 444
selection, 443
sheep, 610
storage, 446–447
types, 438–443
uses, 451–452
Forestry, 64–66, 208
Forests. *See also* Trees; Woodlots
acreage, 196
biodiversity, 213
characterization, 196
China's, 29
composition, 196–197
fires, 216
health, improving, 513
natural resources and, 201–202
North American regions, 197–201
products, 196
resources, 500
water cycle and, 201
wildlife, 201–202, *227*, 229
Formulations, 283
4-H Clubs, 105
Freemartin, 564
Free water, 147
Freeze-drying, 666–667
Frost, 407
Fruit flies, 654
Fruit production
careers, 400
fertilization, 407–408
harvest, 411
planning, 403–406
planting, 406–407
soil factors, 404
soil/site preparation, 406
tree pruning, 409–410
Fruits. *See also specific crops*
categories, 400
characterization, 311–312
climate factors, 403–404
commodities, 646
common, 402
crop pollination, 404
fertility, 404
foods from, 670
growth patterns, 405–406
ripening gene, 408
small-bush, 402
trees, 400–402
true, 311
vine, 402
washing, 662
Fry, 241

Fuchsias, 458
Fungi
benefits, 389
characterization, 262
soil, 177
Fungicides
characterization, 279
chemical structure, 280
classification system, 280
eradicant, 280
protectant, 279
Furrows, 363
Futures
buying/selling, 704–705
definition, 704
description, 703–704
positions, 705

G

Galls, 512
Gardeners, 367
Gardenias, 458
Gardens. *See* Home gardens
Gasoline, 127, 152
Geese, Canadian, *227*
Generation, 47
Genes
definition, 47, 558
disappearing, 98
expression, 558
fish, 249
fruit ripening, 408
function, 559
heterozygous, 559
homozygous, 559
horse color, 622
lean beef, 613
location, 559
mapping, 50
nonsense, 517
splicing, 50
traits, 50
traits and, 560
Genetic engineering. *See also*
Biotechnology
benefits, 36
in breeding, 562
definition, *10*
foods, 39, 690
pest resistant plants, 276
process, *53*
Genetics
animal improvement, 560–561
careers in, 48, 555
code, 48–51
definition, 47
Mendelian, 47–48
principles, 558–560
Genus, 313
Geraniums, 458
Germination
air and, 337
chamber, *191*
hydroponics, 191
light and, 337
temperature, 337
water and, 336

Gills, 242
Girl Scouts of American, 105–106
Glacial deposits, 166
Glands, 681
Glidden, Joseph, 34
Global markets, 703–705
Global warming
  cause, 131
  debate over, 132–134
  precautions, 134
Goats
  breeds, 612
  economics, 611
  history, 611
  origins, 611
  production practices, 612–613
  types, 612
Golden rice, 429
Golf courses. *See also* Turfgrass
  maintenance supervisor, 492
  playability level, 477
Graders, 640, 693
Grades
  beef, 698–699
  quality, 699, *700–701*
  sheep, 699–700
  swine, 699
  yield, 700
Gradient effect, 243
Grading up, 564–565
Grafting, 345–346
Grain handler, 426
Grains
  byproducts, 533
  commodities, 645
  drying, 433
  foods from, 670–671
  types, 416–420
Grants, 718
Grape vines, *410*
Grasscycling, 487
Grasses
  identifying, *481*
  planting, 445–446
  turf (*See* Turfgrass)
  types, 440–443
  waterways, 156–157
Gravitational water, 147
Grazing, 215, 442
Greenhouse effect, 130
Greenhouses, 374
Green Revolution, 38
Grooming, 468
Groundskeepers, 479
Groundwater, 155
  pollution, 155
  safeguarding, 157
  types, 147–148
Group plantings, 501
Growth
  habit, 400
  intravaginal, 480
  patterns, 405–406
  tree, 209–210
  types, 503
Guard cells, 308
Guide dogs, *585*
Gully erosion, 155
Guranine, 49

Guying, 510
Gypsum, 180

H

Hair products, 677
Hand mating, 567
Hardness, wood, 210
Hardwood forests, 204–208
Hartwig, Edgar E., 421
Harvesters, *449*
Harvests
  aspects, 648–649
  definition, 648
  field crops, 433–434
  forage, 446–447
  fruit/nut crops, 411
  hay, 447–449
  machines, 433
  market gardens, 394
  pastures, 447
  silage, 449
  temperatures, 665
  woodlots, 215–217
Hatcheries, 248
Hawaiian Forest, 201
Hay
  cubes, 448–449
  definition, 438
  drying, 448
  harvesting, 447–449
  managers, 444
  maturity levels, *447*
  peanut, 440
  raking, 448
  storage, 449
Healing in, 506
Health. *See also* Diseases
  environmental factors, 539–543
  housing and, 541
  isolation and, 542–543
  maintaining, 538, 547–550
  manure handling, 541–542
  monitoring, 539
  pasture rotation, 543
  paying for, 690
  pest control and, 289–290, 542–543
  sanitation and, 540–541
  signs of, 539
Hedera helix, 300
Heel cuttings, 339, *340*
Heliothis, 284
Hemlock, *203,* 204
Herbicides
  characterization, 274
  chemical families, 275–276
  classification systems, *274*
  contact, 275
  nonselective, 275
  postemergence, 275
  preemergence, 275
  selective, 275
  systemic, 275
Heritability, 561
Hibiscus cannabinus, 450
Hickory, *206,* 208
Hides, 675, 677
Hives, 582
Hoagland solution, 189, *190*

Hogs. *See* Swine
Holt, Benjamin, 34
Homarous spp., 243
Home gardens
  career opportunities, 379
  careers, 368
  cold frames, 371–372
  common crops, 365, 367
  community, 362
  cultivating, 368
  definition, 378
  enterprises, 379
  family's needs, 361
  greenhouses, 374
  harvesting, 370
  hotbeds, 372
  layouts, *364*
  locating, 362
  pest management, 369
  plan, 362
  planting, 363
  raised beds, *364*
  size, 361
  soil preparation, 362–363
  tools, *364*
  watering, 369
  weeding, 368–369
Homes
  middle/upper class, 22–23
  modest incomes, 20–21
  poor, 20
Honeybees
  Africanized, 252–253
  biocontrol role, 583
  domestication, 582
  economics, 582
  history, 582
  pollination by, *405*
  production practices, 584–585
Hoof products, 677
Hormones, 525
Horn products, 677
Horseback riding
  attire, 626
  basic gaits, 627
  English equitation, 625, 627
  saddles, 625, 628
  safety rules, 628–629
  Western horsemanship, 627
Horses
  age, *633*
  anatomy, 622
  associated careers, 620
  breeding, 629–631
  breeds, 619–624
  care, 631–632
  color of, 622
  economics, 618–619
  fossil records, 618
  grooming, *631*
  history, 617–618
  origins, 627–628
  restraining, 550
  types, 619–624
  wild, 619
Horticulture
  bioengineering and, 41
  careers, 64
  description, 13

Host, secondary, 545
Host resistance, 267
Hotbeds, 372
House, William, 531
Housing, 541
Humane Slaughter Act of 1958, 674
Human Genome Project, 557
Humans
  environment and, 26
  intelligence, 5
  life expectancy, 682
  life spans, 31–32
  population, 31–33
  wildlife and, 221, 224–225
Humidity, 667
Hunger, 23
Hunting, 122–123
Hutch, 580–581
Hybridoma, 624
Hybrids, 564
Hydrocarbons, 126
Hydrocooling, 395
Hydroponics
  aeration, 189–190
  careers in, 189
  classroom, 188–191
  definition, 185
  film cans, *191*
  future of, 191
  laboratory, 188–191
  maintenance, 190–191
  nutrient solutions, 189–190
  requirements, 186–187
  seeds for, 191
  solution culture, 188–189
  systems, 185–186
Hygroscopic water, 148
Hyphae, 262

I

Ice-minus, 51
Ictalurus spp., 245
Imbibition, 336
Impatiens, 41, 458–459
Improvement activities, 84, *86, 97*
Inbreeding, 565–566
Indicator species, 151
Industrial technology, 7
Inflorescence, 480
Influenza, avian, 517
Infusions, 548
Injections, 548
Inorganic fertilizers, 181–183
Insecticides
  botanical, 278
  characterization, 277
  classification systems, 278
  production, 279
  synthetic, 278–279
Insects
  adaptation, 258
  anatomy, 258
  beneficial, 258, 373
  Colorado potato beetle, 52
  definition, 255
  development, 260–261
  economic impact, 258
  entomophagous, 259

environment and, 26–27
  feeding damage, 258
  fire ants, 40
  male sterilization, 266
Instars, 260
Insulin, 52, *53*
Integrated pest management. *See also*
      Pesticides
  aspects, 9–10
  biological control, 267
  chemical control, 269
  cultural control, 267–268
  definition, 9–10
  ecosystem manipulation, 264–265
  environmental impact, 291
  history, 263
  host resistance, 267
  key pests, 263–264
  male sterilization, 266
  mechanical control, 269
  monitoring, 266
  physical control, 269
  principles, 263–266
  regulatory control, 266
  strategies, 266–269
  threshold levels, 265–266
Intelligence, gift of, 5
Intergovernmental Panel on Climate
      Change, 133
Internet marketing, 730
Internet sales, 724
Internode, 304
Internships, 84
Interviews, *93*
Intestine products, 681
Intravaginal growth, 480
Inventors, 34
Inventory reports, 737
In vitro mating, 568
Involuntary muscles, 521
Ions, 323, 325
IPCC. *See* Intergovernmental Panel on
      Climate Change
IPM. *See* Integrated pest management
Irradiation, 654, 662
Irrigation
  definition, 147
  field crops, 430–431
  history, 430
  turfgrass, 489
  types, 391–392
  vegetable production, 391–392

J

Jefferson, Thomas, 34
Jockeys, 620
Jog, 627
Johnson grass, 442–443

K

Kenaf, 450
Kentucky bluegrass
  characterization, 441
  cultivars, 490
  fertility requirements, 488
  uses, 482–483
Knowles, Donald, 624

Kosher slaughter, 677
Kreb cycle, 576

L

Labor, migratory, 648–649
Lacustrine deposits, 166
Lakes, wildlife, 233–234
Lambs. *See also* Sheep
  grades, 699–700
  production practices, 610–611
  quality grades, *701*
  retail cuts, 679
Land. *See also* Soils
  cultivation, 415
  demand for, *142*
  erosion, 153–155
  function, 146–147
  ownership, 122–123
  as reservoir, 147
  as resource, 141–142
  sinkholes, 169
  -water relationship, 147–148
Land capability maps
  classes, 170–171
  description, 170
  function, *171*
  subclasses, 171–172
  units, 172
  use of, 172
Landscape architects, 503
Landscape maintenance technician,
      479
Landscaping. *See* Plantscaping
Lasso, 276
Lawn-care services, 477
Lawn mowers, *134*
Lawns. *See* Turfgrass
Layering, 343–344
Lead, 126–127, 235
Leadership
  definition, 102
  development opportunities, 105–110
  importance, 102–104
  at meetings, 115–119
  speaking skills, 110–115
  traits, 104–105
Leaves
  characterization, 306
  compound, *309*
  cuttings, 340–342
  function, *310*
  internal structure, 306
  margins, 306, *307*
  mold, 163
  parts, 306, 307
  scorch, 327
  shape/form, 306
  simple, *309*
  types, 306
Legumes
  examples, 184
  nitrogen fixation, 184
  seeds, planting, *445*
  types, 438–439
Lenticels, 304
Lespedeza, 440
Lesquerella, 434
Lethal dose, 284

Life expectancy, 682
Life spans, 31–32
Light
  for foliage, 470
  in hydroponic systems, 187–188
  intensity, 318
  measuring, 465–466
  for ornamentals, 463, 465–466
  seed germination, 337
Lime, 156
Limestone, standard ground, 180
Liming, 387
Lin, Alexander, 665
Linebreeding, 565
Linseed oil, 422
Liquids, 548
Listeria, 56
Literacy, 82
Living conditions. *See* Environment
Loamy soils, 174
Loans
  add-on, 720–721
  amortized, 721
  comparison chart, *722*
  long-term, 716–717
  production, 717
  seeking, 721, 723
  short-term, 717
  signature, 717
  simple-interest, 720
Loblolly pine, *203*
Lobsters, 243
Loess deposits, 166
Lolium perenne, 483
Long-term loans, 716–717
Lope, 627
Loss-leader pricing, 691
Lumber
  definition, 196
  measure formula, *197*
  seasoning, 217
Lymph glands, 521
Lysine, 527

M

Macronutrients, 322
Mad cow disease. *See* Creutzfeldt-Jakob
  disease
Magnesium, 327
Malaria-resistance, 268
Manage, definition, 103
Management
  definition, 710
  influence, 710
  performance, 710–711
  woodlot, 212
Manure, 184, 541–542
Maple, *205,* 207
Mapping, 50
Maps. *See* Land capability
Marginal burn, 327
Marketing
  advertising, 688–690
  approaches to, 687–688
  careers, 693
  commissions, 698
  considerations, 687
  cycles, 702–703

fees, 698
global, 703–705
Internet, 724, 730
procedures, 698
promotion, 688–690
promotions, 730
retail, 692–694
strategies, 691–692
techniques, 696–698
trends, 702–703
vegetables, 394–395
video, 695
wholesale, 651–652, 694–698
Markets
  auction, 695
  export, 705–706
  farmers', 694
  gardening, 378, 383
  roadside, 694
  terminal, 694–695
Marshes, *241*
Mastitis reduction, 40
Mating. *See* Breeding
Maturity, crop, 648
McCormick, Cyrus, 34
Meat. *See* specific types
Meat cutters, 693
Mechanization
  careers, 67
  companies, 34–35
  inventors, 34
  technologies, 33
  undeveloped countries, 35
Media. *See* Plant-growing media
Medical research, 581, 596
Mediterranean fruit fly, 654
Meetings
  conducting, 116
  officers, 115
  procedures, 115, 117–118
  requirements, 115–116
Meiosis, 559
Membranes, semipermeable, 320
Mendel, Gregor Johann, 47–48, 558
Merozoites, 624
Mesophyll, 307
Methane
  global warming and, *132*
  origins, 134
  use, 134–135
Microbes, 52–53
Micronutrients, 322–323
Microorganisms, 46, 648
Microwave ovens, 669–670
Middlemen, 693
Migratory labor, 648–649
Milk
  characterization, 592
  goats, 39, 611, 612
  nutrients, 672–673
  processed products, 673–674
  production, 516
  products, 672–674
  secreting glands, 40
  shelf life, 665
  water content, 672
Minerals
  characterization, 528–529
  function, *661*

in hydroponic systems, 187
matter, 173
in milk, 673
nutritional role, 661
sources, 530–531
Mississippi River Delta, 153–154
Mistletoe, 301
Mitosis, 558–559
Mix well, 180
Moist-heat cooking, 669
Moisture, soil, 387
Molt, 243
Monoclonal antibodies, 39
Monocotyledons, 304, 379
Monogastric system, 526
Montana Range Days program, 234
Mosquitoes, 268
Moss, peat, 164
Motions, 118–119
Mowers, 447–449
Mulch, 156
Mulching
  application, 510
  characterization, 508
  vegetable production, 392
Mules, 624
Muscular system, 521
Mutualism, 222
Mycelium, 262

N

Natural resources
  careers, 66
  forests and, 201–202
  renewable, 6–7
Natural Resources Soil Conservation
  Service, 158, 172
Natural service, 566
Neighborhoods, influence, 23–24
Nematodes, 177, 262–263
Nervous system, 524
Netting, *492*
Net worth, 733
Newcastle disease, 517
Nitidulid, 509
Nitrates, 184, 245
Nitrites, 245
Nitrogen
  fixation, 184
  function, 326
  ions, 323
  nonprotein, 530
  total ammonial nitrogen, 245
Nitrous oxides
  characterization, 126–127
  emissions, 135
  global warming and, *132*
Nomenclature, 505–506
Nonsense gene, 517
North American Forests, 197–201
North Appalachian Experimental
  Watershed, 386
Northern Coniferous Forest, 198
Northern Hardwoods Forest, 198, 200
No-till planters, 445–446
NRCS. *See* Natural Resources Soil
  Conservation Service
Nurse crops, 443

Nut production
  careers, 400
  disease control, 410
  fertilization, 407–408
  harvest, 411
  pest control, 410
  planning, 403–406
  planting, 406–407
  soil/site preparation, 406
  tree pruning, 409–410
Nutricultures, 185
Nutrients
  classes, 527–529
  deficiencies, *182, 324, 532*
  essential, 322–323, 325
  milk, 672–673
  plant (*See* Plant nutrients)
  soil, 321–322
  sources, 529–531
  supplemental, 225
Nutrition
  basic needs, 659–662
  carbohydrates and, 659
  careers in, 520, 667
  definition, 519
  fats and, 659–660
  food groups, 661–662
  health and, 520–521
  minerals and, 660
  proteins and, 660
  values, 38, 664
  vitamins and, 660
  water and, 660
Nutritionists, 667
Nuts
  characterization, 312
  climate factors, 403–404
  foods from, 670
  growth patterns, 405–406
  types, 403

O

Oaks, *205,* 207
Oats, 419
Obesity, 520
Oil meals, 533–534
Oils. *See also* Field crops
  commodities, 645–646
  cosmetic, 434
  crops, 427
  foods from, 671–672
  as fuel, 423
Oilseed crops, 420–422
Orchard grass, 441
Orchards, 406–407
Organic fertilizers, 181
Organic foods, 10, 289
Organic framing, 391–392
Organic matter, 156, 176, 387
Organisms
  fingerprinting, 56
  in soil, 166, 176
  water/soil benefits, 148
Organophosphate, 279
Ornamentals
  aspects, 455
  associated careers, 463
  definition, 13

design (*See* Plantscaping)
  development, 455
  drainage, 467–468
  as economic indicator, 454–455
  fertilizing, 468
  flowering, 458–459, 469–470
  foliage, 459–462, 470
  grooming, 468
  growing, 465–470
  light needs, 463, 465–466
  pest control, 454–455
  phototropism, 466
  propagation, 469
  selecting, 462–464
  temperature needs, 463–464, 467
  uses, 464–465
  water needs, 467
Osmosis, 241–242, 320
Outcrossing, 566
Ova. *See* Eggs
Ovens, 669–670
Ovulation, 563
Oxidative deterioration, 667
Oxygen
  in aquaculture, 242
  consumption, 135
  dissolved, 244
  in hydroponic systems, 187
  measuring, 244
Ozone, 129, 132

P

Pacific Coast Forest, 200
Packers, 640, 693
Palisade cells, 307
Paper birch, *205*
Parasites
  characterization, 267
  classification, 545–546
  control of, 544
  external, 546
  horse-associated, 624
  internal, 544, 545
Parasitism, 222
Parent material, 166
Parliamentary procedures, 115, 117–118
Particles, soil, 148
Particulates, 129
Parturition, 563
Pastures
  crops, 438
  harvesting, 447
  mating, 567
  renovating, 445–446
  rotation, 543
PCA. *See* Production Credit Associations
PCBs. *See* Polychlorinated biphenyl
Peanut hay, 440
Peanuts, 420, 422
Pears, Bradford, 502
Pears, diseased, *255*
Peat moss, 164, 363
Pedigree, 568
Pelts, 578
Penetration pricing, 691
Percolation, 167
Perennial ryegrass, 483
Perlite, 164–165

Permeable soils, 168
Perryman, Lance, 624
Personal interests, 85, *89*
Pesticides
  breaking down of, *279*
  career areas, 277
  classification, 282
  common names, 283
  controversy, 273
  directions, 284
  disposal, 286
  environmental impact, 28, 290–293
  EPA registration numbers, 284
  exposure, limiting, 287
  formulations, 283
  fungicides, 279–280
  general-use, 282
  in groundwater, 157
  health concerns, 289–290
  herbicides, 274–277
  history of, 273–274
  hygiene, 288
  ingredients, 283
  inorganic, 274
  insecticides, 277–279
  labels, 280–282
  lethal dose, 284
  limitation notice, 286
  manufactures' address, 284
  misuse, 281, 284
  movement of, *291*
  natural toxins and, 275
  net contents, 283
  organic, 274
  precautionary statements, 285
  protective gear, 287–288
  reentry information, 284
  resistance to, 269
  restricted-use, 282
  risk management, 286–288
  safer, 284
  safety, *293*
  signal words, 283–284
  spray, 129
  storage, 286, 288–289
  symbols, 283–284
  trade names, 281
  vapor drift, 290
  water quality and, 152
Pest management
  biological, 432
  chemical, 433
  cultural, 432
  field crops, 431–433
  forage, 446
  fruit production, 410–411
  genetic, 432–433
  good practices, 255
  health and, 542–543
  integrated (*See* Integrated pest management)
  mechanical, 432–433
  nut production, 410–411
  ornamentals, 471
  trees, 513
  vegetable production, 392–393
Pests
  definition, 129, 255–256
  garden, 369

Pests—*continued*
  insects, 258–260
  key, 263–264
  management (*See* Integrated pest
    management)
  population equilibrium, 264–265
  resistance, 354
  resurgence, 269
  weeds, 256–258
  woodlots, 215
Petals, 311
Pet care, 585–587
Petiole tests, 180
pH
  definition, 177
  nutrition and, *326*
  scale, *177, 325*
  soil, 325
Phenoxys, 276
Pheromone, 266
Phloem, 302, 308
Phosphorus, *324, 326*
Photodecomposition, 276
Photosynthesis
  factors, 318
  improving, 137
  process, 135, *136,* 317–318
Phototropism, 466
pH tests, 180
Physiology, animal, 520
Physiology, plant, 444
Pigs. *See* Swine
Pills, 548
Pine trees. *See specific species*
Placement agreement, *96*
Plan, definition, 102–103
Planters, 430, 445–446
Plant-growing media
  amending, 177–185
  decomposed, 163
  definition, 162
  hydroponic, 185–191
  liming, 177–178
  types, 162–165
Plant hardiness zone map, 501
Plant nutrients
  absorption, 187
  deficiencies, *182, 324*
  essential, 322–323, 325
  excess, *327*
  function, 10–11
  functions, 326–327
Plant reproduction
  asexual propagation, 338–343
  definition, 334
  methods, 334
  process, 335
  sexual propagation, 335–337
Plants. *See also* Trees
  air and, 321
  annuals, 365
  aquatic, 240, 242–243
  balled, 506–507
  bare-rooted, 506
  beets, 41
  biennials, 365
  border, 501
  bracts, 310
  breeding, 339

burlapped, 506–507
cell parts, *317*
coarse-textured, 472
composition, 298–299
container-grown, 508
cuttings, 338–342
defoliators, 255
diseases, 261–263
diversity, 310
exotic flowers, 41
explorers, 41
fertilization, 322
-fish, relationship, 242
flowers, 298–299, 310–311
food storage, 327–329
fruits, 311–312
gas exchange, 319
genetic manipulation, 51
germination, 191
group, 501
growth, 156
growth requirements, 188–186
hardiness, 501
hydroponic education, 191
indoor (*See* Ornamentals)
inflorescence, 480
leaves, 306–310
microscopic, 52–53
necessities, 300
new varieties, 330
nomenclature, 505–506
nuts, 312
pathologists, 261
perennials, 367
perfecting, 352
performance, improving, 35–36, 37
photosynthesis, 135, 137, 317–318
physiology, 317, 322
potatoes, 37
residue, 156
respiration, 135, 319
rice, 41
root systems, 298, 300–303
science careers, 2–3
seasonal, 361
selective breeding, 47
shape, 505
soils and, 320, 325
soybeans, 37
specimen, 461–462, 501
stems, 303–306
taxonomy, 312–313
tomato, 38
transpiration, 319–320
use, 354
vegetables, 312
water absorption, 147–148, 321–322
wild, hunters, 296–297
Wisconsin Fast, 191
Plantscaping
  accent, 472
  balance/scale, 472–473
  careers, 473–474
  colors in, 472
  definition, 470
  design aspects, 470–473
  design process, 473
  form, 472
  maintenance, 473

sequence, 472
texture, 472
trees in, 499
Plowing, 387
Plugging, 495
Poa pratensis
  characterization, 441
  cultivars, 490
  fertility requirements, *488*
  uses, 482–483
Poinsettia, 454–455
Poisoning, lead, 235
Poison ivy, 300
Pollen, 311
Pollination, 404, 582
Pollution
  air, 126–130, 135
  ground water, 155
  particulate, 129
  water, 151, 227
  wetland, 231
Polychlorinated biphenyl, 53–54
Polygastric system, 525
Ponderosa pine, *203, 204*
Ponds
  constructed, 246
  roll overs, 246
  wildlife, 233–234
Ponies, 623
Poplar, *205,* 207
Populations
  age profiles, 32
  China, 32–33
  dynamics, 226
  growth rate, 30
  life spans, 31–32
  patterns, changes, 31–33
  wildlife, *221*
  world's, 19
Porcine somatotropin, 51
Pork. *See also* Swine
  byproducts, 646
  commodities, 646
  retail cuts, 679
Potassium, *324,* 327
Potassium nitrate, 323
Potatoes, 37, 348
Poultry
  classes, 574
  coccidiosis in, 40–41
  commodities, 647
  definition, 573
  digestive system, 526–527
  diseases, 517
  domestication, 572
  economics, 573
  history, 572
  processing, 676
  production practices, 575–576
  small breeds, 37–38
  types, 573–575
  uses, 573–575
Powders, 548
Prawns, 245–246
Precipitate, 325
Precipitation, 147
Precooling, 395
Predation, 222–223
Predators, 222, *223*

Preservation, 663, 665–668
Presiding officer, 115
Prestige pricing, 691
Prey, 222
Pricing, 690–691, 714
Procambarus spp., 243, 245
Processed meats, 675
Processing
  agriculture, 63–64
  beef, 675
  careers in, 63–64
  crops, 650
  dairy products, 673–674
  food, 663, 665–668
Processors, 640
Produce managers, 693
Production, enterprise, 83
Production Credit Associations, 719
Production loans, 717
Profit maximization, 691–692
Progeny, 47
Progesterone, 563
Promissory note, 720
Promotions, 688–690, 730
Propagation, 469
Proteins
  in alfalfa, 447
  characterization, 528
  nutritional role, 660
  rich food, 636–637
  sources, 529
Protozoa, 177, 545
Pruning
  definition, 511
  purpose, 409–410
  requirements, 513
  root, 506
  techniques, 512
  timing, 512
  woodlot, 212
PST. *See* Porcine somatotropin
Public lands, 122
Public speaking. *See* Speeches
Pulse, taking, 549
Purebreeding, 564
Purify, definition, 150
Pyrus calleryana, 502

Q
Quaking aspen, *205*
Quality grades, 699, *700–701*
Quarantine Act of 1912, 266
Quarantines, 266
Quincy, Edmund, 34

R
Rabbits
  breeds, 578–579
  domestication, 576–577
  economics, 577
  fur, 578
  history, 576–577
  production practices, 580
  in research, 597
  types, 578–580
  uses, 578–580
Radioactive dust, 128–129

Radon, 127–128
Ranchers, 599, 620
Rangeland forage, 442
Ranger manager, 444
Rape. *See* Canola
Reaper, 34
Recombinant DNA, 429
Records, financial, 733–734, 736–737
Red canary grass, 441
Red clover, 445–446
Red Fescue, 483
Red-spotted ladybug, 509
Refrigeration, 665
Registration papers, 564
Render, 605
Rendering insensible, 674
Renewable resources, 6–7, *8*
Repotting, 469
Reproduction. *See also* Breeding
  female system, 563
  male system, 562–563
  plant, 334–343
  problems, 563
Research
  medical, 581, 596
  projects, 82, 84
  rabbits, 597
Residual soil, 166
Resistance
  host, 267
  pesticides, 269
  pests, 354
Resources
  inventory, 90, *91*
  land, 141–142
  management, 249
  renewable, 6–7, 8
  substitution, 716
Respiration
  in aquaculture, 242
  definition, 135
  plant, 319
  rates, 549
  seed, 337
Respiratory system, 522
Résumés, 90, *92*
Retailer, 640
Retail marketing, 692–694
Rhizomes, *305*, 328, 480
Rhus radicans, 300
Rice
  characterization, 420
  golden, 429
  hybrids, 41
  importance, *420*
Riparian areas, *232*
Risk management, 286–288
Roadside markets, 694
Robertson, William, 275
Rocky Mountain Forest, 201
Roll overs, 246
Root crops, 426
Roots
  adventitious, 300
  caps, 302
  cells, 302–303
  cuttings, 342
  description, 298
  elongation, 302

epidermis, 320
fibrous, 302
food storage, 328
hairs, 302–303, 320–321
pruning, 506
soil space, 300
structure, *303*
systems, 298–301
taproot, 300–301
tissues, 302
turfgrass, 478–479
types, 300
Rootstocks, 401, 403
Rotary mowers, 447
Rotenone, 278
Roughage, 525, 534–535
Roundworms, 545
Row crop planters, 429–430
Rubber production, 38
Ruminants, 525–526
Rye, 419–420

S
SAEP. *See* Supervised agricultural
  experience programs
Safety
  biotechnology, 55
  food, 641–642
  pesticides, *293*
  riding, 628–629
Safflower, 422
Sales, direct, 696
Salinity, 246
Salinity gradient, 240–241
Salmon, 241
Salmonids, 245
Salvage harvesting, 214
Sandy soils, 174
Sanitation, 540–541, 641–642
Scarify, 336
Schultes, Richard, 296
Scorch, 327
Seafood. *See* Fish
Seasons, 381
Seaweeds, 241
Secchi disc, 244
Secretaries, 115
Seedbeds, 428, 444
Seeding rates, *493*
Seedlings, 390, 507
Seeds
  cotton, 422
  crops, 425–426
  culm, 480
  food storage, 329
  forage, 445–446
  germination, 336–337
  growing from, 389
  labels, *493*
  legume crops, 426
  legumes, planting, *445*
  planters, 429–430
  planting, 390
  post-harvest, 426
  propagation, 335
  respiration, 337
  scarify, 336
Seed-tree cutting, 214

Selective breeding, 47, 568–569
Self-employment, 737–738
Semen, 567–568
Semipermeable membrane, 320
Sepals, 311
Septic systems, 153
Sewage disposal, *21*
Sexual propagation, 335–337
Shackles, 674–675
Sheep. *See also* Lambs
  breeds, 610
  byproducts, 677
  commodities, 646
  economics, 609–610
  embryo co-culture, 566
  forage, 610
  history, 609
  Merino, 556
  origins, 609
  slaughter, 675–676
  types, 610
Sheep Experiment Station, 516
Sheet erosion, 155
Shelf life, 665, 668
Shellfish, 243, 245, 247
Shelterwood cutting, 214
Shepherds, 609
Shih, Franklin, 636–637
Shoot systems, 479–481
Short-term loans, 717
Shrimp, 245–246
Shrinkage, wood, 210
Shrubs. *See* Trees
Signature loans, 717
Silage
  acids from, 443
  definition, 438
  description, *47*
  fermentation, 450–451
  harvesting, 449
  spoilage, 449
  storage, 449
Silent Spring, 28, 263
Silos, *449*
Silviculture, 212
Simple, 343
Simple-interest loans, 720
Sinkholes, 169
Sires, 565
Sitka spruce, *203*, 204
Skeletal system, 521, *522, 523*
Skills
  lists of, *87–88*
  plan/record, *98*
  profile, 85
Skimming, 691
Slash and burn farming, 154
Slaughter
  beef, 674–675
  cattle, 699
  hogs, 676
  kosher, 677
  poultry, 676
Slope, 166–167
Small Business Association, 720
Sod crops, 185
Sodding, 493–494
Sod-production farms, 477

Softwoods, 202–204
Soils. *See also* Land; Plant-growing
    media
  acidity, 325
  aggregates, 156, 175–176
  alkalinity, 177, 180–181, 325
  amendments, 491
  classification, 168, 170–173
  clay, 174–175
  climate, 166
  composition, 165–168
  conditioning, 362–363
  conservation, 155–158
  decompser, 177
  definition, 163
  deposits, 166
  drainage, 167
  elements, 325
  erosion, 152, 155–157
  fertilizing, 181–184
  function, 320–321
  gypsum, 180
  horizons, 166, 173
  hydrology, conservation, 150
  loamy, 174
  location, 166
  mechanical analysis, 174
  modification, 491
  moisture, 387
  nutrients in, 321–322
  organic matter, 156, 176
  organisms in, 166, 176, 177
  origins, 165–168
  parent material, 166
  particles, 148
  particle size, *173*
  percolation, 167
  permeable, 168
  pH, 177
  pores, 148
  profile, 166, *167*, 173
  residual, 166
  sampling process, 178–180
  sandy, 174
  saturation, 147
  science, 8–9
  sponge qualities, 147
  structure, 175–176
  sulfur, 180
  tests, 180
  texture, 173–175
  time factors, 168
  topography, 166–167
  watering, 151
  weathering, 168
  wet, 167
Solution cultures, 185
Sorghum, 420
Southern forests, 200
Southern pine, 204
Southern Weed Science Laboratory, 434
Soybeans
  characterization, 420
  film types, 636
  pest-damage index, *266*
  secrets of, 37
  uses, *421*
Space exploration, 358–359

Spades, 507
Spawning, 241
Specialty crops, 422–425
Species, definition, 313
Species, indicator, 151
Specimen plants, 461–462, 501
Speeches
  competitions, 111
  effective, 110–111
  evaluating, *114*
  extemporaneous, 111–112
  giving, 113, 115
  outlines, 112
  planning, 112–113
  prepared talks, 111
  types, 111–112
Spencer, Roy, 133
Sperm, 555
Spermatogenesis, *559*
Sphagnum, 164
Split carcass, 675
Split-vein cuttings, 343
Spoilage, 648
Sprigging, 494
Sprinkler irrigation, 391, 430–431
St. Augustine grass, 486
Staking, 510
Stalks, *324*
Stamen, 311
Standard ground limestone, 180
Starch, 659
Stems
  cross section view, *305*
  cuttings, 338–339
  definition, 303
  food storage, 328–329
  herbaceous, 303
  key parts, *303*
  parts, 304
  turfgrass, 479–480
  types of, 303–304
Stenotaphrum secundatum, 486
Sterility, 563
Sterilization, 266, 348–439
Stiffness, wood, 211
Stimulant crops, 427
Stolonizing, 494
Stolons, 480
Stoma, 308
Stomachs, four-chamber, 526
Storage
  cool, 371
  field crops, 434
  food, 327–329, 663, 665–668
  forage, 446–447
  fruit crops, 411–412
  hay, 449
  nut crops, 412
  silage, 449
  vegetable crops, 370, 395–396
  warm, 370–371
Streams, management, 231
Stream-side forestation, 232
Strength, wood, 211
Strip cropping, 156
Student interests survey, *89*
Subirrigation, 392
Succulents, 459, 461

Sugar
  as additive, 668
  beets, 41
  cane, 424
  characterization, 423
  crops, 427, 646
  function, 659
  maple, *205*
Sulfur
  characterization, 126
  function, 327
  as insecticide, 277
  soil, 180
Sunflowers, 422
Sunlight effects, 130–131
Superior oils, 278
Superstores, 651–652
Supervised agricultural experience
      programs
  American Star Awards, 81
  classroom instruction and, 80
  dairy industry, 90
  description, 78
  enterpreneurship and, 83
  exploratory, 82
  goal setting, *94–95*
  implementing, 90–97
  improvement activity plan, *97*
  opportunities for, 85–90
  placement, 83–84
  placement agreement, *96*
  plan refining, 95, 96
  purpose, 79–80
  resource inventory, 90
  selecting, 90–97
  skills plan/record, *98*
  types of, 82–84
Supply, definition, 687
Supply, principle, 714
Surface irrigation, 391–392, 431
Swathers, 447–448
Sweet clovers, 439
Sweetgum, *205*, 207
Swine. *See also* Pork
  breeds, 606, 608
  death, cause of, 551
  economics, 605–606
  grades, 699
  history, 605
  origins, 605
  production practices, 607, 609
  quality grades, *701*
  slaughter, 676
  species, 605
  types, 606–608
  viruses, 517
Sycamore, *206*, 208
Symptoms, 262
Syringe, 540–541

T

Tall fescue, 483, 484
Tallow, 39–40
TAN. *See* Total ammonial nitrogen
Taxonomy
  binomial system, 313
  definition, 312

  Latin use in, 312–313
  plant, 301
T-budding, 345–346
Technology
  definition, 7
  high, 7–8
  industrial, 7
Temperatures
  aquaculture, 245
  average annual, 24
  global patterns, 132–133
  harvest, 665
  in hydroponic systems, 188
  lakes/ponds, 234
  for ornamentals, 463–464, 467
  photosynthesis and, 318
  seafood, 666
  seed germination, 337
  taking, 549
  vegetable storage, 372
Terminal bud, 304
Terminal markets, 694–695
Testes, 562
Testosterone, 562
Tetraethyl lead gasoline, 127
Thymine, 49
Tidewater, 143
Tillage
  equipment, *491*
  forage, 444
  minimum, 428
  turfgrass, 480
  types, 428
Time factors, 168, 512
Timothy, 441
Tip layering, 343
Tissue cultures
  advantages, 347
  aseptic technique, 350–354
  description, 346
  laboratory, 350
  preparing, 349
  roots, 302
  sterile media for, 347–439
  storage, 349–350
Tobacco, 424–425
Tomatoes
  genetically engineered, 38
  processing, 649–650
  rotten, 650
Topography, 24, 166–167
Total ammonial nitrogen, 245, 246
Toughness, wood, 211
Toxins, 275
Trade names, 283
Traits
  definition, 49
  genes and, 560
  selecting, 561
Transgenic animals, 52
Transpiration, 319–320
Transplants, 389, 390
Transponders, 594
Transporting, 650–651
Trees. *See also* Lumber; Wood
  biocontrol, 509
  characterization, 196
  citrus blight, 329

  classification, 197
  color changes, 318
  cutting, 214
  disease control, 410
  diseases, 513
  dwarf, 401
  fertilizing, 510–511
  fruit, 400–402
  gall infestation, 512
  geographic location, 501–502
  growth, 209–210
  growth habit, 400
  growth types, 503
  guying, 510
  hardwood, 204–208
  indoor, 460
  in landscapes, 499
  major parts, *209*
  mulching, 508, 510
  names, 505–506
  nuts, 403
  obtaining, 506–508
  pest control, 410
  pests, 513
  planting, 508
  production, 410
  products, 198
  pruning, 212, 409–410, 511–513
  rings, *211*
  rootstocks, 401, 403
  semidwarf, 401
  shape, 505
  shrubs *vs.*, 501
  site location, 502
  size, 196, 504
  softwood, 202–204
  spades, 507
  staking, 510
  standard, 401
  uses, 500–501
  water absorption, 210
Triazines, 276
Trickle irrigation, 392
Tropical Forest, 201
Trout
  farmed, *245*
  oxygen needs, 244
  whirling disease, 248
Truck cropping. *See also* Vegetable
      production
  careers in, 383
  definition, 378
  site selection, 382–383
  soil preparation, 387–389
True clovers, 439
Tuber crops, 427
Tubers, 304
Turbidity, 244
Turf farming, *65*
Turfgrass
  adaptation zones, *482*
  associated careers, 479
  cool-season, 478, 483–483, 485
  cultivation, 486–498
  development, 478
  establishment, 489–494
  fertilization, 488–489
  fertilizers, 487

Turfgrass—*continued*
  growth, 478
  industry, 477
  irrigation, 489
  mixing, 484
  mowing, 487–488
  mowing heights, *487*
  netting, *492*
  planting, 492–494
  plugging, 495
  poor quality, *486*
  root system, 478–479
  seeding, 492–493
  selection, 490
  shoot system, 479–481
  site preparation, 490–492
  sodding, 493–494
  species, 481–483, 485–486
  sprigging, 494
  stolonizing, 494
  tillers, 480
  warm-season, 478, 485–486
Turkeys. *See* Poultry

U

United Nations Framework Convention
    on Climate Change, 133–134
United States Department of
    Agriculture
  food safety oversight, 641–642
  organic food standards, 289
  research projects, 98
  shortage study, 15
  wild plant harvesting, 296–297
Universal solvent, 143–144
Urinary system, 524–525
USDA. *See* United States Department of
    Agriculture

V

Vaccinations, 550
Vapor drift, 290
Variety, 313
Vascular bundles, 304
Veal, 699
Vectors, 255
Vegetable production
  cultivation, 390–393
  harvesting, 394
  hydroponics, 393
  marketing, 394
  organic, 391
  planting, 389–390
  site size, 383
  soil preparation, 387–389
  storage, 395–396
Vegetables. *See also specific crops*
  characterization, 312
  classification, 379–381
  commodities, 646
  cool season, 381
  edible parts, 381
  foods from, 670
  growing season, 381
  harvesting, 370
  hothouse, 164
  ideal temperatures, 371–372

planting guide, 366
  seeds, 425–426
  storage, 370–371
  warm season, 381
  washing, 662
Vegetative bud, 304
Vegetative spreading, 480
Ventilation systems, 128
Vertebrates, 221
Vertical integration, 696
Veterinarians, 520, 540, 551
Veterinary skills, 547
Video merchandising, 695
Vines, 460
Viruses, 262
Viscera, 675
Vitamins
  characterization, 529
  function, *660*
  in milk, 673
  nutritional role, 660
  sources, 530–531
Volatilization, 276
Voluntary muscles, 521

W

Warp, wood, 210
Waste disposal
  animal, 605
  biotechnology and, 53–54
  sewage, *21*
Water. *See also* Irrigation
  animal component, 527
  biological function, 528
  brackish, 241
  capillary, 148
  conserving, 148, 150–153
  control projects, 11–12
  cultures, 185
  drainage, 167
  drinking, 21
  free, 147
  fresh, 143
  gravitational, 147
  ground (*See* Groundwater)
  hardness, 244
  hydroponic, 187
  hygroscopic, 148
  imbibitions, 336
  importance, 142
  -land relationship, 147–148
  management, 227
  in milk, 672
  natural flow, 240–241
  nutritional role, 661
  ocean, 143
  for ornamental, 465–466
  oxygen content, *244*
  pesticides in, 152
  plants absorption, 321–322
  plant uptake, 300
  pollution, 151, 227
  potable, 143
  purity levels, 143, 146
  quality, 150–153
  quality, testing, 244
  runoff, 151–152
  salt, 143

soil leaching, 168
  stresses on, *145*
  tree absorption, 210
  turbidity in, 244
  as universal solvent, 143–144
Water cycles
  description, 146
  forests and, 201
  function, *240*
Watershed, 147
Waterways, 156–157
Weeding, 368–369
Weeds. *See also specific species*
  annual, 256
  biennial, 256
  definition, 256
  nonselective control, 490
  noxious, 258
  perennial, 256–257
  vegetable gardens, 390–391
Weight, wood, 210
Wetlands
  constructed, 153, 158
  pollution, 231
  wildlife, 227, 230–231
Wheat
  characterization, 418
  processing, 649
  varieties, 354
Whirling disease, 248
White ash, *206*
Whiteflies, 471
White oak, *205*
White pine, *203, 204*
Whitney, Eli, 34
Wholesale marketing, 694–698
Wholesalers, 640, 651
Wildlife
  biologists, 229
  characteristics of, 221–222
  commensalism, 223
  declines, *221*
  dynamics, 226
  on farms, 226–227
  forests, 229
  future, 235–236
  humans and, 221, 224–225
  lakes practices, 233–234
  mutualism, 222
  parasitism, 222
  ponds practices, 233–234
  predation, 222–223
  relationships, 222–225
  supplemental feeding, 225
  wetlands, 227
Wildlife management
  classifications, 225–227
  farm practices, 228
  forest practices, 229–230
  stream practices, 231–234
  wetland practices, 230–231
Wisconsin Fast Plants, 191
Wood
  bending strength, 211
  decay resistance, 211
  easy of working, 210
  hardness, 210
  nail holding, 211
  painting holding, 210–211

properties, 210–211
shrinkage, 210
stiffness, 211
surface, 212
toughness, 211
warp, 210
Woodlots
fire, 215
growing, 212–213
harvest cutting, 213–214
harvesting, 215–217
pests, 215

protection, 214–215
restocking, 212
Wool, 609
Worms. *See* Nematodes
Wormwood, 434
Wrappings, 667–668

X

Xylem, 308
Xylem cells, 302

Y

Yeast, 177
Yellow poplar, *205*
Yield grades, *700*

Z

Zoysia japonica, 484, 485–486